Polymer Chemistry

Sebastian Koltzenburg
Michael Maskos
Oskar Nuyken

Polymer Chemistry

Springer

Sebastian Koltzenburg
Functional Biopolymers
BASF SE, GMM/B - B001
Ludwigshafen, Germany

Oskar Nuyken
Garching, Germany

Michael Maskos
Fraunhofer ICT-IMM
Mainz, Germany

Translated by Karl Hughes

Forewords by Rolf Mülhaupt and Krzysztof Matyjaszewski

ISBN 978-3-662-56998-6 ISBN 978-3-662-49279-6 (eBook)
DOI 10.1007/978-3-662-49279-6

Softcover re-print of the Hardcover 1st edition 2017

Cover illustration: With kind permission by Gurit

Printed on acid-free paper

This Springer imprint is published by Springer Nature
The registered company is Springer-Verlag GmbH Germany
The registered company address is: Heidelberger Platz 3, 14197 Berlin, Germany

Foreword

Today, synthetic polymers can be found everywhere and are used in nearly every device. All computer chips used in our desktops, laptops, smartphones, or tablets are enabled by polymers used as photoresistors in microlithographic processes or as organic light-emitting diodes. Cutting-edge biomedical applications require polymers for tissue or bone engineering, drug delivery, and tubing and containers for intravenous delivery of medications. The interior of every automobile is almost entirely made from polymers, and they are also used for automobile body parts and under-the-hood applications. New lightweight and strong nanocomposite polymeric materials have enabled energy-efficient Dreamliner and A380 aircrafts. The construction industry uses polymers as insulating materials, sealants, adhesives, and coatings. Many new applications require smart polymers that respond to external stimuli, which can be used in sensors, shape memory materials, and self-healing systems. Thus, polymers are perhaps the most important materials in our society today, and their annual production exceeds 200 million tons. Although approximately 50% of chemists in the USA, Japan, and Western Europe work with polymers, polymer education has not yet reached the appropriate level, and many of those chemists neither take advantage of the unique properties of polymeric materials nor fully comprehend the synthetic pathways to control precisely macromolecular architecture.

Polymer Chemistry by Koltzenburg, Maskos, and Nuyken covers all aspects of polymer science, starting with fundamental polymer physical chemistry and physics, including all classical and modern synthetic techniques, and ending by reviewing various applications including more specialized uses in energy, environment, biomaterials, and other advanced fields. The authors present the material in 22 chapters in a very lucid and attractive way and identify the most important references for each chapter. This textbook is expected to be very helpful for all beginners in polymer science and also for more experienced polymer scientist.

I read the book with great interest and believe that it will become an excellent introductory polymer science textbook for senior undergraduate and graduate students.

Krzysztof Matyjaszewski
J.C. Warner University Professor of Natural Sciences
Carnegie Mellon University,
Pittsburgh, PA, USA
Fall 2015

Foreword to the German Edition

Small molecules such as drugs and food ingredients prolong human life, whereas macromolecules as versatile structural and functional materials are essential for a high quality of life, making "high tech" products available to all mankind. Originally developed as substitutes for natural materials such as ivory, silk, and natural rubber, highly versatile modern synthetic polymers can readily be tailored and processed to meet the diverse needs of our society and our modern technologies. Nowadays polymeric materials and systems are indispensable in modern life. The wide spectrum of polymer applications spans from food packaging to construction, textiles, automotive and aerospace engineering, rubbers, paints, adhesives, and system-integrated functional polymers indispensable in electronics, flexible microsystems, and energy and medical technologies. Their unique versatility in terms of tailored property profiles, ease of processing, application range, and recycling, coupled with their outstanding resource, ecological, energy, and cost efficiency, is not met by any other class of materials. Today polymeric materials play an important role in sustainable development. The success of polymer development and the high demand for polymer products by the rapidly growing world population are reflected by the surging polymer production capacity, which today exceeds 300 million tons per annum. Following the Stone Age, the Bronze Age, and the Iron Age, in the twenty-first century we are now living in the Polymer Age.

Anyone interested in sustainable development become increasingly confronted with synthetic and natural macromolecules and their applications. In almost every facet of modern technology and in innovative problem solutions, the development of engineering plastics and functional polymers plays a key role. In order to convert macromolecules into useful materials and sustainable products it is essential to understand the basic correlations between molecular polymer design, polymer technology, processing, applications, and sustainability. Unlike most conventional textbooks with a rather narrow focus on traditional chemistry and physics, this textbook clearly goes beyond the limits of the individual disciplines and presents a fascinating view of the challenges and prospects of modern interdisciplinary polymer sciences and engineering and their impact on modern technologies. It is obvious that all three authors bring to bear their profound experiences in teaching and research covering the broad field of polymer science and engineering. Furthermore, they are highly skilled in didactics and have succeeded in pointing out the relevance of tailoring polymers with respect to polymer applications. This team of three authors has successfully merged their complementary skills, own experiences, and different points of view. In 22 chapters the authors have impressively managed to present a comprehensive view of the extensive and rapidly developing fields of polymer sciences and technology in an appealing and easy-to-read format. In addition to covering the fundamentals of polymer chemistry and physics, the book describes the synthetic methods and polymer analytics as well as the technological aspects which are essential for tailoring polymeric materials, including multicomponent and multiphase polymer systems. Moreover, illustrated by carefully selected examples of specific applications, this text book gives an excellent view on the hot topics in polymer sciences as well as on environmental aspects of polymers, recycling, bio-based polymers, and modern research trends. The result of this remarkable and successful three-author co-production is a methodically well-conceived and easy-to-read textbook that serves as a desk book reference for polymer scientists, engineers, educators, and students. This textbook represents a valuable source of information for those who are already familiar with science. Because of its clear and didactic style, this textbook represents an excellent choice of reading for those who are approaching the subject of polymer sciences and engineering for the first time. The length of the book is more than adequate to cope with the complexity and breadth of this highly diversified and interdisciplinary field. Despite its high density of valuable information, this textbook reads well and I am firmly convinced that it is certain to become one of the key reference textbooks in the field of polymer sciences.

Rolf Mülhaupt
Freiburg
June 2013

Acknowledgement

The success of the German edition, the encouragement of colleagues in many countries, and the honor of winning the 2015 prize (Literaturpreis der Chemischen Industrie) from the Organization of German Chemical Industry (Verband der Chemischen Industrie VCI) persuaded us to translate our textbook into English to make it more available to a broader international audience.

The authors would like to thank the team of translators from the language department of the TU Munich, led by Mr Karl Hughes, for their dependable cooperation and continuous willingness to discuss suggestions for alteration.

The contribution of Dr Stephen Pask has been particularly valuable, both in terms of language and his specialist knowledge. Thanks to his expertise and tireless unflagging commitment and innumerable discussions in which Karl Hughes and his team were constantly involved, we now have an English text which, we believe, contains numerous improvements compared to the German edition.

We would like to thank Dr Kyriakos A. Eslahian, Dr Thomas Lang, and Jonas Schramm for translating and redrawing the figures and for suggesting and making corrections where necessary. We would like to express our special gratitude to Christoph Bantz who read the final version of the entire document and provided patient and constructive criticism.

We also owe thanks to many interested and critical readers of the German edition who have contacted us to point out typos and mistakes. In this respect we would specifically like to name our colleagues Prof Dr André Laschewsky, University of Potsdam and Prof Dr Ulrich Ziener, University of Ulm.

A special thank you goes to our sponsors for this translation project: the "Fonds der Chemischen Industrie" and the group of macromolecular chemistry of the Gesellschaft Deutscher Chemiker (GDCh).

Thanks are also expressed to Springer-Verlag, especially Ms Merlet Bencke-Braunbeck and Dr Tobias Wassermann, for their support during this project.

Finally, as with the German edition, we have enjoyed continued encouragement and support from our families, for which we can never thank them enough!

Sebastian Koltzenburg
Michael Maskos
Oskar Nuyken
January 2016

Contents

Introduction and Basic Concepts

S. Koltzenburg et al., *Polymer Chemistry*, DOI 10.1007/978-3-662-49279-6_1

Among the many areas of chemistry, polymer science is a comparatively new field. The empirical use of polymeric materials made from natural substances has been documented for centuries; however, only the pioneering work of the late Hermann Staudinger (1926), a Nobel laureate, in the 1920s provided the basis for a systematic understanding of this class of materials. In the decades since then, polymer science has developed to become both technically demanding and industrially extremely important. In particular, polymer science is characterized by its interdisciplinary nature:

- Most technologically relevant macromolecules[1] are based on a carbon backbone and thus belong in the realm of *organic chemistry*.
- Approximately half of all polymers produced today are synthesized using *organometallic* catalysts.
- A description of the behavior of both solid polymers and their solutions is now based on well-established *physical* and *physicochemical theories*.
- Because macromolecules are often used in the area of classical materials, processing and molding of polymers is an essential step in the production of finished products. Thus, *engineering science* is also important. In medical technologies, polymers are used in highly specialized applications, such as artificial heart valves, eye lenses, or as materials for medical devices.

Last but not least, as well as the vast and significant use of synthetic polymers, macromolecules are of crucial *biological* importance. Undoubtedly the most important polymer in the world—without which human existence would not be possible—is DNA. Without its polymeric nature, DNA could not fulfill its essential role as the memory molecule of living systems. If the molecules were not linked to a polymeric strand, DNA would be nothing more than a mixture of four different bases with no defined structure and therefore without biological function. In addition to the millions of tons of natural rubber processed annually, further examples of biopolymers essential to life include proteins that catalyze chemical reactions as enzymes, form membranes, or act as antibodies differentiating between friend and foe.

This chapter deals with the basic concepts and definitions of polymer science and especially the most important question that a natural scientist can ask: "Why?" In particular, why should one take an interest in this field? It is shown that polymers constitute a class of materials that not only make an essential contribution to the existence of life in the form of biological macromolecules, but without which, thanks to their myriad technical applications, our modern daily life would be no longer conceivable.

1.1 Polymers: Unique Materials

Even if we restrict ourselves to the field of non-biogenic, traditional materials, macromolecules are a material class of unparalleled versatility. However, the range of properties covered by polymeric materials is much broader than that of traditional materials. Thus, for example:

---

1 Originally, a distinction was made between macromolecular substances and polymers. This differentiation has become unnecessary. In this book, these terms are used congruently.

- Glass fiber reinforced plastics can have tensile strengths that rival, e.g., steel, whereas other polymers such as polyurethane foams can be used as soft cushions or mattresses.
- Most plastics are electrical insulators, but highly conjugated polymers have also been synthesized with specific conductivities of the same order of magnitude of those of highly conductive metals (Naarmann and Theophilou 1987).
- The density of porous polymeric materials can be varied across a very wide range. In particular, from polymer foams such as Styrofoam®, extremely lightweight articles can be produced.[2]
- The melting point of polymers can also be greatly modified by varying the macromolecular architecture. Some polymers can be physically described as highly viscous melts even at room temperature, whereas other polymers have melting points of several hundred degrees Celsius, and can be heated to red heat or sintered. Of course, the temperature range of the melting or softening point is critical for the temperature at which a material can be used or processed. On the one hand, a high melting point allows a high service temperature but requires a lot of energy to process the molten material into the final shape. For many materials in everyday life, which are only used at room temperature, a low melting point is an advantage because they can be processed much more resource-efficiently than materials with a high melting point. Here, too, the unrivaled variability that polymers offer is often a decisive and advantageous factor.

Because of their great versatility and their resulting unique material properties, synthetic polymeric materials have become indispensable in our daily lives. Many familiar applications can only be realized using macromolecular materials:

- The electrical and electronics industries in their current form are difficult to envisage without polymers. This statement includes seemingly trivial applications such as the sheathing for electric cables—no other non-polymeric substance class provides materials that are both flexible and at the same time act as electrical insulators. Even in technically much more demanding applications, such as the manufacture of solar cells, LEDs, or integrated microchips, polymers play a crucial role, e.g., as etching masks, protective coatings, dielectrics, or fiber optics.
- The modern automobile would also be unthinkable without polymers. All motor vehicles manufactured today are covered with a polymer layer—the so-called *clearcoat*. In addition, polymers, from which, for example, the tires, dashboard, seat cushions, and bumpers are constructed, make a major contribution to reducing the weight of the vehicle, thus limiting the fuel consumption.
- The construction industry has also benefited enormously from this relatively young class of materials. Polymers in the form of insulating foams reduce the energy consumption of buildings, serve as conduits for water supply and sanitation, and provide a weather-resistant alternative to the use of exterior wood.
- As packaging, polymers are now irreplaceable, especially for food packaging or as shock absorbing material for goods in transit.

Polymers find applications not just as classical *materials* but also as, mostly soluble, *active ingredients* and *functional additives.* As such, they often go unnoticed because

---

2 Low-density materials such as metal foams or ceramic aerosols can also be produced from non-polymeric materials; these, however, do not have the same breadth of application in everyday life as polymers.

they are not the actual material but rather, often in relatively small amounts, responsible for the appearance of something. Thus, polymers can be found in modern detergents, cosmetics, or pharmaceutical products. They are also used in water treatment and paper production. In the latter capacity, macromolecules as *functional polymers* are discussed in detail in ▶ Chap. 19.

1.2 Definition of Terminology and Basic Concepts

In the following section a brief introduction to the basic concepts of polymer science is given.

1.2.1 Fundamentals

The term *polymer* refers by definition to molecules formed from a number of building blocks, called *monomers*, usually connected by covalent bonds. The prefix "poly" comes from the Greek word for "many" whereas the Greek prefix "mono" means "single" and refers here to a single block. In the synthesis of many polymers, monomers are linked together in the same manner to form a single chain consisting of covalently connected repeating units (see Fig. 1.1).

There is no definitive limit on the number of repeating units required to meet the definition of the polymer. In general, it is stipulated that the number n, also referred to as the *degree of polymerization*, must be sufficiently high that the physicochemical properties of the resulting molecule no longer change significantly with each addition of a further repeating unit. This definition is not exact. Macromolecules that are composed of relatively few repeating units do not meet this definition and the term *oligomers* ("oligo" = "few") is used for such molecules.

One example of a polymerization reaction is the reaction of ethene to form polyethylene (see Fig. 1.2). In this reaction, the C=C double bond of the ethylene is converted into a single bond.

From the definition of the term polymer, it follows that in principle any chemical molecule that can form two (or more) bonds can be used as a monomer for the synthesis of macromolecules. This allows a huge variety of accessible structures which barely set a limit to the imagination of the synthetic chemist.

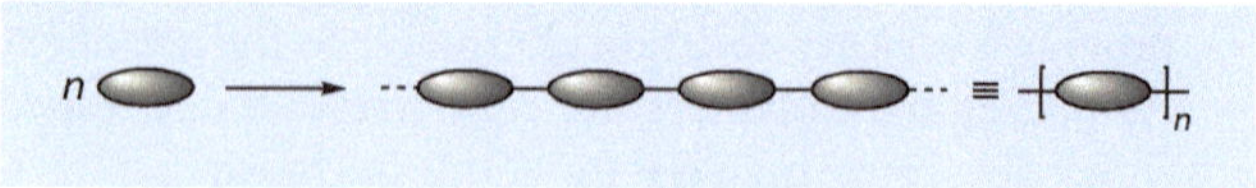

Fig. 1.1 Schematic structure of a polymer of n repeating units

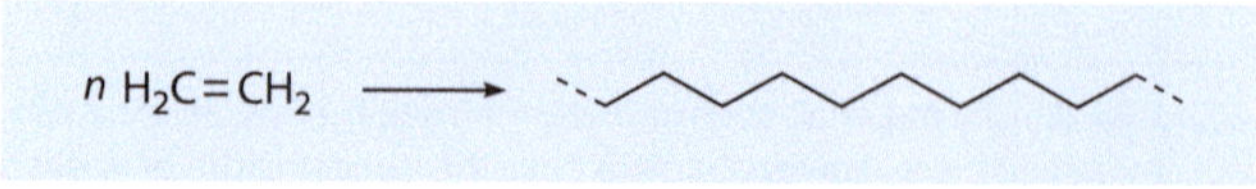

Fig. 1.2 Polymerization of ethene to polyethylene

As already mentioned in the introduction to this chapter, the properties of polymers can be varied within a broad range. For the control of these properties, a huge array of adjustments is available to the polymer chemist. The most important are:

- Type of monomers
- The chemical bond between the repeating units—for example, ether vs amide bonds
- Degree of polymerization
- Architecture of the chain—for example, linear or cross-linked
- Incorporation of chemically different monomers along the polymer chains (*copolymerization*)
- Sequence of monomers in a copolymerization—for example, alternately or in long sequences which consist of only one type of monomer (▶ see Sect. 1.3.3)
- Specific interactions between the components of the polymer chain, e.g., hydrogen bonding or dipole–dipole interactions

In addition to these essential questions, many other factors, such as admixtures (*additives*) and material processing, also have an influence on the properties of macromolecules. The aim of this book is to provide, against the background of an almost infinite variety of possible polymer structures, an overview of the essential principles that can be used for the selective synthesis of structures with desired properties.

1.2.2 Polyreactions

In the following, a brief overview of the basic possibilities for the synthesis of polymers (*polyreactions*) is given. These can be classified according to various criteria.

Depending on the manner in which the polymer chains are constructed in the course of the polyreaction of the monomers, a distinction can be made between step-growth and chain-growth reactions.

Step-Growth Reactions

This polymerization process can, in principle, be applied to all organic compounds which have two functional groups capable of forming a chemical bond. Classic examples of this are ester, amide, or urethane bonds (◘ see Fig. 1.3).

The resulting polymers here are referred to as polyesters, polyamides, or polyurethanes. Details on the nomenclature can be found in ▶ Sect. 1.2.3.

Chain-Growth Reactions

In chain-growth reactions, the polymerization can ensue by an addition to a polymerizable group, especially an olefinic double bond, or by the opening of a ring. The essential criterion for chain growth is the existence of a (usually high energy and unstable) active particle, which is able to add to a monomer unit and thereby transfers its active character to the newly incorporated repeating unit. This leads—as with a falling row of dominoes—to a chain reaction in which the growing chain continuously adds additional monomer units until no more monomer is available or side reactions occur.

Vinyl compounds can often be polymerized by a chain-growth mechanism. Here, the double bond is converted into two single bonds (◘ see Fig. 1.4). Because, in the case of carbon, two single bonds have less enthalpy than one double bond, the reaction is exothermic.

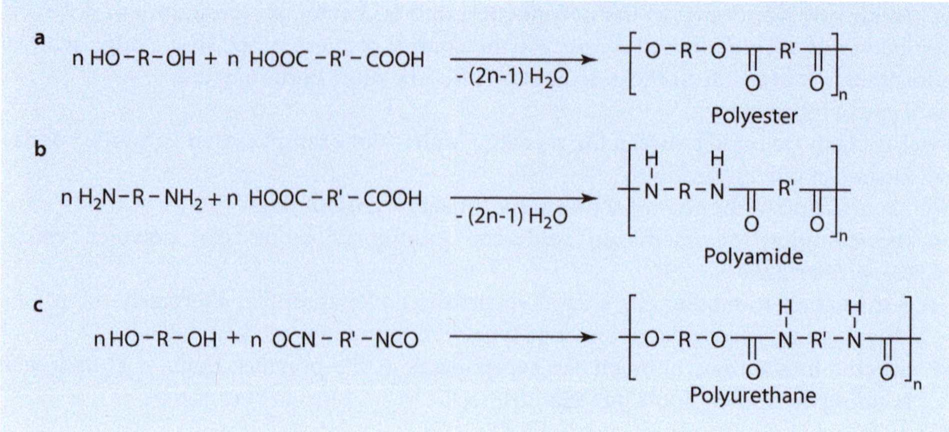

Fig. 1.3 Creation of (**a**) polyester, (**b**) polyamide, and (**c**) polyurethane

Fig. 1.4 Polymerization of an olefinically unsaturated compound

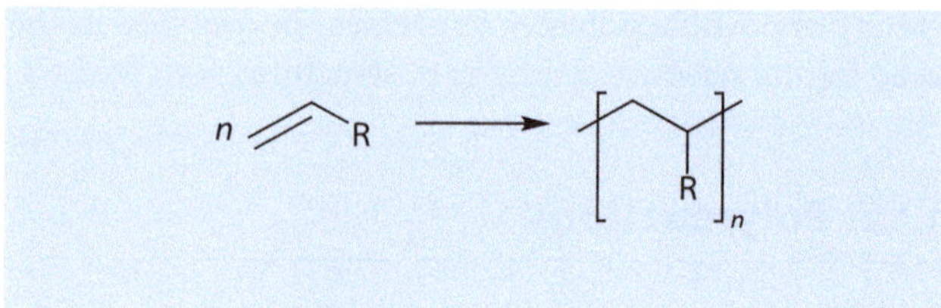

Fig. 1.5 Ring-opening polymerization using the example of ethylene oxide

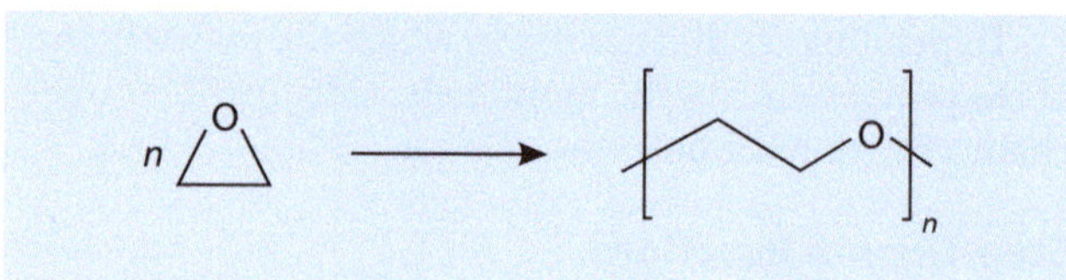

Ring-opening polymerizations also generally involve a chain-growth mechanism (see Fig. 1.5). Here, the driving force is the release of the ring strain. Thus, for instance, three membered rings, such as ethylene oxide (oxirane), undergo facile polymerization.

As with small molecule organic chemistry, chain-growth reactions are classified according to whether they involve radical, positively, or negatively charged reactive species. In these cases, one refers to a *radical*, *cationic*, or *anionic* polymerization, respectively. Another important class of chain-growth polymerizations are those involving transition metal compounds, referred to as *transition-metal catalyzed (catalytic)* or *coordinative* polymerizations.

It follows from Figs. 1.3–1.5 that there are three basic possibilities for the formation of a covalent chemical bond as the result of a polyreaction:

- Addition to a multiple bond
- Elimination of (mostly low-molecular) fragments
- Opening of a ring

These are referred to as *polyaddition*, *polycondensation*, or *ring-opening polymerization*, respectively.

The latter differentiation makes a stringent subdivision of the different types of polymerizations difficult, in particular in the field of chain reactions. Thus, for example, ring-opening polymerization is possible by both catalytic and ionic mechanisms. Likewise, they can occur both via step-growth reactions and chain-growth reactions. However, there is little value in making this distinction between polycondensation and polyaddition and this is not being pursued further here.

This difficulty has been accounted for in this book by the inclusion of a separate chapter on ring-opening polymerization (▶ see Chap. 12), and the basic mechanisms are discussed in the preceding chapters.

A detailed overview of methods for synthesizing macromolecules can be found in ▶ Chaps. 8–12.

1.2.3 Nomenclature of Polymers

The basic principle of polymer nomenclature rests on the designation of the monomer or repeating unit set in brackets, in conjunction with the prefix "poly." For example, poly(styrene) is produced from styrene. Whether one uses the name of the monomer or the repeating unit is rather arbitrary. Thus, for example, the polymerization product of styrene is mostly referred to as poly(styrene), whereas that of ethene, after the repeating unit, usually as poly(ethylene). For reasons of clarity, the brackets are generally omitted with polymers that consist only of a single monomer type (so-called *homopolymers*), and are only used with *copolymers*, i.e., polymers composed of at least two chemically different monomers (▶ see Sect. 1.3.3).

In the formula, the repeating unit is placed in square brackets (◘ see Fig. 1.6). For more complex structures it has also become common practice to indicate with dotted lines that the molecule is further extended (◘ see Fig. 1.9).

If the repeating unit is further altered after polymerization by a chemical reaction (▶ see Chap. 15), the resulting polymer is often designated according to a formal monomer, but one that was not used in the polymerization. For example, the product of hydrolysis of polyvinyl acetate is referred to as polyvinyl alcohol, even though it is not accessible via the direct polymerization of the (unstable) vinyl alcohol (◘ see Fig. 1.7).

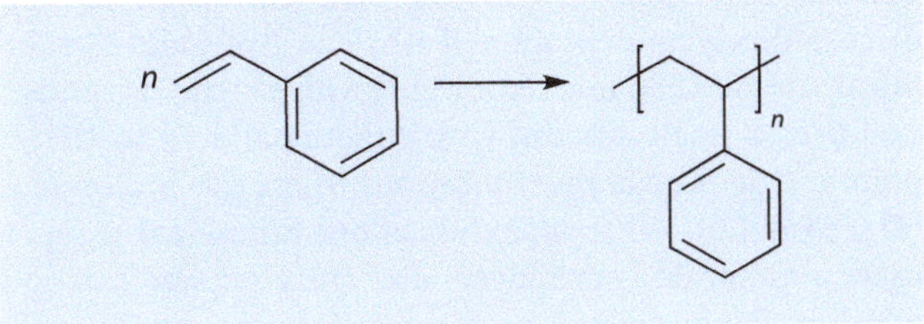

◘ **Fig. 1.6** Polymerization of styrene to polystyrene

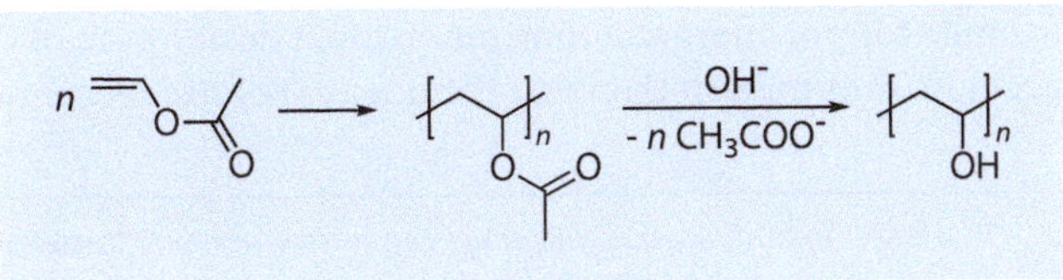

◘ **Fig. 1.7** Synthesis of polyvinyl alcohol from polyvinyl acetate

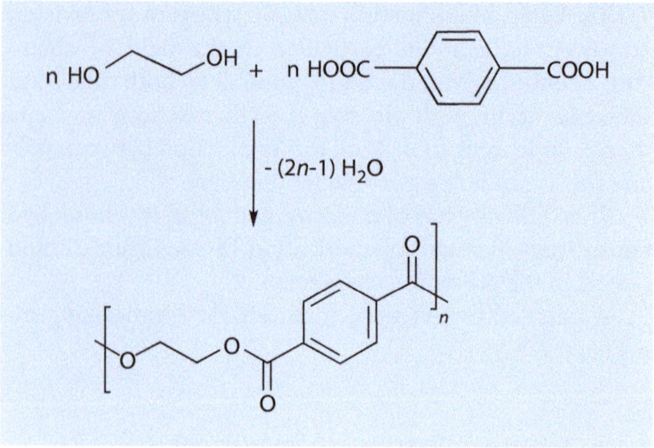

Fig. 1.8 Synthesis of polyethylene terephthalate (PET) from ethylene glycol and terephthalic acid

For polymers resulting from a reaction of a dicarboxylic acid with a diol, common names, based on the name of the corresponding ester, have become usual. For example, the polyester resulting from the complete esterification of ethylene glycol with terephthalic acid (see Fig. 1.8) is referred to as polyethylene terephthalate (abbreviated PET).

1.3 Polymer Architectures

An overview of the different structural principles according to which a polymer can be built is provided below. Several structural variants need to be distinguished:

- Unbranched (linear) and branched polymers
- Isomerism of the repeating units
- Copolymers

1.3.1 Linear and Branched Macromolecules

In addition to their chemical diversity, polymers can also differ substantially from each other in their molecular architecture. The result, as can be readily seen, of the above-discussed polycondensation of ethylene glycol and terephthalic acid (see Sect. 1.2.3), is a linear, unbranched macromolecule with exactly two ends.[3] However, if one was to add, for example, a small amount of trifunctional acid to the monomer mixture, a branching point can develop at those locations where this molecule is incorporated into the polymer (see Fig. 1.9). When an excess of one functional group, for example the OH function, is used, a molecule is produced after the complete reaction of the other functional group which has only alcohol functions as terminal groups which cannot sustain further polycondensation—that is, the polymerization comes to a halt.

In this quite simple architecture it already becomes clear that there are limits to a plain formula for polymers and that, for an exact description of the molecule, one requires more accurate information than just the name(s) of the monomer(s).

3 To a lesser extent, macrocyclic rings can form when one terminal acid group reacts with the terminal alcohol group of the same molecule.

■ **Fig. 1.9** Schematic representation of (**a**) a linear polymer chain and (**b**) a branched macromolecule

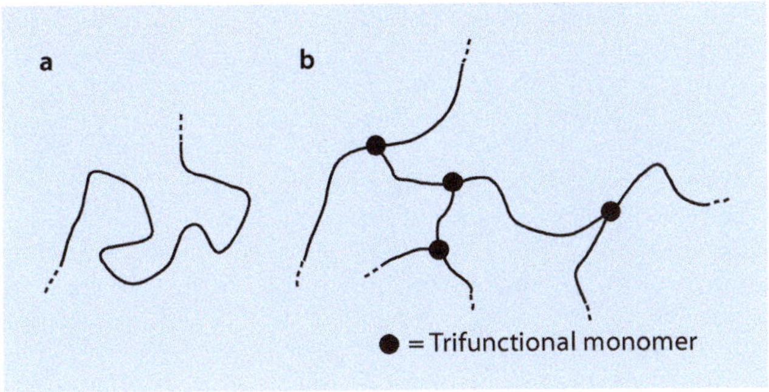

■ **Fig. 1.10** Coexistence of linear, terminal and branched segments in a highly branched polymer (schematic)

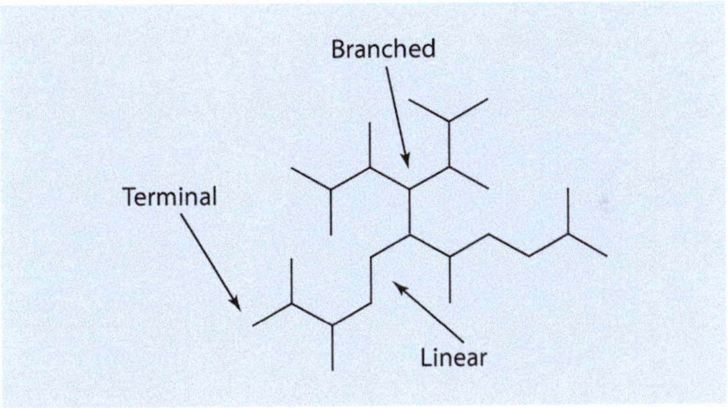

If the number of branches in the macromolecule is increased, one eventually reaches a class of so-called highly branched polymers. These molecules can be notionally divided into linear segments, branches, and so-called *terminal* end segments (■ see Fig. 1.10). An example of such a highly branched polymer is polyethyleneimine, discussed in ▶ Chap. 19.

The greater the ratio of branched to linear segments, the greater the degree of branching of the material. Ideally, there is a perfectly branched, symmetrical macromolecule which grows outward from a central point, resembling a family tree, until the spatial density of branches on the surface of the (generally in good approximation, spherical) molecule is so high that further growth is no longer possible (■ see Fig. 1.11). Because of the similarity with trees, such molecules are referred to, after the Greek word "dendron" for "tree," as *dendrimers* (Buhleier et al. 1978).

Because of the branched polymer's strong outward increase in segment density, its interior contains cavities which can be used for the storage of guest molecules such as dyes or agents. However, the loading density is quite low because of the small number of these cavities. In addition, the synthesis of such perfectly branched systems is only possible in multistep reaction sequences and therefore very time consuming, so dendrimers have, up to now, achieved only limited technical significance (▶ see Chap. 8).

Although highly branched or dendritic polymers represent individual, discrete macromolecules, the synthesis of *polymer networks* leads to the formation of a covalent, branched network within which all of the molecules available in the system are incorporated.

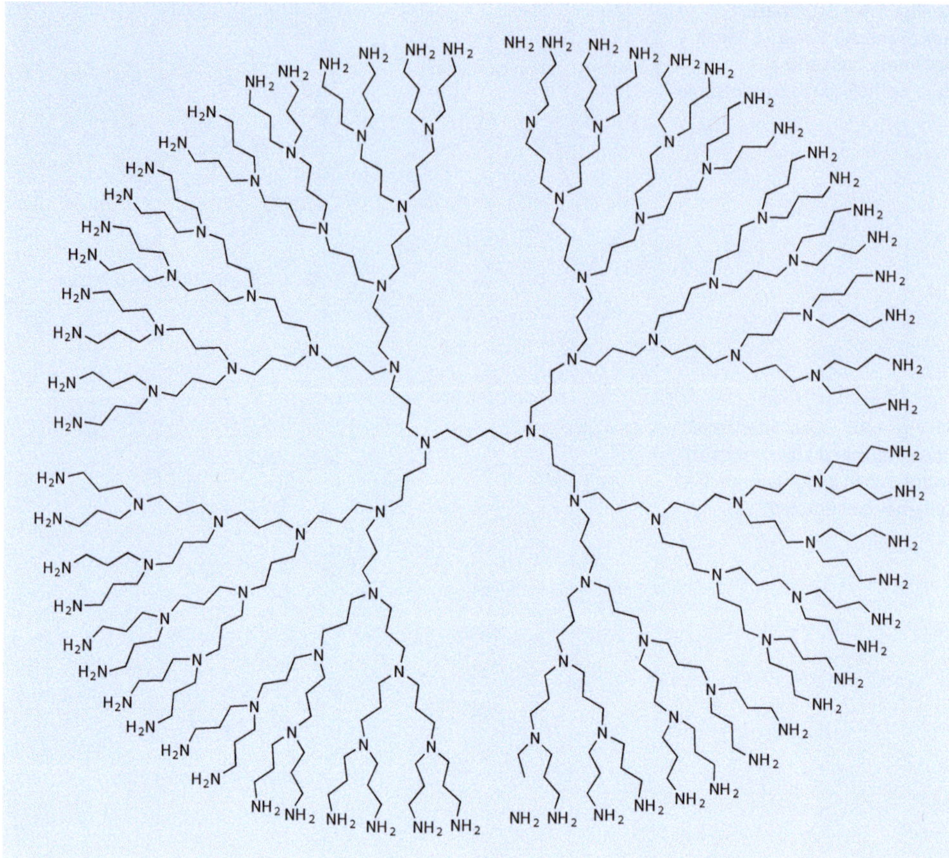

Fig. 1.11 Schematic representation of a dendrimer

Such a network emerges, for example, if higher-functional monomers (i.e., those with more than two reactive groups) are caused to react and the reaction exceeds a certain level of monomer conversion (► see Sect. 8.3.4). Such a network can be described as a single molecule which occupies the entire reaction space (see Fig. 1.12).

1.3.2 Isomerism in Polymers

As with small molecules, macromolecules can also exist as isomers. Here, the following cases, which are particularly important from the viewpoint of the polymer chemist, are *structural-*, *stereo-*, and *conformational isomerism*.

Structural Isomerism

With *structural isomerism* (also referred to as *constitutional isomerism*), the nature of the covalent linkage of the atoms with each other differs, similar to the low-molecular isomers acetone and propionaldehyde. For example, polyvinyl alcohol and polyethylene oxide (see Fig. 1.13) are structurally isomeric polymers.

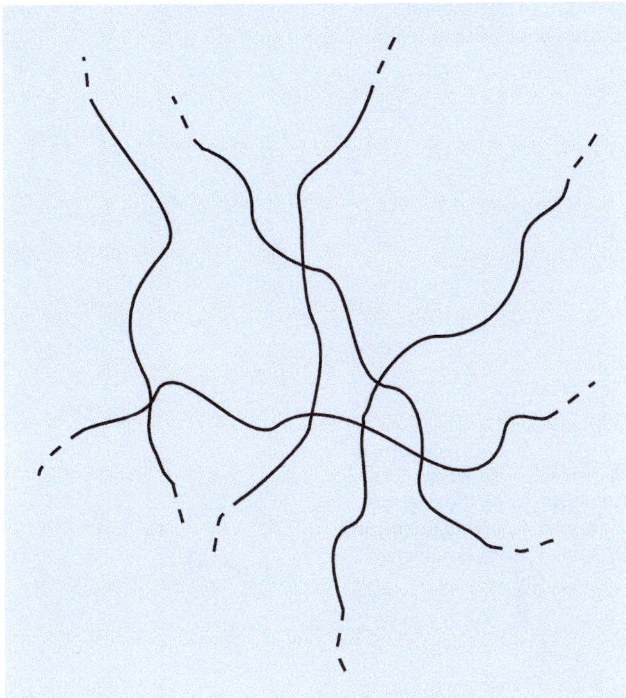

Fig. 1.12 Schematic representation of a polymer network

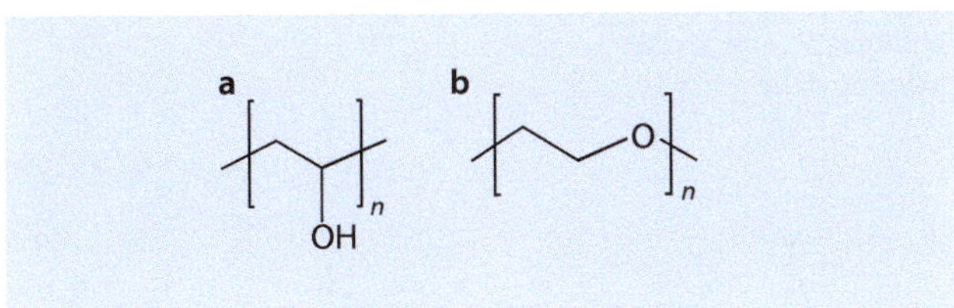

Fig. 1.13 Structural isomerism using the example of (**a**) polyvinyl alcohol and (**b**) polyethylene glycol

An important case of structural isomerism occurs in the polymerization of conjugated dienes such as isoprene (2-methylbut-1,3-diene). During the polymerization of this monomer, as discussed in more detail in ▶ Sect. 14.10, each monomer unit can be added to the macromolecule in different ways. These are shown in Fig. 1.14.

The isomers shown rarely occur as pure forms in practice. Most often the polymers formed are composed of different structural variations within the same polymer chain. In addition to the isomers shown here, 1,4-polyisoprene also exists as *cis* and *trans* isomers (see below).

Stereoisomerism

In analogy to small molecule organic substances, constitutionally identically macromolecules may differ in their spatial form. Notably relevant for polymers are *cis-trans* isomers and enantiomers.

The *cis-trans* isomerism is particularly important for polymers with a double bond in the main chain. The simplest example of this are *cis-* and *trans*-1,4-polybutadiene (Fig. 1.15).

As in the field of low-molecular organic chemistry, they differ in their physicochemical properties. This effect is generally very pronounced in macromolecules. *cis*-1,4-

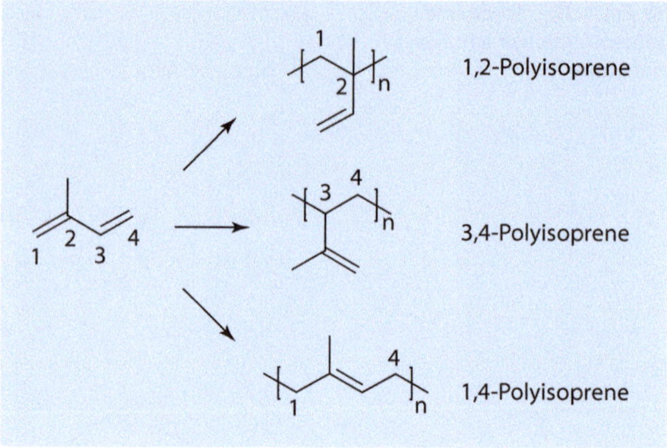

Fig. 1.14 Structural isomers of polyisoprene

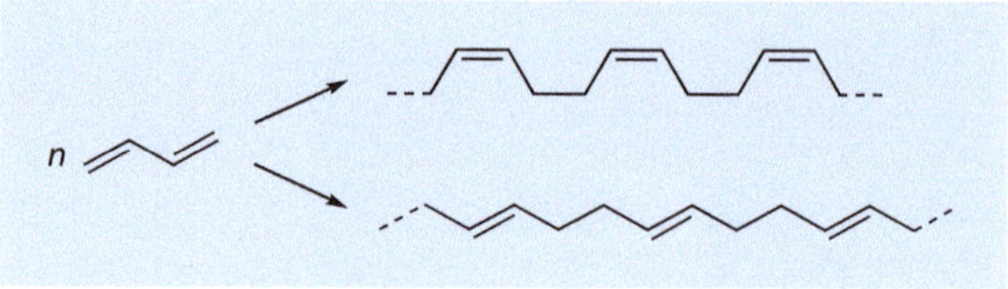

Fig. 1.15 *cis-trans* isomerism using the example of (**a**) *cis*-1,4-polybutadiene and (**b**) *trans*-1,4-polybutadiene

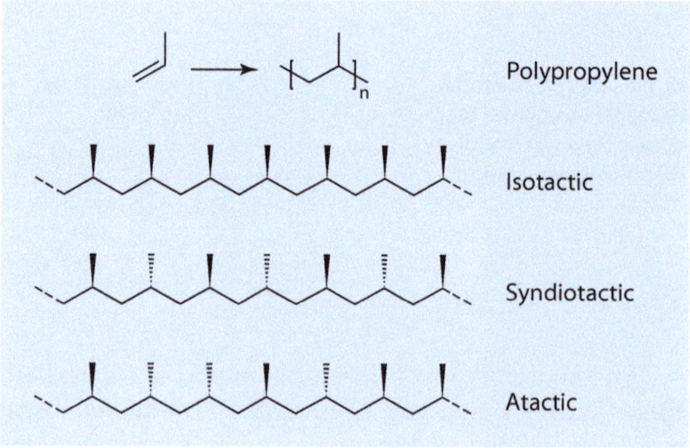

Fig. 1.16 Stereoisomerism using the example of polypropylene: isotactic, syndiotactic, and atactic polypropylene

Polybutadiene (butadiene rubber), for example, is an amorphous, sticky material, whereas the *trans* isomer is a (partially) crystalline material with a high melting point. The situation is similar for the *cis* and *trans* isomers of isoprene (▶ see Sect. 14.10.3). The *cis* isomer is a soft and, in its non-vulcanized state, sticky material which is known as natural rubber, whereas the *trans* isomer, which is referred to as gutta-percha, is a non-sticky and much harder material.

A technically very important case of a polymer exhibiting enantiomerism is polypropylene (PP). In the polymerization of propene (▶ see Chap. 11) a macromolecule is created whose main chain—the so-called polymer backbone—could formally be placed in a plane, whereas the methyl groups of the C_3 carbon point upwards or downwards relative to this plane (◘ see Fig. 1.16). Depending on whether the stereochemistry at the C_2 atom

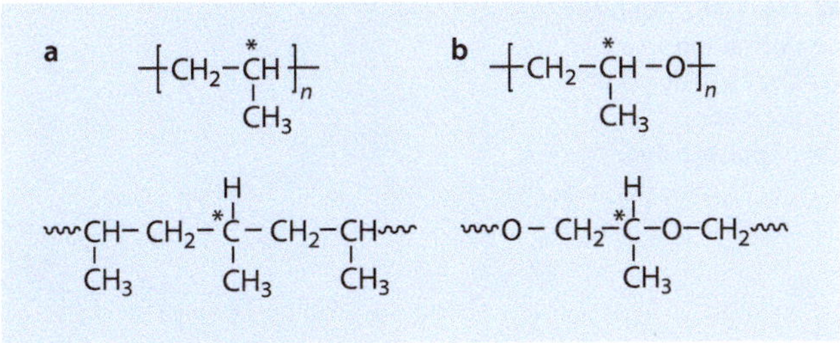

Fig. 1.17 Pseudo-asymmetry using the example of (**a**) polypropylene compared to optical activity in (**b**) polypropylene oxide

is regularly (*stereoregular*) or statistically configured, different stereoisomers result, which are referred to as isotactic, syndiotactic, or atactic polypropylene.

Polypropylene is referred to as isotactic polypropylene (it-PP) if all of its tertiary carbon atoms are configured identically. Syndiotactic polypropylene (st-PP) has an alternating configuration of its methyl groups along the chain, whereas in atactic polypropylene (at-PP) the spatial orientation of the methyl groups is irregular.

The three types of polypropylene mentioned above are diastereomers. Whereas it-PP is a highly crystalline material with a melting point of ca. 160 °C, at-PP cannot crystallize because of its irregular structure and it is an amorphous, sticky material. The control of stereochemistry is thus of considerable importance (► see Chap. 11).

However, in contrast to substances composed of small molecules, stereoregular polyolefins are usually not optically active. This is because the two polymeric chains attached to the chiral center are very large and thus indistinguishable from a physicochemical point of view. A mirror plane can thus be placed approximately through the chiral center. This is also referred to as *pseudo-asymmetry*. In contrast to this, no mirror plane can be placed through the molecule of polypropylene oxide (see Fig. 1.17) because of the additional oxygen atom. Therefore, this macromolecule can exhibit optical activity if one succeeds in producing a stereoregular product and separating the stereoisomers.

Conformational Isomerism

Even with macromolecules, individual molecular groups can rotate around a single bond. Consider the case of polyethylene as the simplest organic polymer. The *trans* conformation is approximately 4 kJ/mol more favorable than the *gauche* conformation. Therefore, the all-*trans* conformation is the enthalpically most favorable conformation. However, the number of possible *trans* and *gauche* conformations of *n* single bonds is 3^n, i.e., a very large number for polymers. For this reason, an all-*trans* conformation is enthalpically favorable, but extremely unlikely, with the result that this conformation is entropically unfavorable. In addition, the enthalpy difference of 4 kJ/mol, which corresponds to about 1.5 times $k_B T$ (k_B: Boltzmann constant, T: absolute temperature) at room temperature, is not particularly large. From simple considerations, it follows that at room temperature the *trans* configuration is only approximately 1.6 times more likely than the *gauche* configuration. A significant number of bonds therefore exist in the *gauche* configuration, and the molecule adopts a spatial arrangement quite different from the extended form, which resembles rather a ball of wool (see Fig. 1.18).[4] Details of this are discussed further in ► Chap. 2.

4 Recent work shows that for isolated chains this picture becomes even more complex: additional forces such as dispersion forces have to be considered, leading to substantial chain-length-dependent influence on the chain architecture; see, e.g., Lüttschwager NOB, Wassermann TN, Mata RA, Suhm MA (2013) The Last Globally Stable Extended Alkane. Angew. Chem. Int. Ed. 52:463–466.

Fig. 1.18 All-*trans* conformation (**a**) and "ball of wool" conformation (**b**) of polyethylene

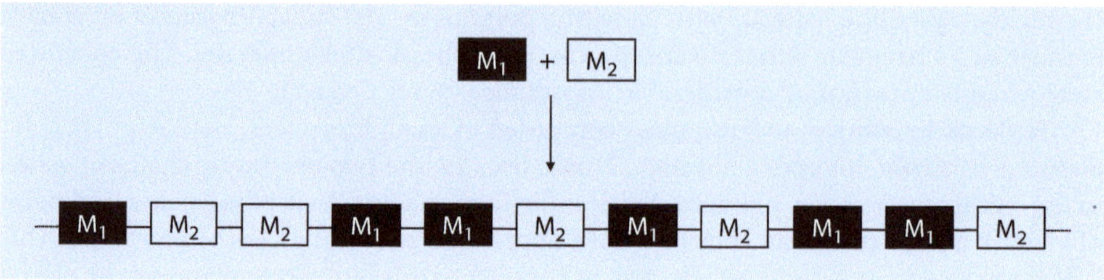

Fig. 1.19 Statistical copolymer consisting of two monomers M_1 and M_2

1.3.3 Copolymers

Copolymers are polymers composed of at least two chemically different monomers. In principle, with copolymerization of different monomers in various quantities, an infinite variety of different macromolecules can be synthesized. This chemical diversity is increased by the various possibilities available for incorporating the *comonomers* into the chain as well as their sequences. These are discussed below.

Statistical Copolymers

If two chemically different monomers M_1 and M_2 are polymerized in a random, statistically determined sequence to each macromolecule, this is referred to as a *statistical copolymer* (see Fig. 1.19).

Here, the incorporation of the monomers into the polymer main chain is determined by the relative reactivity of the monomers and obeys statistical laws (see Chap. 13). The copolymer is referred to as poly(M_1-*stat*-M_2)[5].

---

5 The expression of *random copolymers* (used mostly incorrectly) is often found instead of statistical copolymers. The term *statistical copolymer* should, however, be preferred.

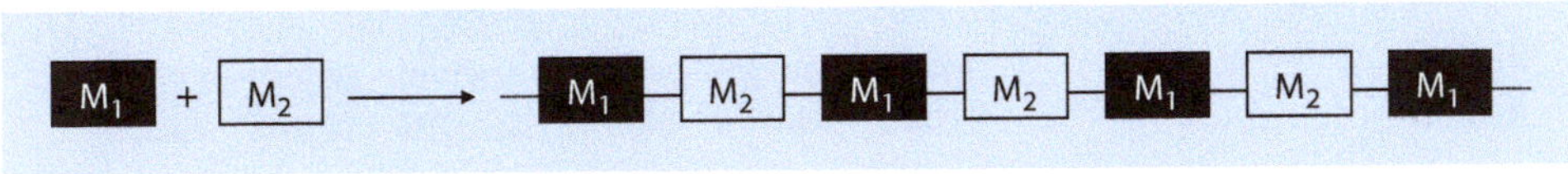

Fig. 1.20 Alternating copolymer consisting of two monomers M_1 and M_2

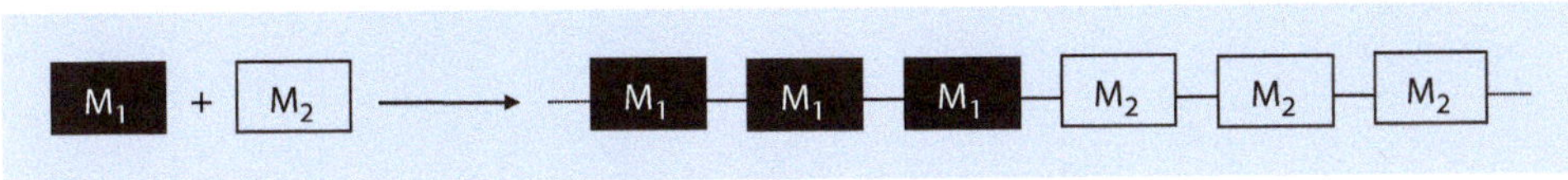

Fig. 1.21 Block copolymer consisting of two monomers M_1 and M_2

Alternating Copolymers

With alternating copolymers, two monomers M_1 and M_2 are built into the polymer main chain in regular alternation. This results in the structure shown in Fig. 1.20.

These polymers are referred to as poly(M_1-*alt*-M_2). Such polymers result from the polymerization of monomers that can react with one another but not with themselves. One example is the polycondensation of a diol and a diacid (see Fig. 1.3). Alternating copolymers result from vinyl polymerizations only when the reactivity of the individual monomers with the other monomer is much greater than with itself. Details are discussed in ► Sect. 13.3.

Block Copolymers

In block copolymers, the comonomers are arranged along the polymer backbone in blocks of consecutive identical monomers (see Fig. 1.21).

Depending on the number of blocks in the polymer backbone, one refers to, for example, di-, tri-, or multiblock copolymers. The nomenclature for these macromolecules is PolyM_1-*block*-PolyM_2, as, for example, in polystyrene-*block*-polybutadiene. Colloquially, the designation Poly(M_1-*block*-M_2) has become common; e.g., poly(styrene-*block*-butadiene).

Graft Copolymers

Polymers in which a side chain of M_2 is linked to the main chain of M_1 are called *graft copolymers* of the monomers M_1 and M_2 (see Fig. 1.22). These macromolecules are referred to as PolyM_1-*graft*-PolyB. Here, PolyM_1 is the main chain, the so-called *graft stock*, and PolyM_2 is the grafted side chain.

Copolymers for which only the monomers are to be designated but in whose name no particular copolymer architecture is to be specified are generally referred to as Poly(M_1-*co*-M_2).

The following chapters present a deeper discussion of the properties and the synthesis of polymeric materials. This is structured as follows:

- ► Chapters 2 and 3 present the size and shape of polymers, as well as methods for determining their molar mass
- In ► Chapters 4–7, the properties of solid polymeric materials are discussed
- ► Chapters 8–13 are devoted to the synthesis of polymers
- ► Chapters 14–17 deal with important classes of polymers, chemical reactions of polymers, and procedures for the manufacture and processing of macromolecules

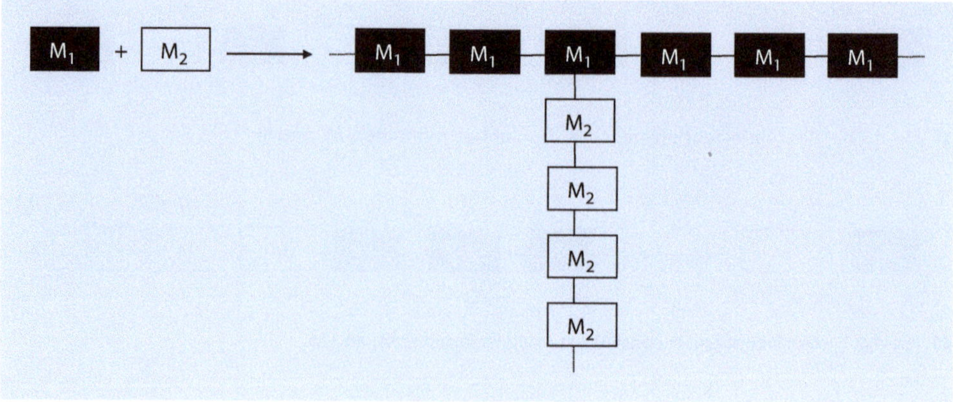

Fig. 1.22 Graft copolymer consisting of two monomers M_1 and M_2

- ▶ Chapters 18–20 are dedicated to special classes of materials (elastomers, functional polymers, and liquid crystalline polymers)
- The textbook concludes with some environmental aspects of synthetic macromolecules (▶ see Chap. 21) and a brief presentation of current trends in polymer chemistry (▶ see Chap. 22)

References

Staudinger H (1926) Die Chemie der hochmolekularen organischen Stoffe im Sinne der Kekuléschen Strukturlehre. Ber Dtsch Chem Ges 59:3019–3043

Naarmann H, Theophilou N (1987) New process for the production of metal-like, stable polyacetylene. Synth Met 22:1–8

Buhleier E, Wehner W, Vögtle F (1978) "Cascade"- and "nonskid-chain-like" syntheses of molecular cavity topologies. Synthesis 1978:155–158

Polymers in Solution

S. Koltzenburg et al., *Polymer Chemistry*, DOI 10.1007/978-3-662-49279-6_2

Polymers, especially when compared with the monomers from which they are built, have a number of special properties. For example, polymers such as starch and polypropylene oxide are much less soluble in water than their monomers, glucose and propylene oxide. Another observation is that many polymers absorb solvents or water without themselves dissolving. Thus, cotton socks, for instance, absorb water without disintegrating when they are washed in a washing machine. To explain and to be able to describe such properties, this chapter is devoted to a description of the polymeric chain structure and the consequences thereof for polymers in solution. Furthermore, the thermodynamics of polymer solutions are discussed and compared with those of small molecules to develop an understanding of the differences in solubility mentioned above.

2.1 Chain Models

The structure of a polymer and the characteristics that can be derived from it can best be visualized by using a chain model. In this model the repeating units are the chain links, which are (in the simplest linear case) connected together to form a chain. In the case of a polymer chain in which all elements are connected in a *trans*-conformation, the simplified image in Fig. 2.1 emerges.

The "*contour length*," l_{cont}—the complete length of a chain with n links of length l, including all chain elements— is

$$l_{cont} = n\,l \tag{2.1}$$

If one considers solely the three most probable conformational possibilities of every $-CH_2-$ entity relative to its direct neighbors, two *gauche*- and one *trans*-conformations, this already gives 3^{n-1} different conformations for the complete polymer chain. Only one of these is the *all-trans*-conformation; thus its actual occurrence is highly improbable. Despite this, statistical methods are used to describe the dimensions of polymer chains as realistically as possible and thereby predict their behavior.

The so-called *Gaussian chain* is a random arrangement of the segments following the erratic flight or *random walk* model and assuming free and unimpeded rotation. Every repeat unit is thus connected to the next by an arbitrary angle. Figure 2.2 shows schematically the basic features of this model.

A statistical treatment is based on the size of the mean end-to-end distance $\vec{R}$.[1] It follows from this that

$$\vec{R} = \sum_{i=1}^{n} \vec{l}_i \tag{2.2}$$

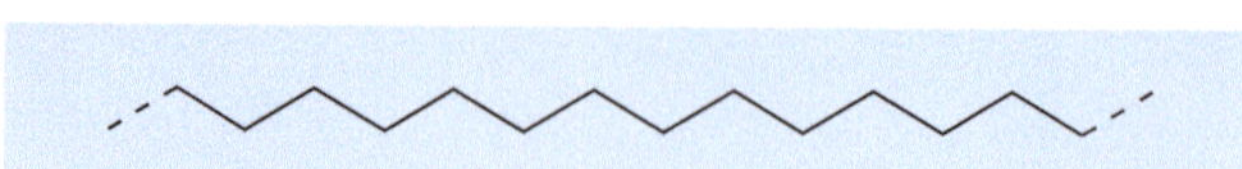

Fig. 2.1 Model of a polyethylene chain in which the $-CH_2-$ units are symbolized by kinks. The basis of this representation is an *all-trans*-conformation (bond angle = 109.47°)

1 For reasons of consistency with the most popularly used nomenclature, the symbol R is used for this variable; needless to say, it should not be confused with the universal gas constant R.

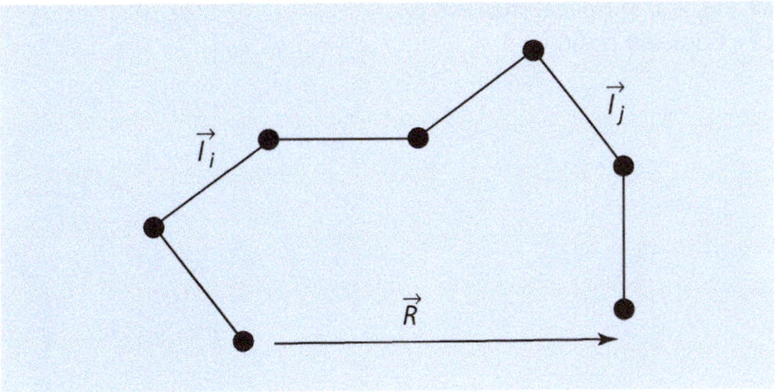

Fig. 2.2 The Gaussian chain model with different repeating units (e.g., $\vec{l}_i$ and $\vec{l}_j$), which are described by vectors according to their spatial position. $\vec{R}$ is a measure of the distance between the chain ends

If a chain end is randomly placed at the point of origin of a one-dimensional coordinate system, the probability $P(x,n)$ of finding the other end of the chain with n links of the length l at a distance x in the interval dx can be described by a Gaussian distribution. This corresponds to the observation that, in the case of random walk, all directions are equally probable. Thus it holds (without proof) that

$$P(x,n) = \sqrt{\frac{2}{nl^2\pi}} \exp\left(-\frac{x^2}{2nl^2}\right) dx \tag{2.3}$$

Essentially, (2.3) includes the exponential term of a Gauss function as well as a pre-exponential factor. The latter ensures that the integral of this probability distribution is 1 for all values of x because the probability of finding the second chain end somewhere is obviously 1 or 100 %.

In three dimensions, this produces the following result (again without proof):

$$P(x,y,z,n) = \left(\frac{2\pi}{3} nl^2\right)^{-\frac{3}{2}} \exp\left(-\frac{3}{2}\frac{x^2+y^2+z^2}{nl^2}\right) dx \cdot dy \cdot dz \tag{2.4}$$

This expression corresponds to the probability of finding the other chain end in a cubic element of volume $dx \cdot dy \cdot dz$. However, if we observe the model not in the Cartesian system but within spherical coordinates to establish the probability of finding the chain end in a spherical shell of a sphere of thickness dr at a distance r from the center of the sphere (Fig. 2.3), it follows that

$$P(r,n) = \left(\frac{2\pi}{3} nl^2\right)^{-\frac{3}{2}} \exp\left(-\frac{3}{2}\frac{r^2}{nl^2}\right) 4\pi r^2 dr \tag{2.5}$$

To be able to make statistically significant assertions, we calculate (comparable to an error calculation) using the corresponding squared values. This can be visualized by looking at $P(r)dr$ (Fig. 2.4).

Fig. 2.3 Spherical shell model of a Gaussian chain

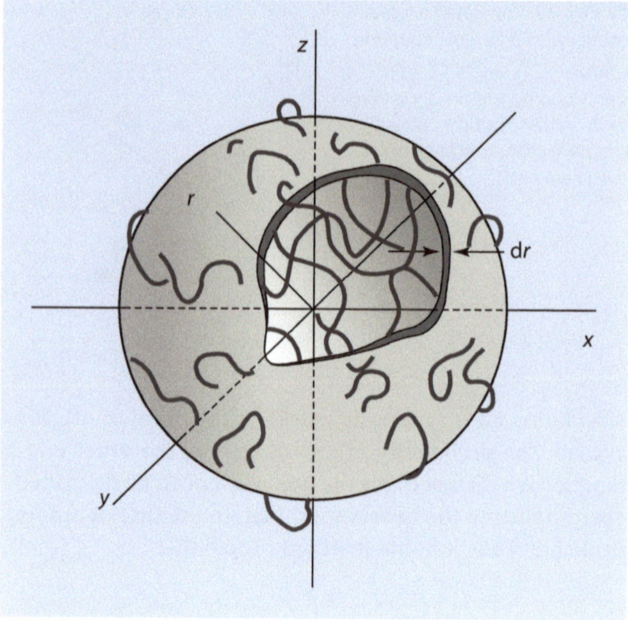

Fig. 2.4 Representation of $P(r)$ dr *for* $l = 0.25$ nm. (*Dashed line*: $n = 100$, *solid line* $n = 1000$)

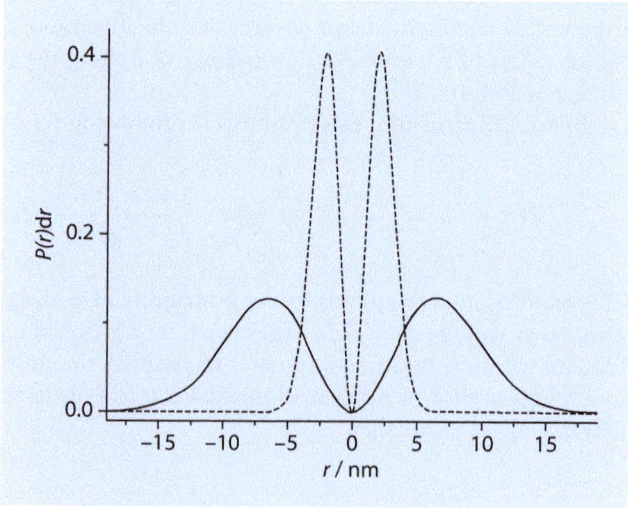

In this case, a simple average would yield zero, which of course would not reflect reality, because both chain ends cannot be in the same place—i.e., the origin of the coordinate system—at the same time.

As can be seen in Fig. 2.4, the segment's density tends to zero at the origin. It initially increases outwards before decreasing again at very large distances, a progression that would be intuitively expected.

As the number of chain links n tends to infinity, the mean square end-to-end distance is

$$\left\langle R^2 \right\rangle = \int_0^\infty P(r,n) r^2 dr \tag{2.6}$$

The solution of the integral in (2.6) yields (without proof)

$$\left\langle R^2 \right\rangle = nl^2 \tag{2.7}$$

When $\langle R^2 \rangle$ is compared to the contour length l_{cont}, it becomes clear that the coiling of the polymer chain has an enormous effect on the dimensions of the chain. If we consider, for example, a polyethylene chain with a repeat unit length $l = 0.25$ nm and 10,000 units, the contour length, as given by (2.1), $l_{cont} = 2500$ nm, and the mean end-to-end distance according to the *random walk* model, $\sqrt{< R^2 >} = 25$ nm. Thus, $\sqrt{< R^2 >}$ is a factor of 100 shorter than l_{cont}.

A more realistic description of a polymer chain compared to that given by a Gaussian chain can be obtained if both the fixed connecting angles and the limited and restricted rotation around the bonds are taken into account. These considerations lead to an expansion of the polymer coil as compared to the Gaussian chain.

If we consider a fixed bond angle θ for the square of the mean square end-to-end distance in accordance with the so-called *valence bond model* (VBM), it follows (without proof) that

$$\left\langle R^2 \right\rangle_{VBM} = nl^2 \frac{1 - \cos\theta}{1 + \cos\theta} \tag{2.8}$$

In the case of a carbon–carbon main chain, a bond angle of 109.47° yields

$$\left\langle R^2 \right\rangle_{VBM} = 2nl^2 \tag{2.9}$$

Thus, taking a fixed bond angle into account leads to a doubling of the mean square end-to-end distance. If one also takes into account the restricted rotation at the angle Φ with non-independent potential, i.e., the preference of *trans* and *gauche* states, one arrives at the so-called *Rotational-Isomeric-State-Model* (*RIS*) (Flory 1953; Painter and Coleman 2009). In this model, the mean square end-to-end distance (without proof) is

$$\left\langle R^2 \right\rangle_{RIS} = nl^2 \left(\frac{1 - \cos\theta}{1 + \cos\theta} \right) \left(\frac{1 + \langle \cos\Phi \rangle}{1 - \langle \cos\Phi \rangle} \right) \tag{2.10}$$

Thus, compared to (2.8), taking account of the *trans* and *gauche* "wells of potential," which are expressed by the statistical mean value of the cosine Φ, leads to a further expansion of the chain as compared to the *valence bond model.*

If these differing values of the mean square end-to-end distance are considered in relation to the Gaussian chain, one can derive fundamental information without having to revert to the physically more correct, but mathematically considerably more complex, models. However, with the current models a generally valid description of the dimensions of real polymer chains in undisturbed states is not possible.

An important value is the *characteristic ratio* c_∞, which describes the expansion and therefore the stiffness of the real polymer as compared to the Gaussian chain (2.10). The influence of solvents is not considered and existing interactions are considered to be mutually neutralized (*unimpeded dimensions*). The larger this characteristic ratio (and thus the corresponding square of the mean square end-to-end distance $\langle R^2 \rangle_{real}$ of the real polymer), the more rigid the polymer.

$$\frac{\langle R^2 \rangle_{real}}{n \cdot l^2} \equiv c_\infty \tag{2.11}$$

Of course the assumption of free bond angles makes the Gaussian chain model a gross simplification of real polymers. By avoiding a more detailed structure, the model can serve as a simplification. Where the characteristic ratio is known from experiments, such as size exclusion chromatography (SEC) under θ conditions (▶ Chap. 3), the model is perfectly adequate to give an approximate description of real polymers. In a development of this model by Kuhn, the length l_k no longer represents that of a repeat unit, but is generally longer. Common to both the real chain and the so-called *Kuhn chain*, which is constructed of $n_k < n$ segments of length $l_k > l$, are the contour length l_{cont} and the real chain end-to-end distance. ◘ Figure 2.5 shows the model schematically.

Thus, l_{cont} is given by

$$l_{cont} = n_k \cdot l_k = n \cdot l \tag{2.12}$$

If one considers the real chain and compares the Kuhn model with the Gaussian chain, one obtains

$$\frac{\langle R^2 \rangle_{real}}{n \cdot l^2} = \frac{\langle R^2 \rangle_k}{n \cdot l^2} = \frac{n_k \cdot l_k^2}{n \cdot l^2} = \frac{l_k}{l} \cong c_\infty \tag{2.13}$$

The example of polyethylene ($n > 100$, $l = 0.25$ nm) provides a useful illustration. As a simplification, one assumes that each segment exists independently of its neighboring segment

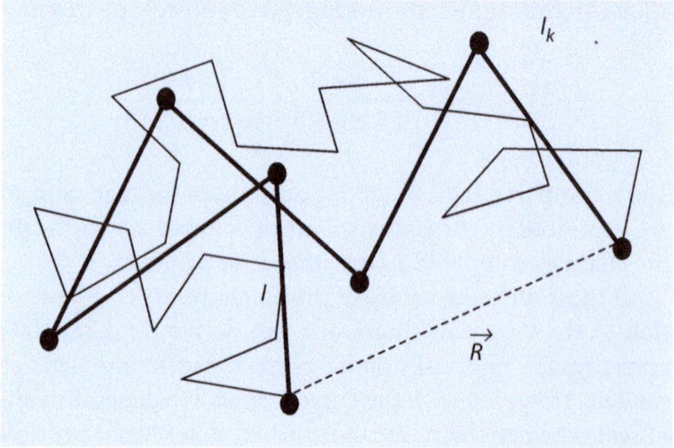

◘ **Fig. 2.5** Schematic representation of the Kuhn model and the corresponding real chain. *Bold* = segment length l_k (Kuhn model), *not bold* = the elements of the real chain

in the *gauche* or *trans* state. In this case, the *gauche-trans* energy levels are statistically independent (unconnected) and it follows from theoretical calculations that

$$\frac{\left\langle R^2 \right\rangle_{VBM}}{n \cdot l^2} = 3.7 \tag{2.14}$$

However, experiments yield

$$\frac{\left\langle R^2 \right\rangle_k}{n \cdot l^2} = 6.7 \tag{2.15}$$

Thus, the mutual influence (coupling) of the conformations cannot be ignored. Only a model considering the linked and restricted rotation is capable of describing the expansion of the real polymer coil compared to the Gaussian chain. With (2.13) the Kuhn length $l_k = 1.68$ nm.

In other words, it is possible to calculate the mean square end-to-end distance from the contour length and the Kuhn length directly:

$$\left\langle R^2 \right\rangle_k = l_k l_{cont} \tag{2.16}$$

2.2 Chain Stiffness

An additional unit of measure for the flexibility of a polymer bundle is the *persistence length* l_p. The model describes a chain with an infinite number of bond vectors $\vec{l}_i$, each of identical length. The projection length of each of the bond vectors $\vec{l}_j$ with $j > i$ onto $\vec{l}_i$ is given by the average of all conformations, $lcos\theta_{i,j}$. This can be graphically interpreted as the portion of all $\vec{l}_j$ that "point" in the direction of $\vec{l}_i$ and can thus be used as a measure of the correlation of the orientation of two independent segments of the polymer chain. Here $\theta_{i,j}$ denotes the angle between the two bond vectors. The so-called *persistence length* is defined as the sum of all the bond vector projections over all conformations on the first vector $\vec{l}_i$:

$$l_p = l \sum_{j=i+1}^{\infty} \left\langle \cos \theta_{i,j} \right\rangle \tag{2.17}$$

The initial summands in this expression make a finite contribution to the persistence length. After a certain separation between the bond vectors $\vec{l}_i$ and $\vec{l}_j$ they become independently orientated in space. Thus $\langle cos\theta_{i,j} \rangle$ converges for these bond vectors to zero and the persistence length becomes a finite value defined by those segments immediately following $\vec{l}_i$. For a stiff polymer, l_p assumes a large value as a result of the considerably restricted rotation. For a statistical coil, l_p has a small value. The persistence length is thus a measure for the stiffness of the coil.

For the limiting case of a bond angle of $\theta = 180°$, the macromolecule forms a straight chain and the mean end-to-end distance is equivalent to the contour length. A mathematical simplification suggests applying the angle complementary to 180°. For $\theta = 180°$ this is $\theta' = 0°$. Although there are real polymers which have a small value for θ', the angle is always greater than zero. An example of this is a double-strand DNA helix. To describe the characteristics of this chain, considerably fewer and correspondingly longer Kuhn segments would be required than would be necessary for a more flexible polymer.

At this stage, the fundamental difference between the contour length l_{cont} and the persistence length l_p should be noted. The persistence length basically depends on the chain stiffness of the polymer in question. Hence, it is a material characteristic closely associated with the chemical composition of the polymer; it does not depend on the molar mass of the particular chain. For this reason alone is it possible, for example, to define the (single) characteristic persistence length of a polymer such as polyethylene. In contrast to this, the contour length is very heavily dependent on the number of repeat units making up the polymer—as can be immediately seen from (2.1); a doubling of the molar mass leads to a doubling of the contour length. Therefore in specific cases it can be that the persistence length is greater than the contour length for a very stiff polymer with a relatively small molar mass, even if this seems to be in conflict with (2.17) (persistence length as projection of all bond vectors onto the first).

The so-called *wormlike-chain-model* (*WLC*) is a simplified way of considering the chain stiffness of semi-flexible polymers. This mathematical model was first described in 1949 (Kratky and Porod 1949). The persistence length l_p was defined as

$$l_p = \frac{2l}{\theta'^2} \tag{2.18}$$

and the auxiliary parameter p :

$$p \equiv l_p / l_{cont} = \left(2 / n\theta'^2\right) \tag{2.19}$$

This auxiliary parameter allows a dimensionless description. This demonstrated that for the mean square end-to-end distance, it holds (without proof) that

$$\left\langle R^2 \right\rangle_{WLC} = l_{cont}^2 \left[2p - \left(\frac{1}{n}\right) - 2p^2 \left(1 - \exp\left(-\frac{1}{p}\right)\right)\right] \tag{2.20}$$

It should be borne in mind that the persistence length is a mathematical paraphrase of the stiffness of the polymer chain. In limiting cases, it tends towards zero (completely flexible chains) or towards infinity (completely stiff rods). Given this, two limiting cases can be distinguished. In the case of extremely stiff chains with a persistence length greater than the contour length, it holds that $p \gg 1$. As a visual example, one could imagine uncooked spaghetti (which could be even longer without doubling up as a chain does). If we imagine spaghetti of immense length, the description would segue into the second case, as presented below. The exponential function can then be expressed as a Taylor development and ignored after the second term. This gives the following relationship between the end-to-end chain distance and contour length:

$$\left\langle R^2 \right\rangle_{WLC} \approx l_{cont}^2 \left[2p - 2p^2 \left(\frac{1}{p} - \frac{1}{2p^2}\right)\right] = l_{cont}^2 \tag{2.21}$$

With this function, stiff chains, for example, rods can be described.

In the second case, we are concerned with molecules that are relatively stiff but have such a large number of segments that the contour length far exceeds the persistence length. An example of this is DNA, mentioned above. DNA is very stiff and has an extremely high contour length. For such molecules $p \ll 1$ which means that p^2 is much smaller than p and, because n is extremely large, $1/n$ is also very small so that (2.20) can be simplified to

$$\left\langle R^2 \right\rangle_{WLC} = 2pl_{cont}^2 = 2l_p l_{cont} \tag{2.22}$$

Making use of the results of the Kuhn chain statistics (2.16) and assuming that $\left\langle R^2 \right\rangle_{WLC} = \left\langle R^2 \right\rangle_k$:

$$l_k = 2l_p \tag{2.23}$$

This restates that the persistence length and the Kuhn length are a measure of the stiffness of polymer chains.

2.3 Entropy Elasticity

The Gaussian chain is the (simplified) state of equilibrium for a chain and therefore represents the energy minimum. As a consequence, to expand a Gaussian chain, energy must be introduced. By stretching the chain we bring it closer to the *all-trans*-conformation, which is the most favorable for flexible chains with regard to the enthalpy, but the most unfavorable formation in terms of entropy. As the number of possible conformations is less in stretched states than before, the entropy decreases. If one takes the Gaussian model (with n chain links of length l) as a basis, the change in entropy ΔS during expansion can (without proof) be expressed with the assistance of the mean end-to-end distance R as follows:

$$\Delta S(R) \propto \frac{3}{2} k_B \left(1 - \frac{R^2}{nl^2}\right) \tag{2.24}$$

Herein, k_B denotes Boltzmann's constant. It can be seen that the further the distance R deviates from the result of the Gaussian model, the greater the loss of entropy associated with chain expansion.

With use of the Gibbs–Helmholtz equation

$$\Delta G = \Delta H - T\Delta S, \tag{2.25}$$

it follows that

$$\Delta G(R) - \Delta H \propto -\frac{3}{2} k_B T + \frac{3}{2} k_B T \frac{R^2}{nl^2} \tag{2.26}$$

In practice, it can be assumed, as a reasonable approximation, that the contribution of the change in enthalpy ΔH to the total free enthalpy ΔG during the stretching of a polymer chain at room temperature (or above) is considerably less than the value $T \cdot \Delta S$ associated with the change in entropy. As a result, ΔH can be ignored for the following consideration as a first approximation.

The returning force F of an elongated coil results from the change in the free energy at the absolute temperature T as a function of R. Consequently, for small deflections we find that

$$R \ll l_{cont} = n \cdot l;$$
$$F = \frac{\partial \Delta G}{\partial R} \propto \frac{3k_B T}{nl^2} R \tag{2.27}$$

The resilience of a polymer chain, which in macroscopic terms corresponds to the resistance of a polymeric material to deformation, depends on the entropy of the chain and is thus proportional to the mean end-to-end distance between chain ends.

2.4 Thermodynamics of Polymer Solutions

In chemical terms, a polymer melt is composed of identically structured molecules. Moreover, density fluctuations are small enough to be ignored. The chemical environment a molecule "sees" is therefore the same throughout the whole melt. Such a situation can be described by an averaged, so-called *mean field*. In this scenario, the dimensions of the polymer are not affected by interactions, for example, with neighboring, chemically dissimilar molecules.

In contrast, in dilute polymer solutions, the density, chemistry, and especially the chemical potential of the system fluctuate as a function of location. In such dilute solutions, the polymer coils are separated by "empty," solvent filled spaces. Moreover, as elucidated in ▶ Sect. 2.1, the segment density within the polymer coils is not homogeneously distributed. In contrast to the melt, a "non-mean field" state prevails. As a result, interactions are present that can influence the conformation of the polymer. To understand the behavior of polymer solutions, it is necessary to consider their thermodynamics.

A polymer solution consists of the dissolved polymer and at least one solvent. Thus, in the simplest case we are dealing with a two-component system. A system consisting of a greater number of components might be a solvent mixture or a solution of more than one polymer.

2.4.1 Ideal and Real Solutions

The characteristic ratio has already been introduced in ▶ Sect. 2.1. It can also be called the ratio of the *undisturbed dimension* of a real polymer coil in relation to the ideal Gaussian chain. The undisturbed dimension is derived directly from the molecular structure of the polymer. This state is the one most probable in the melt, in which the polymer molecules are only surrounded by essentially similar molecules. If, however, the polymer is dissolved in a solution, additional interactions are present that can influence its shape. In a good solvent, the interactions between the polymer and the solvent are favored over the interactions of the individual components. As a result, the polymer coil expands. In contrast, in a poor solvent the coil collapses. The transition from a good to a poor solvent, a so-called *θ-solvent*, presents a special case. In such solvents the interactions between the polymer segments are equally prevalent as those between the polymer and solvent molecules. For any one set of substances this state is normally only present at a specific temperature, the so-called *θ-temperature*. Details of this are discussed further below.

The thermodynamic properties of a polymer solution are determined by the interaction between the solvent and the dissolved substance, the so-called solvent quality. The solvent quality varies considerably for the same solvent, depending on the dissolved polymer. For example, water is a very good solvent for polyethylene oxide (PEO), whereas polystyrene is barely soluble in water. To understand the thermodynamics of polymer solutions, just as with small molecule solutions, one needs to consider the Gibbs–Helmholtz equation:

$$\Delta G^m = \Delta H^m - T\Delta S^m \tag{2.28}$$

Here ΔG is the Gibbs free enthalpy, ΔH the enthalpy, T the absolute temperature, and ΔS the entropy. The index m refers to the respective values for the mixture.

For small molecule mixtures, depending on the quality of the solvent, the following limiting cases can be identified.

Ideal Solution

The special case of an ideal solution arises if the process of mixing with the substance being dissolved does not lead to any change in enthalpy (heat) and the entropy of the mixture is expressible using the simple statistical approach of the so-called *ideal entropy of mixing*. The latter is derived in detail further below. Thus

$$\Delta H^m = 0 \text{ and} \tag{2.29}$$

$$\Delta S^m = \Delta S^m_{ideal} \tag{2.30}$$

Athermal Solution

During the production of an athermal solution, no heat of reaction is observed $(\Delta H^m = 0)$. However, the entropy change deviates from the statistically derived ideal entropy of mixing ΔS^m_{ideal}, the deviation being denoted by ΔS^m_{excess}. Thus

$$\Delta H^m = 0 \text{ and} \tag{2.31}$$

$$\Delta S^m = \Delta S^m_{ideal} + \Delta S^m_{excess} \tag{2.32}$$

Regular Solution

The regular solution is associated with a heat of solution ΔH^m_{excess}, but the change in entropy is ideal:

$$\Delta H^m = \Delta H^m_{excess} \text{ and} \tag{2.33}$$

$$\Delta S^m = \Delta S^m_{ideal} \tag{2.34}$$

Real (Irregular) Solution

Finally, for real, sometimes called irregular solutions, a heat of solution and a deviation from the ideal entropy of mixing can be observed:

$$\Delta H^m = \Delta H^m_{excess} \quad \text{and} \tag{2.35}$$

$$\Delta S^m = \Delta S^m_{ideal} + \Delta S^m_{excess} \tag{2.36}$$

θ Conditions

In this special case, the mixture behaves in a pseudo-ideal fashion. Under these conditions, the free enthalpy of mixing is zero; the enthalpy of mixing and the term including the temperature and the entropy of mixing are equal:

$$\Delta H^s = T\Delta S^m \tag{2.37}$$

The effect that the solvent quality has on the polymer coil's dimensions is shown schematically in Fig. 2.6.

In the following section we concern ourselves with describing those regular solutions which are especially important in practice.

2.4.1.1 Solutions of Lower Molecular Substances

A lattice model is suitable for describing a solution of a small molecule (Fig. 2.7).

If it is assumed that the respective values of molecular and cellular size are identical, the respective volume fraction φ_i is equal to the corresponding mole fraction, given by the respective number N_i:

$$\varphi_i = \frac{V_i}{V_1 + V_2} = \frac{N_i}{N_1 + N_2} \tag{2.38}$$

Here the index i can either describe the solvent (index 1) or solute (index 2). For the case shown in Fig. 2.7, the ideal entropy of mixing can be determined from the number of possible combinations of positions W:

$$\Delta S^m_{ideal} = k_B \ln W \tag{2.39}$$

Statistically, W amounts to (without proof)

$$W = \frac{N!}{N_1!N_2!} \tag{2.40}$$

Because the sum of N is determined by N_1 and N_2, with the aid of the Sterling approximation

$$\ln N! \approx N \ln N - N \tag{2.41}$$

and the following simplification can be made:

$$\Delta S^m_{ideal} = k_B N\left(\ln N - \frac{N_1}{N}\ln N_1 - \frac{N_2}{N}\ln N_2\right) \tag{2.42}$$

For 1 mol (corresponds to N_A molecules, where N_A is Avogadro constant) with the introduction of the volume fraction and the universal gas constant $R = k_B N_A$, it follows that

$$\Delta S^m_{ideal} = R\left(\ln N - \frac{\varphi_1 N}{N}\ln \varphi_1 N - \frac{\varphi_2 N}{N}\ln \varphi_2 N\right) \tag{2.43}$$

$$\Delta S^m_{ideal} = R\left(\ln N - \varphi_1 \ln \varphi_1 - \varphi_1 \ln N - \varphi_2 \ln \varphi_2 - \varphi_2 \ln N\right) \tag{2.44}$$

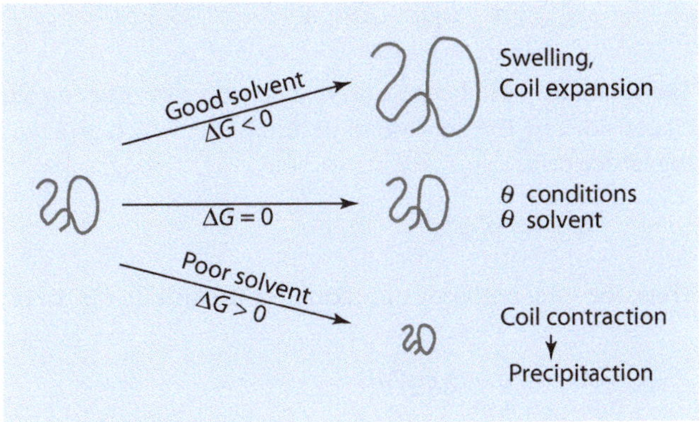

Fig. 2.6 Schematic representation of polymer coils in solvents of different quality

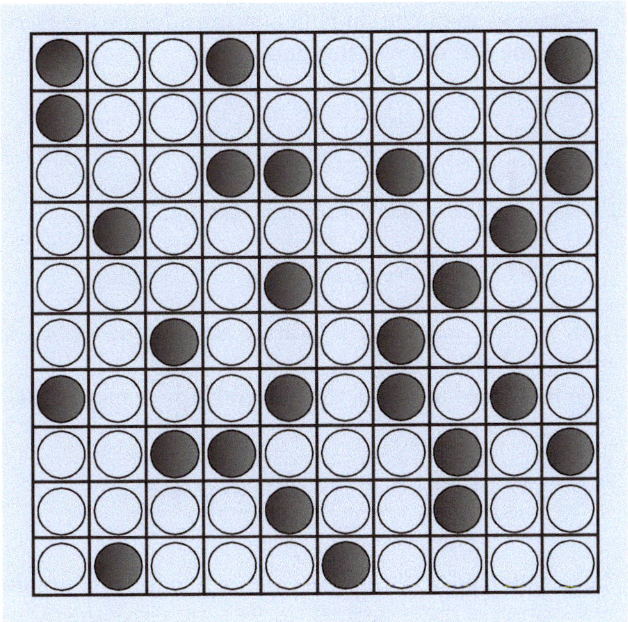

Fig. 2.7 Lattice model to describe a solution of a solute (*dark points*) in a solvent (*light points*). Apart from the contents the lattice cells are identical

Because the sum of the volume fractions equals 1:

$$\Delta S^{m}_{ideal} = -R\left(\varphi_1 \ln \varphi_1 + \varphi_2 \ln \varphi_2\right) \tag{2.45}$$

Furthermore, because φ_1 and φ_2 are always less than 1 for mixtures, the ideal entropy of mixing is always positive and the entropy of mixing always favors the mixture of substances. This should be obvious by considering entropy as a measure of the disorder of a system).

To calculate enthalpies, we take account of the pair-interactions W_{ij} between the neighboring cells in the lattice. To create two contacts between molecules 1 and 2 (with the corresponding enthalpy of interaction W_{12}), the interactions between a pair of solvent molecules ($\rightarrow W_{11}$) and a pair of solute molecules ($\rightarrow W_{22}$) must be ruptured. The enthalpy change for this process is given by

$$\Delta W = 2W_{12} - \left(W_{11} + W_{22}\right) \tag{2.46}$$

The total number of contacts N_{12} between solvent and solute in the lattice can be obtained by considering the coordination number z, which is given by the number of the neighboring lattice cells:

$$N_{12} = N\varphi_1\varphi_2 z \tag{2.47}$$

Thus, the total enthalpy of mixing for the mixture is given by

$$\Delta H^m = \frac{z}{2} N\varphi_1\varphi_2 \Delta W \tag{2.48}$$

The factor 1/2 results from the definition of ΔW (in relation to two interaction contacts). This expression can be simplified by introducing the *Flory–Huggins interaction parameter* χ. From the definition of this parameter

$$\chi = \frac{z}{2k_B T}\Delta W, \tag{2.49}$$

it follows for the molar enthalpy of mixing:

$$\Delta H^m = RT \cdot \varphi_1\varphi_2\chi \tag{2.50}$$

Thus, positive values for χ indicate an endotherm (= unfavorable) interaction between solvent and solute.

With the aid of (2.33), the free enthalpy of mixing ΔG^m can be formulated as

$$\Delta G^m = RT\left(\underbrace{\varphi_1 \ln\varphi_1 + \varphi_2 \ln\varphi_2}_{\textit{Entropy term}} + \underbrace{\varphi_1\varphi_2\chi}_{\textit{Enthalpy term}}\right) \tag{2.51}$$

The variation of ΔG^m with the volume fraction of one component is symmetrical over the composition of the system (Fig. 2.8), i.e., changing the indices for the solute and solvent does not change (2.51).

Figure 2.8 shows plots of the free enthalpy of mixing as a function of the composition for various values of χ.

From Fig. 2.8 it can be deduced that mixtures with smaller values of χ are stable, because for small values of χ the free enthalpy of the mixing is always less than the free enthalpy of the pure components. That is, mixing the components leads to a reduction of the free enthalpy and is thus energetically favorable. This becomes obvious by considering that large, positive values of χ would indicate a strongly endothermic mixing process.

For smaller values of χ, the free enthalpy of mixing is always negative. For $\chi > 2$ the free enthalpy of mixing goes through a maximum and the curve forms a "hump" for solutions of equal volume fractions. For such solutions there are alternative compositions to the left and the right of this maximum with lower energy levels, and such systems begin to separate in the range of equal volume fraction compositions—the systems are unstable. This effect increases with increasing values of χ.

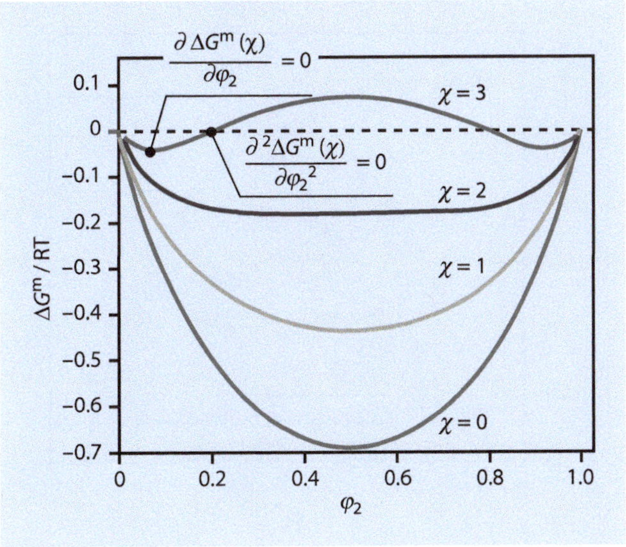

Fig. 2.8 Plot of the free enthalpy of mixing as a function of the composition for various values of the Flory–Huggins

If one observes, for example, a system at $\varphi_2 = 0.5$ for $\chi = 3$, a separation takes place into two phases with $\varphi_2 = 0.07$ and $\varphi_2 = 0.93$ corresponding to the minima of the energy curve. A special case can be observed for mixtures that are between the minimum and the inflexion point of the free enthalpy curve. Because the curve here also increases to the left, these systems are *metastable* and additional (activation) energy is required to induce the process of phase separation. This phenomenon can be compared with the processes of nucleation and growth during crystallization.

The influence χ has on the stability limits, stable–metastable–unstable, can be directly determined by plotting ΔG^m as a function φ for different values of χ and determining the positions of the minima and inflection points. To this end, the minima and inflexion points of the free enthalpy curves in Fig. 2.8 are joined for different values of χ and transposed to give Fig. 2.9.

The transposition of the minima results in the so-called *binodal*, which separates the stable area from the metastable one. The transposition of the inflexion points yields the so-called *spinodal*, which separates the metastable from the unstable area. The critical point is the point where binodal and spinodal curves converge.

2.4.1.2 Polymer Solutions

If one considers polymer solutions, the description in the lattice model according to Flory and Huggins initially changes because monomers are linked together to form the polymer chains (Fig. 2.10).

The mathematical description of polymer solutions in analogy to that for small molecules is given in (2.51), analogously to the lower molecular substances discussed so far. For the free enthalpy of mixing ΔG^m, it holds without proof:

$$\Delta G^m = RT\left(\underbrace{\varphi_1 \ln \varphi_1 + \frac{\varphi_2}{X_N} \ln \varphi_2}_{\text{Entropy term}} + \underbrace{\varphi_1 \varphi_2 \chi}_{\text{Enthalpy term}}\right) \tag{2.52}$$

2

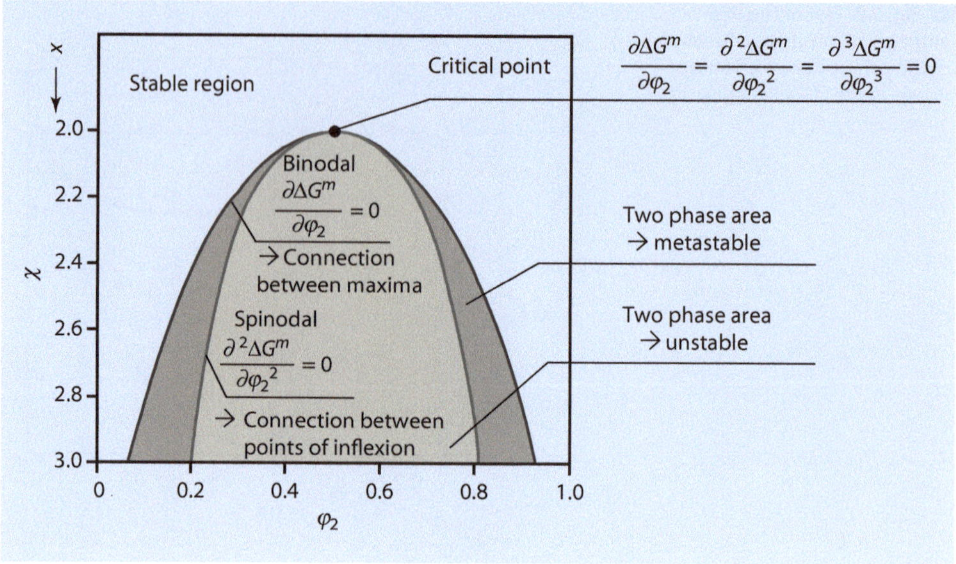

Fig. 2.9 Binodal and spinodal curves for small molecule mixtures

Fig. 2.10 Lattice model representing a solution of a polymer (*dark circles*) in a solvent (*light circles*). Apart from their contents the lattice cells are identical

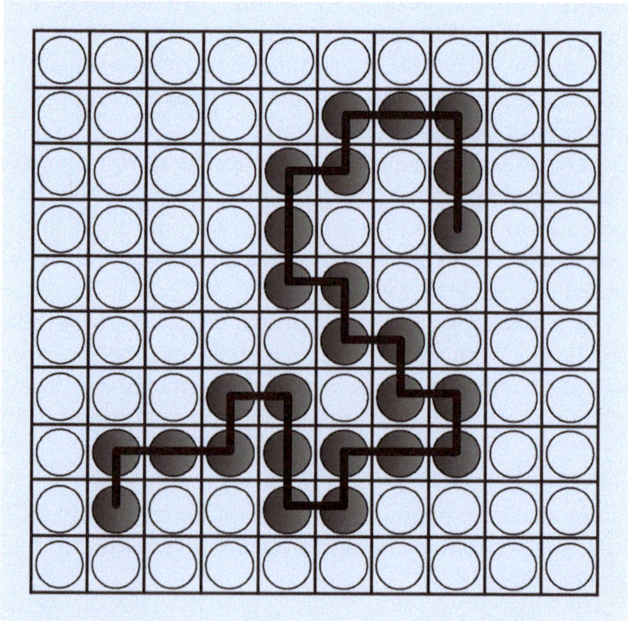

In this model, X_N denotes the relation of the size of the polymer molecule to that of the solvent molecule. Assuming that the solvent molecules occupy the same volume as the repeat units in the polymer, X_N corresponds to the degree of polymerization of the polymer being studied.

It is striking that only the entropy term is different for polymer solutions (cf. (2.51)). Descriptively, the entropy gain is lower when diluting a polymer solution than when diluting

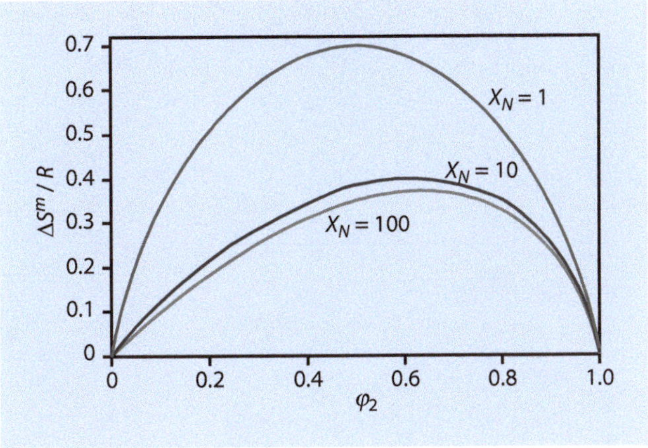

Fig. 2.11 Entropy of mixing as a function of solution composition for polymer solutes of different chain lengths

a corresponding amount of a small molecule solution because the polymer's building blocks cannot arbitrarily spread themselves over the lattice cells. This means that, compared to a small molecule solution, the dissolution of a polymer is always less favorable with respect to its entropy because of its smaller entropy of mixing whereas the enthalpy of mixing does not change considerably to a first approximation (i.e., for regular solutions) whether it is a small molecule or a polymer being dissolved.

Figure 2.11 shows the curve of the entropy of mixing as a function of solution composition for polymer solutes of different chain lengths.

In contrast to the corresponding curves for solutions of small molecules, (2.52) is no longer symmetrical with respect to the two components; i.e., exchanging the component indices would produce a different and new equation. This is also reflected in Fig. 2.11. The curves are no longer symmetrical about a vertical mirror axis running through a 1:1 mixture and the maximum shifts to higher polymer volume fractions for $X_N > 1$ as the size of the polymer increases. The forms of the curves of the free enthalpy of mixing as a function of φ_2 for different values of χ are shown in Fig. 2.12

As in Fig. 2.8, the curves of ΔG^m for large values of χ have 2 maxima. However, the minimum at low polymer concentration (small value of φ_2) is only weakly developed (see insert in Fig. 2.12). The curves in Fig. 2.12, in contrast to those in Fig. 2.8, are also unsymmetrical.

If, in analogy to Fig. 2.9, the binodals for various chain lengths are extrapolated, Fig. 2.13 can be obtained.

From this representation a critical point can be recognized—that the composition at which a miscibility gap is observed for the first time shifts to ever smaller polymer concentrations as the degree of polymerization increases. Moreover, the miscibility gap approaches a critical χ-parameter of $\chi_c = 0.5$. This has two significant implications.

The curves in Fig. 2.13 document that polymers of greater molar mass dissolve less easily than those of smaller molar mass; the solubility of polymers decreases with increasing chain length. The miscibility gap broadens and starts at lower values of χ. For polymers with very high molar mass, a miscibility gap is already evident at a χ-parameter value of 0.5. For such polymers the required entropy of mixing even a much smaller (unfavorable) enthalpy of mixing is necessary to compensate for the reduced (favorable) entropy of mixing and thus to trigger phase separation. The reason for the decreasing solubility of polymers as their degree of polymerization increases is the *entropy* of the system.

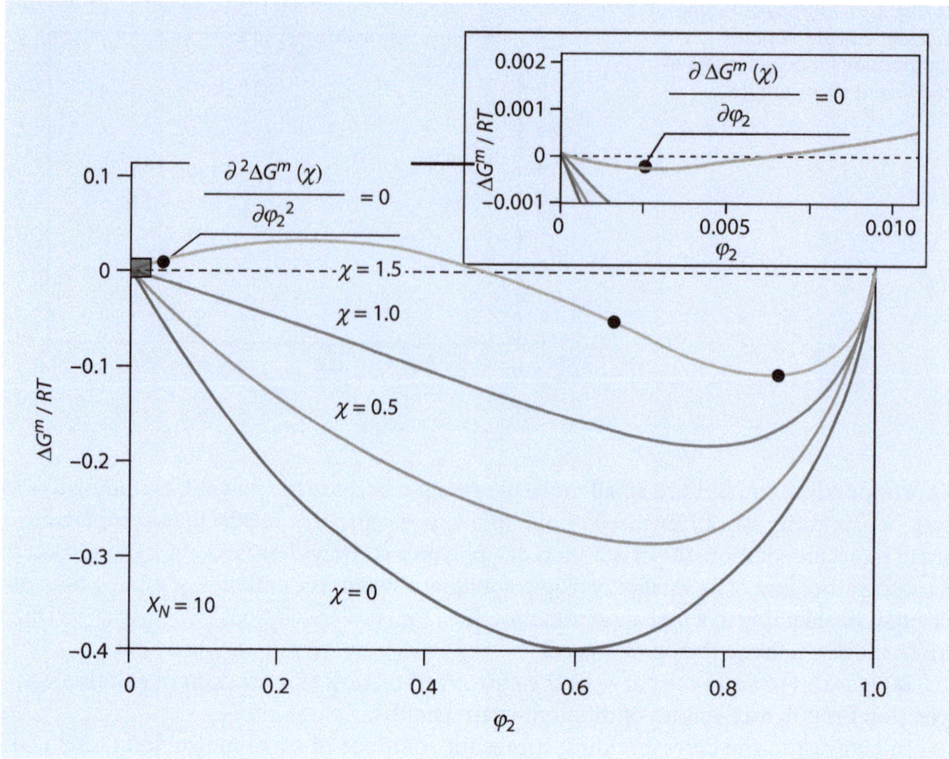

Fig. 2.12 Curves of the free enthalpy of mixing a polymer and a solvent as a function of φ_2 for different values of χ at length $\chi_N = 10$

Furthermore, it follows from this that at high degrees of polymerization the separation leads to a solvent swollen polymer gel and an almost polymer-free solvent phase. Take, for example, a chain with a degree of polymerization of 1000 (we are considering a single lattice cell!) and an interaction parameter χ of 1. In Fig. 2.13 a straight line for $\chi = 1$ parallel to the x-axis intersects the binodal almost at the ordinate ($\varphi_2 = 0$) and again at φ_2 of ca. 0.5. This implies that such a mixture separates into two phases, one containing polymer soaked with 50% solvent (in reality a polymer gel) and the other an almost polymer-free solvent phase. With the aid of the asymmetry of the curves, this explains why some polymers absorb large quantities of a solvent without dissolving—think of cotton socks in the washing machine.

For a system with $X_N = 100$ at $\chi = 1.0$ and a mid-range composition of $\varphi_2 = 0.2$, the phase separation results in two phases with polymer volume fractions of $\varphi_2 = 0.01$ and $\varphi_2 = 0.5$, respectively (Fig. 2.13). From the mass balance it is clear that the volumes of the two phases are not equal. The first phase (containing very little polymer) fills approx. 61% of the total system volume whereas the polymer-rich phase occupies the remaining 39%. These figures are obtained from the "difference" between the resulting phase compositions and the original composition, $\varphi_2 = 0.2$. This can be generalized: the phases whose composition after phase separation is most similar to the overall composition occupy the largest volume. The figures of 61 % and 39 % in our example can be calculated from the difference

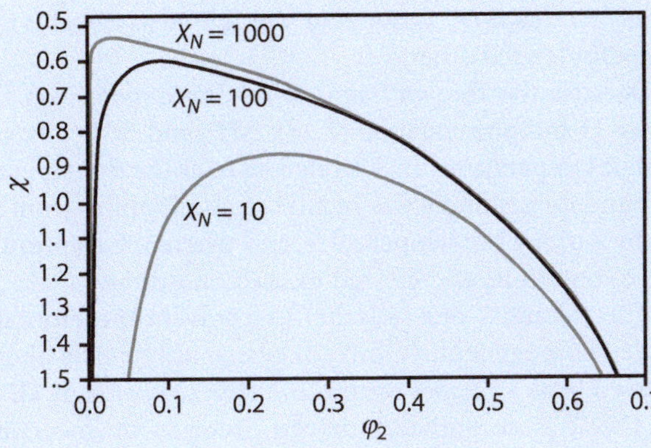

Fig. 2.13 Binodal curves for solutions of polymers with different chain lengths

between the phase composition and the overall composition ($\Delta\varphi_2 = 0.2 - 0.01 = 0.19$ or $0.5 - 0.2 = 0.3$) relative to the difference between the compositions of the two phases ($\Delta\varphi_2 = 0.5 - 0.01 = 0.49$). Because of the formal analogy to the principle governing mechanical levers, this rule of separation is also referred to as the lever rule.

The Flory–Huggins interaction parameter is a central measure for understanding the solubility of a polymer in a specific solvent. A negative interaction parameter indicates that the polymer and the solvent molecules attract each other to a greater extent than they attract their own kind. The polymer dissolves easily. On account of the good solvation, the polymer coil expands compared to the undissolved state.

An interaction parameter of $\chi = 0$ corresponds to an athermal solution. Graphically speaking, this means that all the interactions taking place in the medium are equal, and it does not matter for the enthalpy of the system whether the interacting neighbors are solvent molecules or polymer segments.

When the Flory–Huggins interaction parameter becomes positive, the enthalpy favors phase separation rather than solution. If, however, a degree of miscibility remains, this is because of the contribution from the entropy of the system. As the entropy contributions in polymer systems are much smaller than those with small molecule solutes, even a slightly positive (unfavorable) enthalpy of mixing suffices to prevent the polymer from dissolving. The properties of polymer solutions are therefore strongly dependent on the value of the interaction parameter χ.

If the interaction parameter has a value greater than 0.5 (for extremely high molar masses), the polymer becomes more compact in the solution. The reason for this is a net rejection between polymer and solvent molecules. The limiting case, $\chi = 0.5$, for polymers with a high degree of polymerization reflects the transition between a homogeneous mixture and separation into two phases. As the interaction parameter increases, attraction between polymer and solvent is reduced, interaction becomes more repulsive in nature, and the polymer precipitates.

Additionally, for $\chi = 0.5$ (for the limiting case of infinite molar mass) there is a special case. As mentioned above, this is the value of χ at which a miscibility gap is observed for the very first time. At this point, the contributions from the enthalpy and entropy of mixing compensate each other and the solution acts as an ideal solution, i.e., *quasi-ideal*. As already stated, this condition is referred to as the θ state.

If we compare this to the ideal solution ($\Delta H^m = 0$ and $\Delta S^m = \Delta S^m_{ideal}$), the difference lies in the fact that ΔH^m is not zero and ΔS^m also deviates from ΔS^m_{ideal}, but both terms simply compensate each other.

Because the free enthalpy of mixing depends on the temperature according to the Gibbs–Helmholtz equation (2.28), ΔH^m and ΔS^m_{excess} usually only compensate each other at a single temperature. This is referred to as the θ-temperature. From (2.28), conditions can be found for instances where ΔH^m and ΔS^m differ from zero and ΔS^m_{ideal}, respectively, usually by varying the temperature, and where the solution behaves in a quasi-ideal manner. These conditions are referred to as θ conditions.

The solubility of a polymer in a solvent therefore depends on the temperature. If a single (homogeneously mixed) phase only exists above a certain temperature, this is referred to as an *upper critical solution temperature* (UCST). Here the separation below the UCST is an enthalpy-driven process—at lower temperatures the contribution of the enthalpy of mixing to the free enthalpy of mixing is greater than the effect of entropy (2.28).

However, if a solution is only thermodynamically stable below a certain temperature, the separation that occurs at this temperature must, by analogy, be driven by entropy. When a thermodynamically stable phase of the polymer–solvent-mixture occurs below a critical temperature, this is referred to as the LCST (*lower critical solution temperature*). This temperature denotes the maximum temperature at which the polymer still dissolves.

Formally, this behavior can be explained by interpreting the Gibbs–Helmholtz equation and considering the signs of the enthalpy and entropy terms. Assuming a negative enthalpy of mixing and a normal, positive entropy of mixing, the solution is completely miscible at all temperatures. An LCST occurs when the entropy and enthalpy of mixing take on negative values. As previously stated, because the entropy of mixing is usually positive, this case is rare and is only observed for a few polymers. If the signs of both the entropy and enthalpy of mixing are positive, the system has a UCST. This is often the case in reality. If the enthalpy of mixing is positive and the entropy of mixing negative, the system is not miscible at any temperature.

For real mixtures the situation is not so trivial, as both enthalpy and entropy of mixing are not monotonically dependent on the temperature. A detailed investigation shows that all polymer–solvent mixtures should display a UCST as well as an LCST. However, these are not always observable by experiment. This can often be explained by the UCST and LCST occurring at temperatures above or below the melting or the boiling point of the solvent.

The theory discussed here can be further modified by taking the temperature dependence of χ and other possible influences into account. Such refinements are not included in the present introductory description of the subject.

An approach comparable to the Flory–Huggins model for describing the characteristics of solutions is offered by the *Hildebrand solubility parameter*. This is based on a consideration of *coherence energy densities*, which are not discussed further here (Hildebrand and Scott 1950).

Finally, a few limitations of the Flory–Huggins-model are briefly discussed:

- The lattice model assumes that both the mixture's components (solvent and repeat unit) are of the same size and that the overall volume is composed additively from both components; i.e., the mixing process does not lead to a volume contraction or expansion.

- The model only takes account of combinatory entropy. It is based on a statistical mixture of the chains and segments. This makes sense for concentrated solutions but poses difficulties in the case of dilute solutions.
- Likewise, the model only accounts for non-polar molecules; the influence of, for example, hydrogen bonding between the polymer and the solvent is not considered.

In spite of these limitations, the Flory–Huggins theory serves as an excellent basis for a semi-quantitative understanding of the theory of polymer solutions. In detail, this subject is extremely complex.

The Flory–Huggins parameter can be determined by using various experimental methods (▶ Sect. 3.2.4). Such methods include osmosis as well as the enthalpy of vaporization (making use of solubility parameters) and group contribution calculations based on the results. Because the measurement of the heat of vaporization of a polymer is impractical, swelling experiments are an alternative which is noted here. The reduced solubility of polymers compared to the monomers from which the polymers are composed results from an entropy effect. Knowing this, it is intuitive that the entropy of mixing of two polymers is again considerably reduced. This leads to the important realization that two polymers are generally immiscible.

References

Flory PJ (1953) Principles of polymer chemistry. Cornell University Press, Ithaca, NY

Hildebrand JH, Scott RL (1950) The solubility of non-electrolytes, 3rd edn. Reinhold Publishing Company, New York

Kratky O, Porod G (1949) Röntgenuntersuchung gelöster Fadenmoleküle. Rec Trav Chim Pays-Bas 68:1106–1123

Painter PC, Coleman MM (2009) Essentials of polymer science and engineering. DEStech Publications, Lancaster, PA

Polymer Analysis: Molar Mass Determination

S. Koltzenburg et al., *Polymer Chemistry*, DOI 10.1007/978-3-662-49279-6_3

A key parameter for macromolecular substances is their molar mass or degree of polymerization. This chapter focuses on the question of how to describe mathematically and measure experimentally the molar mass of polymers and their molar mass distribution. A number of the methods presented have been developed exclusively for this purpose and are thus not commonly found in laboratories unconcerned with polymer chemistry. In addition to the molar mass, we discussed in ▶ Chap. 2 that the polymer size (e.g., of the entangled polymer coil in the melt or in solution) depends on the degree of polymerization. Therefore some methods addressing this question are also presented.

Experimental methods for defining molar mass can be divided into two basic categories: those referred to as "absolute methods," the results of which can be directly converted to a molar mass, and a second group of methods, the "relative methods," from which the results need to be calibrated with samples of known molar mass to infer the molar mass of the unknown sample.

In what follows, the theory on which the methods are based is briefly introduced before the most important absolute and relative methods are discussed in ▶ Sects. 3.2 and 3.3.

3.1 Definition of Molar Mass Parameters

Small molecules can usually be characterized by a particular molar mass. All the molecules of a sample of such a compound have the same molar mass. Anomalies are only created by different isotopes, for example, in chlorinated compounds, and are normally not considered further. The average value of the isotopic distribution is defined as the molar mass. This simple method cannot be used for polymers. As described in Chap. 8 and subsequent chapters, most polymerization methods result in a mixture of molecules with different chain lengths, each chain, in some cases, being made up of a very large variety of monomer units. The molar mass of a polymer can also be defined by an average value, but for a broad distribution the average can be calculated in several different ways. This can be demonstrated by looking at a simple example.

Example

Assume that a polymer sample contains three separate molecules, two of which have a molar mass of 10,000 g/mol and the third has a molar mass of 20,000 g/mol. If the average molar mass of this sample is required, the "intuitive" answer is 13,333 g/mol. Mathematically, the weights of the individual molecules are weighted according to their number and added together, and the sum is divided by the total number of molecules. Thus,

$$\begin{aligned} \text{Mean} &= \frac{2 \cdot 10\ \text{kg/mol} + 1 \cdot 20\ \text{kg/mol}}{2+1} \\ &= \frac{2}{3} \cdot 10\ \text{kg/mol} + \frac{1}{3} \cdot 20\ \text{kg/mol} \\ &\approx 13.3\ \text{kg/mol} \end{aligned} \tag{3.1}$$

Generally, the number or amount n_i of the molecules that have a molar mass M_i is multiplied with this molar mass M_i, and the sum of these products divided by the total number of molecules or the total amount of the sample. Alternatively, the product of the molar mass M_i and the mole fraction ($x_i = n_i / \Sigma n_i$) can be used. The "average" calculated from this formula is called the "number average" M_n:

$$M_n = \frac{\sum_i n_i \cdot M_i}{\sum_i n_i} = \frac{\sum_i m_i}{\sum_i n_i} = \sum_i \left(\frac{n_i}{\sum_i n_i} \cdot M_i \right) = \sum_i x_i \cdot M_i \tag{3.2}$$

where m_i is the total mass of all molecules with a molar mass M_i. This is given by the product of the number of molecules with a molar mass M_i, n_i, and their individual molar mass M_i.

In the following the index i is omitted from the summation (sigma) characters; all the sums are valid over all i.

In the preceding equations, all molecules were weighted according to their *number* and these products used for building an average (thus the term "number average", M_n). This average is the one often used in daily life and has thus become intuitive. However, other averages are just as mathematically valid and indeed provide different information about the distribution of the individual items in the sample. To illustrate this, the sample is not considered in terms of the number of molecules in the sample—as chemists mostly do—but in terms of the mass of each molecule type (◘ Fig. 3.1).

From a material scientist's point of view, the situation for the sample shown in ◘ Fig. 3.1 is that half of the material consists of polymers with a molar mass of 10,000 g/mol and the other half of polymers with a molar mass of 20,000 g/mol. From this perspective it makes sense to assume a mean value of ((10+20)/2) = 15,000 g/mol. In analogy to the formula for the number average molar mass, this so-called *weight average* molar mass M_w can be calculated as follows.

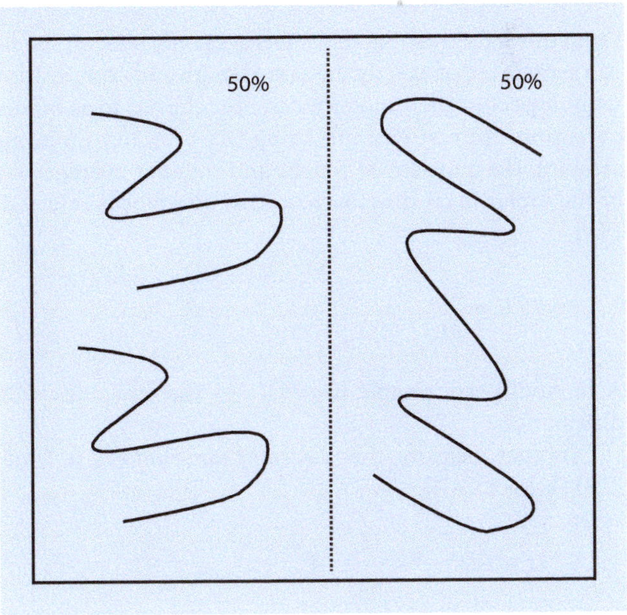

◘ **Fig. 3.1** Sketch showing a polymer sample emphasizing the weight fractions (for an explanation see text)

The mass of all molecules with the molar mass M_i is given by the product of n_i and their individual mass M_i. Thus, if in (3.2) the amount of the molecules is replaced by their mass m_i or the mole fraction with the mass fraction w_i:

$$M_w = \frac{\sum m_i \cdot M_i}{\sum m_i} = \sum\left(\frac{m_i}{\sum m_i} \cdot M_i\right) = \sum w_i \cdot M_i \tag{3.3}$$

where w_i is the mass fraction of the molecules with a molar mass M_i, that is, their mass divided by the total mass of the sample.

As the mass of the molecules in the sample is given by the sum of the products of the amount of substance and molar mass, one can also write

$$M_w = \frac{\sum (n_i \cdot M_i) \cdot M_i}{\sum (n_i \cdot M_i)} = \frac{\sum n_i \cdot M_i^2}{\sum n_i \cdot M_i} \tag{3.4}$$

In analogy with the above, it is possible to obtain further averages, although the derivations are not as straightforward as those for the number and weight averages. The centrifugal average M_z is especially important and is defined by the following equation:

$$M_z = \frac{\sum n_i \cdot M_i^3}{\sum n_i \cdot M_i^2} = \frac{\sum w_i \cdot M_i^2}{\sum w_i \cdot M_i} \tag{3.5}$$

The term "centrifugal average" is derived from this average, being that determined from ultracentrifuge measurements (▶ Sect. 3.2.5).

By definition, the mean molar mass increases in the following order:

$$M_n \leq M_w \leq M_z \tag{3.6}$$

From this it follows that these averages only take on the same value if *all* of the molecules of a sample are of exactly the same length and composition. Such samples, homogeneous with respect to their molar mass, are referred to as *monodisperse*. In practice, this is the exception, one rare example being DNA. In non-monodisperse or so-called *polydisperse* samples, the quotient of weight and number averages is a measurement of the breadth of the molar mass distribution. This quotient is referred to as the polydispersity index *PDI*:

$$PDI = \frac{M_w}{M_n} \tag{3.7}$$

A monodisperse sample has $PDI = 1$. The larger the *PDI*, the broader the molar mass distribution.

Another measure, the *molecular non-uniformity U* of a sample, can be defined as in (3.8) (▶ Sect. 9.4.3):

$$U = PDI - 1 = \frac{M_w}{M_n} - 1 \tag{3.8}$$

Alternatively, the breadth of a molar mass distribution can also be characterized by its variance σ^2:

$$\sigma^2 = \frac{\sum n_i (M_i - M_n)^2}{\sum n_i} \tag{3.9}$$

or its absolute standard deviation $\sqrt{\sigma^2}$.

A relative standard deviation, normalized with respect to M_n, $\sqrt{\sigma^2} / M_n$, can also be defined:

$$\frac{\sqrt{\sigma^2}}{M_n} = \sqrt{U} \tag{3.10}$$

From these definitions it can be calculated that even for a polymer with a narrow molar mass distribution, that is, $PDI = 1.04$, the relative standard deviation is still substantial: 20 %.[1]

The number of repeating units in a polymer chain is referred to as the *degree of polymerization P*. If the end groups are neglected, *P* equals the quotient of the molar mass of the polymer chain and the molar mass of the repeating unit M_{RU}. As with the molar mass, the degree of polymerization of a polymer sample usually has the same (identical) distribution. Thus, the number and weight average degrees of polymerization can be defined as in (3.11) and (3.12), respectively:

$$P_n = \frac{M_n}{M_{RU}} \tag{3.11}$$

$$P_w = \frac{M_w}{M_{RU}} \tag{3.12}$$

1 The relation $\sqrt{\sigma^2} / M_n = \sqrt{U}$ is a result of the following calculation:

$$\frac{1}{M_n^2}\sigma^2 = \frac{1}{M_n^2}\frac{\sum n_i (M_i - M_n)^2}{\sum n_i} = \frac{1}{M_n^2}\frac{\sum n_i \left(M_i^2 - 2M_i M_n + M_n^2\right)}{\sum n_i}$$

$$= \frac{1}{M_n^2}\left[\frac{\sum n_i M_i^2}{\sum n_i} - 2M_n \frac{\sum n_i M_i}{\sum n_i} + M_n^2 \frac{\sum n_i}{\sum n_i}\right]$$

$$= \frac{1}{M_n^2}\left[\frac{\sum n_i M_i^2}{\sum n_i} - M_n^2\right] = \frac{1}{M_n^2}\frac{\sum n_i M_i^2}{\sum n_i} - 1 = \left(\frac{\sum n_i}{\sum n_i M_i}\right)^2 \frac{\sum n_i M_i^2}{\sum n_i} - 1$$

$$= \frac{\sum n_i}{\sum n_i M_i}\frac{\sum n_i M_i^2}{\sum n_i M_i} - 1 = \frac{M_w}{M_n} - 1 = U$$

3

3.2 Absolute Methods

3.2.1 End Group Analysis

If polymers always include a certain number ξ of end groups, regardless of their size, and if they do not contain rings or have branched chains, the molar mass of a polymer can be calculated by quantitatively determining its end groups. The basic idea is to analyze the amount of substance making up the end groups in a certain mass of polymer. The quantitative analysis of the end groups can be carried out chemically or spectroscopically, depending on whether the end groups have a well-defined spectroscopic characteristic or are susceptible to a selective chemical reaction. A typical example of end group analysis using a chemical method is a titration. Thus, a known mass m_2 of a polymer is titrated with the measured amount of reagent n_R necessary to complete the reaction with all the end groups. In this case, $m_2 = \Sigma n_i M_i$ and $n_R / \xi = \Sigma n_i$, and from the quotient of these two values the number average molar mass of the polymer can be calculated:

$$M_n = \frac{m_2}{n_R / \xi} \tag{3.13}$$

Example

In the end group analysis of a polyester with exactly one acid group per chain, 0.7 mL of a KOH solution with a concentration of 0.1 mol/L are consumed for the titration of $m_2 = 1$ g of polymer. From (3.13) the number average molar mass of the polyester is

$$M_n = \frac{m_2 \cdot \xi}{n_R} = \frac{1\,\mathrm{g} \cdot 1}{0.7 \cdot 10^{-3}\,\mathrm{L} \cdot 0.1\,\mathrm{mol/L}} \approx 14,300\,\mathrm{g/mol} \tag{3.14}$$

With increasing molar mass, this method becomes increasingly inaccurate because of the concurrent decreasing concentration of the end groups so that simple titration methods can only be used for end group analysis of molar mass up to a maximum of ca. 20,000 g/mol. Spectroscopic methods such as UV-, IR-, or ^{1}H-NMR-measurements are also frequently employed for the quantitative determination of the end groups and for the definition of the number average molar mass. ^{14}C- or fluorescence labeling of the end groups leads to increased sensitivity and thus to an increased measurement range for the method.

End group analysis is sometimes referred to as the *functional equivalence method.* An important precondition for the applicability of this method is an exact knowledge of the well-defined end groups.

For the special case where the molar mass is determined from an analysis of the end groups by ^{1}H-NMR, the weight of the polymer sample and that of the end groups are not required. In this case, the intensity of the ^{1}H-NMR-signals resulting from the end group are proportional to the amount of substance of the corresponding protons and these signals can be compared with those identified as resulting exclusively from the repeating unit of the chain. If the number of protons of the end group signal is given by ζ_E and the number of protons in each repeating unit by ζ_{RU}, from the corresponding intensities, I_E and I_{RU} respectively, the number average degree of polymerization is given by

$$P_n = \frac{I_{RU} / \zeta_{RU}}{I_E / \zeta_E} \tag{3.15}$$

It is worth mentioning MALDI-TOF-mass spectrometry as an additional method and because this is now widely used for polymer analyses it is discussed separately in Sect. 3.2.7.

3.2.2 Colligative Properties

If a substance is dissolved in a solvent, the chemical potential of the solvent decreases. This leads to changes in, for example, the vapor pressure, the freezing and boiling points, and the osmotic pressure of the solvent. These properties are called the colligative properties of the solution (from the Latin *colligatus* meaning *bound together.*) In a highly diluted state, changes in the colligative properties are approximately proportional to the volume fraction or the concentration of the solute. If such a dilute solution is separated from pure solvent by a membrane through which the solvent but not the solute can pass, an equilibrium can only be established by changing another thermodynamic parameter, such as the temperature or pressure, of either the solution or the pure solvent.

Such a difference allows the mole fraction of the solute and, if the weight of the solute is known, its molar mass to be determined. Changes in the colligative properties of a solution are proportional to the number of dissolved particles per volume, not their size nor, provided there is no chemical reaction between the solute and the solvent, their chemistry. Thus, the molar mass determined from the measurement of colligative properties is the number average molar mass M_n.

Boiling Point Elevation and Freezing Point Depression

If a non-volatile substance which isn't incorporated into the solvent's crystal structure when the latter freezes is dissolved in a solvent, the decrease of the chemical potential of the solvent causes the phase diagram boundaries between the liquid and the gas phase and that between the solid and the liquid phase to shift in such a way that the liquid area increases. Thus, at any pressure, the liquid-gas boundary moves to higher temperatures and the liquid-solid boundary moves to lower temperatures (Fig. 3.2). At a constant temperature the vapor pressure decreases, whereas at constant pressure the boiling temperature is elevated to a temperature T_b and the melting temperature is lowered to a temperature T_f.

The solution boils when the chemical potential of the solvent in the solution is the same as that of the chemical potential of the solvent in the gas phase:

$$\mu_1(gas) = \mu_1^{pure}(fl) + RT_b \cdot \ln x_1 \tag{3.16}$$

Here and in the following, μ_1^{pure} (fl) is the chemical potential of the pure liquid and the subscript 1 indicates the solvent. We explicitly presume that the solute (hereafter denoted with the index 2) does not evaporate and therefore the solvent vapor is present in pure form. Because the sum of the mole fraction of the solvent and that of the solute is 1, $x_1 = 1 - x_2$, so that

$$\ln(1-x_2)=\frac{\mu_1(gas)-\mu_1^{pure}(fl)}{RT_b}=\frac{\Delta_{vap}G_m}{RT_b} \tag{3.17}$$

where $\Delta_{vap}G_m$ is the molar free enthalpy of vaporization for the solvent. With

$$\Delta_{vap}G_m=\Delta_{vap}H_m-T_b\cdot\Delta_{vap}S_m \tag{3.18}$$

(3.17) can be rearranged to give

$$\ln(1-x_2)=\frac{\Delta_{vap}G_m}{RT_b}=\frac{\Delta_{vap}H_m}{RT_b}-\frac{\Delta_{vap}S_m}{R} \tag{3.19}$$

For the pure solvent, $x_2=0$ and the boiling temperature is T_b^{pure}. Because $\ln 1=0$, (3.19) for the pure solvent becomes

$$0=\frac{\Delta_{vap}G_m}{RT_b^{pure}}=\frac{\Delta_{vap}H_m}{RT_b^{pure}}-\frac{\Delta_{vap}S_m}{R} \tag{3.20}$$

Subtracting (3.20) from (3.19) gives

$$\ln(1-x_2)=\frac{\Delta_{vap}H_m}{R}\cdot\left(\frac{1}{T_b}-\frac{1}{T_b^{pure}}\right) \tag{3.21}$$

With $x_2 \ll 1$, the approximation $\ln(1-x_2)\approx -x_2$ can be made and

$$x_2=\frac{\Delta_{vap}H_m}{R}\cdot\left(\frac{1}{T_b^{pure}}-\frac{1}{T_b}\right) \tag{3.22}$$

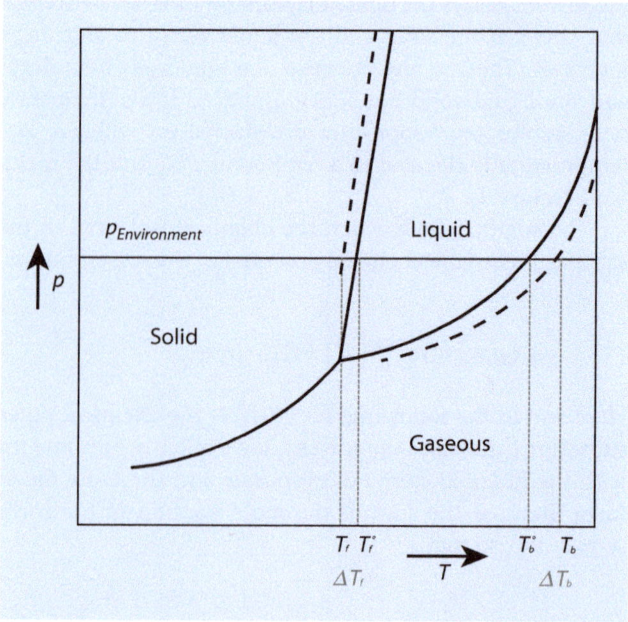

Fig. 3.2 Phase diagram showing an increase of the boiling point (ΔT_b) and a decrease of the freezing point (ΔT_f) as the result of adding a solute; solution (*dashed line*) and pure solvent (*full line*)

Additionally, for dilute solutions T_b does not significantly differ from T_b^{pure} and the term in parentheses can be approximated by

$$\frac{1}{T_b^{pure}} - \frac{1}{T_b} = \frac{T_b - T_b^{pure}}{T_b^{pure} \cdot T_b} \approx \frac{\Delta T_b}{T_b^{pure^2}} \tag{3.23}$$

Here the change in the boiling point $T_b - T_b^{pure} = \Delta T_b$ is what is measured so that it is easier to rearrange the approximate form of (3.22) to give

$$\Delta T_b = \frac{R \cdot T_b^{pure^2}}{\Delta_{vap} H_m} \cdot x_2 \tag{3.24}$$

Interpreting (3.24):

- The boiling point increases because the enthalpy of vaporization is positive
- The increase (for solutions of non-volatile solutes) is directly proportional to the mole fraction of the solute
- The increase is greater, the higher the boiling temperature and the lower the enthalpy of vaporization

Using this principle, the molar mass of the solute can be determined. To this end, the mole fraction of the solute in (3.24) by the respective amounts of the components n_i is

$$x_2 = \frac{n_2}{n_1 + n_2} \approx \frac{n_2}{n_1}, \quad \text{because} \quad n_2 << n_1 \tag{3.25}$$

With the molar mass of the solvent and solute, M_1 and M_2, the total mass m and the approximation $n_1 \approx m / M_1$, it follows that

$$\Delta T_b = \frac{R \cdot T_b^{pure^2} M_1}{\Delta_{vap} H_m} \cdot \frac{n_2}{m} \equiv K_{bp} \cdot \frac{n_2}{m} \tag{3.26}$$

Here, n_2/m is the number of moles of the solute in the solution and is also referred to as the *molality* of the solution. K_{bp} is the *ebullioscopic constant* for the solvent and exclusively dependent on the properties of the solvent. K_{bp} can be seen as an empirical constant, a characteristic of a solvent, and is often known. If this is the case, by measuring the change in the boiling point for different concentrations of solute the molar mass of the solute can be determined. If a weighed amount of polymer m_2 is the solute, the molar mass measured in this way is number average M_n:

$$M_n = K_{bp} \cdot \frac{m_2}{m} \frac{1}{\Delta T_b} \tag{3.27}$$

The freezing point depression can be treated in analogy to the boiling point elevation. In this case a solid solvent (1) is considered in equilibrium with the liquid solvent with the solute (2). Mentioned above, it is assumed that the solute is not included in the crystal structure of the solid solvent.

At equilibrium:

$$\mu_1(solid) = \mu_1^{pure}(fl) + RT \cdot \ln x_1 \tag{3.28}$$

By making use of the identical argumentation as for the boiling point elevation, one obtains

$$\Delta T_f = T_f^{pure} - T_f = \frac{R \cdot T_f^{pure^2}}{\Delta_{fus} H_m} \cdot x_2 = \frac{R \cdot T_f^{pure^2} M_1}{\Delta_{fus} H_m} \cdot \frac{n_2}{m} \equiv K_{fp} \cdot \frac{n_2}{m} \tag{3.29}$$

The ΔT_f is defined in (3.29) inversely to the definition of ΔT_b given above and is the freezing point depression. Here, too, K_{fp} is an empirical constant specific to the solvent, which is referred to as the *cryoscopic constant* and $\Delta_{fus} H_m$ is the molar enthalpy of melting. The index "*fp*" stands for *freezing point.*

3.2.3 Membrane Osmometry

Another important method for determining the number average molar mass is osmometry, which is based on the change in the osmotic pressure of a polymer solution as compared to the pure solvent.

Osmometry is an absolute method which, in the same way as the boiling point elevation and the freezing point depression, relies on colligative properties. An osmotic pressure is created when the solution and solvent are separated by a semipermeable membrane, which ideally is only permeable for the solvent molecules, not for the solute. If that part of the apparatus which contains the solution is sealed and connected to a manometer, the solvent passes through the membrane until a pressure difference of Π is built up, which is as large as is necessary for the chemical potentials to be equal on both sides of the membrane (■ Fig. 3.3). The hydrostatic pressure is given by the height change in the manometer Δh.

In contrast to the boiling and freezing point measurements, carried out at a constant pressure, the measurement of the osmotic pressure is carried out at a constant temperature and this is important for the mathematical treatment of the results.

At equilibrium at a constant temperature, the chemical potential of the pure solvent at a pressure p is equal to the chemical potential of the solvent of a solution whose mole fraction is x_1 at a pressure $p+\Pi$:

$$\mu_1^{pure}(p) = \mu_1(x_1, p+\Pi) \tag{3.30}$$

The different terms $\mu_1(x_1, p+\Pi)$ need to be examined more closely. The free enthalpy and thus the chemical potential is a function of state. To derive $x_1 = 1 - x_2 \quad p+\Pi$ from $x_1 = 1$, P one can conceptually increase the pressure and then add solute 2. Because $d\mu/dp = V_m$, it follows for the chemical potential:

$$\begin{aligned} \mu_1(x_1, p+\Pi) &= \mu_1^{pure}(p+\Pi) + RT \cdot \ln x_1 \\ &= \mu_1^{pure}(p) + RT \cdot \ln x_1 + \int_p^{p+\Pi} V_m^1 dp' \end{aligned} \tag{3.31}$$

where $V_m{}^1$ is the molar volume of the solvent and p' is an auxiliary variable for the pressure to facilitate integration. Combining (3.30) and (3.31) and rearranging gives

$$-RT \cdot \ln x_1 = \int_p^{p+\Pi} V_m^1 dp' \tag{3.32}$$

If the molar volume is constant, with

$$-RT \cdot \ln(1 - x_2) \approx RTx_2, \tag{3.33}$$

it follows that

$$RT \cdot x_2 = V_m^1 \Pi \tag{3.34}$$

With $x_2 \approx n_2 / n_1$ (dilute solution) and $n_1 V_m^1 = V$, one obtains a variant of the van't Hoff equation:

$$\Pi V = n_2 RT \tag{3.35}$$

However, for solutions of large molecules such as polymers, deviations from this equation often arise even at low concentration. To account for such deviations, real solutions are described, e.g., by a *virial equation*. For the case in hand, this is a polynomial of the concentration. Hereby the nonlinear portions of the concentration, which according to the derivation above should not need to be accounted for, are mathematically included. In this virial equation the quotient n_2/V is expressed by the quotient of the mass concentration c_2 and the molar mass M_2 and

$$\Pi = RT\left(\frac{1}{M_2} c_2 + A_2 c_2{}^2 + A_3 c_2{}^3 + \ldots\right) \tag{3.36}$$

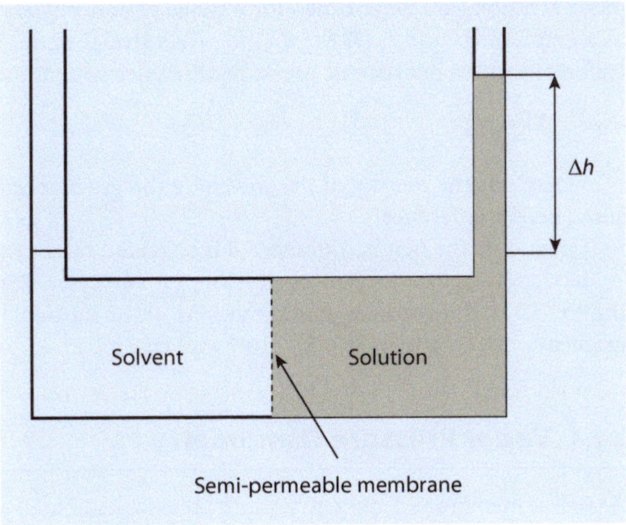

Fig. 3.3 Principle of an osmometer (Pfeffer Cell)

By taking measurements at various concentrations and extrapolating the results to $c_2 = 0$, the molar mass can be obtained from

$$\lim_{c_2 \to 0}\left[\frac{\Pi}{c_2}\right] = RT\frac{1}{M_2} \tag{3.37}$$

The value obtained is the number average molar mass.

If, rather than a simple extrapolation which ignores the nonlinear terms, the results are analyzed, one obtains a measure of the non-ideality of the system.

The parameters A_2, A_3, ... are referred to as the *virial coefficients* of the osmotic pressure. With the aid of the algebraic sign one can determine whether the solvent can be considered as "good" (positive sign) or "poor"(negative sign). With this method a relationship can be established to the experimental determination of the Flory–Huggins parameter (► Chap. 2) for dilute solutions. Thus, it holds (without proof) that:

$$A_2 = \frac{1}{M_{RU}\rho_2}(1-\chi) \tag{3.38}$$

where M_{RU} is the molar mass of the repeating unit, ρ_2 the polymer density (or $1/\rho_2$ the specific volume of the polymer), and χ the Flory–Huggins parameter.

Thus, with membrane osmometry the osmotic pressure between a polymer solution and the pure solvent is measured. This can be accomplished by using a simple manometer or, more usually, by direct pressure measurement in a closed system.

This absolute method can be used successfully in the molar mass range of 10^4 to10^6 g/mol. The lower limit is given by the quality of the membranes, which usually become permeable for smaller molecules. The upper limit is because of Π being proportional to x_2 (Π being inversely proportional to M_2). Thus, the effect being measured becomes smaller as the size of the molecules increases and at some point the margin of error makes reliable measurement impossible.

Example

In a membrane osmometer, the pressure differences between pure water and polyethylene oxide solutions are determined for a range of concentration. The results are then plotted as a graph of Π/c_2 vs c_2 (◻ Fig. 3.4) to give a straight line. Standard osmometers are often calibrated with a hydrostatic pressure difference so that the osmotic pressure

$$\Pi = \rho g h \tag{3.39}$$

where ρ is the density of the solvent, g the gravitational acceleration, and h the measured height difference.

Often only the height difference h is provided rather than the osmotic pressure Π.

Extrapolation to $c_2 = 0$ results in $\lim_{c_2 \to 0}\left[h/c_2\right] = 256\ \mathrm{cm^4\ g^{-1}}$, from which, according to (3.37), the medium molar mass $M_n = 103\ \mathrm{kg\ mol^{-1}}$ can be determined. From the gradient one can obtain $A_2 = 8.7 \cdot 10^{-4}\ \mathrm{mol\ cm^3\ g^2}$.

3.2.4 Vapor Pressure Osmometry

As well as membrane osmometry, vapor pressure osmometry is also an important method for determining the molar mass of polymers and this method is now discussed in more detail.

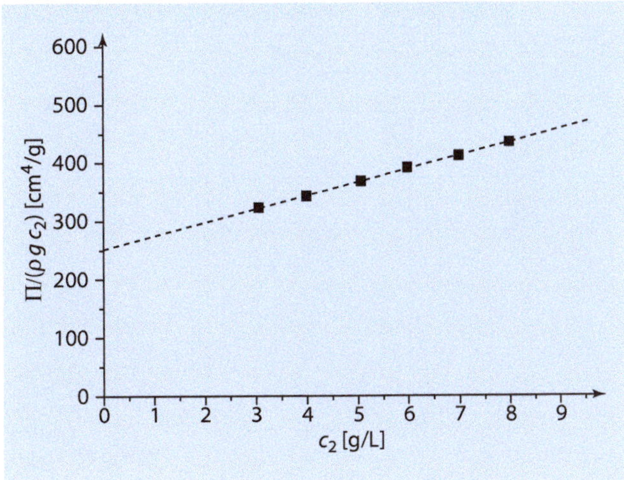

Fig. 3.4 Plot of the quotient of osmotic pressure and the polymer concentration vs polymer concentration, normalized with the density of the solvent ρ and the acceleration produced by gravity g. The extrapolation to $c_2 = 0$ gives from the intercept with the vertical axis.
$\lim_{c_2 \to 0}\left[\Pi/(\rho \cdot g \cdot c_2)\right] = \lim_{c_2 \to 0}\left[h/c_2\right] = RT/M_n$

As explained in ▶ Sect. 3.2.3, the vapor pressure of a solution is less than that of the pure solvent. Thus, if, in a closed chamber, both a compartment of pure solvent and one containing a solution are side by side in the same gas volume, solvent diffuses from the compartment with pure solvent (higher vapor pressure) to that of the solution (lower vapor pressure) via the gas phase. As the solvent condenses into the solution the heat of vaporization warms the solution, leading to an increase in its vapor pressure. The diffusion of solvent continues until the solution attains the same vapor pressure as the pure solvent.

Technical Setup

The floor of a tempered measuring cell (Fig. 3.5) is covered with solvent and equilibrated to ensure a constant vapor pressure. In the vapor chamber of the measuring cell there are two thermal sensors (e.g., thermistors). One drop of solution is applied to one of these temperature sensors, e.g., with a syringe and a drop of solvent to the other. As the vapor pressure of the drop of solution is less than the vapor pressure of the pure solvent, solvent condenses on the drop of solution. The temperature difference ΔT between the two sensors is measured.

To simplify, we can assume that this experiment corresponds to a boiling point elevation experiment—conducted with the partial pressure of the solvent—and that the experimentally determined temperature difference is evaluated (in which the boiling point temperature is to be replaced with the medium temperature of the chamber).

However, there is no real state of equilibrium inside the measuring chamber. A fixed state prevails in which material transfer and heat conduction balance each other. Moreover, the temperature of the chamber is often so distant from the boiling point that the assumption that the enthalpy of vaporization is independent of temperature is not always justifiable. These effects can be corrected, at least empirically, so that it is possible to determine reliably the medium value of the molar mass M_n of an unknown sample by calibrating the device as indicated by the manufacturer or by calibrating with a substance of known molar mass. With this method, it is also necessary to measure a concentration series and to extrapolate to c_2 (or m_2/m_1) = 0. Using a calibration constant K which has to be determined for each apparatus and each solvent:

$$M_n = K \cdot \lim_{m_2/m_1 \to 0}\left[\frac{m_2/m_1}{T - T^0}\right]$$

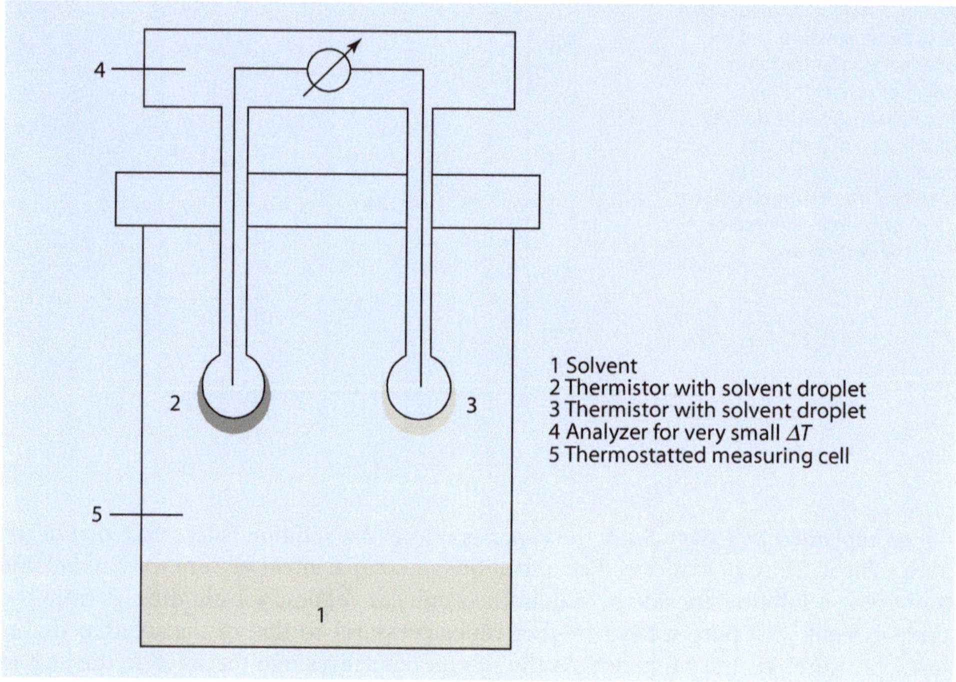

Fig. 3.5 Simplified sketch of a vapor pressure osmometer

or

$$M_n = K \cdot \lim_{m_2/m_1 \to 0} \left[\frac{m_2 / m_1}{\Delta T} \right] \tag{3.40}$$

This method also yields a number average molar mass and is less time consuming than membrane osmometry. It also requires less sample and is suitable for molar masses (M_2) less than 10^4 g/mol. Although the exact recording of absolute temperatures is difficult, it is comparatively easy to measure even minute temperature differences. Thus, vapor pressure osmometry is also suitable for polymers of high molar mass.

3.2.5 Ultracentrifuge

The ultracentrifuge (UC) is a centrifuge which rotates at very high speeds and was originally developed by Svedberg for his research on inorganic and organic colloids (Svedberg and Pedersen 1940). In a solid rotor made of light alloy, two cuvettes are embedded opposite each other, normally within a few centimeters of the axis of rotation, Their top and bottom walls consist of two parallel glass plates and the two sides are parallel to radial lines originating from the axis of rotation to prevent the development of convection patterns. In most cases a powerful optical imaging of the cuvette inside the rotor during the centrifugal process is integrated into the ultracentrifuge.

With an ultracentrifuge, essentially three types of experiment are possible, which allow conclusions about the shape, conformational changes, and size distribution of dispersed particles or dissolved macromolecules. (Maechtle and Boerger 2006).

3.2.5.1 Analysis of Sedimentation Velocity

Polymers usually have a density greater than that of smaller molecules. This can be easily explained by considering that, during polymerization, the distances between the monomer molecules determined by van der Waals forces are replaced by (shorter) covalent bonds. In the ultracentrifuge, a centrifugal force F_Z is imposed on a dispersed particle or a dissolved polymer chain and this force is dependent on the difference in density between that of the particle and that of the surrounding liquid ($\rho_2 - \rho_1$), its volume V_2, and the rotational speed of the rotor:

$$F_Z = V_2 \cdot (\rho_2 - \rho_1) \cdot \omega^2 x \tag{3.41}$$

ω	Angular velocity
x	Distance to the axis of rotation

The particle is accelerated radially outwards by the centrifugal force. The resulting movement induces a frictional force F_R acting opposite to the centrifugal force and dependent on the shape and size of the particle. It is usually also proportional to the viscosity of the surrounding fluid η and the velocity of the particle, dx/dt:

$$F_R = f \cdot \frac{dx}{dt} \tag{3.42}$$

Because the shape of the particle is usually unknown, in the first instance the shape dependency is expressed by an unknown friction factor *f*. If the particles are spheres, the friction factor f_{sphere} is proportional to the hydrodynamic radius R_h of the particle according to Stokes' law and the viscosity of the surrounding liquid:

$$f_{Sphere} = 6\pi\eta R_h \tag{3.43}$$

In the centrifuge, the particle is accelerated until the centrifugal force is balanced by the frictional force and the particle moves outwards with a constant velocity. If the particles are all the same size and shape they all move with the same velocity, and it is possible to observe a 'front' moving outwards leaving a solution free of particles or polymer molecules behind it and moving with a constant speed from the surface of the liquid towards the bottom of the cuvette (◘ Fig. 3.6).

The movement of the front can be determined, e.g., by so-called *Schlieren optics* which can detect even small changes in the refractive index and thus also in the concentration (dc/dx) (◘ Fig. 3.6). As soon as the particle has reached a constant velocity the forces are in equilibrium, $F_z = F_R$, so that

$$V_2 \cdot (\rho_2 - \rho_1) \cdot \omega^2 x = f \cdot \frac{dx}{dt} \tag{3.44}$$

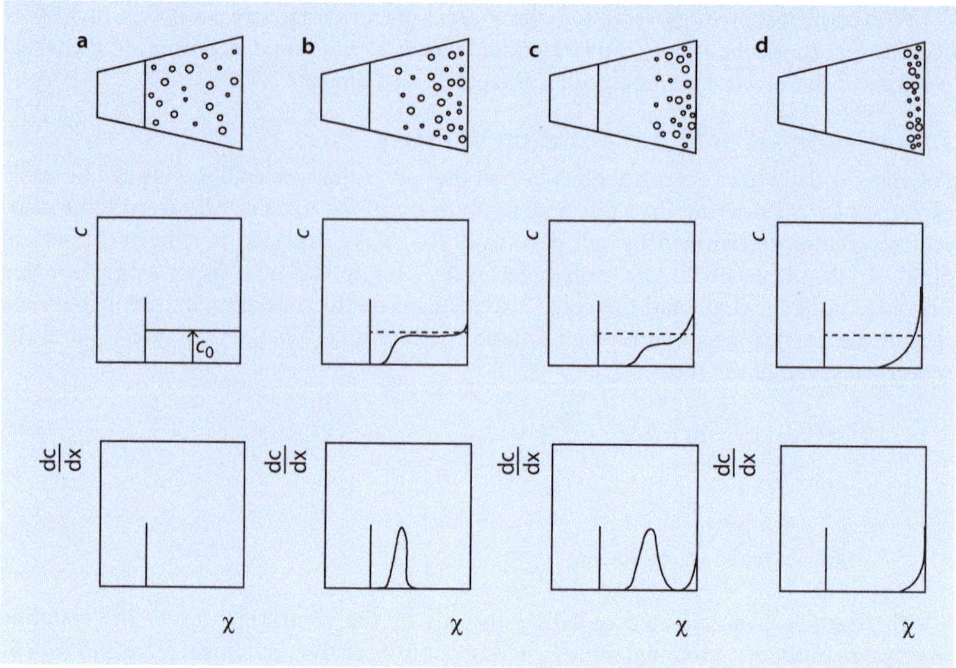

Fig. 3.6 Four characteristic stages of sedimentation in the cuvette of an ultracentrifuge. (**a**) Uniformly distributed particles. (**b**) Formation of a maximum in the concentration gradient. (**c**) Movement of the gradient towards the base of the cuvette. (**d**) "Front" reaches the base of the cuvette

The ratio of the sedimentation velocity, dx/dt, and the strength of the centrifugal acceleration, $\omega^2 x$, is denoted as the sedimentation coefficient S, so that

$$S = \frac{dx/dt}{\omega^2 x} = \frac{V_2 \cdot (\rho_2 - \rho_1)}{f} \tag{3.45}$$

The sedimentation coefficient has the dimensions of time (s). In honor of the pioneer of centrifugal technology, the sedimentation coefficient is usually expressed in the unit 'Svedberg,' $= 10^{-13}$ s (abbreviated by the symbol S). Without the knowledge of the shape, size, or density of the particle, this experiment alone does not allow an assertion extending beyond the sedimentation coefficient. In many biopolymers, this assertion suffices and the sedimentation coefficient becomes the identifier.

Notwithstanding, the diffusion coefficient D of the particle can be determined, e.g., via dynamic light scattering (▶ Sect. 3.2.6.2), which is linked to the friction factor f by the Einstein relation:

$$D = \frac{k_B T}{f} \tag{3.46}$$

If the density ρ_2 of the particle is also known, its volume V_2 can be derived from its molar mass M_2 and the Avogadro's constant N_A by using

$$V_2 = \frac{m_2}{\rho_2} = \frac{M_2}{N_A \cdot \rho_2}, \tag{3.47}$$

and the molar mass can be obtained from

$$S = V_2 \cdot (\rho_2 - \rho_1) \cdot \frac{1}{f} = \frac{M_2}{N_A \rho_2} \cdot (\rho_2 - \rho_1) \cdot \frac{D}{k_B T} \tag{3.48}$$

$$M_2 = \frac{RT}{D} \cdot S \cdot \left(\frac{\rho_2}{\rho_2 - \rho_1} \right) \tag{3.49}$$

Both the sedimentation coefficient S and the diffusion coefficient D depend on the concentration so that measurements at different concentrations and the extrapolation of the measured values to $c_2 = 0$ are necessary.

For spherical particles for which the density is independent of the molar mass (e.g., inorganic particles or non-swollen polymer particles but not swollen polymer entanglements), the volume is proportional to the cube of the radius. Thus, for such spherical particles:

$$S = \frac{4}{3}\pi \cdot R_h^3 \cdot (\rho_2 - \rho_1) \cdot \frac{1}{6\pi\eta R_h} \Leftrightarrow R_h = \sqrt{\frac{9}{2} \frac{\eta S}{\rho_2 - \rho_1}} \tag{3.50}$$

According to (3.50), a radius can be calculated which a solid sphere with an equivalent sedimentation coefficient would have irrespective of the actual shape of the particle. This radius is called the *hydrodynamic radius.*

Because the molar mass M_2 of non-swollen spherical particles is obtained by multiplying the density ρ_2 with the molar volume $\left(V_2^m = \frac{4}{3}\pi R_h^3 \cdot N_A \right)$ using $M_2 = \frac{4}{3}\pi R_h^3 \cdot N_A \cdot \rho_2$, it follows for such particles that

$$M_2 = 9\sqrt{2}\pi \rho_2 N_A \left(\frac{\eta S}{\rho_2 - \rho_1} \right)^{\frac{3}{2}} \tag{3.51}$$

For particles or polymer chains having a distribution of molar mass, not all of the particles at the same distance to the rotor have the same sedimentation velocity. Thus, a sharp front which moves outwards without alteration is not observed. In a mixture of multiple, easily distinguishable sub-groups of uniform particles (e.g., protein mixtures), concentration stages can be recognized and evaluated individually. For particles or polymer chains with a broad molar mass distribution, the border between the 'particle-free' and 'particle-containing' volumes becomes more diffuse within the cuvette as the duration of the centrifugation increases. In this case, an average position for the border area can be determined and an average molar mass determined from its movement. Depending on the method used to determine the average border area, the number, weight, or centrifugal average molar mass is determined; all three average values can be determined from ultracentrifuge measurements. This is a considerable advantage of this method over the other methods discussed above.

3.2.5.2 Measurement at Thermodynamic Equilibrium

In a centrifugal force field, a particle has a potential energy E which is dependent on its distance from the rotor, its volume, and the difference between its density and that of the surrounding fluid:

$$E(x) = -\frac{1}{2} V_2 \cdot (\rho_2 - \rho_1) \cdot \omega^2 x^2 \tag{3.52}$$

If the ultracentrifuge is rotated at a velocity at which the difference between the potential energy of a particle at the bottom of the cuvette and that of a particle in the solution is of the magnitude of the product of the Boltzmann constant k_B and the temperature T, the particles are not all spun outwards but become arranged in the cuvette according to a Boltzmann distribution, so that the ratio of the concentration $c(x)$ at some distance x from the rotor axis to the concentration c_0 at a given point x_0 (e.g., the cuvette floor) is defined by

$$\frac{c(x)}{c_o} = \exp\left\{ -\frac{E(x) - E(x_o)}{2k_B T} \right\} \tag{3.53}$$

$$\frac{c(x)}{c_o} = \exp\left\{ -\frac{V_2 \cdot (\rho_2 - \rho_1) \cdot \omega^2 \left(x_o^2 - x^2 \right)}{2k_B T} \right\} \tag{3.54}$$

A descriptive analogy to this dependence is the barometric formula, which describes the dependence of the density of our atmosphere on the height above sea level. However, this is with the boundary conditions that, in this case, it is not the particles in solution but rather the gas molecules in empty space which are involved and that the potential energy is, in this case, a linear and not a quadratic function of the distance from the axis (height).

From (3.54) it follows that a graph of $\ln(c(x)/c_0)$ vs $\omega^2 x^2/(2k_B T)$ should result in a straight line with a gradient of $-V_2 \cdot (\rho_2 - \rho_1)$.

If the densities of the particle and the surrounding liquid are known, the molar mass can be calculated from

$$M_2 = \frac{\left[V_2 \cdot (\rho_2 - \rho_1) \right] \cdot N_A \rho_2}{(\rho_2 - \rho_1)} \tag{3.55}$$

If the sedimentation coefficient S is known from an analysis of the sedimentation velocity, the friction factor f (from (3.45)), the hydrodynamic radius R_h (from (3.43)), and the diffusion coefficient D (from (3.46)) can be calculated from

$$f = \frac{V_2 \cdot (\rho_2 - \rho_1)}{S} \tag{3.56}$$

$$R_h = \frac{V_2 \cdot (\rho_2 - \rho_1)}{6\pi\eta S} \tag{3.57}$$

$$D = \frac{S \cdot k_B T}{V_2 \cdot (\rho_2 - \rho_1)} \tag{3.58}$$

Compared to the analysis of the sedimentation velocity, the analysis of the sedimentation equilibrium has the advantage that, although the densities of the particle and the surrounding solution still need to be known to calculate the molar mass, the diffusion coefficient is no longer required. However, it has the disadvantage of being far more time consuming. For the measurement of sedimentation velocity, rotor speeds of up to 70,000 rpm can be selected. A single experiment suffices and one doesn't even need to wait for equilibrium to be established. Equilibrium measurements usually require a variation of rotation speeds in the region of a few thousand rpm and waiting until equilibrium is established in order to approach incrementally a sedimentation equilibrium which can be evaluated.

According to (3.54), a semi-logarithmic graph of $\ln c$ vs x^2 should result in a straight line. However, if the particles being analyzed are not of a single molar mass, the graph deviates from a straight line. In such cases it is possible to obtain number, weight, and centrifugal averages using appropriate equations. For more details, the interested reader is referred to the specialist literature. (e.g., Maechtle and Boerger 2006).

3.2.5.3 Sedimentation Equilibrium in a Density Gradient

When a liquid mixture of two low molar mass substances, e.g., an aqueous cesium chloride solution, is centrifuged in an ultracentrifuge at a high rotation speed, the substance of lower density moves to the volume facing the axis of rotation and the substance of higher density moves to the outer volume. As described in ▶ Sect. 3.2.5.2, a concentration gradient of the two low molar mass substances develops. If additional particles or polymer chains are present in the mixture, they move to that part of the cuvette where there is no difference between their density and that of the surrounding solution. The position of the polymers in the density gradient can be determined by analyzing the refractive index at different points with suitable imaging optics (e.g., Schlieren optics), and thus the density of particles can be determined.

As far as it is justified to assume that the two low molar mass substances used in this experiment do not influence the density of the particle or the polymer, the density determined with this method can be used to define the molar mass in conjunction with the method of sedimentation equilibrium already described (▶ see (3.55)).

A particularly nice example of such an experiment is shown in ◘ Fig. 3.7. The photographs of the UV spectra taken during a density gradient ultracentrifuge experiment show that the genetic material in a living cell stays essentially unchanged during 'normal' metabolic processes. During cell division exactly half of the genetic material is transferred to each of the daughter cells and in each new cell the missing half is regenerated. For these experiments, microorganisms were grown in a culture medium containing nutrients enriched with the isotope ^{15}N as the sole nitrogen source. These cells were then transferred to a culture medium which only contained the lighter isotope ^{14}N as its nitrogen source and allowed to continue to multiply. At regular intervals organisms were removed, their genetic material extracted, and their density determined in an ultracentrifuge. The blackening in the bands seen in the figure is a measure of the concentration of genetic material present in the cuvette, the position of a band reflecting the density of that material (increasing from left to right). From top to bottom of the figure, the duration spent by the organisms in the new culture medium increases. It is clear that the density does not decrease with time but instead three discrete bands form: the original genetic material with 100% ^{15}N-labeling (right hand band), a middle band with 50% ^{15}N-labeling, and a third band (left hand band) which no longer contains any ^{15}N.

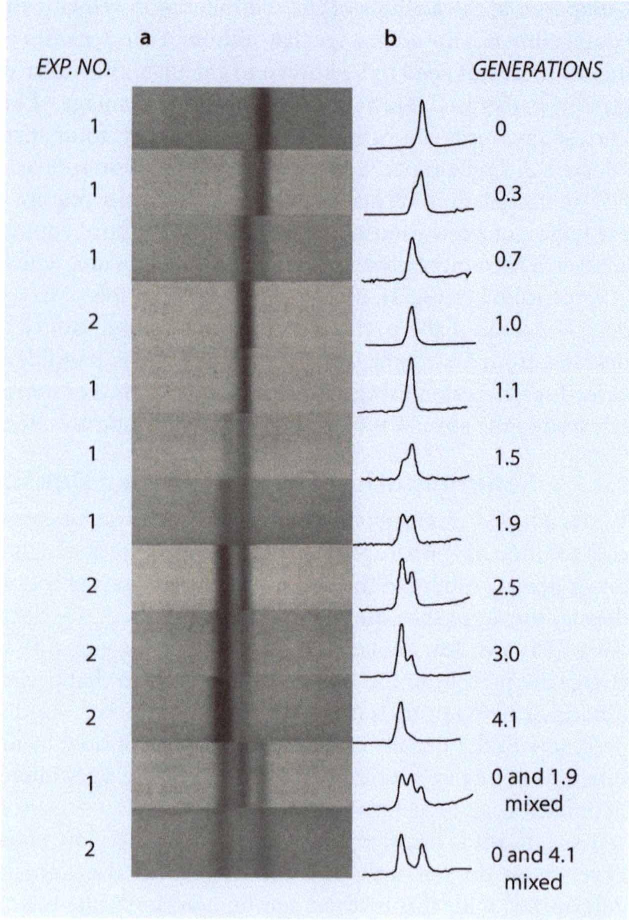

Fig. 3.7 UV-absorption measurements of DNA-bands of bacterial lysates after density gradient ultracentrifugation at varying times after addition of an excess of a ^{14}N-substrate to a ^{15}N-labeled cell culture. The degree of isotopic labeling of a DNA-species corresponds with the relative position of the band and lies between completely labeled and unlabeled DNA, as can be seen in the lowest photograph. Further details can be found in the literature (Meselson and Stahl 1958)

3.2.6 Light Scattering

To describe electromagnetic radiation in a dielectric it is assumed that the dielectric consists of polarizable entities stimulated and made to oscillate by an electromagnetic wave so that these in turn emit electromagnetic radiation of a frequency identical to the initial wave. The waves radially expanding from these individual sources of induced radiation interfere with each other. If the polarizable entities are all identical and equally distributed on a longitudinal scale of at least 1/20 of the wavelength, the interference of the induced waves forms a wave which leads to constructive interference with the incident wave. Thus, stimulation by an even wave also leads to an even wave or, in other words, a ray of light shines through this medium without deflection. However, in polymer solutions the segments of the macromolecules are not homogeneously distributed within the solution (▶ Chap. 2). In this case, components can develop from the interference of the induced radiation which deviates from the original direction of the incident wave. For a polymer solution, the polarizable entities are the repeating units of macromolecules distinguishable by their polarizability from their surroundings (the solution). On the other hand, pure solvents also display a certain scattering behavior. This is caused by local fluctuations in density and the associated inhomogeneities. In solutions of dissolved substances, additional density variations occur.

Thus, an uneven distribution of polarizable entities can originate from the molecules or particles being larger than 1/20 of the wavelength of the incident radiation used and having a characteristic shape or from the molecules being unevenly distributed. A deflection of electromagnetic radiation by large objects of defined shape or from a regular arrangement of objects is usually referred to as *diffraction*. It is usually distinguished by sharp intensity maxima in characteristic directions. Examples of this are the opalescence of opal, moonstone, and other semi-precious stones as well as the iridescent colors of mother of pearl, insect wings, some liquid crystals, and modern effect pigments. A deflection of light by randomly arranged molecules, however, is called *scattering*. Examples of this are the blue color of the cloudless sky or the softening of sunlight by mist.

As the intensity of the scattered light depends on the shape and distribution of the particles or molecules of interest, one can draw conclusions concerning shape, distribution, and their molar mass from the angular dependence of this intensity

3.2.6.1 Static Light Scattering

In static light scattering, the time-average intensity as a function of the observation angle is measured and time-averaged information about the form and distribution of the molecules is obtained.

A polymer solution is illuminated by a light beam (Fig. 3.8). The intensity of the scattered light I_s is determined as a function of the observation angle θ. If a light beam of intensity I_0 is sent through a medium, it is reduced by heat development and scattering. Primary and scattered light are distinguishable by their intensity, the direction of propagation, and their state of polarization.

The following section is concerned with the derivation of the relevant light scattering equations. For information concerning the experimental application and the evaluation of measured data, please refer to ► Sect. "Experimental Execution and Evaluation of the Measurements".

For reasons of length and complexity of the theoretical basics of light scattering, the essential results of the derivation are again briefly summarized in ► Sect. "Experimental Execution and Evaluation of the Measurements" so that the following detailed discussion is not absolutely necessary for understanding the subsequent sections.

Theoretical Basics

To begin with, it is first assumed that the stimulation occurs by a plane wave and that this interacts with a scattering center (e.g., an isolated gas molecule or particle) which is situated in empty space. Additionally, it is assumed that the scattering centers are considerably

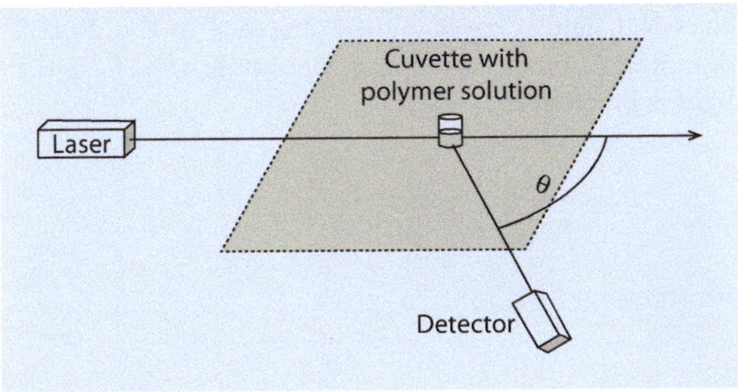

Fig. 3.8 Principle of a light scattering analyzer. Incident and observation directions cross in the cuvette and span a plane

smaller than the wavelength of the incident light so that their shape has no influence on the process and no interactions occur which could force them into preferred pattern. The scattering centers can thus be described as being randomly distributed and without short range order.

The stimulating plane wave produces an electrical field E at the site of the scattering center, which oscillates with the frequency ν:

$$E = E_0 \cdot \cos(2\pi\nu \cdot t) \tag{3.59}$$

E	Electrical field of the light-wave
E_0	Amplitude of the electrical field
ν	Frequency of the incident light
t	Time

The intensity I_0 of the incident wave, i.e., the time average of the ratio of the strength of the radiation and the irradiated area, equals

$$I_0 = c\varepsilon_0\langle E^2\rangle = c\varepsilon_0 {E_0}^2 \left\langle \cos^2(2\pi\nu \cdot t)\right\rangle = c\varepsilon_0 \frac{{E_0}^2}{2} \tag{3.60}$$

ε_0	Permittivity of a vacuum
c	Speed of light

The angular brackets $\langle \dots \rangle$ denote the formation of the average value over a longer time.

If the electromagnetic wave encounters a scattering center, a dipole moment μ is induced by the electrical field of the light-wave E, whose strength is proportional to its polarizability α:

$$\mu = \alpha \cdot \mathrm{E} \tag{3.61}$$

This induced dipole moment oscillates with the frequency of the incident light-wave:

$$\mu = \alpha \cdot E_0 \cdot \cos(2\pi\nu \cdot t) \tag{3.62}$$

The oscillating dipole creates an electromagnetic wave which spreads radially from the dipole with velocity c. This electromagnetic wave contains components of both electric and magnetic fields, which are perpendicular to each other and to the direction of propagation. At a large enough distance from the dipole, the strength of the electrical field is given by the distance to the dipole, r, and the angle between the direction of propagation and the dipole axis, ϑ. If the field strength of the scattered light is given by E_s:

$$E_s = \frac{1}{4\pi\varepsilon_0 c^2} \cdot \frac{1}{r} \sin\vartheta \cdot \frac{d^2\mu}{dt^2} \tag{3.63}$$

Substituting (3.62) in (3.63) gives

$$E_s = -\frac{\pi \cdot \alpha \cdot \nu^2}{\varepsilon_0 c^2} \cdot \frac{1}{r} \sin \vartheta \cdot E_0 \cos(2\pi\nu \cdot t) \tag{3.64}$$

The detector of the light scattering apparatus registers the intensity, i.e., the ratio of the strength of the light impacting on the detector and its area, or, more exactly, the time average of this intensity. This intensity is identical to the product of $\varepsilon_0 c$ and the square of the wave's electrical field strength. Thus, the intensity of the scattered light is given by

$$I = \varepsilon_o c \cdot \left\langle \left[-\frac{\pi \cdot \alpha \cdot \nu^2}{\varepsilon_0 c^2} \cdot \frac{1}{r} \sin\vartheta \cdot E_0 \cos(2\pi\nu \cdot t) \right] \right\rangle^2 \tag{3.65}$$

$$I = \left(\frac{\pi \cdot \alpha \cdot \nu^2}{\varepsilon_0 c^2} \cdot \frac{1}{r} \sin\vartheta \right)^2 \cdot \varepsilon_o c E_0{}^2 \cdot \left\langle \cos^2(2\pi\nu \cdot t) \right\rangle \tag{3.66}$$

$$I = \left(\frac{\pi \cdot \alpha \cdot \nu^2}{\varepsilon_0 c^2} \cdot \frac{1}{r} \sin\vartheta \right)^2 \cdot \varepsilon_o c E_0{}^2 \cdot \frac{1}{2} \tag{3.67}$$

The intensity of the stimulating wave equals $1/2\varepsilon_o c E_0{}^2$ and the relation ν/c is identical to the reciprocal wavelength $1/\lambda_0$. Thus, for the ratio of scattered to incident intensity, I/I_0,

$$\frac{I}{I_o} = \frac{\pi^2 \cdot \alpha^2}{\varepsilon_0{}^2 \lambda_0^4} \cdot \frac{1}{r^2} \sin^2 \vartheta \tag{3.68}$$

In a light scattering experiment, the illuminating light beam and the connecting line from the scattering volume to the detector form a plane. The stimulated wave can now take two marked polarization directions relative to this plane. If the irradiated wave is polarized vertical to the plane (Fig. 3.9a), then the angle between the direction of observation and the dipole axis ϑ is always 90°, regardless of the angle θ between the direction of irradiation and that of observation. As a result, for the case of vertical polarization $I_\perp$:

$$\frac{I_\perp}{I_o} = \frac{\pi^2 \cdot \alpha^2}{\varepsilon_0{}^2 \lambda_0^4} \cdot \frac{1}{r^2} \tag{3.69}$$

If the irradiated wave is polarized parallel to this plane (Fig. 3.9b), then $\vartheta = 90° - \theta$. In this case (scattering intensity for the case of parallel (horizontal) polarization $I_{\|}$):

$$\frac{I_{\|}}{I_o} = \frac{\pi^2 \cdot \alpha^2}{\varepsilon_0{}^2 \lambda^4} \cdot \frac{1}{r^2} \cos^2 \theta \tag{3.70}$$

If the scattering center is stimulated by non-polarized light, this corresponds to a simultaneous stimulation in both polarization directions and the scattered intensity corresponds to the average of (3.69) and (3.70):

$$\frac{I}{I_o} = \frac{\pi^2 \cdot \alpha^2}{\varepsilon_0{}^2 \lambda_0^4} \cdot \frac{1}{r^2} \cdot \frac{1 + \cos^2 \theta}{2} \tag{3.71}$$

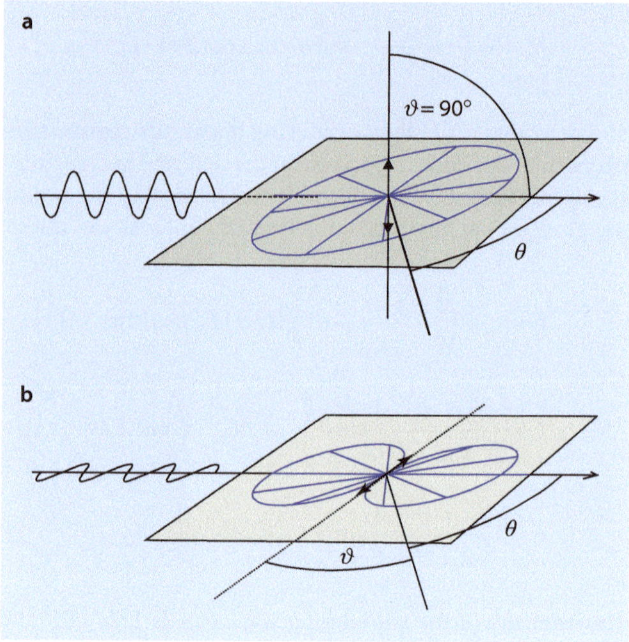

Fig. 3.9 Dependence of the intensity of the scattered light on the observation angle θ. (**a**) Vertical polarization. (**b**) Parallel (horizontal) polarization

If it is assumed that

- Not only one, but N scattering centers are present in the stimulated volume V under observation
- The scattering centers are small enough to exclude interference occurring within the scattering centers
- The distances between the scattering centers are large enough
- The centers are distributed arbitrarily in the volume

then the scattering processes can be considered to be mutually independent. In this case, the intensity of the scattered light is given by (3.71) multiplied by the number of scattering centers

$$\frac{I}{I_o} = \frac{\pi^2 \cdot \alpha^2}{\varepsilon_0{}^2 \lambda_0^4} \cdot \frac{1}{r^2} \cdot \frac{1+\cos^2\theta}{2} \cdot N \tag{3.72}$$

If this is divided by those parameters that solely depend on the apparatus (r, θ, and V), one obtains the so-called Rayleigh ratio R_θ, which describes, for example, the scattering of light through dilute gases:

$$R_\theta \equiv \frac{I}{I_o} \frac{r^2}{\left(1+\cos^2\theta\right)} \frac{1}{V} = \frac{\pi^2 \cdot \alpha^2}{2\varepsilon_0{}^2 \lambda_0^4} \cdot \frac{N}{V} \tag{3.73}$$

Three conclusions can be drawn from (3.73):

- The intensity of the scattered light increases linear to the concentration (number/volume) of the scattering centers, N/V
- The intensity is proportional to the polarizability α of the scattering centers
- The intensity is inversely proportional to the wavelength λ_0

The last point should be familiar from daily life: the sky or faraway mountains appear to be blue because the atmosphere preferentially scatters the short-wavelength rays from the sun laterally and the scattered radiation is superimposed by our perception onto distant objects. If one looks directly at the sun, for example during sunset, it appears to be yellow or red because the scattering preferentially reduces the shorter wavelength parts of the sun's light.

With a knowledge of the polarizability of the scattering centers and their concentration (number), and therefrom the corresponding mass concentration, the molar mass of the centers can be calculated. This is possible, for example, with dilute gases, for which it can be assumed that each atom is a mutually independent scattering center. However, in this case, (3.73) is more likely to be employed to determine the polarizability of the individual molecules or atoms from a knowledge of their number concentration.

If light scattering is not applied to the scattering centers in a dilute gas but rather to a polymer solution, the situation is different in two respects:

1. The electrical field polarizes not only the scattering center, but also its surroundings
2. It cannot be assumed that the scattering centers are mutually independent

To understand the first difference, the influence of a non-conducting medium on an electrical field needs to be considered. If a dielectric is exposed to an external field E_0, it is polarized and inside it a diminished field E_i prevails:

$$E_i = \frac{E_0}{\varepsilon_r} \tag{3.74}$$

Here the relative permittivity ε_r is a material-specific and, usually, frequency-dependent constant. According to Maxwell, in non-conducting materials, this is equal to the square of the refractive index $\tilde{n}$:

$$\varepsilon_r = \tilde{n}^2 \tag{3.75}$$

When electromagnetic radiation encounters a non-conductor, the waves penetrate it. The radiation retains the same frequency and its propagation velocity is reduced to a speed $v = c / \tilde{n} = c / \sqrt{\varepsilon_r}$ relative to the speed of light c in a vacuum.

Conceptually, a polymer solution can be divided into closely packed small units of volume $dV = V/N_{SC}$, whose number N_{SC} equals that of the scattering centers. It is assumed that fluctuations in the composition lead to each of these individual elements having a relative permittivity $\varepsilon_r + \delta\varepsilon_r$, which deviates from the permittivity of its surroundings by a small amount, $\delta\varepsilon_r$. An electrical field which interacts with this small volume causes it to be polarized so that, observed from a suitably large distance, it appears as a simple dipole moment whose moment is proportional to the difference of its permittivity and that of its surroundings:

$$\mu = \varepsilon_0 \cdot \delta\varepsilon_r dV \cdot E \tag{3.76}$$

This means that each of these small volumes can act as scattering centers and that α^2 in (3.73) can be replaced by $(\varepsilon_0 \delta\varepsilon_r dV)^2$. Considering a larger number of scattering centers, the system has to retain the freedom that the permittivity of each and every scattering center may differ from those of its neighbors. Thus, an average value for these deviations needs to be employed and α^2 replaced by $\varepsilon_0^2 \left\langle \delta\varepsilon_0^2 \right\rangle dV^2$. Thus the intensity of the scattered light becomes a function of the variance of the relative permittivity. Next, the dependency of the relative permittivity on composition needs to be defined.

$$\delta\varepsilon_r^{\;2} = \left(\frac{\partial\varepsilon_r}{\partial c_2}\right)^2 \left\langle \delta c_2^{\;2} \right\rangle \tag{3.77}$$

However, it is rather time-consuming to determine relative permittivities at the frequencies of the incident light directly. Experimentally, it is considerably easier to determine changes in the refractive index with concentration $\partial \tilde{n}/\partial c_2$ so that for convenience the following equation is employed:

$$\left\langle \delta\varepsilon_r^{\;2} \right\rangle = \left(\frac{\partial\varepsilon_r}{\partial \tilde{n}}\right)^2 \left(\frac{\partial \tilde{n}}{\partial c_2}\right)^2 \left\langle \delta c_2^{\;2} \right\rangle \tag{3.78}$$

From (3.75) one obtains

$$\frac{\partial\varepsilon_r}{\partial \tilde{n}} = 2\tilde{n} \tag{3.79}$$

Inserting this into (3.73) gives

$$R_\theta = \frac{\pi^2 \cdot 4\tilde{n}^2 \varepsilon_0^{\;2} \left(\frac{\partial \tilde{n}}{\partial c_2}\right)^2 \left\langle \delta c_2^{\;2} \right\rangle (dV)^2}{2\varepsilon_0^{\;2} \lambda_0^4} \cdot \frac{N_{SC}}{V} \tag{3.80}$$

$$R_\theta = \frac{\pi^2 \cdot 2\tilde{n}^2 \left(\frac{\partial \tilde{n}}{\partial c_2}\right)^2}{\lambda_0^4} \cdot \left\langle \delta c_2^{\;2} \right\rangle dV \tag{3.81}$$

From (3.81) it can be seen that the intensity of the scattered light is a function of the variance of the concentration of the dissolved component. At first glance it may appear irritating that the intensity is also dependent on the arbitrarily selected size of the small scattering volume elements (dV). However, as shown below, this apparent dependency is cancelled out by the variance of the composition increasing as the size of the elements decreases. In almost all areas of chemistry it is an admissible simplification to assume that the density of a gas or the composition of a solution within a given volume is identical and independent of any chosen point within the volume. However, for an understanding of light scattering and polymer solutions it is essential to go into more detail. A strictly even distribution of molecules can only be achieved if they are somehow periodically arranged but this would necessitate allocating the molecules to precisely defined positions in space and would only allow a limited number of distribution possibilities. The highest number of distribution possibilities of molecules in space and thus the largest entropy is attained if we assume that we can find a molecule at every position of this space with the same probability (independent of the proximity of another molecule). Such a purely random distribution results in local variations in density and composition. These may be small compared to the average density or composition, but they can be exactly described using statistical methods. We now turn again to the concept that the polymer solution is made up of many, closely packed small volume elements dV whereby the number concentration or the molar concentration of the molecules or particles contained in each single sub-volume can be described by a normal distribution whose breadth increases with decreasing size of the sub-volume.

To obtain the variance of the deviation of the concentration from the average concentration $\left\langle \delta c_2^2 \right\rangle$—the product of the square of the deviation δc_2^2 and the probability of this deviation $p(\delta c_2)$—must be integrated over all possible deviations:

$$\left\langle \delta c_2{}^2 \right\rangle = \int p(\delta c_2)\ \delta c_2{}^2\ d\delta c_2 \tag{3.82}$$

We receive the probability $p(\delta c_2)$ of a deviation via the corresponding Boltzmann distributions of the changes of the free enthalpy of mixing $\delta \Delta G^{\mathrm{m}}(\delta c_2)$ inferred by the deviation. To make the following equations easier to understand, G is used rather than $\Delta G^{\mathrm{m}}(\delta c_2)$:

$$p(\delta c_2) = \frac{\exp\left\{-\dfrac{\delta G}{k_B T}\right\} d\delta c_2}{\int \exp\left\{-\dfrac{\delta G}{k_B T}\right\} d\delta c_2} \tag{3.83}$$

Because all the combined compositions of all the sub-volumes must equal the average composition, the composition of some volumes deviates positively from the average composition whereas others deviate negatively. To simplify, it is assumed that for every sub-volume that deviates from the average composition by the amount δc_2 there exists another sub-volume that deviates from the composition by the opposite amount, $-\delta c_2$. The free enthalpy of mixing of each of these volumes can be written as a Taylor series which is terminated here after the second element. Thus, combining the deviations from the two volumes considered one obtains

$$\delta G = \frac{1}{2}\left[\frac{\partial G}{\partial c_2}\cdot \delta c_2 + \frac{1}{2}\frac{\partial^2 G}{\partial c_2{}^2}\cdot \delta c_2{}^2 + \frac{\partial G}{\partial c_2}\cdot(-\delta c_2) + \frac{1}{2}\frac{\partial^2 G}{\partial c_2{}^2}\cdot(-\delta c_2)^2\right] \tag{3.84}$$

$$\delta G = \frac{1}{2}\frac{\partial^2 G}{\partial c_2{}^2}\cdot \delta c_2{}^2 \tag{3.85}$$

Here the first two terms in the square bracket refer to the first sub-volume and the last two refer to the second sub-volume with opposite deviation. The factor ½ in front of the bracket arises because the change of the free enthalpy for each sub-volume is considered but the terms in parentheses are the sum of the free enthalpies for two sub-volumes.

The two linear terms in the parentheses cancel each other so that, to a first approximation, the deviation of the free enthalpy from the average value is proportional to the square of the deviation.

$$p(\delta c) = \frac{\exp\left\{-\frac{1}{2}\frac{\partial^2 G/\partial c_2{}^2}{k_B T}\cdot \delta c_2{}^2\right\} d\delta c_2}{\int \exp\left\{-\frac{1}{2}\frac{\partial^2 G/\partial c_2{}^2}{k_B T}\cdot \delta c_2{}^2\right\} d\delta c_2} \tag{3.86}$$

Thus, the function $p(\delta c_2)$ is described by a normal distribution with a variance given by

$$\left\langle \delta c_2{}^2 \right\rangle = \frac{k_B T}{\partial^2 G/\partial c_2{}^2} \tag{3.87}$$

The next step requires the second derivative of the free enthalpy as a function of concentration. To obtain this it, is assumed that the free enthalpy of the system is given by the sum of the chemical potentials of the two substances μ_1 and μ_2 multiplied by their amounts, N_1 and N_2, respectively:

$$G = \mu_1 N_1 + \mu_2 N_2 \tag{3.88}$$

The derivative with respect to the concentration of the dissolved substance is given by

$$\frac{dG}{dc_2} = N_1 \frac{d\mu_1}{dc_2} + \mu_1 \frac{dN_1}{dc_2} + N_2 \frac{d\mu_2}{dc_2} + \mu_2 \frac{dN_2}{dc_2} \tag{3.89}$$

According to the Gibbs–Duhem equation it holds that

$$N_1 \frac{d\mu_1}{dc_2} + N_2 \frac{d\mu_2}{dc_2} = 0 \Leftrightarrow \frac{d\mu_2}{dc_2} = -\frac{N_1}{N_2}\frac{d\mu_1}{dc_2} \tag{3.90}$$

Thus, the sum of the first and third terms on the right hand side of (3.89) equals zero. Furthermore, for the scattering center at constant volume, dV, one gets:

$$\frac{dc_2}{dN_2} = \frac{d}{dN_2}\frac{M_2 N_2}{dV} = \frac{M_2}{dV}\frac{d}{dN_2}N_2 = \frac{M_2}{dV} \Leftrightarrow \frac{dN_2}{dc_2} = \frac{dV}{M_2} \tag{3.91}$$

$$\frac{dc_2}{dN_1} = \frac{d}{dN_1}\frac{M_2 N_2}{dV} = \frac{M_2}{dV}\frac{d}{dN_1}N_2 = \frac{M_2}{dV}\frac{d}{dN_1}\frac{dV - \overline{V_1} N_1}{\overline{V_2}} \tag{3.92}$$

$$\frac{dc_2}{dN_1} = -\frac{\overline{V_1}}{\overline{V_2}}\frac{M_2}{dV} \Leftrightarrow \frac{dN_1}{dc_2} = -\frac{\overline{V_2}}{\overline{V_1}}\frac{dV}{M_2} \tag{3.93}$$

Thus, from (3.89)–(3.92):

$$\frac{dG}{dc_2} = \left(\mu_2 - \frac{\overline{V_2}}{\overline{V_1}}\mu_1\right)\frac{dV}{M_2} \tag{3.94}$$

Differentiating (3.94) with respect to c_2 yields

$$\frac{d^2G}{d{c_2}^2} = \left(\frac{d}{dc_2}\mu_2 - \frac{\overline{V_2}}{\overline{V_1}} \frac{d}{dc_2}\mu_1 \right) \frac{dV}{M_2} \tag{3.95}$$

Applying the Gibbs–Duhem equation (3.90) once more, one obtains

$$\frac{d^2G}{d{c_2}^2} = -\left(\frac{N_1}{N_2} + \frac{\overline{V_2}}{\overline{V_1}} \right) \frac{dV}{M_2} \frac{d\mu_1}{dc_2} = -\frac{N_1\overline{V_1} + N_2\overline{V_2}}{N_2\overline{V_1}} \frac{dV}{M_2} \frac{d\mu_1}{dc_2} \tag{3.96}$$

$$\frac{d^2G}{d{c_2}^2} = -\frac{N_1\overline{V_1} + N_2\overline{V_2}}{N_2 M_2} \frac{dV}{\overline{V_1}} \frac{d\mu_1}{dc_2} = -\frac{1}{c_2} \frac{dV}{\overline{V_1}} \frac{d\mu_1}{dc_2} \tag{3.97}$$

Assuming that the chemical potential is dependent on the concentration, concentration-dependent activity constants could be inserted. However, when deriving the theory for osmometry, an alternative approach was used. Thus, deviation from ideal behavior was described with the aid of virial coefficients. The same approach can be made in this case. From the equation

$$\mu_1 = \mu_1{}^\circ + RT \cdot \overline{V_1} \left(\frac{1}{M_2} \cdot c_2 + A_2 \cdot {c_2}^2 + A_3 \cdot {c_2}^3 + \ldots \right), \tag{3.98}$$

it follows that

$$\frac{d\mu_1}{dc_2} = RT \cdot \overline{V_1} \left(\frac{1}{M_2} + 2A_2 \cdot c_2 + 3A_3 \cdot {c_2}^2 + \ldots \right) \tag{3.99}$$

and

$$\frac{d^2G}{d{c_2}^2} = \frac{dV}{c_2} RT \left(\frac{1}{M_2} + 2A_2 \cdot c_2 + 3A_3 \cdot {c_2}^2 + \ldots \right) \tag{3.100}$$

If (3.100) is combined with (3.87) one obtains

$$\left\langle \delta c^2 \right\rangle = \frac{c_2}{N_A \left(\frac{1}{M_2} + 2A_2 \cdot c_2 + 3A_3 \cdot {c_2}^2 + \ldots \right)} \frac{1}{dV} \tag{3.101}$$

Substituting (3.101) into (3.80) yields

$$R_\theta = \frac{4\pi^2}{\lambda_0^4} \tilde{n}^2 \left(\frac{\partial \tilde{n}}{\partial c} \right)^2 \frac{1}{N_A} \cdot \frac{c_2}{\left(\frac{1}{M_2} + 2A_2 \cdot c_2 + 3A_3 \cdot (c_2)^2 + \ldots \right)} \tag{3.102}$$

The value of (3.102) can be more easily appreciated by introducing an optical constant K, whereby

$$K = \frac{4\pi^2}{\lambda_0^4 N_A} \tilde{n}^2 \left(\frac{\partial \tilde{n}}{\partial c} \right)^2, \tag{3.103}$$

so that (3.102) becomes

$$K \frac{c_2}{R_\theta} = \frac{1}{M_2} + 2A_2 \cdot c_2 + 3A_3 \cdot (c_2)^2 + \ldots \tag{3.104}$$

Not only does (3.104) corroborate the expectation that the ratio of the scattering intensity and concentration should increase proportional to the size of the dissolved molecules; it also permits the quantitative determination of the molar mass of the scattering particles. Because solutions are rarely ideal, considerable dilution is necessary to ensure that the non-ideal effects are negligible. For very dilute solutions (3.104) reduces to

$$K \cdot \lim_{c_2 \to 0} \left(\frac{c_2}{R_\theta} \right) = \frac{1}{M_2} \tag{3.105}$$

In practical experiments, the signal/noise ratio deteriorates with concentration so that the scattering intensities are determined for a range of concentrations and the results fitted to the above polynomial equation and this extrapolated to c_2=0. In most cases, a straight line fitted to the data points at lowest concentrations suffices for a useful result. This method not only gives the molar mass of the dissolved substance; by determining the osmotic pressure from the virial coefficient, information about the quality of the solvent can also be obtained.

If the polymer does not have a singular molar mass, it is interesting to consider the following rearrangement of (3.105):

$$1 = K \cdot \lim_{c_2 \to 0} \left(\frac{c_2 M_2}{R_\theta} \right) \tag{3.106}$$

For a polymer with a distribution of molar mass, the contribution of each of the components to the scattering intensity can be considered as additive so that

$$1 = K \cdot \lim_{\sum c_i \to 0} \left(\frac{1}{R_\theta} \sum c_i M_i \right) = K \cdot \lim_{\sum c_i \to 0} \left(\frac{\sum c_i}{R_\theta} \cdot \frac{\sum c_i M_i}{\sum c_i} \right) \tag{3.107}$$

$$1 = K \cdot \lim_{\sum c_i \to 0} \left(\frac{\sum c_i}{R_\theta} \frac{\sum m_i M_i}{\sum m_i} \right) = K \cdot \lim_{\sum c_i \to 0} \left(\frac{\sum c_i}{R_\theta} M_w \right) \tag{3.108}$$

$$1 = K \cdot \lim_{c_2 \to 0} \left(\frac{c_2}{R_\theta} M_w \right) = K \cdot \lim_{c_2 \to 0} \left(\frac{c_2}{R_\theta} \right) \cdot M_w \tag{3.109}$$

Thus, static light scattering does not yield a number average molar mass as the other methods described above, but a weight average molar mass.

Up to this point it has been assumed that the dissolved molecules are small compared to the wavelength of the employed light. For particles whose size is larger than $\lambda/20$ it is necessary to take into account that radiation is emitted from various regions of the particle and that the different portions of radiation interfere with each other (Fig. 3.10). Scattering in the same direction as the incident light, i.e., at a scattering angle of approximately zero, all components of this scattered light are in phase and interfere constructively. Lateral scattering leads to phase shifts between the different components and the scattered light is reduced *depending on the angle.*

The dependency of the scattering intensity on the scattering angle is specific to the shape of the molecule or particle. This can be expressed by multiplying the intensity of the scattered light with an angle-dependent factor, the, so-called form factor.

At least for spherically symmetrical objects it is relatively easy to derive the form from the dependence of scattering intensity on the scattering angle. If two waves of the same amplitude E and identical wavelength λ originate from two points offset against each other in the direction of the direction of propagation x by a line segment S, they can be added together using the rules of addition for trigonometric functions to give a wave whose amplitude depends on the offset S according to

$$E\sin\left[\frac{2}{\,}\ x-\frac{S}{2}\right]+E\sin\left[\frac{2}{\,}\ x+\frac{S}{2}\right]=2E\sin\left[\frac{2}{\,}x\right]\cdot\cos\left[\frac{2}{\,}\frac{S}{2}\right] \qquad (3.110)$$

The ratio of the amplitude at negligible offset, E^*, and the interference reduced amplitude, E, is given by

$$\frac{E}{E^*}=\cos\left[\frac{2}{\,}\frac{S}{2}\right] \qquad (3.111)$$

The geometry of light scattering by a large object can be explained as shown in (Fig. 3.11a). It is observed that all waves which originate from a plane along a single line parallel with

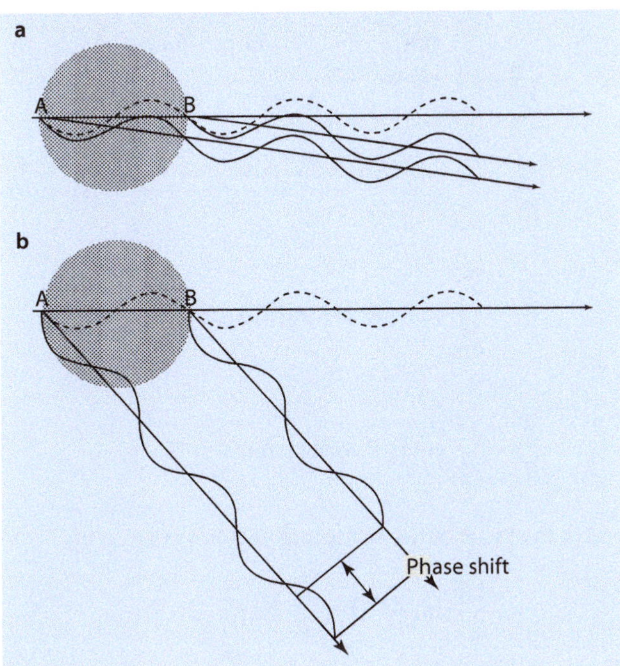

Fig. 3.10 Scattered light in different regions because of a "large" object. (**a**) Negligible phase difference where the scattering angle is close to zero. (**b**) Dependent on the angle and no longer negligible when the scattering angle is considerably larger than zero. This leads to destructive interference so that the scattering intensity is reduced and dependent on the scattering angle

3

half the observation angle (dashed line), are in phase—the line segment $\overline{PO}$ has the same length as the line segment $\overline{PO'}$. Waves that originate from a plane lying parallel to this plane, but shifted by a distance Δx (e.g., the dotted line), compared to waves which originate from the first plane, are phase shifted by an amount S:

$$S = \overline{O'Q} + \overline{QO''} = d\sin(\theta/2) + d\sin(\theta/2) = 2d\sin(\theta/2) \tag{3.112}$$

The simplest symmetrical, spherical object is a hollow sphere with extremely thin walls and uniform density (symbolized by the gray circle in Fig. 3.11b). Such a hollow sphere can be divided into rings lying parallel to a plane which cuts the center of the hollow sphere (represented by the dashed line in Fig. 3.11b) and halves the observation angle. Two of these rings (symbolized by dotted lines in Fig. 3.11b) at an equal distance in front of or behind this plane have an identical circumference. The interference of the radiation scattered by each pair of rings leads to a radiation which is in phase with the radiation scattered by the plane represented by the dashed line. Thus, the amplitude of the radiation originating from each of the pairs can be initially determined and this can then be integrated over all rings. Each pair of rings contributes to the scattering radiation with an intensity proportional to the ratio of the area of the ring-pair to the total area of the sphere, multiplied by the reduction caused by interference according to (3.110). If the radius of the sphere is designated as R and the half apex angle of a cone connecting the center of the sphere with the ring, as α, then the circumference of the ring is given by $2\pi R\sin\alpha$, its breadth by $Rd\alpha$, and its area by $R^2\sin\alpha d\alpha$. As the geometry is mirror-symmetrical, the integration needs only to be made over one hemisphere. The surface area of a hemisphere is equal to $2\pi R^2$. Therefore for a hollow sphere:

$$\frac{E}{E^*} = \int_{\alpha=0}^{\pi/2} \frac{\cos\left[\frac{2\pi}{\lambda}\frac{S}{2}\right] \cdot 2\pi R\sin\alpha \cdot Rd\alpha}{2\pi R^2} = \int_{\alpha=0}^{\pi/2} \cos\left[\frac{2\pi}{\lambda}\frac{S}{2}\right]\sin\alpha \cdot d\alpha \tag{3.113}$$

$$\frac{E}{E^*} = \int_{\alpha=0}^{\pi/2} \cos\left[\frac{2\pi}{\lambda}\frac{2\cdot 2R\cdot\cos\alpha\cdot\sin(\theta/2)}{2}\right]\sin\alpha d\alpha \tag{3.114}$$

$$\frac{E}{E^*} = \int_{\alpha=0}^{\pi/2} \cos\left[qR\cdot\cos\alpha\right]\sin\alpha d\alpha \tag{3.115}$$

$$\frac{E}{E^*} = \int_{\alpha=0}^{\pi/2} \cos\left[qR\cdot\cos\alpha\right]\sin\alpha d\alpha \tag{3.116}$$

where the value of the scattering vector q is given by

$$q = \frac{4\pi\tilde{n}}{\lambda_0}\sin(\theta/2) = \frac{4\pi}{\lambda}\sin(\theta/2) \tag{3.117}$$

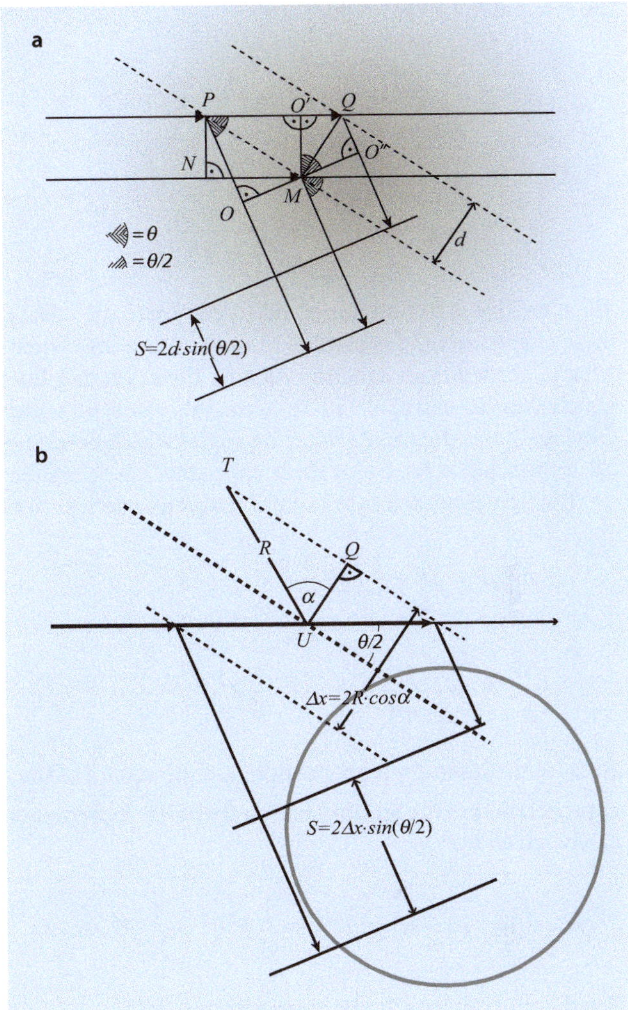

Fig. 3.11 (**a**) Geometrical demonstration of phase differences as a function of scattering angle. (**b**) Alternative approach to the geometric relationships (see text for explanation)

Although the integral in (3.116) looks quite complicated with its nested trigonometric functions, it can be solved relatively easily by introducing the substitution

$$a = qR \cdot \cos\alpha \tag{3.118}$$

Because

$$\frac{da}{d\alpha} = -qR \cdot \sin\alpha \Leftrightarrow d\alpha = -\frac{1}{qR \cdot \sin\alpha} da \tag{3.119}$$

and applying the integration limits

$$a(\alpha = 0) = qR, \quad a\left(\alpha = \pi/2\right) = 0 \tag{3.120}$$

the integral in (3.116) reduces to

$$\frac{E}{E^*} = \int_{a=qR}^{0} \cos a \cdot \sin\alpha \left(-\frac{1}{qR \cdot \sin\alpha} \right) da = -\frac{1}{qR} \int_{a=qR}^{0} \cos a \cdot da \tag{3.121}$$

$$\frac{E}{E^*} = -\frac{1}{qR}\left(\sin 0 - \sin qR\right) = \frac{\sin qR}{qR} \tag{3.122}$$

The objective is to find a method to eliminate the effects of interference. It is known that these effects become negligible in the limiting case when the angle of scattering is zero so what is needed is an equation relating the scattering intensity and the angle of scattering which can be extrapolated to zero, the value of which cannot be directly measured. Obviously, it is the small scattering angles which need to be examined. In the limiting case, qR approaches zero so that three approaches are possible.

The first approach is to expand $\sin qR$ as a series truncated after the third element:

$$\sin qR = qR - \frac{1}{3!}(qR)^3 \tag{3.123}$$

$$\frac{E}{E^*} = \frac{qR - \frac{1}{3!}(qR)^3}{qR} = 1 - \frac{1}{6}R^2q^2 \tag{3.124}$$

Because the intensity is proportional to the square of the amplitude ($I = |E|^2$), the second approach makes the approximation that qR is extremely small so that the form factor $P(q)$ can be given by

$$\frac{I}{I^*} = \left(1 - \frac{1}{6}R^2q^2\right)^2 = 1 - \frac{1}{3}R^2q^2 + \frac{1}{36}R^4q^4 \approx 1 - \frac{1}{3}R^2q^2 = P(q) \tag{3.125}$$

For the third approach, the approximation $1/(1 - \mathrm{x}) \approx 1 + \mathrm{x}$ is made so that

$$\frac{I^*}{I} = \frac{1}{P(q)} \approx 1 + \frac{1}{3}R^2q^2 \tag{3.126}$$

The transition from a hollow sphere to any symmetrical spherical solid is now simple. The sphere can simply be built up from concentric spheres in which the local concentration of polymer segments, $c(R)$, is a function of the radius, and integrate over the radii:

$$\frac{1}{P(q)} \approx 1 + \frac{1}{3}q^2 \frac{\int_{R=0}^{\infty} R^2 c(R) 4\pi R^2 dR}{\int_{R=0}^{\infty} c(R) 4\pi R^2 dR} \tag{3.127}$$

The quotient of integrals in (3.127) is called the square of the radius of gyration r_G. Here the radius of gyration—as is the norm in polymer science—is defined by the average square distance from the center of gravity of the object. However, in engineering it is the norm to define the radius of gyration by the average square distance from a defined axis of rotation. It is important not to confuse these two definitions.

By inserting the radius of gyration into (3.127) one obtains

$$\frac{1}{P(q)} \approx 1 + \frac{1}{3} q^2 r_G^2 \tag{3.128}$$

It can be shown that (3.128) is also valid for spherically asymmetrical objects, e.g., cylinders, ellipsoids, or discs, but the mathematical proof is not given here. The "trick" of observing pairs of planes at equal distances from the center of the object and limiting observations to small scattering angles serve as a logical explanation that all further observations of the interferences lead simply to additive combinations of the corresponding intensities. In this way, the scattering of a number of different objects can be treated as the scattering by uniform objects of an averaged shape. The sum of asymmetrical spherical objects scattered in space with no predominant orientation act, on average, as symmetrical spherical objects with identical radii of gyration.

With the form factor $P(q)$, in analogy to (3.104) and (3.105):

$$K \frac{c_2}{R_\theta} = \frac{1}{M_2 P(q)} + 2A_2 \cdot c_2 + 3A_3 \cdot (c_2)^2 + \ldots \tag{3.129}$$

and

$$K \cdot \lim_{c_2 \to 0} \left(\frac{c_2}{R_\theta} \right) = \frac{1}{M_2 P(q)} \tag{3.130}$$

Inserting (3.128) leads to

$$K \frac{c_2}{R_\theta} \approx \frac{1}{M_2} \left(1 + \frac{1}{3} q^2 r_G^2 + \ldots \right) + 2A_2 c_2 + 3A_3 (c_2)^2 + \cdots \tag{3.131}$$

$$K \cdot \lim_{c_2 \to 0, q \to 0} \left(\frac{c_2}{R_\theta} \right) = \frac{1}{M_2} \tag{3.132}$$

Equations (3.128) and (3.131) are the quintessence of this section. To eliminate the effect of interference by extrapolating to a zero scattering angle, a graph of Kc_2/R_θ vs q^2 or $\sin^2 (\theta/2)$ gives a straight line ably suitable for extrapolation to $\theta = 0$. For polydisperse samples, light scattering yields a weight average molar mass M_w.

Experimental Execution and Evaluation of the Measurements

As discussed above, with this method the light scattered by a polymer solution is measured at different observation angles. The refraction increment $d\tilde{n}/dc_2$ is determined from separate measurements. To eliminate non-ideal effects, measurements are made over a range of concentrations.

From the results of the previous section, the following equation, the so-called *Zimm equation*, can be derived for vertically polarized incident light using (3.73). This equation describes the intensity of light scattered by polymer solution:

$$\frac{4\pi^2 \cdot \tilde{n}^2}{N_A \cdot \lambda_0^4} \cdot \left(\frac{d\tilde{n}}{dc_2}\right)^2 \cdot \frac{c_2}{R_\theta} = \frac{1}{M_w}\left(1 + \frac{16\pi^2\,\tilde{n}^2}{3\lambda_0^2} \cdot \sin^2\left(\theta/2\right) \cdot r_G^2\right) + 2A_2 \cdot c_2 \tag{3.133}$$

For the purpose of clarity, the definitions of the symbols are listed again here:

$\tilde{n}$	Refractive index of the solvent
N_A	Avogadro's constant
λ_0	Wavelength of light in vacuum
$d\tilde{n}/dc_2$	Refractive index increment
c_2	Concentration of the dissolved substance
R_θ	Experimentally defined Raleigh ration (see below)
M_W	Weight average molar mass
θ	Scattering angle
r_G	Radius of gyration of the polymer chain
A_2	Second virial coefficient

The Raleigh ratio of the sample—the actual experimental measurement result of the light scattering experiment—is usually obtained by subtracting the measured scattering intensities of the solvent I_1 from those of the solution I_2; additionally, the Raleigh ratio is normalized using the ratio of a standard R_{stand} (nowadays generally toluene) determined at the wavelength being employed:

$$R_\theta = \frac{I_2 - I_1}{I_1} R_{stand} \tag{3.133a}$$

Furthermore, the parameters dependent on the device and the solvent are combined in the optical constant K (cf. (3.103)):

$$K = \frac{4\pi^2}{\lambda_0^4 N_A} \tilde{n}^2 \left(\frac{\partial \tilde{n}}{\partial c}\right)^2 \tag{3.133b}$$

If the scattering vector (cf. (3.117)) is also employed:

$$q = \frac{4\pi\tilde{n}}{\lambda_0} \sin(\theta/2) = \frac{4\pi}{\lambda} \sin(\theta/2), \tag{3.133c}$$

The Zimm equation (cf. (3.104)) is now much simpler:

$$\frac{Kc_2}{R_\theta} = \frac{1}{M_w}\left(1 + \frac{1}{3} \cdot q^2 \cdot r_G^2\right) + 2A_2 \cdot c_2 \tag{3.134}$$

To obtain the molar mass of the substance being analyzed from light scattering data, the Raleigh ratio R_θ at various concentrations and angles needs to be determined and

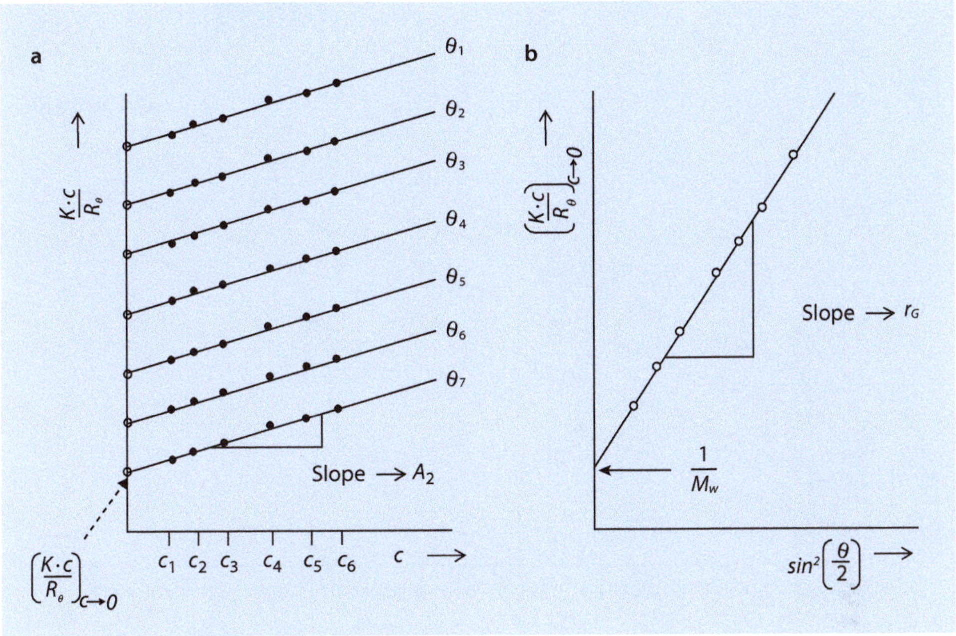

Fig. 3.12 Twofold extrapolation for a set of light scattering data. (**a**) Extrapolation to a zero concentration $c_2 = 0$. (**b**) Then to a scattering angle $\theta = 0$. c Polymer concentration c_2, indices indicate different polymer concentrations

extrapolated to $\theta = 0$ and $c_2 = 0$; the sequence of the procedure is not important. This procedure is summarized in Fig. 3.12.

As a "by-product" of this evaluation, the gradients of the linear extrapolations result in the radius of gyration r_G and the second virial coefficient A_2. The radius of gyration determined in this way is the z-average of the squares of the radii of gyration.

With the aid of modern data evaluation programs, the visualization of the light scattering data as a perspective projection of a three-dimensional diagram is trivial and customizing a plane directly onto the light scattering data according to (3.131) is easily achieved. However, in this case it is easy to lose sight of any deviations of the experimental data from the expected linearity. A good compromise is the so-called Zimm plot which comes close to the visual appearance of the 3D projection but allows an exact graphical evaluation (Fig. 3.13). In this diagram, Kc_2/R_θ is plotted against the sum of $\sin^2(\theta/2)$ and the polymer concentration c_2, whereby the latter is multiplied by a more or less randomly selected constant k to simplify the extrapolation. The constant k is chosen in such a way that the data points are equally distributed across the diagram but its value does not influence the result. All data points that belong to different angles but the same concentration c_{2i} are connected by one line. On this line, the position at which the lateral position is $k \cdot c_{2i}$ is marked. This corresponds to an extrapolation to $\sin^2(\theta/2) = 0$. The resulting dots are connected by a further line, which is extended to the ordinate. The gradient of this line is $2A_2 \cdot k$, its intercept with the ordinate is identical to an extrapolation to $c_2 = 0$ and $\sin^2(\theta/2) = 0$ and has the value of $1/M_w$. The same procedure is then executed again in reverse. A line is drawn through all dots that belong to a given angle θ_i, but to different concentrations, on which the lateral positions of $\sin^2(\theta/2)$ are

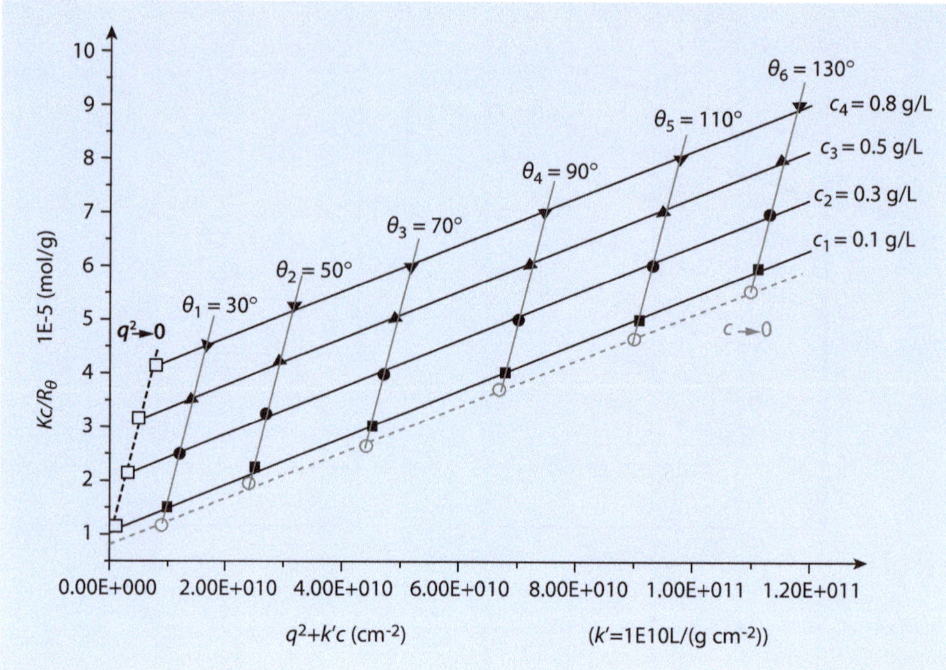

Fig. 3.13 Zimm-diagram from light scattering measurements on a polymer in toluene. The extrapolations yield the molar mass $M_w = 121$ kg/mol, the radius of gyration $r_G = 124$ nm, and the second virial coefficient $A_2 = 2.2 \cdot 10^{-4}$ mol cm^3 g^{-2}. c polymer concentration c_2, indices refer to the different polymer concentrations

marked. These markings are identical to an extrapolation onto c_2=0 at a constant scattering angle. A line is again drawn through the resulting dots and extended to the ordinate. The gradient of this line equals $r_G^2 \cdot (4\pi / \lambda)^2 (3M_W)$, its intercept with the ordinate should be the same as that for the previously drawn line and has the value of $1/M_W$.

Thus, not only information about the molar mass of the solute can be obtained from the Zimm plot; the radius of gyration and the second virial coefficient of the osmotic pressure can also be determined. The static light scattering thus offers with a single method the molar mass and also information about the size of the polymer coil in the solution as well as information about the quality of the solvent.

The scattering intensity is measured, e.g., with a photomultiplier or a photo diode. For light scattering measurements the sample solution is irradiated with incident light in a cylindrical cuvette. The intensity of the scattered light is then measured at selected scattering angles θ. In many cases, argon ion lasers (λ_0 =488 nm), helium-neon lasers (λ_0 =633 nm), or NdYAG solid state lasers (λ_0 =532 nm) are used at light sources.

Typically, solutions with concentrations between 0.1 and 0.5 g/L are measured. Interference from impurities, especially from dust particles, must be carefully prevented and the measured values are compared to those of the dust-free solvent.

3.2.6.2 Dynamic Light Scattering

In contrast to static light scattering, in which the scattered light is measured over relatively long time periods and time averaged values of the scattering intensity obtained, for dynamic light scattering the fluctuation of the scattering with time is measured—the "noise." Because of the complexity of this measurement method we only elaborate the basic principles here.

As shown in Fig. 3.8, the incident and the scattered light beams form a plane. The detector therefore only determines the scattering intensity of those scattering centers which are in the volume defined by the overlap of the incident and the scattered light beams. This volume is comparatively small. The diffusion of the polymers in the scattering volume leads to fluctuations in intensity over time. Resolution of these intensity fluctuations with respect to time gives a measure of the speed at which the polymers diffuse through the solution and thus also within the scattering volume. Thus, this method allows direct determination of the diffusion coefficient of the dissolved macromolecules which has been discussed in connection with the ultracentrifuge (▶ Sect. 3.2.5.2).

For further information the interested reader is referred to the specialist literature (e.g., Schaertl 2007).

3.2.7 MALDI-TOF-MS

3.2.7.1 General

The **M**atrix **A**ssisted **L**aser **D**esoption **I**onization **T**ime **o**f **F**light **M**ass **S**pectroscopy (MALDI-TOF-MS) is a fast and sensitive absolute method for determining both the number average and weight average molar masses. It has a special status in the analysis of polymers of biological origin. In an ideal case, molar masses of <300,000 g/mol can be measured with an accuracy of ±0.01% (Pasch and Schrepp 2003).

3.2.7.2 Basics

For the measurement the polymer is dissolved in a solution with an approx. 1000-fold excess of an organic matrix. Typical matrix molecules are presented in (Fig. 3.14).

After that the solvent is evaporated and, as the matrix crystallizes, the polymer molecules become isolated. This isolation is an important prerequisite for the success of the molar mass determination with MALDI-TOF-MS. In a high vacuum (10^{-6} to 10^{-8} mbar) a short laser pulse is directed onto the sample via special optics. Typically, nitrogen lasers with a wavelength of $\lambda = 337$ nm are used. The matrix strongly absorbs at the wavelength of the laser and the energy absorbed leads to an explosive phase transition of the matrix and the macromolecules into the gas phase.

As well as facilitating the absorption of the energy, the matrix also induces ionization. Electrically charged macromolecules are created by proton transfer in the gas phase. Alternatively, a cation-generating salt (typically a silver, sodium or potassium salt) is often added to the original solution, especially for measurements on synthetic polymers.

The ions generated are then accelerated in an electrostatic field with a strength of 10 to a few 1000 V/mm (Fig. 3.15). The mass analysis follows with the aid of a time-of-flight analyzer (Fig. 3.15). For this purpose, the time is determined which the ions need to

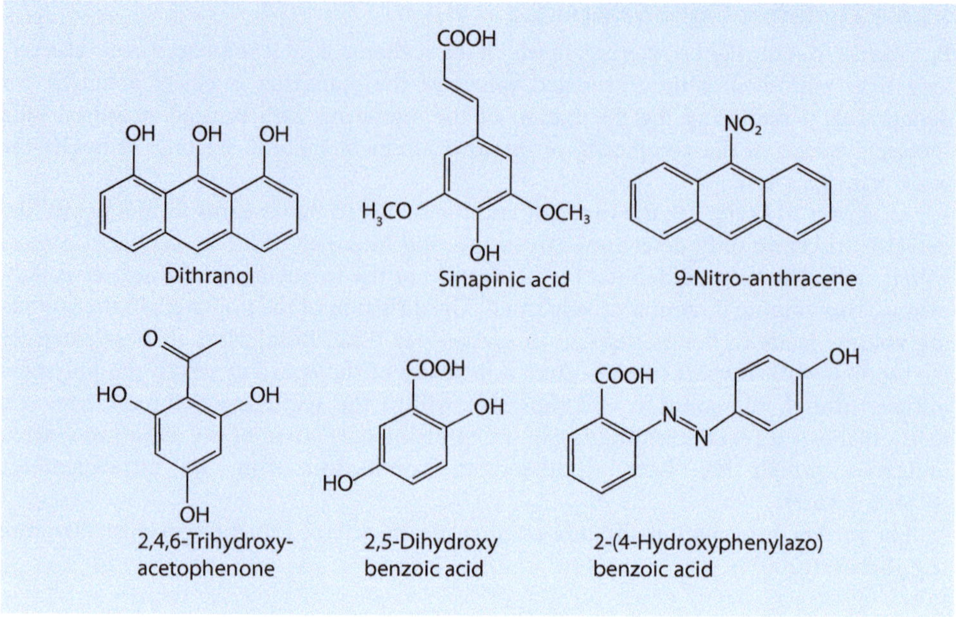

Fig. 3.14 Typical matrix molecules. Dithranol (also known as anthralin and Cignolin) is shown in its enol-form which is in tautomeric equilibrium with the keto-form

travel across a defined, field-free flight distance between the accelerating electrode (3) and the detector (4). From the flight time, the relation of mass to electric charge of the molecules can be determined. As the MALDI-TOF-MS typically leads to singularly charged ions (the detailed mechanisms are still being discussed), the result is the corresponding molar mass.

Applying the voltage U, according to the principle of energy conservation, gives for the kinetic energy of the ions E_{kin}:

$$E_{kin} = \frac{m \cdot \bar{v}^2}{2} = z \cdot e \cdot U \tag{3.135}$$

m	Mass of the particles
$\bar{v}$	Average particle velocity
z	Ion charge number
e	Elementary charge
U	Velocity voltage

The velocity reached results from

$$v = a\, t_{acc} \tag{3.136}$$

a	Acceleration
t_{acc}	Time of acceleration

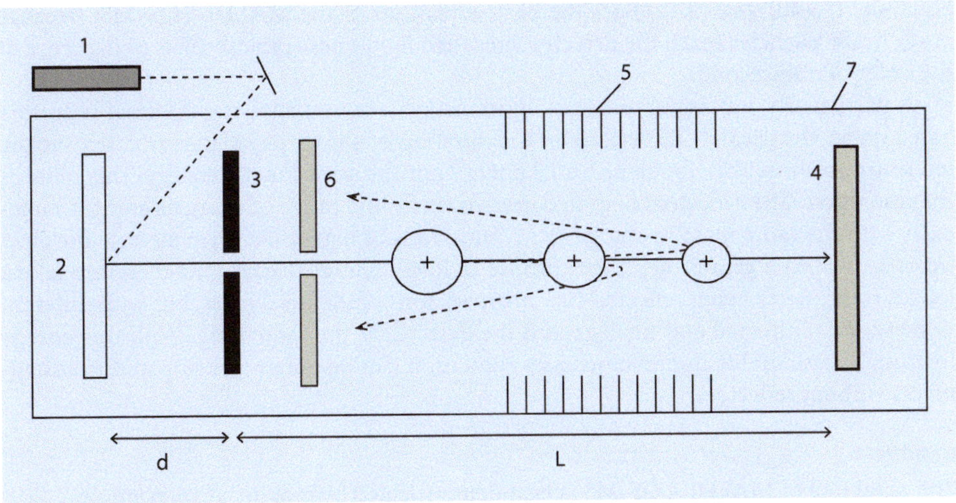

Fig. 3.15 Setup of a time-of-flight analyzer. (1) Laser, (2) sample holder, (3) accelerating electrode, (4) linear-detector, (5) reflectron, (6) reflectron-detector, (7) vacuum chamber, *d* acceleration section, *L* field-free flight distance (linear), (*dashed line*) flight routes (for the function of the optional reflectron, ion mirrors – see text)

With an acceleration distance d where

$$d = \frac{a}{2} t_{acc}^2 \tag{3.137}$$

the time of acceleration is

$$t_{acc} = d\sqrt{\frac{2\, m/z}{Ue}} \tag{3.138}$$

The resulting duration of the time taken t_{drift} to cross the field-free distance L is given by

$$t_{drift} = \frac{L}{v} = L\sqrt{\frac{m/z}{2Ue}} \tag{3.139}$$

For the total time of flight:

$$t = t_{acc} + t_{drift} = const \cdot \sqrt{\frac{m}{z}} \tag{3.140}$$

from which, after rearrangement one obtains:

$$\frac{m}{z} = const' \cdot t^2 \tag{3.141}$$

3

Equations (3.140) and (3.141) are the basic equations of the MALDI-TOF-MS. Because $m \sim t^2$, heavy particles reach the detector later than light ones. Typical times of flight are of the order of milliseconds.

A particularly successful mass resolution is achieved by applying optional reflectors based on an electrical field. Because of the duration of a laser pulse, ions which leave the ion source have neither the same initial energy nor the same kinetic energy. This reduces the resolution. The reflectors help to compensate for the range of energies and for molecules with the same mass to charge ratio. Thus, ions of higher energy penetrate the electrostatic field to a greater depth than those of lower energy. Faster ions therefore take a longer route. After being reflected (*ion mirrors*), ions of identical mass, but with different velocity, are redirected and finally reach the detector at the same time. With this energy focusing, considerable improvements in resolution can be obtained compared to instruments without reflectors.

Example

The capability of MALDI-TOF-MS is best demonstrated by looking at an example.

▣ Figure 3.16 shows the analysis of a polystyrene from a dithranol (▣ Fig. 3.14) matrix containing a silver salt as cation generating agent.

The peaks are easily identified. They can be assigned to individual chains of different mass (in this case, z equals 1). The molar mass of the individual chains are separated by the mass of a single styrene unit. From these peaks and knowledge about the cation generating salt, it is possible to determine the molar mass of the end groups. Furthermore, by integration over all the peaks, an average molar mass can be obtained. When determining the average molar mass it is important to note that the higher molar mass ions may be underrepresented, especially for broader distributions, because of peak broadening and effects dependent on the matrix and the cation generating salt so that the average value can be distorted.

3.3 Relative Methods

Molar mass determinations by methods which do not involve direct, physical-chemical correlation which can be mathematically explained between the actual measurements and the molar mass of the sample are referred to as relative methods. Such methods require a calibration by samples of known molar mass. It is important for this calibration that, among other things, it is known how the molar mass of the calibration standard has been measured and which average value has been determined. Precision demands that calibration should only be carried out using strictly monodisperse standards. However, in practice, such standards are rarely available. It is worth remembering here that even a sample with a polydispersity index of 1.04, which is seen in standard literature as a narrow distribution and may even be considered as monodisperse, still has a molar mass standard deviation of 20% (► Sect. 3.1) so that M_n and M_w are not identical. Thus, if the calibration involves samples whose molar mass was determined by a method which generates, for example, number average M_n, the calibration is not the same as that obtained using sam-

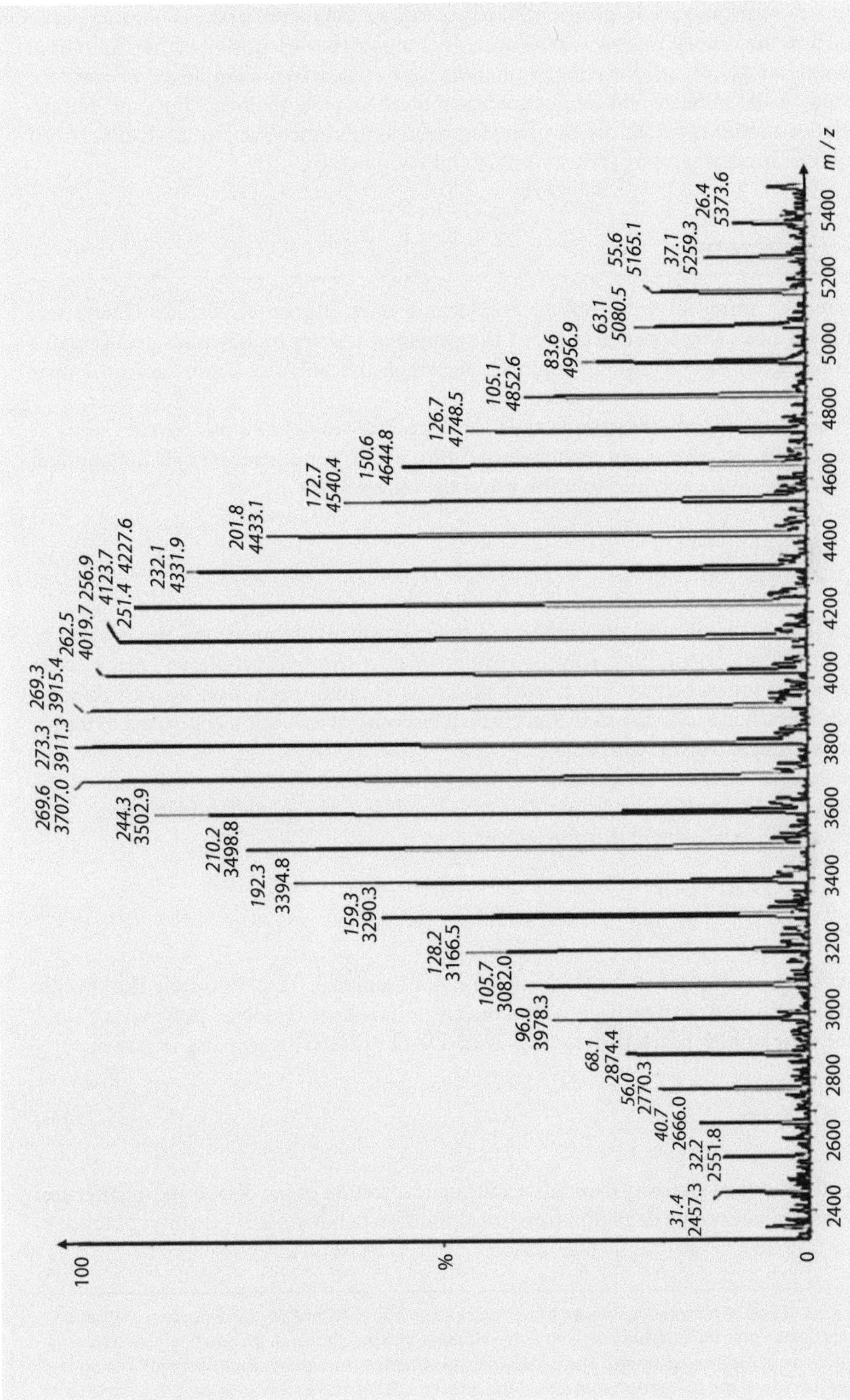

Fig. 3.16 MALDI-TOF-MS-spectrum of a polystyrene. The relative abundance (normalized to the most abundant peak) is plotted against the mass/charge ratio (*m/z*). The *numbers* above the peaks are the relative abundance and the corresponding *m/z* value

ples whose weight average is known.[2] Strictly speaking, the nature of the calibration samples and how their molar mass was determined belongs to the description of the calibration. Such problems do not arise if strictly monodisperse standards are employed because for such samples the number and weight average molar mass are identical. The most important relative methods which are of greatest practical relevance are size exclusion or gel permeation chromatography (SEC or GPC) and viscometry.

3.3.1 Viscometry

Compared to pure solvents, polymer solutions have higher viscosities. This effect depends on both the concentration of the polymer and its molar mass and remains measurable even for very dilute solutions in which the polymer chains are no longer entangled.

Before evaluating the measured values, it is necessary to define a few terms.

The relative viscosity η_{rel} of a polymer solution with a concentration c_2 is the quotient of the viscosity of the polymer solution η and the pure solvent η_0:

$$\eta_{rel} = \frac{\eta}{\eta_0} = \frac{t}{t_0} \tag{3.142}$$

This quotient is identical to the quotient of the experimentally measured times t and t_0, which a solution or the pure solvent require to flow through capillaries, respectively (details can be found below). The relative viscosity, as can be seen from the definition, is dimensionless. It is a measure of the increase in viscosity of a solution compared to that of the pure solvent at a given temperature.

The specific viscosity (also dimensionless) is a measure for the contribution of the polymer to the viscosity of a solution. It is calculated from the quotient of the *difference* in viscosity and the viscosity of the pure solvent:

$$\eta_{sp} = \frac{\eta - \eta_0}{\eta_0} = \frac{\Delta\eta}{\eta_0} \tag{3.143}$$

The reduced viscosity (of unit mL/g and, in older literature, dL/g) is simply the specific viscosity normalized with respect to the concentration of the dissolved polymer c_2. Thus, it is a measure of how much the viscosity is affected by dissolving, e.g., 1 g of polymer.

$$\eta_{red} = \frac{\eta_{sp}}{c_2} \tag{3.144}$$

Because the reduced viscosity depends on the concentration of the dissolved polymer and is therefore influenced by deviation from ideal solution behavior, it is common practice to

2 One should not be tempted, however, to try to determine both M_n and M_w by using two calibration curves to interpret the numbers derived as the number and weight averages and finally even to determine a polydispersity index. The polydispersity obtained in this way would depend also on the polydispersity of the calibration substance. The errors involved make such an approach of little value.

extrapolate the measured values to zero concentration. The number so obtained is called the Staudinger index $[\eta]$:

$$\lim_{c_2 \to 0} \eta_{red} = [\eta] \tag{3.145}$$

The Staudinger index, also referred to as the intrinsic viscosity, does not vary with concentration as does the reduced viscosity. It is a measure of the volume of the polymer molecules in a dilute solution and has the same units as the reduced viscosity, mL/g (or dL/g).

In the theoretical description of this phenomenon we can view the polymer chains as impenetrable spheres. If we disperse such spheres in a liquid with a viscosity η_0, the viscosity increases to a value η. The specific viscosity for small volume fractions φ_2 of the dispersed spheres can be described with the Einstein equation:

$$\eta_{sp} = \frac{\eta - \eta_0}{\eta_0} = 2.5 \cdot \varphi_2 \tag{3.146}$$

In a dilute solution the volume fraction ($V_1 + V_2 \approx V_1$), depends on the mass concentration c_2, the mass of a single sphere m_2 and its volume v_2 as follows:

$$\varphi_2 = \frac{V_2}{V_1} = \frac{v_2}{m_2} c_2 \tag{3.147}$$

V_2	Volume of all spheres
V_1	Volume of the solvent
v_2	Volume of one sphere
m_2	Mass of one sphere

It is assumed that the polymer chains in the dilute solution are swollen with solvent, i.e., that the quotient $v_2\, c$ is not simply the inverse of the density of the pure polymer but rather a larger value which depending on the quality of solvent.

By combining (3.146) and (3.147), one obtains

$$\frac{\eta - \eta_0}{\eta_0} \frac{1}{c_2} = 2.5 \cdot \frac{v_2}{m_2} \tag{3.148}$$

The term on the left hand side of (3.148) involves only experimentally measured values and is referred to as the reduced specific viscosity (see above). In practice, deviations are observed depending on the concentration of the solution. These can be eliminated by assuming an infinitely diluted solution ($c_2 = 0$, see Eq. 3.145) and use $[\eta]$ instead of the first term.

With (3.148), measuring the viscosity of a polymer solutions yields a measure for the quotient of the volume of a (swollen) polymer chain and its molar mass. For hard, solid spheres, e.g., highly cross-linked, sphere-shaped particles, $v_2 \sim m_2$. Thus, for such particles, the Staudinger index doesn't depend on the particle mass so that only information about the density of the particles but not their mass can be derived.

However, for solvent penetrated coils, Kuhn's square root law can be used whereby the average end-to-end distance, i.e., the effective diameter of the coil d_2, is proportional to $\sqrt{m_2}$ (► Chap. 2).

Thus:

$$v_2 \sim d_2^3 \sim m_2^{3/2} \tag{3.149}$$

From (3.148) and (3.149) one obtains for the molar mass M_2:

$$[\eta] \sim \frac{m_2^{3/2}}{m_2} \sim m_2^{1/2} \sim M_2^{1/2} \tag{3.150}$$

As a general case for the various structures of polymer molecules in solution, the so-called Mark–Houwink equation, (3.151), is used:

$$[\eta] = K_{M,H} \cdot M^a \tag{3.151}$$

$K_{M,H}$ Solution-dependent and polymer-specific constant

a Exponent, between 0.5 (ideal sphere) and 2,0 (stick-like molecules)

The preparation for viscometry measurements is comparatively simple. The time t is experimentally measured, for which a certain amount of a solution ΔV needs to flow through a capillary of a known diameter d. According to Hagen and Poiseuille's Law:

$$\eta = \frac{\pi \cdot g \cdot \Delta h \cdot d^4}{128 \cdot \Delta V \cdot l} \cdot \rho \cdot t \tag{3.152}$$

Here, g stands for the Earth's acceleration produced by gravity, Δh for the difference in height between the upper and lower reservoir of the liquid, l for the length of the capillary, and ρ for the density of the polymer solution. For a specific viscometer the quotient in (3.152) is constant so that a device constant K_G can be defined and (3.152) becomes

$$\eta = K_G \cdot \rho \cdot t \tag{3.153}$$

The device constant can be determined (if not supplied by the device manufacturer) using any pure solvent with a known viscosity and (3.154):

$$K_G = \frac{\eta_0}{\rho_0 \cdot t_0} \tag{3.154}$$

Here the variables indicated with 0 represent the density, viscosity, and time required by the pure solvent to flow through the capillary.

For dilute polymer solutions it can be assumed to a good approximation that the density of the polymer solution does not differ considerably from that of the pure solvent:

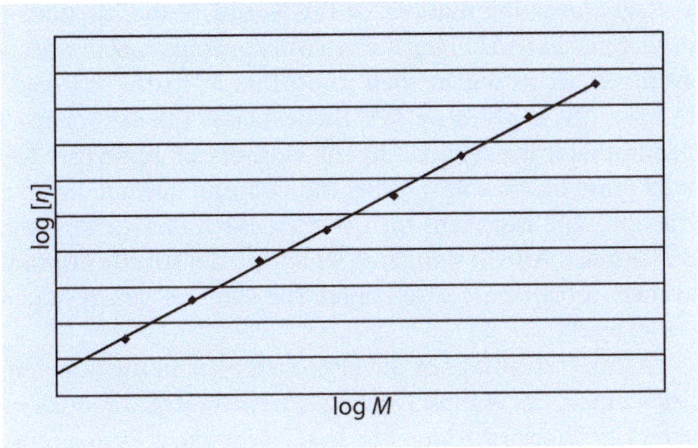

Fig. 3.17 Double logarithmic plot of the Staudinger index vs the molar mass

$$\rho \approx \rho_0 \tag{3.155}$$

so that

$$\eta = \eta_0 \cdot \frac{t}{t_0} \tag{3.156}$$

Thus, for a once calibrated apparatus the measurements are reduced to simply measuring the time for a certain amount of sample liquid. In modern viscometers this is done automatically. To reduce the error, multiple measurements are made. As the viscosity of polymer solutions is largely dependent on the temperature, excellent temperature control of the viscometer is extremely important to obtain reproducible results.

With the Mark–Houwink equation (3.151) a calibration with polymer probes of various molar mass, determined with other (absolute) methods, can be performed. The equation predicts that a double-logarithmic plot of the Staudinger index against the molar mass for different samples results in a straight line with gradient a and an intercept of the ordinate equal to the solvent and polymer dependent constant of the Mark–Houwink equation(Fig. 3.17).

From the value of the exponent a determined with this graph, important conclusions can be drawn about the shape of the dissolved polymer. An exponent of 0.5 indicates a solvent penetrated polymer in a θ-solvent (▶ Chap. 2). In thermodynamically better solvents, higher values of around 0.8 are observed. Even higher values of around 2 indicate that the polymer molecules are stiff or stick-shaped.

If the calibration—as explained in the introduction to ▶ Sect. 3.3—was made with monodisperse standards, the following average for the molar mass from the viscosity measurements is obtained; this is referred to as the viscosity average molar mass:

$$M_\eta = \left(\frac{\sum_i m_i M_i^a}{\sum_i m_i} \right)^{1/a} \tag{3.157}$$

At first glance, (3.157) looks rather complicated, but further consideration makes an interpretation relatively simple. For the number and the weight averages, the molar mass M_i is averaged over the number or the weight of the fraction, respectively. By analogy, for the numerator in (3.157) for the viscosity average molar mass, the individual molar masses are averaged according to their contribution to the viscosity of the solution. According to (3.151), this is given by M^a. This explains the exponent in the numerator of (3.157). The denominator is explained by the viscosity of a polymer solution being proportional to the total mass of the solute. Thus the viscosity average involves the mass of each molar mass fraction. The exponent for the quotient serves to arrive at a correct unit for the average molar mass. Also, if standards which are not strictly monodisperse are employed, a viscosity average molar mass is obtained, the value of which depends on how the calibration was performed.

A great advantage of the viscometry lies in the large range of molar mass which can be determined. For samples having extremely large or small molar mass, the flow rates can be varied by choosing a larger or smaller capillary. Even a molar mass >100 g/mol can be measured with a suitable solvent. The upper measurable molar mass limit is about $20 \cdot 10^6$ g/mol.

Because viscosity measurements to determine the Staudinger index require a concentration series ($c \leq 1$ g/100 mL) to be measured, ca. 250 mg of material is necessary.

As well as exceptionally good thermal control, it should also be noted that the method is extremely sensitive to dust particles and insoluble polymer. The solution should thus be filtered before the measurements and it should be confirmed that the polymer is completely soluble.

3.3.2 Size Exclusion Chromatography (SEC)

So-called *Size Exclusion Chromatography* is also referred to by the acronym SEC. In older literature, and the name *Gel Permeation Chromatography*, abbreviated as GPC, can also be found.

The apparatus preparation is, in principle, rather easy and similar to what is required for HLPC known from small molecule organic chemistry. The sample is introduced into the system with a syringe and carried by the mobile phase through one or more separation columns with the help of a pump. Detection at the exit of the column(s) is achieved by conventional methods, such as refractive index or UV/vis-detectors. However, in contrast to HLPC, the separation does not rely on a different adsorption of molecules to the stationary phase of the separation columns but rather on the different molecular size, more precisely; their different hydrodynamic radii. To achieve this, a porous polymer gel, i.e., a swollen, cross-linked polymer, is employed as the stationary phase. The size of the gel's pores is approximately equal to the size of the polymer molecules which can enter these pores. (◘ Fig. 3.18) shows some electron micrographs of such a gel.

It is essential for the separation by size that the pores of the gel are of different sizes. As a result, a polymer of a certain molar mass (and thus of a certain size) can only enter into a certain proportion of these pores. The smaller the molecule, the larger the number of pores that it can enter during the chromatographic process. Thus, the smaller polymer molecules enter and exit a *larger* number of pores within the column and thus travel a longer distance when inside the column, and they exit the column *later* than larger molecules (◘ Fig. 3.19).

The elution volume V_E, at which the polymer of a certain molar mass exits the separation column, can be described by the following equation:

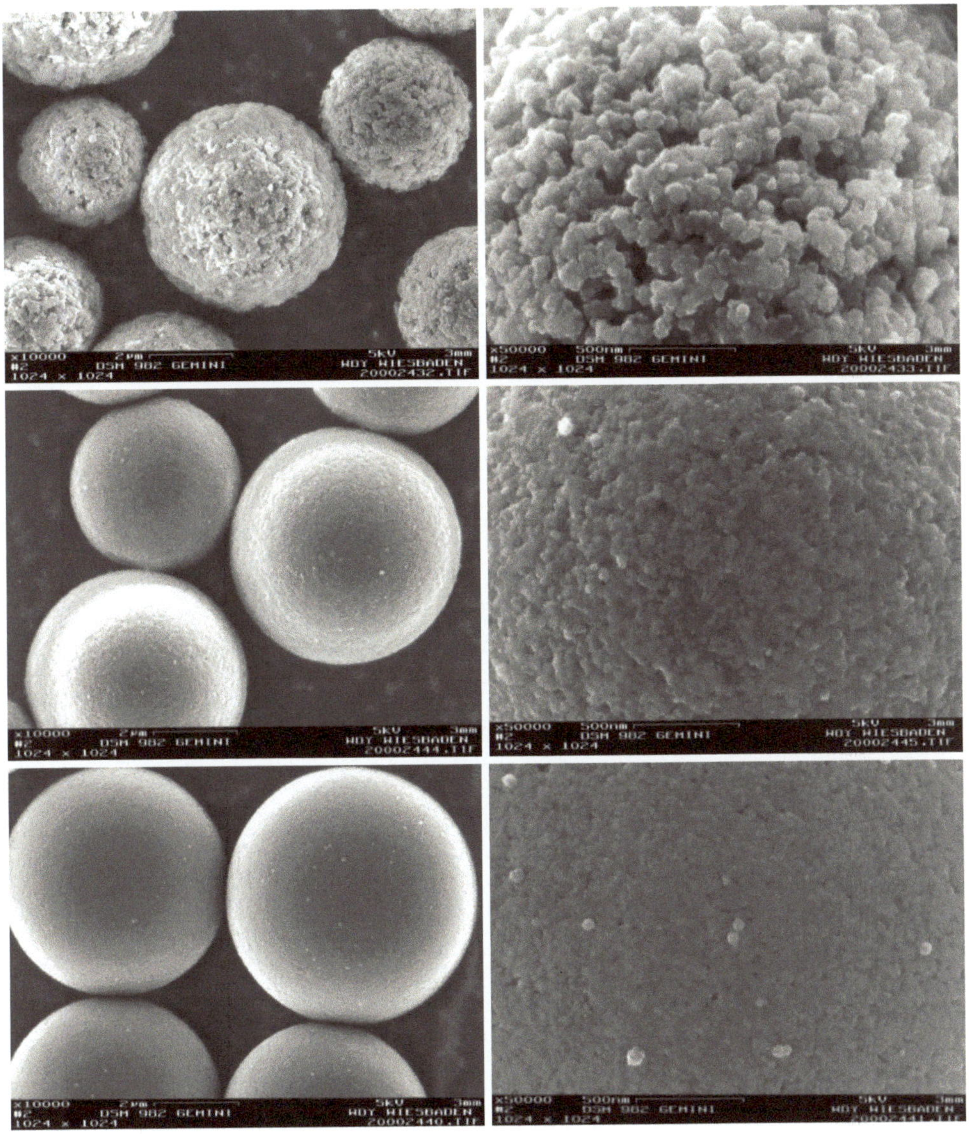

Fig. 3.18 Electron micrographs at various magnifications of SEC-gels with various pore sizes. (By kind permission of Polymer Standards Services GmbH, Mainz)

$$V_E = V_l + V_i \cdot f\left(M_2\right) \tag{3.158}$$

Here, V_l is the volume between the gel particles. V_i is the pore volume inside the gel particles. The course of the function denoted as $f(M_2)$ represents the fraction of the available pores that, because of their size, are able to accept macromolecules of molar mass M_2. It has a value between 0 (very large molecules) and 1 (very small molecules). The form of the function defines the so-called upper and lower exclusion limit. Above a certain molar mass, the molecules are so large that they do not fit into any gel pores ($f(M_2)$=0). Above

3

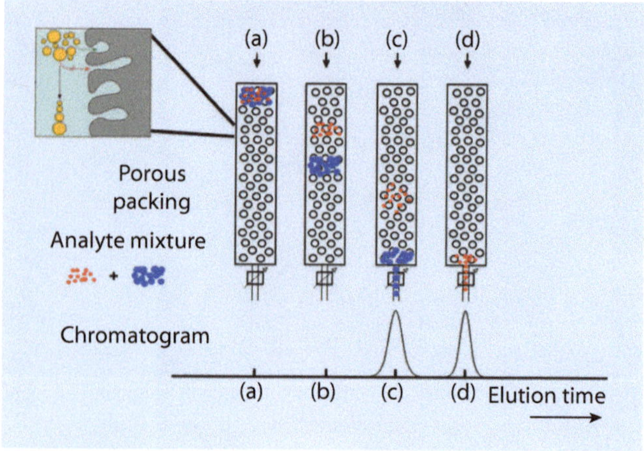

Fig. 3.19 Separation principle of SEC according to polymer size

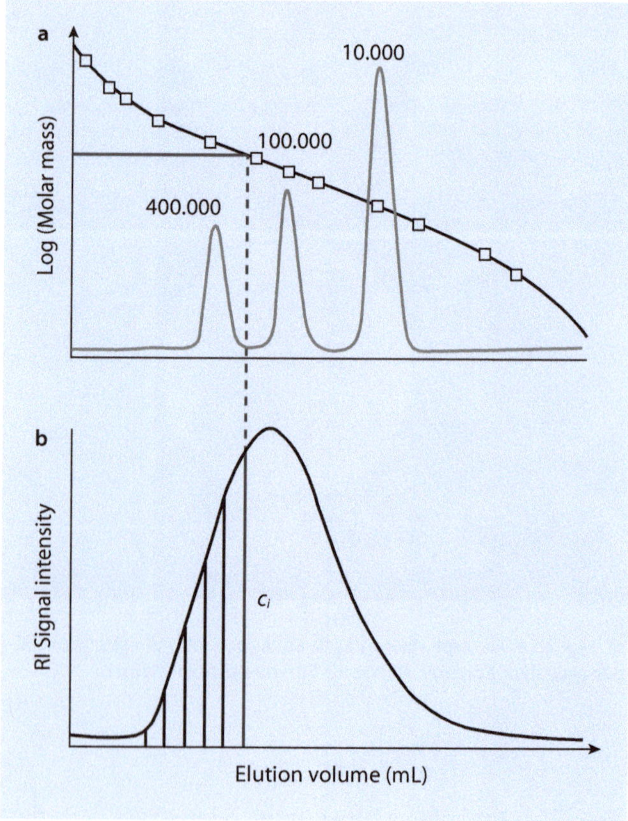

Fig. 3.20 (**a**) SEC calibration with standards having narrow molar mass distributions. (**b**) Determination of the concentration fractions c_i and the corresponding molar masses M_i of a polystyrene sample with a broad molar mass distribution. *RI* refractive index detector

this molar mass, no further separation occurs ($V_E=V_I$). All molecules leave the separation column at the same time at the minimum elution time. In analogy, a minimum molar mass also exists, below which the total pore volume of the stationary phase is available to the molecules ($f(M_2)=1$). Thus for molecules smaller than the lower exclusion limit, no further separation occurs ($V_E = V_I + V_i$). The molecules exit the column simultaneously at a maximum elution time. The function $f(M_2)$ and thus also the exclusion limits depend on the employed column as well as the eluent, the temperature, and the analyte. The calibration of the SEC is usually executed with standards of narrow molar mass distribution (often polystyrene) (Fig. 3.20).

The following equation is usually experimentally determined:

$$\log M = a - b \cdot V_E \tag{3.159}$$

Here a and b depend on the experimental setup and the column's stationary phase. Assigning a molar mass to an elution time (or an elution volume) is normally made simply from a calibration curve measured with standards having narrow molar mass distributions and of known molar mass (relative method). With such a calibration, a molar mass can be assigned to every elution volume. After recording the chromatogram it is normally analyzed by suitable software. With the assumption that the intensity of the detector signal corresponds to the concentration of the macromolecules, which is usually accurate enough, this analysis yields the number of molecules n_i for each molar mass M_i.

From this data set, all the average values can be calculated from the equations discussed at the beginning of this chapter. Furthermore, as well as this numerical information, a graphical representation of the molar mass distribution is directly available. This is especially important if the molar mass of the polymer being analyzed is not evenly distributed around a defined molar mass average but, instead, the molar mass distribution has multiple maxima. As opposed to the so-called *monomodal* samples in which the molar mass distribution has only a single maximum, such samples are referred to as *bimodal* or *polymodal*, depending on the number of maxima.

The separation of the macromolecules in SEC depends on their hydrodynamic volume V_h. This depends on the molar mass in a different way for each polymer. To avoid needing to consider dependence of the physical chemical properties on the particular polymer, the following equation (without proof) can be employed:

$$\log\left(M \cdot [\eta]\right) \propto V_h \tag{3.160}$$

Thus, the hydrodynamic volume of a polymer should be independent of its chemical composition and its microstructure, so that for a given chromatography column and a certain solvent it should be proportional to the logarithm of the product of the Staudinger index and the molar mass. Indeed, this can be confirmed by experiment and a plot of the logarithm of this product vs the elution volume results in a straight line to a good approximation (Fig. 3.21). (Grubisic et al. 1967; Wild and Guliana 1967).

A great advantage of SEC is the speed and simple execution of the measurement. Only ca. 0.1 mg is injected onto the column as a dilute solution. The measurement of different concentrations is unnecessary. The range of molar mass which can be measured depends on the exclusion limits of the separation column. As opposed to most of the methods presented in this chapter, the SEC delivers any desired molar mass average and offers information about the shape of the molar mass distribution. These advantages have led to great importance being placed on the molar mass definition with SEC in practice. However, two key limiting conditions must be kept in mind when using this method:

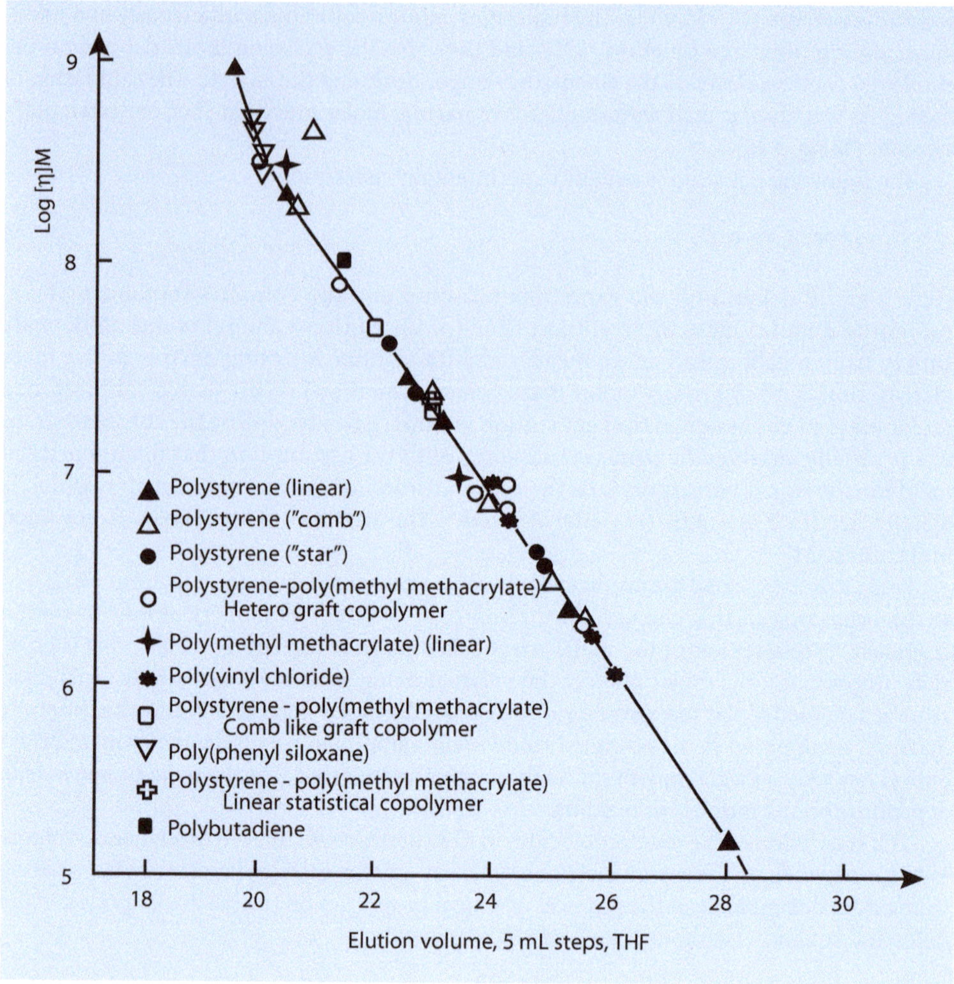

Fig. 3.21 Example of a universal calibration with THF as solvent

1. The interactions between polymer, solvent and column packing should be approximately equal for all combinations, and preferably as minimal as possible. If the polymer interacts with the stationary phase, for example, the separation according to hydrodynamic volume is superimposed by effects common to adsorption chromatography. Choosing a suitable column that meets all the criteria in the best possible way is therefore crucial for meaningful results.
2. Because of the inherent influence of the solvent quality on the effective size of the polymer (see above), the system of calibration standard and solvent has a great influence on the determined molar mass of the measured sample, especially if the polymers being examined differ chemically from the standard. Thus, it is essential to report the standard used. Because this is often polystyrene, the determined molar mass should be given as a polystyrene equivalent molar mass (in the solvent used).

3. In some co-polymerization processes it can happen that some comonomers are preferentially incorporated in polymer molecules having greater or smaller molar mass so that the chemical composition of the polymer is not the same over the molar mass distribution. Such chemical inhomogeneity with respect to the molar mass can indeed make the quantitative evaluation of the detector questionable. In such cases, multidimensional chromatography techniques that combine the SEC with conventional chromatography techniques can provide a solution (Rittig and Pasch 2008).

The combination of static light scattering and SEC and a concentration-dependent detector (RI- or UV/vis) yields absolute molar mass information without the necessity of a calibration and thus provides more information about the sample, simultaneously retaining the advantages of SEC.

The speed and simplicity of SEC and viscometry have made these methods standard practice in industrial polymerization processes where they are routinely employed to verify adherence to product specification where molar mass is involved.

Both SEC and viscometry are relative methods. If an absolute molar mass is required, a characterization by light scattering (if necessary, coupled with SEC) or analytical ultracentrifugation is recommended. These methods are also convenient if no appropriate, inert column material can be found.

For the characterization of comparatively small macromolecules, MALDI-TOF or osmometry can be successfully employed.

References

Grubisic Z, Rempp P, Benoit H (1967) A universal calibration for gel permeation chromatography. Polym Lett 5:753–759

Maechtle W, Boerger L (2006) Analytical ultracentrifugation of polymers and nanoparticles. Springer, Heidelberg

Meselson M, Stahl FW (1958) The replication of DNA in *Escherichia coli*. Proc Natl Acad Sci 44:671–682

Pasch H, Schrepp W (2003) MALDI-TOF mass spectrometry of synthetic polymers. Springer, Heidelberg

Rittig F, Pasch H (2008) Multidimensional liquid chromatography in industrial applications. In: Cohen SA, Schure MR (eds) Multidimensional liquid chromatography, theory and applications in industrial chemistry and the life sciences. Wiley, New York, pp 387–423

Schaertl W (2007) Light scattering from polymer solutions and nanoparticle dispersions. Springer, Heidelberg

Svedberg T, Pedersen KO (1940) The ultracentrifuge. Oxford University Press, Oxford

Wild L, Guliana R (1967) Gel permeation chromatography of polyethylene: effect of long chain branching. J Polym Sci A-2(6):1087–1101

Polymers in Solid State

S. Koltzenburg et al., *Polymer Chemistry*, DOI 10.1007/978-3-662-49279-6_4

The vast majority of all polymers produced in the world are utilized in their solid form—that is, as a classic material. Thus, a discussion of the solid-state properties of polymers, their morphology, and the impact of this and their properties on their applications is an essential part of this book.

Polymers in their solid state can be roughly divided into three categories, of which, in practice, only two—the *amorphous* and the *semicrystalline* states—play a role (◘ Fig. 4.1). The single-crystalline state is of academic importance only.

Many polymers cannot crystallize. They exist in an amorphous state, similar to that found, for example, in window glass. Substances with a low molar mass can also solidify to amorphous solids, particularly complex molecules, such as certain pharmaceuticals. Water can be shock-frozen to an amorphous solid. However, for substances with a low molar mass this state is very unstable. Thus, many low molar mass amorphous phases tend to re-crystallize, that is, transform into a crystalline state corresponding to the thermodynamic equilibrium.

Under certain circumstances, however, polymers can at least partially crystallize. The crystallization of polymers is a very complex process because the long chains must arrange themselves into a defined crystalline structure. Because the intertwining of the polymer coils—so-called *entanglements*—are an obstacle to crystallization, this is a relatively slow process. Additionally, crystallization of the polymer is usually imperfect, resulting in the creation of a partially crystalline or semicrystalline material in which both amorphous and crystalline domains coexist.

Polymeric single crystals can be obtained by performing crystallization from very dilute solutions. However, these are more of an academic interest and without importance for industrial polymer chemistry.

As is known from the analysis of substances with a low molar mass, amorphous and crystalline areas can be easily distinguished, for example, by X-ray spectroscopy. Here, one observes so-called halos from the amorphous phases, whereas the crystalline phases appear as peaks in an X-ray diffractogram. An analysis of the phase transitions upon heating a sample also provides valuable information about whether a sample is amorphous or crystalline. The morphology of a solid greatly influences the properties (e.g., the optical, mechanical, and thermal properties) of products made from polymers.

Before we discuss this in more detail, a number of fundamental physical concepts need to be introduced. The mechanical properties of a solid are measured in terms of its so-called "moduli." These include *Young's modulus* for uniaxial extension, and the *shear modulus* for shear deformation of a test piece (◘ Fig. 4.2).

According to Hooke's law, the uniaxial stress of a test piece is given by

$$\sigma = E \cdot \varepsilon \tag{4.1}$$

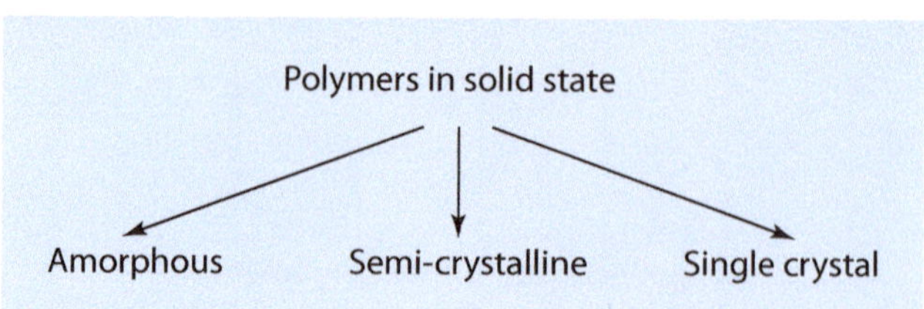

◘ **Fig. 4.1** Schematic representation of the morphologies of polymeric solids

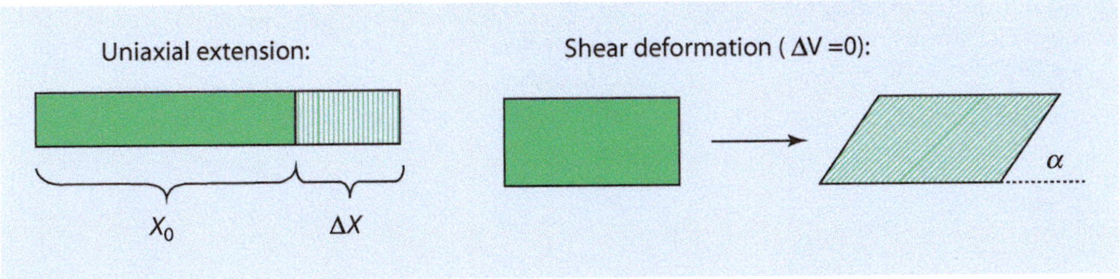

Fig. 4.2 Schematic representation of the shape change under uniaxial stress by Δx or shear stress by a shear angle α

This means that the applied stress σ (force per unit area) is proportional to the elongation of the test piece ε. The proportionality constant E is referred to as the Young's or E-modulus. Whereas, if a test piece is deformed under shear, the stress σ_s—again defined as the force applied to a given area—is proportional to the tangent of the shear angle α. In this case, the proportionality constant is again referred to as a modulus, in this case, the shear modulus G:

$$\sigma_s = G \cdot \tan\alpha \tag{4.2}$$

4.1 Phase Transitions in Polymeric Solids

As mentioned above, amorphous and crystalline solids can be differentiated by their thermal phase transitions. When crystalline substances are heated, a melting point is observed in analogy familiar from low molar mass crystals. At the melting point, an abrupt rearrangement occurs in the material. Below the melting point, there is crystalline order and this disappears during the melting process. This is the reason why the properties such as volume undergo a stepwise change at the melting point (Fig. 4.3).

Thermal transitions of this type are referred to as *first-order phase transitions*; they can be assigned an enthalpy, in this case, the enthalpy of melting. Because of the structural change on melting, the entropy of the melted phase is greater than that of the crystalline phase so that a ΔS can be defined for this transition as well as a corresponding free enthalpy ΔG.

In contrast to this, amorphous solids are as disordered in the solid phase as they are in their melts.[1] Vitreously solidified amorphous states are supercooled, disordered melts or glasses. This means that they undergo no structural change when heated and transitioning to the liquid phase. This process, which corresponds to the melting process of crystalline compounds, is referred to as the *glass transition*, and the associated temperature T_G as the *glass transition temperature*. If we now consider the dependence of, for example, the volume on the temperature (Fig. 4.4), we can see that indeed the thermal expansion, i.e., the slope of this line, is different above and below T_G, but that no abrupt change of volume occurs at the glass transition.

This is because the structure of the material is identical—i.e., disordered—above and below T_G.

1 On closer inspection, there are subtle differences here as well, which are, however, not further discussed in the present book.

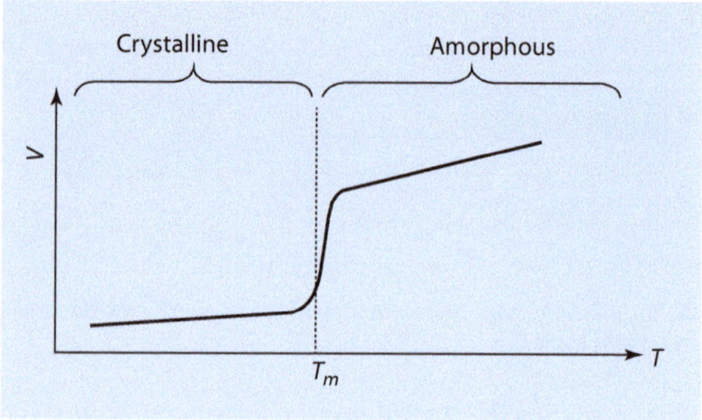

Fig. 4.3 Sketch showing the change of volume with melting. T_m melting temperature

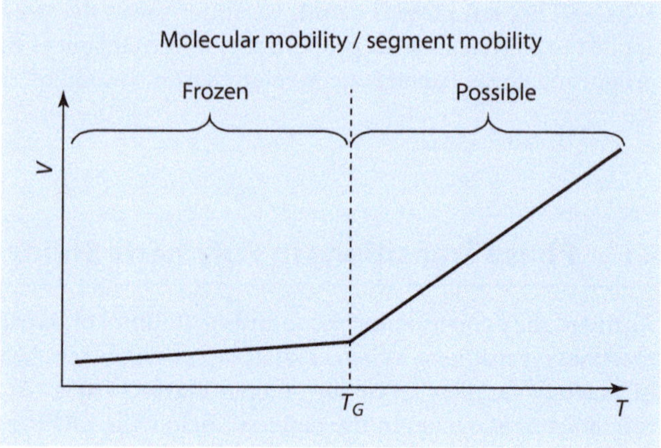

Fig. 4.4 Sketch showing the volume change at a glass transition. T_G glass transition temperature

Transitions of this type are referred to as *second-order transitions*. In contrast to first-order phase transitions, there is no transition enthalpy required here because no enthalpy needs to be invested to break a crystal lattice. Nor is there any substantial change in entropy, because both states are approximately equivalent in structure. As a consequence, no ΔG can be assigned to a second-order transition. However, abrupt changes in other properties, such as heat capacity, are observed. Because the melt generally has more translational degrees of freedom above T_G, stimulation requires more enthalpy and the addition of a larger amount of heat is required to raise the temperature by one degree. Hence, a sharp rise in heat capacity occurs across a glass transition, which can be measured as Δc_p.

Semicrystalline polymers consist, as already mentioned, both of crystalline and amorphous areas. In principle, polymers are, aside from the above-mentioned exceptions, never crystalline to 100 %, so there is a co-existence of crystalline and amorphous domains. Thus, when heating of a semicrystalline or partially crystalline polymer, both a glass transition at a temperature T_G and a melting process at a temperature T_m can be observed (Fig. 4.5).

It is always the case that T_G is lower than T_m. This should be obvious. If it were not so, a substance would, when cooling from the melt, solidify to a glassy solid at the higher T_G and would then have to crystallize below T_G. However, below T_G the molecules are insufficiently mobile to allow the reorientation necessary for crystallization.

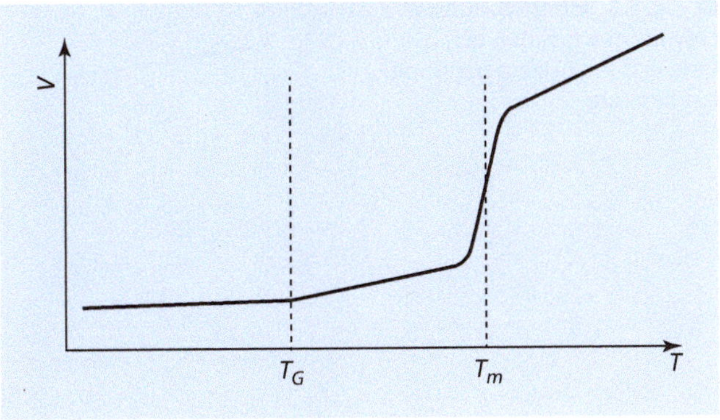

Fig. 4.5 Sketch showing the volume change at the glass transition and the melting temperature of partially crystalline polymers. T_G glass transition temperature, T_m melting temperature

4.2 Methods for the Determination of T_G and T_m

Below, the most important methods for determining the glass transition temperature and the melting temperature are presented. These can be classified according to static and dynamic procedures.

4.2.1 Static Procedures

Static procedures are based on the measurement of a material property, such as density, volume, heat capacity, or refractive index, as a function of temperature. In these procedures, the sample is slowly heated to maintain equilibrium, and the change of the respective intrinsic property is measured. Examples are dilatometry (▶ see Sect. 5.3) and the measurement of the refractive index, where discontinuities in the curves of the volume or refractive index can be observed as a function of temperature at the glass transition temperature (Figs. 4.5 and 4.6, respectively).

However, in general, the preferred method is *differential scanning calorimetry* (DSC). The principle of DSC is that a sample is heated at a constant rate and the heat flow required to maintain this defined rate is measured. The heat flow correlates with the heat capacity c_p of the sample, i.e., the amount of heat necessary to produce a given change in temperature. As mentioned above, we observe an abrupt increase in heat capacity c_p at the glass transition temperature, whereas, at the melting point the latent heat ΔH_m (the heat of melting) is required. Thus, the temperature of the sample does not change during the melting process, even though heat is absorbed. This is valid until the moment when the crystal lattice is completely "broken" and the transition to an isotropic melt has occurred. Formally, the heat capacity tends to infinity during the melting process so that a peak results in the DSC (Fig. 4.7). From the area under this peak the heat of melting for the sample can be calculated. Alternatively, if ΔH_m is known, the degree of crystallization can be determined.

DSC is a method frequently used both in industrial and academic laboratories. Its advantages lie in the very small amount of sample material required (approximately 5 mg). Furthermore, the technology is well established and the available devices are quite robust and easy to use. In particular, sample preparation is also relatively simple. The sample has to be placed in a measuring vessel—usually a disposable aluminum pan.

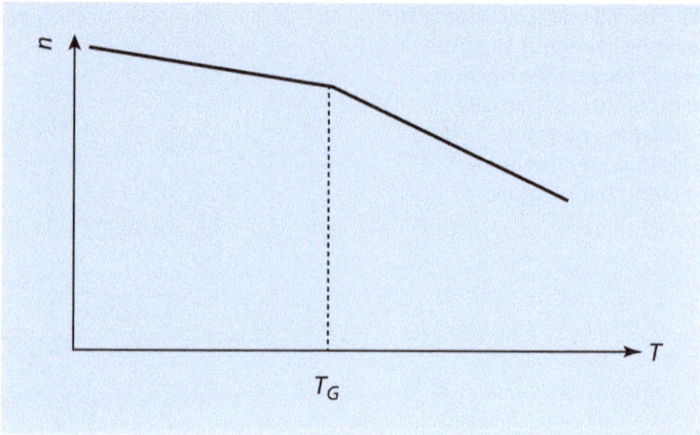

Fig. 4.6 Refractive index of a polymer as a function of temperature. T_G glass transition temperature

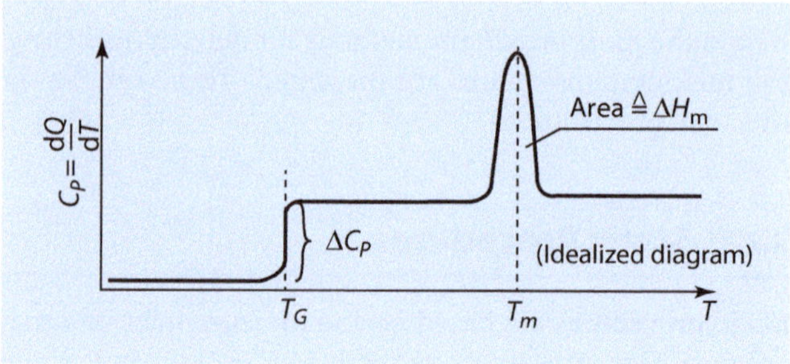

Fig. 4.7 Heat capacity as a function of temperature on heating for semicrystalline polymers as observed with DSC

The disadvantages of DSC lie in its limited sensitivity. Not every phase transition is resolved. Moreover, the measuring curve is sometimes difficult to interpret. Thus, the curve in Fig. 4.7 is an idealized DSC curve rarely generated by real polymers. Measured DSC curves are often not parallel to the axes, but arched or curved, so that it is sometimes hard to discern the individual transitions correctly.

DSC provides the possibility of varying the heating rate and *hysteresis* can be observed for first-order phase transitions. Particularly large hysteresis is observed for the crystallization peak, associated with the transition from a melt to a crystalline phase. This is because crystallization requires the formation of crystal-nuclei. This process requires a certain time so that, at a very high cooling rate, the crystallization temperature can be significantly undershot.

4.2.2 Dynamic Procedures

In *dynamic mechanical thermal analysis* (DMTA), a sample is subjected to periodic sinusoidal, mechanical stress, for example, by a periodic tensile or torsional movement. This is accomplished at a frequency f and with a stress σ. During such measurements it is often observed that the deformation of the material is delayed in time with respect to the applied stress, so that the periodic oscillation of the material and the mechanical excitation are phase-shifted (Fig. 4.8).

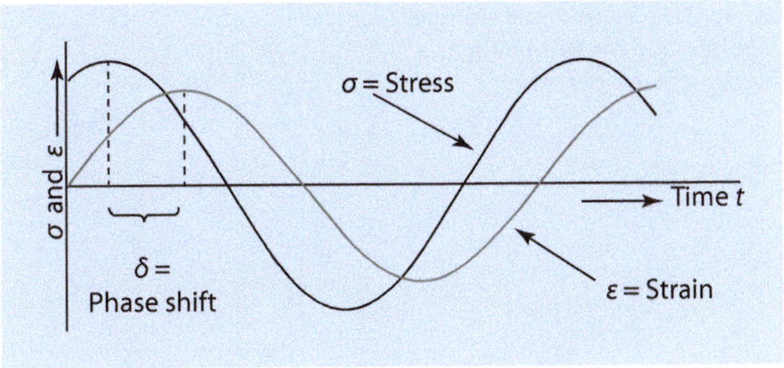

Fig. 4.8 Periodic progression of stress σ and strain ε

The angle of the phase shift is referred to as δ. Such a periodic oscillation can be—as known from mathematics—represented in the form of complex numbers. Thus, we can define the strain ε as

$$\varepsilon = \varepsilon_0 e^{-i\omega t} \tag{4.3}$$

and the stress σ as

$$\sigma = \sigma_0 e^{-i\omega t+\delta} \tag{4.4}$$

In doing so, ε_0 or σ_0 signify the respective amplitudes, ω the angular frequency ($\omega = 2\pi f$), and t the time.

Although Young's modulus represents the ratio of stress and strain for simple systems which obey Hooke's law, a complex E-modulus (E^*) can be defined for these complex cases. The rule is

$$E^* = \frac{\sigma}{\varepsilon} = \frac{\sigma_0}{\varepsilon_0}\left(\cos\delta + i\sin\delta\right) \tag{4.5}$$

The complex modulus E^* thus consists of a real part and an imaginary part, which are referred to as E_1 and E_2 (sometimes also as E' and E''):

$$E^* = E_1 + iE_2 \tag{4.6}$$

E_1 is referred to as the *storage modulus* and E_2 as the *loss modulus*. E_1 is a measure of the energy that can be stored in the system and E_2 a measure of the energy that is lost in the system. The quotient of the moduli E_2/E_1 therefore is given by $\sin\delta/\cos\delta$, i.e., $\tan\delta$—this is designated the *loss factor*, and describes the ratio of dissipated to stored energy.

Without proof, it can be shown that the loss factor can be written as

$$\tan\delta = \frac{E_2}{E_1} = \frac{1}{\omega\tau} \tag{4.7}$$

Here, τ denotes the relaxation time of the system.

If E_1 and E_2 are plotted as a function of temperature for materials going through the glass transition, the typical graph shown in Fig. 4.9 is obtained.

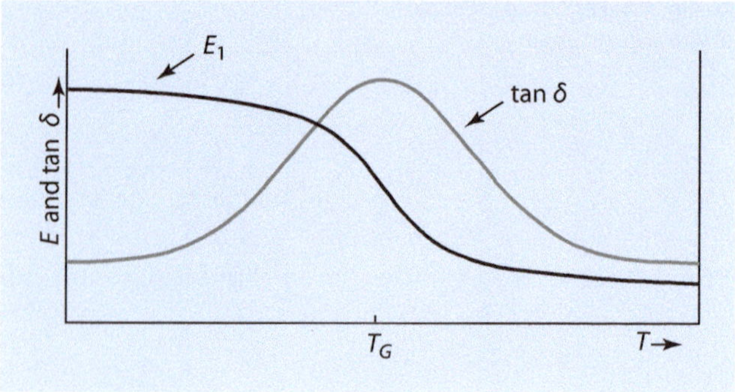

Fig. 4.9 Graph of the storage modulus and the loss factor as a function of temperature

It can be seen that the storage modulus drops abruptly across the glass transition. This becomes obvious by considering that the melt after the glass transition is a liquid and as such is generally not able to store energy. By contrast, the loss factor passes through a maximum at temperatures around the glass transition temperature. A maximum of the loss factor is equivalent to the material being a damping element. This can easily be explained graphically.

In a thought experiment, one imagines a polymer at low temperature as a hard metal spring. At low temperature the polymer chains are in a glassy solid state, they are not very mobile, and relatively large forces are required for small elongations. In other words, the material can be represented by a hard spring with a relatively high modulus.

As this system passes through its glass transition, the chains become more flexible, the material becomes softer, and the modulus of a spring representing the material is low. A polymer above its glass transition temperature can be represented by a soft metal spring.

Both hard and soft springs have their own resonance frequency. As is known from physics, a spring can store a lot of energy in the region of its resonance frequency, and it can be easily excited. With a hard spring, this resonance frequency is relatively high and for a soft spring the frequency is low. However, at their respective resonance frequencies, large oscillations can easily be induced in both springs.

What happens when a material passes through its glass transition? Because polymers always have a certain molar mass distribution, and, in particular, amorphous phases are heterogeneous with respect to their local order, there are both soft and hard springs present in the material close to the glass temperature. Thus, the system no longer finds its resonance frequency and any attempt to store energy in these materials results in a much larger energy dissipation. As a consequence, energy dissipation is observed at temperatures close to the glass transition, and thus a high damping of input energy. In an experiment, one would find that a weight hanging on a hard and a soft spring, arranged in series, cannot be excited to a greater, persistent oscillation.

DMTA measurements are designed to measure the storage and loss moduli of the sample as a function of temperature whereby the glass transition can then be identified from the position of the maximum in the loss modulus curve.

This can be achieved by heating a sample with a constant frequency and measuring the phase shift between the stress and strain curves. However, in practice it is simpler to vary the frequency rather than the temperature. This leads us to the so-called *time-temperature superposition principle*. How are frequency and temperature related?

Let us imagine that we are observing a sample at high temperature. If we put a load on the sample, for example by denting it or pulling on it, then at high temperature the polymer chains have sufficient mobility to respond to the strain by changing their conformation; the material appears soft. The same thing happens when we put a strain on a material at low frequency. The duration of the strain is relatively long and the polymeric material has a relatively long time to counteract the strain. Again, it appears soft. At low temperatures or high strain frequencies the opposite is the case. At low temperatures, the mobility of the material is not sufficient to react to the strain by changing its coil conformation. The material appears hard. At high measuring frequencies, the material does not have enough time to react to the strain and it appears hard.

This behavior—the dependence of the mechanical properties on the duration of the applied strain—is by no means limited to polymers. Compare stepping into a water-filled bathtub with a belly flop from a 5-m springboard. You can probably notice dramatically the dependence of the mechanical effect on your body on the speed of your plunge. Similarly, water behaves differently at a bathtub temperature of 37 °C than at −10 °C. Here, low temperatures cause a similar effect to that of short strain durations.

Thus, there is an analogy between measurements at high frequencies and measurements at low temperatures—and vice versa. Therefore DMTA measurements can either be carried out at a constant measuring frequency when varying the temperature, or—which is more often done technically—the temperature is kept constant and the measuring frequency varied. The data are then converted into a temperature curve.

Figure 4.10 shows the dependence of the storage and loss moduli on the measuring frequency, or more specifically on $\log(1/\omega\tau)$. The maximum of the loss modulus is achieved when this logarithm becomes zero, i.e., when the relaxation time τ is identical to the reciprocal of the measuring frequency. Thus, when the time in which the material can relax is approximately similar to the observation time, the sample is in an intermediate state where the material cannot yet fully relax but the first movements have already started. This state corresponds to the gradual transition from a hard to a soft spring. Exactly here, analogous to the above thought experiment, high damping is observed.

The DMTA suits itself ideally for the detection of very fine phase transitions. This can be shown by the example of polystyrene (Fig. 4.11).

In the upper part of the figure, the dependence of the storage modulus, in this case the shear modulus, on the temperature is shown for polystyrene. One can see a large transition at a temperature of about 100 °C, which corresponds to the glass transition of polystyrene. The curve at lower temperatures is, however, more difficult to interpret. This region is easier to analyze if the logarithm of the loss factor $\tan\delta$ is plotted as a function of the

Fig. 4.10 Graph of storage and loss moduli as a function of frequency

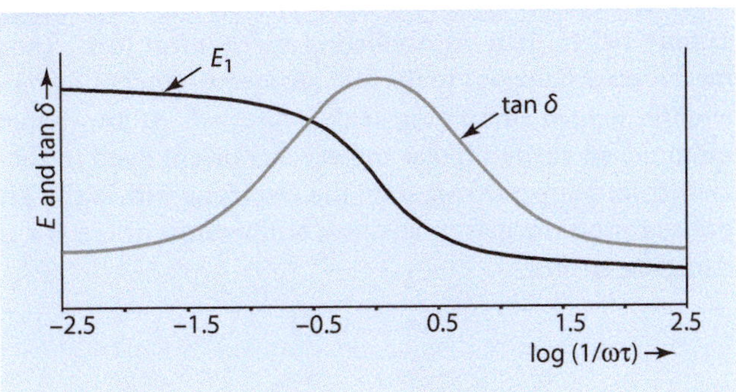

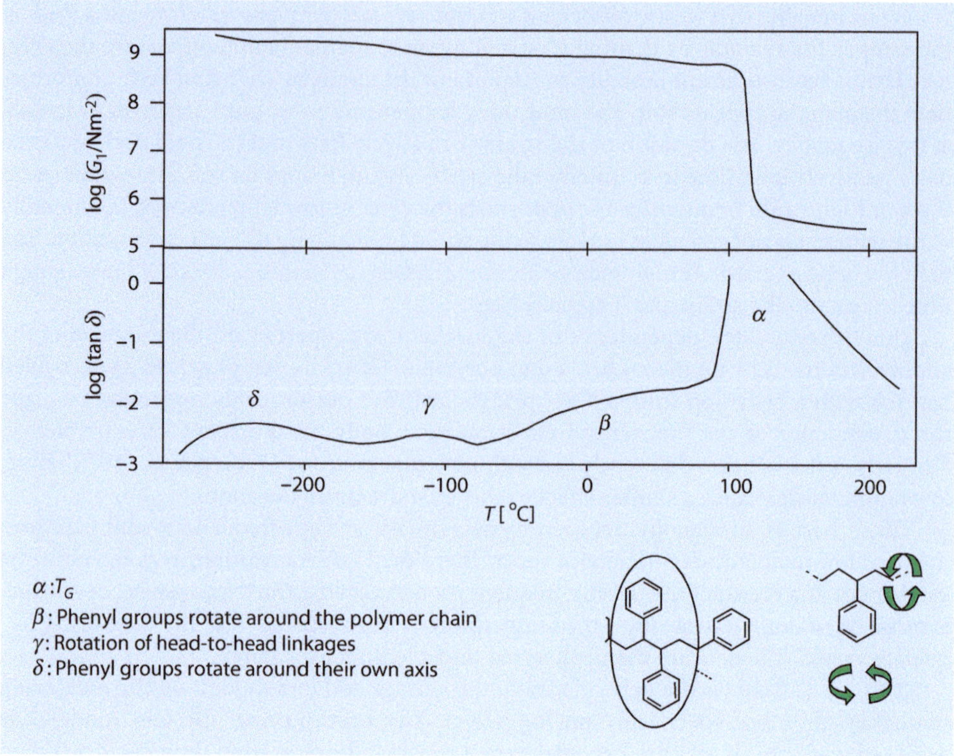

Fig. 4.11 DMTA measurement of polystyrene (Arridge 1975)

temperature (lower part of the diagram). Here we also see a very distinctive maximum at about 100 °C, which can be assigned to the glass transition. However, three additional transitions at lower temperatures can also be identified. These are referred to as the β, γ, and δ transitions. It has been shown that, at the β transition, the phenyl groups stop rotating around the polymer chain. If the sample is cooled below the γ transition, the rotations of head-to-head linkages also become frozen. At the δ transition, the rotation of the phenyl groups around their own axes also ceases. Thus, interestingly, we find further thermal transitions in the material below the main glass transition temperature T_G.

Dielectric thermal analysis (DETA) functions according to principles similar to DMTA. Analogous to DMTA, in DETA an alternating electric field is applied to the sample rather than an oscillating mechanical field. The polarization of the material is measured. Analogous to the DMTA measurements, a phase shift between the polarization and the applied alternating field is observed. At low frequencies or high temperatures, for example, an entire dipolar moiety can orient itself in the electric field. At high frequencies/lower temperatures, only the electrons within the group are displaced. At temperatures around the glass transition temperature there is a non-uniform reaction and high damping results.

In principle, DETA can be used in the same way as DMTA; however, it can only be applied to polymers with dipolar groups. Thus, in combination with DMTA, DETA allows conclusions as to whether dipolar groups are involved in a particular transition. If they are, the transition is observed using both techniques. If not, the transition is only identified using DMTA.

Reference

Arridge RGC (1975) Mechanics of polymers. Clarendon Press, Oxford University Press

Partially Crystalline Polymers

S. Koltzenburg et al., *Polymer Chemistry*, DOI 10.1007/978-3-662-49279-6_5

As a corollary to the basic introduction to polymers in their solid state, this chapter deals with partially crystalline polymers. To begin, we need to ask ourselves which polymers are able to crystallize.

Because crystallization involves the transition to a highly ordered state, the entropy of crystallization is negative, that is, unfavorable. This has to be compensated for by a favorable, and thus likewise negative, crystallization enthalpy. Exothermic crystallization requires polymer chains that can interact energetically enough with one another in their crystallized state. This can happen in two different ways: either through symmetrical chains that allow a high packing density or through strong interactions between the chains. Examples of polymers with symmetrical chains are polyethylene or isotactic polypropylene. By contrast, atactic polypropylene cannot crystallize because of its asymmetrical chain. An example of a polymer with strong interactions (in this case hydrogen bonds) between its chains is the polyamide family, particularly polyamide 6.6.

Polymers with especially flexible chains tend to form random coils, as discussed in ▶ Chap. 2. This is obviously not favorable for crystallization. Therefore, these types of polymers are either amorphous, or crystallization is very slow. Thus, for example, polyisobutene only crystallizes if it is annealed at its ideal crystallization temperature over a long time period.

Crystallization takes place in two stages according to classical crystallization theory, namely, *nucleation* and ensuing growth. For nucleation to take place, crystal nuclei with a critical minimal size first have to form. This process can be induced, for example, by the addition of nucleation agents, and the rate of growth is largely dependent on the temperature.

Both steps significantly influence the morphology of the polymer crystals formed and thus also influence their properties, such as the temperature at which they melt.

5.1 Factors that Influence Melting Temperature

The melting temperature of polymers is not a fixed constant, in contrast to that of small molecules. For polymers it is influenced by various aspects of the polymer composition as well as the thermal history of the material. The significant parameters are discussed in detail from the following points of view:

- Conditions of crystallization
- Chain flexibility
- Symmetry
- Interactions between chains or chain segments
- Tacticity
- Branching
- Molar mass
- Comonomers

5.1.1 Influence of the Conditions for Crystallization

As a rule of thumb, polymers can crystallize in a temperature range from at least 30 °C above the glass transition temperature to at least 10 °C below the melting temperature. At temperatures that are too low, the mobility of the chain segments is insufficient to allow crystallization. At temperatures that are too high, the segments are too mobile for any crystallization to take place.

Because of the entanglement of polymer chains with one another, polymers, in principle, crystallize relatively slowly, especially when compared to most small molecular substances. Thus, it is possible to control the degree of crystallization by adjusting the rate of cooling. As a general rule, very rapid cooling does not give the polymer enough time to form a highly crystalline material. Indeed, by cooling a polymer extremely rapidly it is even possible to quench polymers that would normally crystallize into a completely amorphous state.

Rapid cooling leads to the formation of many crystallization nuclei and thus also to a large number of small crystals. Small crystals have a large surface to volume ratio and thus have a relatively large surface energy. As a result, small crystals are energetically disadvantaged in comparison to large crystals. This leads to their melting at comparatively low temperatures. Rapid cooling thus leads to the melting temperature being reduced by up to 30 °C. It also leads to polymers with a broad melting range because crystals of very different sizes form, each of which have their own unique melting temperature. If the rate of cooling is slower, this effect is less developed and the melting range is narrower and more easily defined.

5.1.2 Influence of Chain Flexibility

The flexibility of the polymer chain has a substantial influence on its crystallization entropy. As has already been stated, the crystalline state is more ordered than the melted state. This loss of entropy is more pronounced the greater the number of conformations that the polymer can adopt in its melted state. This number is, in turn, dependent on the difference in energy between the *trans* and *gauche* conformations (▶ see Chap. 2). If this energy difference is small, the polymer tends strongly to coil up on itself. Thus, the conformational entropy in the melt is large and the change in entropy on crystallization very unfavorable. This is only overcompensated by the crystallization enthalpy at lower temperatures. Polymer chains that have a small difference in energy between their *trans* and *gauche* forms generally have low melting temperatures (◘ Table 5.1).

Because the crystallization enthalpies for high density polyethylene and poly(tetrafluoroethylene) (PTFE) are very similar, the difference in their melting temperatures is a result of the greater tendency of polyethylene to form coils and is thus an entropy effect.

◘ **Table 5.1** Energy difference between polymers in their *trans* and *gauche* forms

Polymer (high-density)	Energy difference (*trans/gauche*)	Melting temperature
Polyethylene	$\Delta\varepsilon = 3$ kJ/mol	$T_m = 137$ °C
PTFE	$\Delta\varepsilon = 18$ kJ/mol	$T_m = 372$ °C

5.1.3 Influence of Chain Symmetry

Asymmetrical elements in the polymer backbone mean that the polymer chain can no longer fit so well into the crystal structure. The resulting disturbance to the organization of the crystals leads to a lowering of the melting temperature. This effect is especially marked for structural units that lead to a kink in the polymer chain. Examples of this are *cis*-double bonds, *cis*-connected rings or aromatic rings, which have a polymer backbone chain linked to them at their *ortho* or *meta* positions (◘ Fig. 5.1).

Symmetry-related disorder leads, for example, to *trans*-polybutadiene being crystalline with a melting temperature of 148 °C, whereas *cis*-polybutadiene is amorphous.

5.1.4 Influence of Interactions Between the Chains

Examples of interactions between polymer chains and/or between different segments of one and the same chain are dipole–dipole interactions and hydrogen bonds (◘ Fig. 5.2). Thus, as mentioned above, polyamide 6.6 forms a strong hydrogen bond network throughout the whole material. In this polymer the alkylene chains between the groups that operate either as donors or acceptors are of equal lengths and an ideal network develops. As a result of this, polyamide 6.6 melts at a temperature of roughly 267 °C (compared to high density polyethylene, which does not build any hydrogen bonds and melts at temperatures as low as 137 °C).

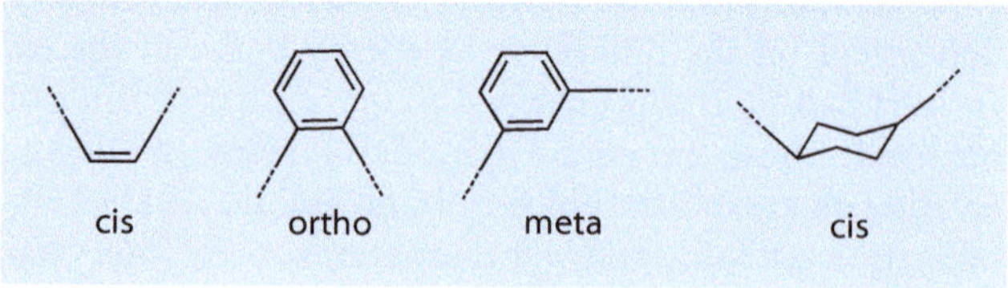

◘ **Fig. 5.1** Chain elements that lead to kinks in the polymer chain

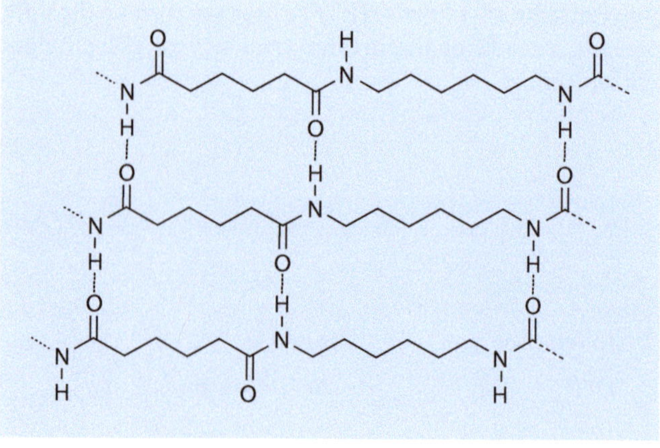

◘ **Fig. 5.2** Hydrogen bond network in polyamide 6.6

5.1.5 Influence of Tacticity

Isotactic polymers usually arrange themselves into helical superstructures so as to minimize the steric interactions between the substituents. In such helical structures, main chain substituents are so arranged as to point away from the helix. The helices themselves are then arranged in the crystal structure in a regular fashion.

In the case of syndiotactic polymers, the main chain substituents are further apart from each other and the main chain takes on a zig-zag conformation (Fig. 5.3).

As a general rule, atactic polymers cannot crystallize because of their irregular structures. Because of this, atactic polystyrene ($R=C_6H_5$) is amorphous, whereas isotactic and syndiotactic polystyrene are crystalline and melt at 240 °C and 270 °C, respectively.

5.1.6 Influence of Branching

A compact, branched side group generally leads to a stiffening of the polymer chain and creates an increase in the melting temperature because of the effects that have been explained above (crystallization entropy) (Fig. 5.4).

In contrast to this, long, flexible side chains interfere with the crystal lattice, and thus cause the melting temperature to sink. An example of this can be found in the isotactic α-polyolefins, where the melting temperature drops from polypropylene to poly-1-pentene (Table 5.2) (Jones 1964).

Initially, the melting temperature continues to drop with increasing side chain length. However, after a certain length, it increases again, slightly. This can be observed for side chains that are longer than eight methyl units long, such as poly-1-decene. The reason for

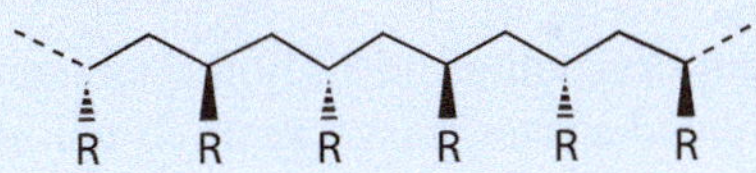

Fig. 5.3 Zig-zag conformation of a syndiotactic polymer

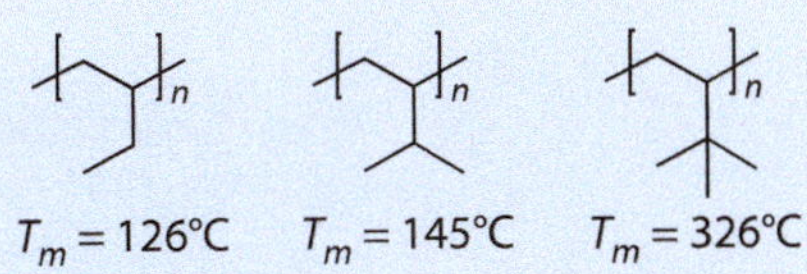

Fig. 5.4 Influence of different branching of side chains on the melting temperature.

Table 5.2 Influence of length of the side chain on the typical melting temperatures of isotactic α-polyolefins

α-Polyolefin	Melting temperature T_m (°C)
Polypropylene	170
Poly-1-butene	126
Poly-1-pentene	80

this is the increasing influence of the side chains to crystallize on the conformation of the main chain.

Branching in the main chain also leads to a lowering of the packing density. This lowers the crystallization enthalpy and thus also the melting temperature. Good examples of this are the various branched polyethylenes. As is shown in ▶ Chap. 14, ethylene can be polymerized, using radical polymerization, to a relatively highly branched material, so-called *low-density* polyethylene (PE-LD). This material has a melting temperature of 110 °C, whereas *linear-low-density*-polyethylene (PE-LLD) produced by transition metal catalysis, which has a lower degree of branching, only starts to melt at a higher temperature of around 122 °C.

5.1.7 Influence of Molar Mass

The ends of the chains are, in comparison to the segments within the chain, relatively mobile. This mobility interferes with the crystal lattice and reduces the degree of crystallization. The resulting drop in melting temperature at low molar masses is a thermodynamic effect.

This thermodynamic effect can, however, be superimposed by kinetic effects, because polymers with very large molar mass crystallize very slowly. It is possible, especially if the rate of cooling is fast, that polymers with high molar masses show lower crystallinity than their analogues of lower molar mass. ◘ Table 5.3 lists the melting temperatures of isotactic polypropylene of differing molar masses (Natta et al. 1964).

5.1.8 Comonomers

Most partially crystalline, linear homopolymers are either characterized by a high melting temperature and a high glass transition temperature or a low melting temperature and a low glass transition temperature. A rule of thumb is that the glass transition temperature (in Kelvin) is roughly 0.7 times the melting temperature. This relationship is also valid for

◘ **Table 5.3** Influence of molar mass (*M*) on melting temperature (T_m) using isotactic polypropylene as an example

M (g/mol)	T_m (°C)
905	93
1050	105
1340	110
2000	114
14,000	154
30,000	170

many small molecule substances in that they can also be converted into an amorphous state, such as pharmaceuticals.

For some applications this relatively simple connection between the melting temperature and the glass transition temperature is undesirable. There are some applications for which a material with a low glass transition temperature but a high melting temperature would be considered desirable. The melting temperature of a polymer (as an empirical observation) generally correlates to its E-module. Most partially crystalline polymers are packed in an ordered fashion and have strong interactions between their polymer chains. These properties give rise to a high module. Thus, rigidity, melting temperature, and glass transition temperature are usually closely related to one another.

In the following paragraph we discuss which polymer structures enable the above-mentioned relationship between T_m and T_G (at least within certain boundaries) to be circumvented. The question is, for example, how to produce a soft material (with a low melting temperature) that nevertheless has a high glass transition temperature. The crucial fundamental idea here is that the crystallinity of a material is dependent on whether the chain segments in the crystal structure can arrange themselves in an ordered manner, whereas the glass–liquid transition temperature is more dependent on how mobile the molecules in the polymer chains are.

Let us consider the interactions between the chains of polyamide 6.6 (◘ Fig. 5.2). The polymer chains take on a regular zig-zag conformation in the solid state. It is immediately apparent from ◘ Fig. 5.2 that because of the regularly ordered structure a very large number of hydrogen bonds can form. This network of hydrogen bonds makes this polyamide very rigid and the melting temperature very high. However, if a certain amount of comonomer is incorporated, for example decane dicarboxylic acid, the regular sequence of hydrogen bonds in the material is significantly disturbed. The resulting disturbance of the crystalline order leads to a drop in both the melting temperature and the module. The glass transition temperature T_G, which is essentially dependent on chain flexibility, is only slightly influenced by the incorporation of this kind of comonomer. Therefore, it is possible, by producing a statistical copolymer, to lower the melting temperature and the module without an equivalent lowering of the glass transition temperature.

One advantage of such materials is their lower processing temperature. As most polymer materials are processed as melts (► see Chap. 17), a lower processing temperature is less aggressive to the material and uses less energy.

The general effect of the statistical incorporation of a comonomer into a polymer can be seen in ◘ Fig. 5.5.

The figure shows the change in the melting temperature and glass transition temperature of two homopolymers poly(A) and poly(B) when their composition is gradually varied from 100% A to 100% B. Because of the disturbance of the crystal structure, the melting temperature drops significantly, even if only relatively small amounts of comonomer are present. The disturbance can be so strong in materials where the comonomers account for half of the composition that the material does not crystallize at all and remains amorphous.

The linear interpolation of the glass transition temperatures shown in ◘ Fig. 5.5 is a simplification and generally only approaches the real case when the comonomers are chemically very similar to one another.

The effect of copolymerization on melting temperature can be approximately (without proof) estimated as follows:

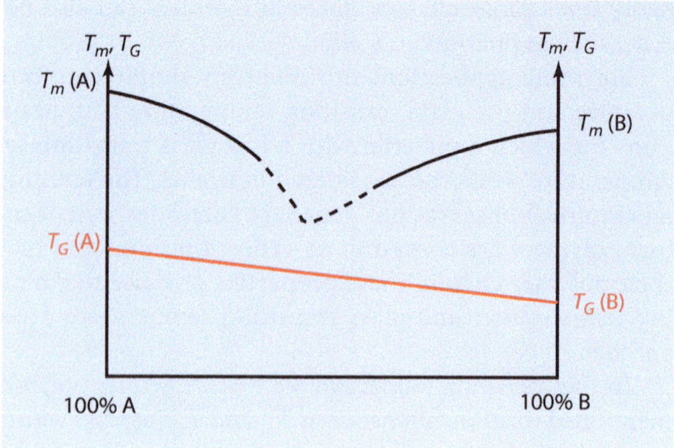

Fig. 5.5 Idealized description of the change in melting temperature (T_m) and glass transition temperature (T_G) as a function of the composition of statistical copolymers of monomer A and B. Further explanations are given in the text

$$\frac{1}{T_{AB}} - \frac{1}{T_A} = -\frac{R}{\Delta H_m} \ln x_A \qquad (5.1)$$

T_{AB}	Melting temperature of the copolymer
T_A	Melting temperature of the homopolymer A
ΔH_m	Melting enthalpy of the homopolymer
x_A	Mole fraction of the homopolymer

A comparison between statistical and block copolymers can be found in ▶ Sect. 7.2.1.

The different variables that influence melting temperature mentioned above are responsible for the slightly different melting temperatures listed in the literature for chemically identical polymers. The differences can be ascribed especially to molar mass, the degree of branching, tacticity, and the thermal history of the material (cooling rate, tempering). Therefore, it is untrue that any one of the listed values is more or less correct than any of the others.

5.2 Morphology of Partially Crystalline Polymers

A 100 % crystalline material is never produced after the cooling period because polymer chains are so entangled in the molten state. Thus, in partially crystalline polymers there are always both crystalline domains and amorphous domains present. Polymer single crystals can only be obtained from very dilute (less than one per thousand) solutions but this is not of technological relevance.

In practice, polymers are often processed from the melt and crystallization takes place from a very viscous system. The entanglements prevent a perfect crystal structure from being formed. This leads to the proportion of the amorphous domain varying in size.

The crystalline areas often form lamellae, typically less than 100 nm thick. The main chain of the polymer is usually orientated orthogonally to the plane of the lamellae, resulting in an "accordion-like" structure.

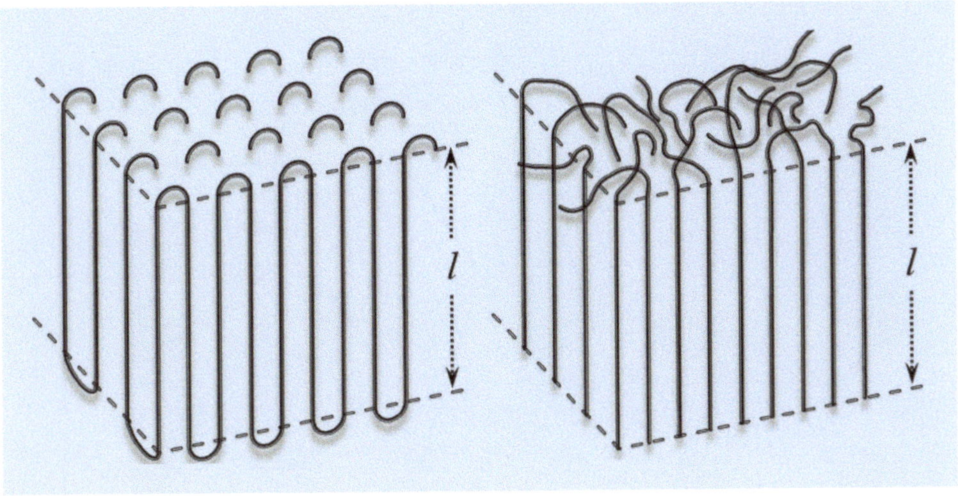

Fig. 5.6 Lamella model for polymer crystallization

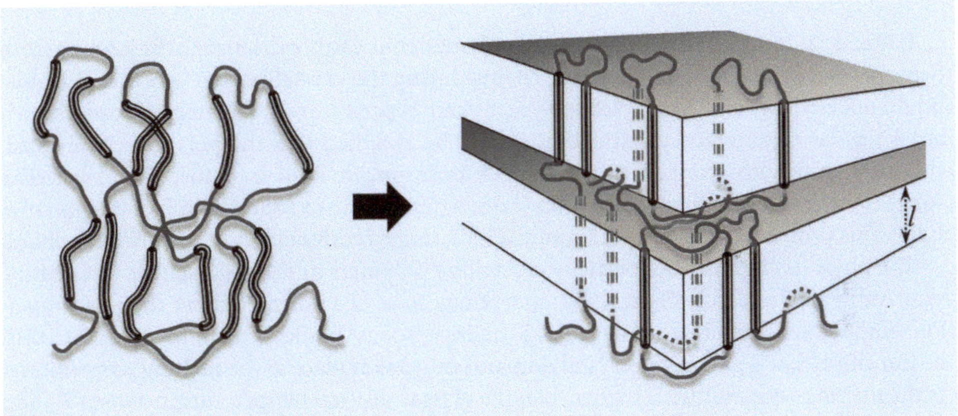

Fig. 5.7 Switchboard-model of polymer crystallization. Segments that have been incorporated in the crystal lamellae are *highlighted*

The question as to how the polymer chains fold at the ends of the lamellae has not been conclusively resolved. There are essentially two models (Figs. 5.6 and 5.7).

According to current hypotheses, the first, regularly folded model (Fig. 5.6) is obtained from a dilute solutions, whereas the second model, also known as the "switchboard model" (Fig. 5.7) is obtained from the melt (Dettenmaier et al. 1980).

The thickness of the crystal lamellae increases with the crystallization temperature. Whereby the number of repeat units in the main chain is, as a general rule, markedly below 100.

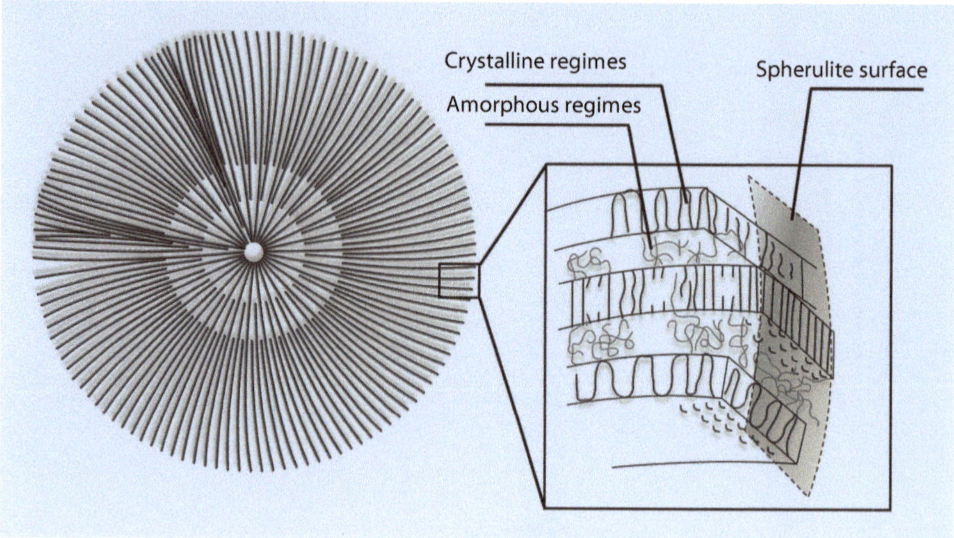

Fig. 5.8 Superstructure (spherulite) of a partially crystalline polymer

It is crucial for crystallization that the polymer coils cannot disengage themselves from their entanglements with neighboring chains during the crystallization process. Thus one should not envisage that a complete polymer chain separates from the melt and orders itself into a regular crystal lattice. Instead, it should be assumed that the polymer chain must, without significantly moving away from its position, fit itself into the crystal lamellae alongside segments of other chains. This is often described as a solidification of the polymer chain. This process itself does not require a long range reorientation of the polymer chain.

The physical chemistry of partially crystalline polymers in the solid state is fascinating. As already explained in ▶ Chap. 4, the amorphous areas of a polymer above the glass transition temperature should be envisaged as a (highly viscous) liquid or melt. In contrast to this, the transition to a liquid in the crystal domains only takes place at the melting temperature. As the melting temperature is higher than the glass transition temperature, between T_G and T_m two phases coexist in partially crystalline polymers, one of which is liquid and one solid. In terms of the right hand sketch in Fig. 5.7, this means that in this temperature range there are solid crystal lamellae surrounded by a liquid, amorphous phase. We are dealing with a stable, two phase system, in which crystallites coexist with a chemically identical melt—and this below the melting temperature. This extremely unusual situation, which, of course, is impossible to replicate with small molecules, leads to outstanding mechanical properties. For example, if strain is applied to the material, such as some kind of impact, the liquid phase dissipates the energy and the material does not break. This explains the ductility of many partially crystalline polymers between T_m and T_G. We are thus dealing with a material that in some respects is similar to a highly viscous liquid. However, of course, the material as a whole cannot flow because, as can be seen in Fig. 5.7, the polymer chains are generally integrated into more than one crystallite, which prevents any macroscopic flow.

The crystal lamellae of partially crystalline polymers often build superstructures, and an example of this is the so-called spherulites (McCrum et al. 1988) with spherical superstructures as shown in Fig. 5.8.

Fig. 5.9 Isotactic polypropylene viewed through a polarizing microscope (dark regions: α-spherulite, light regions: β-spherulite)

As well as the structures shown here, one- or two-dimensional superstructures (fibers or discs) are also observed. These superstructures can be analyzed by a polarizing microscope because they are birefringent (Fig. 5.9).

This image shows isotactic polypropylene. Polypropylene forms two different kinds of spherulites (called α- and β-spherulites) which can be seen in Fig. 5.9.

5.3 Crystallization Kinetics

An analysis and the knowledge of the crystallization kinetics are crucial requirements for a rational processing of a polymer. During the manufacturing process, in which the polymer is generally melted (▶ see Chap. 17), the cooling rate, as has been explained, determines the degree of crystallinity and thus the properties of the resulting material. It is therefore important to know how a polymer crystallizes under any given processing conditions.

The analysis of the crystallization kinetics can, for example, be carried out using a dilatometer (Fig. 5.10), whereby the change in volume of the sample as a function of temperature is measured from the change in the height of a column of mercury (which also serves to contain the sample). For such measurements, the material is tempered above its melting temperature and then rapidly cooled to the desired crystallization temperature. Because the density of the crystals is greater than that of the amorphous domains, a volume contraction accompanies crystallization. By measuring the change in volume with time the desired crystallization kinetics can be determined.

The crystallization of polymers can be described by the Avrami equation:

$$\frac{w_{m,t=t}}{w_{m,t=0}} = \exp\left(-kt^n\right) \tag{5.2}$$

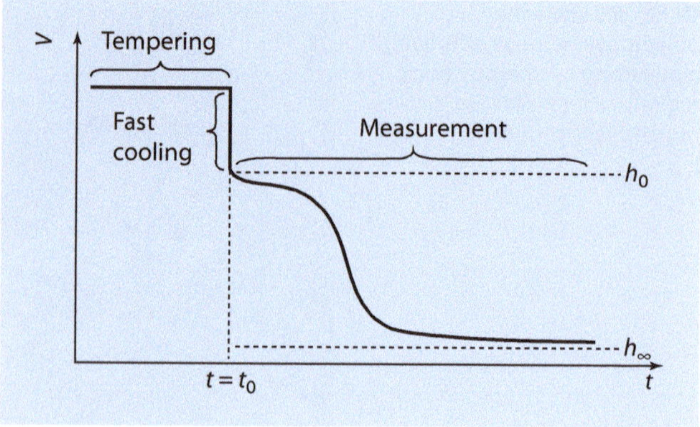

Fig. 5.10 Schematic diagram of the change in volume *V* of a polymer sample over time *t* during crystallization as measured in a dilatometer

Table 5.4 Interpretation possibilities of the Avrami exponent *n*

Growth in the form of....	Slow nucleation ($T \sim T_m$)	Rapid nucleation ($T \ll T_m$)
Fibers	2	1
Discs	3	2
Spherulites	4	3
Complex bundles/coils	6	5

w_m Weight of the non-crystalline melt as a function of time *t*
k Rate constant
n Avrami exponent

This equation describes the weight fraction of the non-crystalline melt at time t relative to the original mass of the melt at time t_0. The relation between these two numbers is a function, which, where $n = 2$, corresponds to a Gaussian distribution.

To evaluate the results of dilatometric measurements, the Avrami equation is usually transformed as follows:

$$\frac{h(t) - h_\infty}{h_0 - h_\infty} = \exp\left(-kt^n\right) \tag{5.3}$$

h_0, $h(t)$, h_∞ Height of the column of liquid at time $t = 0$, t and t_∞

From the value of the Avrami exponent, n, it is possible to predict what kinds of superstructures form in the melt. The correlations are listed in Table 5.4. However, a difference between slow and rapid nucleation must be taken into consideration.

References

Dettenmaier M, Fischer EW, Stamm M (1980) Calculation of small-angle neutron scattering by macromolecules in semicrystalline state. Colloid Polym Sci 258:343–349

Jones AT (1964) Crystallinity in isotactic polyolefins with unbranched side chains. Makromolekulare Chem 71:1–32

McCrum NB, Buckley CP, Bucknall CB (1988) Principles of polymer engineering. Oxford University Press, Oxford

Natta G, Pasquon I, Zambelli A, Gatti G (1964) Dependence of the melting point of isotactic polypropylenes on their molecular weight and the degree of stereospecificity of different catalytic systems. Die Makromolekulare Chemie 70:191–205

Amorphous Polymers

S. Koltzenburg et al., *Polymer Chemistry*, DOI 10.1007/978-3-662-49279-6_6

The fundamental principles of "amorphous" polymers were introduced in ▶ Chap. 4. Some of their particular properties are described in more detail in this chapter.

6.1 Dependence of the Mechanical Properties of Amorphous Polymers on Temperature

To understand the unique behavior of amorphous polymers, let us first consider their behavior when they are heated. The change of the E-module of an amorphous polymer with a large molar mass when heated is shown in ◘ Fig. 6.1.

The polymer chains are frozen at a low temperature (◘ Fig. 6.1, temperature range I); that is, the molecules are immobile. If an applied force acts upon them, they cannot change position, and the E-module, the force needed to stretch the sample, is very large. The material behaves as if it was a glass-like solidified liquid. Even though the polymer is not crystallized, its shape is macroscopically stable over this temperature range; it does not flow. If the deformation force is large enough, the material simply breaks.

The first chain segments start to move as the temperature approaches the glass transition temperature (◘ Fig. 6.1, temperature range II). At this temperature the material can react better to an external force; the polymer chains can rearrange and the material is less brittle. The tenacity, the opposite of brittleness, of the material, increases. The material behavior is similar to that of leather. Therefore, this temperature range is also referred to as the *leathery state*.

The material can, theoretically, be considered as a melt above the temperature T_G. Because of its high viscosity, however, and especially the numerous entanglements of the polymer chains, the material (as long as the temperature is not that high) does not, macroscopically, behave as a liquid. Instead the material appears, at first glance, to be a solid. This changes, however, if mechanical force is applied to the material over a longer period of time. In this case, the polymer chains have enough time to yield to the forces being exerted upon them and to rearrange. The material flows and takes on a different shape.

The amount of time during which a force is applied to a sample determines whether the material irreversibly changes its shape, whether flow actually takes place or not. A single blow or other short application of force does not give the polymer chains time to yield to the forces by flow. After the application of force, the material can, as long as it is not broken, regain its original shape. If, however, the force is applied for a longer time, the material can flow irreversibly. In this temperature range (◘ Fig. 6.1, temperature range III) the material

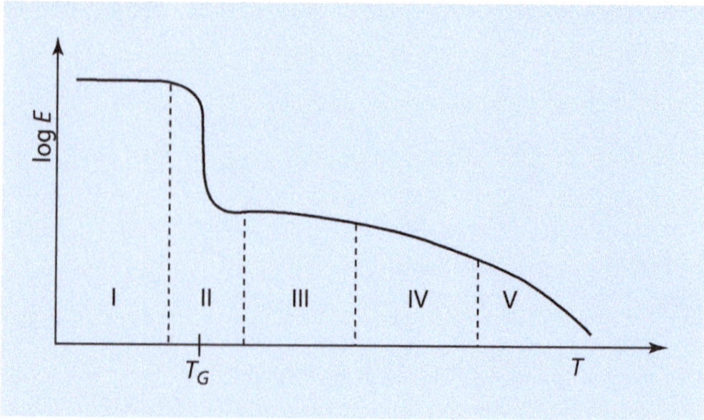

◘ **Fig. 6.1** Change of the E-module of an amorphous polymer with a large molar mass as a function of temperature

exhibits both viscous and elastic mechanical behavior. This type of behavior is called viscoelasticity and is discussed in more detail in ▶ Sect. 6.6. Over temperature range III the material is, theoretically, an extremely viscous melt, called the *rubber-elastic state*. In this state the elasticity of the material still dominates over its viscosity.

If the temperature is increased even further (◘ Fig. 6.1, temperature range IV), the polymer chains in the material gain molecular mobility. As a result of this, the material can flow more easily and its ability to react elastically decreases. That is, the ability of a sample to take on its original shape after a force has been applied decreases.

If a mechanical stress is applied to the material and then released, part of the resulting deformation is restored, whereas another part of it remains. The latter, at a molecular level, involves the polymer chains reacting to the applied force by taking on a different conformation. Because the newly assumed conformation is not the same as that of the undisturbed state, it is less favorable energetically (▶ Chap. 2). By moving relative to one another, the polymer chains can rearrange in an energetically more favorable conformation if the applied stress acts over a longer period of time or the temperatures are higher.

In temperature range IV shown in ◘ Fig. 6.1, we are dealing with two opposing effects: on the one hand, an elastic component that is dependent upon the entanglements of the polymer chains; on the other, a viscous component that is present if the polymer chains have enough energy to move relative to one another, allowing the sample to flow macroscopically. This intermediate stage is referred to as *rubber-elastic flow*.

Temperature range V describes the gradual transition into a viscous melt that does not possess a significant shape memory. The polymer is in a viscous state in this temperature range.

The boundaries between temperature ranges III and V in ◘ Fig. 6.1 are not clearly defined. The transitions from one stage to another are gradual. Elasticity and shape stability decrease as temperature increases, whereas the ability to flow on a macroscopic level increases.

6.2 Amorphous State

In an amorphous, glass-like solid state, a polymer is similar to a plate of frozen pasta. Considerable force is necessary to deform the material. The chain segments are immobile; in some cases there may be some mobility of the side chains. The latter leads to the material being less brittle (i.e., tougher) than, for example, quartz glass. In this state, even strains of a few percent lead to failure—*brittle fracture*. As already explained in ▶ Chap. 4, at temperatures above T_G the material can yield to an external force without breaking. The energy is dissipated by the molecules rearranging and the material is tough rather than brittle.

As a first approximation, in the frozen state the polymer chains are not arranged in any, sort or long range, regular pattern. At temperatures below T_G, movement of the chain segments is not possible and the molecules cannot change their relative positions. The only movements possible are vibrations at an atomic level. If the temperature rises above T_G, the movement of individual segments becomes possible. A transition to a melt state follows in which the material can flow—even if the viscosity is very high.

A macroscopic flow of the material requires that the polymer chains do not have a high packing density. This is, however, generally the case in the amorphous state. There is a certain amount of *free volume* given by the imperfect packing of the irregularly arranged chains. This is the case both above and below T_G, and as a consequence the density of the amorphous material is lower than that of the corresponding crystal. An increase in temperature leads to a thermal expansion of the material, the free volume increases, and the polymer chains become more mobile; the viscosity decreases and the melts flows more easily.

6.3 Glass Transition

The changes in the mechanical properties of polymers at the glass transition can be described by a transition from a hard ($T < T_G$) to a soft ($T > T_G$) spring. A polymer chain surrounded by enough free volume with enough thermal energy to allow significant molecular movement is similar to a soft spring. A molecularly frozen polymer chain below T_G can be considered in terms of a hard spring. From physics it is known that both soft and hard springs store elastic energy. The ability to store energy is greatest at the resonance frequency of the spring, whereby the resonance frequency of a hard spring is higher than that of a soft one—hard springs oscillate more rapidly than soft springs.

An amorphous material is, however, not completely homogenous. Generally it is made up of polymer chains of different lengths, and, additionally, the distribution of the free volume is not identical for each chain. Thus, if a polymeric material is heated, as the temperature approaches that of the glass transition, initially only some of the polymer chains transition from a hard to a soft state. This has a strong influence on the ability of the material to store energy. In this temperature range the material is made up of a mixture of 'springs' of different resonance frequencies. Macroscopically, the whole system does not have a homogenous resonance frequency. The result of this is a strong damping of mechanical energy.

Using a material at temperatures close to the glass transition temperature is problematic because of the radical and erratic variation of the mechanical properties (moduli) (Fig. 6.1).

Other properties of the material which change at the glass transition temperature are:
- Specific volume
- Heat capacity
- Refractive index

These changes in properties can be used experimentally to measure the glass transition temperature.

6.4 Factors that Influence the Glass Transition Temperature

The temperature at which the glass transition occurs is essentially determined by the mobility of the polymer chains. In turn the chain mobility is most strongly influenced by the following factors:
- Chain flexibility
- Steric effects
- Tacticity

- Branching and crosslinking
- Molar mass

6.4.1 Chain Flexibility

Fundamentally, flexible chains are more mobile and lead to a lower glass transition temperature. Examples of flexible chain elements are alkyl, ether, or silicone moieties (Fig. 6.2).

In contrast, chain rigidity is increased by the incorporation of rigid segments such as aromatic rings into the polymer backbone.

Table 6.1 shows the influence of chain flexibility on the glass transition temperature of some selected examples.

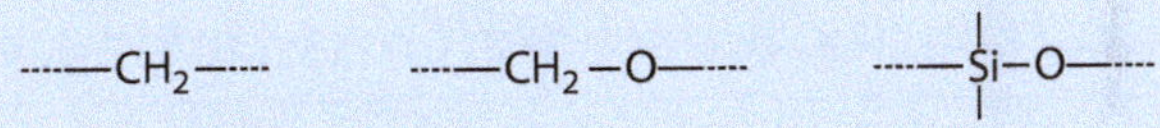

Fig. 6.2 Examples of flexible chain elements

Table 6.1 Glass transition temperature (T_G) of selected polymers having chains of varying flexibility (decreasing from top to bottom)

	T_G (°C)
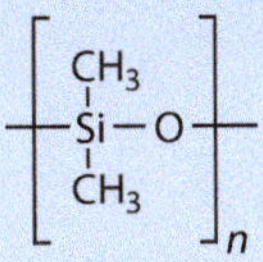	−123
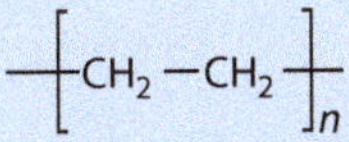	−93
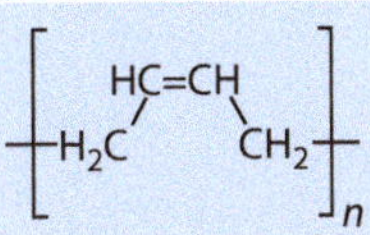	−85
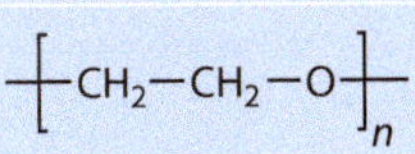	−67
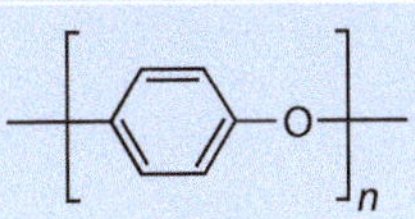	+83

Sterically demanding, stiff segments in the main polymer chain lead to a less mobile polymer chain and the glass transition temperature increases. Thus, polyphenyl ether has a glass transition temperature of +83 °C (bottom line of ◘ Table 6.1).

6.4.2 Steric Effects/Substituents

Flexible side chains increase the available free volume and thus the mobility of the polymer chain. As a result of this the glass transition temperature is lowered (◘ Table 6.2).

Polar substituents can form hydrogen bonds or dipole–dipole interactions which fix the position of the polymer chains and decrease their mobility; the glass transition temperature is increased (◘ Table 6.3).

Additional methyl groups in an α-position relative to the substituent result in a substantial stiffening of the polymer chain and so to an increase in the glass transition temperature of 70–100 °C. A classic example of this is polymethacrylates, which have higher glass transition temperatures than the corresponding polyacrylates.

6.4.3 Tacticity, Branching/Cross-Linking, and Molar Mass

The influence of the configuration or tacticity of polymers from vinyl monomers is only pronounced if there is more than one substituent, such as an additional methyl group, adjacent to the double bond—as is the case for the polymethacrylates mentioned above. Such monomers are, however, usually radically polymerized, whereby the tacticity of the polymer formed cannot be, or is only marginally, controlled. The glass transition

◘ **Table 6.2** Influence of linear alkyl substituents on the glass transition temperature (T_G) of polymers with the structure $(CH_2–CHX)_n$

–X	T_G (°C)
-H	–93
$-CH_3$	–20
$-C_2H_5$	–24
$-C_3H_7$	–40
$-C_4H_9$	–50

◘ **Table 6.3** Influence of polar substituents on the glass transition temperature (T_G) of $(CH_2–CHX)_n$

–X	T_G (°C)
$-CH_3$	–20
-Cl	+81
-CN	+105

temperature of typical polyolefins such as polypropylene is, in contrast, not (or only marginally) dependent on the tacticity because of the missing second substituent.

If the material is cross-linked, the cross-links function as permanent entanglements and chain movement is restricted. At low cross-link densities the glass transition temperature increases almost linearly with the number of cross-links. Materials with a high cross-link density tend to exhibit a broad, ill-defined glass transition.

The polymer chain ends are, in contrast to the other segments along the chain, relatively mobile. Because of this they are able to 'create' comparatively more free volume around themselves. By the same argument as that used above, polymers with lower molar mass have lower glass transition temperatures. This effect can be described by Bueche's empirical equation:

$$T_G = T_G^{\infty} - \frac{const.}{M_n} \qquad (6.1)$$

T_G^{∞} Glass transition temperature for $M_n \rightarrow \infty$

The equation can be applied to molar masses M_n of above ~5000 g/mol. Thus, the glass transition temperature asymptotically increases to a limiting temperature valid for very high (infinite) molar mass.

Theories Describing Free Volume

The term 'free volume' has already been used several times when describing the glass transition temperature. It is assumed that the volume V of a polymer sample is made up of a volume V_0 that is occupied by the polymer molecules and an additional free volume V_f that is available for the rotational and translational movement of the molecules:

$$V = V_0 + V_f \qquad (6.2)$$

V Total volume
V_0 Volume of the chains
V_f Free volume

Both V and V_0, as well as V_f, are dependent on the temperature. For a polymer in its fluid or a rubber-elastic state, the free volume increases with temperature. Visually, the mobility of the molecules increases. However, if the temperature decreases, the free volume shrinks and below a critical temperature—the glass transition temperature—it doesn't change any more; the molecules are immobile.

The main difficulty of theoretically describing the glass state using free volume is that free volume is not an equilibrium state. Basic approaches to a theoretical description of the glass state that should be mentioned are:

- Williams, Landel, and Ferry's kinetic approach (WLF equation)
- Gibbs and DiMarzio's thermodynamic approach
- Gibbs and Adam's combination approach

For more details the interested reader is referred to the corresponding literature (Adam and Gibbs 1965; DiMarzio et al. 1976; Eisele 1990; Williams et al. 1955). Estimating what proportion of the material is free volume is both theoretically and experimentally difficult, and estimates vary according to parameters such as polymer and temperature between 2.5 % and 12 % of the total volume.

6.4.4 Plasticizers

Plasticizers are low molar mass, non-volatile compounds added to a polymer to change its properties. Generally these compounds have a 'lubricating effect' on the polymer chains. The molecules can move more easily when surrounded by low molar mass material. The addition of a plasticizer thus results in an increase in the mobility of the chains and consequently to a lowering of the glass transition temperature. Plasticizers are therefore used so that soft, flexible materials can be obtained from stiff, brittle base polymers.

A technically important material, which often has large quantities of plasticizers added to it, is polyvinyl chloride (PVC). PVC without plasticizer has a glass transition temperature of 81 °C. If 30–40 % plasticizer is added, so-called soft-PVC is obtained, which has a glass transition temperature of <0 °C. Important uses of this material are for the manufacture of raincoats, curtains, and cable sheathing.

The effect of the addition of a plasticizer on the glass transition temperature can be estimated using

$$\frac{1}{T_G} = \frac{w_{Pol}}{T_{G,\,Pol}} + \frac{w_W}{T_{G,\,W}} \tag{6.3}$$

w_{Pol}	Mass fraction of the polymer
w_W	Mass fraction of the plasticizer
$T_{G,\,Pol}$	Glass transition temperature of the polymer
$T_{G,\,W}$	Freezing point or glass transition temperature of the plasticizer
T_G	Glass transition temperature of the plasticized polymer

Water can have a strong plasticizing effect on hydrophilic polymers. In these cases, the properties of the material are very dependent on the surrounding conditions, especially the humidity of the surroundings. This effect, which, for example, can be observed with the biopolymer starch, means that using such materials is a challenge.

6.5 Rheological Behavior of Polymer Melts

The rheological behavior of polymer melts is complex but relevant for understanding amorphous materials at temperatures above T_G as well as for understanding the processing of polymers (▶ Chap. 17), as processing usually involves polymer melts. For this reason, the next part of this chapter deals with *rheology*, the science of flow behavior, and its relevance to polymers.

6.5.1 Newtonian Fluid

According to Newton, the rheological behavior of a simple fluid can be described by

$$\sigma = \eta\left(\frac{dv}{dx}\right) \tag{6.4}$$

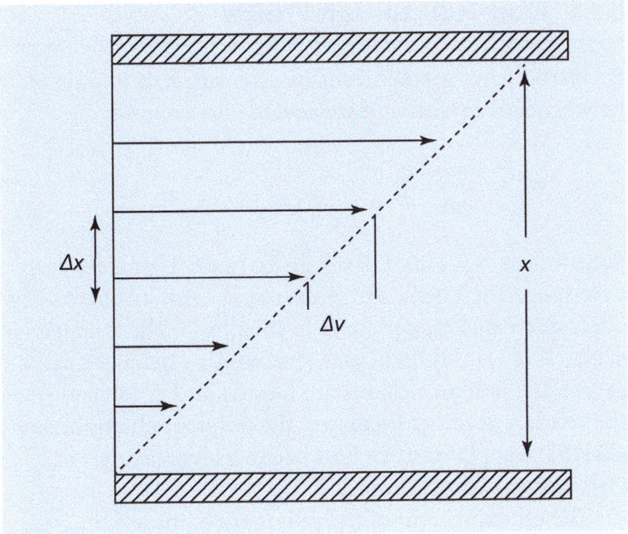

Fig. 6.3 Rheological properties of a Newtonian fluid. The bottom plate remains static and the top plate moves with a velocity *v* and a separation *x* relative to the bottom plate

Equation (6.4) establishes a correlation between the shear rate (dv/dx) and the shear stress σ. For the simplest case, a Newtonian fluid, the shear rate is linearly proportional to the shear stress. The proportionality constant η is the viscosity.

The flow properties of a Newtonian liquid between a static and moving plate are shown schematically in Fig. 6.3.

Two smooth sheets, between which there is a layer of fluid of thickness **x**, are moved relative to one another with a tangential velocity v. The top layer of fluid adheres to the sheet above it and pulls a layer of fluid with it. The friction between the fluid layers leads to a further layer of fluid following the one above it, albeit with a slower speed. The layer of fluid next to the bottom sheet is at rest ($v = 0$). The friction F_R which opposes the movement is proportional to the shear rate dv/dx and the area A (► see (6.5)). The proportionality constant η in this equation is also called the *dynamic viscosity* for this reason.

$$F_R = A \cdot \eta \left(\frac{dv}{dx} \right) \tag{6.5}$$

With the definition of shear stress

$$\sigma = \frac{F_R}{A} \tag{6.6}$$

(6.5) can be transformed into Newton's law of friction (6.4).

6.5.2 Non-Newtonian Fluids

By introducing a coefficient of viscosity n it is possible to take nonlinear relationships between shear stress and shear rate into account:

$$\sigma^n = B\frac{dv}{dx} \tag{6.7}$$

Substituting $n=1$ and $B=\eta$ gives (6.4). Using $n>1$ the rheology of polymers can be described. The coefficient B decreases with an increasing shear rate. The gradient of the curve decreases (upper curve in ◘ Fig. 6.4). The fluid (polymer melt) initially behaves as an elastic body and then begins to flow. This behavior can be explained by considering that, at rest, the polymer chains are looped and entangled so that they resist displacement. As the velocity gradient increases, the polymer chains become disentangled and aligned parallel to the applied stress; flow becomes increasingly easy. This effect is known as structural viscosity or shear thinning.

The entanglement of the polymer chains are the main reason for the high viscosity of polymer melts. However, they are effective when the chains are longer than a certain length. Short chains are unable to form effective entanglements. The minimal chain length needed for entanglements to form and viscosity to be affected is called the critical chain length ζ_c. As a rule of thumb, ζ_c corresponds to main chains consisting of about 100 units. The critical chain length is generally longer for non-polar polymers than for polar polymers. Because of the stronger interactions between the chains, polar polymers tend to be more able to form effective entanglements.

With a plot of log η against log M_n (◘ Fig. 6.5) it is often possible to observe that, below a critical chain length, the logarithm of the viscosity varies with the logarithm of the molar mass with a gradient of 1 (n *in* (6.7) = 1). Above the critical chain length, the viscosity increases roughly with the cube of the molar mass in a way that is roughly proportional to the cube of the molar mass (n *in* (6.7) ≈ 3).

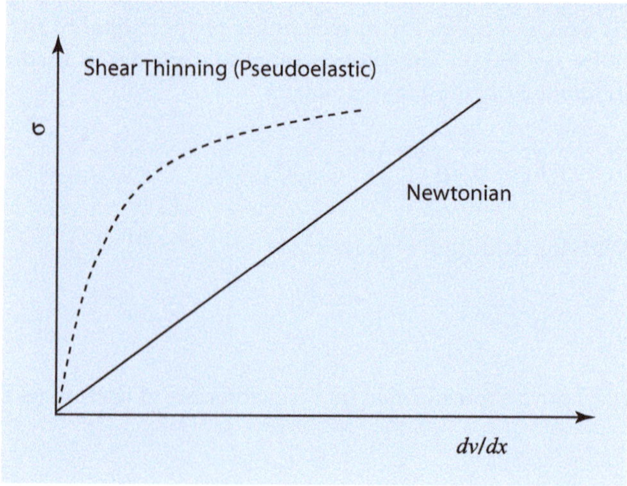

◘ **Fig. 6.4** Comparison of a Newtonian and a structurally viscous fluid

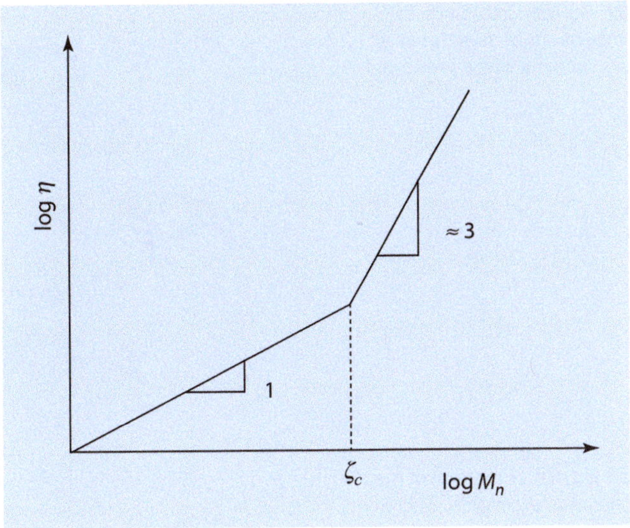

Fig. 6.5 Determining the critical chain length ζ_C above which entanglements of the polymer chains affect the viscosity of a polymer melt

6.5.3 Process of Reptation

Being entangled with other chains, the movement of intact polymer coils through the melt does not take place. Instead, the individual chains have to 'wriggle' around the entanglements. This is effected by the movement of individual chain segments and allows the polymer to be divided into independent kinetic units for a theoretical treatment. The polymer melt flows when these units change their position. They can, for example, move in a crankshaft-like movement in the melt (Fig. 6.6).

This process is called *reptation*. In the reptation model the entanglements are considered as fixed sites in the network which define the shape of a tube through which the polymer chain moves (Fig. 6.7a).

The change in position in the tube takes place by a meandering of a few chain segments. The movement can be visualized by thinking of the way a snake moves, or the way a carpet slides over a smooth surface (Fig. 6.7b).

Rate of the Reptation Process

The time that the polymer chain takes before it leaves the tube created by the entanglements is defined as the relaxation time τ_P and, without proof:

$$\tau_P = \frac{(n \cdot l_0)^2}{2D_t} \tag{6.8}$$

τ_P	Time the polymer requires to leave the 'tube'
n	Number of reptation units in the polymer chain
l_o	Length of a segment (under θ-conditions, ► Chap. 2)
D_t	Diffusion coefficient within the reptation tube

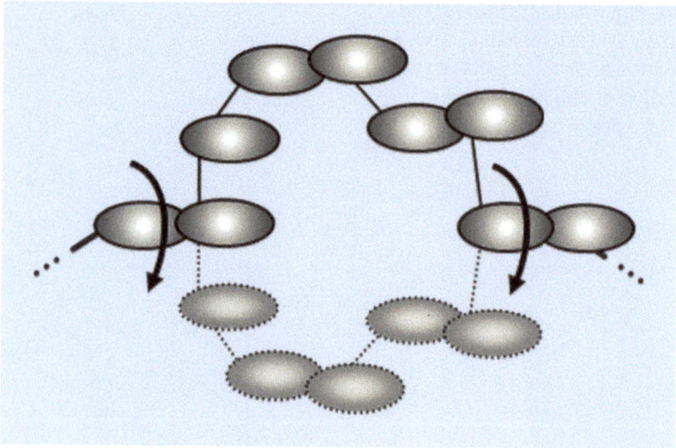

■ **Fig. 6.6** Crankshaft-like movement of the segments of a polymer chain in the melt

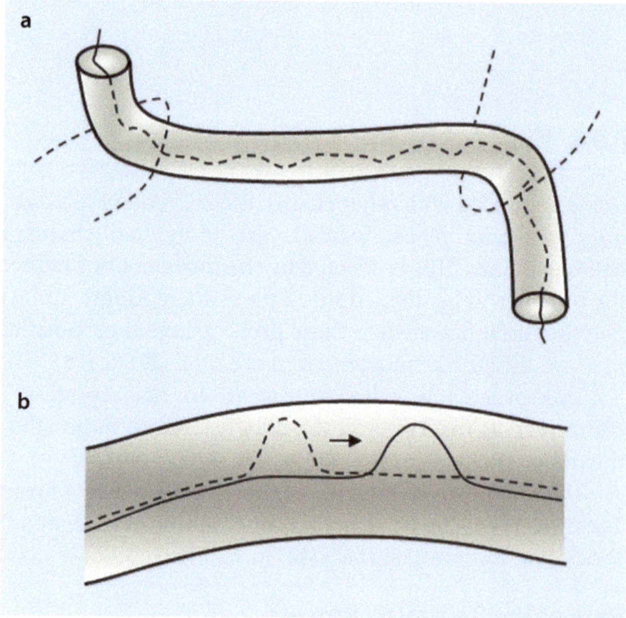

■ **Fig. 6.7** Illustration of the reptation process. (**a**) Definition of the reptation pathway by the entanglements with neighboring chains. (**b**) Meandering movement of a macromolecule

The relaxation time is thus proportional to the square of the chain length and indirectly proportional to the diffusion coefficient D_t of the polymer chain within the tube.

This diffusion coefficient is given by (also without proof)

$$D_t = \frac{k_B T}{f_t} = \frac{k_B T}{n\zeta} \tag{6.9}$$

f_t	Friction factor of the chain
ζ	Friction factor of a segment

By introducing a theoretical relaxation time for the single segments τ_{P0}:

$$\tau_P = \frac{l_0^2}{2D_t} n^2 = \frac{l_0^2}{2D_t n} n^3 = \frac{l_0{}^2 \cdot \zeta}{2k_B T} \cdot n^3 \cong \tau_{P0} \cdot n^3 \tag{6.10}$$

On account of the dependence of the diffusion coefficient on the chain length, if effective entanglements exist, the relaxation time of the polymer chain is proportional to the chain length cubed, consistent with the gradient of the line in ◘ Fig. 6.5 above the critical chain length.

The relaxation time τ_{P0} defined in (6.10) is the hypothetical relaxation time of a polymer chain with only one chain segment and is of the order of 10^{-10} s. Thus the relaxation time for a polymer with 10,000 bonds is of the order of 100 s.

The reptation process can describe the diffusion of a polymer molecule under different conditions. As well as describing the viscous flow process, it can also be used to describe the solution of a polymer molecule at the surface between a solid polymer and a solvent. This model can also be used to describe the welding of polymeric materials (▶ Chap. 17). In both latter cases the polymer diffuses over a boundary layer made up of either a solvent or another polymer.

The viscosity of a polymer chain is temperature dependent and can be described by a simple Arrhenius equation:

$$\eta \propto \exp^{\frac{-E_A}{k_B T}} \tag{6.11}$$

E_A	Activation energy
k_B	Boltzmann constant
T	Absolute temperature

This relationship is valid at temperatures that are at least 100 °C above the glass transition temperature. At lower temperatures the velocity of the reptation process and thus the melt viscosity is not determined by the thermal energy available for movement but rather by the shrinking of the free volume at low temperatures.

6.6 Viscoelasticity

6.6.1 Influence of Time on the Mechanical Behavior

As has already been discussed in ▶ Sect. 6.1, amorphous polymers of large molar mass do not suddenly transition into a low viscosity melt at their glass transition temperature but pass through a range of states in which the material still exhibits a degree of elasticity, i.e., shape memory. The requirement for this is that the molar mass is not too small. A rotation of the individual chain segments around single bonds in the polymer chain is possible at temperatures above the glass transition temperature, as has already been set out in ▶ Sect. 6.1. Thus, if a stress acts on the entangled polymer coil the individual segments can assume an alternative conformation without any bonds being broken. If the length of the macromolecule is longer than the critical chain length introduced in ▶ Sect. 6.5 then effective entanglements exist that counteract the deformation of the sample. Only at temperatures that are considerably above the glass transition temperature do the polymer chains disentangle so fast that no elastic behavior is observed; at such high temperatures the polymer melt behaves almost as a 'normal' low viscosity fluid.

Deformation of a polymer sample at temperatures that are not much higher than the glass transition temperature leads to the polymer chains assuming an energetically less favourable conformation. As a result, as has already been mentioned above, elastic energy is stored in the material. What then happens is largely dependent on the duration of the stress.

Short Term Exertion of Force

If the externally applied force is only applied for a short period of time, then the polymer chain takes on its original energetically favorable conformation again. The material springs back completely and returns to its original shape; the deformation process is totally reversible.

Prolonged Exertion of Force

Alternatively, during a prolonged exertion of force, a reorientation (or at least a partial one) of the polymer chains relative to one another occurs. Because of the externally applied forces, the polymer chains are unable to take on their original entangled conformation. This leads to a movement of the polymer chains relative to one another—the material starts to flow.

The same considerations apply to the temperature at which the exertion of force takes place. The mobility of the polymer chains increases at a higher temperature so that the polymer flows after shorter times under stress than at lower temperatures.

When considering polymers under these conditions, we are dealing with materials that behave, depending on the duration of the force exerted upon them and the temperature, either as viscous fluids or as elastic materials. This characteristic is known as *viscoelasticity*. The range of viscoelasticity is marked by two cases at the extremes of the viscoelastic spectrum: a 'free flowing (and elastic) fluid' and an 'elastic (non-flowing) material.' According to Hooke, for an elastic material follows:

$$\sigma = E \cdot \varepsilon \tag{6.12}$$

σ Stress
E Elastic modulus
ε Elastic extension of the length l by the factor x

From Newton for viscous, free flowing liquids follows:

$$\sigma = \eta \frac{d\varepsilon}{dt} \tag{6.13}$$

η Viscosity

Thus, the resulting stress is proportional to the rate of extension but independent of the extension itself.

These laws can either be visualized in terms of a damper (Newton) or a spring (Hooke). In the case of a Hookean spring, the applied stress σ is proportional to the expansion ε and the elastic energy can be stored in the spring. A damper can be envisaged as the mechanical equivalent of a plate that carries out work in a fluid with viscosity η. In this case the applied stress is independent of the expansion; energy is dissipated (Fig. 6.8).

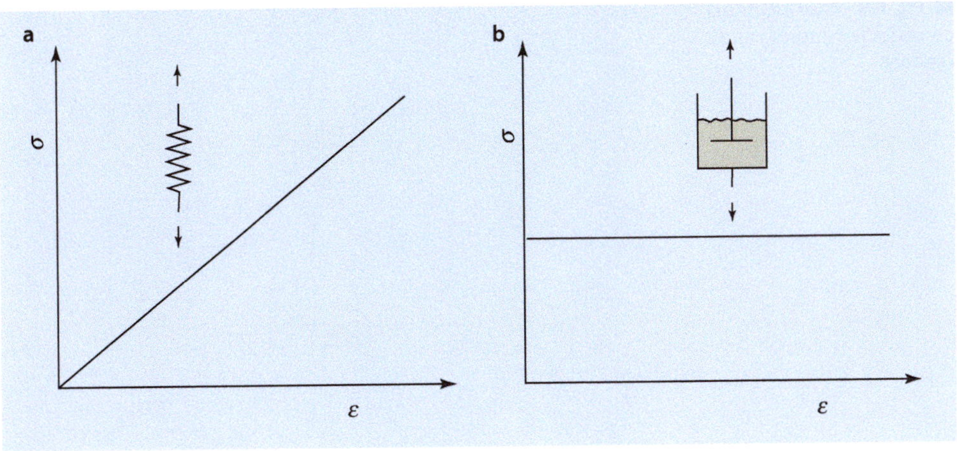

Fig. 6.8 $\sigma = f(\varepsilon)$: (**a**) for an elastic material (storage of energy); (**b**) for a viscous fluid (dissipation of energy)

Neither of these models describes viscoelastic polymers correctly. To describe such systems, combination models are necessary.

6.6.2 The Maxwell Approach

The Maxwell model is a serial arrangement of a spring and a damper (Fig. 6.9).

The total elongation is composed of an elastic (ε_{elast}) and a viscous (ε_{visc}) component in viscoelastic systems:

$$\varepsilon = \varepsilon_{elast} + \varepsilon_{visc} \tag{6.14}$$

Equations (6.12) and (6.13) (Hooke and Newton) can now be transformed as follows:

$$\text{Hooke}: \frac{d\varepsilon}{dt} = \frac{1}{E}\frac{d\sigma}{dt} \quad \text{Newton}: \frac{d\varepsilon}{dt} = \frac{\sigma}{\eta} \tag{6.15}$$

and with (6.14) it follows:

$$\frac{d\varepsilon}{dt} = \frac{1}{E} \cdot \frac{d\sigma}{dt} + \frac{\sigma}{\eta} \tag{6.16}$$

For a constant extension ε, i.e., for $\frac{d\varepsilon}{dt} = 0$ it results (6.17):

$$\sigma = \sigma_0 \exp\left(-\frac{E}{\eta}t\right) \tag{6.17}$$

σ_0	Stress at the time $t = 0$
E	Module of elasticity
η	Viscosity (Fig. 6.11)

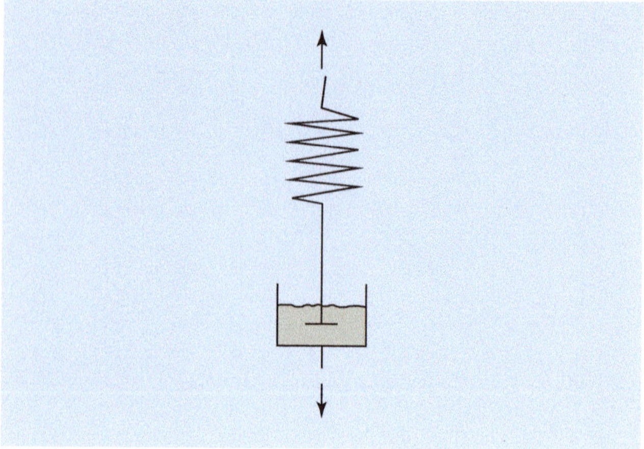

■ **Fig. 6.9** Maxwell model consisting of a spring and a damper

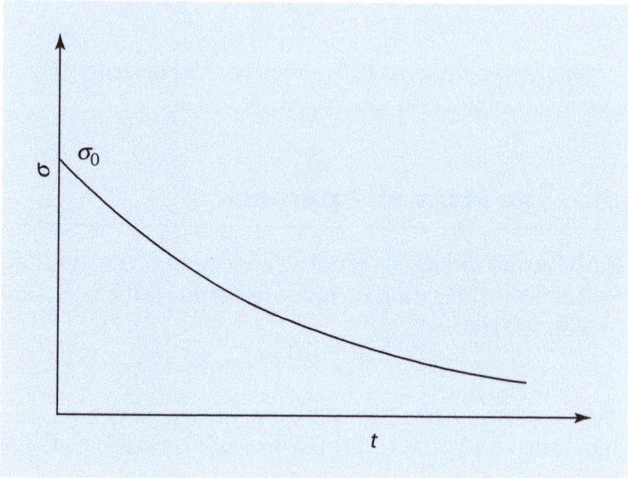

■ **Fig. 6.10** $\sigma = f(t)$ according to the Maxwell model

6

This means that the stress σ, the force necessary to stretch a particular object, decreases exponentially with time (■ Fig. 6.10). The model describes the *stress relaxation* of a viscoelastic system that has suddenly been stretched to a given length and is then fixed. This can easily be demonstrated using a serial arrangement of spring and damper: if the model is stretched quickly, in the first instance, the damper element is not stretched, only the spring, because of the inertia of the damper element. If, at this moment, the applied force is removed just as quickly as it was applied, the spring springs back completely. If the force is applied over a longer period of time—as is the case in the Maxwell model—then the damper is irreversibly stretched, whereas the spring contracts back into its original shape. As a result, over time, the applied tension decreases exponentially to zero.

This simple model describes the behavior of a viscoelastic polymer under constant elongation. If the elongation is maintained over a longer period of time, then the polymer chains have enough time (by moving relative to one another) to rearrange themselves into an energetically more favourable conformation.

The quotient of viscosity η and module of elasticity E is defined as the relaxation time τ in (6.18):

$$\tau = \frac{\eta}{E} \tag{6.18}$$

If $t = \tau$ then the applied stress has decreased to $1/e$ of the initial value.

The Maxwell model therefore describes the time dependency of stress at constant elongation and the relaxation of stress that follows an elongation. From Fig. 6.10 it can be derived that for extremely short periods of elongation the system exhibits an (extremely short) phase of elastic behavior followed by creep or flow. In contrast to what is actually observed for viscoelastic systems, the model does not predict elastic behavior for an extended, observable period. The model also does not predict that elongation under constant stress is time dependent because it assumes that the constant elongation is independent of time.

6.6.3 Voigt–Kelvin Model

Unlike the Maxwell model, the Voigt–Kelvin model defines elongation under constant stress as a function of time. This model can be described mechanically by the parallel arrangement of spring and damper (Fig. 6.11).

In analogy with (6.16), the stress is the sum of the elastic and the viscous components:

$$\sigma = E\varepsilon + \eta \frac{d\varepsilon}{dt} \tag{6.19}$$

When applying a constant tension σ_0 that has an immediate effect, for example by attaching a weight, integration of (6.19) gives

$$\varepsilon = \frac{\sigma_0}{E}\left(1 - \exp\left(-\frac{E}{\eta}t\right)\right) \tag{6.20}$$

The change in elongation as a function of time according to (6.20) is shown in Fig. 6.12.

The form of the curve in Fig. 6.12 can also easily be explained using the mechanical model from Fig. 6.11. If a parallel arrangement of a damper and a spring is stretched by a constant force, initially, the damper prevents an elongation of the system because of its inertia. After an elapse of time an elongation does occur but it is increasingly attenuated by the resilience of the spring and eventually plateaus. Thus, this simple mechanical model describes the delayed reaction of a viscoelastic polymer to an external force caused by the polymer chains taking a certain length of time to rearrange.

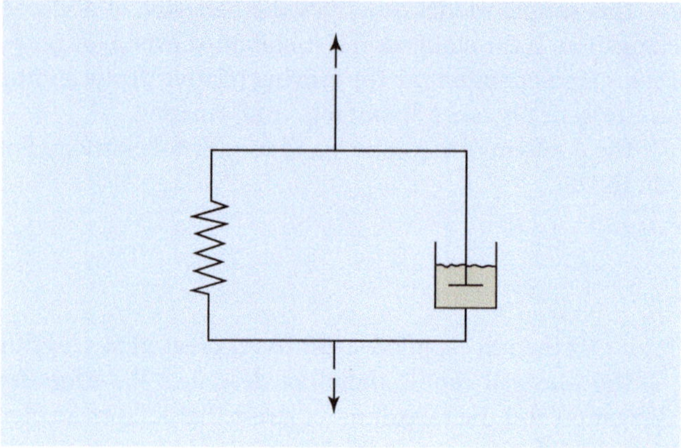

Fig. 6.11 The Voigt–Kelvin model for a spring and a damper

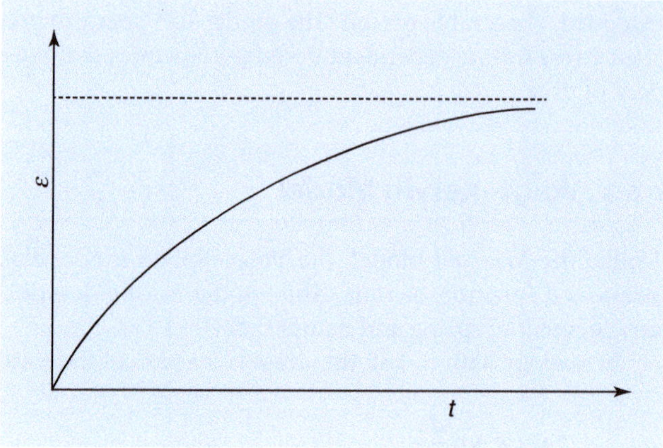

Fig. 6.12 Elongation ε as a function of time t according to the Voigt–Kelvin model

Here too, in analogy with the treatment above, the retardation time τ_R of the Kelvin–Voigt model can be defined by the quotient of η and E:

$$\tau_R = \eta / E \tag{6.21}$$

If $t = \tau_R$ then the elongation ε has increased to 63% $(1 - 1/e)$ of its equilibrium value.

As with the Maxwell model, the Voigt–Kelvin model does not offer a complete description of viscoelastic behavior. It allows for delayed elongation, the delayed response of a viscoelastic polymer to a constant applied force. It does not, however, describe the irreversible flow that occurs in a viscoelastic system.

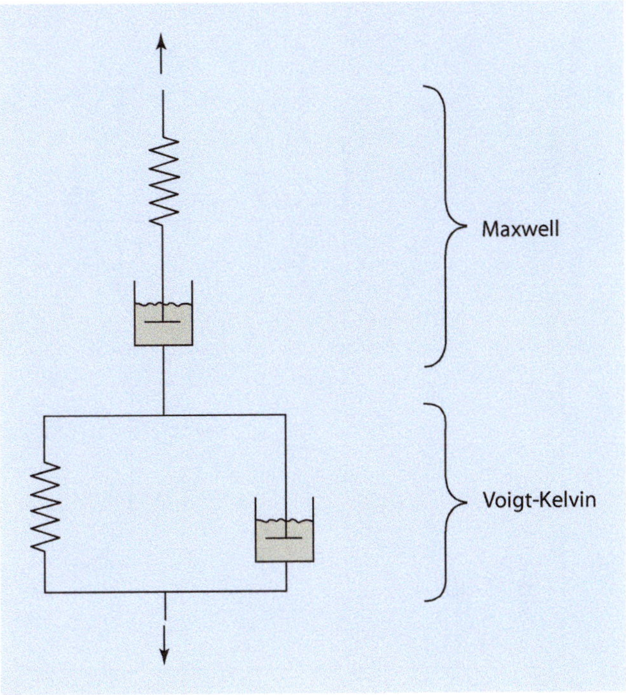

Fig. 6.13 Set-up of springs and dampers in the Burgers model

6.6.4 The Burgers Model

The Burgers model is a combination of the Maxwell model with the Voigt–Kelvin model. Thus, it can be described by the arrangement of two springs and two dampers shown in Fig. 6.13.

To understand the model better, the following describes the reaction of the system if it is suddenly stretched by a certain amount at a certain time t_0, this applied force then being maintained for a given time and then removed. The dependence of the elongation on time is shown in Fig. 6.14. This process can be divided into five phases:

a. From 0 to A no force is being applied to the sample and it is not stretched. This conforms to the arrangement shown in Fig. 6.14a.
b. From A to B the force is suddenly applied to the sample at t_0. In the mechanical model the upper Hookean spring reacts to this force instantaneously. If the applied force is removed just as quickly, then the system springs back completely to point A. The bottom spring cannot follow this quick movement as it is linked to the damper.
c. From B to C the two dampers in the mechanical model only move if the stress is maintained for a finite period of time. The second spring is then stretched. In this case the bottom damper describes the irreversible flow of the model, and the energy that is needed to stretch the upper damper is stored in the spring connected in parallel to it.

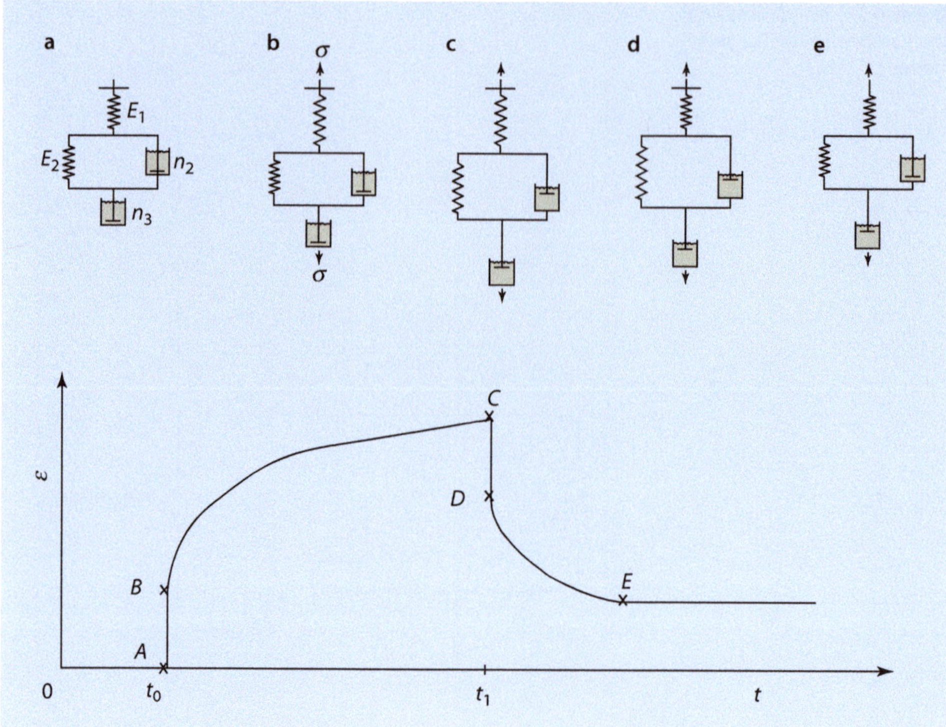

Fig. 6.14 Elongation ε as a function of time t according to Burgers' model

d. From C to D, if the applied force is removed then the upper spring instantaneously springs back to its original position but the lower spring cannot follow this movement.
e. From D to E the second spring also springs back to its original position but with a time delay determined by the relaxation of the damper coupled with it. The bottom damper remains elongated.

The Burgers model describes the behavior of a viscoelastic system relatively well. When a force is applied for a short period of time, for example by a short impact, the entangled polymer chains are unable to rearrange relative to one another. Elastic behavior is observed and a complete return to the original shape takes place. If the force is applied over a longer period of time the polymer chains can detach themselves from one another and can change their relative position, and the material starts to flow. If the applied force is removed a partial return to the original shape takes place but a permanent deformation remains.

To summarize the above, viscoelasticity is characterized by the following phenomena:

1. A time and frequency of the material properties—either elastic or viscous behavior is observed depending on the duration of the applied force
2. Hysteresis, i.e., irreversible processes resulting in the system not returning to its initial state

Important hysteresis effects are:
- Creep or viscous flow
- Stress relaxation
- Time delayed reaction of the material (phase shift) if an oscillating force is applied

The latter is the basis for the dynamic-mechanical-thermal analysis (DMTA) discussed in ▶ Chap. 4.

References

Adam G, Gibbs JH (1965) On the temperature dependence of cooperative relaxation properties in glass-forming liquids. J Chem Phys 43:139–146

DiMarzio EA, Gibbs JH, Fleming PD III, Sanchez IC (1976) Effects of pressure on the equilibrium properties of glass-forming polymers. Macromolecules 9:763–771

Eisele U (1990) Introduction to polymer physics. Springer, Heidelberg

Williams ML, Landel RF, Ferry ID (1955) The temperature dependence of relaxation mechanisms in amorphous polymers and other glass-forming liquids. J Am Chem Soc 77:3701–3703

Polymers as Materials

S. Koltzenburg et al., *Polymer Chemistry*, DOI 10.1007/978-3-662-49279-6_7

Polymers as materials appear in myriad forms. Everyone is familiar with floor coverings made of polyvinyl chloride (PVC) and with Plexiglas windows (polymethyl methacrylate), and the latter's particularly successful version: the roof-top of the Munich Olympic Stadium. Many are equally familiar with the strengthening of polymers by compounding them with glass fiber. Polymers are also increasingly being used in medical applications, for instance as bone and organ prostheses. One can easily imagine that these must meet completely different requirements than, for example, an ordinary PVC tube in a chemical laboratory. These few examples amply demonstrate how diversified and partially contradictory the requirements for a material in its specific application are, and that an ideal material for all applications cannot exist.

The respective arguments for the application of plastics as an alternative to metals, glass, wood, or ceramic are multifarious. Often, weight advantages, thermal insolation, sound insulation, or electrical insulation are foremost. Depending on the size of the production series, polymer articles can have comparatively low manufacturing costs. Thermoplastics also offer a relatively large degree of design freedom. Complicated parts can often be produced from a single piece, for instance, by injection molding (▶ see Chap. 17). Synthetic materials are typically also relatively corrosion-resistant. Last but not least, the comparatively low temperatures at which synthetic materials are usually processed leads to energy and, consequently, cost reduction during processing.

As a result, many polymer researchers spend their time on the development of macromolecules that can be customized to particular requirements. This involves trying to incorporate as many desired attributes as possible into one polymer. Ideally, this would be, for example, a polymer that is as hard as steel, as clear as glass, as light as a feather, as heatproof as quartz, and as cheap as possible.

When searching for a suitable material, some basic requirements need to be considered:

- Mechanical and thermal resilience
- Resistance to chemicals, ultraviolet (UV) radiation, and electrical fields
- Appearance (inherent color/colorability, transparency, and printability)
- Cost of raw materials, production, rework, and assembly
- Weight
- Dimensional stability, and thus form constancy under different conditions
- Resistance to fire

Clearly, deciding on the right material is a complex process. Therefore, only the basic methods of customizing polymers are described in this chapter.

7.1 Fracture Behavior

As the material properties of polymers can be varied over a broad range, more than just one mechanism can lead to material failure. ◘ Figure 7.1 shows two characteristic cases.

The left curve (◘ Fig. 7.1a) shows the dependence of the stress on the elongation for a brittle polymeric material. For stiff materials (with a high Young's modulus) a relatively minor elongation ε already requires a high stress σ. The curve is therefore rather steep. Even a relatively minor elongation leads to fracture. This fracture behavior is referred to as a brittle fracture. It is typical for brittle, rigid materials and is characteristic of polymers

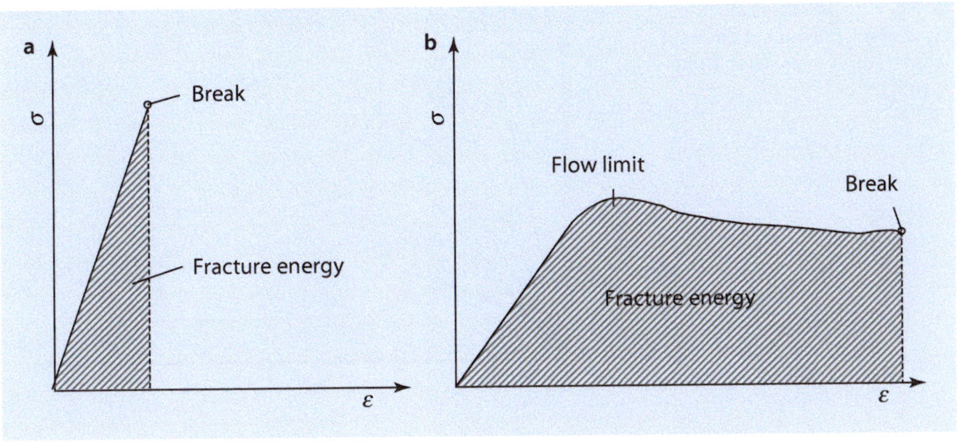

Fig. 7.1 Tension–elongation curves ($\sigma = f(\varepsilon)$) of (**a**) brittle and (**b**) viscoelastic polymers

below their glass transition temperatures. The Young's modulus is given by the initial slope of the curve shown and the energy needed for fracture by the area under the curve until fracture. An everyday example of such behavior is glass.

The other curve (Fig. 7.1b) shows the result of the same test for a tough, viscoelastic material. These materials are generally less stiff than crystalline polymers, or polymers at temperatures beneath their glass transition temperature. Thus, the curve becomes flatter. In such cases the material typically starts to flow under the external stress at a particular degree of elongation ε. This elongation is referred to as the yield point. The shape of the curve after exceeding the yield point is relatively poorly defined. Here, too, the material eventually fractures whereby the mechanism is referred to as ductile fracture. The energy leading to fracture is again given by the area under the curve. An everyday example of such behavior is chewing gum.

The fracture behavior of a polymer is strongly dependent on temperature. That way, for instance, a material that exhibits brittle fracture at low temperature can fracture in a ductile manner at higher temperatures. This is depicted in Fig. 7.2.

In Fig. 7.2, the transition from brittle fracture to ductile fracture is clearly recognizable. At low temperatures, only limited deformation is possible. The material is characterized by a high Young's modulus, i.e., the material is very stiff and a great deal of force is needed to stretch it. On the other hand, the material is able to sustain considerable stress before it breaks. The material becomes softer with increasing temperature (from T_1 to T_6), the Young's modulus sinks, and the curves become flatter. The *strength* of the material, the stress at which the material fractures, tends to decrease.

At even higher temperatures, the material reaches its yield point. Beyond the yield point, the form of the curve is no longer well defined. The material becomes progressively softer, which is reflected in a sinking Young's modulus and relatively flatter curves.

It is important to note that the maximum *toughness* of the material, i.e., the maximum energy required for fracture, is reached at an average temperature T_4. The area below the curve is smaller at lower temperatures because of a different fracture mechanism (brittle fracture). At temperatures above T_4, the curves are, taken as a whole, flatter, so that the integral of the curves are also smaller.

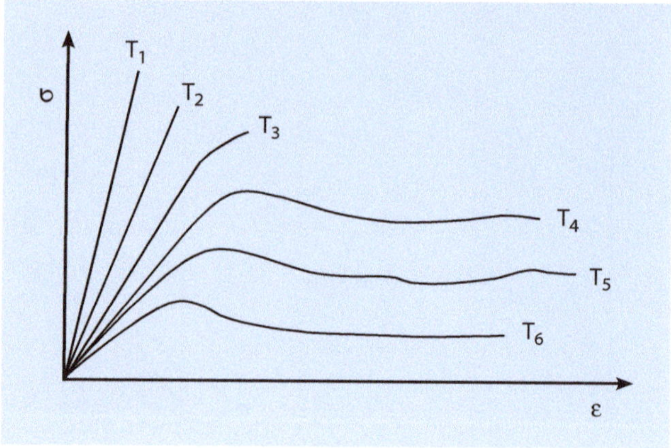

Fig. 7.2 Stress–strain curves for a polymer at different temperatures (increasing from T_1 to T_6)

7

7.2 Tailor-Made Plastics

7.2.1 Mechanical Characteristics

All polymers have approximately the same molecular main chain stability, no matter what their particular structure, because they consist of covalent C–C-, C–O-, C–N-, or C–S-bonds with bond strengths of between 330 and 420 kJ/bond. Important, distinctive differences are produced by intermolecular interactions or cross-links (Fig. 7.3).

Intermolecular interactions between hydrocarbon chains such as in polyethylene and polypropylene are primarily caused by van der Waals forces. Polar groups, such as the nitrile groups in polyacrylonitrile, increase the stiffness and strength of the polymer through dipole–dipole interactions. Hydrogen bonds are the reason for the exceptional strength of polyamides. An interesting variant is the ionic interactions between carboxylate functions of a polymer and bivalent metal ions, such as Zn^{2+}. A considerable improvement in the mechanical characteristics can be achieved through cross-linking.

The mechanical characteristics of thermoplastics can be improved by increasing their molar mass and by avoiding or removing low molar mass moieties or by crystallization. When nonpolar sections along the chain are replaced by polar ones, for instance if a C–C- is replaced by a CO–NH-function (an amide function), the stiffness, strength, and viscosity of the polymer increase. Replacing nonpolar aliphatic groups by polar side groups, for example CI, CN, or OH, works in a similar way. These polar groups do not necessarily need to be part of every repeating unit, but can be introduced into the polymer chain by copolymerizing a nonpolar monomer with a polar monomer in a controllable manner (▶ Chap. 13), which can be seen as fine tuning of the polymer characteristics (Nuyken 1987).

As already explained in ▶ Chap. 2, because of entropy effects, polymers have limited solubility in low molecular solvents. Such effects are even more pronounced when two polymers are mixed so that chemically different polymers are not, as a rule, miscible. For this reason, when dealing with solid block copolymers, the two different polymer blocks usually form separate phases. However, because both blocks are chemically bonded, a macroscopic separation is impossible but the material is heterogeneous at a nanoscale. This can lead to different morphologies depending on the volume ratio of the polymer blocks whereby the minor component forms spherical or rod-like aggregates dispersed in

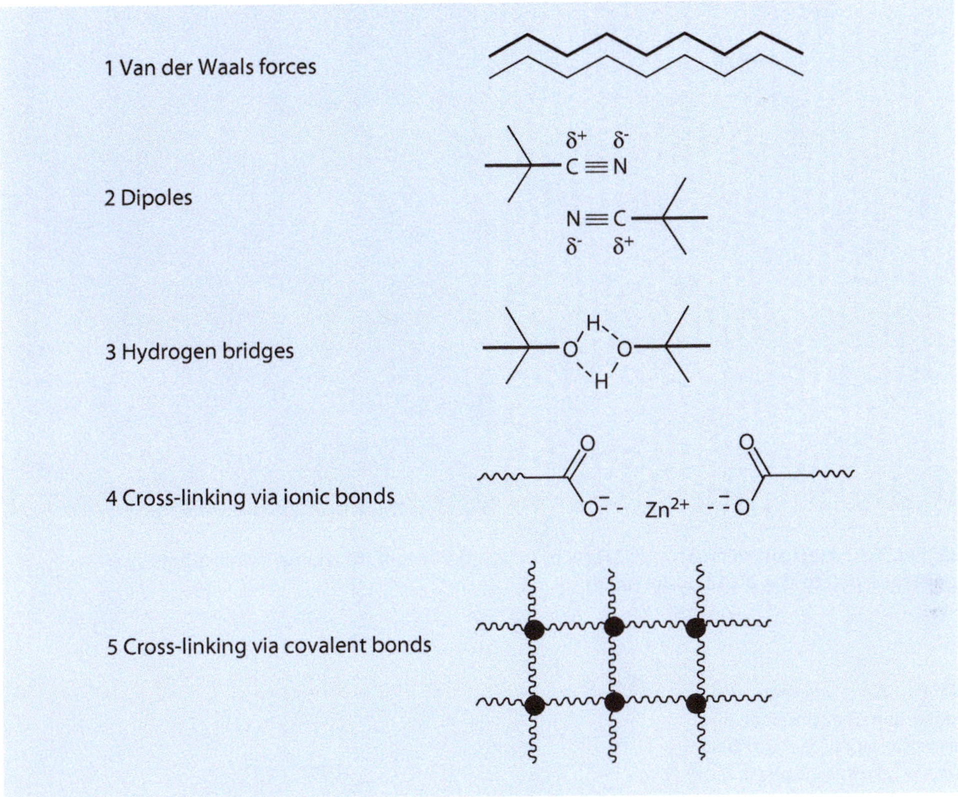

Fig. 7.3 Typical examples of intermolecular interactions or cross-links between polymer chains

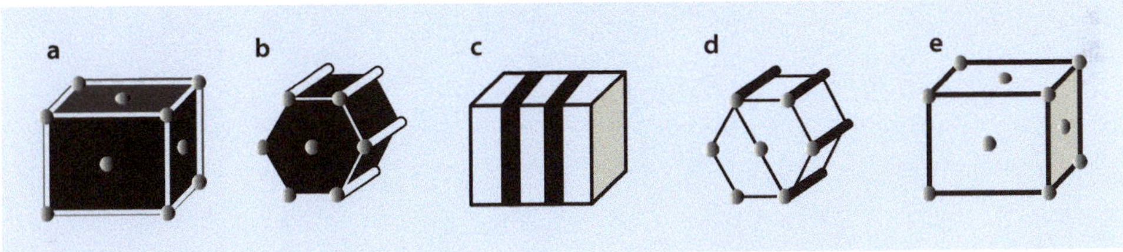

Fig. 7.4 Schematic depiction of the morphologies of block polymers having different compositions (see text)

a matrix of the major component. If both blocks are about equally distributed in the system, lamellae can form. These morphologies are schematically shown in Fig. 7.4.

Spherical aggregates arise for strongly segregated block copolymers when the weight fraction of the one component is less than ca. 15 % (Fig. 7.4a). When the portion is between about 15 % and 35 %, one-dimensional (rod-like) structures are formed (Fig. 7.4b). Between 35 % and 65 %, lamellae can be found (Fig. 7.4c). For even higher portions, the corresponding inverse structures are formed (Fig. 7.4d and Fig. 7.4e). Of course, these percentages should only be taken as a rough guide.

If the driving force for the phase separation of the blocks is relatively low, interpenetrating co-continuous networks can form at certain compositions because of surface area

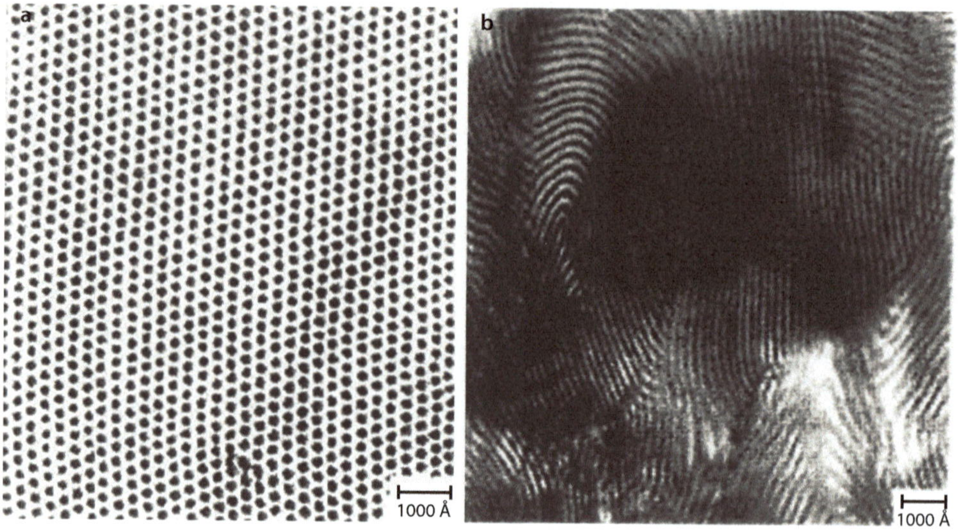

Fig. 7.5 Electron-microscopies (TEM) of (**a**) a rod-like and (**b**) a lamellar morphology (seen perpendicular to the plane of the page)

Fig. 7.6 Schematic depiction of a spherical morphology with a stiff polymer as the continuous phase

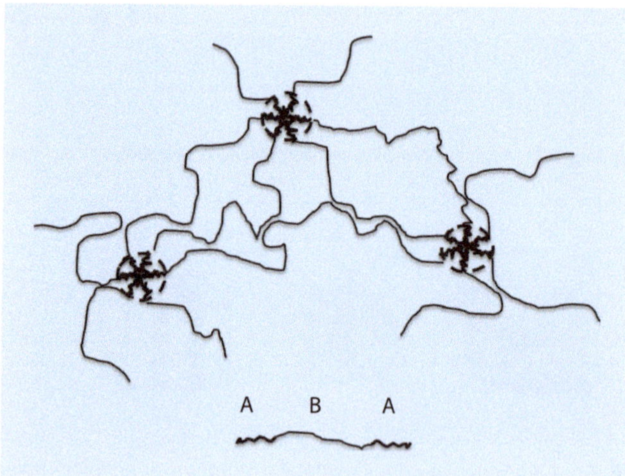

considerations. These are also referred to as *gyroide* phases and lead to a minimum in the surface area between the two phases.

The morphology of block copolymers can be very well visualized by transmission electron microscopy (TEM) (Fig. 7.5).

The mechanical characteristics of block copolymers are very interesting and make them useful for a wide variety of applications. One example would be the spherical morphology as shown in Fig. 7.6.

If we assume that the continuous phase consists of a stiff polymer but that the dispersed spheres are composed of a soft, elastic material, then the combination of these two

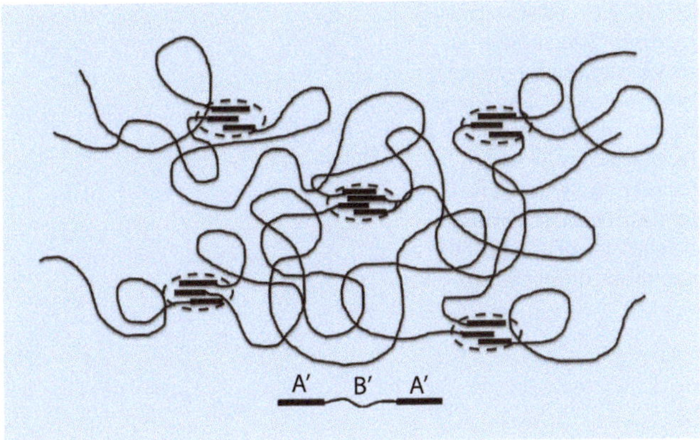

Fig. 7.7 Phase separation in an A′B′A′-block copolymer via physical cross-linking. Example: A′ polystyrene segment, B′ polybutadiene segment

materials exhibits very interesting characteristics: The continuous, so-called *hard phase* makes the material seem stiff macroscopically. Stiff materials are characterized by a high Young's modulus, but often suffer from their brittleness. If a so-called *soft phase* made from a soft, elastic polymer is embedded in such a stiff material, this phase can dissipate energy, for example, impact energy. Thus, the material becomes substantially less brittle. This principle is used industrially and is referred to as *impact modification*. It enables the production of materials that are not only stiff, but tough.

This principle is applied technologically, for instance to styrene copolymers. Non-transparent (opaque), high impact polystyrene (HIPS) consists of a mixture of neat polystyrene with a graft copolymer having a polybutadiene backbone and polystyrene side chains. The incompatibility of these two components leads to phase separation whereby the polystyrene forms the continuous (coherent) phase. Under stress cracks are generated which propagate through the polystyrene phase until they meet a softer, polybutadiene phase where the propagation halts. As a result, the mechanical resilience of the material is considerably greater than that of pure polystyrene, which is relatively brittle at room temperature because of its relatively high glass transition temperature of 100 °C.

One class of *thermoplastic elastomers* (*TPE*) is A′B′A′-block copolymers where A′, for example, is a polystyrene segment and B′ a polybutadiene segment. The incompatibility of these polymers also leads to phase separation. In this case, the continuous phase is generally the elastic polybutadiene (similar materials are also made using polyisoprene as the elastomeric phase). Because of the hard polystyrene domains that are glassy at room temperature, the polybutadiene behaves as if it were cross-linked (Fig. 7.7). However, the physical cross-links are only stable up to the glass transition temperature of the hard phase, in these examples the polystyrene. The polystyrene transitions into a plastic state when heated above its glass transition, and the material becomes malleable.

The thermal characteristics of block copolymers are different from those of homopolymers or statistical copolymers. Because of the phase separation in the solid state, the material has two melting and two glass transition temperatures at which the respective, separate polymer blocks undergo the corresponding phase transitions. These can be observed, for example, by DSC.

However, practically, only two of these four transitions are important. The material starts to flow and loses its form only if the temperature exceeds the highest of the melting temperatures. In other words, it can be thermally strained up to the higher melting

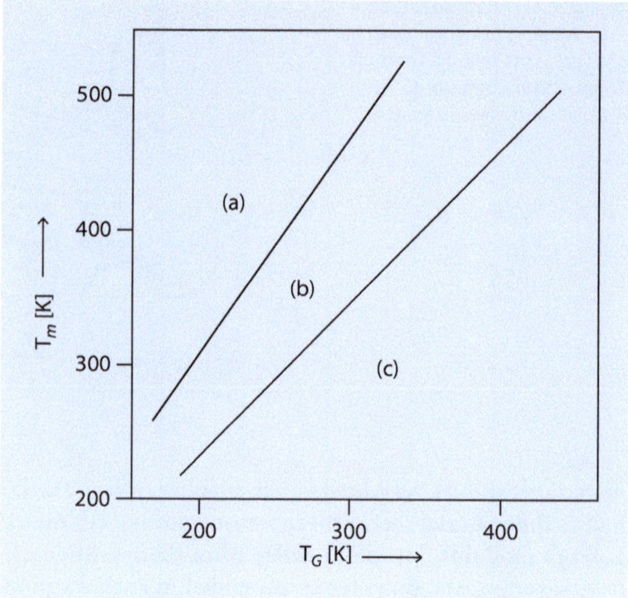

Fig. 7.8 Schematic representation of the relationship between melting temperature T_m and the glass transition temperature T_G for (**a**) block copolymers, (**b**) homopolymers, and (**c**) statistical copolymers having identical or similar chemical compositions

temperature without losing its form. This temperature is relevant for processing, and defines the highest temperature at which the polymer can be used.

On the other hand, if the desired property is impact modification, it is the low glass transition temperature of the soft phase that is critical and this defines the temperature at which the article becomes brittle.

By combining a polymer with a very high glass transition temperature or melting point with a polymer having a low glass transition temperature, a material can be obtained with a broad range of useful impact resistant properties.

In summary, by using the concepts of statistical copolymers (▶ Sect. 5.1.8) and those of block copolymers, the practically relevant melting and/or glass transition temperatures can be adjusted by using the rules of thumb mentioned in ▶ Sect. 5.1.8 over a wide range. This is shown schematically in Fig. 7.8.

Another interesting way to influence the mechanical properties of a polymer is to mix it with a solid filler such as carbon black, graphite, silica gel, glass fiber, or aluminum oxide (▶ Sect. 17.2). If unsaturated polyesters, epoxy resins, phenol formaldehyde resins, and other materials are reinforced with minerals or chopped glass fibers, they can be processed, for example, by injection molding as long as the polymers are not cross-linked.

As an example, the effect of short and long glass fibers on the tensile strength of polyamide and polypropylene is shown in Table 7.1.

The mechanical characteristics of polymer/GF-compounds can be further improved by replacing the fiber by glass mats and three-dimensional structures. However, the properties also depend on the fiber material (diameter, kind of glass), the kind of polymer, and the interaction between the polymer and the fiber. As well as the examples mentioned above, among others, polystyrene, polyoxymethylene, polyester, polyether ether ketone (PEEK), polycarbonate, and epoxy and phenolic resins are also used in such composites. Such composites are perfectly able to compete with metals in terms of their tensile strengths.

Table 7.1 Comparison of the effect of short and long glass fibers (GF) on the tensile strength of polyamide 6 (PA) and polypropylene (PP). (Data kindly provided by G. Heinrich, IPF/Dresden)

Material	Source	Tensile strength (MPa)
PP + GF, short fibers (30 wt% GF)	IPF/Dresden	90
PP + GF, UD[a] (50 wt% GF), 17 mm	IPF/Dresden	884
PP + GF, UD[a] (50 wt% GF), 12 mm	IPF/Dresden	964
PP + GF, UD[a] (60 wt% GF)	Twintex® R PP 60	760
PA + GF, short fibers (30 wt% GF)	IPF/Dresden	183
PP + GF, UD[a] (50 wt% GF)	IPF/Dresden	983

[a]*UD* unidirectional

Alternatives to glass fibers are carbon fibers and SiC fibers. Spectacular applications of such composites are the tail of the Airbus A380, the gripping arm of the Space Shuttle, and various items of sports equipment.

Aramid yarn can be seen as a "self-reinforcing system". The stiff chains of the aromatic polyamide interact via intermolecular hydrogen bonds to give a composite-like system with a very high tensile strength. Here, the stiff aramid yarn corresponds to a high-modulus fiber (for instance, glass or carbon fiber) embedded in soft resin.

7.2.2 Optical Characteristics

Amorphous polymers usually appear optically transparent, whereas crystalline polymers are generally opaque. This is because the light that falls through a crystalline polymer is scattered by the crystal lamellae; it is unable to pass straight through the sample. Only a few crystalline polymers, such as polycarbonate and poly(4-methyl-1-pentene), are transparent because the crystallites are smaller than the wavelength of visible light. In addition, the refractive indices and densities of the crystallites and the amorphous phases are almost identical so that the light is not scattered and they appear transparent.

However, there are exceptions to this basic rule: amorphous polymers naturally appear opaque if they contain chromophores that absorb the incoming light. Crystalline polymers, on the other hand, can be optically transparent if the crystallites they contain are so small that they do not scatter visible light. This can be achieved by strongly increasing the formation of crystal nuclei by adding additives. Hereby, the number of crystallites increases to such an extent that their size becomes limited. If the crystallites are made substantially smaller than the wavelength of the incoming light, no scattering occurs.

Amorphous thermoplastics, such as polystyrene, polymethyl methacrylate, polyvinyl chloride, or cellulose ester, are thus transparent, and used for glasses, frostings, photographic materials, and packaging. However, crystalline polymers, such as polyethylene, polyamides, and polyacetals, are not suitable for glass production. To make such polymers transparent the crystallization has to be prevented. This can be accomplished by, for

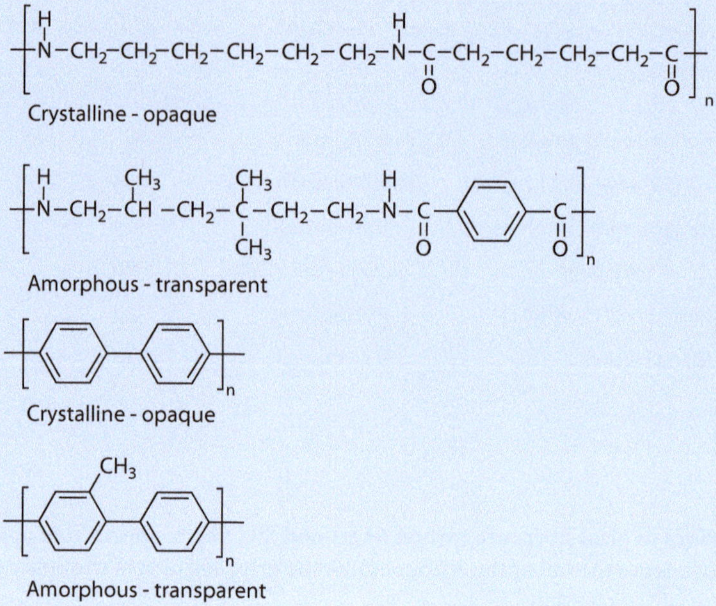

Fig. 7.9 Examples of crystalline and amorphous polymers

example, introducing side chains; CH_3-functions are often sufficient. In Fig. 7.9 crystalline and amorphous polymer structures are compared to each other.

7.2.3 Materials for Lightweight Construction

Synthetic materials are usually specifically much lighter than metals. If the mechanical strength is less critical than a lower specific weight for an article, polymer foams are a suitable solution, for example, for sound and heat insulation, foams have special advantages (▶ Chap. 17). Mechanically more demanding foams, such as furniture cushions, are also available. Polymer foams can be obtained from many polymers, for instance, from polyvinyl chloride, polystyrene, urea-formaldehyde and phenol-formaldehyde resins, poly-urea, as well as from polyurethanes. Foams can also be obtained from polyvinyl acetal and polyethylene.

Polymeric foams can be prepared by mixing at high shear and introducing gas by adding low molecular mass additives which are then allowed to evaporate or chemically. For more details, refer to ▶ Sect. 17.3.

Foaming a natural rubber is successful if a foaming agent, such as soap, is added to an initial volume of a mixture of latex rubber, extender, and vulcanization accelerator. The mixture is then vigorously stirred, expanded, and afterwards solidified using foam stabilizers. In this way, the density of the material can be reduced by up to 90%.

Viscose sponges can emerge when Na-cellulose xanthate is mixed with coarse grained salts, such as Glauber's salt, and coagulated in electrolyte solution. The material

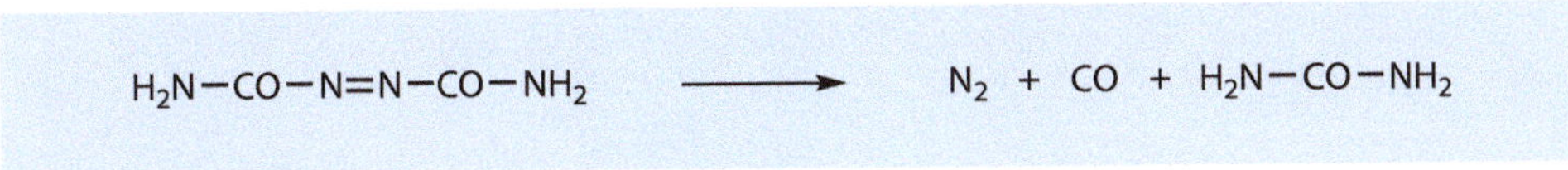

Fig. 7.10 Example of a chemical blowing agent

Fig. 7.11 Reaction of a carboxylic acid with an isocyanate

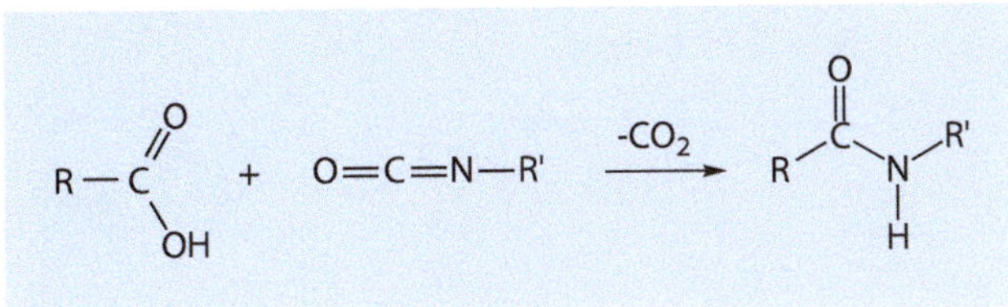

attains it porous structure because of the multiple salt inclusions which are later washed out.

In many cases, chemical agents such as NH_4NO_2, $NH_4Cl + NaNO_2$, azo compounds, sulfohydrazides, or azides that produce nitrogen when heated are employed as blowing agents (Fig. 7.10).

Chemical and physical agents are also used in the production of foams from isocyanates. For instance, the reaction of a carboxylic acid with an isocyanate to an amide, with the elimination of CO_2, is depicted in Fig. 7.11. To enhance foam formation, physical agents, such as volatile solvents, are frequently added to these systems.

The reaction of water with isocyanates to form a polyurethane foam also needs to be mentioned here. More details of this reaction are given in ▶ Sect. 17.3.2.

7.2.4 High-Temperature Materials

One of the biggest disadvantages of plastics compared to other materials such as steel and ceramic is their usually limited shape retention at higher temperatures. Crystalline polymers keep their shape at temperatures approaching their melting point whereas amorphous polymers tend to soften at much lower temperatures and over a wide range of temperatures. Thus, syndiotactic and isotactic polymers which can crystallize can be used at higher temperatures than their atactic or amorphous analogs.

Another way to improve the shape retention of polymers when they are heated is to introduce intermolecular interactions or links (Fig. 7.3) between functional components of the main chains and the side chains. Such functions have a distinct influence on the melting temperature and can be seen as a criterion for estimating heat stability as a first approximation (Table 7.2).

If every second CH_2-group in polyethylene is replaced by O, the polymer becomes more polar, and the melting temperature rises from 137 °C (polyethylene) to 175 °C (polyoxymethylene). CONH-moieties in the main chain of polyamides lead to a further increase of the melting temperature to 255–267 °C. Some side groups, e.g., CH_3, Cl, C_6H_5, or CN, also increase the polymer's melting temperature. In contrast, the melting temperature is decreased by longer alkyl side chains (▶ Chap. 5) (Brandrup et al. 1999). An important means of increasing the temperature at which a polymer can be used is to include aromatic structures in the main chain. Some typical examples are shown in Fig. 7.12.

Table 7.2 Influence of the polymeric structure on the melting temperature (comments ▶ Chap. 5)

Structure		Name	Melting temperature (°C)
$\left[CH_2-CH_2 \right]_n$		Polyethylene	137
$\left[CH_2-O \right]_n$		Polyoxymethylene	174–180
$\left[R-CO-NH-R'-CO-NH \right]_n$		Polyamide	255–267
$\left[CH(A)-CH_2 \right]_n$	$A=CH_3$	Polypropylene	170–180
	$A=Cl$	Polyvinylchloride	180–240
	$A=C_6H_5$	Polystyrene	240–270
	$A=CN$	Polyacrylonitrile	317
$\left[CH(R)-CH_2 \right]_n$	$R=C_2H_5$	Poly(butene-1)	126

Polyimides are synthesized in a two-step process and are particularly well researched (Fig. 7.13).

These polymers are resilient until ca. 500 °C for short periods of time and can be used long-term at temperatures of up to 270 °C.

Another concept for developing heat-stable polymers, which has proved successful, is the systematic replacement of carbon atoms by heteroatoms, such as Si, Sn, Al, Ti, etc. (Fig. 7.14).

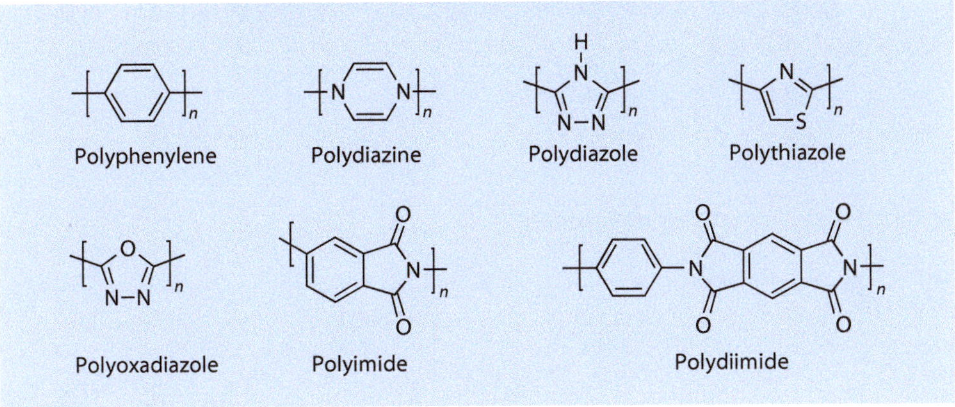

Fig. 7.12 Polymers with aromatic rings in the main chain

Fig. 7.13 Synthesis of polyimides

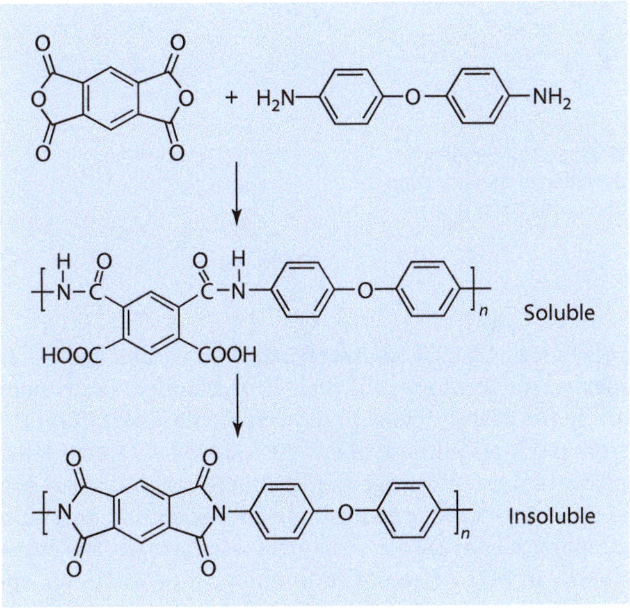

The transition from single-chain polymers to *ladder polymers* (Fig. 7.15) is particularly interesting. Evidently, polymers with a single main chain only sustain a decrease in their molar mass if one bond of the main chain breaks, and thus their mechanical and thermal characteristics deteriorate. In contrast, breaking a single bond in a ladder polymer does not change the molar mass. Indeed, there is a good chance that the breakage heals because the radicals initially formed by the bond breaking are held in close proximity by the rest of the molecule and they can recombine.

Ladder polymers such as those depicted in Fig. 7.15 can have thermal distortion stabilities close to that of quartz. Carbon fibers (▶ Chap. 4) also exhibit excellent heat

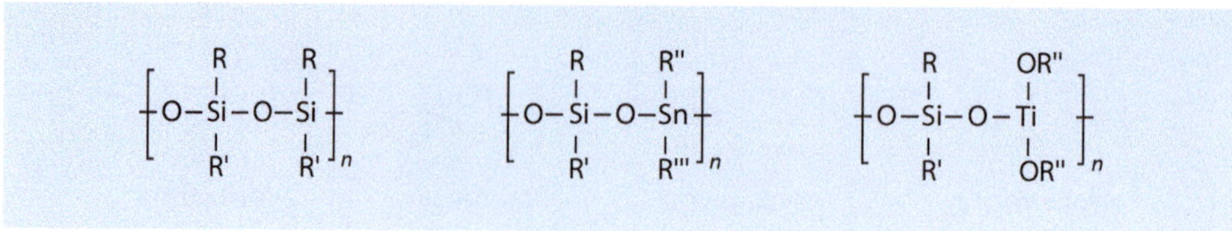

Fig. 7.14 Thermally stable heteropolymers

Fig. 7.15 Siloxane-based ladder polymer

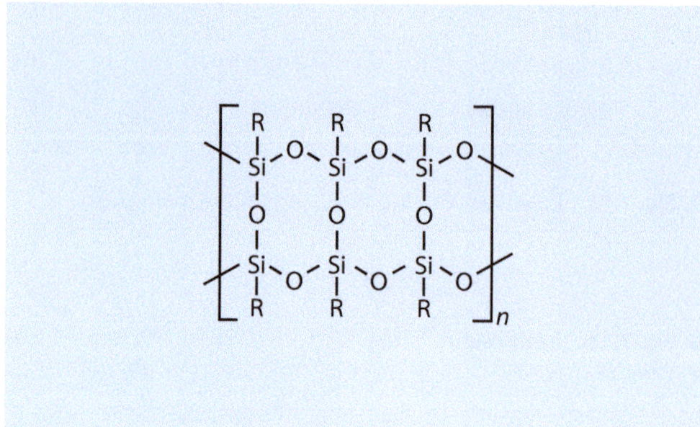

Fig. 7.16 Simplified depiction of the formation of carbon fibers

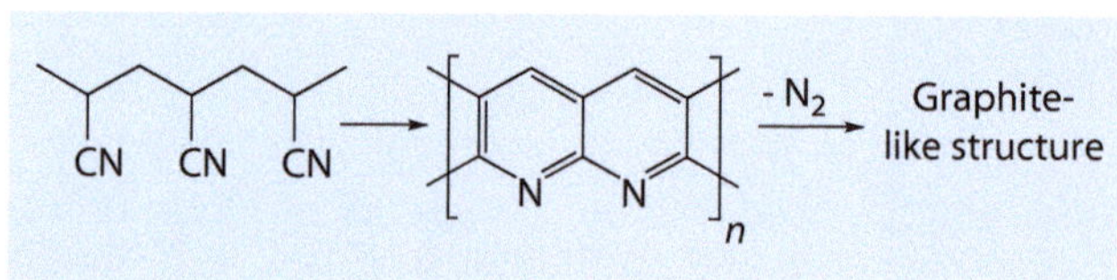

resistance. Crucial characteristics of carbon fibers are their extraordinarily high mechanical strength and their heat stability. Their manufacture is complex, and they are made mainly from polyacrylonitrile fibers (Fig. 7.16). They can also be made from pitch or cellulose fibers. At 150–350 °C an oxidation and partial cyclization takes place. In a second stage the fibers are carbonized at temperatures up to 2000 °C and ever more highly organized structures similar to graphite, which can be considered simplistically as ladder structures, are formed. The strength and low density of carbon fibers can best be exploited as composites with, e.g., epoxy resins or polyesters. They are extensively applied in aircraft construction but also, e.g., in high performance sports equipment.

7.3 Cross-Linked Materials

Connecting several polymer chains together via covalent bonds leads to 3D network structures. These consist theoretically of a single molecule of which the molar mass is essentially infinite. Cross-linking is one of the most important means of modifying the material characteristics of a polymer. Cross-linking can take place either during the synthesis of a polymer by copolymerizing a multi-functional monomer (▶ Chap. 8) or by subsequent reaction of a pre-prepared polymer (▶ Chap. 15).

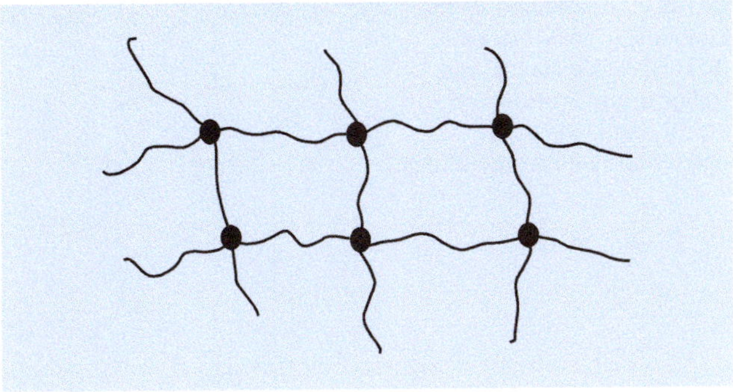

Fig. 7.17 Schematic depiction of a polymeric network

7.3.1 Structure and Application of Networks

A network is characterized by network points of specific functionality f (f=3, 4, ...) connected to other network points by f chains (Fig. 7.17).

Polymeric networks can be soft, rubber-like materials (elastomers, ▶ Chap. 18) but also very hard materials. They can also form swollen gels with solvents. The thermo-reversibly cross-linked thermoplastic elastomers, discussed above, also belong to the class of polymeric networks.

Examples of applications for covalent networks include:

- Vulcanized rubbers (elastomers, for example car tires)
- Composite materials
- Organic coatings, such as car paints
- Separation media (ion exchange resins)
- Prostheses, contact lenses
- Carriers for the controlled release of active ingredients
- Electronic systems for printed circuits
- Biological gels, for example the lens of human eyes

This short list of examples emphasizes the numerous material requirements and particularly that these requirements can differ greatly. To fulfill these requirements the materials need to be specifically adapted.

Generally, one can say that networks can be differentiated from their linear analogues by their insolubility and infusibility.

7.3.2 Mechanical Characteristics of Networks

Cross-linked materials are similar to semi-crystalline systems in some of their properties; neither semi-crystalline nor cross-linked polymers are viscoelastic. A macroscopic flow of the material is either hindered by the crystal lamellae or by the cross-links. The lamellae or cross-links act as anchor points and fix the macroscopic form of the article even for temperatures above T_G. However, the material appears softer and is tougher at temperatures between T_G and T_m because the amorphous phase dissipates impact energy (▶ Sect. 7.2.1).

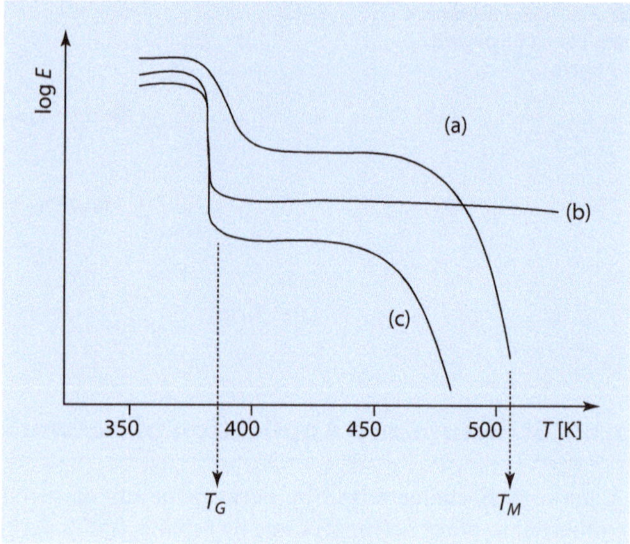

Fig. 7.18 Changes in the E-module of an (**a**) isotactic, (**b**) cross-linked atactic, and (**c**) linear, non-cross-linked atactic (high molecular) polystyrene with temperature

With an increasing number of network points in an amorphous material:

- The relaxation time and module increase
- The yield point rises until it disappears when cross-linking is complete
- The viscous flow and stress relaxation decrease; both these processes also cease when cross-linking is complete

Both crystalline and amorphous polymers can be cross-linked whereby, to a first approximation, cross-linking does not change the morphology of the material.

Figure 7.18 shows the influence of the temperature on the mechanical characteristics of the material using polystyrene as an example. In this diagram isotactic, atactic, and cross-linked, atactic polystyrene are compared with each other.

The curve of the E-module with temperature for atactic polystyrene has already been discussed above. If the material is cross-linked, the material characteristics—for instance, the module—below the glass transition temperature are relatively independent of whether the material is cross-linked or not. The module drops for both materials at the glass transition but for the cross-linked material the module stabilizes at a higher level (to a first approximation) until the material thermally decomposes. This happens because the material cannot flow. The semi-crystalline, isotactic polystyrene also passes through a glass transition but, because only the amorphous part of the material undergoes a glass transition, the module decrease is not as large as that of the amorphous material. Between T_G and T_m the module is determined by the crystalline portion of the material and changes very little with temperature; this also leads to a plateau in the log E vs. T curve. The module only drops again significantly when the melting temperature of the crystalline portion is reached.

The industrially very important, so-called elastomers are more or less soft, cross-linked polymeric materials. Because of their considerable importance they are discussed separately in ► Chap. 18.

7.3.3 Network Synthesis

To form stable polymeric networks, any reaction in which covalent bonds are formed can be used. The number of organic reactions that could be used is nearly infinite. In practice the following inter alia are employed:

- Reactions of epoxides
- Reactions of isocyanates
- Hydrolysis and condensation of alkoxy silanes and hydrosilylation
- Reactions of phenol and formaldehyde
- Reactions of urea, thiourea, and melamine with formaldehyde
- Addition of SH- or NH_2-terminated molecules to C=C-bonds
- Vulcanization of rubber by sulfur, peroxides, or phenol-formaldehyde resins
- Radical copolymerization of poly-unsaturated monomers
- Radical polymerization of α,ω-telechelics

Because the direct synthesis of polymeric networks is independent of the mechanism of the polymerization, some selected reactions are discussed here. The subsequent cross-linking of pre-formed polymers is discussed in Chap. 15.

7.3.4 Typical Cross-Linking Reactions

Multi-functional Monomers

The stoichiometric reaction of bifunctional alcohols with trifunctional isocyanates results in cross-linked polyurethanes (Fig. 7.19).

In a similar way, networks can be formed directly from monomers by radical chain reactions. One example is the copolymerization of styrene with 1,4-divinyl benzene (Fig. 7.20).

When the two vinyl groups of 1,4-divinyl benzene are incorporated into different chains and when their incorporation exceeds a certain critical level, the resulting copolymer consists (ideally) of a single macromolecule network.

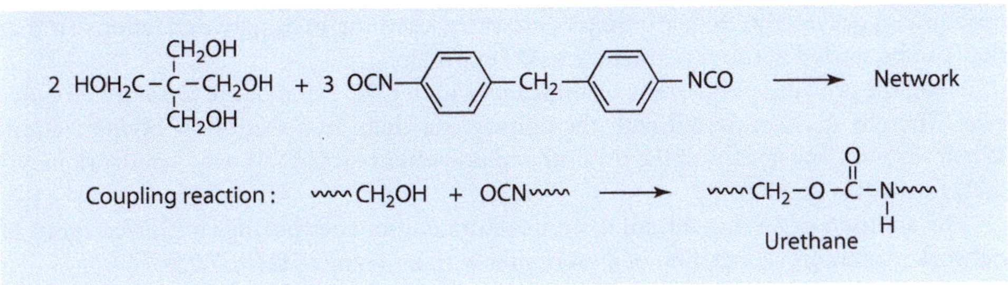

Fig. 7.19 Example of the formation of a network via a step-growth reaction

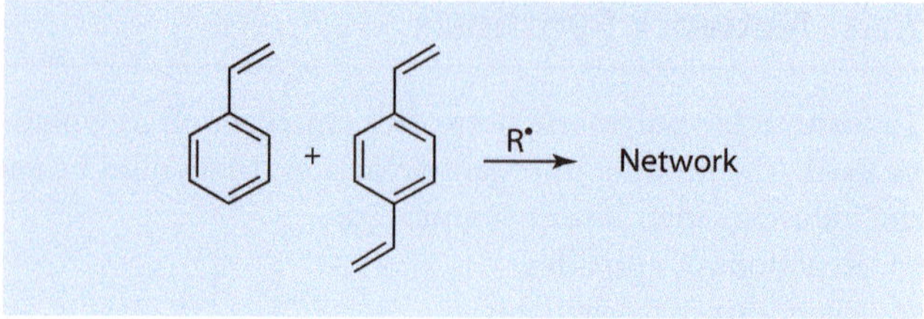

Fig. 7.20 Cross-linking during radical copolymerization

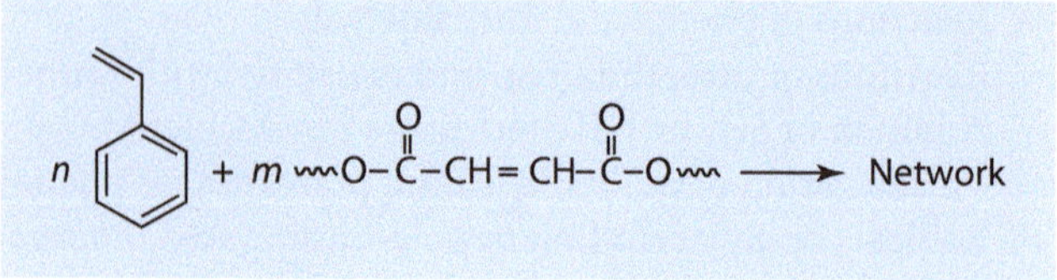

Fig. 7.21 Cross-linking via copolymerization of a monomer with a multi-unsaturated polymer

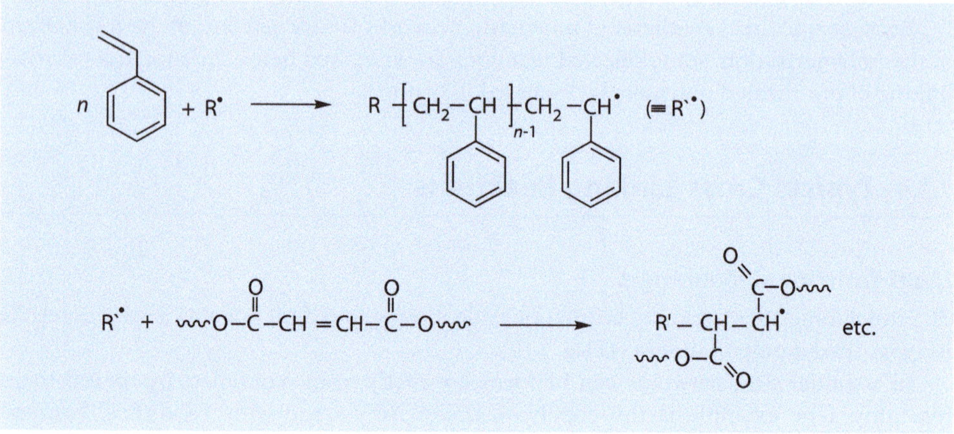

Fig. 7.22 Elementary steps of the reaction of styrene with an unsaturated polyester

Reaction of Monomer with Polymer

A polymer network results from the radical polymerization of styrene in the presence of unsaturated polyesters. Redox initiators are often used for such polymerizations so that they can be started at room temperature (Fig. 7.21).

Here, the growing polystyrene chain reacts with a C=C-bond of the unsaturated polyester. Thereby, it is connected with the polystyrene chain in a sort of copolymerization. When several C=C-groups of the unsaturated polyester react in this way, a network inevitably forms (Fig. 7.22).

The addition of an α,ω-dithiol to an unsaturated polymer belongs to this category of network formation via reaction of a monomer with a polymer (Fig. 7.23).

In place of the thiol in Fig. 7.23, SiH- or NH_2-functionalized molecules can be employed.

A cross-linking reaction frequently used is the stoichiometric hardening of α,ω-diepoxides with amines (Fig. 7.24) or acid anhydrides.

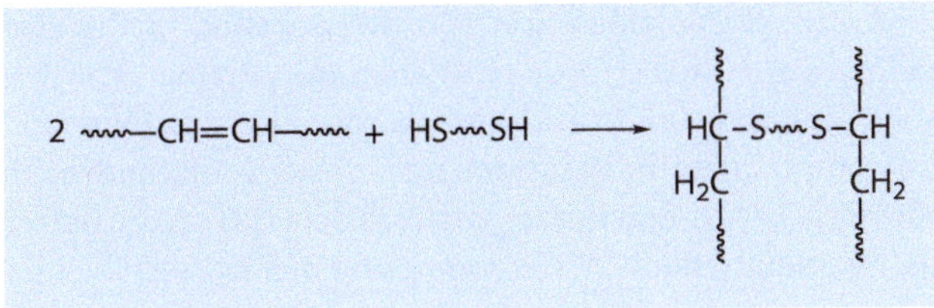

Fig. 7.23 Reaction of an α,ω-dithiol with an unsaturated polymer

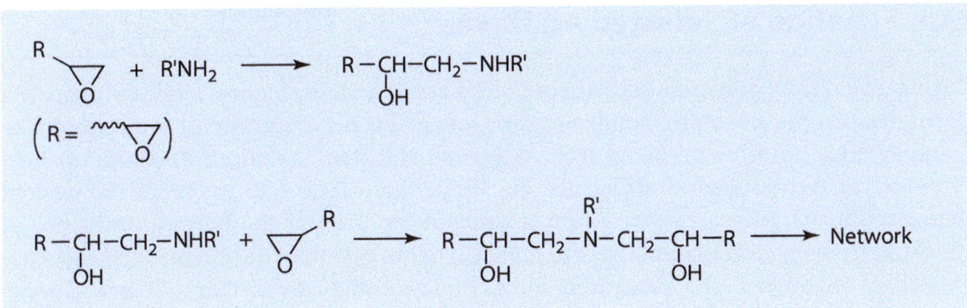

Fig. 7.24 Fundamental step of the cross-linking of an epoxy resin with a primary amine

7.4 Polymer Additives

Polymer additives are substances that improve the characteristics of a polymer, often in very small amounts. They are often necessary to convert a polymer into a useful material. Such additives:

- Improve the processability of the polymer
- Improve the mechanical properties of the polymer
- Reduce the costs
- Modify the surface of the polymer
- Influence the optical characteristics
- Improve the aging resistance of the polymer (▶ Chap. 15)

In addition, there are other additives which induce fire resistance or which act as foaming agents (blowing agents).

7.4.1 Technological Requirements on Polymer Additives

Additives need to be effective and cost effective. They must be compatible with the polymer matrix and not be removed from the polymer during processing. In addition, they should not be hazardous to health. These requirements vary depending on the application. Particularly strict requirements naturally exist for application in polymers for medical engineering or for food packaging.

An inevitable side effect is a deterioration of the electrical insulating effect because additives are usually more polar than the polymer. For the same reason, the absorption of water, and thus the hydrolysis susceptibility, can increase. The polymer and the additives are compounded by solution, solid suspension, and melting, but also by mixing as dry powders. The chosen mixing method depends on the inherent characteristics of the materials, the requirements of the compound quality, and the mixing plant's available equipment.

7.4.2 Function of Selected Additives

During processing, polymers experience both thermal and mechanical stress (▶ Chap. 17). Depending on the structure, bonds break or low molar mass entities are eliminated. For example, HCl can be eliminated from polyvinyl chloride. *Stabilizers* (▶ Chap. 15) help to hinder or repress undesirable reactions. Their main task is to maintain the desired characteristics of polymers over a longer time and to prevent the harmful influence of atmospheric oxygen. Lubricants help to reduce friction between the polymer and the reactor wall, or the inner wall of processing tools. These additives are similar to soap and accumulate on the polymers surface.

Equally important are *UV-stabilizers*, which protect the organic material from the negative effects of visible and UV light (▶ Chap. 15). One of the most effective additives against the negative effects of light is *carbon black*. The mechanical characteristics of the polymers are influenced by *plasticizers* or *reinforcing fillers* (glass fiber, ▶ Chap. 17). Fillers can also be used simply to stretch the polymer and make it more cost-effective, and to modify the polymer's specific weight. The surface characteristics of polymers are improved by *antistatic agents*. In this way, the danger of the electrostatic charge on PVC-floors, for instance, can be reduced. The optical characteristics are influenced by *dyes* and *pigments*, and the above-mentioned *carbon black* also plays a role here.

Depending on their structure, polymers are more or less susceptible to attack by unwanted microorganisms. To prevent this, *fungicides*, for example, are added to the polymer. Because polymers are usually organic substances, their flammability needs to be reduced by adding *fire retardants*. *Blowing agents* are used to transform the polymer into a foam.

The weakening of the polymer by radiation, temperature, oxygen and other reactive gases, water, microorganisms, and by mechanical stress leads to discoloration, the loss of optical clarity, the loss of insulation quality, and/or the deterioration of the mechanical properties. The above-mentioned additives can delay the deterioration, but, as a rule, they cannot prevent it, because they are used up as they act (for example, antioxidants react with oxygen) or slowly diffuse out of the polymer.

7.4.3 "White Carbon Black"

Quartz powder has become a proven filler for casting resins. Casting resins are relatively low molecular, uncross-linked precursors which usually consist of one or two components to which, before being processed (hardened) to cross-linked polymers, an additional reactant and fillers are added. The addition of quartz powder reduces the heat development during the hardening (dilution effect), the volume shrinkage is decreased, and the thermal conductivity is increased. Particularly those fillers that bind chemically to the polymer influence its properties favorably, because a separation of the components is thereby prevented and the problems that often result from the

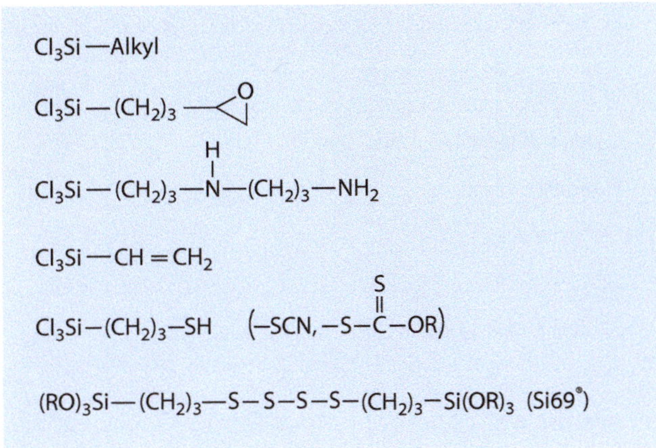

Fig. 7.25 Selected silanes for the modification of silica

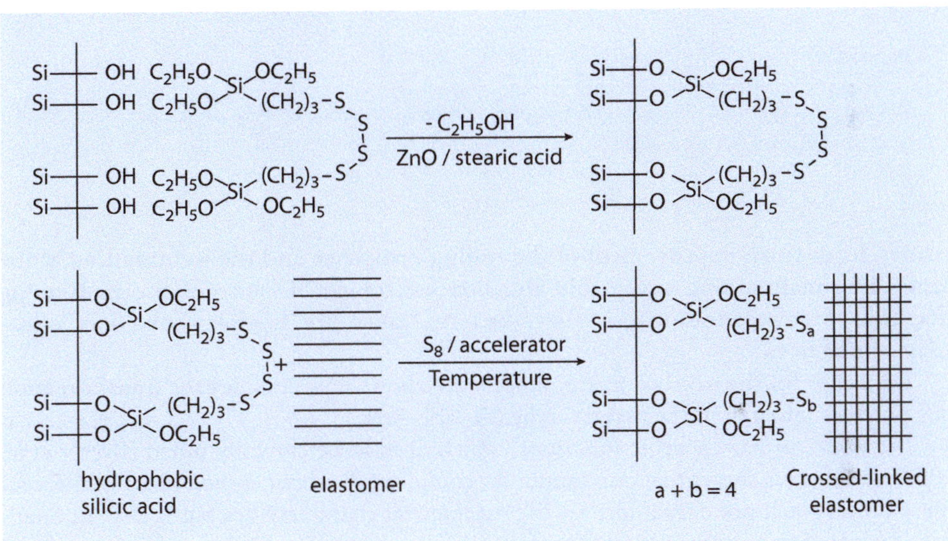

Fig. 7.26 Si69® as a coupling agent between silica and a rubber matrix

interface between the polymer and the filler are reduced. Free OH-groups on the surface of silica, for example, can take part in the hardening process of cast resins. Furthermore, reacting these groups with silanes modifies the silica surface and increases the range of applications where it can be usefully employed. Thus silica can be specifically modified depending on which polymer it is to be compounded with. A selection of modification reagents is listed in Fig. 7.25.

Attempts have been made for quite some time to replace carbon black, at least partially, with quartz, i.e., to develop a "white carbon black." This initially rather unsuccessful research was given a renewed fillip by the large increase in the raw materials for carbon black during the oil crisis. Because of their polarity difference, rubber and silica are incompatible. The surface of the silica can be made hydrophobic by reacting it with coupling agents such as Si69® (Fig. 7.26). Subsequently, the tetrasulfan group of the Si69® reacts with the double bonds of the elastomer matrix. The partial substitution of carbon black as filler by silica, combined with the use of such

Table 7.3 Example of a recipe for the tread of an automobile tire

Components	Weight amount (phr)[a]
Styrene-butadiene rubber (SBR)	70
Butadiene rubber	30
Carbon black	55
SiO_2	49
Specific coupling agent	12
ZnO	6
Stearic acid	2
Vulcanization accelerator	1–3
Sulfur	1–2
Stabilizers	1

[a] *phr* parts per hundred rubber; weight amounts based on the total amount of rubber

silanes, leads to an improvement of the rolling resistance and the wet traction, at the same time maintaining acceptable abrasion resistance. Because of the fuel-saving potential of reduced rolling resistance, the term "green tire" has been coined for silica-silane-filled tires.

The recipe for the tread of an automobile tire shows how complex the transformation of a polymer into a useful material can be (Table 7.3).

The additives have specific functions, which need to be carefully tuned (Eyerer et al. 2008). Table 7.3 shows that one cannot do completely without carbon black in this recipe. Carbon black not only improves the mechanical characteristics but is also an established protection against UV-light (► Chap. 15).

References

Brandrup J, Immergut EH, Grulke EA (1999) Polymer handbook, 4th edn. Wiley, New York

Eyerer P, Hirth T, Elsner P (2008) Polymer engineering. Springer, Berlin

Nuyken O (1987) Antrittsvorlesungen, Bd 2, Johannes-Gutenberg-Universität Mainz, pp 103–139

Step-Growth Polymerization

S. Koltzenburg et al., *Polymer Chemistry*, DOI 10.1007/978-3-662-49279-6_8

The synthetic processes of producing polymers from their monomers can be divided into step-growth and chain-growth polymerization. These two polymer formation reactions (◘ Figs. 8.1 and 8.2) are fundamentally different in their mechanisms, intermediate products, the way the molar mass increases as a function of monomer conversion, and the activation energy of their elementary steps.

8.1 Differences Between Step-Growth and Chain-Growth Polymerizations

The basic characteristics of these two reaction types are illustrated by reference to the synthesis of polystyrene from styrene (◘ Fig. 8.1) and nylon 6.6 from 1.6-hexamethylene diamine and adipic acid (◘ Fig. 8.2).

As already briefly introduced in ▶ Chap. 1, one industrially significant polymerization mechanism involves the transformation of a double bond into two single bonds, as shown in ◘ Fig. 8.1.

Such polymerization of unsaturated compounds can be caused by radicals (▶ Chap. 9), ions (▶ Chaps. 10 and 12), or transition metal catalysts (▶ Chap. 11). During the radical polymerization of styrene, an initiating radical adds to the double bond of monomeric styrene and creates a styryl radical. This in turn can attack another styrene molecule. Because the reaction always leads to a chain terminating with a styryl radical, it can occur several thousand times before it ends. Because of its chain reaction character, such a process is referred to as a chain-growth polymerization. Such chain-growth polymerization have various distinguishing characteristics:

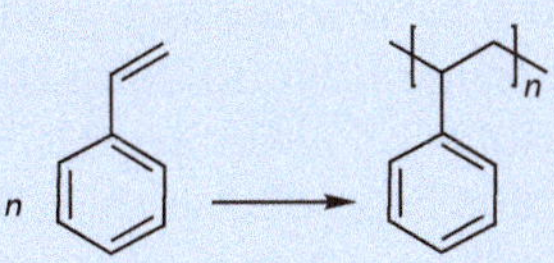

◘ **Fig. 8.1** Polymerization of a double bond using styrene (polystyrene) as an example

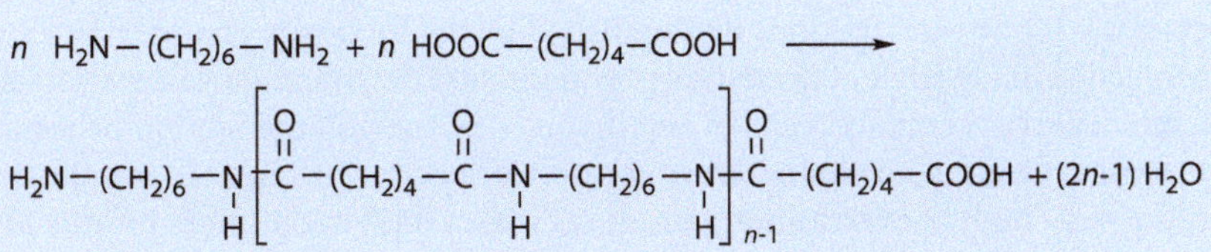

◘ **Fig. 8.2** Synthesis of a polyamide from 1.6-hexamethylene diamine and adipic acid

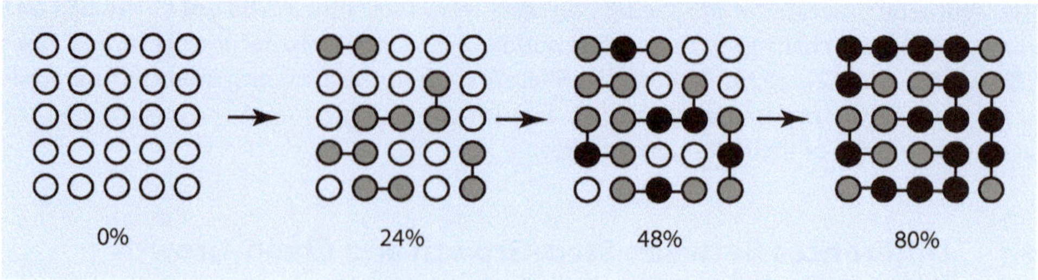

Fig. 8.3 A visualization of a polycondensation at 0 %, 24 %, 48 %, and 80 % functional group conversion. *White balls* represent monomers which have not reacted yet, *gray balls* those monomers which have reacted at one of their two active groups, and *black balls* are those molecules which have reacted at both ends

- It is initiated by a highly reactive particle.
- Only a small portion of the molecules in the process are actively involved in the polymerization process (in this case, the growing styryl terminated radicals). The remaining molecules are unreactive; that is, they can only react with radicals but not with themselves.
- The chain grows quickly (generally within a matter of seconds) to a high molar mass

A typical example of step-growth polymerization is the synthesis of nylon 6.6 from 1.6-hexamethylene diamine and adipic acid (Fig. 8.2). This polymerization is activated by heat, which causes dehydration. There is no chemical initiator. In contrast to chain-growth polymerization, it is impossible to differentiate between active and passive particles. A new monomer results from each individual reaction step and has the same reactivity as the original monomers. Further monomer conversion depends solely on statistical probabilities. All intermediate products are stable and can be isolated. In contrast to chain-growth polymerization, it takes considerable time for polymers with a high molar mass to form. Figure 8.3 illustrates this. Step-growth polymerizations in which a small molecule is a reaction byproduct are also termed "polycondensations."

The example in Fig. 8.3, shows that, even with a conversion of 80 % of the functional groups, there are five "polymer chains" with "degrees of polymerization" of 3, 4, 5, 6, and 7. Not until very high conversions do the oligomers formed condense to chains with a high molar mass. The quantitative details of such processes are given in ▸ Sect. 8.2.

Step-growth polymerization is distinguished by the following characteristics:

- An initiation in the sense of the radical polymerization described above does not take place.
- All molecules are equally reactive and involved in the polymerization process. In particular, there are no chain-carrying reactive species with limited stability.
- Only at very high conversions are products with a high molar mass produced.

Step-growth polymerizations require that the monomers possess chemical functions that can form covalent bonds with each other. These functions can be united in the same molecule, such as ω-hydroxyl carboxylic acid, which can produce a polyester via dehydration (Fig. 8.4). Such a monomer is called an AB-monomer, where A and B represent the reactive groups. Alternatively, two di- or multi-functional molecules can react with each other to form the polymer (Figs. 8.2 and 8.5). Using similar terminology, such monomers are referred to as AA and BB.

A particular characteristic of the step-growth polymerization of two monomers AA and BB is the need to accurately adjust the stoichiometry of the reacting molecules if

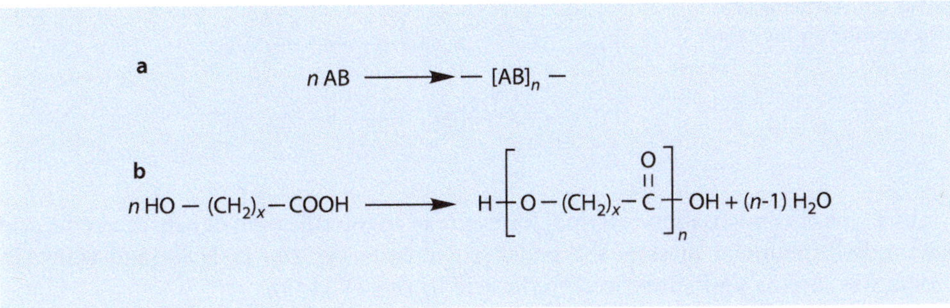

Fig. 8.4 Polymerization of an AB-monomer: (**a**) a general scheme and (**b**) the polymerization of a ω-hydroxyl carboxylic acid to a polyester as an example

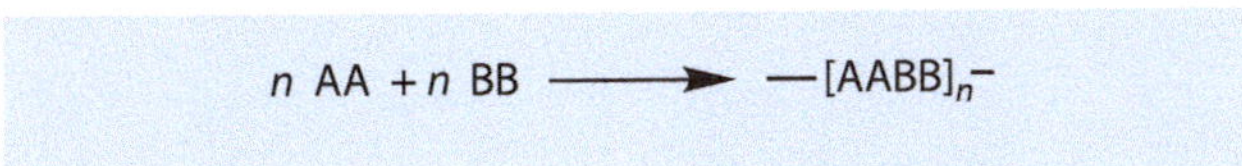

Fig. 8.5 Generic diagram of the polymerization of AA- and BB-monomers

high molar masses are required. For example, if there is an excess of BB monomer, all the A-functionalities react and the oligomers are terminated with B groups, which cannot react with each other.

Given the criteria described here, polymerization reactions can be unambiguously classified into chain-growth and step-growth polymerization.

8.2 Molar Mass, Degree of Polymerization, and Molar Mass Distribution

The most important factors used to characterize macromolecules are the molar mass, the degree of polymerization, and the molar mass distribution. These characteristics depend on the polymerization mechanism of the macromolecules and on the degree of monomer conversion.

8.2.1 Degree of Polymerization and Molar Mass of Step-Growth Polymerizations

During the stepwise transformation of an AB-monomer or of a stoichiometric AA/BB-system into a polymer (Fig. 8.6), the molar mass is calculated by multiplying the molar mass of the components AB, AA and BB respectively, by the number n of repetitions in the polymer molecule:

$$M_{Polymer} = n\left(M_{AB}\right) \tag{8.1}$$

$$M_{Polymer} = n\left(M_{AA} + M_{BB}\right) \tag{8.2}$$

In (8.1) and (8.2), $M_{Polymer}$ represents the molar mass of the polymer, M_{AB} the molar mass of the monomer AB, and M_{AA} and M_{BB} the molar masses of the monomers AA and BB, respectively.

$$n\ \mathrm{AB} \longrightarrow -[\mathrm{AB}]_n-$$

$$n\ \mathrm{AA} + n\ \mathrm{BB} \longrightarrow -[\mathrm{AABB}]_n-$$

Fig. 8.6 A scheme of a step-growth polymerization

In a typical condensation reaction, for example an esterification of a dicarboxylic acid with a diol, the molar mass of the polymer decreases by $(2n-1)\cdot H_2O$ (and with AB-monomers such as a 6-hydroxy carboxylic acid by $(n-1)\cdot H_2O$).

The degree of polymerization, P_n, is identical to the number of repetitions n and is calculated as follows (▶ Sect. 3.1):

$$P_n = \frac{n_0}{n_t} \tag{8.3}$$

n_0 Initial number of molecules
n_t Number of molecules at time t

or:

$$P_n = \frac{c_0}{c_t} \tag{8.4}$$

c_0 Initial concentration of functional groups
c_t Concentration of functional groups[1] at time t

Because at a conversion, p:

$$p = \frac{c_0 - c_t}{c_0} = 1 - \frac{c_t}{c_0} \tag{8.5}$$

or

$$c_t = (1-p)\cdot c_0 \tag{8.6}$$

then P_n is given by

$$P_n = \frac{c_0}{c_0(1-p)} = \frac{1}{1-p} \tag{8.7}$$

This implies that only with high conversion rates (p almost 1) can sufficiently high degrees of polymerization be achieved (Fig. 8.7).

If the concentration of the reactive groups is not stoichiometric, the degree of polymerization can be determined from the following considerations.

1 The concentration of the functional groups is the same as that of the existing molecules; the former is experimentally easier to access.

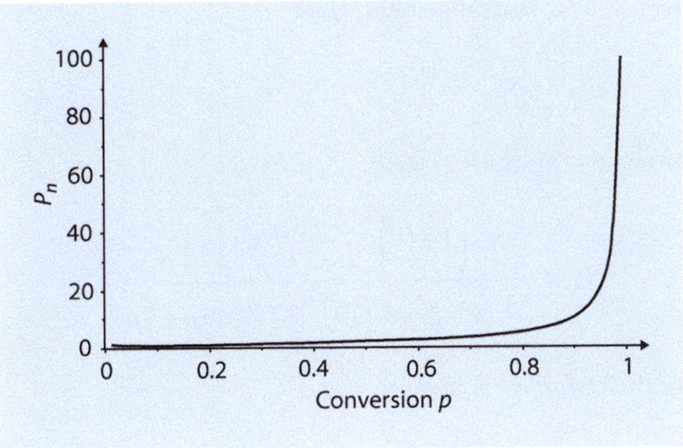

Fig. 8.7 Dependence of the degree of polymerization on conversion for a polycondensation

Because

$$P_n = \frac{n_0}{n_t} \tag{8.3}$$

the ratio of shortfall (A) to excess (B) is defined as the non-stoichiometric factor r:

$$\frac{n_A}{n_B} = r \leq 1 \tag{8.8}$$

n_A Starting number of the functional groups A
n_B Starting number of the functional groups B
$n_{A,t}$ Number of functional groups A at time t
$n_{B,t}$ Number of functional groups B at time t

From this it follows that r is always less or equal to 1.
From (8.3) and (8.8) we obtain

$$P_n = \frac{n_A + n_B}{n_{A,t} + n_{B,t}} = \frac{n_A\left(1+\frac{1}{r}\right)}{n_{A,t} + n_{B,t}} \tag{8.9}$$

Because p represents the conversion of the functions A, unconverted A is given by

$$n_{A,t} = (1-p)\,n_A \tag{8.10}$$

Because the conversion of B is limited by the non-stoichiometric factor r, the number of B functions is given by

$$n_{B,t} = (1-rp)\,n_B \tag{8.11}$$

The total number of functions $n_{A,t} + n_{B,t}$ that have not been converted at time t is

$$n_{A,t} + n_{B,t} = (1-p)\,n_A + (1-rp)\,n_B \tag{8.12}$$

which, after rearrangement, gives

$$n_{A,t} + n_{B,t} = \frac{n_A}{r}(r - 2rp + 1) \tag{8.13}$$

Inserting this in (8.9) yields

$$P_n = \frac{n_A\left(1+\frac{1}{r}\right)}{\frac{n_A}{r}(r-2rp+1)} = \frac{\frac{n_A}{r}(r+1)}{\frac{n_A}{r}(r-2rp+1)} \tag{8.14}$$

or the *Carothers equation*

$$P_n = \frac{r+1}{r-2rp+1} \tag{8.15}$$

8

In the case of exact stoichiometry of the functional groups, i.e., $r = 1$, (8.7) derived above is obtained:

$$P_n = \frac{2}{2-2p} = \frac{1}{1-p}$$

With quantitative conversion of those active groups in the minority ($p = 1$, $r < 1$) it follows that the degree of polymerization is given by

$$P_n = \frac{r+1}{r-2r+1} = \frac{1+r}{1-r} \tag{8.16}$$

Thus, the degree of polymerization can be controlled by the conversion p, but also by the non-stoichiometry r. To limit the molar mass, non-stoichiometry is often induced by intentionally adding monofunctional elements. If [AA] < [BB], the degree of polymerization P_n of the polymer can be adjusted by (8.16). Furthermore, if AA is completely consumed, the polymer molecules are exclusively terminated by B groups which are available for subsequent reactions.

According to (8.16) or ◘ Fig. 8.8, at a non-stoichiometry of $r = 0{,}975$, despite complete conversion ($p = 1$), the degree of polymerization is only 79 ($P_n = 79$). If the conversion is reduced to $p = 0.95$ with otherwise identical conditions, the degree of polymerization decreases to $P_n = 16$.

The Dependence of the Conversion on the Equilibrium Constant K

Because the esterification reaction is a typical organic equilibrium, it is important to know what conversion can actually be achieved. This is determined by the equilibrium constant K, given in molar concentration terms by

$$K = \frac{[RCOOR'][H_2O]}{[RCOOH][R'OH]} \tag{8.17}$$

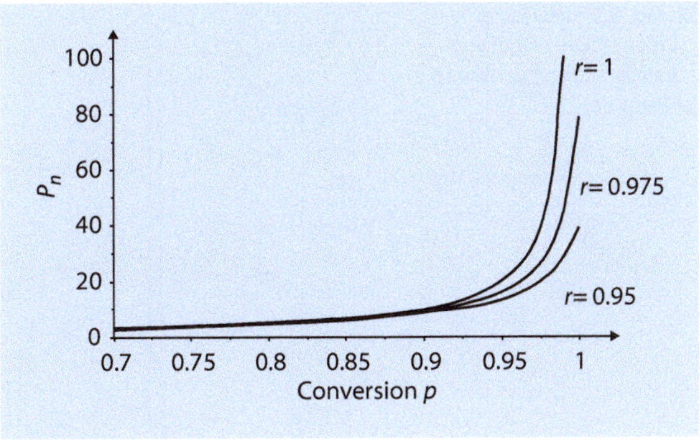

Fig. 8.8 Dependence of the degree of polymerization on conversion at different stoichiometric ratios r for the polymerization of AA- with BB-monomers

with

$$p = \frac{[RCOOR']}{[RCOOH]_0} = \frac{[H_2O]}{[RCOOH]_0} \tag{8.18}$$

and

$$(1-p) = \frac{[RCOOH]}{[RCOOH]_0} = \frac{[R'OH]}{[RCOOH]_0} \tag{8.19}$$

($[RCOOH]_0 = [R'OH]_0$: starting concentration of the educts; see also (8.6))

From (8.17)–(8.19):

$$K = \frac{p[RCOOH]_0 \cdot p[RCOOH]_0}{(1-p)[RCOOH]_0 \cdot (1-p)[RCOOH]_0} = \frac{p^2}{(1-p)^2} \tag{8.20}$$

Solving (8.20) for p yields

$$p = \frac{\sqrt{K}}{1+\sqrt{K}} \tag{8.21}$$

and, at equilibrium

$$P_n = \frac{1}{1-p} = \frac{1}{1-\frac{\sqrt{K}}{1+\sqrt{K}}} \tag{8.22}$$

As an example, for $K=5$ one can calculate $p=0.69$ and $P_n=3$.

Higher degrees of polymerization (at a constant K) can only be reached by shifting the equilibrium to the products side. The system adjusts, creating new products (reactants are consumed) to establish equilibrium (8.17). As a result, higher conversions and higher

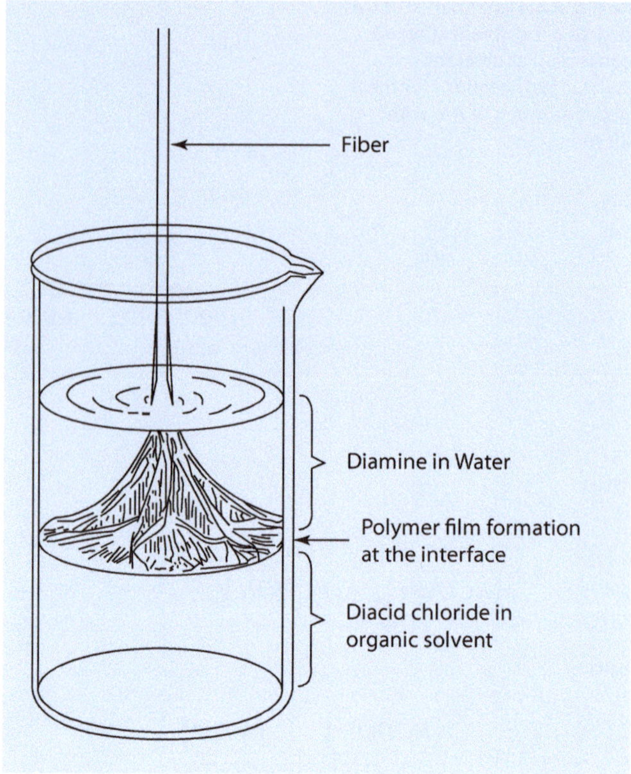

Fig. 8.9 Interfacial condensation of a diacid chloride with a diamine to a polyamide

8

degrees of polymerization can be reached in accordance with the Carothers equation (see (8.15) or (8.7)).

To achieve high degrees of polymerization it is necessary to remove, as completely as possible, the water produced. Another option is to remove the resulting polymer, for instance by interfacial (Fig. 8.9) or precipitation polymerization.

At the interface between the aqueous (containing the diamine) and organic phases (e.g., CCl_4 with diacid chloride) polyamide forms spontaneously and can be continuously removed. The polymeric film in the boundary layer forms a barrier for the reactants so that they only react with the chain ends of the polymers, and thus higher molar masses are achieved than with the more commonly employed melt condensation.

8.2.2 Molar Mass Distribution of Step-Growth Polymerization

The conversion of an AB-monomer into an $[AB]_P$ – polymer involves $(P-1)$ polymerization steps (e.g., esterifications). The probability that an A-group has reacted by time t corresponds to the conversion p $(0 \le p \le 1)$. The probability that an A-group has not reacted is thus $1-p$. The probability that a polymer molecule forms, with exactly P repeating units, requires not only $(P-1)$ reactions of the functional group A but also that one A-group does not react. With this the number n_P of these polymers is given by the product of the individual probabilities (in the following derivation note the distinction between P and p):

$$n_p = p^{p-1} \cdot n_t \tag{8.23}$$

n_P Number of molecules with degree of polymerization P
n_t Number of molecules remaining at time t

The sum of all n_p equals the number of molecules remaining, n_t:

$$\sum n_p = n_t \tag{8.24}$$

(All sums (Σ) in this chapter are considered over all elements, P)
From (8.3) and (8.7) it also follows that

$$n_t = n_0 \cdot (1-p) \tag{8.25}$$

where n_0 is the initial number of molecules.
Applying (8.25) to (8.23) yields

$$n_p = n_0 \cdot p^{P-1} \cdot (1-p) \tag{8.26}$$

For the weight fraction m_P it follows that

$$m_P = \frac{P \cdot n_P}{P_n n_0} = \frac{P n_P}{n_0}(1-p) \tag{8.27}$$

$$\sum m_P = 1 \tag{8.28}$$

Combining (8.26) and (8.27) gives

$$m_P = P \cdot p^{P-1} \cdot (1-p)^2 \tag{8.29}$$

Figure 8.10 shows how the ratio n_P/n_0 and m_P shift with increasing conversion ($p = 0.96$, 0.9875, 0.995) to higher degrees of polymerization P.
With the definition of the number average degree of polymerization, P_n:

$$P_n = \frac{\sum (n_P \cdot P)}{\sum n_P} = \sum \left(P \cdot \frac{n_P}{n_t} \right) \tag{8.30}$$

and from (8.23):

$$P_n = \sum \left(P \cdot p^{P-1} \right) \cdot (1-p) \tag{8.31}$$

can be derived.
According to progression theory:

$$\sum \left(P \cdot p^{P-1} \right) = \frac{1}{(1-p)^2}, \tag{8.32}$$

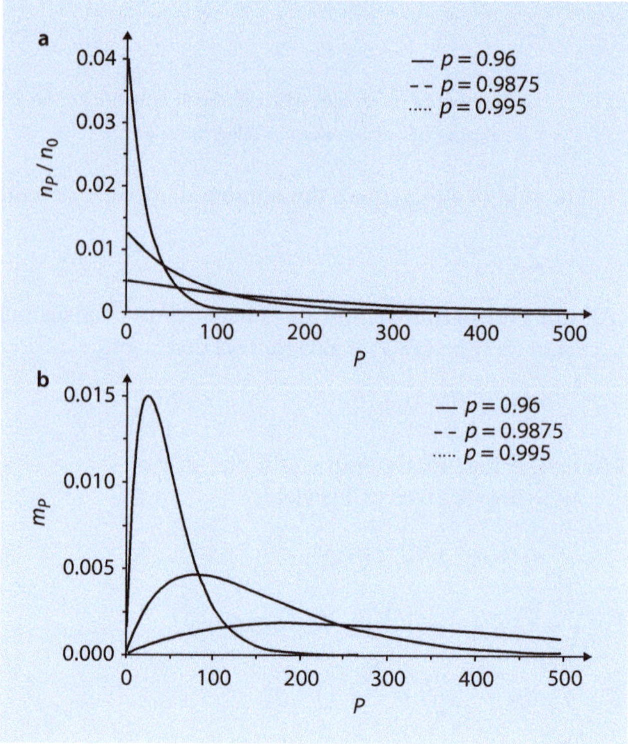

Fig. 8.10 (**a**) n_P/n_0 and (**b**) m_P as a function of the degree of polymerization P. Calculated from (8.25) and (8.28)

so that the familiar Carothers equation (8.7) is found for P_n:

$$P_n = \frac{1}{1-p} \tag{8.7}$$

For the weight average degree of polymerization, P_w:

$$P_w = \sum (P \cdot m_P) \tag{8.33}$$

With (8.27) this gives

$$P_w = \sum \left(P^2 \cdot p^{P-1}\right) \cdot (1-p)^2 \tag{8.34}$$

Applying progression theory:

$$\sum \left(P^2 \cdot p^{P-1}\right) = \frac{1+p}{(1-p)^3} \tag{8.35}$$

We obtain for P_w:

$$P_w = \frac{1+p}{1-p} \tag{8.36}$$

For the polydispersity index (PDI) it follows:

$$PDI = \frac{P_w}{P_n} = \frac{1+p}{1-p} \cdot (1-p) = 1+p \tag{8.37}$$

As $p \to 1$ it follows from (8.37) that the PDI for an ideal step-growth polymerization (no secondary reactions, no ring formation) has the value 2 (8.38):

$$\frac{P_w}{P_n} = 2 \tag{8.38}$$

This PDI is the same as the PDI for radical polymerizations where termination is via disproportionation (▶ Sect. 9.3).

8.3 Linear, Branching, and Crosslinking Step-Growth Polymerizations

In this chapter, the extraction of linear polycondensates from AB-components and the reaction of AA- and BB-monomers, as well as of branched polycondensates from A_xB-components, is explained using the example of polyesters. Additionally, the conditions for crosslinking step-growth polymerization is introduced.

8.3.1 Monomers with Two Different Functional Groups

In general notation, an $[AB]_n$ – polymers originates from n AB-monomers. Corresponding to this schema, a polyester is derived from an ω-hydroxy carboxylic acid (HO–$(CH_2)_x$–COOH). However, the condition $x>4$ must hold because for $x \leq 4$ the competing and alternative reactions shown in ◘ Figs. 8.11–8.13 are possible.

For $x=1$ a lactide forms as the main product (◘ Fig. 8.11).

By means of anionic and cationic initiation or in the presence of special metal compounds (e.g., stannous octanoate), lactides undergo ring-opening polymerization (▶ Chap. 12).

When $x=2$, heating produces mainly acrylic acid via dehydration (◘ Fig. 8.12).

When $x=3$ or $x=4$, lactones are the major products from the corresponding monomers (◘ Fig. 8.13).

In contrast, 6-hydroxyhexanoic acid ($x=5$) can be converted into the corresponding polyesters without any problems (cf. ◘ Fig. 8.4).

8.3.2 Reaction of Two Different Monomers Each Having Two Identical Functional Groups

A typical example of this reaction type, where the respective polymer –$[AABB]_n$– is formed from AA + BB, is shown in ◘ Fig. 8.2. As described earlier, very high conversions and strict adherence to the stoichiometry are required for high molar masses (▶ Sect. 8.2.1).

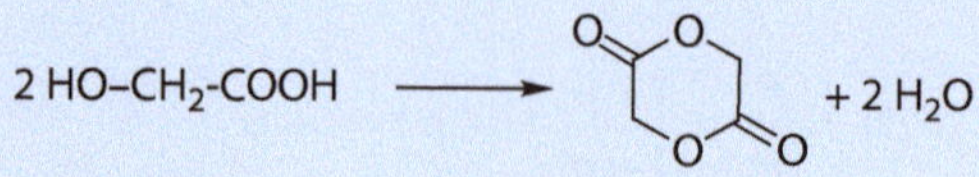

Fig. 8.11 Synthesis of a lactide from hydroxyacetic acid

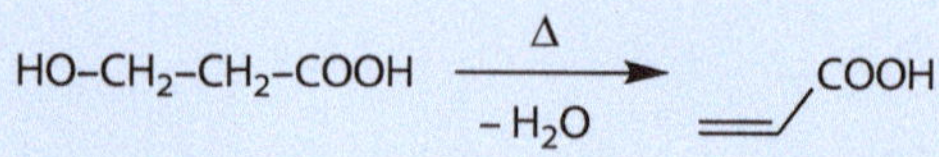

Fig. 8.12 Reaction of 3-hydroxypropionic acid to acrylic acid

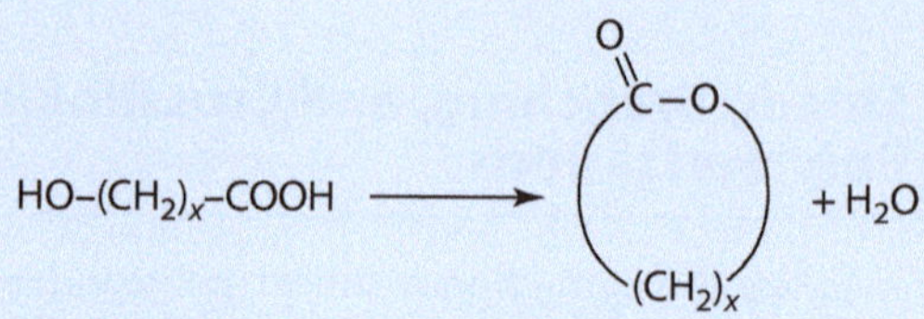

Fig. 8.13 Formation of the corresponding lactones from 4-hydroxybutanoic acid ($x=3$) or 5-hydroxypentanoic acid ($x=4$)

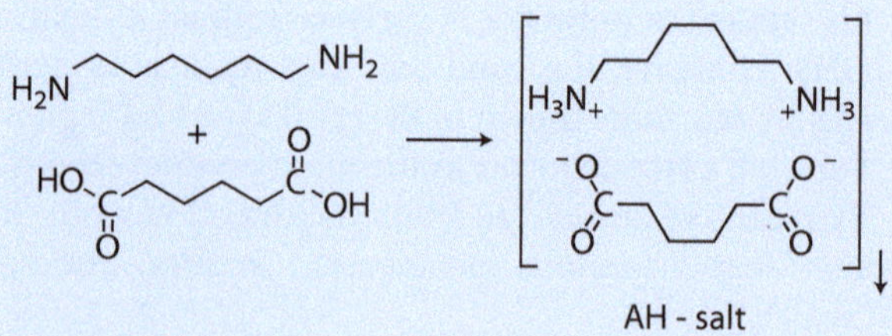

Fig. 8.14 Formation of an AH-salt from 1.6-hexamethylene diamine (H) and adipic acid (A)

The problem of the stoichiometry of the reactants 1.6-hexamethylene diamine (AA) and adipic acid (BB) can be elegantly avoided by mixing the constituents beforehand and converting them into a so-called AH-salt that can be isolated in its solid state (Fig. 8.14). Here, both constituents are always exactly in the molar ratio 1:1.

By fusing in vacuo, the AH-salt can be dehydrated and converted into the corresponding polyamide 6.6 (Fig. 8.15).

Linear aliphatic polyamides are designated by numbering: either by using one number (e.g., polyamide 6 or polyamide 12) which indicates the number of C-atoms in the starting material (AB-monomer) or by using two digits (polyamide 6.6 or polyamide 6.12) when two starting materials are employed (AA and BB). The convention is that the first number gives the number of C-atoms in the diamine and the second number represents the number of C-atoms in the dicarboxylic acid.

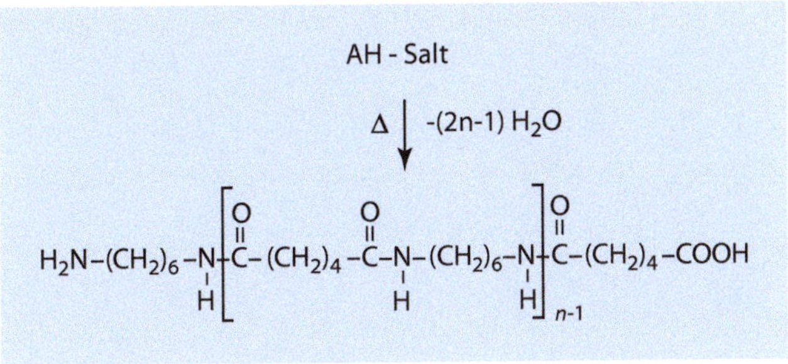

Fig. 8.15 Synthesis of polyamide 6.6 by fusion of the AH-salt

8.3.3 Polymers from A_xB-Monomers ($x \geq 2$)

A_2B-monomers can be converted into hyperbranched polymers through self-condensation. Such polymers have drawn not only scientific but also industrial attention in recent times, particularly as the polymerization of such A_xB-monomers ($x \geq 2$) cannot lead to crosslinking (▶ Sect. 1.3) (Flory 1952; Turner et al. 1993). These hyperbranched polymers are related to the so-called *dendrimers*. However, *dendrimers* have a perfect architecture which hyperbranched polymers do not. The latter are cheaper to produce. Schematically, an A_2B-monomer reacts as illustrated in Fig. 8.16.

The total number of free functionalities A (free A) in the polymer can be calculated from the overall functionality f in the A_xB-monomer ($f = 3$ for A_2B) and the number of links n with the formula (8.39) (Table 8.1).

$$\text{Free A} = \underbrace{(f-1)}_{\substack{\text{Number of}\\ \text{function A}\\ \text{in the monomer}}} \cdot \underbrace{(n+1)}_{\substack{\text{Repeating units}\\ \text{in each}\\ \text{polymer molecule}}} - \underbrace{n}_{\substack{\text{Number of}\\ \text{reacted A}\\ \text{functional groups}}} = (n+1)\cdot f - 2n - 1 \qquad (8.39)$$

The branching coefficient α specifies the probability of a functional group of one branching unit linking with another branching unit and thus, in hyperbranched systems, corresponds to the probability p_A that a randomly chosen A has reacted:

$$\alpha = p_A \qquad (8.40)$$

Because the number of reacted A and B is equal, it follows that

$$p_A(f-1) = p_B = p \quad (f - 1 = x \text{ in } A_xB) \qquad (8.41)$$

Here p_B is the probability that B has reacted and P is the probability of the reaction of A with B. With (8.41), (8.40) can be transformed into (8.42):

$$\alpha = \frac{p}{f-1} \qquad (8.42)$$

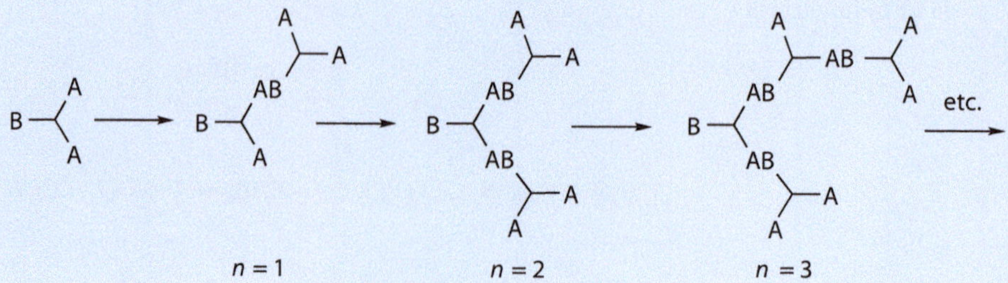

Fig. 8.16 Formation of hyperbranched polymers from A_2B-monomers

Table 8.1 Number of free A per polymer molecule as a function of reacted A, n (A_2B-monomer, $f=3$)

n	Free A
1	3
2	4
3	5
4	6
⋮	⋮
20	22

For an A_2B-monomer ($f=3$) the relation (8.43) can be obtained from (8.42) at complete conversion ($p=1$):

$$\alpha = \frac{1}{2} \ (\text{for } f = 3 \text{ and } p = 1) \tag{8.43}$$

Only when B in A_2B has completely reacted (this practically never happens) is $\alpha = ½$. Because crosslinking only occurs when this limit value is exceeded, the resulting polymer is not cross-linked and therefore soluble and fusible.

It must always be assumed that all functionalities of one type (A or B), independent of the size of the molecule, are equally reactive and thus react statistically, and that no intramolecular rings form. Macromolecules with a functionality B and $(n+1) \cdot f - 2n - 1$ unreacted A-positions then develop from A_nB, where n is the number of links (see (8.39)).

Figure 8.17 shows the result of the self-condensation of two dihydroxybenzoic acid derivatives protected with different protecting groups (TMS protecting group/COCl; CH_3CO protecting group/COOH).

Because of their multifunctionality and good solubility, their low solution viscosities (resulting from the high degree of branching), and their low tendency to crystallization and entanglement formation compared with linear polymers, these polymers have potential applications in catalytic converter technology and medicine as substrates, as well as materials in their own right.

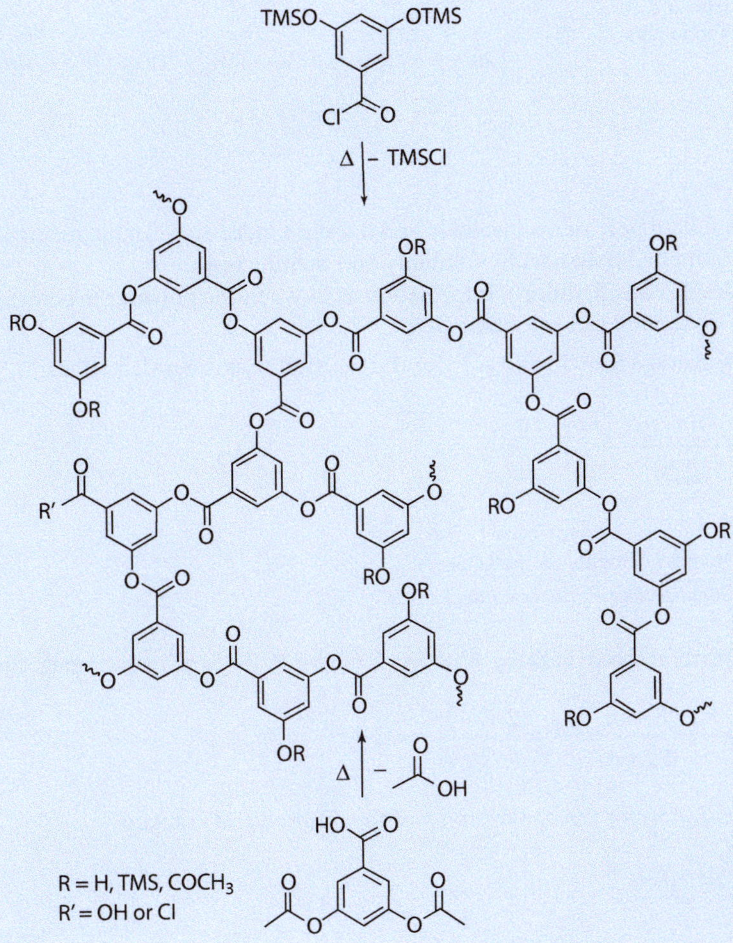

Fig. 8.17 Structure of a hyperbranched polyester based on two different monomers (TMS = trimethylsilyl)

8.3.4 Cross-Linking

The reaction of different monomers with more than two functional groups (e.g., the reaction of trifunctional alcohols with difunctional acids, Fig. 8.18) with suitable stoichiometry of the functional groups eventually results in a three-dimensional network. Such networks have become indispensable materials because of their insolubility and their inability to melt.

Fig. 8.18 Formation of a network from di- and trifunctional monomers

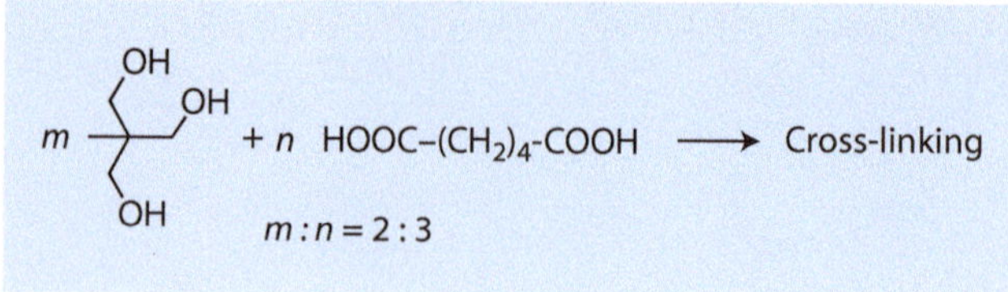

A cross-linked polymer is insoluble and does not melt; they can be distinguished from linear and branched polymers by solubility and melting tests.

From simple considerations, information as to when the polymer becomes a network, the so-called *gel point* can be obtained.

First, an average functionality $\bar{f}$ for the reaction mixture is defined as

$$\bar{f} = \frac{\sum n_j f_j}{\sum n_j} \tag{8.44}$$

n_j	Number of monomer molecules j
f_j	Functionality of the monomer j

For the reaction shown in Fig. 8.18 ($m=2$ and $n=3$), (8.45) gives the average functionality:

$$\bar{f} = \frac{n_m \cdot f_m + n_n \cdot f_n}{n_m + n_n} = \frac{2 \cdot 3 + 3 \cdot 2}{2+3} = 2.4 \tag{8.45}$$

With the definition for the conversion p of the difunctional reactant:

$$p = \frac{2(n_0 - n_t)}{\bar{f} \cdot n_0} \tag{8.46}$$

n_0	Number of molecules (monomers) at time $t=0$
n_t	Number of molecules (monomers + polymers) at time t
$2(n_0-n_t)$	Number of functional groups which have reacted at time t
$\bar{f} \cdot n_0$	Average number of existent functional groups at time $t=0$.

With the definition, given above, for the degree of polymerization P_n:

$$P_n = \frac{n_0}{n_t} \tag{8.3}$$

it follows that

$$p = \frac{2}{\bar{f}}\left(1 - \frac{1}{P_n}\right) \tag{8.47}$$

At the gel point, it holds that $P_n \to \infty$. The conversion required to reach the gel point can be derived as

$$p = \frac{2}{\bar{f}} \tag{8.48}$$

The conversion at the gel point as a function of the functionality of the reaction mixture is given in Table 8.2.

Figure 8.19 clearly shows how the sharp rise of P_n shifts to lower conversions with increasing average functionality of the mixture.

Industrially important networks can be made, for example, from the reaction of diisocyanates with multifunctional alcohols or from reaction of triisocyanates with diols (Fig. 8.20).

By varying the isocyanates and alcohols both with regard to structure and functionality, many and varied network properties can be achieved.

Table 8.2 Conversion required to reach the gel point depending on the functionality of the reaction mixture

$\bar{f}$	P
2	1.0000
2.1	0.9524
2.2	0.9091
2.3	0.8696
2.4	0.8333
3	0.6667
4	0.5000

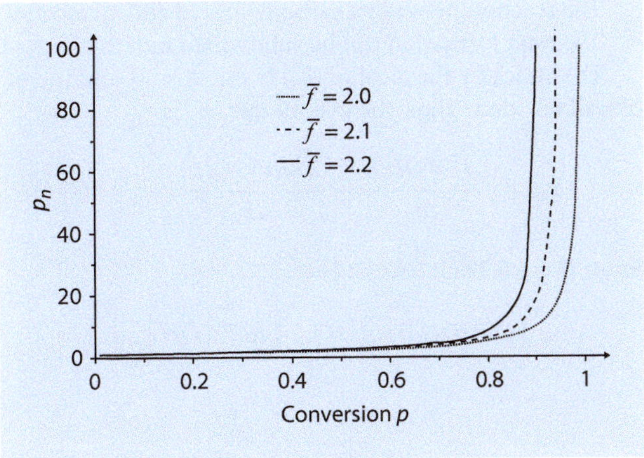

Fig. 8.19 Influence of the functionality $\bar{f}$ on the degree of polymerization P_n

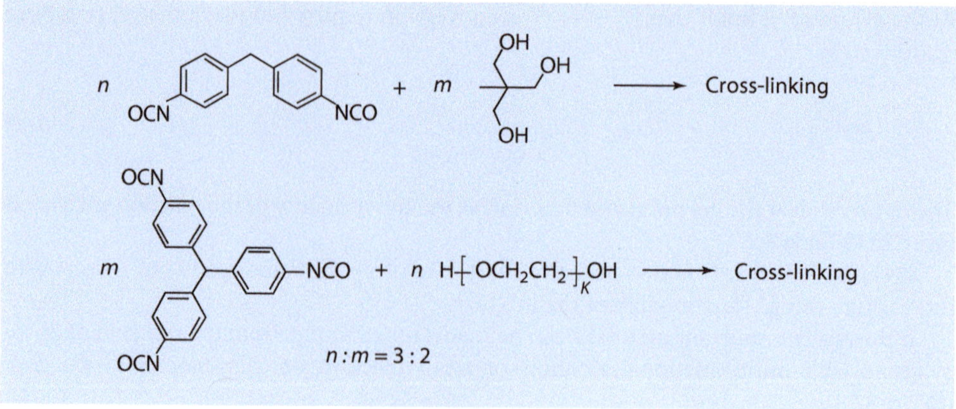

Fig. 8.20 Formation of polyurethane networks

8

Fig. 8.21 Condensation reaction of a carboxylic acid with an alcohol to give an ester

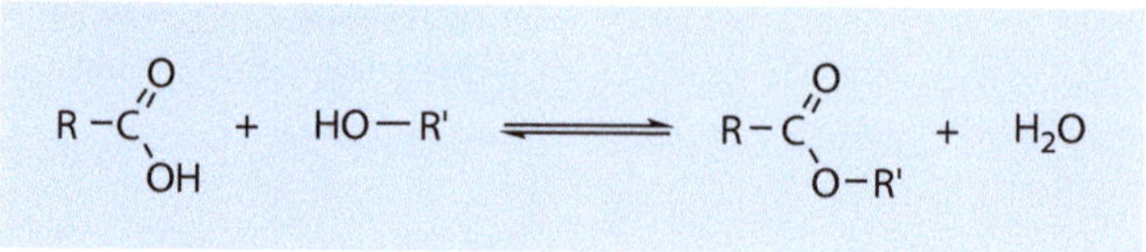

8.4 Kinetics of Step-Growth Polymerization

The fundamental kinetic theories of step-growth polymerization date back to P.J. Flory. These are based on the assumption that the individual reaction steps can be seen as reactions between the functional groups, and that the activity of these groups is independent of the size of the molecules. Using the example of the esterification, the processes and their consequences are explained (Figs. 8.21 and 8.22). The fundamental propositions can also be applied to other step-growth polymerizations.

The reaction between a carboxylic acid and an alcohol results in an ester (Fig. 8.21).

The ester formation can be subdivided into the stages illustrated in Fig. 8.22.

The attack by the alcohol (k_3) is the slowest and therefore determines the speed of the overall reaction. Thus, the overall rate, v_{br} is

$$v_{br} = -\frac{d[R'OH]}{dt} = -\frac{d[RCOOH]}{dt} = k_3\left[RC^+(OH)_2\right][R'OH] \tag{8.49}$$

From Fig. 8.22a it follows that

$$k_1 \cdot [RCOOH][HA] = k_2 \cdot \left[RC^+(OH)_2\right]\left[A^-\right] \tag{8.50}$$

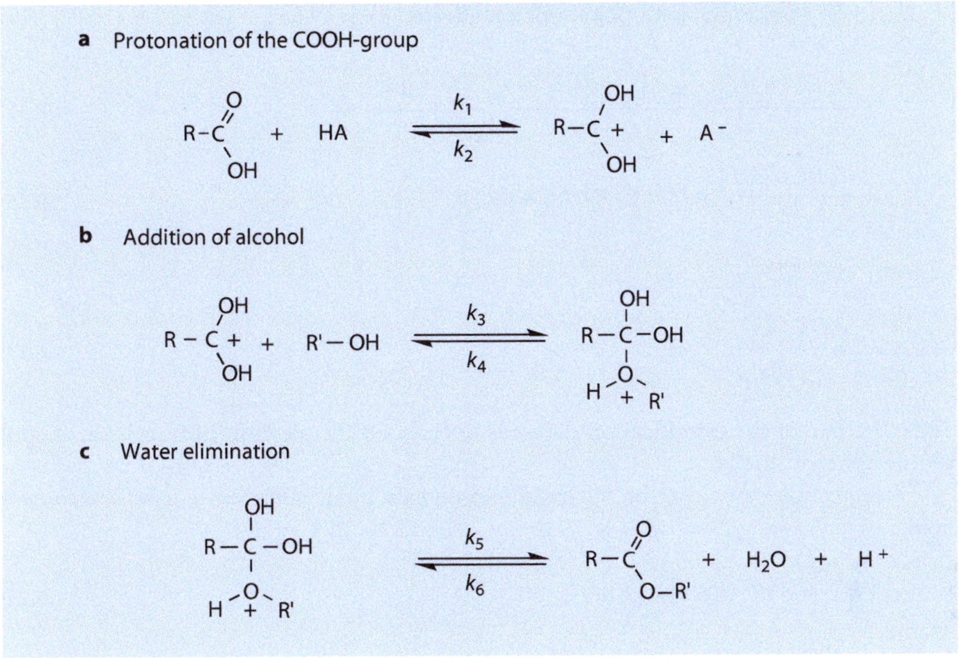

Fig. 8.22 (**a–c**) Mechanism of acid-catalyzed ester formation by reaction of a carboxylic acid with an alcohol

which can be rearranged as

$$\left[RC^{+}(OH)_{2}\right]=\frac{k_{1}}{k_{2}}\cdot\frac{[RCOOH][HA]}{\left[A^{-}\right]} \tag{8.51}$$

The term for the dissociation constant of the acid HA (8.52) can be rearranged to (8.53), and substitution in (8.51) then gives (8.54):

$$K_{\mathrm{HA}}=\frac{\left[H^{+}\right]\left[A^{-}\right]}{[HA]} \tag{8.52}$$

$$\frac{[HA]}{\left[A^{-}\right]}=\frac{\left[H^{+}\right]}{K_{\mathrm{HA}}} \tag{8.53}$$

$$\left[RC^{+}(OH)_{2}\right]=\frac{k_{1}}{k_{2}}\cdot\frac{[RCOOH]\left[H^{+}\right]}{K_{\mathrm{HA}}} \tag{8.54}$$

If (8.54) is combined with (8.49) one obtains (8.55) or (8.56):

$$-\frac{d[RCOOH]}{dt}=\frac{k_1\cdot k_3}{k_2}\cdot\frac{[RCOOH][R'OH]\left[H^+\right]}{K_{\mathrm{HA}}} \tag{8.55}$$

$$-\frac{d[RCOOH]}{dt}=k'[RCOOH][R'OH]\left[H^+\right] \tag{8.56}$$

whereby in (8.57)

$$k'=\frac{k_1\cdot k_3}{k_2\cdot K_{\mathrm{HA}}} \tag{8.57}$$

Thus, the rate of the esterification depends linearly on the alcohol, carboxylic acid, and proton concentrations.

Esterification can either be self-catalyzed or externally catalyzed. For *self-catalysis* it holds:

$$\left[H^+\right]=[RCOOH]=[Kat] \tag{8.58}$$

Furthermore, with precise stoichiometry:

$$[RCOOH]=[R'OH] \tag{8.59}$$

With (8.58) and (8.59), (8.56) becomes

$$-\frac{d[RCOOH]}{dt}=k'[RCOOH]^3 \tag{8.60}$$

For simplification we put

$$[RCOOH]=c \tag{8.61}$$

and (8.60) becomes

$$-\frac{dc}{dt}=k'c^3 \tag{8.62}$$

By separating the variables and subsequent integration one obtains

$$\int_{c_0}^{c_t}\frac{dc}{c^3}=-k'\int_0^t dt \tag{8.63}$$

$$\left[-\frac{1}{2}\left(\frac{1}{c^2}\right)\right]_{c_0}^{c_t}=-\frac{1}{2}\left(\frac{1}{c_t^2}-\frac{1}{c_0^2}\right)=-k't+k'\cdot 0 \tag{8.64}$$

$$\frac{1}{c_t^2}-\frac{1}{c_0^2}=2k't \tag{8.65}$$

It has already been shown that for the conversion p:

$$p = \frac{c_0 - c_t}{c_0} = 1 - \frac{c_t}{c_0} \tag{8.5}$$

or

$$c_t = (1 - p) \cdot c_0 \tag{8.6}$$

Thus, from (8.65):

$$\frac{1}{(1-p)^2 \cdot c_0^2} - \frac{1}{c_0^2} = 2k't \tag{8.66}$$

$$\frac{1}{c_0^2} \cdot \left(\left(\frac{1}{1-p} \right)^2 - 1 \right) = 2k't \tag{8.67}$$

$$\left(\frac{1}{1-p} \right)^2 = 2k't \cdot c_0^2 + 1 \tag{8.68}$$

Using Carothers' equation (▶ Sect. 8.2.1) for the degree of polymerization (8.7) one obtains (8.69):

$$P_n = \frac{1}{1-p} \tag{8.7}$$

$$P_n = \sqrt{2k't \cdot c_0^2 + 1} \tag{8.69}$$

As a first approximation for the case of self-catalysis, the average degree of polymerization increases with the square root of the time:

$$P_n \propto \sqrt{t} \tag{8.70}$$

For *external catalysis*, (8.56) and (8.59) also apply. If $[H^+]$ is replaced by $[Kat]$ then one gets:

$$-\frac{d[RCOOH]}{dt} = k'[RCOOH][R'OH][Kat] \tag{8.71}$$

The catalyst concentration $[Kat]$ and the rate constant k' can be consolidated to a new constant k'':

$$k'[Kat] = k'' \tag{8.72}$$

Thus, assuming the concentration of catalyst remains constant, the third-order, self-catalyzed reaction (8.62) becomes a second-order reaction:

$$-\frac{dc}{dt} = k''c^2 \tag{8.73}$$

By separating the variables and subsequent integration one obtains

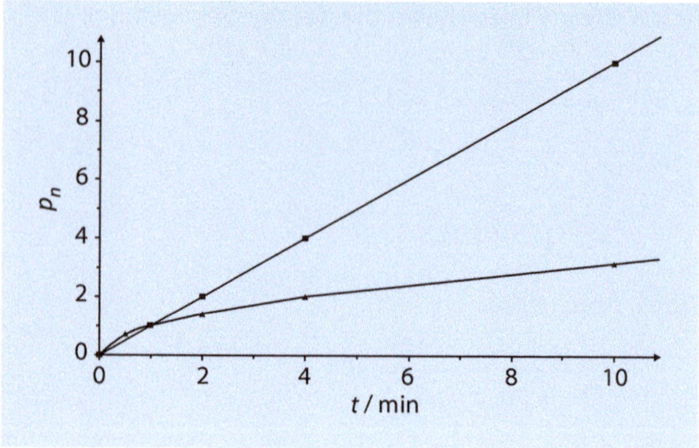

Fig. 8.23 Comparison of an externally catalyzed with a self-catalyzed polycondensation. *Squares* externally catalyzed, *triangles* self-catalyzed

$$-\frac{dc}{c^2} = k''dt \tag{8.74}$$

$$-\int_{c_0}^{c_t} \frac{dc}{c^2} = k''\int_0^t dt \tag{8.75}$$

$$-\left[-\frac{1}{c}\right]_{c_0}^{c_t} = \left[k''t\right]_0^t \tag{8.76}$$

$$+\left(\frac{1}{c_t} - \frac{1}{c_0}\right) = k''(t-0) \tag{8.77}$$

Combining this with (8.6) one obtains

$$\frac{1}{(1-p)\cdot c_0} - \frac{1}{c_0} = k''t \tag{8.78}$$

$$\frac{1}{c_0}\cdot\left(\frac{1}{1-p} - 1\right) = k''t \tag{8.79}$$

$$\frac{1}{1-p} - 1 = k''t \cdot c_0 \tag{8.80}$$

$$\frac{1}{1-p} = k''c_0 t + 1 \tag{8.81}$$

Finally, combining (8.81) with (8.7) one obtains

$$P_n = k''c_0 t + 1 \tag{8.82}$$

Because external catalysis leads to the average degree of polymerization increasing linearly with time, external catalysis is preferred over self-catalysis, where the degree of polymerization is proportional to the square root of time (Fig. 8.23).

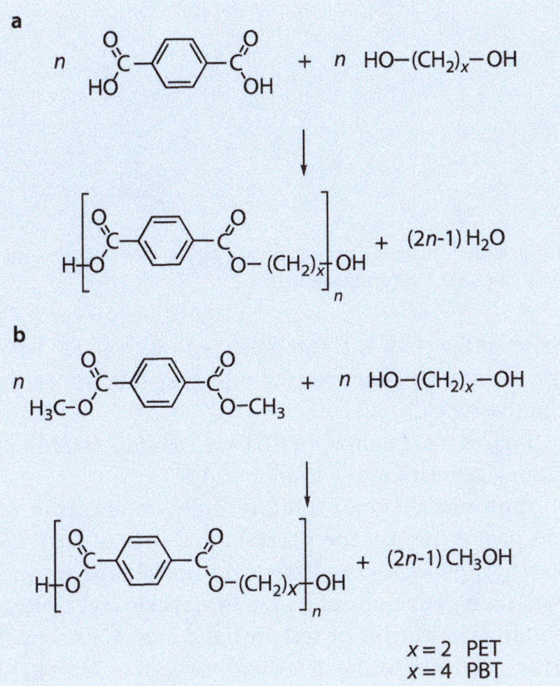

Fig. 8.24 Syntheses of PET and PBT via (**a**) polycondensation of terephthalic acid or (**b**) transesterification of terephthalic acid dimethyl ester with the respective diol

8.5 Typical Polycondensates

In this chapter the most important technological polycondensates, such as polyester, polycarbonates, and polyamides are introduced. Their synthesis, processing, and typical applications are described.

8.5.1 Polyester

Polyethylene terephthalate (PET) and polybutylene terephthalate (PBT) belong to the most important polymers of this class. PET is used as a fiber and in the packaging sector, in particular as plastic bottles. PBT is used for high-quality construction materials, e.g., in the automobile and electrical industries. PET and PBT are available via reaction of terephthalic acid with ethylene glycol or with 1.4-butandiol, respectively, or through transesterification of terephthalic acid dimethyl ester with the respective diol (Fig. 8.24).

Typical trade names for PET are Arnite A (DSM), Rynite (DuPont), and Valox (SABIC).

PET crystallizes very slowly and can be partially crystalline or amorphous depending on the processing conditions.

A partially crystalline PET (PET-C) has medium strength, high stiffness and hardness, but relatively poor impact strength. Areas of application are in electrical (e.g., telephones, computers, switch parts, and spark plugs.) and mechanical engineering (e.g., bearings, gears, and pump parts). Moreover, it is also processed to make zippers, buttons, and furniture fittings. Amorphous PET (PET-A) is transparent at thicknesses up to 5 mm. Above 90 ° C it crystallizes and becomes opaque. Because of its good oxygen and carbon dioxide

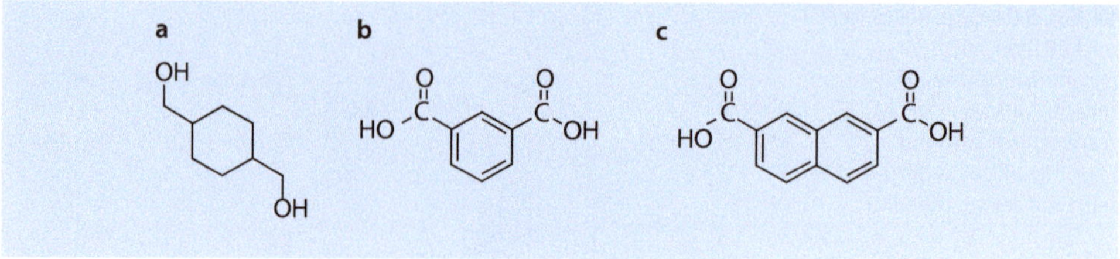

Fig. 8.25 Structural formula of (**a**) cyclohexane-1,4-dimethylol, (**b**) isophthalic acid, and (**c**) naphthalene-2,7-dicarboxylic acid

barrier properties, it is the preferred material for beverage containers. It is also used as a support material for thermal and magnetic typewriter ribbons, photographic film, and adhesive strips.

Typical trade names for PBT are Arnite T (DSM), Crastin (DuPont), Enduran (SABIC), Pocan (Lanxess), and Ultradur (BASF).

8

High dimensional stability, rigidity, abrasion resistance, and low creep deformation properties are the special attributes of PBT. Typical applications include sliding bearings, pulleys, couplings, and household appliances. By inclusion of bulky monomers such as cyclohexane-1,4-dimethylol (cyclohexane-1,4-dimethanol), as well as by partial substitution of terephthalic acid (benzene-1,4-dicarboxylic acid) with isophthalic acid (benzene-1,3-dicarboxylic acid) or naphthalene 2,7-dicarboxylic acid (Fig. 8.25), the crystallization tendency of PET and PBT can be suppressed and their transparency improved. This expands the range of applications to the packaging for hot foods.

8.5.2 Polycarbonates (PC)

Polycarbonates are linear polyesters of carbonic acid. Well-known trade names of polycarbonates are Makrolon (Bayer Material Science), Lexan (SABIC), and Xantar (Mitsubishi). The industrially most important polycarbonate is produced from bisphenol A and phosgene or from diphenyl carbonate (Fig. 8.26).

The reaction of bisphenol A with diphenyl carbonate can be performed as a melt condensation; the crucial reaction here is a transesterification.

The reaction between bisphenol A and phosgene is carried out as an interfacial condensation. In this case, the phosgene is in the organic methylene chloride phase and the bisphenol A is present as the disodium salt in an aqueous alkaline solution. The reaction occurs at the phase boundary. A variant is homogeneous polymerization in pyridine.

By complete replacement of bisphenol A with other bisphenols (Fig. 8.27), important properties of the resulting polycarbonates, such as glass transition temperature and refractive index, can be varied. Polycarbonates from o,o,o′,o′-tetramethyl-bisphenols are very different from those with unsubstituted bisphenol constituents. They have higher glass transition temperatures and lower melt viscosities and hence are easier to process.

Typical fields of application worth mentioning are terminal strips, special plugs, baby bottles, parts for office machines, film and slide cassettes, optical data storage (CD), headlight reflectors, canopies, soundproof walls, and the side and rear windows of trucks and tractors.

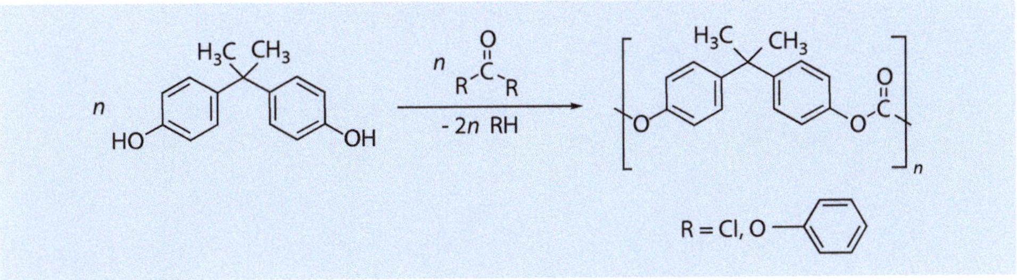

■ **Fig. 8.26** Synthesis of polycarbonates by polycondensation of bisphenol A with phosgene or diphenyl carbonate

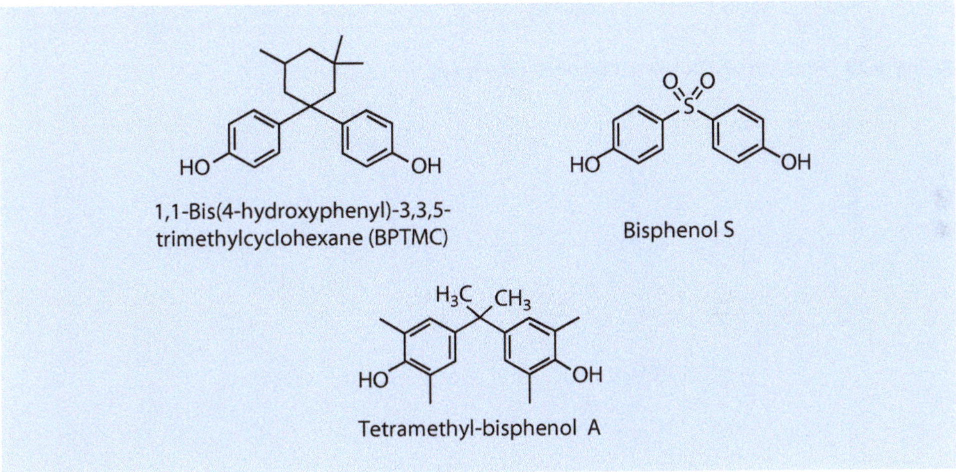

■ **Fig. 8.27** Some alternative bisphenols for the preparation of polycarbonates

8.5.3 Polyestercarbonate (PEC)

By co-condensation of bisphenol A with terephthaloyl dichloride and phosgene, a polyestercarbonate is formed that has a higher heat resistance than the simple polycarbonate from bisphenol A and phosgene possesses. This is because of the partial substitution of CO by –CO–Ar–CO– units (■ Fig. 8.28).

Typical applications for PECs are curling tongs, microwave dishes, headlight reflectors, and bobbins, where operating temperatures of 140–180 °C are possible and at the same time higher strengths are required. Examples of trade names are Lexan PPC (SABIC) and Ardel (BP-Amoco).

8.5.4 Aliphatic Polyamides (PA)

Polyamides can be prepared from a single component AB, e.g., 6-aminohexanoic acid or from the corresponding ε-caprolactam. Alternatively, they can be prepared from two components AA + BB, e.g., 1,6-hexamethylene diamine (1,6-diamino hexane) and adipic acid (butane-1,4-dicarboxylic acid) (■ Fig. 8.29).

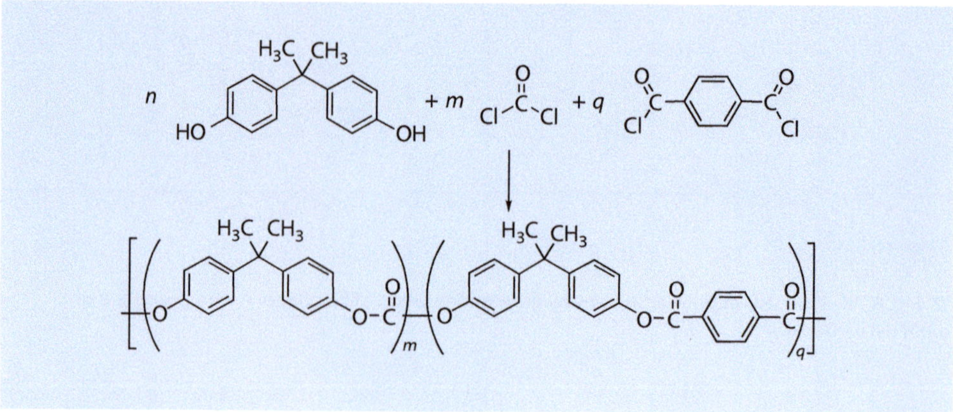

Fig. 8.28 Synthesis of a polyestercarbonate ($m+q=n$)

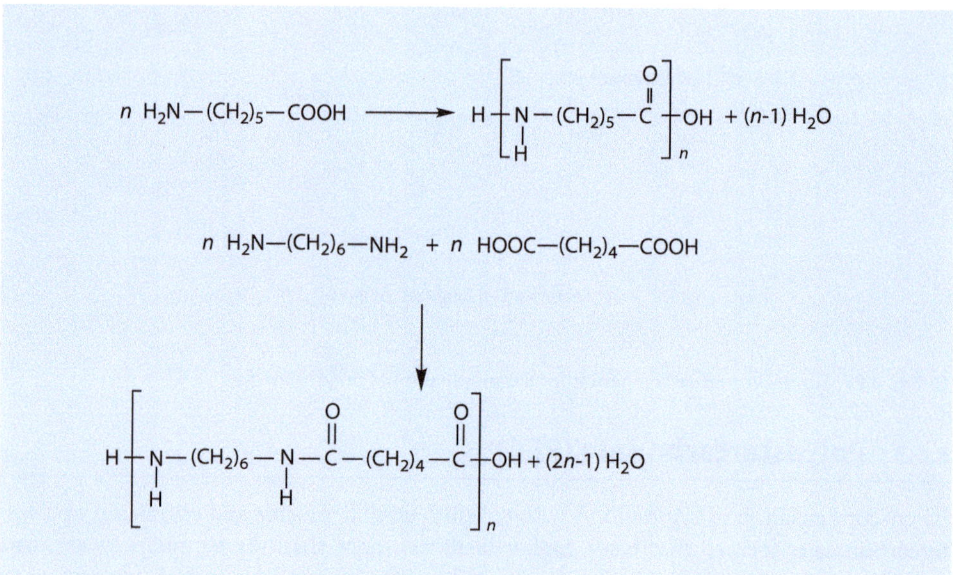

Fig. 8.29 Synthesis of (**a**) polyamide 6 and (**b**) polyamide 6,6

Polyamides are semi-crystalline. Hydrogen bonds between the amide groups of adjacent molecules are the reason for the elevated glass tranition temperatures compared to other non-polar polymers. Their high resilience, low coefficients of friction, and excellent wear properties are valued material properties. Table 8.3 contains selected examples of industrially interesting polyamides.

Polyamides are used in mechanical engineering, e.g., for gears, wheels, and bearings, and in the automotive industry as headlight housings, fuel lines, fenders, and intake pipes. In electrical engineering, they are indispensable as cable and wire sheathing. In the packaging sector, polyamides find a variety of uses as roasting bags, sausage casings, and coatings.

■ **Table 8.3** Selected examples of trade names and manufacturers of various polyamides

Polyamide (PA)	Trade mame (manufacturer)
PA 6	Durethan B (Lanxess), Ultramid B (BASF), Akulon (DSM)
PA 11	Rislan B (Atofina)
PA 12	Grilamid (EMS), Rislan B (Atofina)
PA 4,6	Stanyl (DSM)
PA 6,6	Durethan A (Lanxess), Ultramid A (BASF), Minlon (DuPont)
PA 6,10	Ultramid S (BASF), Zytel (DuPont)
PA 6,12	Vestamid (Evonik), Zytel (DuPont)

The properties of the various types of polyamides are essentially determined by the hydrogen bonds between the amide bonds of neighboring macromolecules. Thus, for example, the melting temperature of elongated PA 6 is 222 °C, whereas the melting temperature of PA 6,6 is almost 40 °C higher, 260 °C. As shown in ■ Fig. 8.30, polymer chains of PA 6,6 in a stretched state can form hydrogen bonds between the chains extremely well. An optimal arrangement can be achieved by moving the chains relative to each other during processing, for example, by drawing, until all the hydrogen bridges "snap" together.

In contrast, the main chain of PA 6 is less symmetric—the mirror plane that is present in PA 6,6 is missing. From ■ Fig. 8.30 it is clear that the optimal formation of hydrogen bonds between the chains is only possible if the chains are facing away from each other. If both chains are facing in the same direction, the number of hydrogen bonds is reduced compared with PA 6.6. However, a complete "turn" of the polymer chains does not occur because of the high viscosity of the system as well as the entanglement of the individual chains (▶ Chap. 6).

8.5.5 Partially Aromatic and Aromatic Polyamides and Polyimides

Polyamides in which aliphatic units are partially replaced by aromatic moieties (■ Fig. 8.31) are known as partially aromatic polyamides. They are superior to purely aliphatic polyamides with respect to their mechanical properties and their heat resistance. Typical commercial products are Ultramid T (BASF) and Durethan T (Lanxess). As a general abbreviation the convention is that, e.g., a polyamide synthesized from 1,6-hexamethylene diamine, adipic acid, and terephthalic acid is called PA 6/6 T.

Simple aromatic polyamides (polyaramides) are formed from aromatic components, e.g., from 1,4-, or 1,3-diaminobenzene and terephthalic acid (or terephthalic acid dichloride) or isophthalic acid (or isophthalic acid dichloride) (■ Fig. 8.32).

These polymers are soluble in concentrated sulfuric acid and can be spun from it. Polyaramide fibers excel in their high rigidity, high impact resistance, high elongation, and good vibration absorption. The best-known applications are in the field of security

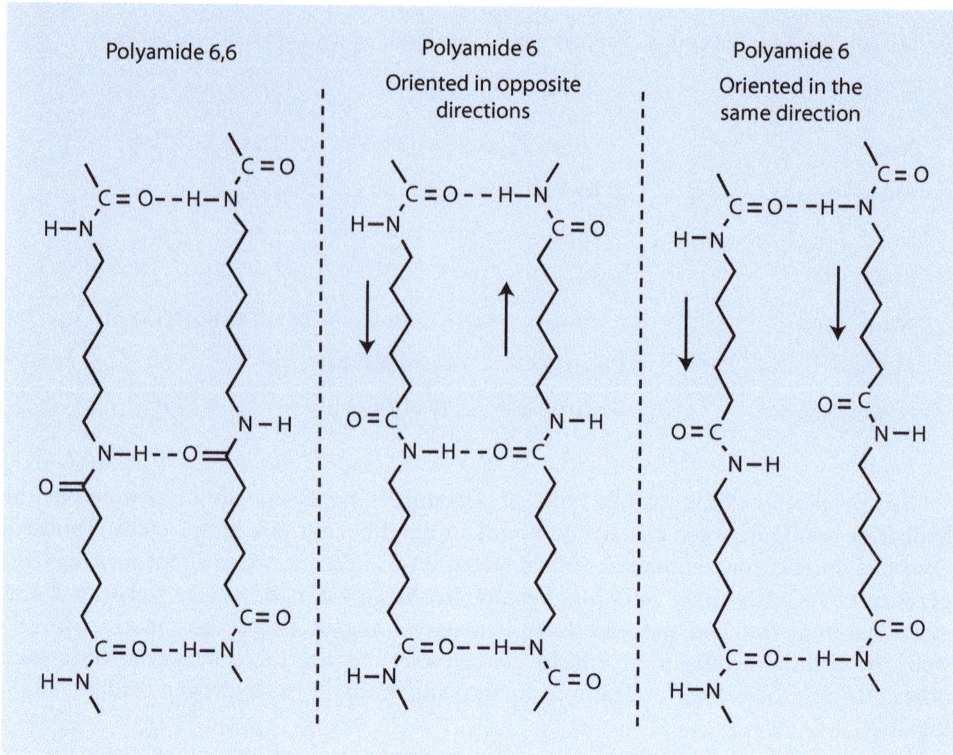

Fig. 8.30 Molecular arrangement and hydrogen bonds in PA 6.6 and PA 6

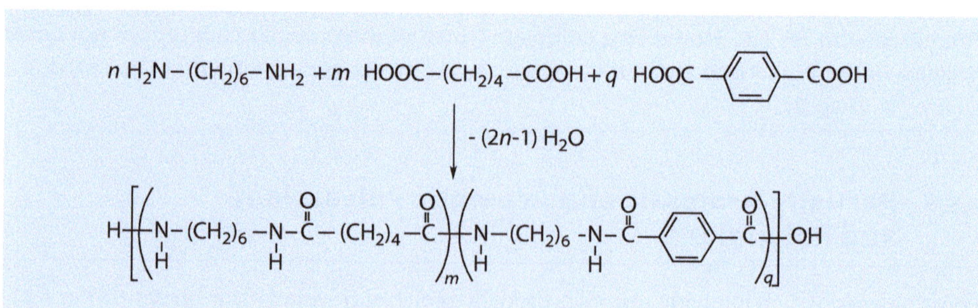

Fig. 8.31 Synthesis of a partially aromatic polyamide ($m+q=n$)

and safety (bullet-proof vests, armor for vehicles), but they are also used as a replacement for asbestos (in brake linings and gaskets) and as reinforcing agents (for fiber-optic cable, rubber materials, and sporting goods).

Because of their structure and the possibility of forming hydrogen bonds, these polymers have lyotropic liquid crystalline properties (Fig. 8.33; ► Chap. 20). This explains the

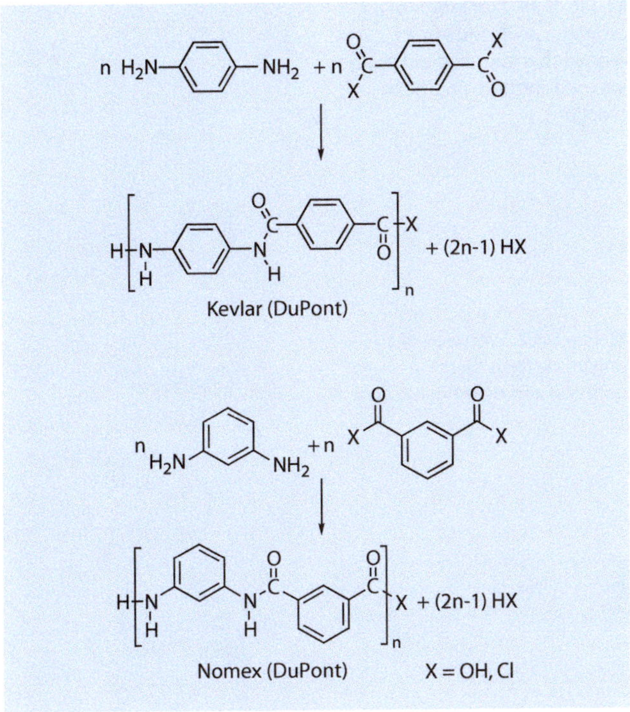

Fig. 8.32 Synthesis of aromatic polyamides

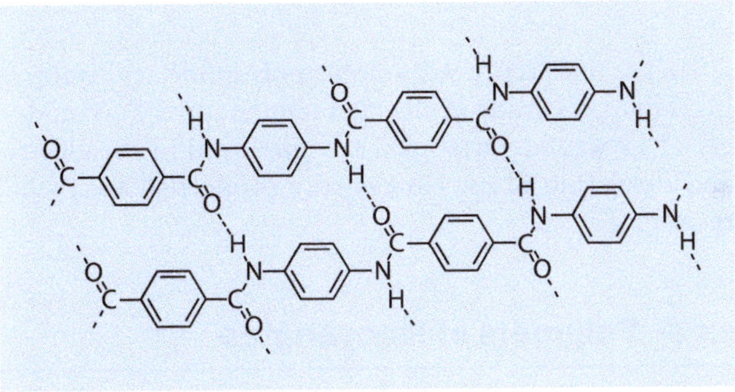

Fig. 8.33 Two neighboring para-polyaramide chains with hydrogen bonds

strong directional dependence of their mechanical properties (high-strength and resistance to bending in the direction of the polymer chains).

The synthesis of aromatic polyamides with an alternating sequence of NH- and CO-functions can be accomplished by employing 4-aminobenzoic acid but such polymers can also be synthesized indirectly via a Schotten–Baumann reaction (Fig. 8.34).

The direct path from 4-aminobenzoic acid to polymer is not useful because of partial decomposition of monomers at higher temperatures.

In contrast to the polyamides, the polyimides contain tertiary nitrogen atoms (imides) in the polymer chain. A polyimide is obtained, for example, by reaction of 1,3-diamino benzene with pyromellitic acid in two steps (Fig. 8.35).

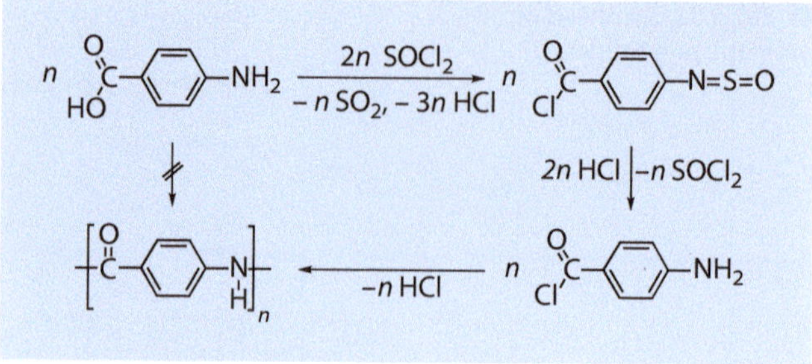

Fig. 8.34 Formation of an aromatic polyamide from 4-aminobenzoic acid indirectly via the Schotten–Baumann reaction

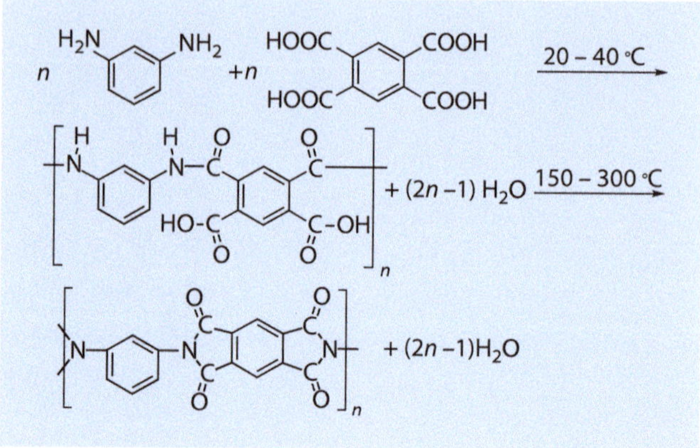

Fig. 8.35 Synthesis of a polyimide from diamino benzene and pyromellitic acid

In the first step, a still soluble polyamide acid is formed, which is converted into the insoluble polyimide at elevated temperature. Polyimides, which are available in a variety of structural variations, are employed in the electronic industry as flexible cables and insulating films. Noteworthy properties are their fire resistance and good heat resistance.

8.5.6 Polymers of Isocyanates

Typical reactions of isocyanates which can be used for a stepwise synthesis of polymers are summarized in Fig. 8.36.

The most commonly used diisocyantes are shown in Fig. 8.37.

From the reaction of trifunctional isocyanates with di- or trifunctional alcohols, amines, or carboxylic acids, cross-linked, i.e., insoluble polymers are formed.

The diisocyanate polymerization reactions are characterized by a wide variety of available components (monomers and oligomers). The reactions often occur at room temperature and without the formation of by-products. By adjusting the stoichiometry, the molar masses can be controlled; the use of monomers with more than two functional groups leads to the formation of crosslinking points (▶ Sect. 8.3.4). The main areas of application are highly elastic foams (mattresses), rigid foams, and rigid and flexible molded parts with a compact outer skin (window frames, skis, steering wheels, shoe soles, etc.).

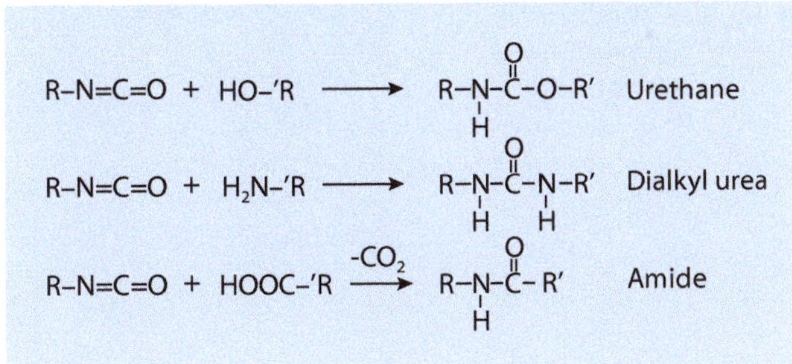

Fig. 8.36 Reaction of isocyanates with alcohols, amines and carboxylic acids

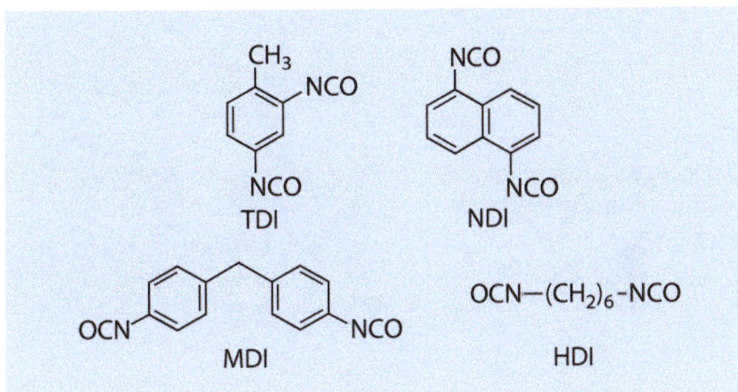

Fig. 8.37 Selection of commonly used diisocyanates (TDI: 2,4-tolyl diisocyanate (often a mixture of 2,4- and 2,6-isomers), NDI: 1.5-naphthyl diisocyanate, MDI: 4,4'-methylene diphenyl isocyanate, HDI: 1,6-hexamethylene diisocyanate)

8.5.7 Polysiloxanes

The complete hydrolysis of dimethyl dichlorosilane results in a mixture of cyclic compounds and hydroxyl-terminated oligomeric dimethyl siloxanes (Fig. 8.38).

By varying the hydrolysis conditions, the proportion of the cyclic compounds ($n=3$, $n=4$) can be controlled. The starting materials for the so-called equilibrium polymerization are cyclic compounds (▶ Chap. 12), whereas polycondensation is only possible with hydroxyl oligomers.

By adding a monchloro trialkyl silane, the molar mass of these polymers can be limited (Fig. 8.39).

α,ω-Dihydroxy polysiloxanes can be crosslinked by addition of trichloroalkyl silane or tetrachlorosilane (Fig. 8.40).

By varying the proportions of the reactants and by using different R, the properties of the networks can be widely varied.

Polysiloxanes are thermally stable and UV resistant. Because of the low intermolecular forces and the rotational capability of the Si-O-Si bond, they have a low glass transition temperatures.

Industrially used polysiloxanes are available as low-viscosity liquids, oils, fats, resins, and rubbers. Oils and fats are used as dielectrics, hydraulic oils, lubricants, and water repellents. Resins are required, for instance, for electrical insulation and protective coatings. Molded and extruded products as well as wire and cable insulation are made from silicone rubber.

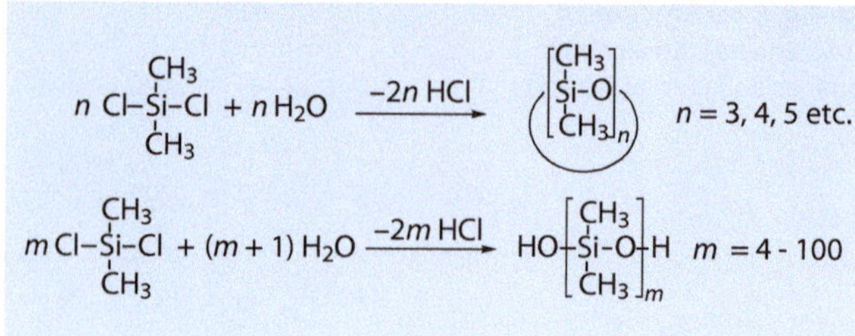

Fig. 8.38 Hydrolysis of dimethyl dichlorosilane

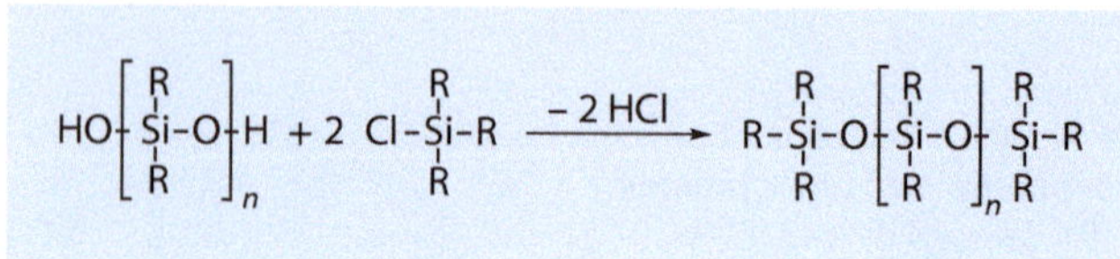

Fig. 8.39 Regulation of molar mass with monofunctional chloroalkyl silanes

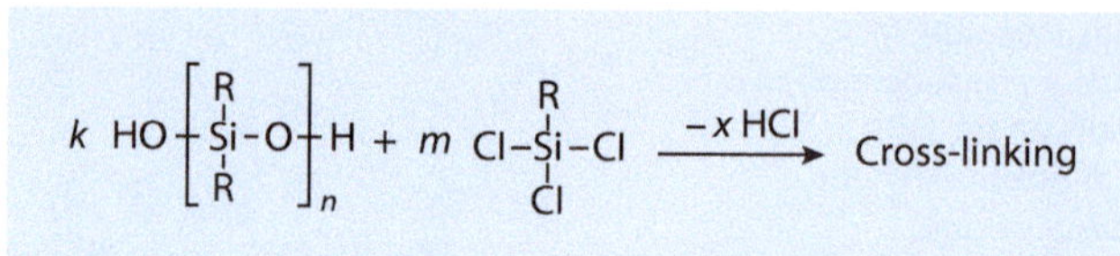

Fig. 8.40 Crosslinking of polysiloxanes with trichloro alkyl silane

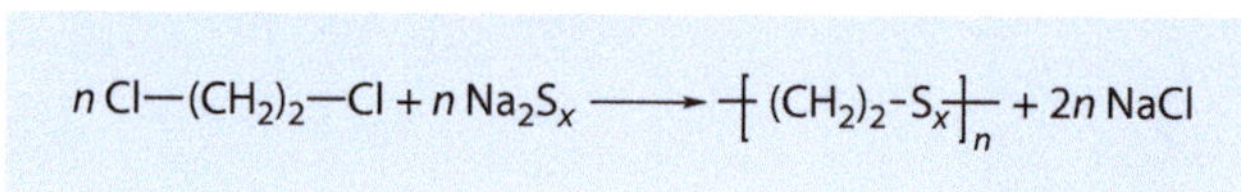

Fig. 8.41 Synthesis of a polyalkylene polysulfide

8.5.8 Selected Specialty Polymers

Here, examples of polymers that are also synthesized via a step-growth mechanism and which have achieved a certain economic importance are described. These include polyalkylene polysulfides, polyarylene sulfides, polysulfones, polyether ketones, and polyphenylene oxides.

8.5.8.1 Polyalkylene Polysulfides

The reaction of aliphatic dihalides with alkali metal polysulfides results in polymers with rubber-like properties. A typical reaction example is presented in Fig. 8.41.

Subsequently, when some of the S_x groups are split by a reducing reaction, liquid oligomers are formed that have SH groups at both ends and are can therefore be described as telechelic[2] (Fig. 8.42).

In the telechelic molecules, S_x bonds remain that have not been split. The degree of division and thus the molar mass of the telechelic polymers can be controlled by the amount of reducing agent used.

By oxidation, for example with PbO_2 or quinone, the SH groups can be converted back into polysulfide bonds. Such materials, which have become known by their trade name, Thiokol, are used in large quantities for seals in the building industry, the automotive

2 Telechelics are oligomers with well-defined functional end groups, with which further chemical reactions are possible.

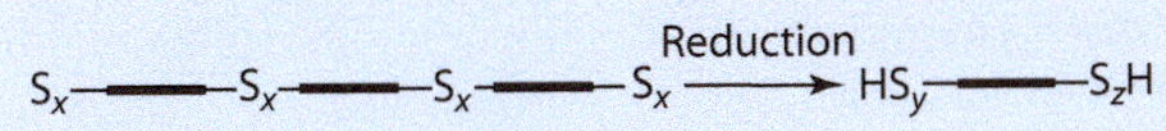

■ **Fig. 8.42** Synthesis of SH-terminated telechelic polymers by splitting the S_x groups

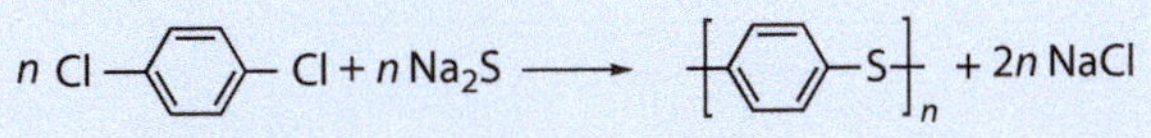

■ **Fig. 8.43** Synthesis of polyphenylene sulfide

industry, and shipbuilding, where they are greatly appreciated for their oxygen and solvent resistance.

8.5.8.2 Polyarylene Sulfides

By reacting 1,4-dichlorobenzene in *N*-methyl pyrrolidone with Na_2S, polyphenylene sulfide (■ Fig. 8.43) can be obtained.

This polymer has become known by its trade name, Ryton (Chevron Phillips), and is very popular because of its exceptional mechanical properties, excellent dimensional stability, and insulating properties. The main fields of application are heat- and corrosion-resistant metal coatings and glass fiber reinforced injection molded parts.

8.5.8.3 Polysulfones

Polysulfones are thermoplastic construction materials that have a high strength, hardness, impact resistance, and chemical resistance. Their main fields of application are in the aerospace and electronics industries and in the household sector.

They can also replace die castings of zinc, bronze, and lead. Polysulfones can be produced via nucleophilic and electrophilic substitution reactions. ■ Figure 8.44 shows the nucleophilic attack of deprotonated bisphenol A on 4,4′-dichlorodiphenyl sulfone, i.e., one of the monomers already contains the sulfone group (■ Fig. 8.44).

Alternatively, polysulfones can be obtained by electrophilic substitution. Here, the sulfone group is formed by a Friedel–Crafts reaction of a sulfonic acid dichloride (■ Fig. 8.45).

Both synthetic methods are very varied because of the possible reactants.

8.5.8.4 Polyether Ketones

Entirely analogous to the polysulfones are the polyether ketones (PEK), which are available either by nucleophilic or by electrophilic substitution reactions (■ Fig. 8.46).

Polyether ketones are resistant to many organic and inorganic chemicals. Applications of PEK can be found in the automotive and aerospace industries and for high voltage as well as medical engineering (reusable, sterilizable, biocompatible, and transparent to X-rays).

8.5.8.5 Polyphenylene Oxides

Polyphenylene oxides (PPO) can be obtained by the oxidative dehydrogenation of 2,6-disubstituted phenols (■ Fig. 8.47).

Poly-2,6-dimethyl-1,4-phenylene oxide is completely miscible with polystyrene. This PPO and PPO/PS blends are used for electrical components. PPO are ideal materials for impellers, valves and flow meters. Surgical devices made of PPO can be easily sterilized in superheated steam. Additionally, PPO is a base material for heat-resistant connectors, insulators, and lamp bases.

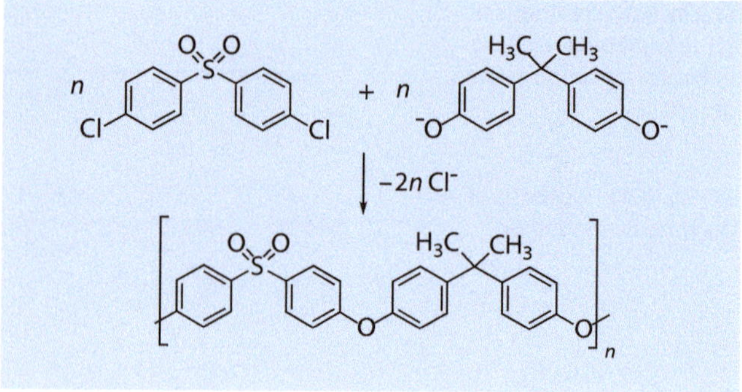

Fig. 8.44 Synthesis of a polysulfone by nucleophilic substitution

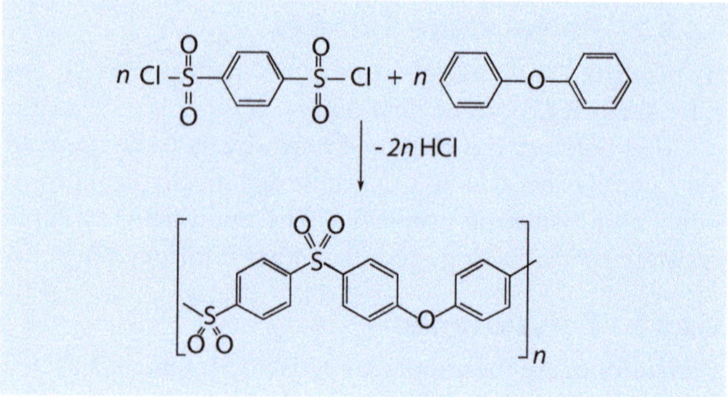

Fig. 8.45 Formation of a polysulfone by electrophilic substitution

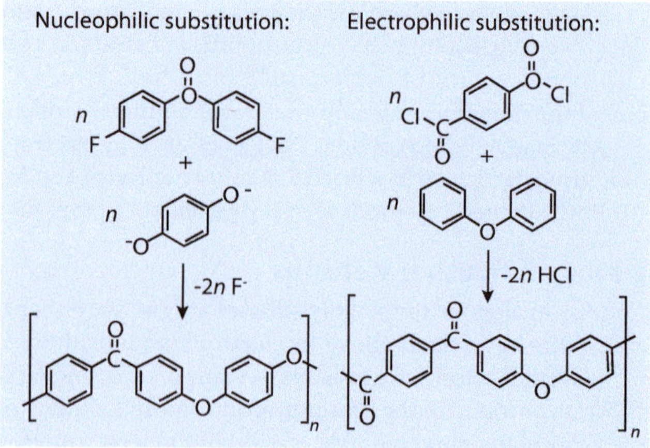

Fig. 8.46 Synthesis of polyether ketones

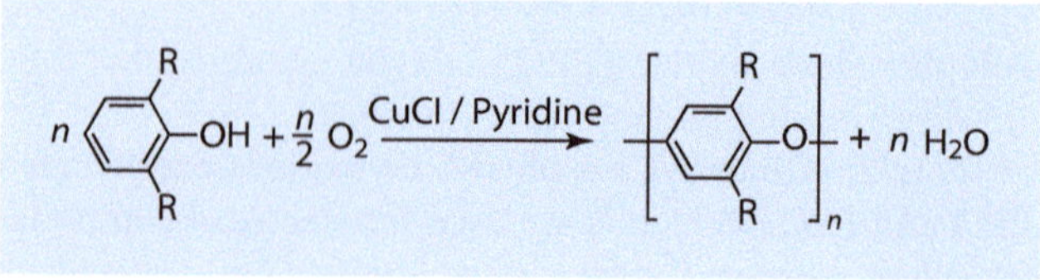

Fig. 8.47 Synthesis of polyphenylene oxide by oxidative dehydrogenation

8

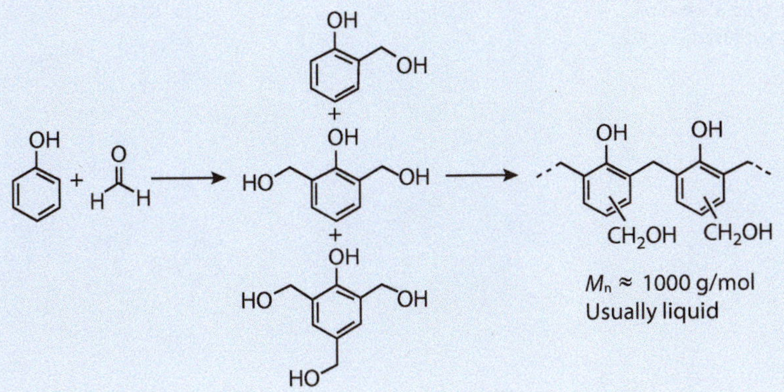

Fig. 8.48 Synthesis of resols (in a basic milieu, ratio of phenol : formaldehyde = 1:1.5)

8.6 Industrially Relevant Crosslinking Systems

As described in ► Sect. 8.3.4, polymers that are built up gradually can be cross-linked by using monomers with an average functionality of $\overline{f}$ > 2, allowing them to be converted into insoluble three-dimensional networks. In this section, systems in which the cross-linking is carried out in two stages by reaction of a prepolymer (first stage) with a cross-linking agent (second stage) are discussed.

8.6.1 Phenolic Resins

Condensation of phenol with formaldehyde can yield low molar mass, soluble, often still liquid intermediate materials (prepolymers). By further reactions (e.g., by heating), these can be converted into networks that are insoluble and do not melt. These materials are among the oldest industrially used polymers. The condensation of phenol with formaldehyde in aqueous solution is pH-dependent. The prepolymers obtained in an alkaline medium are called *resols* (cross-linked products are called resins or *resites*) (Fig. 8.48). The prepolymers prepared in acidic solution are called *Novolac* (Fig. 8.49).

In an acidic environment, an excess of phenol is necessary, otherwise the polymerization cannot be controlled and no stable and storable intermediates are obtained. Novolac is usually solid at room temperature.

Cross-linking of resols is accomplished by heating. At 130 °C methylol phenols simply dehydrate (Fig. 8.50).

By contrast, at 150 °C, methyl groups are predominantly formed as formaldehyde and water are eliminated, (Fig. 8.51).

The crosslinking of Novolac is achieved by adding crosslinking agents, such as hexamethylene tetramine. This hydrolyzes to dimethylol amine and reacts with Novolac as shown in Fig. 8.52.

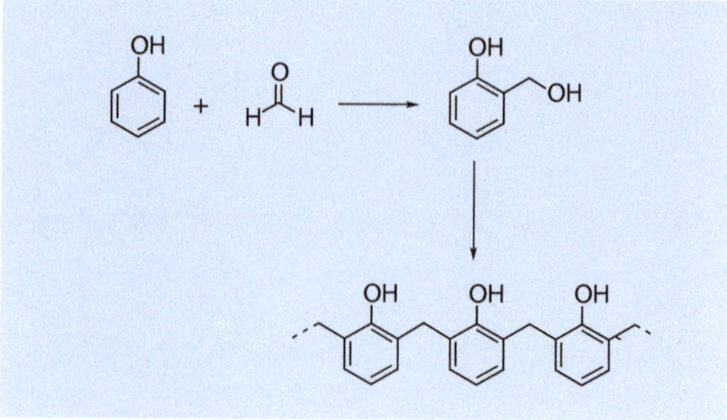

Fig. 8.49 Synthesis of Novolac (at pH ≤ 3, ratio of phenol:formaldehyde = 1:0.8)

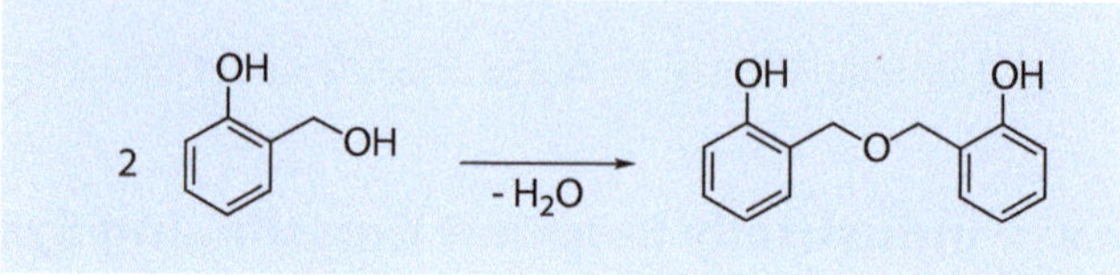

Fig. 8.50 Elementary step in the cross-linking of a resol by dehydration ($T = 130\ °C$)

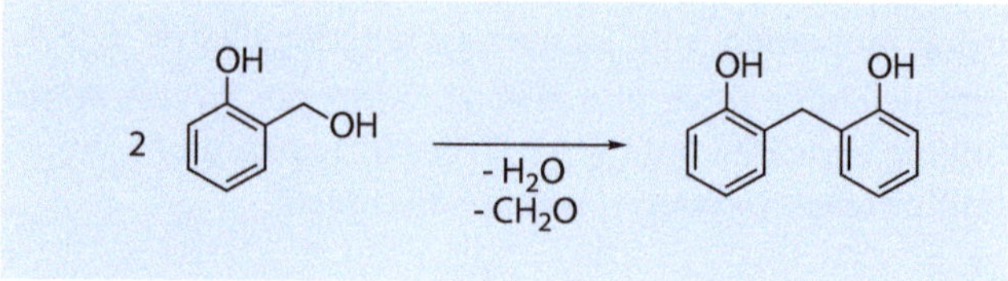

Fig. 8.51 Elementary step in the cross-linking of a resol by elimination of water and formaldehyde ($T = 150\ °C$)

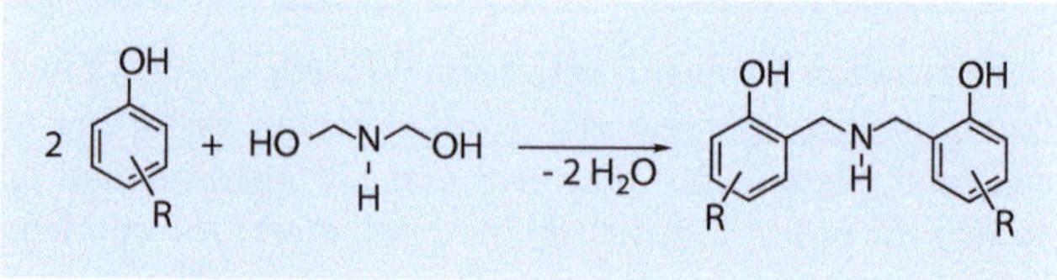

Fig. 8.52 The cross-linking of Novolac with dimethylol amine

Phenolic resins are used mainly as molding compounds with fillers. The hardened (cross-linked) resins are tough construction materials with good resistance to stress cracking. Webs of paper or tissues impregnated with phenolic resins can be processed into laminates.

8.6.2 Urea Resins

The reaction products of formaldehyde with urea (carbamide) are called urea resins. This reaction, which is a Mannich reaction, is pH-dependent.

In a neutral to slightly alkaline environment, well-defined products are created (Fig. 8.53).

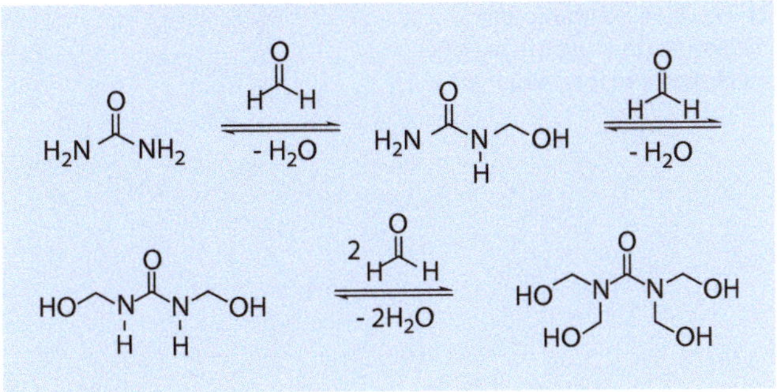

■ **Fig. 8.53** Conversion of urea with formaldehyde to urea resin precursors in a neutral to slightly alkaline environment

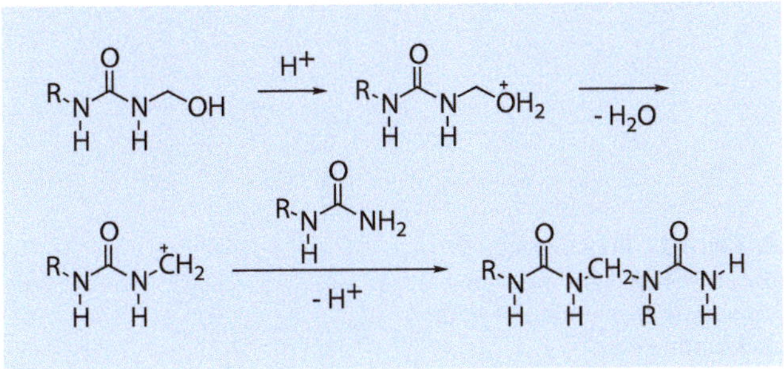

Fig. 8.54 Individual steps of the acid-catalyzed cross-linking of urea resin precursors

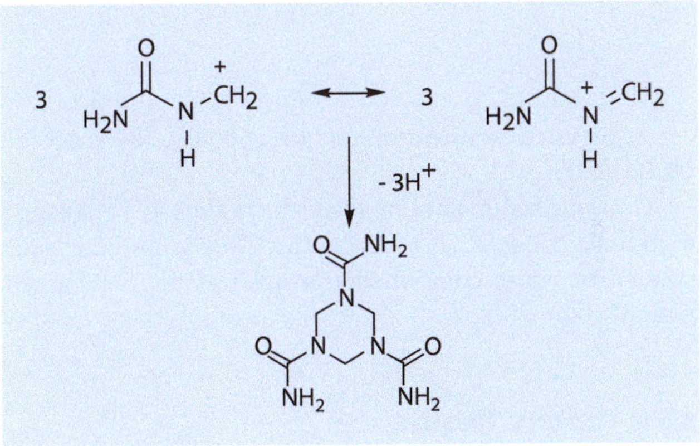

■ **Fig. 8.55** Trimerization of urea in an acidic environment

In an acidic environment, cross-linked structures rapidly form (■ Fig. 8.54).

The cross-linking can be continued to incorporate all the -NH_2 groups. In addition, the carbenium ion can trimerize (■ Fig. 8.55).

Thereafter, the NH_2 groups can react further (as described above) and cross-link. The urea resins are similar in their properties and their applications to phenolic resins.

8.6.3 Melamine Resins

Melamine (1,3,5-triaminotriazine) and formaldehyde react stepwise to yield intermediates which have similar structures to those of urea resins (■ Fig. 8.56).

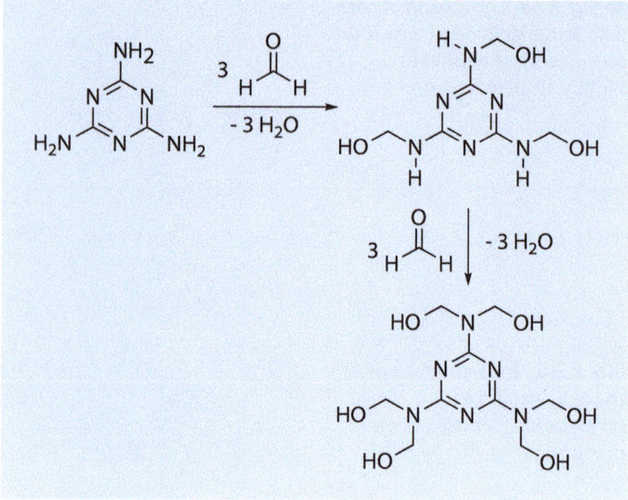

Fig. 8.56 Formation of melamine resin precursors from melamine and formaldehyde

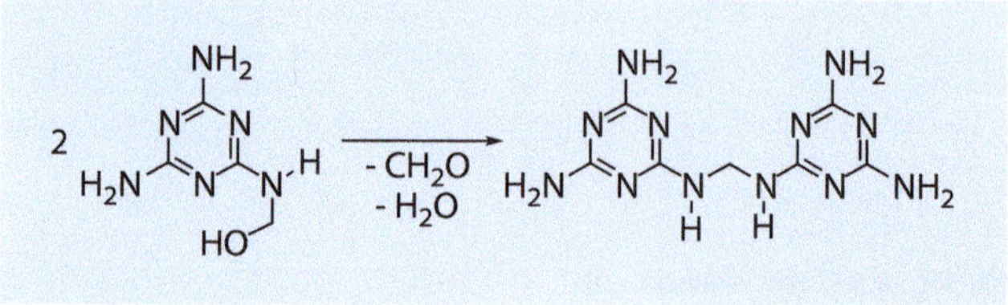

Fig. 8.57 Principle step in the cross-linking of melamine resins with elimination of water and formaldehyde

At elevated temperatures (140–160 °C), the molecules bridge and form a network (Fig. 8.57).

Cross-linked melamine resins have slightly better mechanical and thermal properties than both phenolic and urea resins. They are used especially for the production of light-colored or white components, instead of phenolic and urea resins which have a dark, natural color.

8.6.4 Epoxy Resins

Epoxy resins are produced by the reaction of low molar mass epoxy compounds with phenols (Fig. 8.58).

With a small excess of epichlorohydrin (epichlorhydrin:phenol < 2:1), higher molar mass epoxy resins are formed via multiple repetition of the reactions shown in Fig. 8.58 (Fig. 8.59).

These prepolymers can be converted into networks either by stoichiometric reactions with acid anhydrides (heat or latent curing) or with primary or secondary amines (cold curing), but also catalytically with tertiary amines.

The first step of the cross-linking of the prepolymers is an esterification of the –OH group with an acid anhydride (Fig. 8.60a). In the second step, the resulting –COOH function opens an epoxide ring and connects two chains together in the intermolecular course of the reaction (Fig. 8.60b).

■ **Fig. 8.58** Condensation reaction of epichlorohydrin with bisphenol A

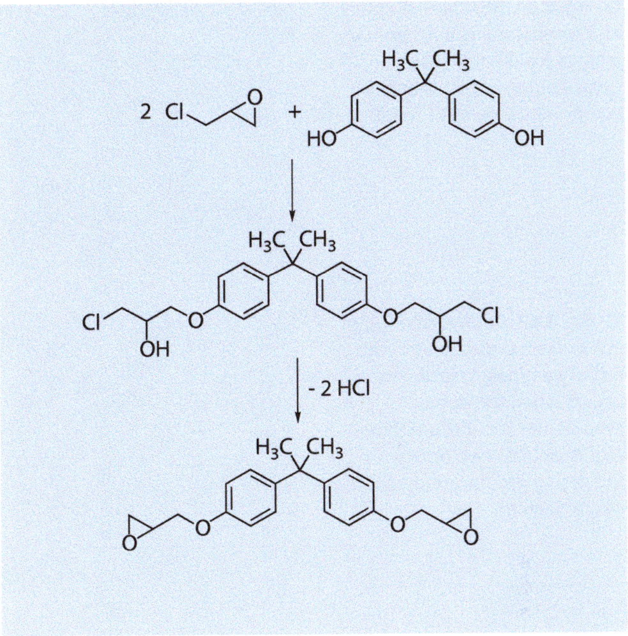

■ **Fig. 8.59** Structural formula of epoxy resin prepolymers

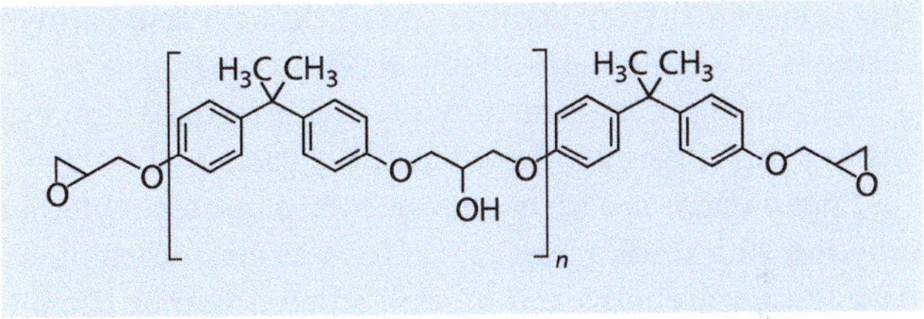

■ **Fig. 8.60** Mechanism of crosslinking of epoxy resin prepolymers with carboxylic acid anhydrides (heat curing)

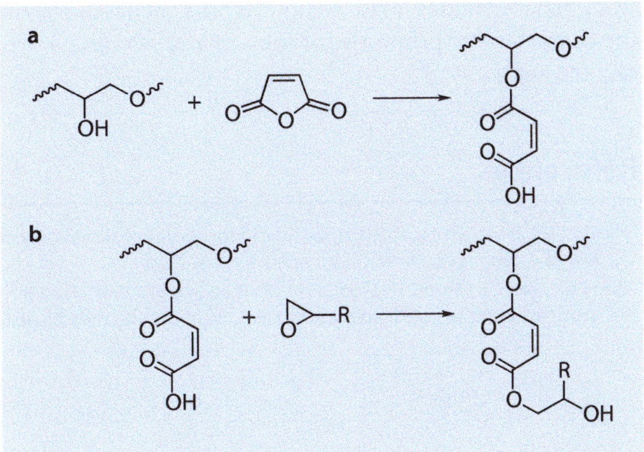

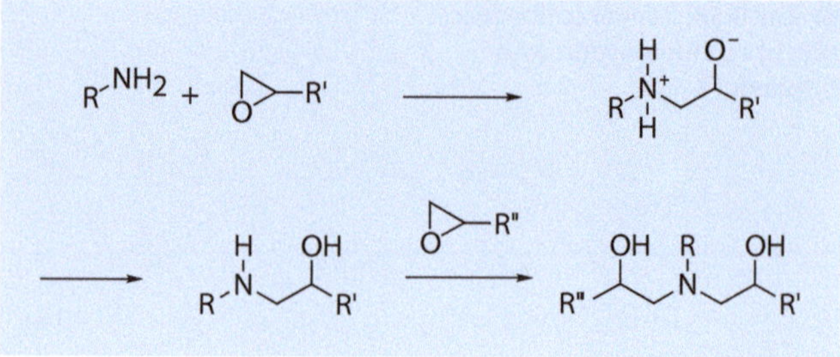

Fig. 8.61 Reaction of epoxy resin prepolymers with primary amines (cold hardening); Cross-linking occurs when the components are multifunctional

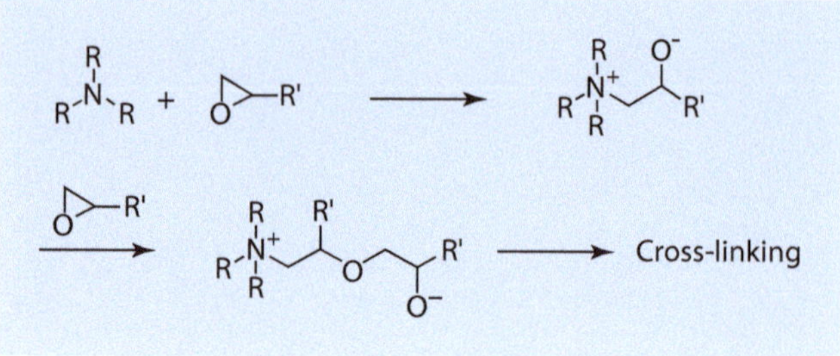

Fig. 8.62 Catalytic curing of epoxy resin prepolymers with tertiary amines. Crosslinking occurs when the epoxy prepolymer is multifunctional, that is, at least two epoxy functional groups are present per molecule

The first step of the curing of the prepolymers with primary amines is the opening of the oxirane ring. In the second step, the resulting secondary amine opens another oxirane ring (most likely from another chain). As with acid hardening (Fig. 8.60), primary amine curing is a stoichiometric reaction (Fig. 8.61).

The curing of oxirane polymers with tertiary amines is a catalytic process. The basic steps of this curing are shown in Fig. 8.62.

Epoxy resins are high-quality, indispensable materials, which are often used in conjunction with reinforcing agents such as glass, carbon, or aramid fibers. The composites have tensile strengths and moduli greater than or equal to steel alloys. Areas of application are in electrical engineering (carrier material for printed circuit boards, printed circuits), the automotive industry (structural and body components), and sports articles (e.g., bicycle frames, skis, hockey, tennis rackets). By varying the starting components, the characteristic properties of these materials can be precisely adjusted to the particular requirements.

References

Flory PJ (1952) Molecular size distribution in three-dimensional polymers, IV Branched polymers containing A-B-Bf-1-type units. J Am Chem Soc 74:2718–2723

Turner SR, Voit BI, Morey TH (1993) All-aromatic hyper branched polyesters with phenol and acetate end groups: syntheses and characterization. Macromolecules 26:4617–4623

Radical Polymerization

S. Koltzenburg et al., *Polymer Chemistry*, DOI 10.1007/978-3-662-49279-6_9

This chapter introduces the fundamental concepts of radical polymerization. Typical monomers, initiators, and transfer and termination reagents are discussed. Furthermore, the kinetic equations, the degree of polymerization, and molar mass distribution are derived.

9.1 Mechanism

During radical polymerization, initially (start) a radical (species with a single electron), which arises from, for example, the decay of peroxides or azo compounds (■ Fig. 9.9), adds to the C=C double bond of a monomer, resulting in a new radical extended by a monomer unit (■ Fig. 9.1). If the radical that was extended by a monomer unit is able to add an additional monomer and to form a macro radical, it is referred to as *chain growth* (■ Fig. 9.2).

This chain reaction is continued until two radicals meet and prohibit the addition of another monomer (■ Figs. 9.3 and 9.13d$_1$) by, for example, forming a covalent bond (*combination*). Alternatively, the two radicals can be individually deactivated by *disproportionation* (■ Fig. 9.13d$_2$).

If the radical character of a macro radical is transferred to another moiety, and if this then becomes able to add an additional monomer, it is referred to as a transfer reaction (■ Fig. 9.4).

These reaction types are discussed in detail in the following sections.

9.1.1 Typical Monomers

■ Figure 9.5 shows the general structure of radically polymerizable monomers, R^1, R^2, and R^3, which are mostly H (■ Table 9.1). R^4 are usually substituents that stabilize the growing radicals.

■ **Fig. 9.1** Reaction of a radical $R^\bullet$ with a monomer $H_2C{=}CHR'$

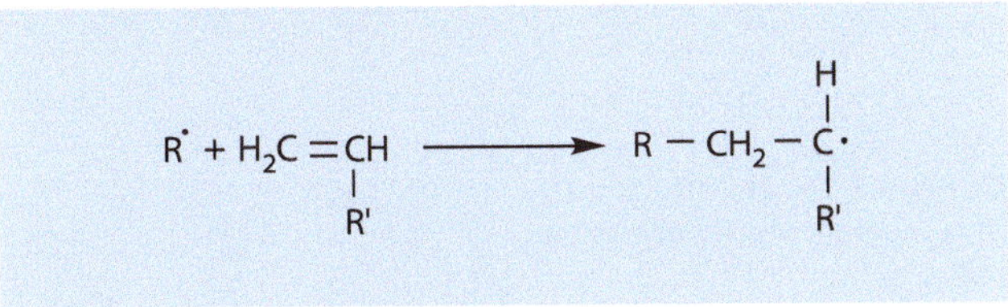

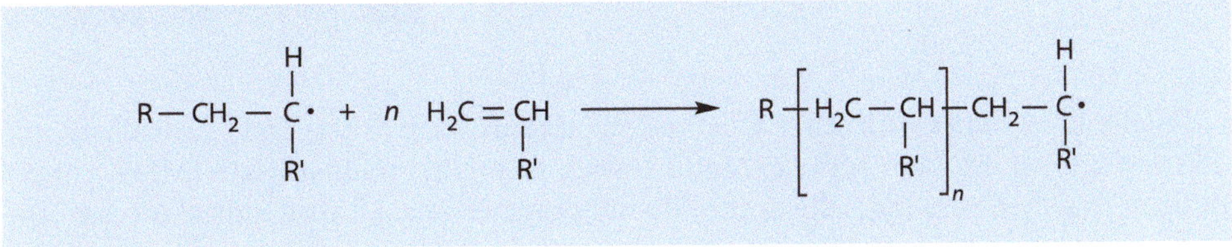

■ **Fig. 9.2** Formation of a macro-radical

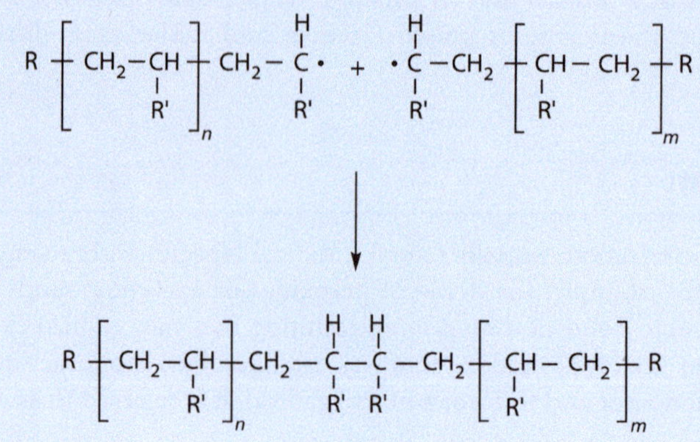

Fig. 9.3 Termination of chain growth by radical combination

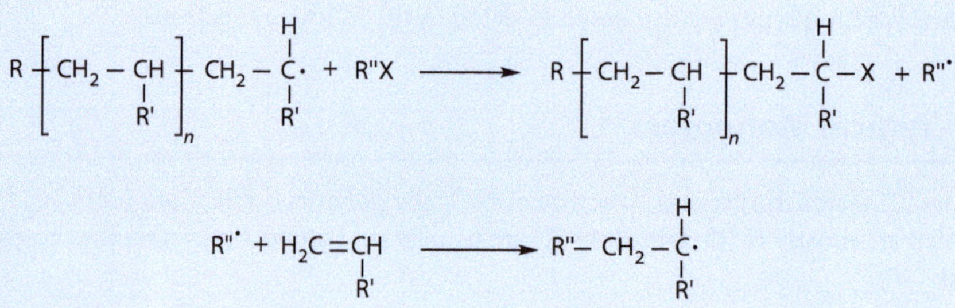

Fig. 9.4 Example of chain transfer

Fig. 9.5 General structures of monomers which polymerize radically

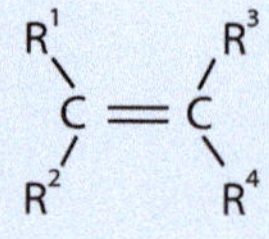

Generally, alkenes with other substitution patterns, such as 1,1- and 1,2-disubstituted alkenes, are , apart from some exceptions such as methyl methacrylate (MMA), more difficult to polymerize radically or they do not polymerize at all via a radical mechanism.

Table 9.1 Typical examples for radically polymerizable monomers of the structure $H_2C{=}CHR'$

Monomer	*R'*	Polymer
Ethene	$-H$	$-[CH_2-CH_2]_n-$
Styrene	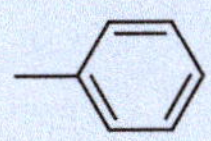	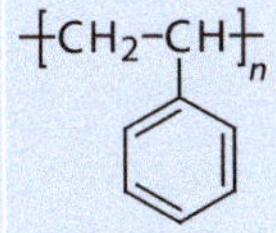
Acrylonitrile	$-C{\equiv}N$	$-[CH_2-CH(CN)]_n-$
Acrylic acid methyl ester	$-C(=O)-O-CH_3$	$-[CH_2-CH(C(=O)-O-CH_3)]_n-$
Acrylic acid	$-C(=O)-OH$	$-[CH_2-CH(C(=O)-OH)]_n-$
Vinyl chloride	$-Cl$	$-[CH_2-CH(Cl)]_n-$
Vinyl acetate	$-O-C(=O)-CH_3$	$-[CH_2-CH(O-C(=O)-CH_3)]_n-$

9.1.2 Radical Sources

Radical polymerizations are triggered by reactive species that have an unpaired electron (radicals). Such radicals can result from, for instance, the influence of light (visible or ultraviolet light) on neutral compounds (Fig. 9.6). Another means of creating radicals involves the photochemical decay of special substances (*photoinitiators*) (Figs. 9.7 and 15.7).

Radicals can also be created by γ- and β-radiation (Fig. 9.8). In such a system, not only radicals but also anions and cations are created, and the respective chain growth reactions can proceed either via radicals, anions, or cations, or the different growing mechanisms can compete with each other.

Compounds that decompose to yield radicals during moderate heating (50–100 °C) are also frequently used as radical sources. The best-known radical sources are organic peroxides and azo compounds; industrially, organic peroxides are most often used (Fig. 9.9).

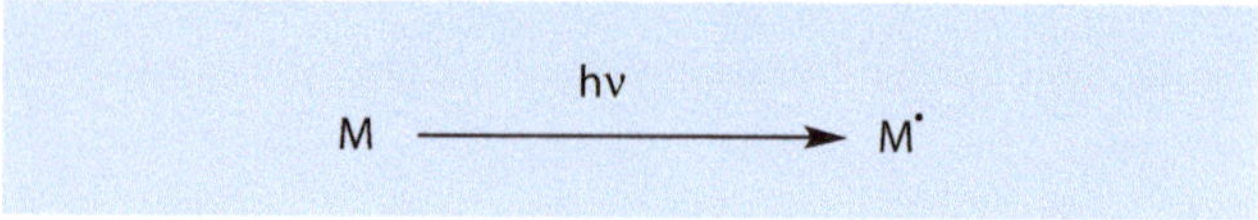

Fig. 9.6 Photochemical activation of a monomer

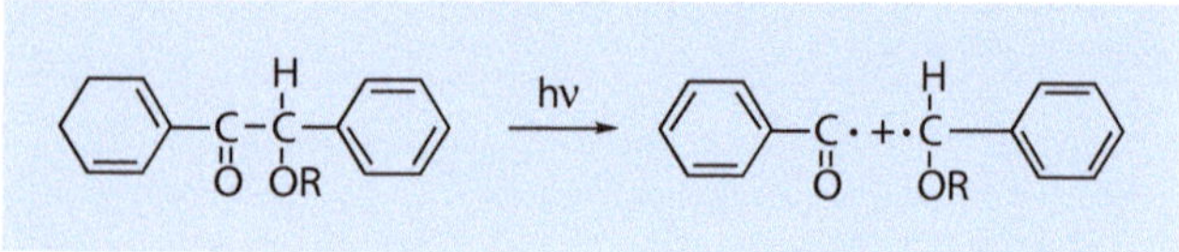

Fig. 9.7 Photochemical decomposition of benzoin ether

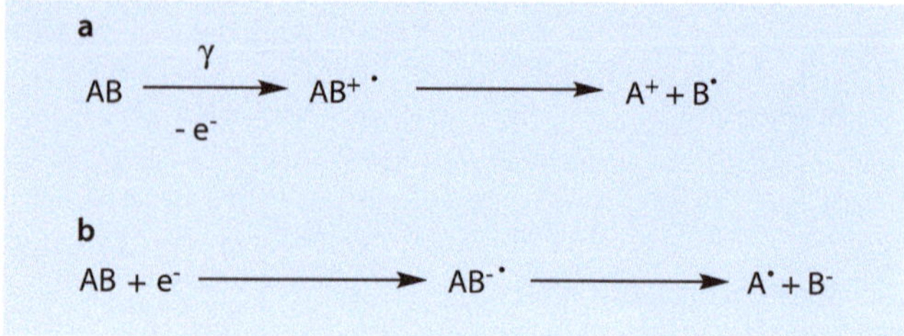

Fig. 9.8 Creation of radicals with high energy radiation. (**a**) γ-Radiation. (**b**) β-Radiation

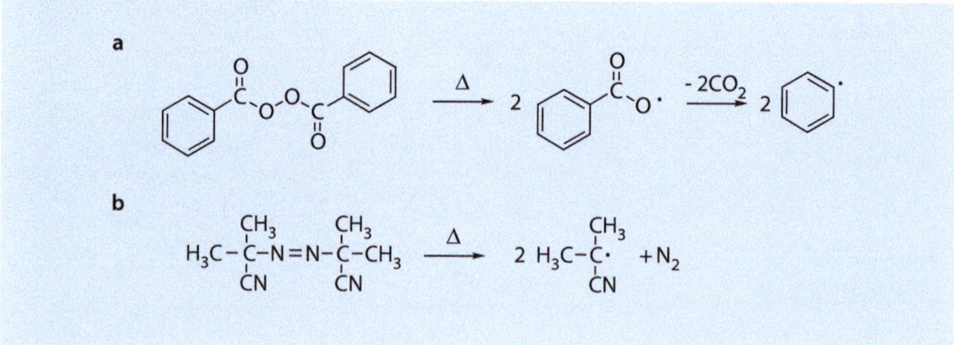

Fig. 9.9 Radicals from (**a**) dibenzoyl peroxide and (**b**) azo-bis-isobutyronitrile (AIBN)

Furthermore, redox reactions that proceed via radical intermediates can be used to initiate a radical polymerization (Fig. 9.10). In the literature, particularly in the patent literature, various combinations of oxidizing agents (e.g., peroxides, chlorates, hypochlorites, and permanganates) and reducing agents (e.g., sulfites, thiosulfates, sulfinic acid, and hydrazine) are suggested for this purpose. The major advantage of such redox systems is their ability to function at lower temperatures. Such systems have become irreplaceable for polymerizations where water is either the solvent or the dispersant.

It is also possible for monomers to polymerize simply by warming, known as *thermal polymerization*. This is, however, restricted to two monomers, namely, styrene and MMA. The list of thermally polymerizable monomers was originally longer. Upon closer investigation, however, with the exception of styrene and MMA, impurities have been identified as the radical source and thus responsible for the polymerization.

Fig. 9.10 Radical formed during a redox reaction (*Fenton reagent*)

$$Fe^{2+} + H_2O_2 \longrightarrow Fe^{3+} + OH^- + OH^{\bullet}$$

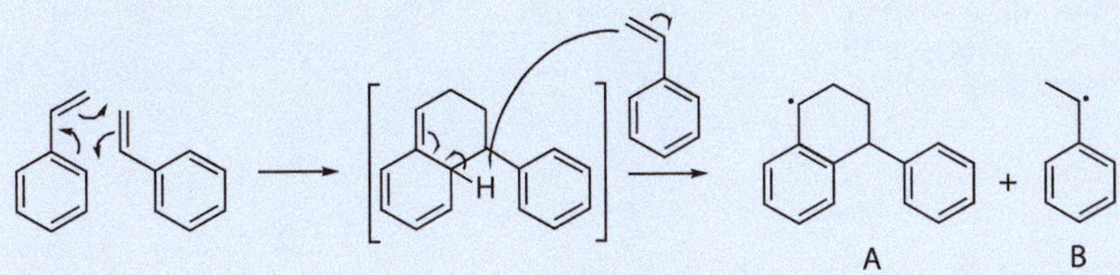

Fig. 9.11 Primary steps of the thermal polymerization of styrene

The thermal polymerization of styrene proceeds via a Diels–Alder adduct from which the radicals *A* and *B* emerge after reaction with an additional styrene molecule and H-transfer. The radicals are then able to initiate radical polymerization (Fig. 9.11).

9.2 Kinetics of Radical Polymerization

On the basis of the most precise mechanism, a radical polymerization can be divided into several individual reaction steps. A kinetic equation can be developed for every step, and from these the overall kinetics can be derived. The key question for the kinetics of radical polymerization is how the monomer concentration changes with time at a given temperature.

9.2.1 Basic Processes

The basic processes—initiation, growth, and termination—can be formally summarized by the scheme in Fig. 9.12. The initiator dissociates into two radicals at a given temperature with the rate constant k_d. These radicals, in the case of AIBN (**a**zo-bis-**i**so**b**utyro**n**itrile) (Fig. 9.9b), sufficiently stabilized by the nitrile group, are reactive enough to add to monomers (Table 9.1) with the rate constant k_i. The new radical $RM^{\bullet}$ adds additional monomers with the rate constant k_p. Depending on the temperature, termination can result from either radical combination (with k_t) or from disproportionation (with k_t') to a saturated chain RM_n^H and an unsaturated chain $RM_m^=$ (Figs. 9.12 and 9.13), or both termination reactions occur simultaneously. With an increase in temperature, termination by disproportionation generally becomes more important because $E_{A,disp.} > E_{A,comb.}$.

In Fig. 9.13 the individual steps—initiator decomposition, initiation, growth, and termination – are illustrated using the example of the AIBN-initiated polymerization of the monomer CH_2=CHR′.

In the following paragraphs, several limiting cases for the kinetics of radical polymerization are discussed.

Fig. 9.12 Formal scheme for a radical polymerization. *I* initiator, *R*• initiator radical, *M* monomer, *RM*• initiator radical extended by a single monomer unit, RM_n• initiator radical extended by *n* monomer units, RM_n^H polymer with a saturated chain end, $RM_m^{=}$ polymer with an unsaturated chain end

Initiator decomposition	I	$\xrightarrow{k_d} 2R^{\bullet}$
Start	$R^{\bullet} + M$	$\xrightarrow{k_i} RM^{\bullet}$
Propagation	$RM^{\bullet} + (n\text{-}1)\,M$	$\xrightarrow{k_p} RM_n^{\bullet}$
Termination	$RM_n^{\bullet} + {}^{\bullet}M_mR$	$\xrightarrow{k_t} RM_{n+m}R$
		$\xrightarrow{k'_t} RM_n^H + {}^{=}M_mR$

9.2.2 *R*• and *P*• are Equally Reactive

With the simplified assumption that all radicals have the same reactivity independent of their chain length ($R^{\bullet} \equiv RM^{\bullet} \equiv RM_n^{\bullet} \equiv P^{\bullet}$), the scheme in Fig. 9.12 can be simplified as follows (Fig. 9.14).

We then have

$$\frac{d[P^{\bullet}]}{dt} = 2k_d[I] - k_t[P^{\bullet}]^2 \tag{9.1}$$

With the assumption of a quasi-stationary state adapted from Bodenstein[1] it follows that

$$\frac{d[P^{\bullet}]}{dt} = 0 \tag{9.2}$$

The radicals are consumed at the same rate at which they are created. For the radical concentration $[P^{\bullet}]$ the following applies:

$$[P^{\bullet}] = \sqrt{\frac{2k_d[I]}{k_t}} \tag{9.3}$$

Because termination by combination and disproportionation cannot be kinetically distinguished, k_t here is the average rate constant for both types of termination.

The monomer is only consumed by the growing chains, and thus the monomer consumption over time is a direct measure of the rate of polymerization v_{br} (▶ see (9.4)). (This assumption is not completely correct because monomer is also consumed by initiation (the first growth step) (Fig. 9.12). However, the amount of monomer consumed in the initial step of chain growth is very small compared with that consumed during polymerization to give chains many hundreds of monomer units long.)

$$-\frac{d[M]}{dt} = k_p[M][P^{\bullet}] \tag{9.4}$$

1 The concept of a quasi-stationary state involves the assumption that the change in the concentration of intermediate products (here *P*•) can be neglected when compared with changes in the concentration of educts (here *M*) and products (here *P*) (Frost and Pearson 1964)

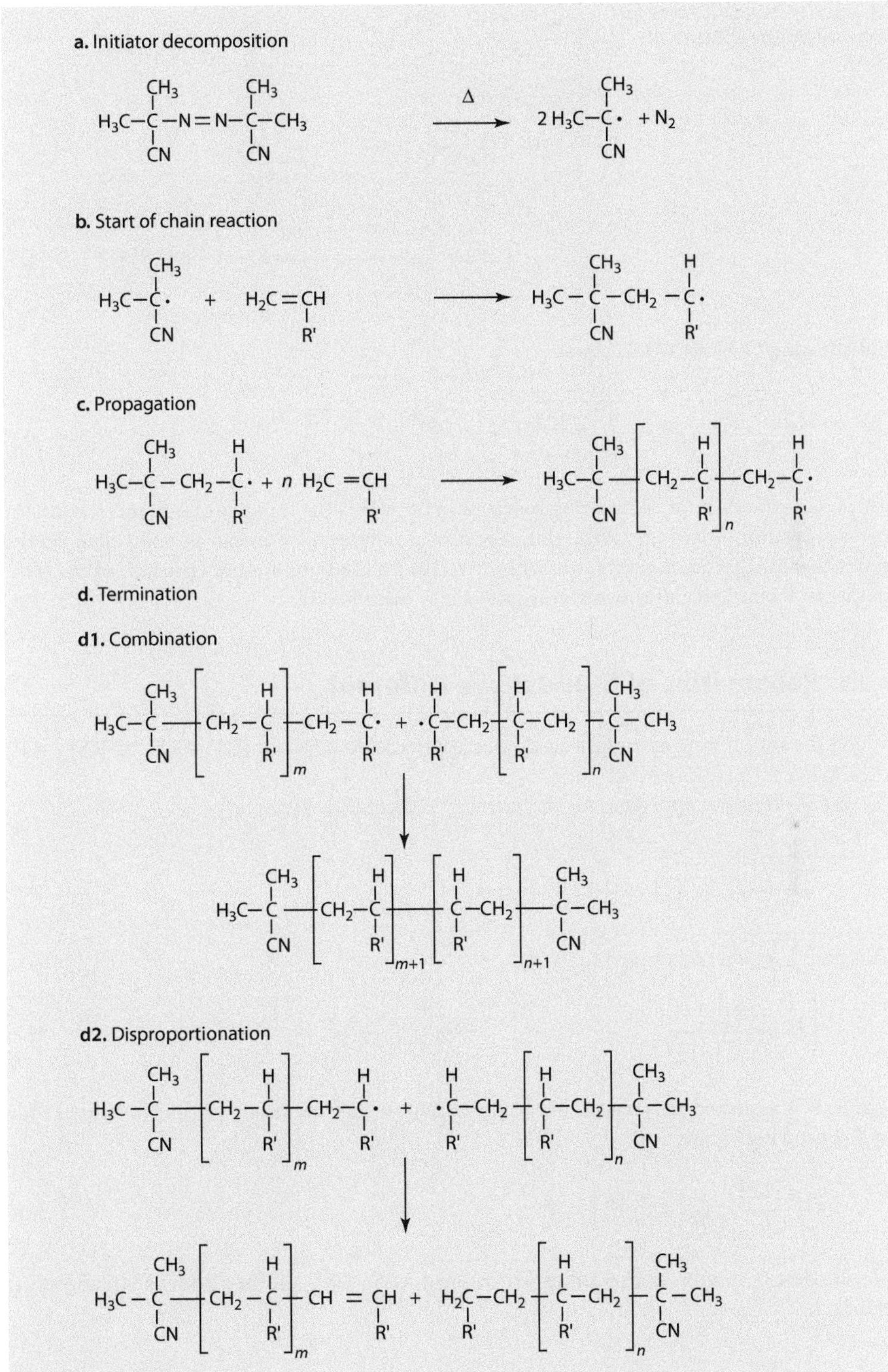

Fig. 9.13 Individual steps of the polymerization of a vinyl monomer CH_2=CHR′ initiated by AIBN (the elements of the last row correspond to RM_n^H and $RM_m^=$ in Fig. 9.12)

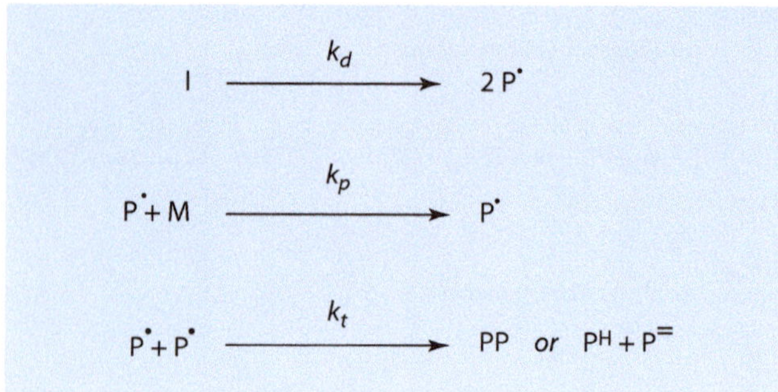

Fig. 9.14 Formal scheme for a radical polymerization with $R^\bullet \equiv P^\bullet$

Substituting (9.3) into (9.4) yields

$$-\frac{d[M]}{dt} = v_{br} = k_p \sqrt{\frac{2k_d[I]}{k_t}} \cdot [M] \tag{9.5}$$

This basic equation for radical polymerization describes the kinetics of polymerization for most vinyl monomers very well. Thus, the rate of polymerization can be controlled via the monomer and initiator concentrations as well as by the temperature (because of the temperature dependence of the rate constant; see ► Sect. 9.2.6).

9.2.3 Reactivities of $R^\bullet$ and $P^\bullet$ are Different

When the simplifying assumptions do not apply and $R^\bullet \neq P^\bullet$, but $RM^\bullet \equiv RM_2{}^\bullet \equiv RM_n{}^\bullet \equiv P^\bullet$, then Fig. 9.14 turns into Fig. 9.15.

The Bodenstein approach for the primary radical $R^\bullet$ is given by

$$\frac{d[R^\bullet]}{dt} = 2k_d[I] - k_i[R^\bullet][M] = 0 \tag{9.6}$$

This can be easily rearranged to

$$[R^\bullet] = \frac{2k_d[I]}{k_i[M]} \tag{9.7}$$

Again, it is assumed to a reasonable approximation that the monomer is consumed only by the growing chains and

$$-\frac{d[M]}{dt} = k_p[M][P^\bullet] \tag{9.8}$$

For the determination of the unknown concentration $[P^\bullet]$, another Bodenstein approach is that

$$\frac{d[P^\bullet]}{dt} = k_i[R^\bullet][M] - k_t[P^\bullet]^2 = 0 \tag{9.9}$$

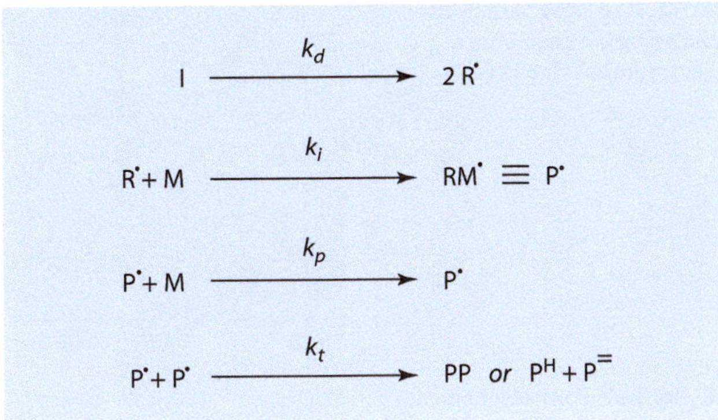

Fig. 9.15 Formal scheme of a radical polymerization with $R^{\bullet} \neq P^{\bullet}$ (the possibility of a reaction between $R^{\bullet}$ and $P^{\bullet}$ into $R\text{–}P$ is not considered)

So the molar concentration $[P^{\bullet}]$ is given by

$$\left[P^{\bullet}\right] = \sqrt{\frac{k_i\left[R^{\bullet}\right][M]}{k_t}} \tag{9.10}$$

If $[R^{\bullet}]$ is replaced in (9.10) by (9.7):

$$\left[P^{\bullet}\right] = \sqrt{\frac{k_i \cdot 2k_d[I][M]}{k_t \cdot k_i[M]}} = \sqrt{\frac{2k_d[I]}{k_t}} \tag{9.11}$$

and for the rate of polymerization after substituting (9.11) into (9.8):

$$-\frac{d[M]}{dt} = k_p\sqrt{\frac{2k_d[I]}{k_t}} \cdot [M] \tag{9.5}$$

In other words, differing reactivities of $R^{\bullet}$ and $P^{\bullet}$ cannot be determined kinetically.

9.2.4 Cage Effect

If one considers that the primary radicals from the decomposition of the initiator $R^{\bullet}$ are trapped by solvent molecules in a cage, it becomes clear why not all initially formed radicals initiate chain growth. Thus, the radicals first have to diffuse out of the cage (k_{diff}) in order to react with a monomer. However, they can also react with another primary radical if still in the cage (Fig. 9.16).

If, in addition, we assume that only the free radicals $R^{\bullet}_{free}$ are able to add to a monomer, the mechanism shown in Fig. 9.17 can be developed.

In this case, there are three unknown radical concentrations, $[R^{\bullet}_{cage}]$, $[R^{\bullet}_{free}]$, and $[P^{\bullet}]$. Consequently, the Bodenstein approach needs to be applied for each of these concentrations.

Cage Radicals $R^{\bullet}_{cage}$

Assuming the reaction process is as given in Fig. 9.17, the concentration of the radicals in the solvent cage $[R^{\bullet}_{cage}]$ changes with time. The radicals $R^{\bullet}_{cage}$ form in pairs in the cage, which means that, when a radical is present, the second exists inevitably in close proximity and does not need to diffuse into such a position. Thus, the recombination of radicals in

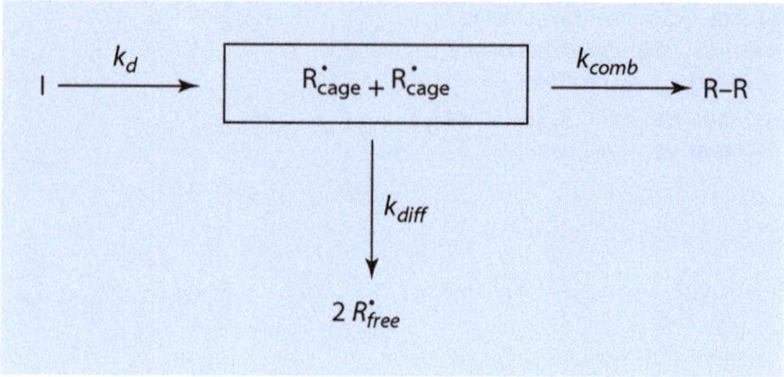

Fig. 9.16 Loss of radicals caused by the cage effect. *Frame* symbol for the cage

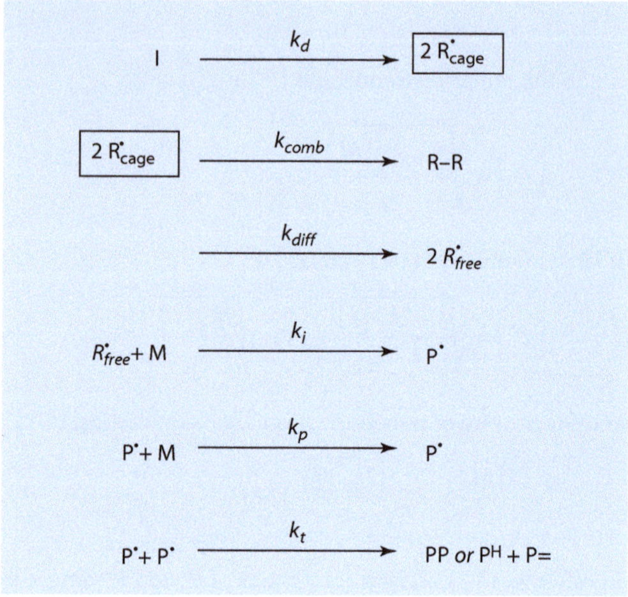

Fig. 9.17 Formal scheme for a radical polymerization including the cage effect

the solvent cage is of first order in radical concentration (Fig. 9.16) (9.12). In the equation the superscript 1 is included for emphasis:

$$\frac{d\left[R^{\bullet}_{cage}\right]}{dt} = 2k_d\left[I\right] - k_{comb}\left[R^{\bullet}_{cage}\right]^1 - k_{diff}\left[R^{\bullet}_{cage}\right] = 0 \tag{9.12}$$

Thus, it follows that

$$\left[R^{\bullet}_{cage}\right] = \frac{2k_d\left[I\right]}{k_{diff} + k_{comb}} \tag{9.13}$$

Free Radicals $R^{\bullet}_{free}$: Formation and Reaction

For the change of $[R^{\bullet}_{free}]$ with time, it follows from Fig. 9.17 that

$$\frac{d\left[R^{\bullet}_{free}\right]}{dt} = k_{diff}\left[R^{\bullet}_{cage}\right] - k_i\left[R^{\bullet}_{free}\right]\left[M\right] = 0 \tag{9.14}$$

Simple conversion yields

$$\left[R^{\bullet}_{free}\right]=\frac{k_{diff}\left[R^{\bullet}_{cage}\right]}{k_i\left[M\right]} \tag{9.15}$$

With (9.13) it follows that

$$\left[R^{\bullet}_{free}\right]=\frac{k_{diff}}{k_{diff}+k_{comb}}\cdot\frac{2k_d\left[I\right]}{k_i\left[M\right]} \tag{9.16}$$

Polymer Radicals $P^{\bullet}$: Fate and Definition of the Net Reaction Rate

The polymer radicals $P^{\bullet}$ are formed through the reaction of $R^{\bullet}_{free}$ with monomers and are destroyed by chain termination reactions. Hence, the following applies for the change of the radical concentration $[P^{\bullet}]$ with time:

$$\frac{d\left[P^{\bullet}\right]}{dt}=k_i\left[R\right]\left[M\right]-k_t\left[P^{\bullet}\right]^2=0 \tag{9.17}$$

and

$$\left[P^{\bullet}\right]=\sqrt{\frac{k_s\left[R^{\bullet}_{free}\right]\left[M\right]}{k_t}} \tag{9.18}$$

Using the definition of $[R^{\bullet}_{free}]$ from (9.16) gives

$$\left[P^{\bullet}\right]=\sqrt{\frac{k_i\left[M\right]}{k_t}\cdot\frac{k_{diff}}{k_{diff}+k_{comb}}\cdot\frac{2k_d\left[I\right]}{k_i\left[M\right]}} \tag{9.19}$$

Equation (9.19) can be simplified by introducing the efficiency f of the overall process producing polymerization active radicals:

$$f=\frac{k_{diff}}{k_{diff}+k_{comb}} \tag{9.20}$$

$$\left[P^{\bullet}\right]=\sqrt{\frac{f\cdot 2k_d\left[I\right]}{k_t}} \tag{9.21}$$

The efficiency of the initiator f can take values between 0 and 1. For AIBN, values of $f=0.7$ are usually found, which means that that portion of the primary radicals already destroyed in the cage should not be neglected.

Combining (9.21) and (9.8) an expression for the overall rate of polymerization can be obtained:

$$-\frac{d\left[M\right]}{dt}=k_p\cdot\sqrt{\frac{f\cdot 2k_d\left[I\right]}{k_t}}\cdot\left[M\right] \tag{9.22}$$

and thus

$$v_{br} = -\frac{d[M]}{dt} \propto [M]^1 \tag{9.23}$$

$$v_{br} = -\frac{d[M]}{dt} \propto \sqrt{[I]} \tag{9.24}$$

The rate of polymerization v_{br} increases linearly with the monomer concentration (the highest possible $[M]$ is the molar concentration of the pure monomer) and with the square root of the initiator concentration.

9.2.5 Experimental Determination of the Rate of Polymerization

To determine the rate constants in (9.22), polymerizations are studied at low conversions ($[M]$ and $[I]$ are essentially constant so that $[M] \approx [M]_0$ and $[I] \approx [I]_0$) by first varying $[M]_0$ while keeping $[I]_0$ constant and then varying $[I]_0$ while keeping $[M]_0$ constant (◘ Fig. 9.18).

The rate of polymerization can be determined gravimetrically from the resulting polymer, because this is equal to the quantity of the monomer consumed. For this purpose, the polymer yield at a given time needs to be isolated. This is best achieved by adding the polymer solution drop by drop into a suitable precipitant agent.[2] The precipitated polymer is filtered, dried, and weighed and the weight converted into the amount of monomer polymerized. Gravimetry is a universal method for measuring conversion but it is time-consuming because only a single experimental data point is obtained for each experiment.

Alternatively, especially low conversions can be followed in a dilatometer (◘ Fig. 9.19). A dilatometer is a vessel of a known volume to which a capillary is attached. During polymerization, the total volume generally decreases because polymers usually have a higher density than monomers.[3] The volume reduction ΔV with time can be followed very exactly in the capillary. If the densities of the monomer and polymer being studied are known, the overall rate of polymerization can very easily be determined from the ΔV over time (for a sufficiently large range of monomer concentration $[M]$).

In specific cases, not only IR-, UV-, and ^{1}H-NMR-spectroscopy but also gas chromatography and other analytical methods can be used to follow the decrease in monomer concentration or the increase in polymer concentration.

2 For polystyrene methanol is a suitable non-solvent. Extensive information about solvents and non-solvents can be found, for example, in Brandrup and Immergut (1989).

3 During the polymerization, the Van der Waals distances between monomer molecules are replaced by covalent bonds which are shorter so that the polymer usually has a greater density than the monomers.

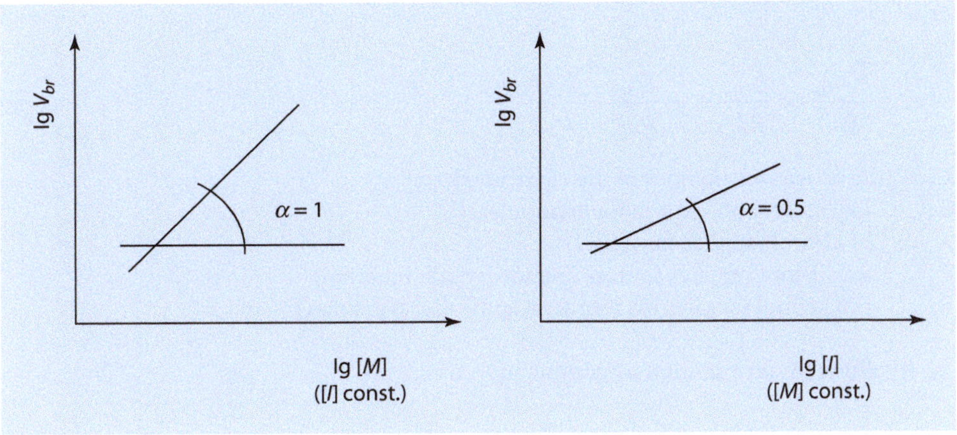

Fig. 9.18 Determination of the reaction order with respect to [M] and [I]

Fig. 9.19 Measuring the rate of polymerization using a dilatometer

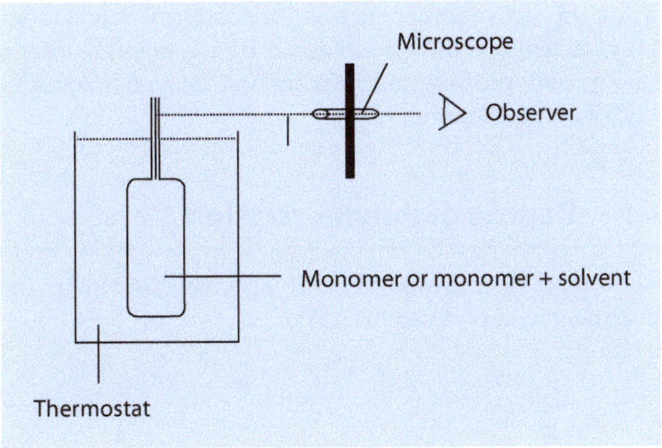

9.2.6 Temperature Dependence of the Overall Rate of Polymerization

For the overall rate of polymerization (as already derived above):

$$v_{br} = k_p \cdot \underbrace{\sqrt{\frac{f \cdot 2k_d}{k_t}}}_{T-dependent} \cdot \underbrace{[M] \cdot \sqrt{[I]}}_{T-independent} \tag{9.22}$$

According to Arrhenius, if the efficiency f is assumed to be temperature independent:

$$\frac{d \ln v_{br}}{dT} = \frac{d \ln k_p}{dT} + \frac{1}{2} \cdot \frac{d \ln k_d}{dT} - \frac{1}{2} \cdot \frac{d \ln k_t}{dT} \tag{9.25}$$

or

$$\frac{E_{A,v_{br}}}{RT^2} = \frac{E_{A,p}}{RT^2} + \frac{1}{2} \cdot \frac{E_{A,d}}{RT^2} - \frac{1}{2} \cdot \frac{E_{A,t}}{RT^2} \tag{9.26}$$

Using the activation energies of the elementary steps:
$E_{A,p} \approx 20$ kJ/mol (applies for many monomers)
$E_{A,d} \approx 120$ kJ/mol (applies for AIBN)
$E_{A,comb.} \approx 0$ kJ/mol (applies for termination by combination)
$E_{A,disp.} \approx 10$ kJ/mol (applies for termination by disproportion)

It follows for termination by combination that

$$E_{A,v_{br}} \approx 20\,\text{kJ/mol} + \frac{1}{2} \cdot 120\,\text{kJ/mol} - 0\,\text{kJ/mol} = 80\,\text{kJ/mol} \tag{9.27}$$

If the rate constants for two temperatures, for example 80 °C and 90 °C, are calculated using this activation energy and the Arrhenius equation, it can be seen that the rate of polymerization almost doubles for this temperature increase of 10 °C. Thus, in addition to the concentration of monomer and initiator, temperature is a third crucial variable for the rate of polymerization.

9

9.3 Degree of Polymerization

The degree of polymerization P_n specifies how many monomer moieties make up one polymer chain (▶ Chap. 3, (3.11)):

$$P_n = \frac{M_n}{M_M} \tag{3.11}$$

M_n	number-average molar mass
M_M	molar mass of the monomer (equal to M_{RU}: molar mass of a repeating unit)

Similar to the mean values of molar mass M_n and M_w, it is also appropriate to distinguish between the different mean values, e.g., P_n and P_w, for the degree of polymerization.

9.3.1 Kinetic Chain Length and Degree of Polymerization

The kinetic chain length v is a measure of the average number of monomer molecules which polymerize per initiating radical before chain growth is terminated. Thus, v is determined by the ratio of the probabilities of chain growth p_p and the probability of the chain termination p_t or by the ratio of the rate of growth v_p to the rate of termination v_t:

$$v = \frac{p_p}{p_t} = \frac{v_p}{v_t} = \frac{k_p[M][P^\bullet]}{k_t[P^\bullet]^2} = \frac{k_p[M]}{k_t[P^\bullet]} \tag{9.28}$$

From (9.3)

$$\left[P^{\bullet}\right]=\sqrt{\frac{2k_d\left[I\right]}{k_t}} \quad (9.3)$$

and it follows that

$$\nu=\frac{k_p\left[M\right]}{\sqrt{k_t\cdot 2k_d\left[I\right]}} \quad (9.29)$$

By rearranging

$$v_{br}=-\frac{d\left[M\right]}{dt}=k_p\left[M\right]\left[P^{\bullet}\right] \quad (9.8)$$

to obtain an expression for $[P^{\bullet}]$ and substitution into (9.28), the kinetic chain length as a function of the overall rate of polymerization can be obtained:

$$\nu=\frac{k_p^2\cdot\left[M\right]^2}{k_t\cdot v_{br}} \quad (9.30)$$

With (9.30), it is now possible to determine the ratio k_p^2/k_t as both the kinetic chain length ν and the overall rate of polymerization v_{br} can be determined experimentally with considerable accuracy.

There is a close correlation between the kinetic chain length and the degree of polymerization. Although the nature of the termination reaction (combination or disproportionation) is not considered when discussing the kinetic chain length, it is important for the degree of polymerization. For termination solely via disproportionation, the degree of polymerization P_n equals the kinetic chain length. On the other hand, a termination via combination joins two chains together so that the degree of polymerization is twice that for termination via disproportionation.

Termination by Disproportionation

$$P_n=\nu \quad (9.31)$$

and

$$P_n=\frac{k_p\left[M\right]}{\sqrt{k_t\cdot 2k_d\left[I\right]}} \quad (9.32)$$

Termination by Combination

$$P_n=2\nu \quad (9.33)$$

or

$$P_n=2\cdot\frac{k_p\left[M\right]}{\sqrt{k_t\cdot 2k_d\left[I\right]}} \quad (9.34)$$

Table 9.2 Comparison of measured degree of polymerization P_n (osmotically determined) with the calculated kinetic chain length for the bulk polymerization of styrene at 50 °C (Henricí-Olivé and Olive 1969)

[AIBN] · 10^2 (mol/L)	$v_{Br} \cdot 10^5$ (mol/(L s))	ν	2ν	P_n
0.061	0.66	3600	7200	6800
0.122	0.96	2650	5300	5500
0.305	1.52	1670	3340	3570
1.83	3.57	655	1310	1380

The comparison of the measured degree of polymerization P_n with the calculated kinetic chain length v for styrene leads to the conclusion that termination of polymerization occurs via combination (Table 9.2).

From the definition of the degree of polymerization P_n (9.32) and (9.34) it follows that to increase the degree of polymerization, either the monomer concentration must be increased or the initiator concentration decreased. The greatest possible monomer concentration is the concentration of the pure substance. The effect of the temperature is more complex and is discussed in more detail in Sect. 9.3.2.

9

9.3.2 Temperature Dependence of the Degree of Polymerization

The defining equation for the degree of polymerization P_n (9.32) can be separated into temperature dependent and temperature independent terms:

$$P_n = \underbrace{\frac{k_p}{\sqrt{2k_t \cdot k_d}}}_{T-dependent} \cdot \underbrace{\frac{[M]}{\sqrt{[I]}}}_{T-independent} \tag{9.32}$$

If one disregards the volume change of the system caused by a change in temperature, then the monomer and initiator concentrations are independent of temperature. For the effect of temperature on the degree of polymerization P_n, Arrhenius equations for the rate constants k_p, k_t, and k_d give

$$\frac{d \ln P_n}{dT} = \frac{d \ln k_p}{dT} - \frac{1}{2} \cdot \frac{d \ln k_t}{dT} - \frac{1}{2} \cdot \frac{d \ln k_d}{dT} \tag{9.35}$$

$$\frac{E_{A,P_n}}{RT^2} = \frac{E_{A,p}}{RT^2} - \frac{1}{2} \cdot \frac{E_{A,t}}{RT^2} - \frac{1}{2} \cdot \frac{E_{A,d}}{RT^2} \tag{9.36}$$

With the individual activation energies $E_{A,p}$, $E_{A,t}$ and $E_{A,d}$, which were already used for the calculation of $E_{A,v_{br}}$ (9.27), a value of –40 kJ/mol ((9.37) and (9.38)) can be determined for E_{A,P_n} (termination by combination):

$$E_{A,P_n} \approx 20\,\text{kJ/mol} - \frac{1}{2} \cdot 0\,\text{kJ/mol} - \frac{1}{2} \cdot 120\,\text{kJ/mol} \tag{9.37}$$

$$E_{A,P_n} \approx -40\,\text{kJ/mol} \tag{9.38}$$

From this it follows that the degree of polymerization decreases with increasing temperature (negative E_{A,P_n}). On simple reflection we see that qualitatively identical results can be inferred because higher temperatures lead to a faster decomposition of the initiator and therefore more radicals are formed per unit time. As a result, given a constant monomer supply, the chains formed are shorter.

Thus, temperature is a third critical variable, in addition to the monomer and initiator concentrations, for the average degree of polymerization. In addition, it is possible to control the degree of polymerization by adding transfer agent to the reaction mixture.

9.3.3 Regulation of Degree of Polymerization by Transfer

Reagents which can assume the radical nature of a growing chain and then start a new chain growth are referred to as *regulators*, *chain transfer agents*, or simply *transfer agents*. Monomers (Fig. 9.20 and Table 9.3), polymers (Fig. 9.23), and solvents (Figs. 9.21 and 9.22), and also intentionally added substances such as mercaptans and aldehydes, can serve as chain transfer agents (Table 9.4).

Transfer to Monomer

$M^\bullet$ is able to start a new polymer chain by adding an additional monomer unit. A characteristic of transfer agents is that they terminate an individual chain but they do not affect v_{br} because a new polymer chain starts from the small radical molecule formed.

The general approach for determining the degree of polymerization from the rate of the transfer reaction is

$$P_n = \frac{v_p}{v_t + \sum v_{tr}} \tag{9.39}$$

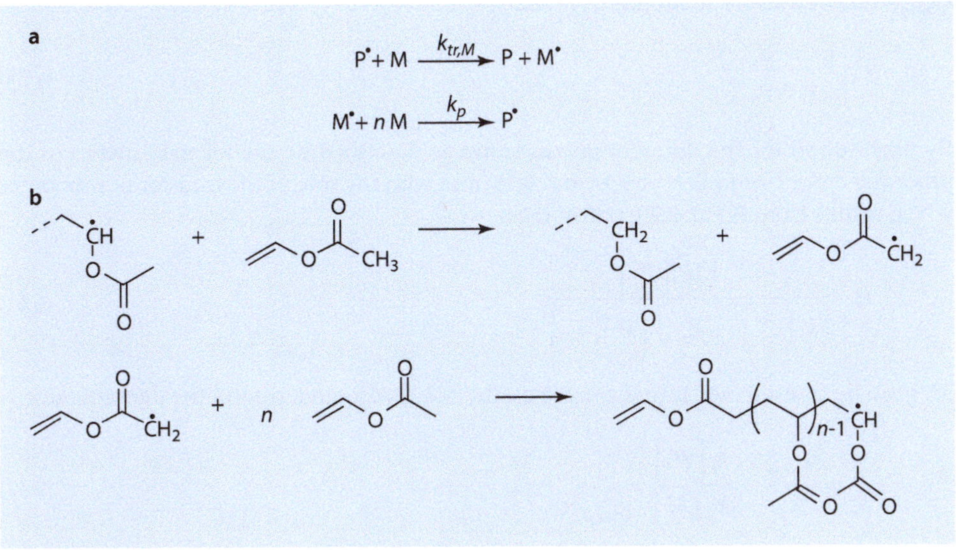

Fig. 9.20 Chain transfer to monomer: (**a**) general scheme and (**b**) vinyl acetate as an example

Table 9.3 Transfer coefficients C_M for selected monomers (Brandrup and Immergut 1989)

Monomer	$C_M \cdot 10^4$
Acrylamide	0.5
Acrylonitrile	0.26
Ethylene	0.4–4.2
Styrene	0.3–0.6
Vinyl acetate	1.75–2.8
Vinyl chloride	10.8–16

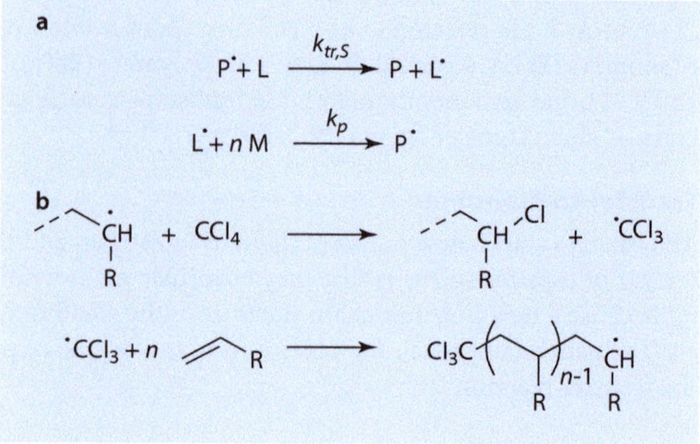

Fig. 9.21 Chain transfer to solvent: (**a**) general scheme and (**b**) CCl_4 as an example

where v_{tr} represents the rate of the transfer reaction and v_p and v_t represent the growth and termination rates, already introduced above. If the only transfer is transfer to monomer ($v_{tr,M}$ = rate of transfer to monomer) then

$$P_n = \frac{v_p}{v_t + v_{tr,M}} \tag{9.40}$$

By substitution for the different process rates as described by the formal kinetics of the processes described in Figs. 9.14 and 9.15, and with the rate of the transfer to monomer which results from Fig. 9.20, one obtains

$$P_n = \frac{k_p[M][P^\bullet]}{k_t[P^\bullet]^2 + k_{tr,M}[M][P^\bullet]} \tag{9.41}$$

This can be transformed into the more useful (9.42) without a sum in the denominator:

$$\frac{1}{P_n} = \frac{k_{tr,M}}{k_p} + \frac{k_t[P^\bullet]}{k_p[M]} = \frac{1}{P_{n,0}} \tag{9.42}$$

where $P_{n,0}$ is the degree of polymerization when no transfer or other chain length limiting reactions apart from transfer to monomer occur.

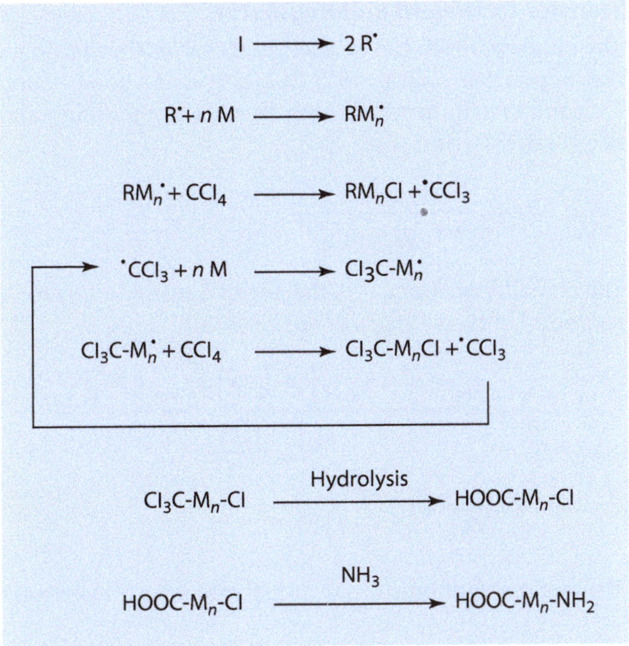

Fig. 9.22 Reaction scheme to form terminally functionalized polymers by chain transfer to CCl_4 solvent and subsequent hydrolysis followed by reaction with ammonia

Table 9.4 Transfer constants C_S for selected solvents and regulators for the polymerization of styrene at 60 °C (Brandrup and Immergut 1989)

Solvent	$C_S \cdot 10^4$
Benzene	0.023
Toluene	0.125
Isopropyl benzene	0.82
Chloroform	3.4
Carbon tetrachloride	110
Carbon tetrabromine	22 000
n-Butyl mercaptan	210 000

Substitution of (9.3) for $[P^\bullet]$ gives

$$\frac{1}{P_{n,0}} = \frac{k_{tr,M}}{k_p} + \frac{\sqrt{2k_d \cdot k_t}}{k_p} \cdot \frac{\sqrt{[I]}}{[M]} \tag{9.43}$$

Thus a plot of $1/P_n$ against $\sqrt{[I]}/[M]$ allows the transfer coefficient $C_M = k_{tr,M}/k_p$ to be determined from the intercept.

The transfer to monomer is intrinsic, inevitable, and is often called *direct chain transfer*. Compared to acrylamide, it plays a significant role for monomers such as vinyl acetate and vinyl chloride (Table 9.3).

Transfer to Solvent and Regulator

The same approach can be used as above for transfer to monomer (Fig. 9.20). The reaction of polymer radicals with CCl_4 serves as a good example (Fig. 9.21).

For a system in which both transfer to monomer and transfer to solvent take place, (9.39) can be written as

$$P_n = \frac{v_p}{v_t + v_{tr,M} + v_{tr,S}} \tag{9.44}$$

The new additive term, $v_{tr,S}$, the rate of transfer to solvent appears in the denominator. The reciprocal of (9.44) yields (9.45) or (9.46):

$$\frac{1}{P_n} = \frac{k_t\left[P^{\bullet}\right]^2 + k_{tr,M}\left[P^{\bullet}\right][M] + k_{tr,S}\left[P^{\bullet}\right][S]}{k_p\left[P^{\bullet}\right][M]} \tag{9.45}$$

$$\frac{1}{P_n} = \frac{1}{P_{n,0}} + C_S \cdot \frac{[S]}{[M]} \tag{9.46}$$

The rate constant ratio $k_{tr,S}/k_p$ is referred to as transfer coefficient C_S:

9

$$C_S = \frac{k_{tr,S}}{k_p} \tag{9.47}$$

A graph of $1/P_n = f([S]/[M])$ has the coefficient C_S as the gradient. In Table 9.4, some transfer coefficients for several solvents and the polymerization of styrene at 60 °C are compiled.

More important than the absolute numbers in this table, which not only depend on the monomer but also on the temperature, is the trend.

Mercaptans are thus highly effective regulators. Even very small amounts of mercaptan suffice to reduce drastically the molar mass of the resulting polymers. The kinetic analysis and reaction scheme are analogous to that for transfer to solvent (Fig. 9.21).

An interesting application of transfer to solvent is the polymerization of styrene into CCl_4 which yields oligomers with a CCl_3 head group and a Cl terminal group. From these oligomers α,ω-amino acids can be synthesized (Fig. 9.22).

Transfer to Polymer

Transfer to polymer leads to a branching of the main polymer chain and thus, in contrast to those transfer reactions discussed already, to an increase in molar mass (Fig. 9.23).

9.4 Molar Mass Distribution

The macromolecules formed by chain growth have different chain lengths resulting from the kinetics of the processes which formed them (statistical sequence of start, growth, termination, and transfer). Thus, they are a mixture of molecules with a distribution of molar mass. This distribution has an important influence on their physical and material characteristics.

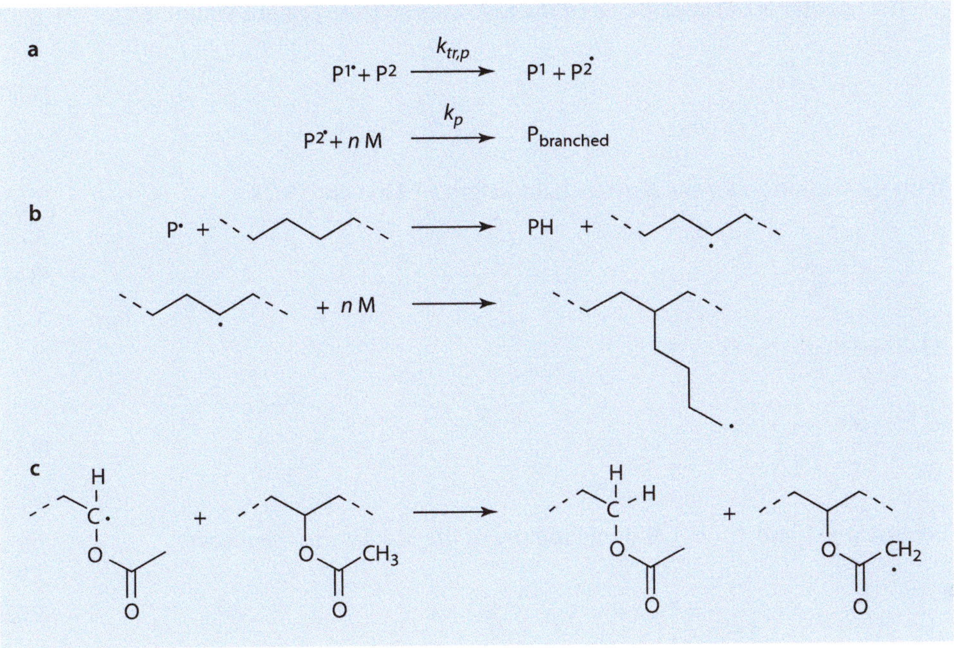

Fig. 9.23 Chain transfer to polymer: (**a**) general scheme, (**b**) polyethylene, and (**c**) polyvinyl acetate as examples

9.4.1 Molar Mass Distribution Resulting from Termination by Disproportionation

The numerical frequency n_P of those molecules that show a specific degree of polymerization P is proportional to the statistical probability Π_P that these molecules are formed:

$$n_P = const \cdot \Pi_P \tag{9.48}$$

n_P Numerical frequency of the molecules with degree of polymerization P
Π_P Statistical probability that the molecules are formed

Molecules with the degree of polymerization P result if propagation can continue unimpaired P times. Reactions that hinder propagation are transfer and termination. Thus, the probability α for a single propagation step is as follows:

$$\alpha = \frac{v_p}{v_p + v_t + \sum v_{tr}} \tag{9.49}$$

α Probability for a single growth step
v_p Rate of propagation
v_t Rate of termination
v_{tr} Rate of transfer

The transfer is not considered in the following so (9.49) can be simplified to

$$\alpha = \frac{v_p}{v_p + v_t} \tag{9.50}$$

With the definition for the kinetic chain length ν (9.51) and (9.28)

$$\nu = \frac{v_p}{v_t} \tag{9.51}$$

it follows that

$$\alpha = \frac{1}{1 + \frac{1}{\nu}} \tag{9.52}$$

Because $\nu \gg 1$ and $1/\nu \ll 1$ and making use of the MacLaurin's sequence:

$$\text{for } x \ll 1 : \frac{1}{1+x} = 1 - x + x^2 - x^3 \ldots \approx 1 - x \tag{9.53}$$

9

Thus:

$$\alpha = 1 - \frac{1}{\nu} \tag{9.54}$$

For the probability Π_P that the propagation continues consecutively P times so that molecules with the degree of polymerization P are formed, the following applies:

$$\Pi_P = \alpha_1 \cdot \alpha_2 \cdot \alpha_3 \ldots \cdot \alpha_P = \alpha^P \tag{9.55}$$

The combination of (9.48) and (9.55) yields (9.56):

$$n_P = const \cdot \alpha^P \tag{9.56}$$

From this it can be derived that

$$m_P = const \cdot M_P \cdot \alpha^P = const \cdot M_M \cdot P \cdot \alpha^P = A \cdot P \cdot \alpha^P \tag{9.57}$$

m_P	Frequency of the molecules in terms of mass (mass fraction) with the degree of polymerization P
M_P	Molar mass of the molecules with the degree of polymerization P
M_M	Molar mass of the monomers

From (9.57):

$$A = M_M \cdot const \tag{9.58}$$

For the general case:

$$\int_0^\infty m_P dP = 1 \tag{9.59}$$

so that

$$A \cdot \int_0^\infty P \cdot \alpha^P dP = 1 \tag{9.60}$$

or

$$A = \frac{1}{\int_0^\infty P \cdot \alpha^P dP} \tag{9.61}$$

The general solution of the integral is

$$\int_0^\infty P^n \cdot \alpha^P dP = \frac{n!}{(-\ln \alpha)^{n+1}} \tag{9.62}$$

Substitution of (9.62) into (9.61) (with $n = 1$) gives

$$A = \ln^2 \alpha \tag{9.63}$$

and for m_P (9.57):

$$m_P = \ln^2 \alpha \cdot P \cdot \alpha^P \tag{9.64}$$

This mass distribution function for the product of a polymerization where termination is via disproportionation gives the weight fraction of molecules with a degree of polymerization P (Schulz 1939; Flory 1953) (Fig. 9.24).

9.4.2 Molar Mass Distribution with Termination by Combination

The polymer molecules resulting from a termination via combination termination are composed of two polymer radicals. The degree of polymerization P can be obtained by the combination $1 + (P - 1) = P$; $2 + (P - 2) = P$ … $(P - 2) + 2 = P$ and so on. Consequently, the number of combinations leading to the degree of polymerization P is $P/2$. The frequency of the xth combination $n_{P,x}$, is proportional to the product of the frequencies of the single chain lengths x and $P - x$:

$$n_{P,x} = const' \cdot \alpha^x \cdot \alpha^{P-x} = const' \cdot \alpha^P \tag{9.65}$$

$n_{P,x}$	Numerical frequency of the xth combination that leads to the degree of polymerization P
α^x	Probability of the formation of molecules with the degree of polymerization x
α^{P-x}	Probability of the formation of molecules with the degree of polymerization $(P - x)$

Fig. 9.24 Mass distribution function for $\nu = 10$, calculated from (9.63) and (9.69). *I* Termination by Disproportionation, *II* Termination by Combination

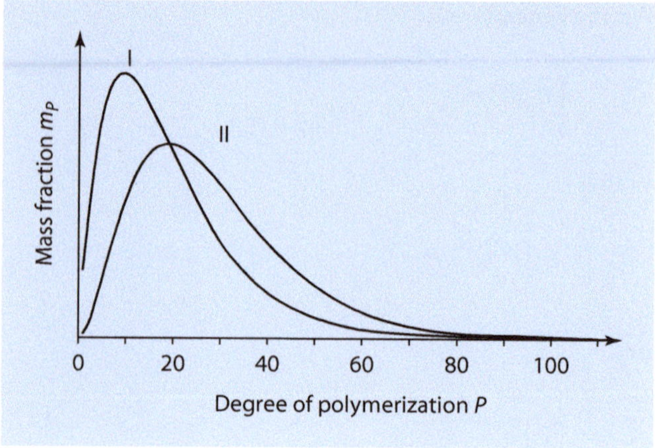

The numerical frequency n_P of the degree of polymerization is the sum of the individual frequencies $n_{P,x}$:

$$n_P = \sum_{x=1}^{x=\frac{P}{2}} n_{P,x} = const' \cdot \frac{P}{2} \cdot \alpha^P \tag{9.66}$$

In analogy to (9.57):

$$m_P = M_P \cdot const' \cdot \frac{P}{2} \cdot \alpha^P = M_M \cdot P \cdot const' \cdot \frac{P}{2} \cdot \alpha^P = A' \cdot P^2 \cdot \alpha^P \tag{9.67}$$

From (9.59) and (9.67) one obtains

$$\int_0^\infty m_P dP = 1 = A' \cdot \int_0^\infty P^2 \cdot \alpha^P dP \tag{9.68}$$

and with (9.62)

$$A' = -\frac{\ln^3 \alpha}{2} \tag{9.69}$$

So that with (9.67):

$$m_P = -\frac{\ln^3 \alpha}{2} \cdot P^2 \cdot \alpha^P \tag{9.70}$$

In Fig. 9.24, the values of m_P for termination by disproportionation (I) and combination (II) with an assumed kinetic chain length of $\nu = 10$ are plotted as a function of the degree of polymerization using (9.64), (9.70), and (9.54).

9.4.3 Average Degree of Polymerization and Non-Uniformity

The number-average P_n (9.71) and weight-average P_w (9.72) degrees of polymerization can be derived from the mass distribution functions and the equations for the number and weight-averages of the degrees of polymerization:

$$P_n = \frac{1}{\int_0^\infty \frac{m_P}{P} dP} \quad (9.71)$$

P_n Number-average of degree of polymerization

$$P_w = \int_0^\infty m_P \cdot P \cdot dP \quad (9.72)$$

P_w Weight-average of degree of polymerization

Derivation of the (9.71) and (9.72):
The number-average molar mass is defined as follows according to ► Chap. 3:

$$M_n = \frac{\sum n_P \cdot M_P}{\sum n_P}$$

From $M_P = M_M \cdot P$ and $m_P = n_P \cdot P$:

$$M_n = \frac{\sum \frac{m_P}{P} \cdot M_M \cdot P}{\sum \frac{m_P}{P}}$$

For the number-average degree of polymerization P_n it follows from $M_n = P_n \cdot M_M$ that

$$\frac{M_n}{M_M} = P_n = \frac{\sum m_P}{\sum \frac{m_P}{P}}$$

In integral notation one obtains

$$P_n = \frac{\int_0^\infty m_P dP}{\int_0^\infty \frac{m_P}{P} dP}$$

Because, by definition, $\int_0^\infty m_P dP = 1$ (see (9.59)), this results in (9.71).

Similarly, starting from the equation for the weight-average molar mass M_w:

$$M_w = \frac{\sum n_P \cdot M_P^2}{\sum n_P \cdot M_P}$$

One obtains (9.72) for the weight-average degree of polymerization P_w (9.72).

Termination by Disproportionation

Substituting (9.64) into (9.71) yields the number-average of degree of polymerization P_n:

$$P_n = \frac{1}{\ln^2 \alpha \int_0^\infty \alpha^P dP} \tag{9.73}$$

Using (9.62) it follows that

$$P_n = -\frac{1}{\ln \alpha} \tag{9.74}$$

For $\alpha \leq 1$, i.e., for large molar masses:

$$-\ln \alpha = 1 - \alpha \tag{9.75}$$

With (9.54) and (9.74) and (9.75) one obtains

$$P_n = \nu \tag{9.76}$$

From (9.72) and (9.64):

$$P_w = \ln^2 \alpha \int_0^\infty P^2 \cdot \alpha^P dP \tag{9.77}$$

With (9.62) it now follows that

$$P_w = -\frac{2}{\ln \alpha} \tag{9.78}$$

and substitution of (9.75) and (9.54) into (9.78) yields

$$P_w = 2\nu \tag{9.79}$$

With these equations for P_n (9.76) and P_w (9.79) and the definition of the non-uniformity U (9.80), one obtains (9.81):

$$\mathrm{U} = \frac{P_w}{P_n} - 1 \quad (9.80)$$

$$U = \frac{2\nu}{\nu} - 1 = 1 \quad (9.81)$$

Thus, if polymerization is terminated by disproportionation, the non-uniformity $U = 1$.

Termination by Combination

A similar treatment can be applied for polymerizations terminated by combination. Thus, substitution of (9.70) into (9.71) for n yields

$$P_n = \frac{1}{\int \left(-\frac{\ln^3 \alpha}{2} \cdot \frac{P^2}{P} \alpha^P dP \right)} \quad (9.82)$$

From (9.62), (9.75), and (9.54) it follows that

$$P_n = -\frac{2}{\ln \alpha} = 2\nu \quad (9.83)$$

P_w can then be derived from (9.70) and (9.72):

$$P_w = \int \left(-\frac{\ln^3 \alpha}{2} P^3 \alpha^P dP \right) \quad (9.84)$$

Finally, using (9.62), (9.75), and (9.54) yields for P_w (9.85) and for U (9.86):

$$P_w = -\frac{3}{\ln \alpha} = 3\nu \quad (9.85)$$

$$U = \frac{3\nu}{2\nu} - 1 = 0.5 \quad (9.86)$$

If termination of polymerization takes place by combination, the non-uniformity of the polymer is given by (9.86)—$U = 0.5$.

Thus, the non-uniformity of the product of a polymerization terminated by combination is lower than that for polymers terminated by disproportionation. Because termination is generally not exclusively by one or other mode, the experimentally determined values for U lie between these limits. However, much higher values for U can be observed caused by transfer reactions.

In the scientific literature, the terms polydispersity and *polydispersity index PDI* (9.87) are often used instead of non-uniformity *U*:

$$PDI = \frac{P_w}{P_n} = U + 1 \tag{9.87}$$

9.5 Controlled Radical Polymerization (CRP)

The synthesis of block copolymers, i.e., of polymers with longer, alternating sequences of at least two different monomers, by the sequential addition of the different monomers, was a domain of ionic polymerization until 1985 (▶ Chap. 10). In the following sections methods a method of synthesizing such polymers via radical polymerization is introduced.

9.5.1 Comparison of Living and Radical Polymerization

A polymerization is considered as living when the rate of initiation is significantly greater than the rate of propagation and, additionally, when termination and transfer reactions are absent. The average degree of polymerization P_n can be determined at complete monomer conversion and for an initiator which is 100 % effective from the following relation:

$$P_n = \frac{[M]_0}{[I]_0} \tag{9.88}$$

$[M]_0$	Initial monomer molar concentration
$[I]_0$	Initial initiator molar concentration

Essentially ideal polymerizations, as living ionic polymerization are, yield products with narrow molar mass distributions (Poisson distribution; $M_w/M_n < 1.1$, ▶ Sect. 10.2.6). The active lifetime of the growing chains is "infinite"—i.e., these chains only stop growing when all the monomer is consumed. Additional monomer leads to additional chain growth and the molar masses increase proportional to conversion. Adding an alternative, skillfully selected monomer results in a block copolymer.

The reaction schemes for radical polymerization (initiation, propagation, termination, and transfer, Sect. 9.2.1) suggest that living radical polymerization should not exist. New chains are constantly started and growing ones terminated. The molar mass distributions are broad. As discussed above (▶ Sect. 9.4), $M_w/M_n \geq 2$ results if termination is by disproportionation and $M_w/M_n \geq 1.5$ if termination is by combination . Usually, it is therefore not possible to obtain block copolymers by radical polymerization simply by sequentially adding monomers (▶ Sect. 10.2.2.2).

In ◘ Table 9.5 the most important characteristics of living polymerization and of free radical polymerization are compared.

It would be of great scientific and technical value if free radical polymerizations were living. This would entail the ubiquitous irreversible termination and transfer reactions being curtailed or eliminated and the development of a suitable rapid, quantitative initiation so

Table 9.5 Characteristic differences between living polymerization and typical free radical polymerization

Living polymerization	Free radical polymerization
Block copolymers by sequential monomer addition	Block copolymers not possible
$M_w/M_n \approx 1.1$	$M_w/M_n \geq 1.5$ or ≥ 2.0
P_n increases linearly with conversion	P_n greatest at the start; decreases with increasing conversion because $P_n \sim [M]_t$
Sensitive to contaminations, extreme conditions, often $T < 0\,°C$	Insensitive to contaminations, often $T > 70\,°C$
Demanding, expensive technology, technically only used for specialties	Inexpensive technology, often used for technical processes

that all the chains grow simultaneously. If one compares the essential reaction for the formation of the polymer chain (v_p) with that responsible for the termination (v_t), it is apparent that both are dependent on the radical concentration but with different reaction orders:

$$v_p \sim \left[P^{\bullet}\right] \quad \text{first-order} \tag{9.89}$$

$$v_t \sim \left[P^{\bullet}\right]^2 \quad \text{second-order} \tag{9.90}$$

By reducing the radical concentration $[P^{\bullet}]$, the relative rate of the second-order reaction is reduced more than that of the first-order reaction. The rate of polymerization does decrease but not as much as the undesired termination reaction.

A reduction of $[P^{\bullet}]$ can be achieved by adding a stable radical $T^{\bullet}$ which reacts with the active end of a growing chain $P^{\bullet}$. By adding energy, $T^{\bullet}$ can be split off again so that the chain can continue growing. $T^{\bullet}$ itself should not be able to react with the monomer. This reaction between $P^{\bullet}$ and $T^{\bullet}$ should also be reversible (Fig. 9.25).

Free $P^{\bullet}$ can add additional monomers but, after every few growth steps, reacts with $T^{\bullet}$ and is transformed into the thermolabile P-T-species.

The concentration of the active chains is very small:

$$\frac{\left[P^{\bullet}\right]}{\left[P-T\right]} \leq 10^{-5} \tag{9.91}$$

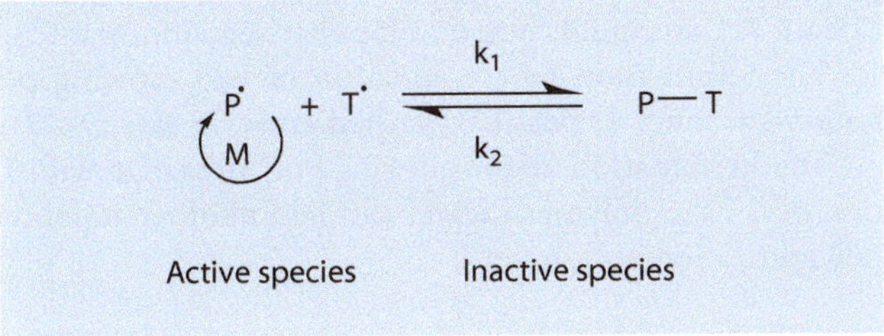

Fig. 9.25 Equilibrium between the active species $P^{\bullet}$ and the *dormant* species *P-T*

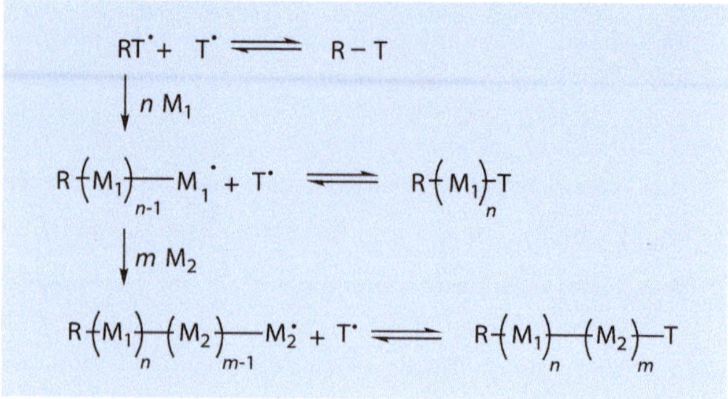

Fig. 9.26 Synthesis of block copolymers by controlled radical polymerization

If the exchange between dormant and active species takes place quickly enough, the active species only add a few monomer units and, if the next activation follows statistical rules, a linear increase in the degree of polymerization with conversion, typical of a living polymerization, can be observed. Through this reversible chain termination, this polymerization differs significantly from a true living polymerization (no termination), but also from a typical free radical polymerization (irreversible chain termination) and it is therefore better known as *controlled radical polymerization* (CRP). An important goal of CRP is to enable the synthesis of block copolymers (Fig. 9.26).

9

The characteristics of a CRP are:

- Linear increase in the degree of polymerization with the conversion
- Low *PDI* ($PDI \leq 1.3$)
- High level of end functionalization (here end group *T*)
- Possibility of the synthesis of block copolymers
- Typical free radical polymerization insensitivity to contamination and water

The possibility of obtaining block copolymers that are inaccessible by ionic polymerization, as well as the insensitivity of the growing species to contamination and water, are obvious advantages of CRP. Prerequisites for the success of a CRP are a low concentration of $P^{\bullet}$ and a reversible chain termination with $T^{\bullet}$ which cannot start chains by itself. In the following Sects. 9.5.2, 9.5.3, 9.5.4, and 9.5.5, important CRP methods are introduced.

9.5.2 Dissociation-Combination-Mechanism

The dissociation combination mechanism involves the propagation of the polymer chains being interrupted by conversion to a stable radical $T^{\bullet}$ which is unable to trigger a radical polymerization by itself, as already shown in Fig. 9.25 (*p*ersistent *r*adical *e*ffect, PRE; Fig. 9.27), an equilibrium develops between "dormant" species *P-T* and growing radicals $P^{\bullet}$. The termination by combination of two growing polymer radicals ($P^{\bullet}$) should be reduced as much as possible (dashed arrow in Fig. 9.27).

The "termination" of the growing polymer chain with $T^{\bullet}$ is reversible. After the separation of $T^{\bullet}$, the polymer radical can add another monomer (Fig. 9.29, example of the polymerization of styrene).

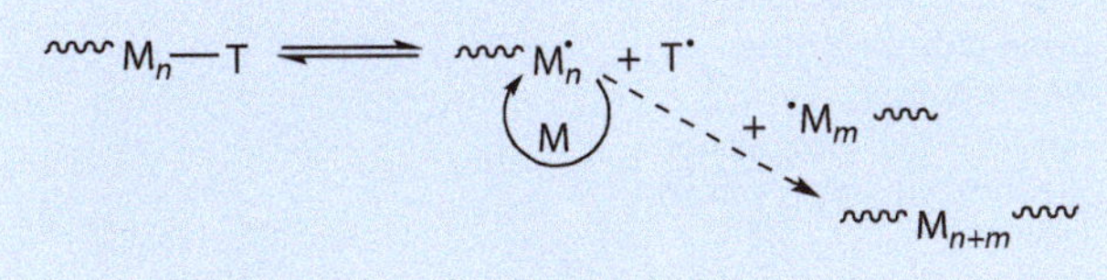

Fig. 9.27 Controlled radical polymerization by dissociation-combination-mechanism

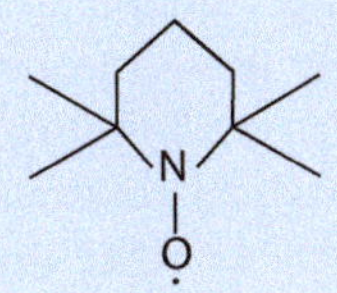

Fig. 9.28 Structural formula of TEMPO

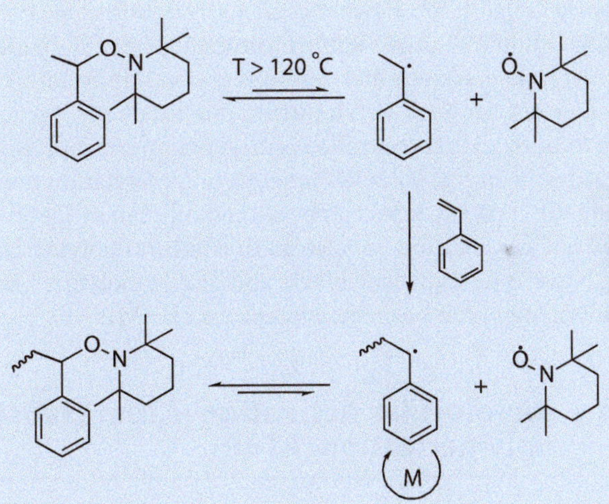

Fig. 9.29 Alkoxy amine as initiator of the controlled radical polymerization of styrene

An important class of materials for T• are the nitroxide radicals (Solomon et al 1985, Georges et al. 1993, Hawker et al. 2001, Benoit et al. 1999, 2000) and TEMPO ((2,2,6,6-**Tetramethylpiperidin-1-yl)oxyl)** (Fig. 9.28) is the example which has been most often studied.

The initiation of CRP can take place with common initiators such as benzoyl peroxide and azobisisobutyronitrile in the presence of nitroxide radicals. It can, however, be initiated by special alkoxy amines (Fig. 9.29).

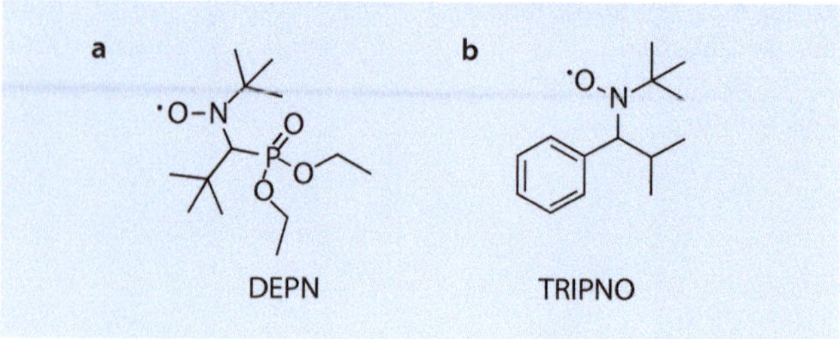

Fig. 9.30 Examples of *N*-oxide-radicals for the NMP method. (**a**) *N*-*t*Bu-*N*-[1-**d**iethyl**p**hosphono-(2,2-dimethylpropyl)]-**n**itroxide (DEPN). (**b**) 2,2,5-**Tri**methyl-4-**p**henyl-3-aza-hexan-**n**itr**o**xide (TRIPNO)

Using CRP, polymers such as polystyrene, with a narrow molar mass distribution, can be synthesized. By these means, block copolymers can also be obtained (styrene/substituted styrene, styrene/methyl methacrylate, styrene/*N*-alkyl acrylamide, etc.). However, controlled radical polymerization occurs significantly slower than typical free radical polymerization. Another disadvantage is that, with the exception of TEMPO, alkoxy amines are not commercially available and these end groups are, for example, for the synthesis of telechelics, not easily converted into other interesting functional end groups.

9

Controlled radical polymerization in the presence of alkoxy amines can be found in the literature as NMP-method (**N**-oxide **m**ediated **p**olymerization) (Solomon et al. 1985; Georges et al. 1993; Hawker et al. 2001; Benoit et al. 1999, 2000).

Figure 9.30 shows some more examples of *N*-oxide-radicals.

As well as styrene and various acrylates, monomers polymerizable using this method include acrylamides, acrylonitrile, and butadiene. Methacrylates form sterically demanding radicals so that controlled radical polymerization with such monomers via NMP is not possible. Whereas TEMPO requires polymerization temperatures of $T \geq 120$ °C, TRIPNO and DEPN (Fig. 9.30) can be successful even at $T \geq 80$ °C. The NMP-method can be carried out as a solution but also as an emulsion polymerization (▸ Sect. 16.5).

Numerous statistical, block, and graft copolymers, as well as star and highly branched (co)polymers can be created by means of NMP.

9.5.3 Bimolecular Activation (Atom Transfer Radical Polymerization, ATRP)

At almost the same time as CRP was being developed, in 1995, Matyjaszewski and Sawamoto described radical polymerizations that involve a bimolecular initiation reaction and have become known as **a**tom **t**ransfer **r**adical **p**olymerizations (ATRP) (Fig. 9.31).

As the Y (in Fig. 9.31) Cu(I)-(Wang and Matyjaszewski 1995), Ru(II)-(Kato et al. 1995), and also Fe(II)-, Ni(II)-, and Ni(0)-complexes can be employed. One well known example of ATRP is shown in Fig. 9.32.

The list of monomers that are polymerizable by means of ATRP includes styrene, various substituted styrenes, (meth-)acrylates, (meth-)acrylamides, acrylonitrile, and vinyl pyridine.

Typical initiators are depicted in Fig. 9.33.

Some of the diverse structures of the ligands which have been used are shown in Fig. 9.34.

■ **Fig. 9.31** Scheme of a controlled radical polymerization through bimolecular activation

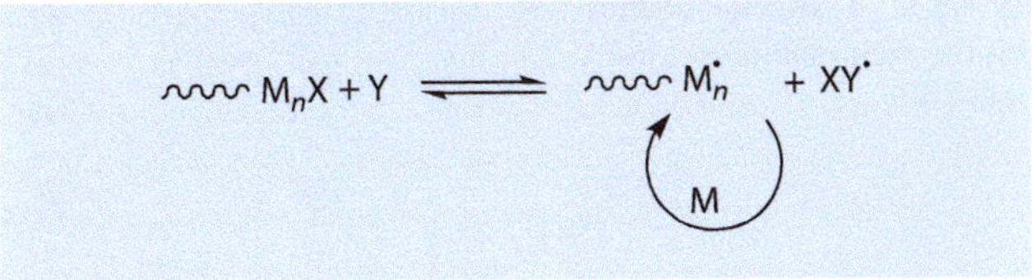

■ **Fig. 9.32** ATRP of styrene, initiated with 1-bromoethyl benzene and Cu(I)-bis-(bipyridine)

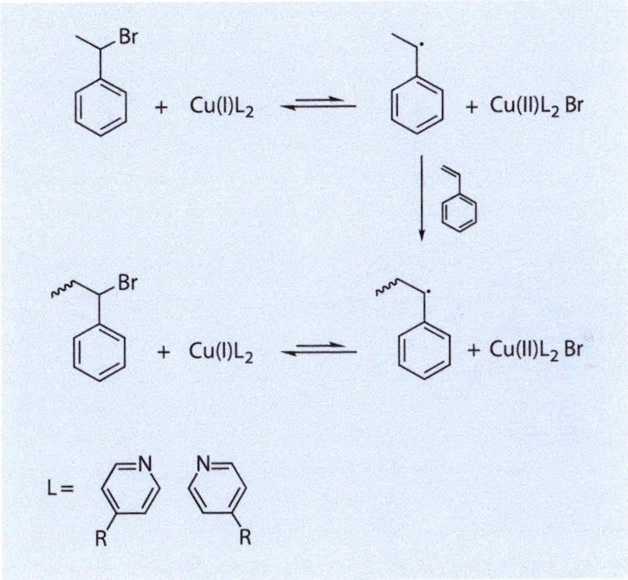

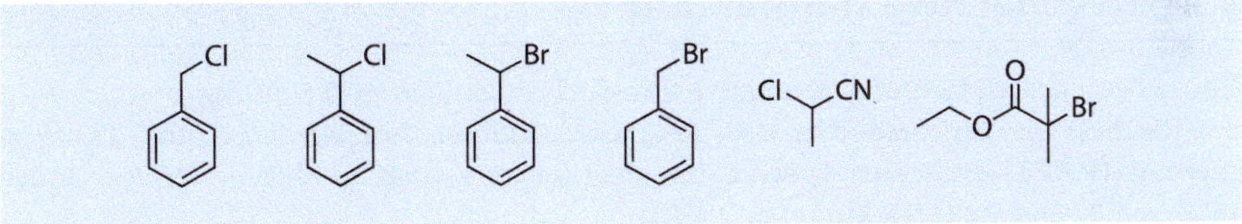

■ **Fig. 9.33** Typical initiators for ATRP

Because the end group (e.g., X = Br) remains throughout the polymerization, it can be easily used for the synthesis of block copolymers (■ Fig. 9.35). To this end, monomer M_2 is added after the polymerization of monomer M_1 is complete.

ATRP can be applied to many radically polymerizable monomers. The reaction conditions are comparable to those for typical free radical polymerization but the polymerization temperatures for ATRP are significantly higher than normal for free radical polymerization. The molar masses are controlled by the $[M]/[I]$-ratio. Furthermore, polymers with narrow molar mass distributions can be obtained. The color of many metal complexes as well as their toxicity and the difficulty of their removal are disadvantages. Because here, as well as with NMP (► Sect. 9.5.2), the concentration of the active species must be very low, the rate of polymerization is comparatively slow despite the higher reaction temperatures.

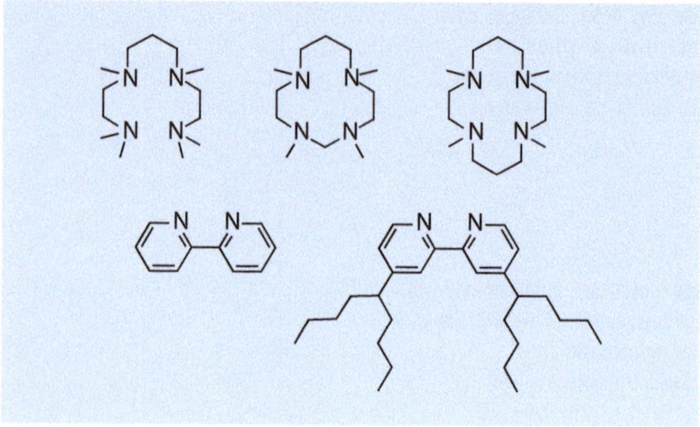

Fig. 9.34 Selected ligands for the metal complexes used for ATRP

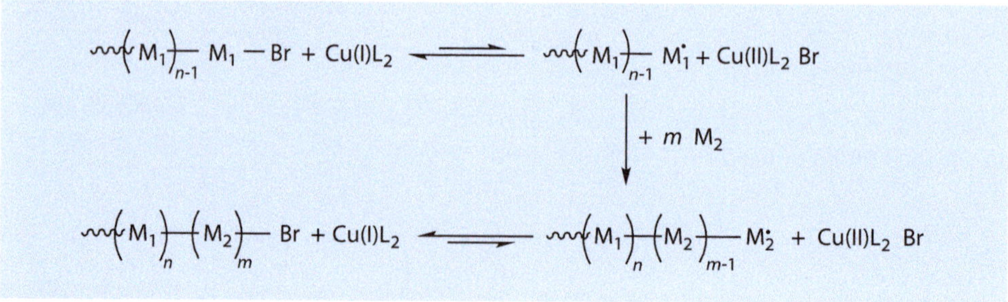

Fig. 9.35 Synthesis of poly(M_1-*Block*-M_2) by ATRP

9.5.4 Degenerative Transfer (RAFT)

The general description of degenerative transfer is illustrated in Fig. 9.36.

The best known method is the "***Reversible Addition Fragmentation and Transfer***" method (RAFT) in which specific transfer reagents such as dithioester are added (Figs. 9.37 and 9.38) (Delduc et al. 1988).

If transfer occurs quickly and the transfer reagent is used in sufficient concentration ($[Transfer\ agent] > 10 \cdot [I]$), a free radical polymerization can relatively easily become a controlled radical polymerization.

RAFT has the following advantages compared to ATRP:

- No metals
- Milder conditions

Similar to ATRP, RAFT can be used for many vinyl monomers. However, complex research is required to find an appropriate dithioester for each monomer. Furthermore, colored and odor-intensive byproducts are common. Dithioester end groups are less easily converted into other interesting end groups and they are not particularly stable at higher temperatures or with respect to light.

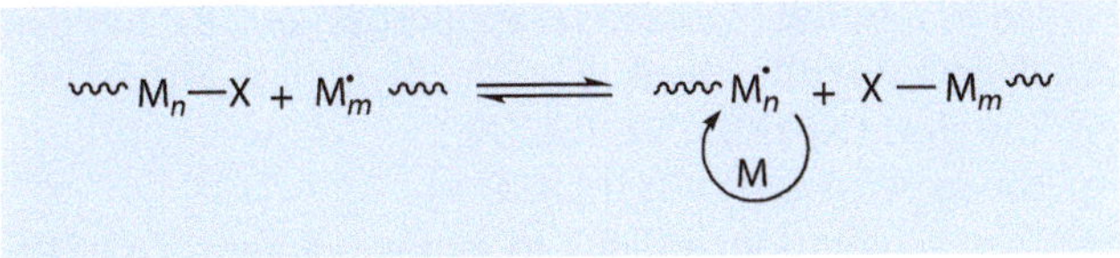

Fig. 9.36 Controlled radical polymerization via degenerative transfer

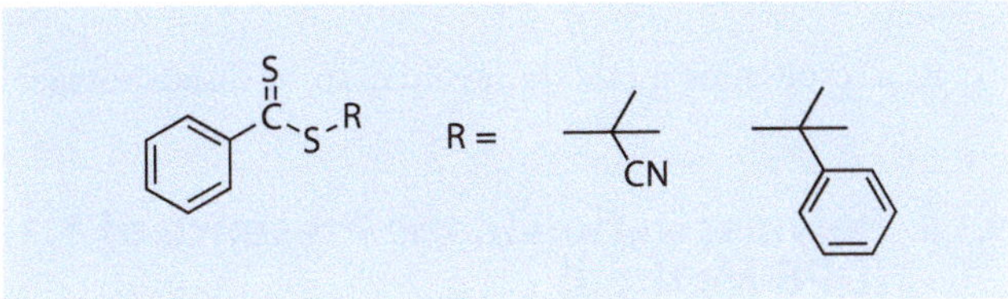

Fig. 9.37 Structure of the dithioester used for RAFT

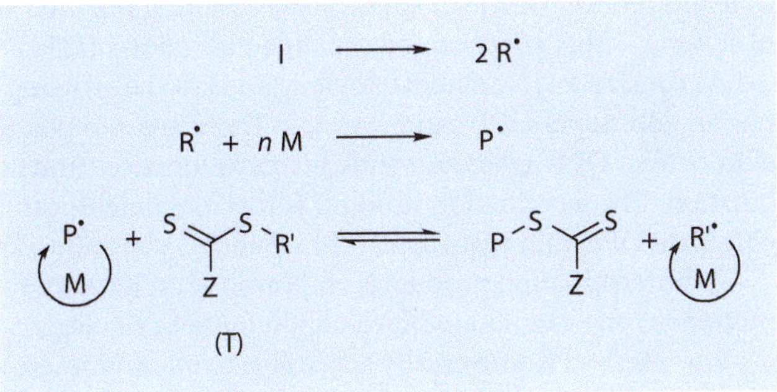

Fig. 9.38 Controlled radical polymerization by RAFT

9.5.5 Comparison of NMP, ATRP, and RAFT

Every CRP-system has advantages and disadvantages. The advantage of the NMP-system is that it is a purely organic system and that it is applicable for many monomers including acids. Disadvantages are the high reaction temperature and the lack of availability of the *N*-oxide-precursors as well as the difficult conversion of the end groups into more attractive groups. Furthermore, the *N*-oxide-radicals need to be used at the same concentration as the initiator.

For ATRP, only catalytic amounts of the metal complex transfer agents are required and many monomers are polymerizable using this method. The end groups (X = Br) are easily converted. Additionally, many potential initiators have simple structures and are commercially available. The principal disadvantages of ATRP are the color, the toxicity, and the difficulty of removing the metal residues.

RAFT is equally universally applicable as ATRP. However, only a few dithioesters are commercially available. Other disadvantages are the odor, color, and toxicity of the products, byproducts, or their decomposition products.

The industrial application of CRP has the following advantages:

- The technology is simple
- Block copolymers can be made by radical polymerization
- The molar masses are easily controllable and the molar mass distributions are narrow

Until now, the commercial application of CRP has been limited to niches because the disadvantages still prevail:

- Too slow (NMP, ATRP, RAFT)
- Too expensive (NMP, RAFT)
- Toxic, coloration, difficult to remove catalysts (ATRP)
- Colored products (ATRP, RAFT)
- Odor-intensive products (RAFT)

As a result, new methods devoid of these disadvantages are still being looked for.

9.5.6 Polymerization in the Presence of 1,1-Diphenylethylene (DPE-Method)

1,1-**Diphenylethylene** (DPE) is noted because even traces retard the rate of polymerization and reduce the molar mass of the resulting polymer (e.g., during the polymerization of styrene.). This effect can be explained as follows (◘ Fig. 9.39).

As soon as a 1,1-diphenylethylene adds to the growing chain, neither the monomer M nor an additional DPE can be added. There are two possibilities for this radical: either it eliminates a DPE again and adds further monomer units or it simply terminates by combination. The termination product is thermolabile; it can break open at the original link, eliminate DPE, and then, as described above, continue to propagate.

If the termination product is employed as an initiator for the polymerization of another monomer, one can obtain block copolymers in an elegant way (◘ Fig. 9.40).

This method is universally applicable to all monomers that are radically polymerizable (Raether et al. 2001; Wieland 2003) and block copolymers are easily synthesized from more than two different monomers.

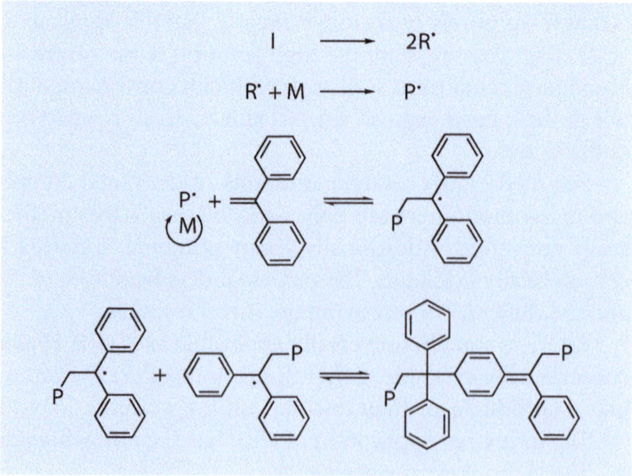

◘ **Fig. 9.39** Effect of 1,1-diphenylethylene on radical polymerization

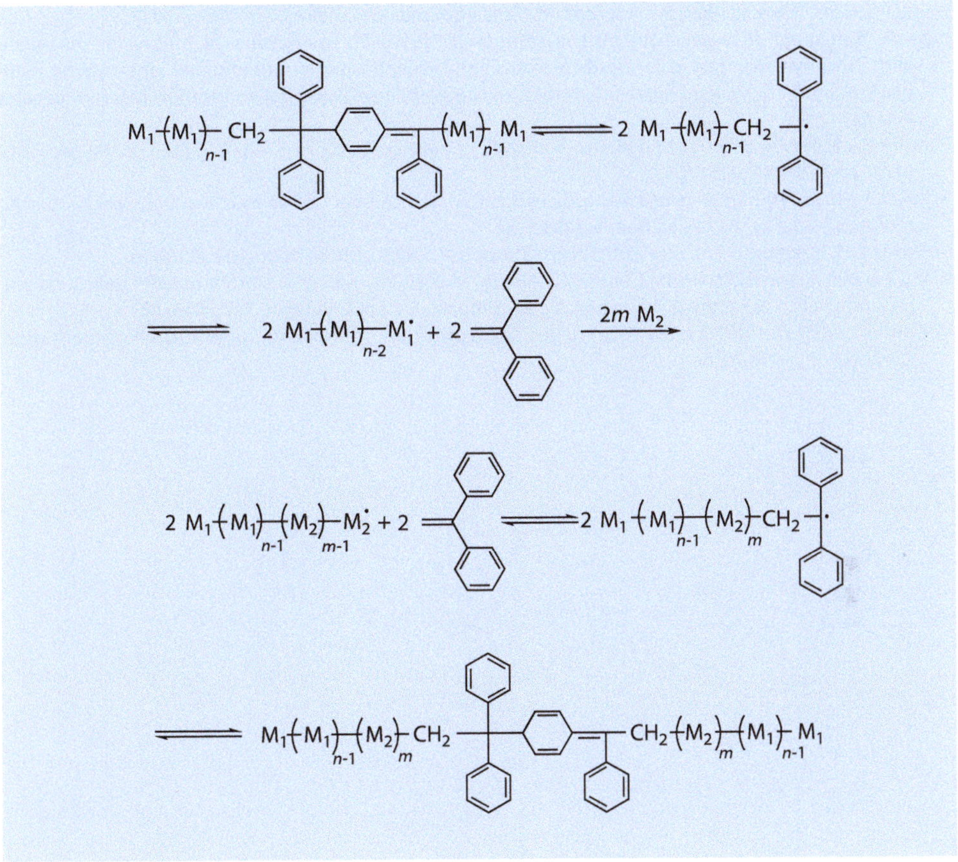

Fig. 9.40 Synthesis of a block copolymer from M_1 and M_2 by the DPE-method

References

Benoit D, Chaplinski V, Braslau R, Hawker CJ (1999) Development of a universal alkoxyamine for "living" free radical polymerization. J Am Chem Soc 121:3904–3920

Benoit D, Grimaldi S, Robin S, Finet J-P, Tordo P, Gnanou Y (2000) Kinetics and mechanism of controlled free-radical polymerization of styrene and *n*-butylacrylate in the presence of an acyclic beta-phosphonylated nitroxide. J Am Chem Soc 122:5929–5939

Brandrup J, Immergut EH (1989) Polymer handbook, 3rd edn. Wiley, New York

Delduc R, Tailhan C, Zard SZ (1988) A convenient source of alkyl and acyl radicals. J Chem Soc Chem Commun: 308–310

Flory PJ (1953) Principles of polymer chemistry, vol VIII. Cornell University Press, New York

Frost AA, Pearson RG (1964) Kinetik und Mechanismus homogener chemischer Reaktionen, Verlag Chemie, Weinheim

Georges MK, Veregin RPM, Kazmaier PM, Hamer GK (1993) Narrow molecular weight resins by a free-radical polymerization process. Macromolecules 26:2987–2988

Hawker CJ, Bosman AW, Harth E (2001) New polymer synthesis by nitroxide mediated living radical polymerizations. Chem Rev 101:3661–3688

Henricí-Olivé G, Olive S (1969) Polymerisation. Verlag Chemie, Weinheim
Kato M, Kamigaito M, Sawamoto M, Higashimura T (1995) Polymerization of methyl methacrylate with the carbon tetrachloride/dichlorotris-(triphenylphosphine)ruthenium(II) methylaluminum bis(2,6-di-tert-butylphenoxide) initiating system: possibility of living polymerization. Macromolecules 28:1721–1723
Raether B, Modery B, Braun F, Brinkmann-Rengel S, Christie D, Haremza S. WO 0144327 (2001) BASF AG, Inv.; Chem Abstr 135:61763
Schulz GV (1939) The kinetics of chain polymerization, V. The effect of various reaction species on the multimolecularity. Z Physik Chemie B43:25–46
Solomon DH, Rizzardo E, Cacioli P (1985) Eur Pat Appl EP135280. Chem Abstr 102:221335q
Wang J-S, Matyjaszewski K (1995) Controlled/"living" radical polymerization, atom transfer radical polymerization in the presence of transition metal complexes. J Am Chem Soc 117:5614–5615
Wieland P (2003) Das DPE-System—Ein neues Konzept für die kontrollierte radikalische Polymerisation. Dissertation, TU München

Ionic Polymerization

S. Koltzenburg et al., *Polymer Chemistry*, DOI 10.1007/978-3-662-49279-6_10

Ionic polymerization, similar to radical polymerization, involves a chain mechanism in which either cations (cationic polymerization) or anions (anionic polymerization) are the active centers. Because the solution is electroneutral, the number of active centers and their counterions is identical. Whether a monomer with a carbon-carbon double bond can be polymerized anionically or cationically depends on the electron density of the double bond, which in turn depends on the substituents. If the substituents induce a donor effect (OR, NR_2, $C_6H_4{-}CH_3$), then the monomer favors a cationic polymerization. By contrast, monomers with acceptor-substituents (CN, COOR, $CONR_2$) can be anionically polymerized. Because the growing chains are identically charged they cannot terminate the reaction by, for example, combining with one another which means they remain active. A complete lack of termination reactions is known as a *living polymerization*. The chains can be extended by the addition of more monomers. Block copolymers can be synthesized by adding other monomers capable of polymerization. Additionally, the molar masses can be controlled using the ratio $[M]:[I]$.

10.1 Cationic Polymerization

Although some of the first polymerizations observed by chemists were probably cationic polymerizations, it was not until it was discovered that isobutene, trioxane, cyclic ethers such as oxirane and tetrahydrofuran, for example, as well as cyclic siloxanes, could be cationically polymerized that a general interest developed in this type of polymerization.

A special feature of cationic polymerization is its high selectivity toward monomers with donor functions (Fig. 10.1).

As is the case for radical polymerization, cationic polymerization has two distinguishable parts, initiation and propagation. A termination of the reaction caused by two growing chains reacting with one another, a typical feature of radical polymerization, is not possible (Fig. 10.2). There are, however, reactions that result in a termination of chain growth, as discussed in ► Sect. 10.1.4.2.

As the active species (mostly carbocations) used in cationic polymerization are very reactive, the solvent needs to be chosen with special care and it should be extremely clean; even traces of water or methanol should be avoided. Preferred solvents are CH_2Cl_2, CH_3Cl, $CHCl_3$, benzene, and toluene. The polymerizations are usually carried out at low temperatures because of the reactivity of the species, which increasingly take part in undesirable

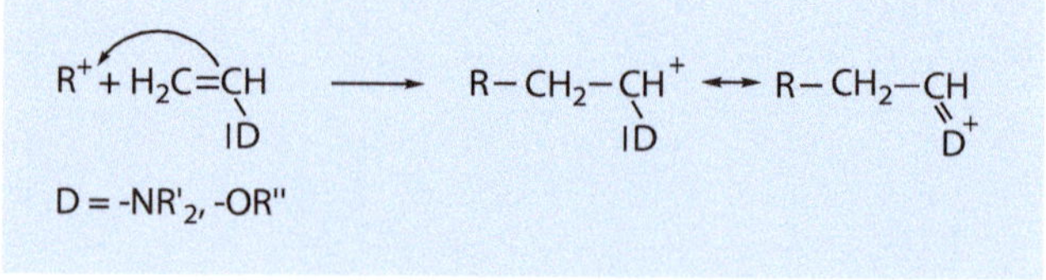

Fig. 10.1 Addition of a carbocation onto a nucleophilic monomer. *R*, *R'*, *R''* e.g., alkyl

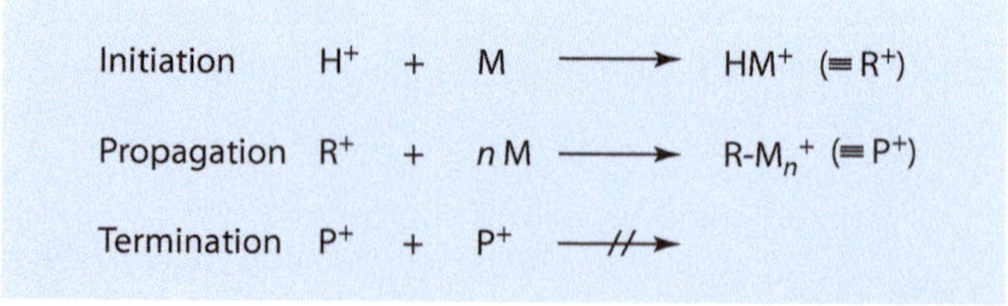

Fig. 10.2 Basic steps in a cationic polymerization

side reactions, such as Friedel–Crafts reactions or deprotonation, at higher temperatures. It is also advisable to polymerize in a high-vacuum glass apparatus or in a dry box. As an alternative, Schlenk techniques can be used.

10.1.1 Reaction Mechanism

Cationic polymerization can be described using the equation shown in Fig. 10.3. Vinyl monomers (e.g., isobutene, vinyl ether), carbonyl compounds (e.g., formaldehyde, acetaldehyde), and heterocyclic compounds (e.g., tetrahydrofuran, trioxane) can be cationically polymerized. The growth mechanisms for each of these types of monomers are discussed in the following paragraphs.

In the growth step of the polymerization of a vinyl monomer the growing chain has a terminal positive charge $\sim M_n^+$, and these coordinate with an additional monomer molecule to yield a chain with an additional monomer and a terminal charge $\sim M_{n+1}^+$. The counterion A^- ensures the electroneutrality of the system. The higher the nucleophilicity of the carbon-carbon double bond, the easier it is for cationic polymerization to take place (Fig. 10.4).

The growth step of the polymerization of carbonyl compounds is shown in Fig. 10.5. Polymerization takes place because of the nucleophilic attack of the carbonyl function on the charged chain end.

The polymerization of heterocyclic compounds, which occurs via ring-opening polymerization, is shown in Fig. 10.6.

Some cyclic monomers containing oxygen as the hetero atom, which are especially suitable for cationic polymerization, are shown in Fig. 10.7.

Fig. 10.3 Propagation in cationic polymerization

$$\sim\!M_n^+\ A^- + M \longrightarrow \sim\!M_{n+1}^+\ A^-$$

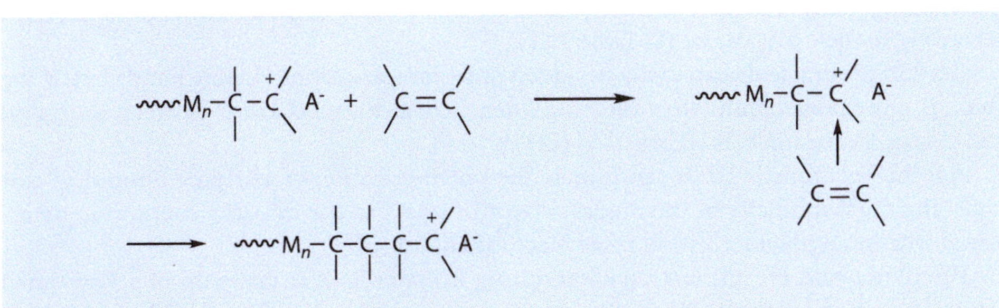

Fig. 10.4 Detailed propagation mechanism of the cationic polymerization of a monomer with a C=C double bond

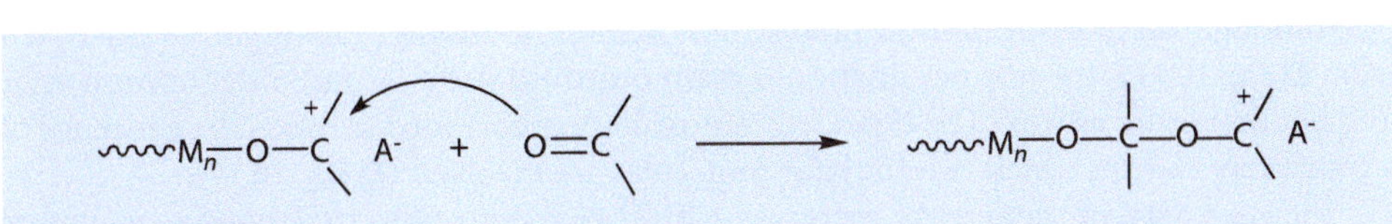

Fig. 10.5 Propagation mechanism of the cationic polymerization of a monomer with carbonyl group

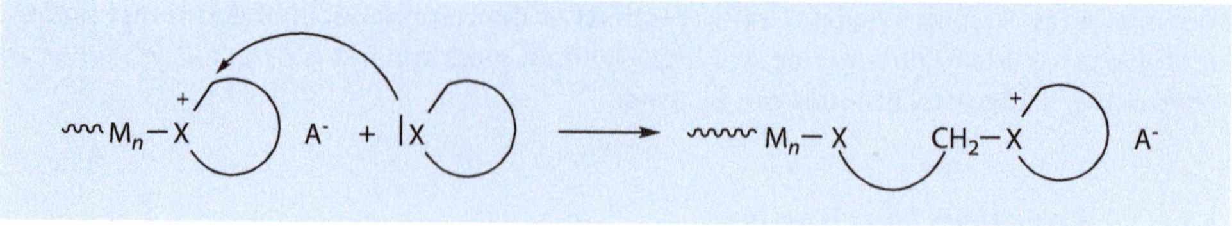

■ **Fig. 10.6** Nucleophilic attack of a heterocyclic monomer on a cationically growing chain

■ **Fig. 10.7** Some cyclic monomers containing oxygen as the hetero atom, suitable for cationic polymerization

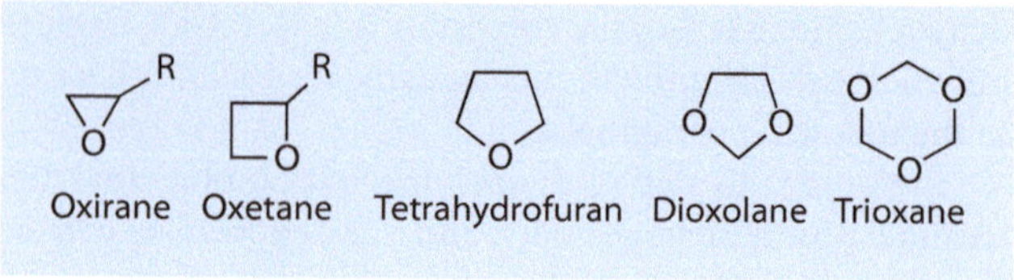

As well as those cyclic monomers shown in ■ Fig. 10.7, aziridines, thiiranes, phosphazenes, siloxanes, and oxazolines can all also be cationically polymerized via ring-opening polymerization (▶ Sect. 12.3).

10

10.1.2 Initiators

Numerous Brønsted and Lewis acids are available for the initiation of cationic polymerization. The choice of initiator depends to a large extent on the monomer to be polymerized.

10.1.2.1 Brønsted Acids

Brønsted acids with a suitable acidity can be used as initiators for cationic polymerization (■ Fig. 10.8). They can be classified either in acetic acid or acetonitrile as the solvent according to their pK_a values (■ Table 10.1).

Certain proton acids can easily be added onto carbon-carbon double bonds but, if the anion is more nucleophilic than the monomer, a covalent bond forms between the anion and cation, for example, if HCl is used (■ Fig. 10.9).

Another example is HI; it can initiate the polymerization of aziridine but not of oxirane. The nucleophilicity of the monomer with respect to the initiator counterion determines whether a polymerization takes place (■ Fig. 10.10).

Perchloric acid $HClO_4$ is a very interesting initiator as it is made up of a very small cation and a non-nucleophilic anion. However, a water-free handling of this acid is very difficult (and important) as even the smallest contamination of the substance can result in an explosion. For this reason it is often only used diluted with CH_2Cl_2 as a solvent. However, care must be taken, even with the diluted acid.

Although the polymerization rate is first order in monomer (as might be predicted from ■ Fig. 10.11), the rate of initiation is often observed to be second order in monomer and not first order as would be expected. A possible explanation for this is the existence of a transition state, in which two initiator molecules are involved (■ Fig. 10.11).

The reactivity of monomers increases with their nucleophilicity: styrene < α-methyl styrene < 4-methoxy styrene.

Fig. 10.8 Dissociation of a Brønsted acid

$$HA \rightleftharpoons A^- + H^+$$

Table 10.1 pK_a-values for a selection of Brønsted acids suitable as initiators for cationic polymerization

Brønsted acid		pK_a-Wert	
Name	Formula	In CH_3COOH	In CH_3CN
Trifluoromethane sulfonic acid	HO_3SCF_3	4.7	2.6
Perchloric acid	$HClO_4$	4.9	1.6
Hydrobromic acid	HBr	5.6	5.5
Sulfuric acid	H_2SO_4	7.0	7.3
Hydrochloric acid	HCl	8.4	8.9
Methane sulfonic acid	HO_3SCH_3	8.6	8.4
Trifluoro acetic acid	$HOOCCF_3$	11.4	10.6

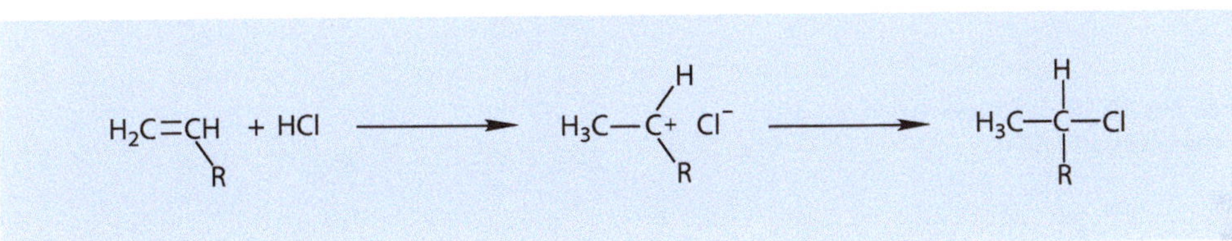

Fig. 10.9 Addition of HCl to an olefin

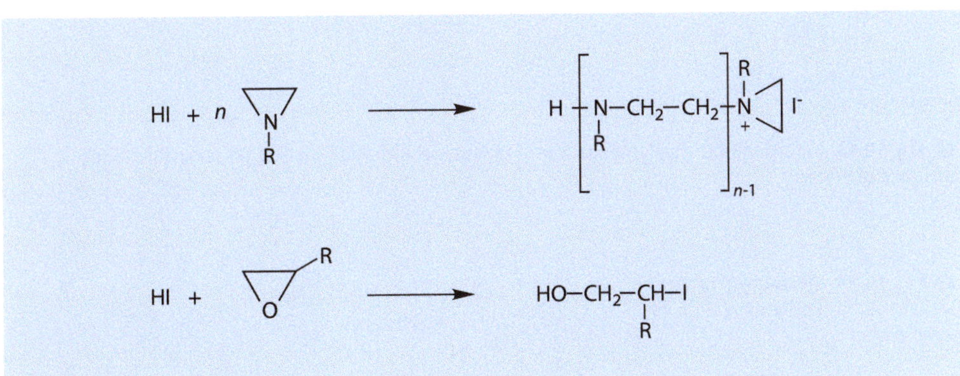

Fig. 10.10 Different reactions of HI with aziridine (polymerization) and oxirane: ring-opening and addition of the counterion

10.1.2.2 Lewis Acids

Typical Lewis acids that can be used for initiating a cationic polymerization are, e.g., $AlCl_3$, BF_3, $TiCl_4$, $SnCl_4$, and $SbCl_5$. The self-dissociation of some of them has been examined (Fig. 10.12).

The initiation of a polymerization proceeds occurs according to Fig. 10.13.

Iodine is also able to initiate cationically the propagation of certain monomers (e.g., vinyl ether) (Fig. 10.14).

The most common and also the most economically significant cationic initiation makes use of the concept of coinitiators, e.g., $TiCl_4/H_2O$ (Fig. 10.15).

Alternative methods for obtaining initiators for cationic polymerization are listed in Fig. 10.16.

Iodonium and sulfonium salts are interesting initiators which, when irradiated in the presence of donors (DH), form protons which can then initiate a polymerization of vinyl or heterocyclic monomers (Fig. 10.17).

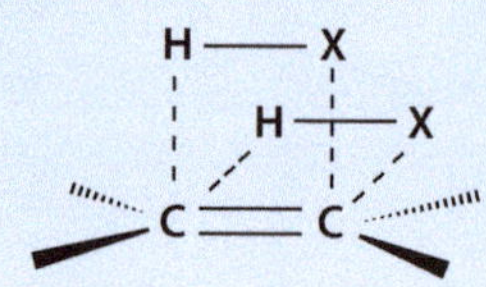

Fig. 10.11 Transition state for the reaction of HX with an olefin. *X* halogen

$$2\,AlCl_3 \rightleftharpoons AlCl_2^+ + AlCl_4^-$$

Fig. 10.12 Self-dissociation of the Lewis acid $AlCl_3$

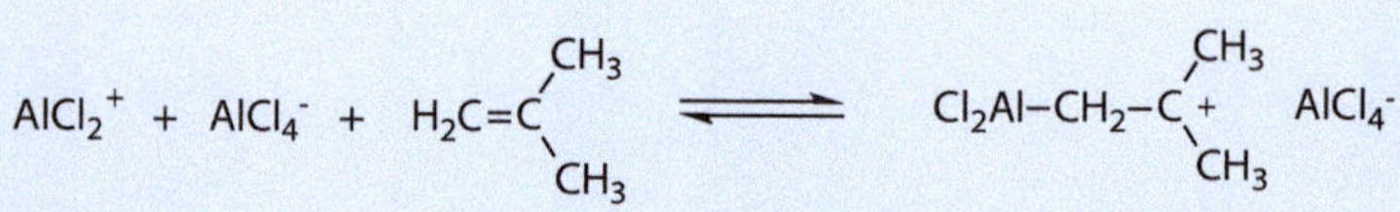

Fig. 10.13 Initiation of the polymerization of isobutene with $AlCl_3$ via self-dissociation (in the absence of water)

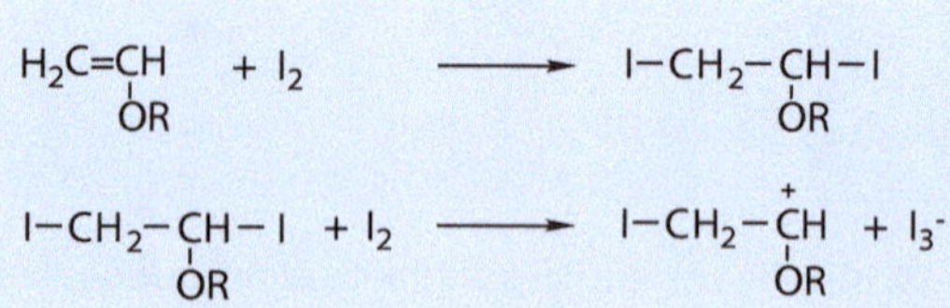

Fig. 10.14 Mechanism of the iodine initiated polymerization of vinyl ethers

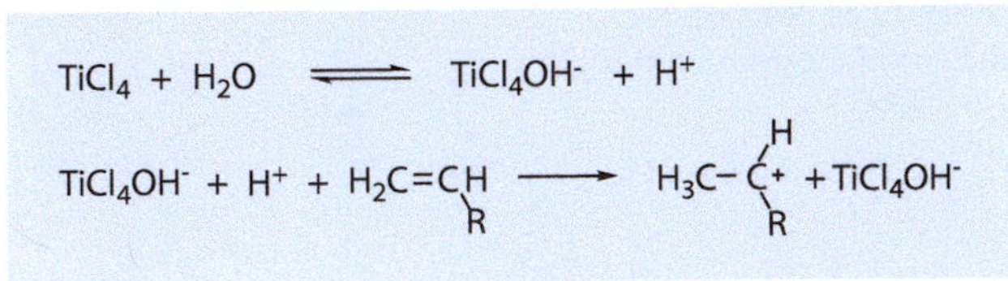

Fig. 10.15 Initiation with a Lewis acid and coinitiator taking $TiCl_4/H_2O$ as an example

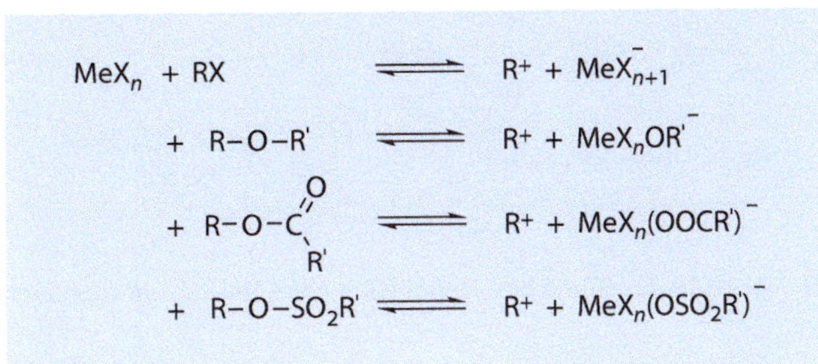

Fig. 10.16 Possibilities for synthesizing R^+. *Me* metal, *X* halogen, *R, R'* e.g., CH_3

$$Ar_3S^+ SbF_6^- \xrightarrow{h\nu} [Ar_3S^+ SbF_6^-]^* \longrightarrow Ar^{\bullet} + Ar_2S^{\bullet+} SbF_6^-$$

$$Ar_2S^{\bullet+} + SbF_6^- + DH \longrightarrow Ar_2HS^+ + SbF_6^- + D^{\bullet}$$

$$Ar_2HS^+ + SbF_6^- \longrightarrow Ar_2S + H^+ + SbF_6^-$$

Fig. 10.17 Mechanism of the formation of protons from sulfonium salts in the presence of a proton donor (DH). *Ar* aromatic ring

These types of initiators are easy to synthesize, structurally flexible, and thus can be adapted to a range of light wavelengths. They have been successfully used in printing inks, various types of coating materials, and for the production of multi-layered OLEDs.

Initiators such as HI/I_2, $HI/Bu_4N^+ClO_4^-$, or $RCO^+SbF_6^-$ are especially suitable for controlling the cationic polymerization of specific monomers (▶ Sect. 10.1.5).

10.1.3 Chain Growth During Cationic Polymerization

The chain carriers are often carbon cations which propagate via an electrophilic addition to the monomer (Fig. 10.18).

The counterion can exist as an ion pair with the active chain end, as a solvent separated ion pair, or as a free ion (Fig. 10.19).

The considerable reactivity of the carbocations can lead to alternative, irregular structures along the polymer chain because of, for example, rearrangements (Figs. 10.20 and 10.21).

In the cationic polymerization of heterocycles, propagation is via onium ions and ring-opening (Fig. 10.22 and ▶ Chap. 12. The different types of ring-opening polymerizations are dealt with in more detail in ▶ Chap. 12. Only the mechanistic details essential for a general understanding of cationic polymerization are discussed here.)

An example of this type of reaction is the polymerization of 2-methyl-2-oxazoline (Fig. 10.23).

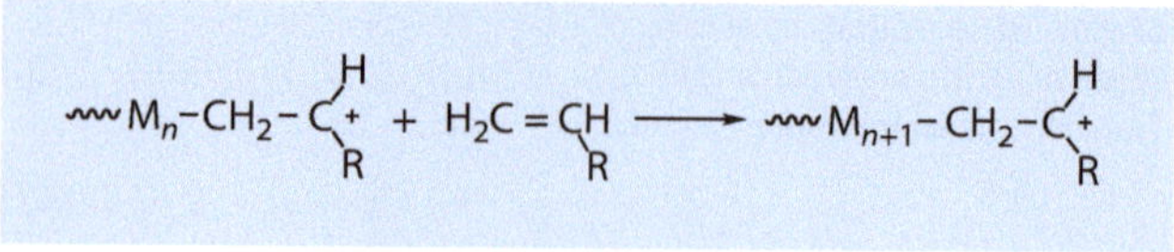

Fig. 10.18 Electrophilic addition of the growing chain to a monomer

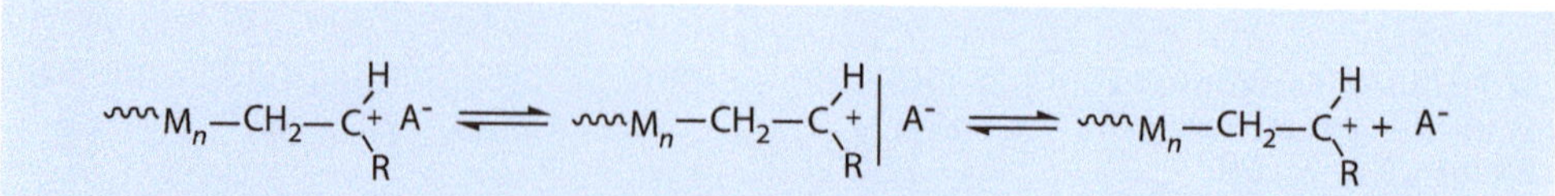

Fig. 10.19 Equilibria between the ion pairs, the solvent separated ion pairs, and the free ions

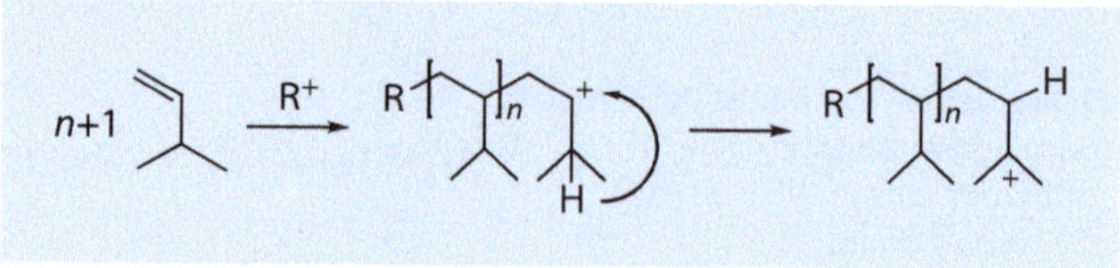

Fig. 10.20 Irregular structures resulting from hydride transfer during the polymerization of 3-methylbut-1-ene

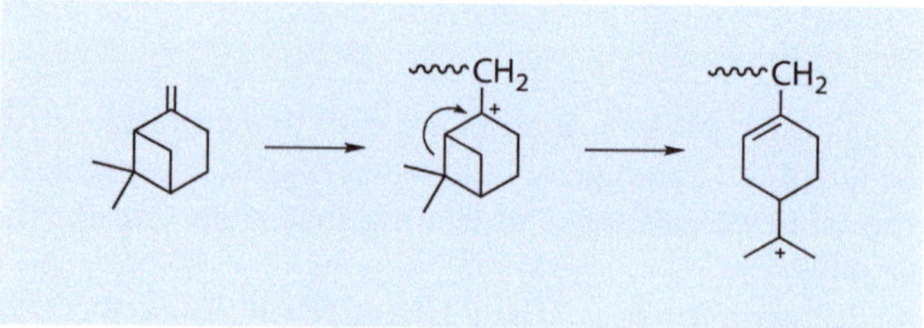

Fig. 10.21 Rearrangement during the cationic polymerization of *β*-pinene

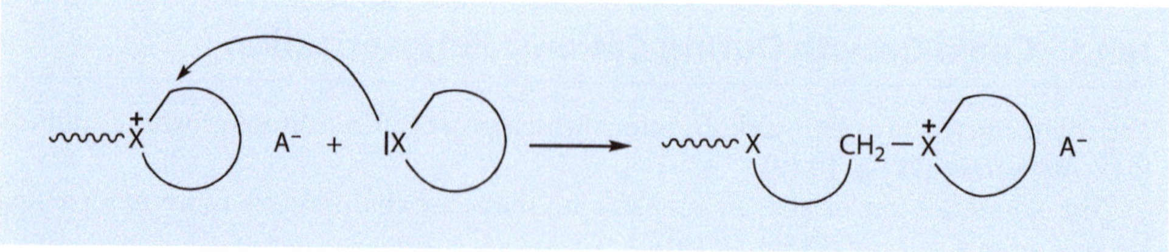

Fig. 10.22 Ring-opening and chain extension via the reaction of an onium ion with a heterocyclic monomer

The polymerization of lactams (most often the seven-atom ring lactam) is interesting and involves a so-called *activated monomer* in the transition stage (Fig. 10.24).

According to this, mechanism propagation is via a monomer protonated at the lactam-nitrogen. However, it has also been suggested that the initial step is the protonation of the carbonyl oxygen.

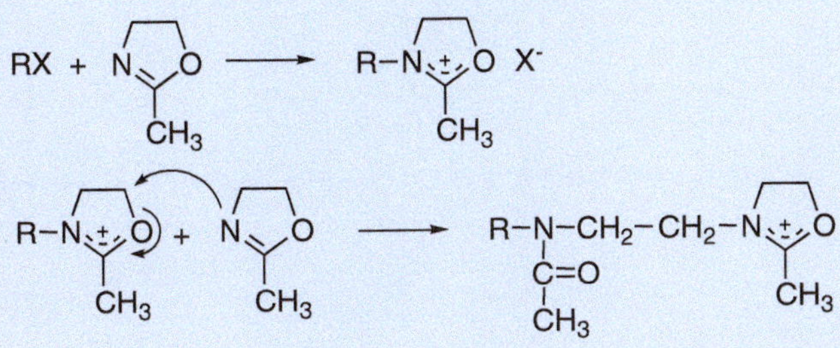

Fig. 10.23 Polymerization of 2-methyl-2-oxazoline initiated by RX (e.g., CH_3Br)

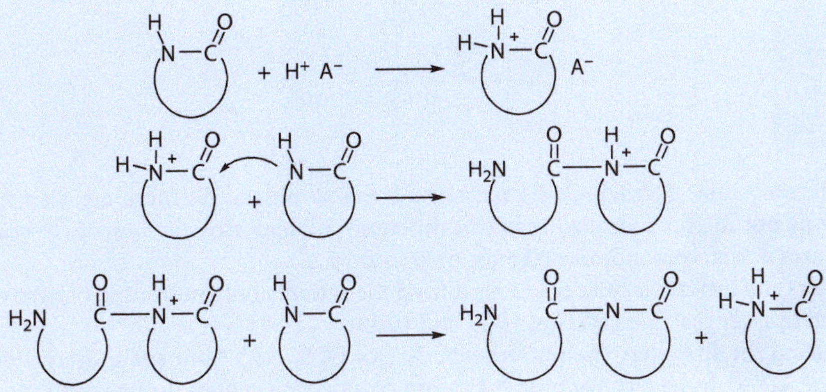

Fig. 10.24 Lactam polymerization involving *activated monomers*

10.1.4 Transfer and Termination

As has already been explained at the beginning of the chapter, the growing chains cannot terminate by combining with one another. Because of the reactivity of cationic chain ends there are a large number of reactions which can lead to chain termination or the transfer of the charge from one chain to another. These reactions are discussed in the following paragraphs (► Sects. 10.1.4.1 and 10.1.4.2).

10.1.4.1 Transfer Reactions

A transfer reaction that often occurs during the cationic polymerization of vinyl monomers is a β-H-elimination (Fig. 10.25).

If aromatic groups are present the intramolecular Friedel–Crafts reaction is another transfer reaction worth considering (Fig. 10.26).

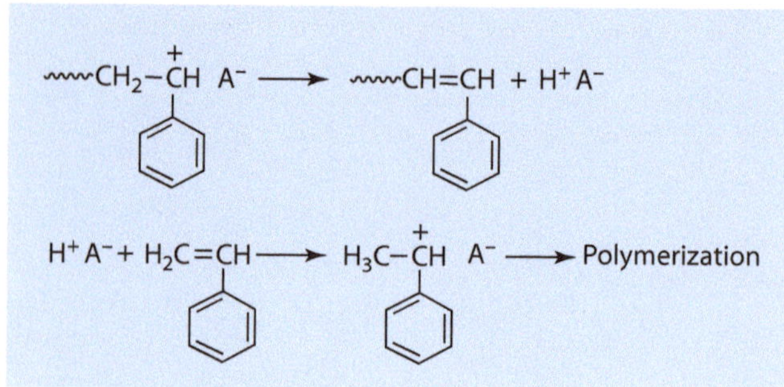

Fig. 10.25 Transfer via β-H-elimination. Cationic polymerization of styrene as an example

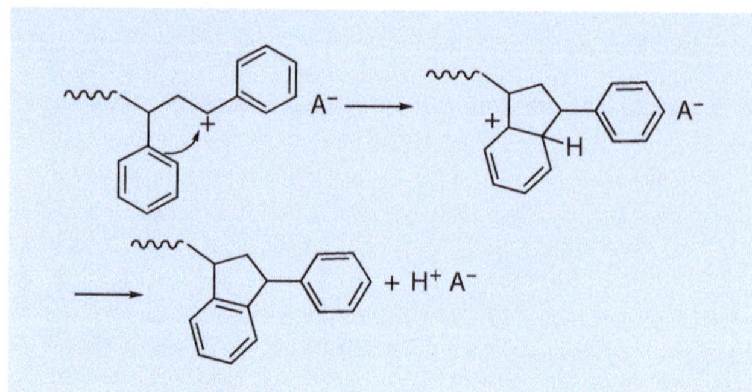

Fig. 10.26 Transfer by Friedel–Crafts alkylation (formation of indane)

Intermolecular Friedel–Crafts reactions are also observed. These are also transfer reactions but in this case also result in different polymer structures, namely branched structures or macromonomers (Figs. 10.27 and 10.28).

Inter- and intramolecular reactions during the cationic polymerization of heterocycles are also *transfer* reactions (Figs. 10.29 and 10.30).

During the first step, the nucleophilic X (X=O, S, NR) from somewhere along the chain reacts with the electrophilic α-C-atom of an active center. If remaining monomer reacts with such a branched cation at the α-C-atom the reaction yields a neutral linear chain and a charged species which continue to grow.

When such reactions are intramolecular so that rings are formed, then this is referred to as *back-biting* (Fig. 10.30).

The result of this reaction is a depolymerization with the formation of rings of varying size.

10.1.4.2 Termination Reactions

Because of the reactivity of the active species, there are many possibilities for termination during cationic polymerization. These reactions do not necessarily destroy the active center. It is often sufficient for the cation to rearrange into an inactive species (Fig. 10.31). For example, a hydride shift can lead to an inactive species caused by steric considerations.

If *anion splitting* occurs, which is also occasionally referred to as the decomposition of the anion, then termination occurs through the reaction of the active cation with part of the counterion (Fig. 10.32).

Termination most frequently occurs, however, because of remaining impurities, such as H_2O (Fig. 10.33).

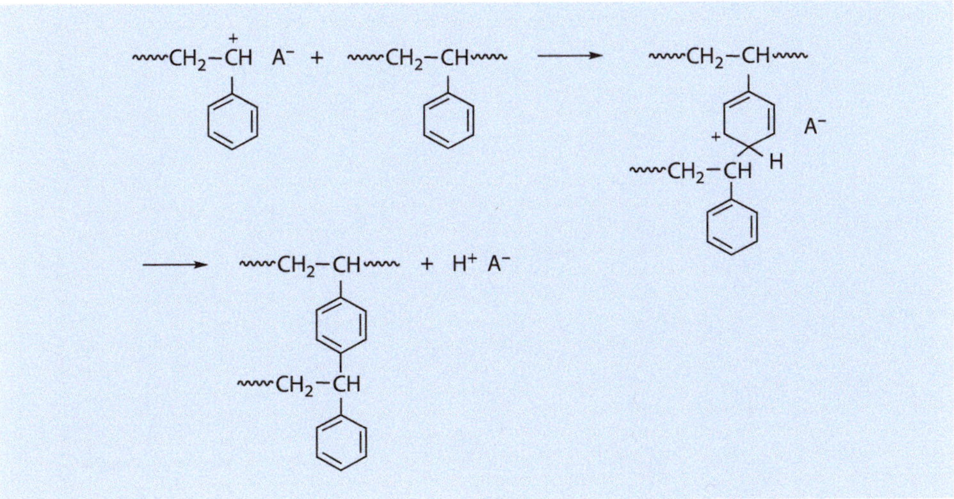

Fig. 10.27 Transfer by intermolecular Friedel–Crafts reactions to form branched polymers

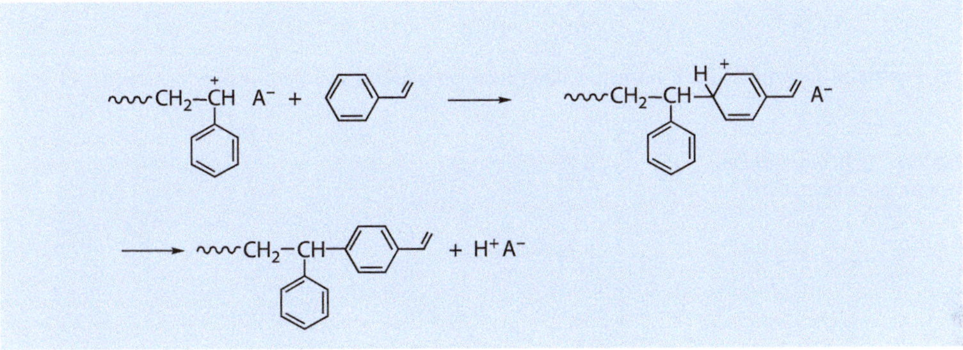

Fig. 10.28 Transfer by an intermolecular Friedel–Crafts reaction to yield a macromonomer

10.1.5 Living and Controlled Cationic Polymerization

Characteristic of living polymerization are the rapid initiation and the absence of transfer and termination. As a result, the molar mass of the polymer should increase in a linear fashion with conversion. Because of the reactivity of the carbenium ions, the above requirements are difficult to meet for the cationic polymerization of vinyl compounds, for example for the technically important polymerization of isobutene. However, it has still proved possible to develop systems which can be called controlled polymerizations.

The best known example of such a system is the INIFER-polymerization of isobutene (Kennedy and Smith 1979, 1980). An *ini*tiator-trans*fer*-reagent such as 1,4-bis (chloroisopropyl)benzene reacts with BCl_3 to become a carbenium ion which is stabilized by neighboring aromatic compounds; however, the equilibrium is shifted far towards the neutral component (Fig. 10.34).

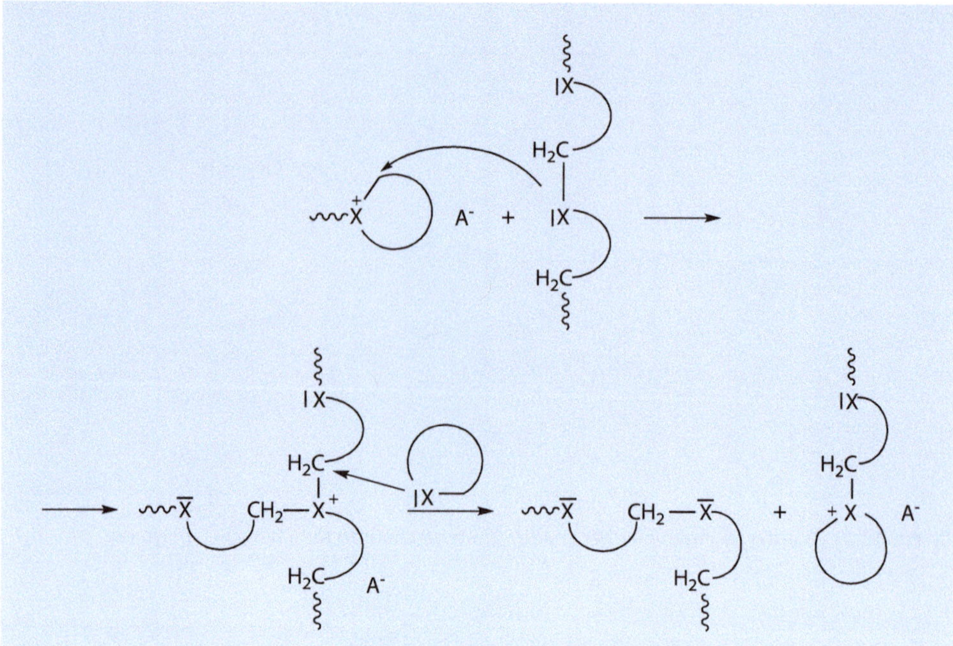

Fig. 10.29 Intermolecular transfer reactions between growing and inactive polymer chains

Fig. 10.30 *Back-biting*: Intramolecular transfer

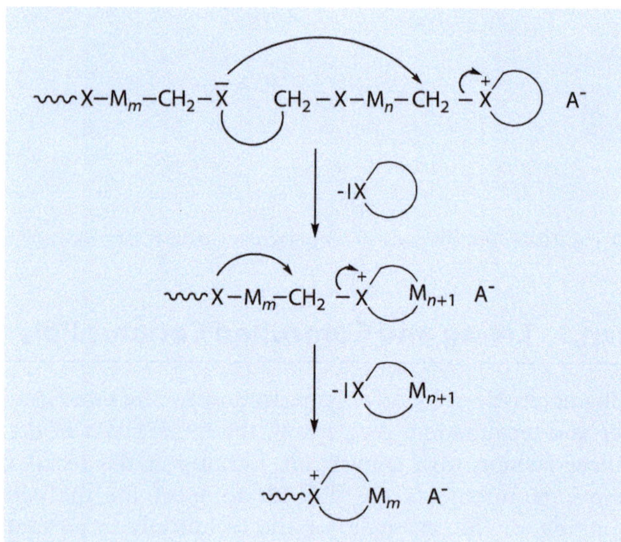

Fig. 10.31 Termination by hydride shift

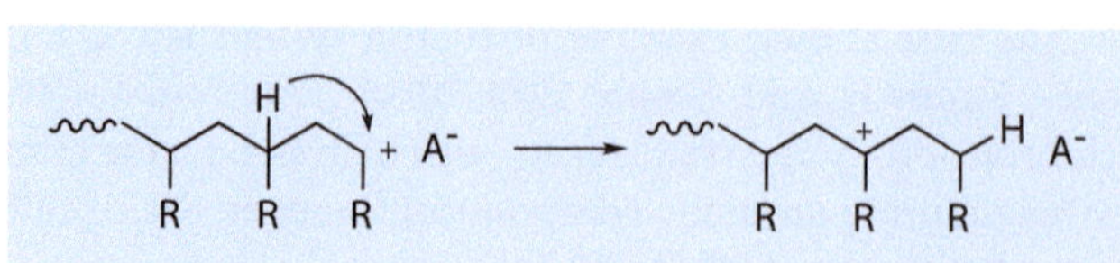

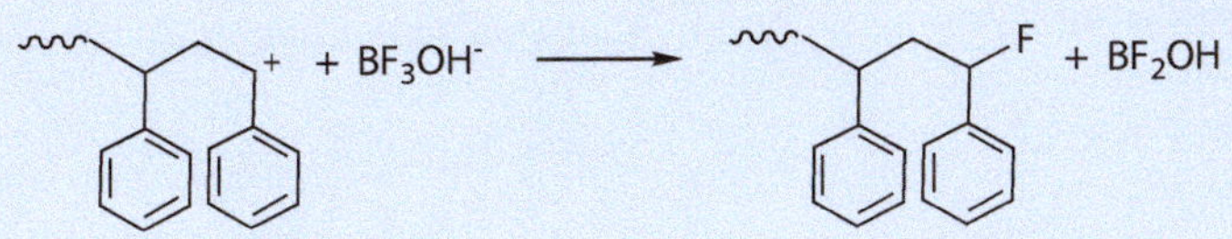

Fig. 10.32 Termination via *anion splitting*

Fig. 10.33 Termination by impurities or added terminating agents, e.g., H_2O

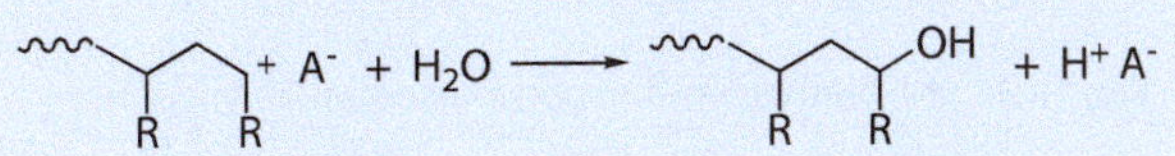

The cation (II, Fig. 10.34) formed during initiation using BCl_3 is able to add isobutene (Fig. 10.35).

The growing chain can then be terminated either by *anion splitting* or by a transfer reaction (reaction between III and I) (Fig. 10.36).

A Cl-terminated polyisobutene (IV) is produced by the reaction of III with BCl_4^-. This reaction is reversible but IV usually reacts from the second isopropyl chloride group adjacent to the aromatic ring with BCl_3—the second arm is initiated (Fig. 10.37).

The species VI can be produced from V by the addition of isobutene (Fig. 10.38).

Finally, VI can terminate by *anion splitting* or by reacting with I to become Cl-terminated (Fig. 10.39).

The latter reaction continues until all I and similar functionalities (IV) have been consumed; polymerization continues until all the monomer has been used up. According to this scheme, macromolecules of controllable molar mass and interesting functional end groups (in this case Cl), so-called *telechelic macromolecules*, can be synthesized.

Using the INIFER mechanism as an example, it was possible to demonstrate that it is unnecessary for the charged chain ends to retain their activity or identity. The desired goal (telechelic macromolecules, block copolymers) can also be achieved by making use of the reversibility of the termination step. The same concept of reversible termination step is also crucial for controlled radical polymerization (▶ Sect. 9.5).

Vinyl ethers can be polymerized in a controlled manner by HI/I_2 or HI/NR_4ClO_4 as the initiator (Fig. 10.40); polymers with adjustable molar mass ($P_n = [M]/[I]$), narrow molar mass distribution ($M_w/M_n \leq 1{,}2$), and with controlled functional end groups can be synthesized in this way.

Block copolymers can be synthesized simply by the sequential addition of different vinyl ethers. If the same polymerization is initiated by Brønsted acids, such block copolymer syntheses are not successful.

By using the bi-functional adduct of the addition of hydrogen iodide to a divinyl ether as an initiator, telechelic macromolecules can be obtained (Fig. 10.41).

Heterocyclic monomers such as tetrahydrofuran can also undergo controlled cationic polymerization (Fig. 10.42).

Particularly worth noting in terms of controlled cationic polymerization are the oxazoline monomers (Fig. 10.23). An amphiphilic block copolymer can be obtained simply by adding a different monomer (e.g., $R' = n\text{-}C_9H_{19}$) after conversion of the first monomer (e.g., $R' = CH_3$) is complete (Fig. 10.43).

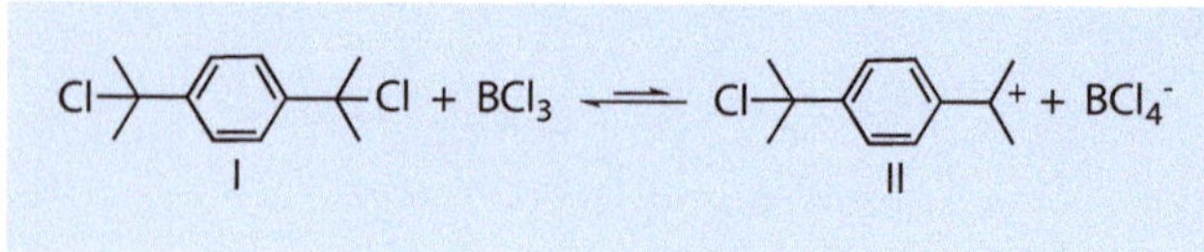

Fig. 10.34 INIFER-mechanism: formation of the initiating cation

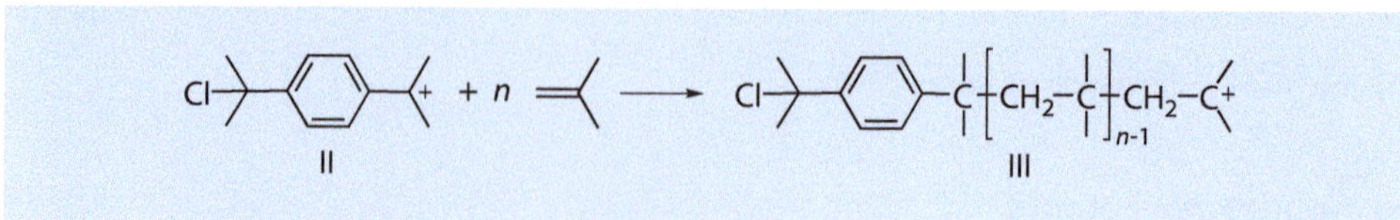

Fig. 10.35 INIFER-mechanism: polymerization of isobutene

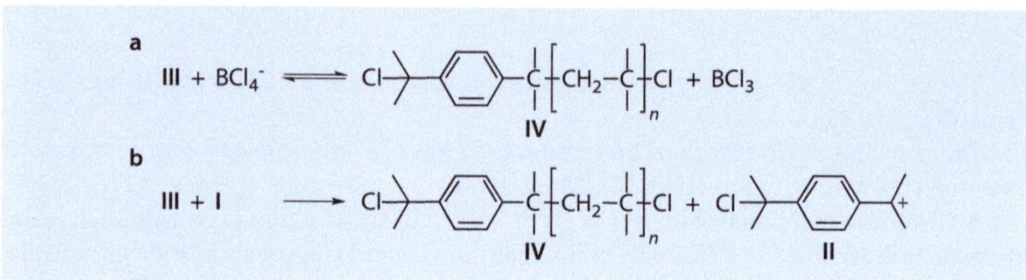

Fig. 10.36 INIFER-mechanism: termination of propagation by: (**a**) anion splitting or (**b**) the initiator

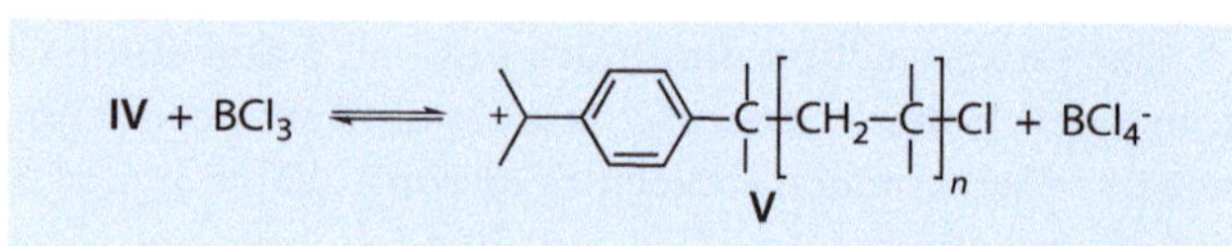

Fig. 10.37 INIFER-mechanism: activation initiating of the second arm of a chloro-terminated polymer

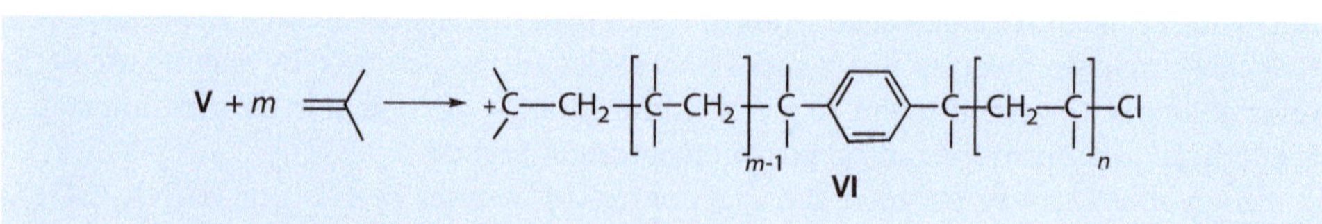

Fig. 10.38 INIFER-mechanism: growth of the second polymer arm

The suitable choice of initiator RX and termination reagent opens further possibilities for variation. As a result of this, molecules such as fluorophilic surfactants with hydrophobic end groups can be synthesized (Fig. 10.44).

For the example in Fig. 10.44, $C_8F_{17}CH_2CH_2\text{–}SO_3CF_3$ was used as an initiator and an alkyl-substituted piperazine as a terminating agent.

The controlled polymerizations described in this section result in macromolecules with well defined molar mass, narrow molar mass distributions, and well defined functional end groups; this is an elegant versatile method for synthesizing telechelic macromolecules and block copolymers.

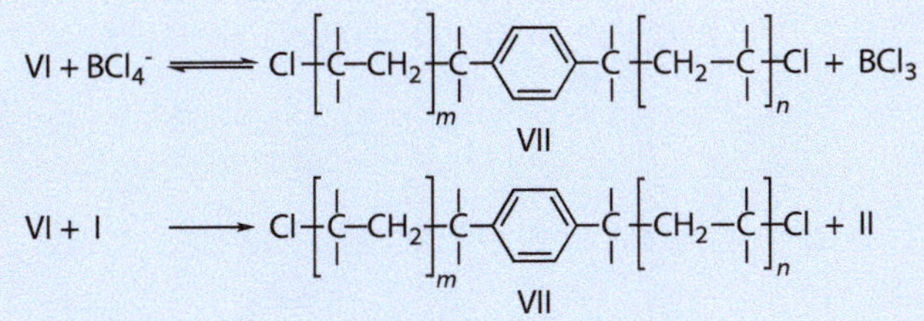

Fig. 10.39 INIFER-mechanism: reactions which lead to α,ω-dichloro polyisobutene

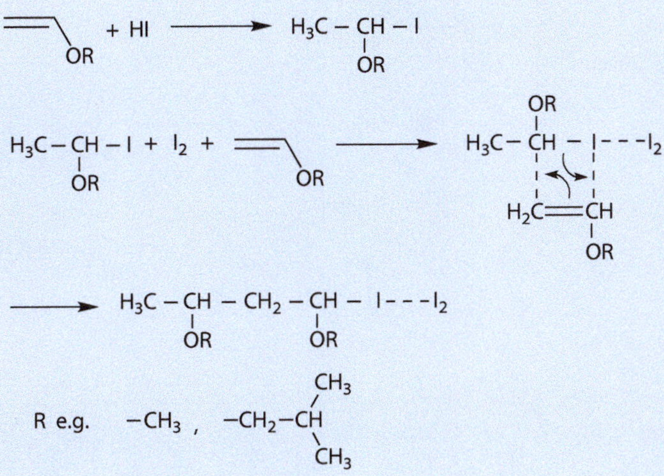

Fig. 10.40 Controlled polymerization of vinyl ethers with HI/I_2

10.1.6 Kinetics of Cationic Polymerization

In contrast to radical or anionic polymerization, there are no coherent, generalized rules for cationic polymerization. This is, apart from other things, because of the many different types of active centers—e.g., free ions, solvated ions, ion pairs, and ion aggregates. Often the influence of transfer reactions is disregarded, the effect of the solvent is underestimated, and unwarranted assumptions of quasi-stationarity are made. Additionally, the extreme reactivity of the active centers, their low concentrations, and their sensitivity to ubiquitous impurities often lead to the real mechanism often not being properly understood.

In contrast, the polymerization of isobutyl vinyl ether (IBVE) initiated with 1-(2-methyl propoxy)-1-iodoethane and tetrabutyl ammonium perchlorate (TBAP) as a co-initiator ($(R'I)/(n\text{-}Bu)_4NClO_4$) is relatively well defined.

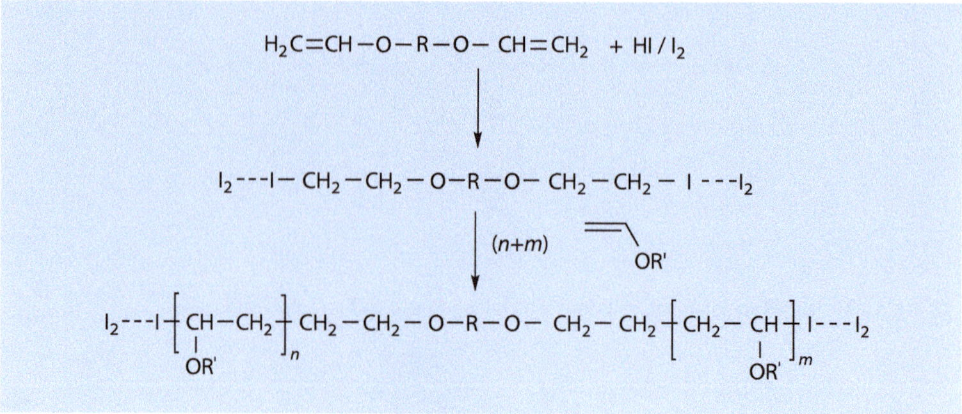

Fig. 10.41 Synthesis of telechelic vinyl ethers

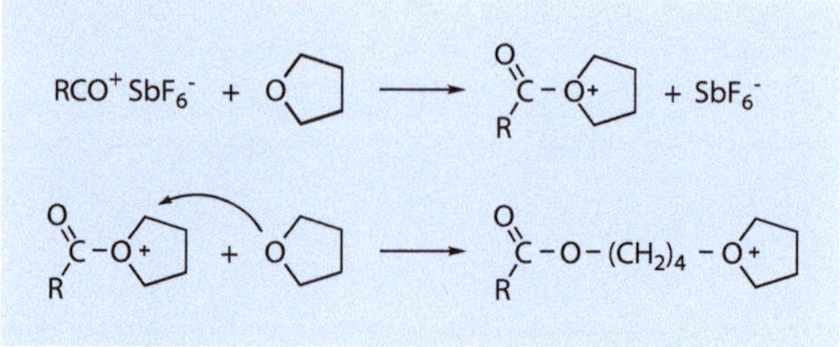

Fig. 10.42 Controlled polymerization of tetrahydrofuran

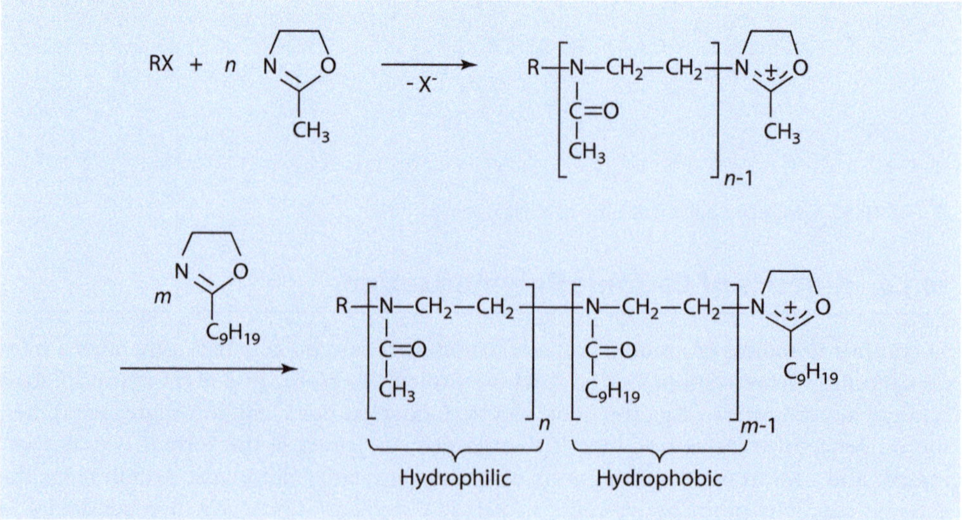

Fig. 10.43 Synthesis of an amphiphilic block copolymer. *Hydrophilic* $COCH_3$ side chain, *hydrophobic* COC_9H_{19} side chain

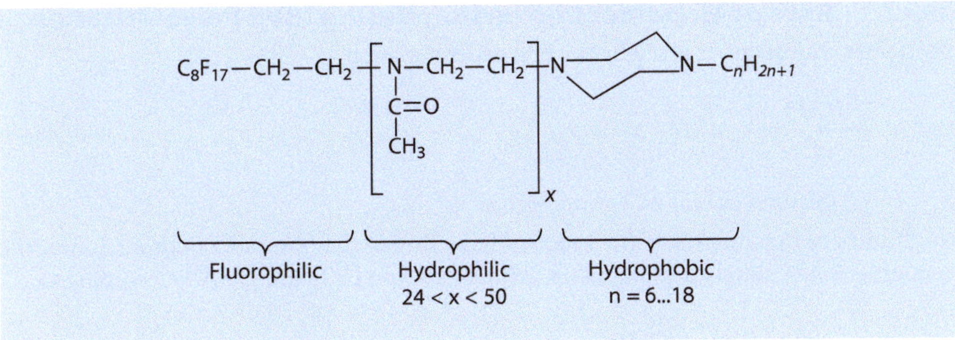

Fig. 10.44 Polymeric fluorophilic tenside with lipophilic terminal groups

As can be seen in Fig. 10.45, the substitution of I^- by ClO_4^- results in an active species, $R'^+ClO_4^-$. The addition of monomer (M) to this leads to the active species, $P^+ClO_4^-$, which is in equilibrium with the PI (see ▶ Sect. 9.5 to appreciate the similarity to controlled radical polymerization). Kinetic measurements such as those described in the following paragraphs (▶ Sects. 10.1.6.1 and 10.1.6.2) corroborate this mechanism.

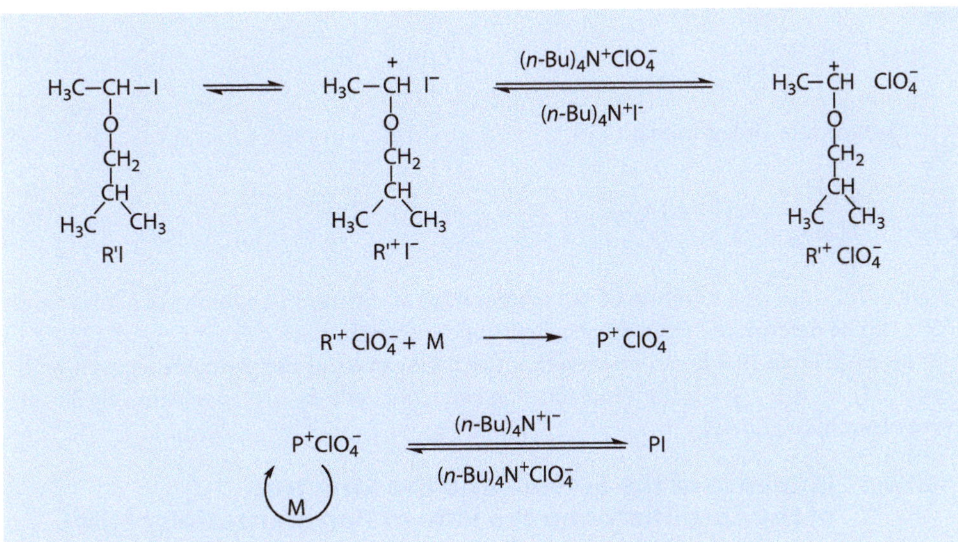

Fig. 10.45 Mechanism of the polymerization of isobutyl vinyl ether (M), initiated with 1-(2-methyl propoxy)-1-iodoethane (R′I)/TBAP

10.1.6.1 Rate of Monomer Conversion During IBVE Polymerization

Monomer conversion is given by the first order equation

$$-\frac{d[M]}{dt} = k_p' \cdot [M] \quad (10.1)$$

k_p' Measured rate constant of propagation

Studies of the kinetics of the reaction have shown that the rate of conversion of the monomer is not simply dependent on $[M]$, but also on $[R'I]$ and $[TBAP]$ according to

$$k'_p = k_p \cdot [R'I]^{\alpha} \cdot [TBAP] \quad (10.2)$$

k_p "True" rate constant of propagation

α is called the coefficient of interaction and experimentally $\alpha = 1$ $(0.9 < \alpha < 1.1)$.

Taking the expression for the conversion of monomer to polymer p:

$$p = \frac{[M]_0 - [M]}{[M]_0} \quad (10.3)$$

or

$$1 - p = \frac{[M]}{[M]_0} \quad (10.3a)$$

(10.1) gives, after integration:

$$\ln\frac{[M]}{[M]_0} = \ln(1-p) = -k'_p \cdot t \quad (10.4)$$

Plots of $\ln(1-p)$ as a function of the temperature at different temperatures allows values for k_p' to be determined from the gradients of the straight lines.

From Table 10.2 it can be seen that the molar mass of the products is given by the ratio $[M]_0 : [R'I]_0$ ($p = 100\%$) and that the rate constants k_p' are approximately linearly proportional to $[TBAP]_0$.

10.1.6.2 Influence of the Solvent and the Structure of the Co-initiator on the Rate of Polymerization of IBVE

The rate of polymerization of IBVE increases significantly with the polarity of the solvent. At −40 °C the polymerization of IBVE is 300 times faster in methylene dichloride than in toluene.

Both the co-initiator cation (Table 10.3) and its anion (Table 10.4) affect the rate of polymerization.

The rate of polymerization of IBVE is reduced by a factor of 17 by changing $R = C_2H_5$ to $R = n\text{-}C_8H_{17}$ or $n\text{-}C_{12}H_{25}$ in R_4NClO_4.

Although the rate of polymerization of IBVE is similar with the anions PF_6^- and ClO_4^-, it is noticeably slower with BPh_4^-.

Table 10.2 Polymerization of IBVE, initiated with R′I/TBAP—influence of $[M]_0$, $[R'I]_0$, $[TBAP]_0$ and temperature on the rate constants $k_w{}'$ and on M_n

T (°C)	$[M]_0$ (mol/L)	$[R'I]_0$ (10^{-2}) mol/L	$[TBAP]_0$: $[R'I]_0$	k_p' (10^{-3} s^{-1})	M_n (g/mol)	M_w/M_n
−15	1.0	0.61	0.49	0.64	16,400	1.12
−15	1.0	0.57	1.75	1.55	17,200	1.10
−15	1.0	0.57	2.63	2.48	17,600	1.12
−30	0.5	2.50	0.10	0.14	2,100	1.10
−30	0.5	2.50	0.50	0.83	2,000	1.08
−30	0.5	2.50	1.00	1.42	2,100	1.10
−30	0.5	2.50	2.00	4.39	1,900	1.12

Table 10.3 Polymerization of IBVE at −15 °C in methylene dichloride, initiated with R′I/R_4NClO_4—influence of the coinitiator R on the rate of polymerization ($[M]_0 = 1.0$ mol/L, $[R'I]_0 = 0.57 \cdot 10^{-2}$ mol/L, $[R_4NClO_4]_0 = 0.3 \cdot 10^{-2}$ mol/L)

R	CH_3	C_2H_5	n-C_4H_9	n-C_6H_{13}	n-C_8H_{17}	n-$C_{12}H_{25}$
k_p' [10^{-4} s^{-1}]	Poorly soluble	5.89	6.37	2.10	0.28	0.28

Table 10.4 Polymerization of IBVE at −15 °C in methylene dichloride, initiated with R′I/R_4NClO_4—influence of the coinitiator A^- on the rate of polymerization ($[M]_0 = 1.0$ mol/L, $[R'I]_0 = 0.57 \cdot 10^{-2}$ mol/L, $[R_4NClO_4]_0 = 0.3 \cdot 10^{-2}$ mol/L)

A^-	BPh_4^-	BF_4^-	PF_6^-	ClO_4^-
k_p' [10^{-4} s^{-1}]	0.05	1.30	4.97	6.37

This trend is in accordance with the *special salt effect* described for nucleophilic substitution reactions in the presence of tetraalkyl ammonium salts (Winstein et al. 1961).

In summary it can be stated that:

- IBVE polymerization, initiated with R′I and TBAP, is a living polymerization with a reversible termination step
- The rate of polymerization increases linearly with $[M]$, $[R'I]$ and $[TBAP]$
- The rate of polymerization increases linearly with the polarity of the solvent

Even from this simple and comparatively well-defined example it is clear that the kinetics of cationic polymerization are rather complex. Thus, it becomes clear why there are, as yet, no generalized kinetic schemes for cationic polymerization and also that the results obtained with one monomer cannot simply be used to describe another system.

10.2 Anionic Polymerization

Already in the early part of the twentieth century butadiene was being polymerized anionically in hydrocarbons with a dispersion of sodium in hydrocarbon solvents (*BUNA*). This process could, however, not compete technologically against the radical polymerization of butadiene and was soon abandoned. The seminal work of M. Szwarc on anionic living polymerization paved the way for telechelic macromolecules, block copolymers, and polymers with diverse architectures (Szwarc 1956).

Similar to cationic polymerization, a high selectivity towards monomers is characteristic of anionic polymerization. In contrast to cationic polymerization, anionic polymerization works best with monomers having acceptor groups adjacent to their double bond (Fig. 10.46).

The sensitivity of the active species demands great care when choosing and purifying the solvents to be employed. Solvents often chosen are tetrahydrofuran and aliphatic and aromatic hydrocarbons. Traces of water and alcohols should be avoided as these deactivate the growing chains. Solvents such as CH_2Cl_2 and $CHCl_3$ also lead to chain termination (Fig. 10.47).

The polymerizations are generally carried out at temperatures below room temperature to facilitate control of the otherwise rapid reactions and to suppress side reactions.

The absence of termination and transfer steps allows living polymerization and enables polymers with narrow molar mass distributions ($M_w/M_n < 1{,}1$), well-defined functional end groups (telechelic structures), and block copolymers to be synthesized. Telechelic polymers can be obtained by deliberately stopping the polymerization with specific termination reagents (Fig. 10.48).

10

10.2.1 Reaction Mechanism

Anionic polymerization takes place according to the scheme given in Fig. 10.49.

The active, negatively charged chain reacts with an additional monomer and is thus extended by one monomer unit. K^+ represents the positively charged counterion, gener-

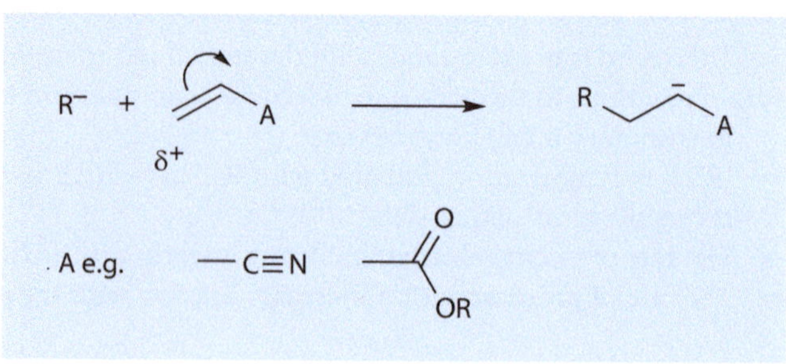

Fig. 10.46 Addition of a nucleophile to an electrophilic monomer

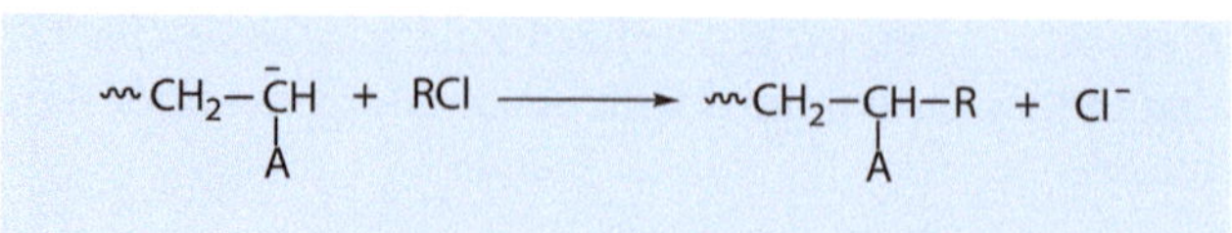

Fig. 10.47 Chain termination via nucleophilic substitution of the halogen in RCl by a polymeric anion

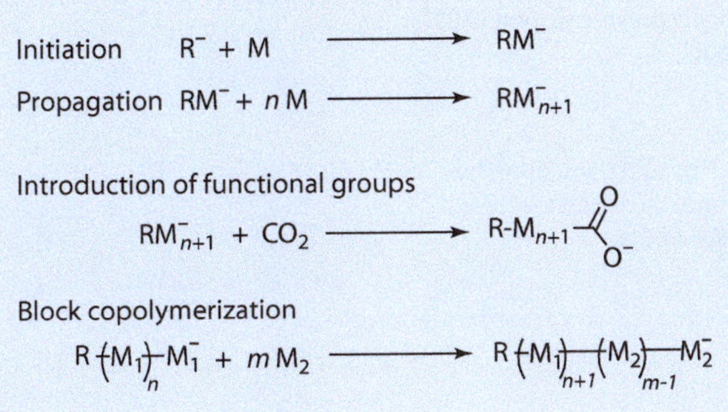

Fig. 10.48 Reactions of anionic species with monomer and terminating reagents, here: CO_2. *M*, M_1, M_2 monomers

$$\sim M_n^- \ K^+ + M \longrightarrow \sim M_{n+1}^- \ K^+$$

Fig. 10.49 Propagation during anionic polymerization. K^+ counterion, cation

ally a metal cation such as Li^+. The similarity of propagation in the anionic polymerization of vinyl compounds with the same steps in cationic or radical polymerizations makes detailed discussion here unnecessary.

In addition to vinyl compounds with acceptor groups, such as acrylic esters, acrylonitriles, and vinyl ketones mentioned above, butadiene, styrene, and vinyl pyridine can also be anionically polymerized. It is also possible to polymerize heterocyclic compounds such as oxirane, thiirane, and lactam anionically (▶ Sect. 12.4).

10.2.2 Initiation

The initiation of anionic polymerization takes place either by the direct addition of a nucleophile to the carbon-carbon double bond of a monomer or by an electron transfer reaction. Both mechanisms are dealt with in the following paragraphs (▶ Sects. 10.2.2.1 and 10.2.2.2). Individual paragraphs deal with the so-called *group-transfer* polymerization (▶ Sect. 10.2.2.3) and the polymerization of lactams, of Leuchs' anhydride, and of acrylamides (▶ Sect. 10.2.2.4), and a further paragraph deals with multifunctional initiators (▶ Sect. 10.2.2.5).

10.2.2.1 Nucleophilic Initiators

Covalent or ionic metal amides such as $NaNH_2$ and $LiN(C_2H_5)_2$, alkoxides, hydroxides, cyanides, phosphines, and amines can all be employed as initiators. Organic metal compounds such as *n*-C_4H_9Li (Fig. 10.50) and Grignard compounds such as PhMgBr are especially popular.

Lithium alkyl and aryl compounds are often used as these are easily soluble in hydrocarbons.

Neutral nucleophiles, which generate zwitterionic growing chain ends, are not so frequently used as initiators for anionic polymerization (Fig. 10.51).

The choice of initiator depends on the monomer to be polymerized. Monomers such as styrene and 1,3-butadiene require more nucleophilic initiators such as amide or alkyl

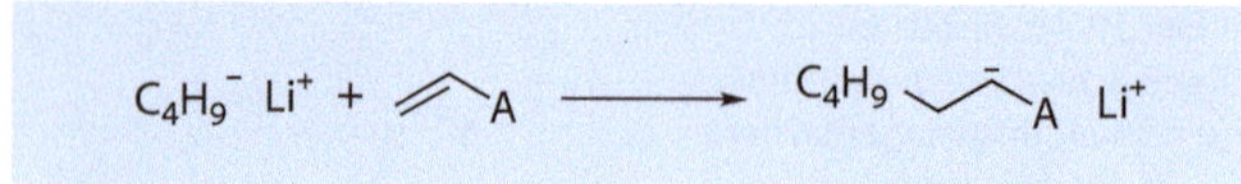

Fig. 10.50 Initiation of anionic polymerization with *n*-BuLi

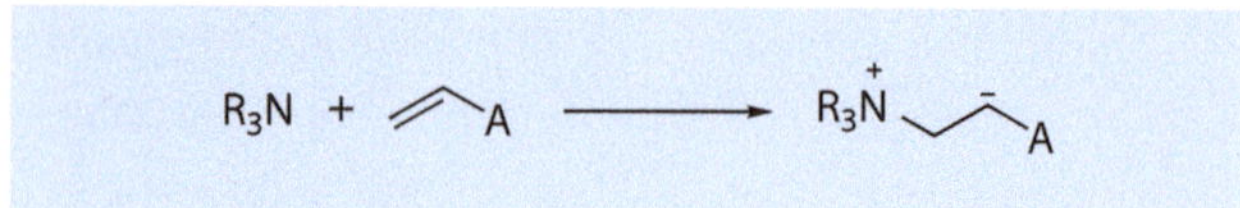

Fig. 10.51 Initiation by a neutral nucleophile. *A*, e.g., COOR or CN

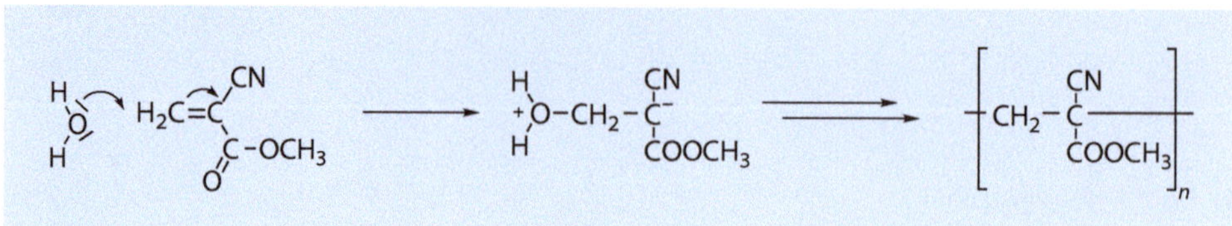

Fig. 10.52 Water as initiator for the anionic polymerization of 2-cyano-acrylic acid methyl ester

carbonate ions. Less nucleophilic initiators such as alkoxides can be used for the polymerization of acrylonitrile, methyl vinyl ketone, and methyl methacrylate. The polymerization of a monomer with two electron-withdrawing substituents, e.g., methyl cyanoacrylate, can be initiated by very weak nucleophiles such as Br^-, CN^-, amines, phosphines, and even water ("Superglue") (Fig. 10.52).

10.2.2.2 Electron Transfer

The reaction of sodium with naphthalene results in a radical ion (Fig. 10.53) which can be used as an initiator for the polymerization of styrene and other anionically polymerizable monomers.

The color changes spontaneously when the polymerization of styrene is initiated by a solution of naphthalene radical ions in tetrahydrofuran; the dark green solution turns a deep cherry red (Fig. 10.54).

The styrene radical ion dimerizes to a dianion (Fig. 10.55).

Because of resonance stabilization I (Fig. 10.55) is formed. The alternative conceivable products of a radical combination, II and III, are either not stabilized at all (II) or significantly less stabilized (III) and therefore do not form.

The dianion I grows by adding additional monomer and becomes a macromolecule with a negative charge at both ends. These dianions can be converted into molecules with interesting functional groups at both ends, telechelics, by reaction with specially chosen termination reagents (Fig. 10.56).

If stored with suitable precautions (protection against moisture), living polystyrene anions, for example, can be kept for longer periods. The polymerization can then be resumed either by further addition of monomer or, if a different monomer is introduced, a block copolymer can be produced (Fig. 10.57).

The synthesis of ABA block copolymers with A = polystyrene and B = 1,4 polybutadiene is commercially very important. The polymerization of styrene is started using BuLi. Then butadiene is added to yield a simple A-B block copolymer. After the addition of SiR_2Cl_2, two growing anions are linked via an SiR_2 bridge produced by the

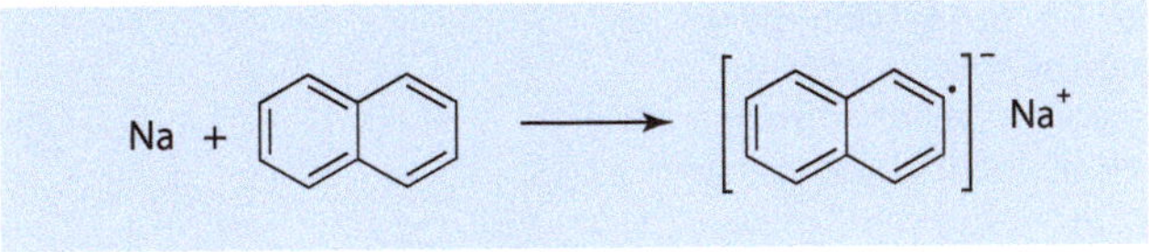

■ **Fig. 10.53** Formation of a radical anion from the reaction of sodium with naphthalene

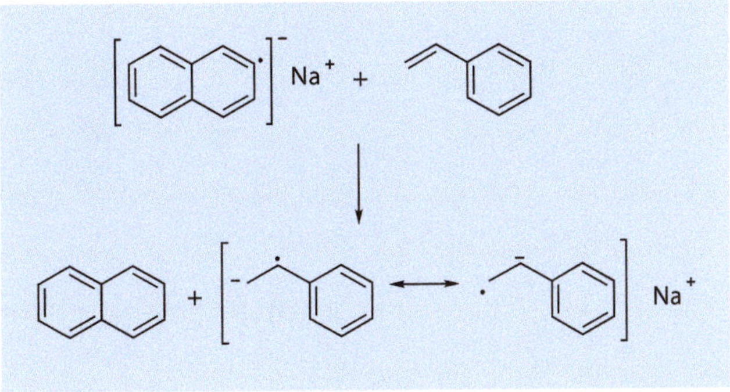

■ **Fig. 10.54** Sodium naphthalene as an initiator for the polymerization of styrene

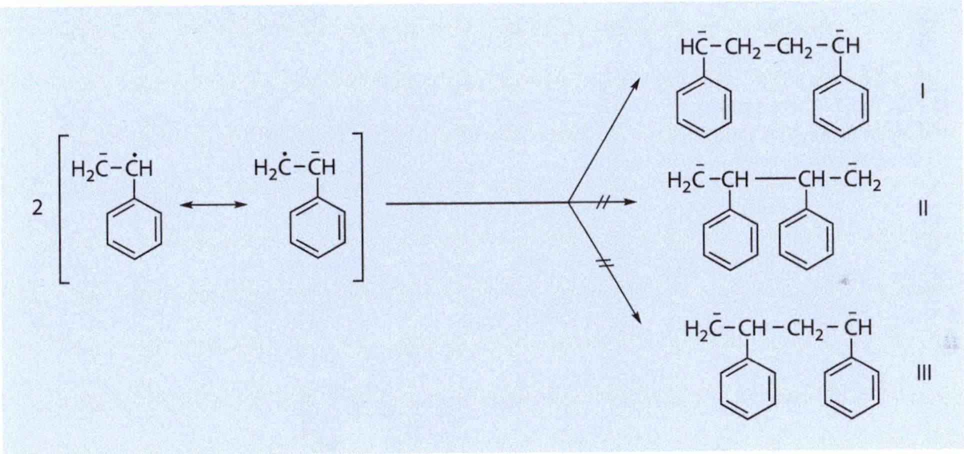

■ **Fig. 10.55** Dimerization of a styryl-radical anion to the corresponding dianion

nucleophilic substitution of the halogens to yield an A-B-A block copolymer (■ Fig. 10.58).

Theoretically it would be possible to synthesize the A-B-A block copolymer simply by adding styrene monomer to the living A-B block copolymer but the lack of reactivity of the butadienyl anion (allyl anion) with respect to styrene prohibits this route.

10.2.2.3 Group Transfer Polymerization

Group transfer polymerization (GTP) is shown schematically using the example of methyl methacrylate in ■ Fig. 10.59. A silylated acetal of dimethyl ketene (1-methoxy-2-methyl-1-trimethylsilyl oxypropene) capable of initiating the controlled anionic polymerization

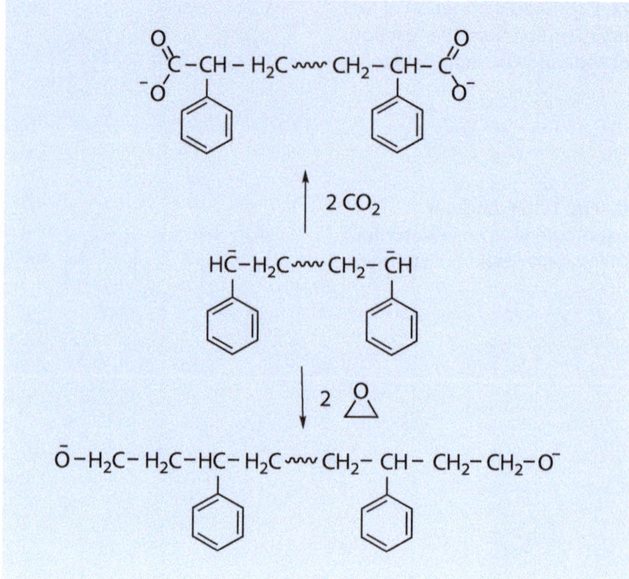

Fig. 10.56 Synthesis of telechelic polystyrene by reacting the growing polymeric dianion with CO_2 or oxirane

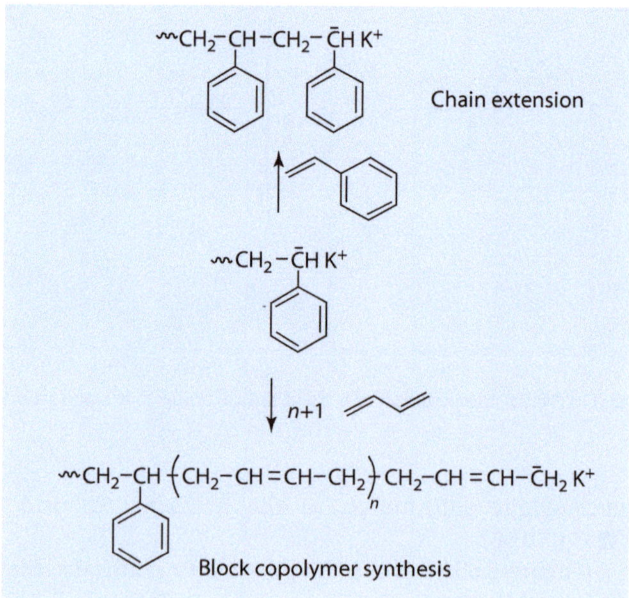

Fig. 10.57 Chain extension and block copolymer synthesis via living polymerization

of methyl methacrylate in the presence of fluoride anions (as NR_4F), $ZnCl_2$, or AlR_2Cl is used as an initiator (Fig. 10.59).

In contrast to the conventional anionic polymerization of MMA, e.g., initiated by BuLi, GTP is especially suited for the synthesis of PMMA with a narrow molar mass distribution.

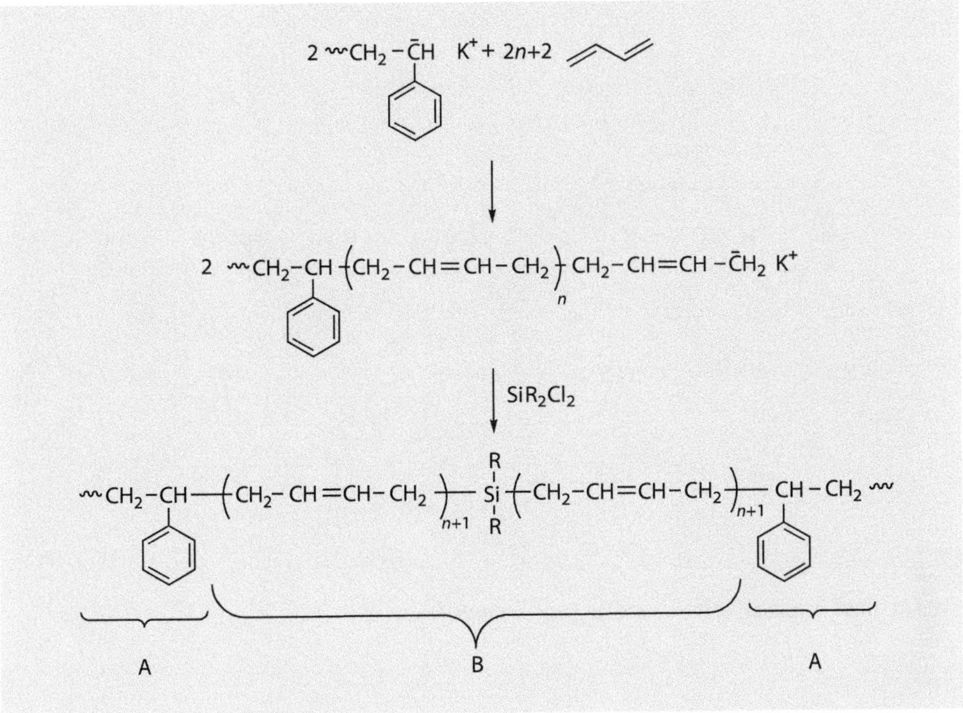

Fig. 10.58 Synthesis of an ABA-block copolymer from a living anionic polymerization by a controlled termination with the bifunctional termination reagent (SiR_2Cl_2). *R*, e.g., CH_3

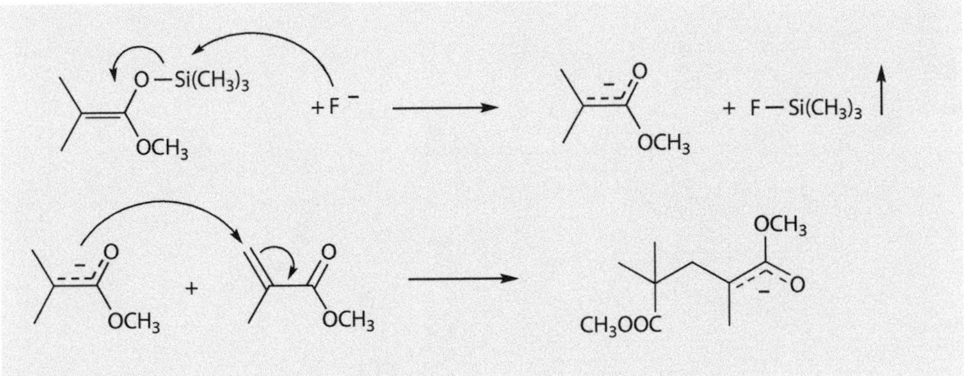

Fig. 10.59 Initiation of the group-transfer-polymerization of methyl methacrylate

10.2.2.4 Polyamides from Lactams, Leuchs' Anhydride, or Acrylamide

Lactams and Leuchs' anhydride polymerize via an *activated monomer* mechanism (▶ Sect. 12.4.3 and Figs. 10.60 and 10.61).

Polyamide-2 is produced during this process (▶ see Sect. 8.3.2 for the polyamide nomenclature) and CO_2 is eliminated with each propagation step (▶ Sect. 12.4.3).

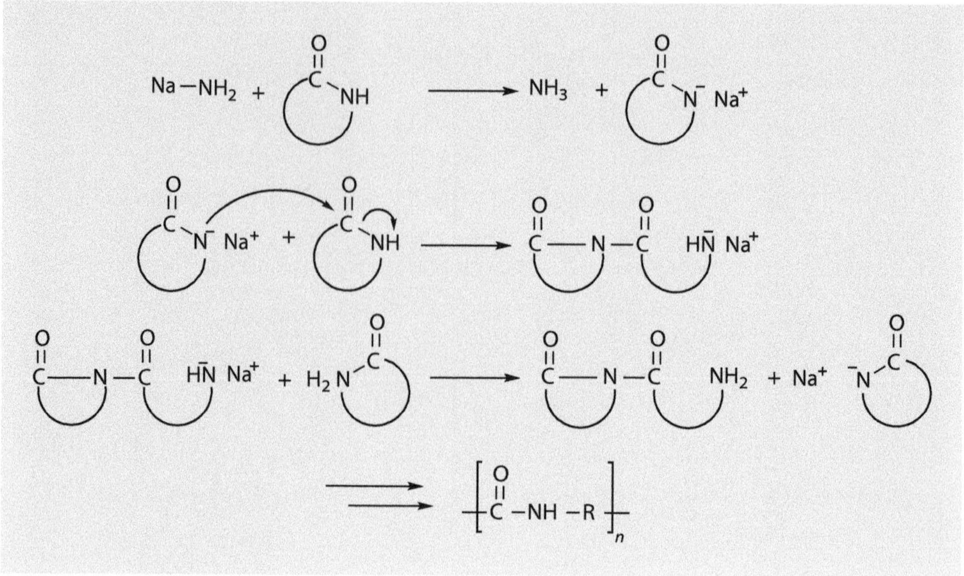

Fig. 10.60 Initiation of the anionic polymerization of lactams

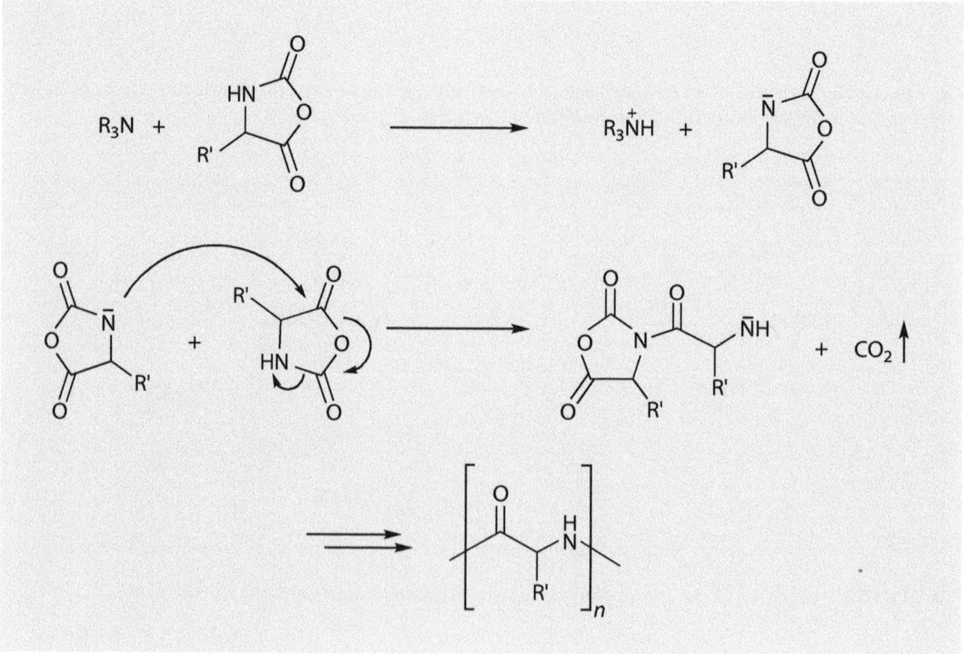

Fig. 10.61 Initiation of the anionic polymerization of Leuchs' anhydride

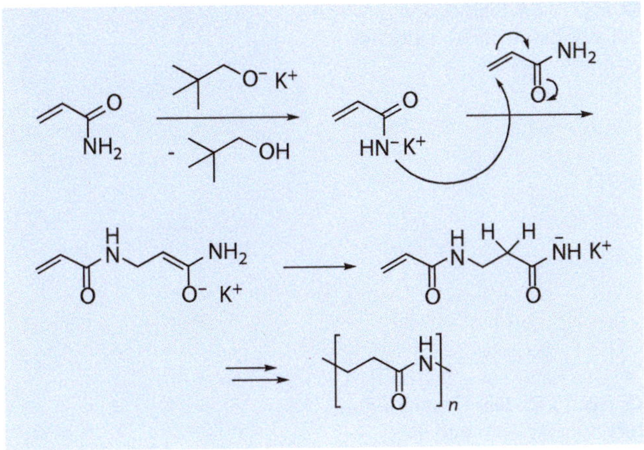

Fig. 10.62 Synthesis of polyamide-3 from acrylamide

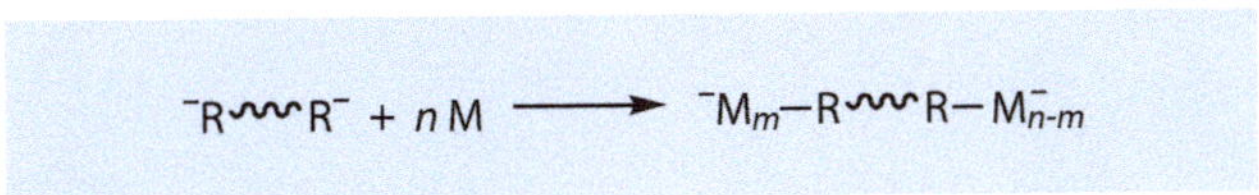

Fig. 10.63 Initiation with a bifunctional initiator

A polyamide-3 instead of the expected polyacrylamide is produced if a strong base is present (Fig. 10.62).

10.2.2.5 Bi- and Multifunctional Initiators

Bifunctional initiators that lead to chain growth in two directions are suitable for the synthesis of telechelics, and multifunctional initiators are used to synthesize star polymers (three or more arms) (Fig. 10.63).

The synthesis of a bifunctional initiator from styrene was discussed in ▶ Sect. 10.2.2.2. By analogy, a similar initiator can be synthesized from 1,1-diphenylethylene (Fig. 10.64).

A selection of bifunctional initiators which can be synthesized by combining the appropriate olefins with BuLi are shown in Figs. 10.65 and 10.66.

However, the synthesis of multifunctional initiators is not trivial. A more elegant solution is to couple growing chains with a multifunctional termination reagent (Fig. 10.67).

10.2.3 Role of Termination Reactions and Reagents in Anionic Polymerization

Reactions that deactivate the growing chain so that no further monomer can be added to it are described in this section.

10.2.3.1 Polymerization Without Termination

Termination reactions, such as the combination which can be observed during radical polymerization, do not occur during living ionic polymerization because identically charged species cannot terminate one another. The anionic polymerization of non-polar

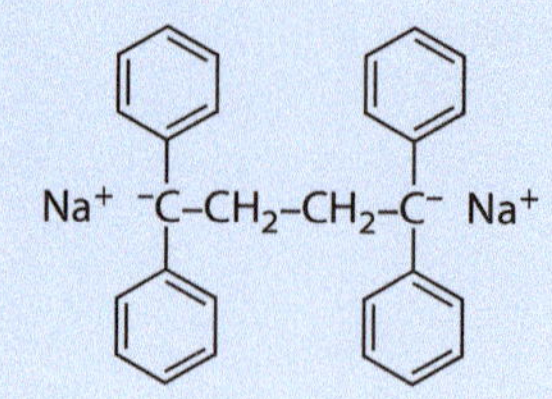

■ **Fig. 10.64** Bifunctional initiator based on 1,1-diphenyl ethylene

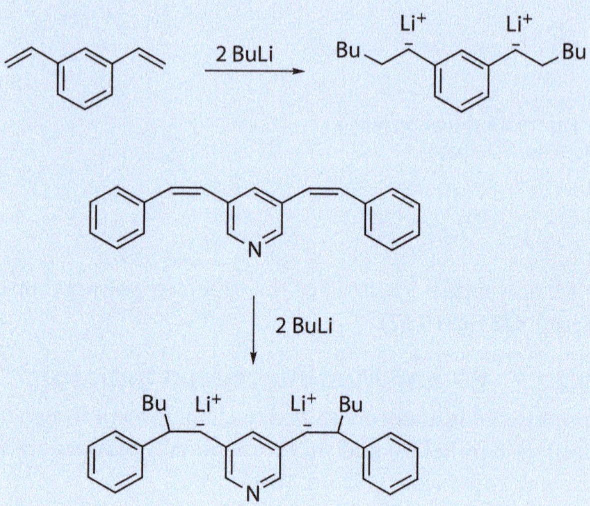

■ **Fig. 10.65** Bifunctional initiators from diolefins and BuLi

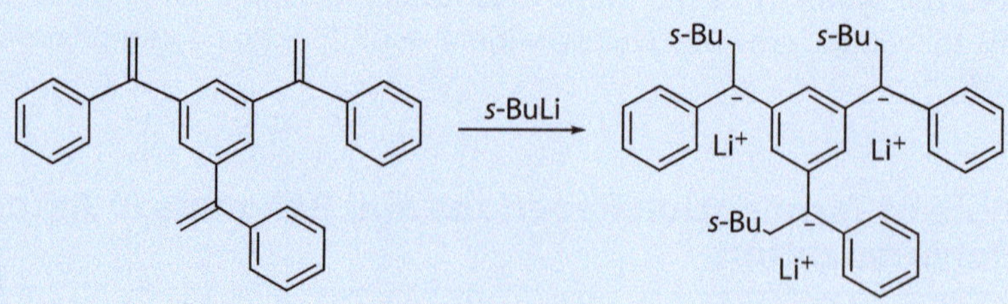

■ **Fig. 10.66** An example of a trifunctional initiator

monomers such as styrene and 1,3-butadiene takes place without termination because the growing chain ends are relatively weak nucleophiles and are thus only rarely terminated by proton abstraction from, for example, the solvent or already formed polymers. A simple collapse of the counterions is also not observed (■ Fig. 10.68).

Fig. 10.67 Synthesis of a three-arm polymer by controlled termination

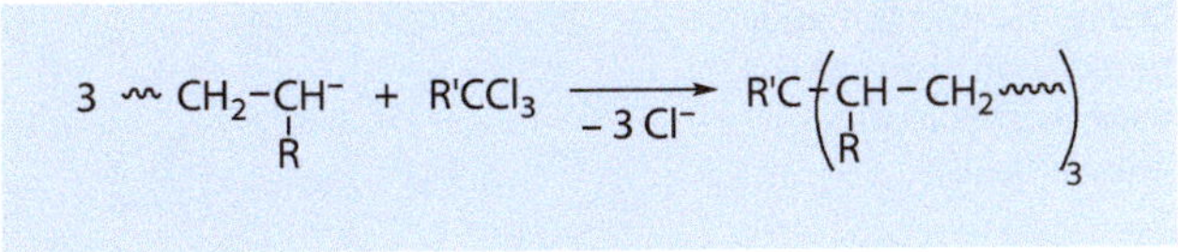

Fig. 10.68 Collapse of the counterion does not occur

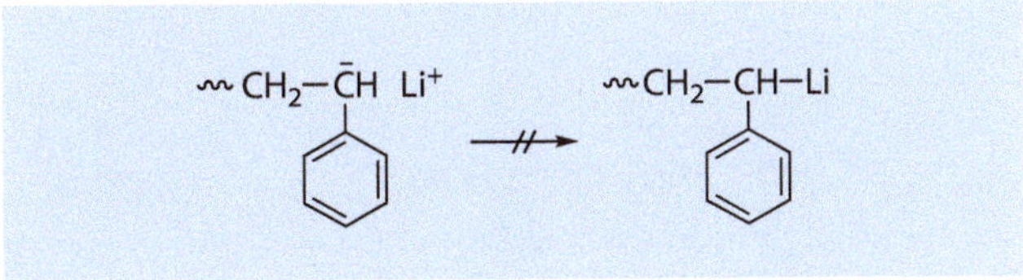

Fig. 10.69 Termination of an anionic polymerization by water

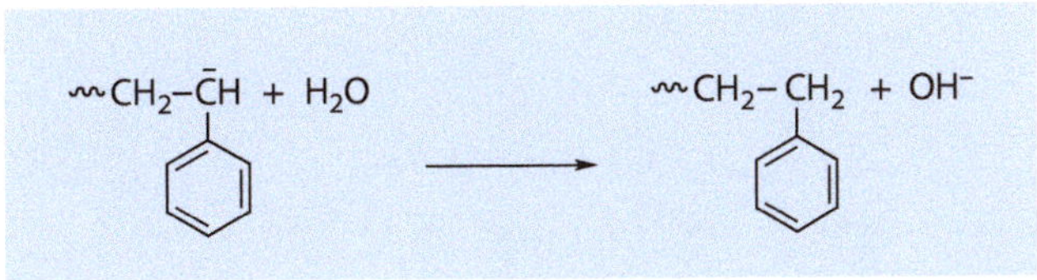

If propagation continues until all the monomer has been consumed, the growing chain remains active provided no transfer takes place. Because of this, this type of polymerization is referred to as living polymerization. These polymeric carbanions are often colored because of conjugation with the substituent R of the last monomer unit; R = Ph, for example, is cherry red. A fading of the color would be an indication for a termination reaction.

10.2.3.2 Termination by Contamination and Deliberately Added Reagents

Elaborate experimental techniques (including, for example, high vacuum or Schlenk techniques and carefully purified and dried solvents) are indispensable for synthesizing good quality products; moisture on the glass surfaces is often sufficient to have a negative effect (Fig. 10.69).

By analogy with the reaction shown in Fig. 10.56, the active chain ends can be converted into interesting functional groups if solid CO_2, oxiranes, or halogen-containing termination reagents are added rather than water. The addition of oxirane can result in the formation of block copolymers. This can, however, be avoided by adding only a stoichiometric amount of the oxirane and rapidly quenching with water. As well as the introduction of –COOH and –OH end groups, the introduction of amino groups may also be desired. One way of producing $-NH_2$ terminated polystyrene is shown in Fig. 10.70.

10.2.3.3 Hydride Elimination

Over a longer time period of time (days or weeks), a decrease in the carbanion concentration in a solution of polystyryl anions in THF is often observed despite the absence of termination reagents. This is because of hydride elimination taking place (Fig. 10.71).

The first step is a hydride elimination. The reaction of the resulting CH = CH(Ph) chain ends with another carbanion yields a stable and thus unreactive allyl anion. NaH can, if there is still monomer present, initiate new chain growth.

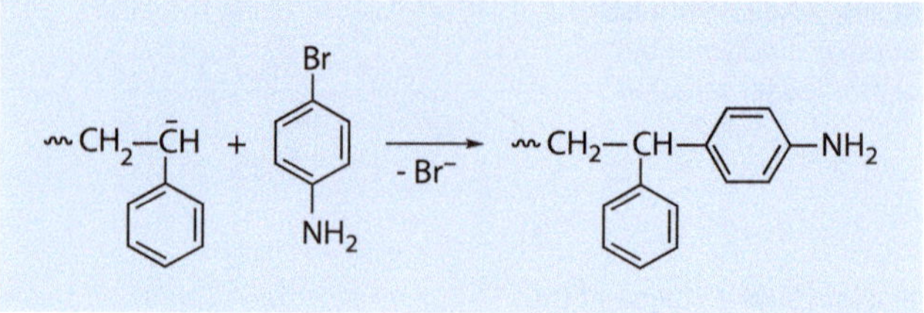

Fig. 10.70 Intentional termination of an anionic polymerization with 4-bromoaniline

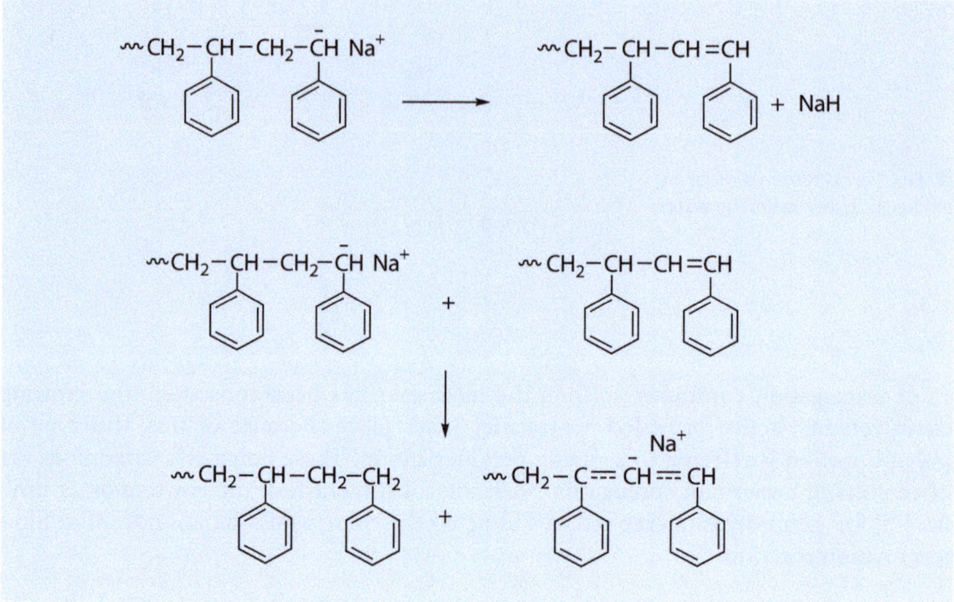

Fig. 10.71 Formation of an unreactive 1,3-diphenyl allyl anion

10.2.3.4 Termination Reactions During the Anionic Polymerization of Methyl Methacrylate

Three different termination reactions can be observed during the anionic polymerization of methyl methacrylate:

1. A reaction of the initiator with the functional group of the monomer leading to the less reactive alkoxide (Fig. 10.72)
2. A nucleophilic reaction of the polymeric anion with the monomer (Fig. 10.73)
3. An intramolecular back-biting reaction (Fig. 10.74)

These termination reactions can be avoided, for the most part, by using initiators that are less nucleophilic, such as 1,1-diphenyl hexyl lithium (instead of *n*-BuLi), lower temperature (approx. –70 °C), and polar solvents (ether rather than hydrocarbons).

10.2.4 Kinetics of Anionic Polymerization

In contrast to cationic polymerization, the proposed mechanisms and the experimental results from kinetics studies with vinyl monomers can be explained by a single consistent model. In the following paragraphs the kinetics of polymerizations without termination

Fig. 10.72 Nucleophilic substitution of the OCH_3-group of the monomer by R

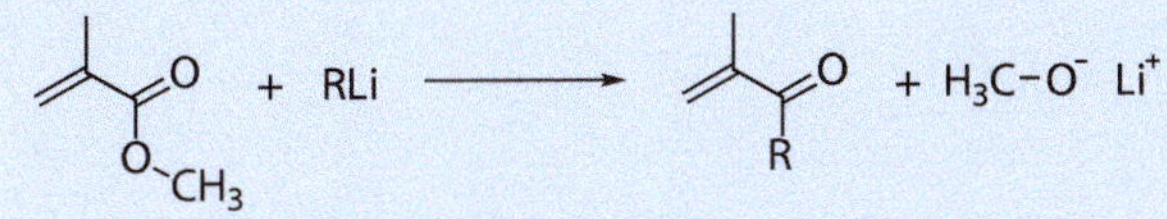

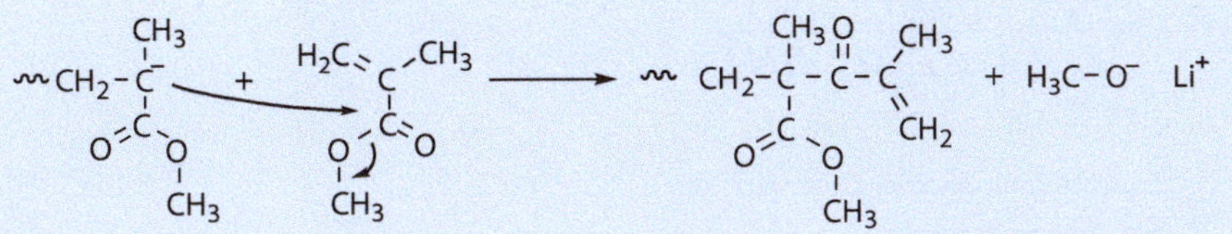

Fig. 10.73 Reaction of the polymeric anion with the monomer

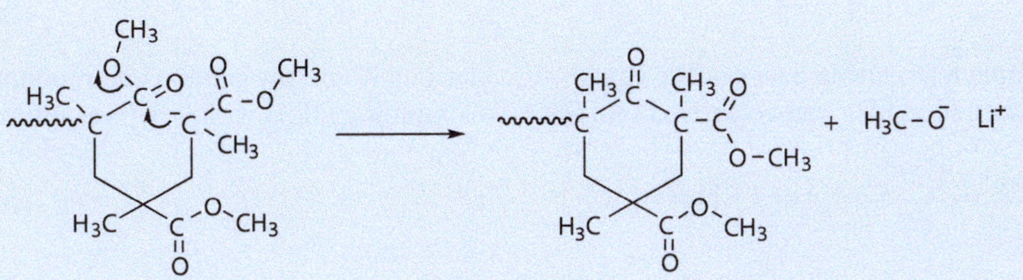

Fig. 10.74 Ring closure as the result of intramolecular back-biting reaction

Fig. 10.75 Polymerization-depolymerization equilibrium. k_p rate constant of propagation, k_{dp} rate constant of depolymerization

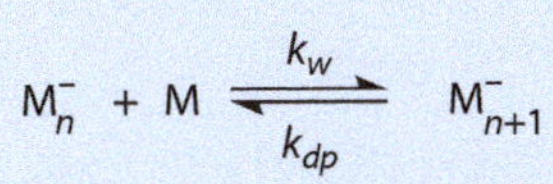

(▶ Sect. 10.2.4.1), those in which free ions and ion pairs compete with one another (▶ Sect. 10.2.4.1), and, lastly, those which take place in non-polar solvents (▶ Sect. 10.2.4.2) are discussed.

The monomer is consumed by propagation and regenerated by depolymerization (Fig. 10.75).

All active species M^- are present at the beginning of polymerization because of the rapid initiation and very quick start in comparison to growth. The following holds for the change in monomer concentration over time:

$$-\frac{d[M]}{dt} = k_p[M][M_n^-] - k_{dp}[M_{n+1}^-] \quad (10.5)$$

k_p Rate constant of propagation

k_{dp} Rate constant of depolymerization

If the rates of polymerization and depolymerization are equal:

$$\frac{d[M]}{dt} = 0 \tag{10.6}$$

and

$$K = \frac{k_p}{k_{dp}} = \frac{[M_{n+1}^-]}{[M_n^-][M]} = \frac{1}{[M]_e} \tag{10.7}$$

$[M]_e$ Equilibrium monomer concentration
K Equilibrium constant

With the assumption that

$$[M_{n+1}^-] = [M_n^-] \tag{10.8}$$

a simple relationship between the equilibrium constant K and the equilibrium monomer concentration $[M]_e$ can be derived (see (10.7)). Combining (10.5) with (10.7) results in

$$-\frac{d[M]}{dt} = k_p[M_n^-]([M]-[M]_e) \tag{10.9}$$

Separation of the variables yields the following:

$$-\int_{[M]_0}^{[M]} \frac{d[M]}{[M]-[M]_e} = k_p[M_n^-]\cdot\int_0^t dt \tag{10.10}$$

Integration of (10.10) gives

$$\left[-\ln([M]-[M]_e)\right]_{[M]_0}^{[M]} = \left[k_p[M_n^-]t\right]_0^t \tag{10.11}$$

$$\ln\frac{([M]_0-[M]_e)}{([M]-[M]_e)} = k_p[M_n^-]t \tag{10.12}$$

The rate constant k_p can be determined from the gradient of a plot of ◘ Fig. 10.76.

In some cases the experimentally determined values for k_p increase inversely proportional to the initiator concentration (◘ Fig. 10.77). This is an indication of a complication, namely a competition between free ions and ion pairs as the initiating species. If both of these are present at the same time then, with increasing dilution, the equilibrium shifts and the concentration of free ions increases (► Sect. 10.2.4.1).

The addition of $NaB(Ph)_4$ (*Kalignost*), which is completely dissociated in tetrahydrofuran, results in the equilibrium between the free ions and ion pairs of the initiator being shifted towards the side of the ion pairs because of an increase in the concentration of sodium ions (◘ Fig. 10.78).

Fig. 10.76 Plot to determine the propagation rate constant k_p

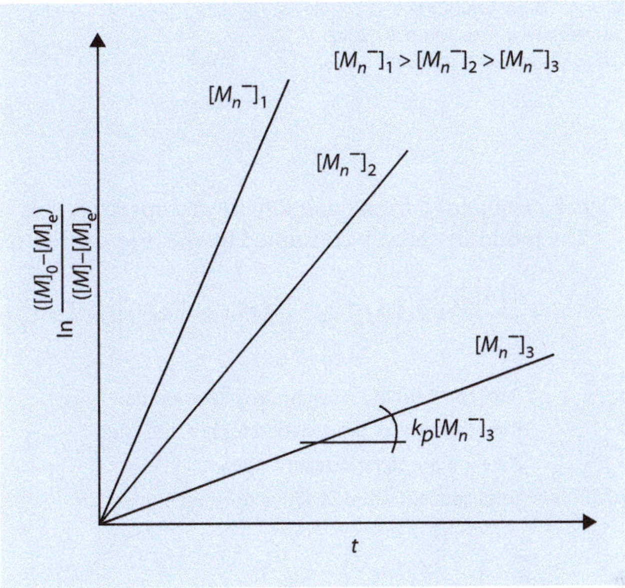

Fig. 10.77 Typical plot of the propagation rate constants experimentally determined for a polymerization involving both free ions and ion pairs (e.g., sodium naphthalene styrene in THF)

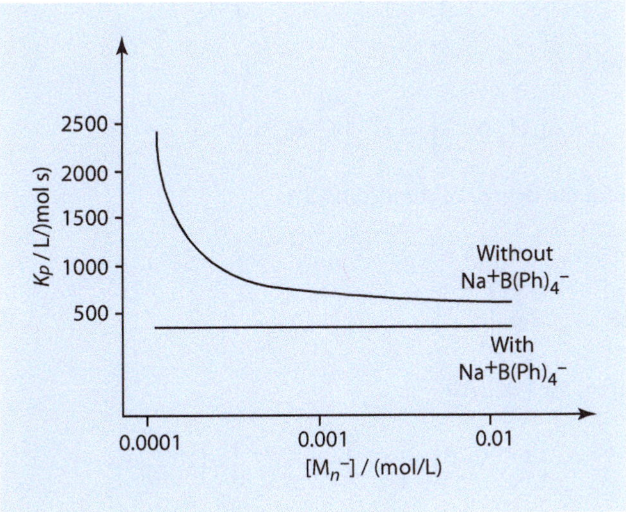

10.2.4.1 Competitive Growth of Two Different Species

During ionic polymerization it is to be expected that both the initiator and the growing chains, are part of the equilibria between the ion pairs and the free ions, depending on the polarity of the solvent and the concentration of the active species (Fig. 10.78).

The following demonstrates how the coexistence of two different active polymerization species can be taken into account kinetically.

Considering only the equilibrium between the free ions and ion pairs, for the dissociation constant K_D with $[Na^+] = [M_n^-]$:

$$K_D = \frac{[Na^+][M_n^-]}{[M_n^- Na^+]} = \frac{[M_n^-]^2}{[M_n^- Na^+]} \tag{10.13}$$

Fig. 10.78 Equilibria between free ions, ion pairs, and the undissociated species

$$M_n^- + Na^+ \rightleftharpoons M_n^-\ Na^+ \rightleftharpoons M_n\text{–}Na$$

Free ions — Ion pair — Undissociated species

K_D can be obtained from conductivity and spectroscopic measurements.

The monomer is only consumed by the propagation reaction so that

$$-\frac{d[M]}{dt} = k_\pm\left[M_n^- Na^+\right][M] + k_-\left[M_n^-\right][M] = \overline{k_p}\left[C^-\right][M] \qquad (10.14)$$

$k_\pm$	Rate constant of propagation for the ion pair
k_-	Rate constant of propagation for the free ions
$\overline{k_p}$	Average propagation rate constant
$[C^-]$	Total concentration of active polymer anions

$$\overline{k_p} = \frac{k_\pm\left[M_n^- Na^+\right]}{\left[C^-\right]} + \frac{k_-\left[M_n^-\right]}{\left[C^-\right]} \qquad (10.15)$$

with

$$\left[M_n^- Na^+\right] = \left[C^-\right] - \left[M_n^-\right] \qquad (10.16)$$

and the degree of dissociation α:

$$\alpha = \frac{\left[M_n^-\right]}{\left[C^-\right]} \qquad (10.17)$$

then for $\left[M_n^- Na^+\right]$:

$$\left[M_n^- Na^+\right] = \left[C^-\right] - \alpha\cdot\left[C^-\right] = \left[C^-\right](1-\alpha) \qquad (10.18)$$

Combining (10.13), (10.17), and (10.18) gives for K_D:

$$K_D = \frac{\alpha^2\cdot\left[C^-\right]}{(1-\alpha)} \qquad (10.19)$$

As the degree of dissociation α is usually significantly smaller than 1, $1-\alpha$ is close to 1 and

$$\alpha = \sqrt{\frac{K_D}{\left[C^-\right]}} \qquad (10.20)$$

If α is small then the total concentration of the growing chains is essentially identical to the concentration of the ion pairs and

$$\left[M_n^- Na^+\right] = \left[C^-\right] \tag{10.21}$$

Thus, with (10.15), (10.17), and (10.21) one obtains for $\overline{k_p}$:

$$\overline{k_p} = k_\pm + k_- \alpha \tag{10.22}$$

By inserting (10.20) into (10.22) one obtains

$$\overline{k_p} = k_\pm + k_- \sqrt{\frac{K_D}{\left[C^-\right]}} \tag{10.23}$$

From a series of experiments with varying initiator concentration a plot of the mean propagation rate constant as a function of $1/\sqrt{\left[C^-\right]}$ (◘ Fig. 10.79) allows the determination of $k_\pm$ (from the intercept) and k_- (from the gradient, if K_D is known).

As discussed in ▶ Sect. 10.2.4.1, the addition of $NaB(Ph)_4$ completely suppresses the dissociation of the initiator and the growing chains (Gl. (10.13) and Gl. (10.21)). From (10.13), (10.17), and (10.22), and with $[M_n^- Na^+] = [C^-]$ (because of the shift in equilibrium produced by $NaB(Ph)_4$), one obtains

$$\overline{k_p} = k_\pm + k_- \cdot \frac{K_D}{\left[Na^+\right]} \tag{10.24}$$

A plot k_p vs. $1/[Na^+]$ allows $k_\pm$ and k_- to be determined using (10.24) (◘ Fig. 10.80).

Thus, with two alternative equations, all three constants K_D, $k_\pm$, and k_- can be determined. K_D can also be obtained independently from conductivity measurements.

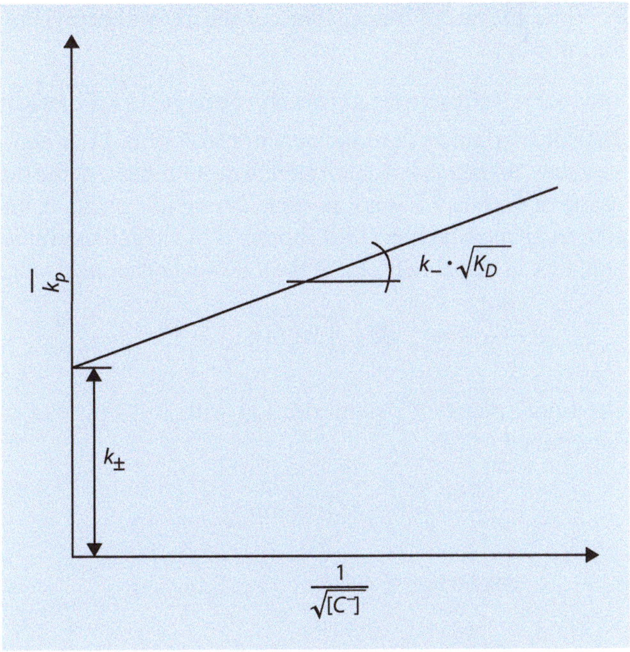

◘ **Fig. 10.79** Plot to determine the rate constants $k_\pm$ and k_-

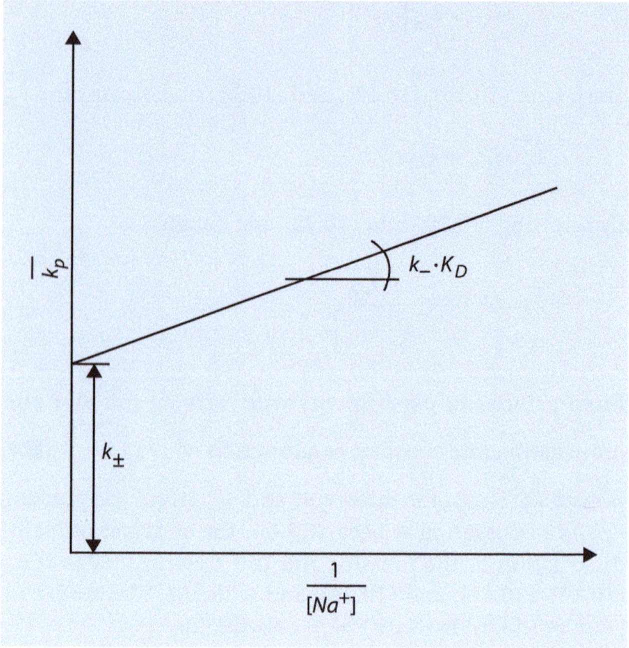

Fig. 10.80 Determine the rate constants $k_\pm$ and k_- from experiments with varying Na^+-concentration, e.g., by adding sodium tetraphenylborate (Kalignost)

The constants $k_\pm$ and k_- differ considerably. For the polymerization of styrene at 25 °C, initiated with sodium naphthalene in THF, the values

$$k_\pm = 80\,\mathrm{L}/(\mathrm{mol}\bullet\mathrm{s})$$
$$k_- = 65{,}000\,\mathrm{L}/(\mathrm{mol}\bullet\mathrm{s})$$

have been determined.

Thus, the rate of propagation for the free polystyryl ions is significantly faster than that for ion pairs.

10.2.4.2 Polymerization in Non-polar Solvents

The polymerization of monomers initiated by BuLi in non-polar solvents such as hexane is very slow. Initiation and chain growth occur initially at the same time so that an exact determination of the kinetics is not possible during this phase. If, however, the initiator is completely converted into oligomers (BuLi becomes M_nLi) and this initial solution is used for polymerization, we can empirically obtain the following relationship for monomer conversion:

$$-\frac{d[M]}{dt} = k_p[M]\sqrt{[M_nLi]} \tag{10.25}$$

The initial phase of polymerization with BuLi instead of M_nLi follows the empirically determined rule:

$$-\frac{d[BuLi]}{dt} = k_{st}[M]\cdot\sqrt[6]{[BuLi]} \tag{10.26}$$

k_{st} Start rate constant

These equations can be derived from the equilibria for the initiator and the growing chains given in Figs. 10.81 and 10.82), respectively.

Equation (10.26) can be analyzed experimentally by converting the remaining BuLi *n*-butane by adding water. The *n*-butane can then be determined, for example, by gas chromatography.

10.2.5 Ceiling Temperature

The rate of polymerization of α-methyl styrene decreases with increasing temperature and above a temperature of 61 °C polymerization ceases completely. The reason for this is the reversibility of the propagation reaction, and the relative rates of polymerization and depolymerization (Fig. 10.83).

From Fig. 10.83 one can develop for the conversion of monomer with time:

$$-\frac{d[M]}{dt} = k_p [M][M_n^-] - k_{dp} [M_{n+1}^-] \tag{10.5}$$

Again, making the assumption that

$$[M_n^-] = [M_{n+1}^-] \tag{10.8}$$

at equilibrium and with (10.5):

$$-\frac{d[M]}{dt} = (k_p [M] - k_{dp})[M_n^-] = 0 \tag{10.5a}$$

The temperature at which $k_p[M] - k_{dp} = 0$ and polymerization stops is referred to as the *ceiling temperature* (T_c) (Fig. 10.84). This equilibrium can be determined for anionic polymerization because, in ideal cases, no termination takes place. However, principally, this equilibrium is independent of the type of polymerization.

Fig. 10.81 Equilibrium between BuLi and the hexamer $(BuLi)_6$

$$(BuLi)_6 \rightleftharpoons 6\ BuLi$$

Fig. 10.82 Equilibrium between M_nLi and the dimer $(M_nLi)_2$

$$(M_nLi)_2 \rightleftharpoons 2\ M_nLi$$

Fig. 10.83 Polymerization–depolymerization equilibrium

$$M_n^- + M \underset{k_{dp}}{\overset{k_p}{\rightleftharpoons}} M_{n+1}^-$$

Fig. 10.84 Kinetic determination of the ceiling-temperature

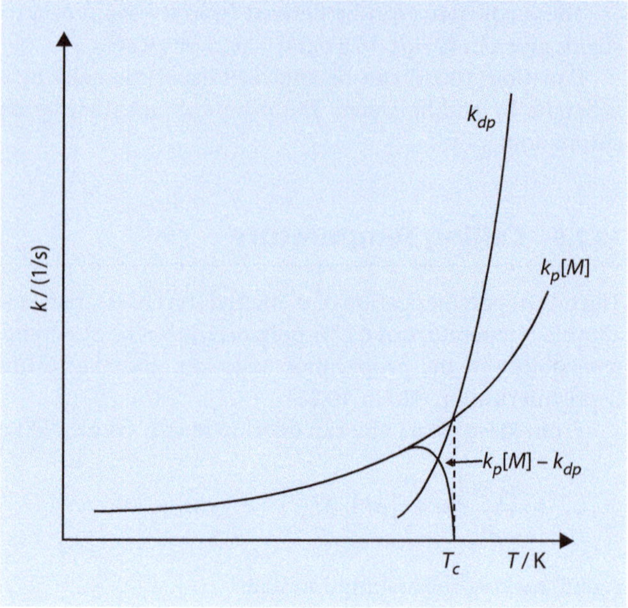

Depolymerization has a larger activation energy than polymerization so that the increase of k_{dp} with increasing temperature is greater than that of k_p . To have the same dimensions, k_{dp} is compared to k_p x $[M]$ in Fig. 10.84. This does not change the basic assertion about the different temperature dependence of both partial reactions in Fig. 10.83.

As for every other chemical reaction, the basic thermodynamic equation also applies to the conversion of monomer into polymer:

$$\Delta G^0 = \Delta H^0 - T\Delta S^0 \tag{10.27}$$

As well as this, the following applies for the free enthalpy of reaction ΔG:

$$\Delta G = \Delta G^0 + \mathrm{R}T \ln K \tag{10.28}$$

ΔG^0	Molar free enthalpy of polymerization
ΔH^0	Molar enthalpy of polymerization
ΔS^0	Molar entropy of polymerization
T	Reaction temperature
R	Gas constant

At equilibrium:

$$\Delta G = 0 \tag{10.29}$$

and from (10.28):

$$\Delta G^0 = -\mathrm{R}T \ln K \tag{10.30}$$

Combining the equilibrium constant K:

$$K = \frac{\left[M_{n+1}^{-}\right]}{\left[M_{n}^{-}\right]\left[M\right]} \tag{10.7}$$

with the assumption

$$\left[M_{n+1}^{-}\right] = \left[M_{n}^{-}\right] \tag{10.8}$$

yields the very simple relationship

$$K = \frac{1}{\left[M\right]} \tag{10.31}$$

Inserting (10.31) into (10.30) gives

$$\Delta G^{0} = \mathrm{R}T\ln\left[M\right] \tag{10.32}$$

which, with (10.27), gives

$$\Delta H^{0} - T\Delta S^{0} = \mathrm{R}T\ln\left[M\right] \tag{10.33}$$

Thus, for the ceiling temperature:

$$T_{c} = \frac{\Delta H^{0}}{\Delta S^{0} + \mathrm{R}\ln\left[M\right]} \tag{10.34}$$

From (10.34) it can be seen that every monomer concentration is associated with a different ceiling temperature. A 1 M solution is usually viewed as the standard concentration in physical chemistry and at this concentration the ceiling temperature is given by

$$T_{c} = \frac{\Delta H^{0}}{\Delta S^{0}} \tag{10.35}$$

However, conventionally, ceiling temperatures are only discussed for polymerizations of pure monomers. The concentration c for pure α-methyl styrene is ($\rho = 0.940$ g/mL, $M = 118$ g/mol):

$$c = \frac{n}{V} = \frac{m}{M \cdot V} = \frac{\rho}{M} = \frac{940\ \mathrm{g/L}}{118\ \mathrm{g/mol}} = 7.966\ \mathrm{mol/L} \tag{10.36}$$

Living anionic polymerization is especially well suited for experimentally determining ceiling temperatures as the polymerization-depolymerization equilibrium is not disturbed by termination reactions. The polymerizations are carried out at different temperatures ($T < 60$ °C), allowing equilibrium to become established. The amount of polymer is then determined and from this the monomer conversion can be calculated. Solving (10.33) for $\ln[M]$ gives the basis for determining the ceiling temperature (■ Fig. 10.85):

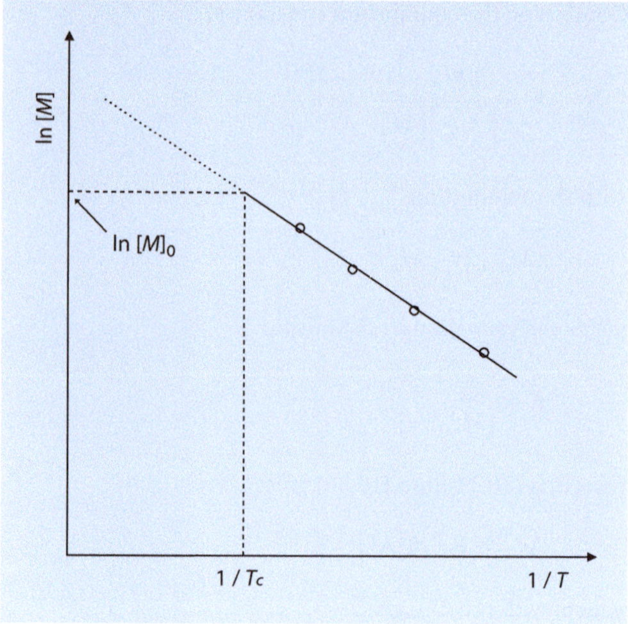

Fig. 10.85 Thermodynamic determination of the ceiling temperature. T_c ceiling temperature, $[M]_0$ concentration of pure monomer

10

$$\ln[M] = \frac{\Delta H^0}{RT} - \frac{\Delta S^0}{R} \tag{10.37}$$

By plotting $\ln[M]$ against $1/T$ (Fig. 10.85), $\Delta H^0/\mathrm{R}$ can be obtained from the gradient and $\Delta S^0/\mathrm{R}$ from the intercept. Both ΔH^0 and ΔS^0 are negative. ΔH^0 is negative as a higher-energy monomer (π-bond) is converted into a lower-energy polymer (σ-bond). Negative values for ΔS^0 are the result of the loss of degrees of freedom for the polymer compared to the monomer.

From (10.34) it can be seen that at every polymerization temperature a minimum monomer concentration is necessary for polymerization to take place (Table 10.5).

10.2.6 Molar Mass Distribution from an Ideal (Living) Anionic Polymerization: Poisson Distribution

The term ideal anionic polymerization is used if the following conditions are met:

1. Every chain starts to grow at the same time, i.e., at time $t=0$ all active centers have identical degrees of polymerization, either $P=0$ or $P=1$. (This implies that the chains are initiated very rapidly and that the rate of initiation is much faster than the rate of propagation so that the time required for initiation is negligible.)
2. The number of growing chains remains constant during polymerization, i.e., there is no chain termination.
3. The chains are constructed out of a series of kinetically identical monomer units (Fig. 10.86).
4. Growth continues until the monomer has been completely converted or until it is intentionally stopped.

Table 10.5 Limiting concentration for a selection of monomers at a polymerization temperature of 25 °C and the ceiling temperatures for the pure monomers

Monomer	$[M]_c$ (mol/L)	T_c (pure monomer) (°C)
Vinyl acetate	$1 \cdot 10^{-9}$	–
Methyl acrylate	$1 \cdot 10^{-9}$	–
Styrene	$1 \cdot 10^{-6}$	310
Methyl methacrylate	$1 \cdot 10^{-3}$	220
α-Methyl styrene	2.2	61

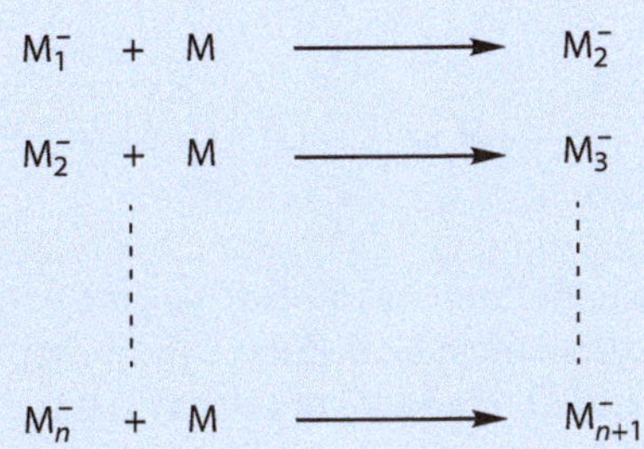

Fig. 10.86 Propagation of an ideal anionic polymerization

5. The chains should only grow in one direction.
6. There is no chain transfer.

The molar mass distribution deviates considerably from those distributions derived above (► Sect. 9.5) as demonstrated below.

The change of the molar concentration of the active species with *a single* monomer unit per molecule ($P=1$) with time is given by

$$\frac{d\left[M_1^-\right]}{dt} = -k_p[M]\left[M_1^-\right] \tag{10.38}$$

The change of the molar concentration of the active species with *two* monomer units per molecule ($P=2$) with time is given by

$$\frac{d\left[M_2^-\right]}{dt} = k_p[M]\left[M_1^-\right] - k_p[M]\left[M_2^-\right] \tag{10.39}$$

By analogy, for $[M_n^-]$:

$$\frac{d\left[M_n^-\right]}{dt} = k_p[M]\left[M_{n-1}^-\right] - k_p[M]\left[M_n^-\right] \tag{10.40}$$

With every propagation step, monomer is consumed so that

$$-\frac{d[M]}{dt} = k_p [M] \sum_{n=1}^{\infty} \left[M_n^-\right] \tag{10.41}$$

According to condition (2) (above) the sum of all the active species is constant:

$$\sum_{n=1}^{\infty} \left[M_n^-\right] = const. = \left[C^-\right] \tag{10.42}$$

From (10.41) and (10.42):

$$-\frac{d[M]}{dt} = k_p [M]\left[C^-\right] \tag{10.43}$$

or

$$-\int_{[M]_0-[C^-]}^{[M]_t} d[M] = k_p \left[C^-\right] \int_0^t [M]\,dt \tag{10.44}$$

(One monomer unit has already reacted with each initiator molecule to become M_1^- at time $t=0$ so these must be deducted from $[M]_0$.)

Solving the left hand side of (10.44) yields

$$\frac{-\left([M]_t - [M]_0 + \left[C^-\right]\right)}{\left[C^-\right]} = k_p \int_0^t [M]\,dt \tag{10.45}$$

The equation for the degree of polymerization P_n during an anionic polymerization, i.e., when not all the monomer has been converted to polymer ($[M]_t \neq 0$), is

$$P_n = \frac{[M]_0 - [M]_t}{\left[C^-\right]} \tag{10.46}$$

$$P_n - 1 = \frac{[M]_0 - [M]_t - \left[C^-\right]}{\left[C^-\right]} \tag{10.47}$$

From (10.45) and (10.47) if follows that

$$\frac{[M]_0 - \left[C^-\right] - [M]_t}{\left[C^-\right]} = P_n - 1 = k_p \int_0^t [M]\,dt \tag{10.48}$$

Introducing the kinetic chain length ν, which denotes the number of monomer units which have been added to the chain started by the anion M_1^-:

$$\nu = P_n - 1 \tag{10.49}$$

and from (10.48) one obtains

$$dv = k_p[M]dt \tag{10.50}$$

$$\frac{dv}{dt} = k_p[M] \tag{10.51}$$

From (10.51) and (10.38):

$$-\frac{d[M_1^-]}{[M_1^-]} \cdot \frac{1}{dt} = \frac{dv}{dt} \tag{10.52}$$

or

$$d[M_1^-] = -[M_1^-]dv \tag{10.53}$$

Similarly, from (10.39) and (10.50):

$$d[M_2^-] = [M_1^-]dv - [M_2^-]dv \tag{10.54}$$

and for the trimer:

$$d[M_3^-] = [M_2^-]dv - [M_3^-]dv \tag{10.55}$$

Thus, for an n-mer anion, from (10.40) and (10.50):

$$d[M_n^-] = [M_{n-1}^-]dv - [M_n^-]dv \tag{10.56}$$

Integration of (10.53) yields

$$\int \frac{d[M_1^-]}{[M_1^-]} = -\int dv \tag{10.57}$$

Solving (10.57) yields

$$\ln[M_1^-] = -v + const. \tag{10.58}$$

When $t=0$, $\nu=0$ and the constant can be calculated:

$$const. = \ln[M_1^-] \tag{10.59}$$

Because $[M_1^-]=[C^-]$ (when $t=0$), i.e., one monomer unit has already been added to the initiator:

$$const. = \ln\left[C^-\right] \tag{10.60}$$

By insertion into (10.59) one obtains for $t \neq 0$ and $[M_n^-] \neq [C^-]$:

$$\ln\left[M_1^-\right] = -\nu + \ln\left[C^-\right] \tag{10.61}$$

so that for $[M_1^-]$:

$$\left[M_1^-\right] = \left[C^-\right] \cdot e^{-\nu} \tag{10.62}$$

Insertion of (10.62) into (10.54) results in

$$d\left[M_2^-\right] = \left[C^-\right] \cdot e^{-\nu} d\nu - \left[M_2^-\right] d\nu \tag{10.63}$$

Integrating (10.63) by the method of integrating factors:

$$e^{\nu} \cdot d\left[M_2^-\right] = \left[C^-\right] d\nu - \left[M_2^-\right] \cdot e^{\nu} d\nu \tag{10.64}$$

whereby

$$e^{\nu} d\nu = d\left(e^{\nu}\right) \tag{10.65}$$

and substituting (10.65) into (10.64) gives

$$\left[C^-\right] d\nu = e^{\nu} d\left[M_2^-\right] + \left[M_2^-\right] d\left(e^{\nu}\right) \tag{10.66}$$

Using the integration rule (total differential):

$$x \cdot y = \int x \cdot dy + \int y \cdot dx \tag{10.67}$$

Equation (10.66) can be integrated ($x = e^{\nu}$; $dx = e^{\nu}\, d\nu = d(e^{\nu})$; $y = [M_2^-]$; $dy = d[M_2^-]$) and transformed into (10.68):

$$\left[C^-\right] \cdot \nu + \text{const.} = e^{\nu} \cdot \left[M_2^-\right] \tag{10.68}$$

const. = 0, because $[M_2^-]$ and $\nu = 0$ at $t = 0$ so that

$$\left[M_2^-\right] = \frac{\left[C^-\right] \cdot \nu}{e^{\nu}} \tag{10.69}$$

As deduced above (see (10.55)) for trimers:

$$d\left[M_3^-\right] = \left[M_2^-\right] d\nu - \left[M_3^-\right] d\nu \tag{10.70}$$

which, with (10.69) gives

$$d\left[M_3^-\right]=\frac{\left[C^-\right]\cdot v\cdot dv}{e^v}-\left[M_3^-\right]dv \tag{10.71}$$

After multiplication by e^v one obtains

$$e^v\cdot d\left[M_3^-\right]+\left[M_3^-\right]\cdot e^v dv=\left[C^-\right]\cdot v\cdot dv \tag{10.72}$$

By analogy with (10.66), after integration one obtains

$$\left[M_3^-\right]\cdot e^v=\left[C^-\right]\cdot\frac{v^2}{1\cdot 2}+const. \tag{10.73}$$

$const=0$, as at $t=0$, $[M_3^-]=0$ and $\nu=0$

$$\left[M_3^-\right]=\left[C^-\right]\cdot\frac{v^2}{2!\cdot e^v} \tag{10.74}$$

Extrapolating this approach one obtains for $[M_n^-]$:

$$\left[M_n^-\right]=\left[C^-\right]\cdot\frac{v^{n-1}}{(n-1)!\cdot e^v} \tag{10.75}$$

Thus, the mole fraction of n-mer anions M_n^- of the total ions$[C^-]$ generates a *Poisson* distribution :

$$\frac{\left[M_n^-\right]}{\left[C^-\right]}=\frac{v^{n-1}}{(n-1)!\cdot e^v} \tag{10.76}$$

The weight fraction of the n-mer anion m_P of the total weight m is then

$$w_P=\frac{m_P}{m}=\frac{\frac{\left[M_n^-\right]}{\left[C^-\right]}\cdot n\cdot M_0}{(v+1)\cdot M_0}=\frac{v^{n-1}\cdot n}{(n-1)!\cdot(v+1)\cdot e^v} \tag{10.77}$$

Using this equation and the definition of the weight average, one obtains for the weight average of the degree of polymerization P_w:

$$P_w=\sum_n w_P\cdot n=\sum_n\frac{v^{n-1}\cdot n^2}{(n-1)!\cdot(v+1)\cdot e^v}=$$

$$=\frac{v^2+3v+1}{v+1} \tag{10.78}$$

By solving this equation using the series theory one obtains

$$P_w = \sum_n \frac{\nu^{n-1} \cdot n^2}{(n-1)! \cdot (\nu+1) \cdot e^\nu} = \frac{\nu}{(\nu+1) \cdot e^\nu} \sum_n \frac{\nu^{n-2} \cdot n^2}{(n-1)!} =$$
$$= \frac{\nu}{(\nu+1) \cdot e^\nu} \cdot \left(\nu + 3 + \frac{1}{\nu}\right) \cdot e^\nu = \frac{\nu^2 + 3\nu + 1}{\nu + 1}$$

The definition of the non-uniformity of a molar mass distribution (▶ Sect. 9.4.3) is

$$U = \frac{P_w}{P_n} - 1 \tag{9.80}$$

which (if $P_n = \nu + 1$ (10.49)) can be expanded to

$$U = \frac{\nu^2 + 3\nu + 1}{(\nu+1) \cdot (\nu+1)} - 1 = \frac{\nu}{(\nu+1)^2} = \frac{P_n - 1}{P_n^2} = \frac{1}{P_n} - \frac{1}{P_n^2} \tag{10.79}$$

Under ideal conditions, polymers with very narrow molar mass distributions are usually obtained during anionic polymerization. This is why anionic polymerization is used, for example, for the synthesis of polymer standards, e.g., for calibrating size exclusion chromatography (▶ Sect. 3.3.2).

10

From (10.79) it follows that non-uniformity decreases with an increase in the degree of polymerization.

From ◘ Fig. 10.87 and (10.79) it can be seen that a polymer from an anionic polymerization still contains chains with degrees of polymerization between $n = 30$ and $n = 80$ even when its inconsistency is $U = 0.02$, despite the fact that it has a considerably narrower molar mass distribution than a polymer from a radical polymerization with an identical kinetic chain length ($\nu = 50$).

At this point, it is important to draw the readers' attention to a comparison between the frequency distribution of the individual polymer fractions $x_i(P_i)$ and the weight fraction distribution $w_i(P_i)$, both as functions of the degree of polymerization P_i. The basic concept beneath the former is the mole fraction and for the latter the weight fraction. For a Poisson distribution, in analogy with (10.76) and (10.77) one gets:

$$x_i(P_i) = \frac{(P_n - 1)^{P_i - 1}}{\Gamma(P_i)} \exp(1 - P_n) \tag{10.80}$$

$$w_i(P_i) = x_i(P_i) \frac{P_i}{P_n} \tag{10.81}$$

$x_i(P_i)$ describes the mole fraction of those polymer chain with a degree of polymerization P_i, $w_i(P_i)$ the corresponding weight fraction, P_n the mean degree of polymer-

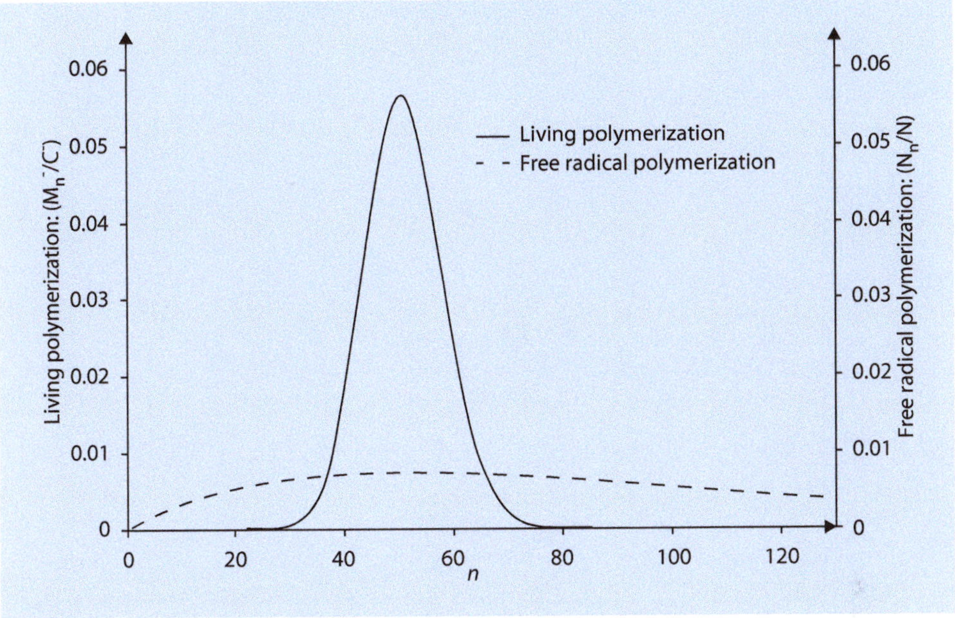

Fig. 10.87 Comparison of the number average distribution of n-mers from an anionic and a free radical polymerization. Anionic polymerization: M_n^- number of active chains with n repeating units, C^- total number of active chains; radical polymerization, N_n number of polymers with n repeating units, N number of all polymer chains; for both cases $\nu = 50$

ization (number average), and Γ the gamma function with the auxiliary variable k given by

$$\Gamma(P_i) = \int_0^\infty k^{P_i - 1} \exp(-k)\, dk \tag{10.82}$$

It should also be noted that for natural numbers:

$$(P_i - 1)\,! = \Gamma(P_i) \tag{10.83}$$

On the other hand, Schulz Zimm distributions are usually obtained from free radical polymerizations:

$$x_i(P_i) = \frac{\zeta^{\zeta+1}}{P_n^{\zeta}\Gamma(\zeta+1)} P_i^{\zeta-1} \exp\left(\frac{-\zeta P_i}{P_n}\right) \tag{10.84}$$

ζ describes the degree of coupling, i.e., the number of chains that grew independently and are terminated by combination:

$$\zeta = \frac{P_n}{P_w - P_n} \tag{10.85}$$

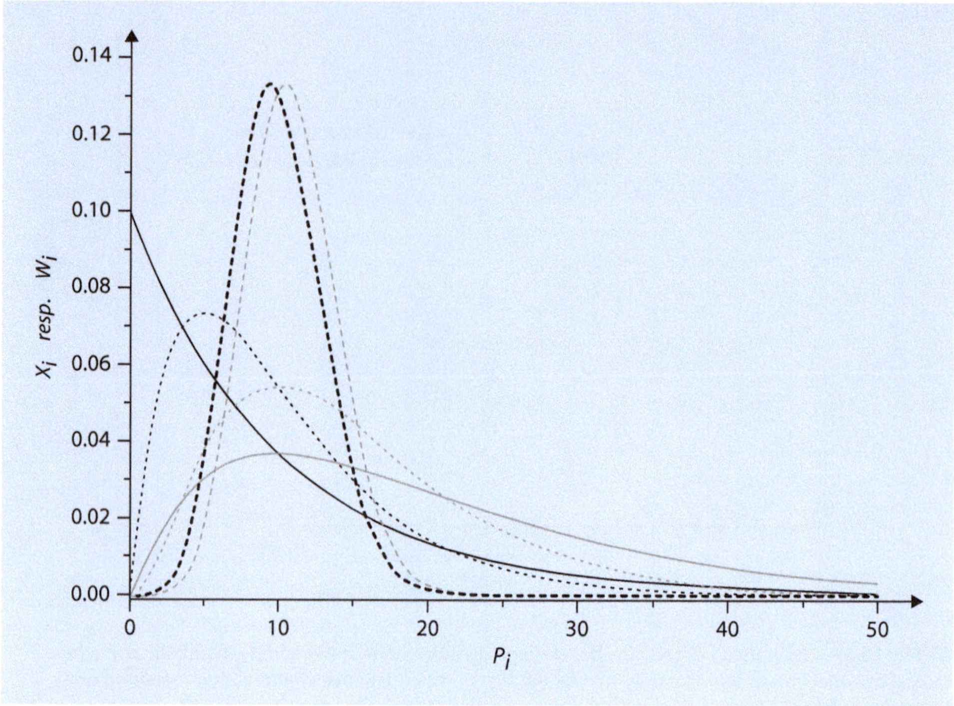

Fig. 10.88 Distribution functions with $P_n = 10$. *Dashed line:* Poisson-distribution, *full line:* Schulz–Zimm distribution with $\zeta = 1$, *dotted line:* Schulz–Zimm distribution with $\zeta = 2$. (black: mole/number fraction distributions x_i, gray: weight fraction distributions w_i)

P_w describes, as above, the weight average degree of polymerization. Where termination is solely by disproportionation $\zeta = 1$ (the resulting distribution is also called a Schulz Flory distribution). Where termination is by recombination $\zeta = 2$. Examples of the graphs of these functions are shown in Fig. 10.88.

References

Kennedy JP, Smith RA (1979) New telechelic polymers and sequential copolymers by polyfunctional initiator-transfer agents (inifer). I. Synthesis and characterization of α, ω-di(tert-chloro)polyisobutylene. Polym Prepr 20:316–319

Kennedy JP, Smith RA (1980) New telechelic polymers and sequential copolymers by polyfunctional initiator-transfer agents (inifer). III Synthesis and characterization of poly (α-methyl styrene-*b*-isobutylene-α-methyl styrene). J Polym Sci Polym Chem Ed 18:1539–1546

Szwarc M (1956) "Living" polymers. Nature 178:1168

Winstein S, Klinedinst P, Clippinger E (1961) Salt effects and ion pairs in solvolysis and related reactions, XXI Acetolysis, bromide exchange, and the special salt effect. J Am Chem Soc 83:4986–4989

Coordination Polymerization

S. Koltzenburg et al., *Polymer Chemistry*, DOI 10.1007/978-3-662-49279-6_11

Nearly half of all polymers produced worldwide are produced by catalytic polymerization reactions carried out in the presence of transition metal compounds. Coordination of a monomer to a metal center is a crucial step in the catalytic cycle. Therefore, these polymerizations are referred to as "coordination polymerizations". Especially polypropylene and a large proportion of polyethylene are produced in this way. Therefore, this chapter deals with the fundamental principles of this industrially enormously important but also academically interesting and multifaceted field of chemistry.

11.1 Polymerization of α-Olefins

In contrast to many other unsaturated compounds, simple α-olefins cannot be polymerized (with the exception of ethene) via the radical or ionic polymerizations described in the previous chapters. This is because of the chain carrying the active center—for example, a radical—preferring to abstract a hydrogen atom or an ion from the carbon atom adjacent to the double bond. This results in an allyl species, which is so stable that no other monomer can be added because this would lead to the resonance stabilization being lost. An example of this is the reaction of propene with a source of radicals in Fig. 11.1. The reaction with corresponding anions or cations takes place in an analogous manner.

The polymerization of α-olefins only became possible after Karl Ziegler's discoveries at the Max Planck Institute for Carbon Research in Muelheim an der Ruhr (Ziegler et al. 1955).

11.2 Ziegler Catalysts

Karl Ziegler, who was conducting research on the oligomerization of ethene by aluminum alkyls, discovered by accident that transition metals could massively influence this reaction (Fig. 11.2). He noticed that traces of nickel prevented the oligomerization of

Fig. 11.1 Reaction of propene with a radical

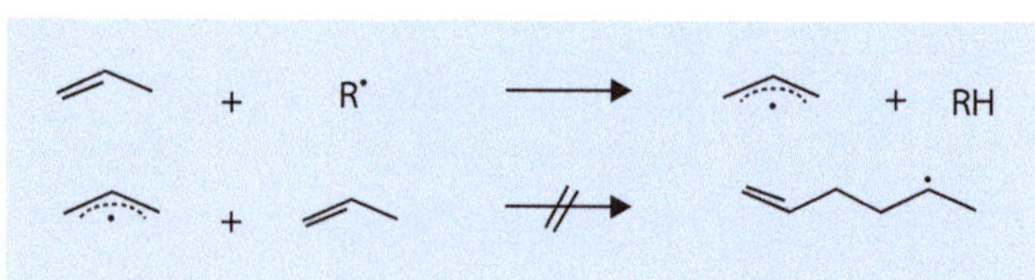

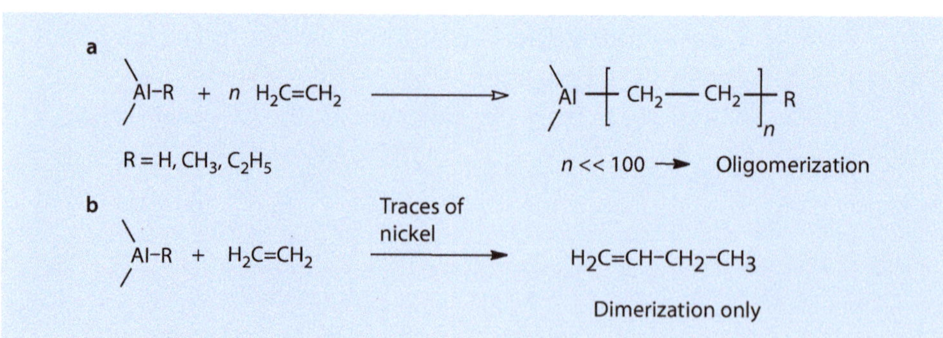

Fig. 11.2 (**a**) Ziegler's "Aufbaureaktion" (Chain extension reaction). (**b**) The nickel effect

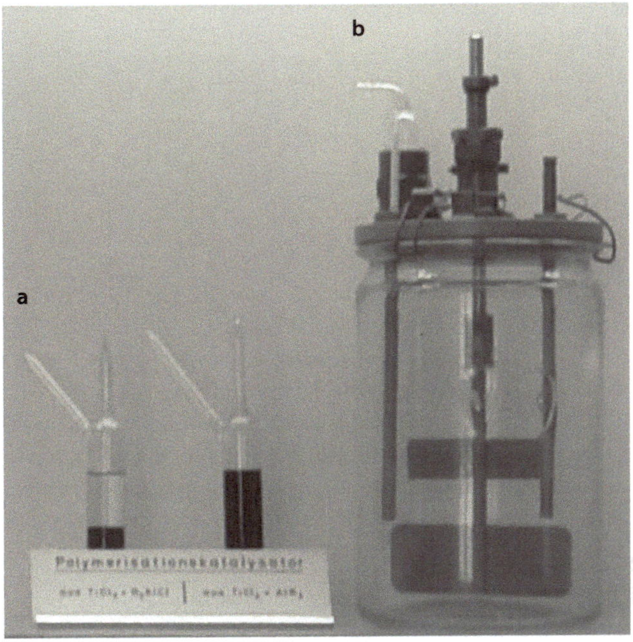

Fig. 11.3 (**a**) Glass containers with two catalysts. (**b**) Ziegler's polymerization reactor (reproduced here with kind permission from Professor Dr. G. Fink)

Table 11.1 Examples of Ziegler catalysts and cocatalysts (Et = C_2H_5)

Catalyst	Formula	Cocatalyst	Formula
Titanium tetrachloride	$TiCl_4$	Triethyl aluminum	$AlEt_3$
Zirconium tetrachloride	$ZrCl_4$	Diethyl aluminum chloride	$AlEt_2Cl$
Titanium trichloride	$TiCl_3$	Ethyl aluminum dichloride	$AlEtCl_2$
Titanium tetraisobutylate	$Ti(O\text{-}i\text{-}C_4H_9)_4$	Diisobutyl aluminum hydride	$AlH(i\text{-}C_4H_9)_2$
Vanadium oxytrichloride	$VOCl_3$		

ethene and that in the presence of nickel only dimers are produced. This effect came to be known as the *nickel effect* and spurred Ziegler on to analyze more closely the influence of transition metals on oligomerization. During his research he discovered that a mixture of titanium compounds, such as titanium tetrachloride together with aluminum alkyl compounds, could spontaneously polymerize ethene to macromolecular polyethylene at atmospheric pressure and room temperature.

At that time this procedure was revolutionary because a typical radical polymerization of ethene, involving a very active primary radical at the growing chain end, required extreme reaction conditions: a temperature of 300 °C and a pressure of up to 2000 bar. In contrast to this, the polymerization using Ziegler catalysts takes place at ambient conditions. To highlight this, Karl Ziegler used a converted preserving jar for presentation purposes (Fig. 11.3).

The reaction takes place with organometallic compounds, formally transition metals, and requires a second component, an aluminum alkyl, as a so-called cocatalyst. Examples are given in Table 11.1.

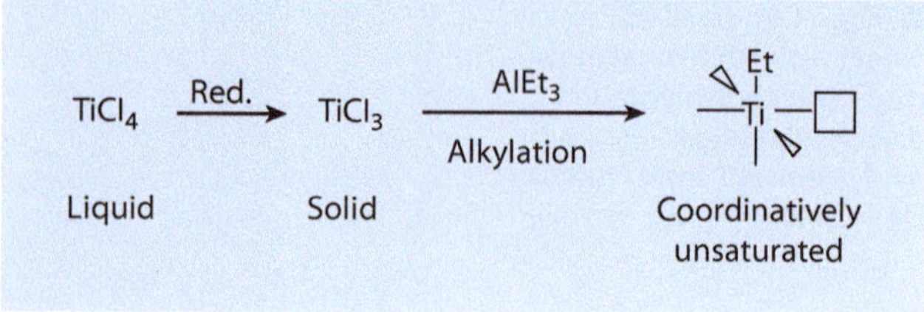

Fig. 11.4 Initiation process for Ziegler catalysts

Heptane or toluene is used as a solvent. The use of metal alkyls dictates strictly anhydrous conditions. Alternatively, traces of water can be scavenged by a surplus of aluminum alkyls.

The active catalysts discussed above are insoluble in hydrocarbon solvents and can thus be described as *heterogeneous* catalysts.

The following model is the generally accepted one for Ziegler catalysis and is discussed in the following using titanium tetrachloride as an example. The catalytic cycle can be divided into the basic steps—initiation, growth, chain transfer, and termination—analogous to other chain growth reactions, such as radical polymerization.

11.2.1 Initiation

Because of the reducing effect of aluminum compounds, the titanium(IV) chloride is reduced and a solid titanium-containing species is formed.[1] On the surface of the solid particles the titanium centers are alkylated by the aluminum alkyl. During this process one of the chloride ions is swapped for an ethyl group. Some of the metal centers on the surface are coordinatively unsaturated and have a free coordination point. The metal atoms are the catalytically active centers (Fig. 11.4).

11.2.2 Growth

An ethene molecule coordinates with the metal atom via the π-electrons of the double bond as the first part of the growth step. Then an insertion of the ethene molecule over a four-center-four-electron-transition state involving the metal carbon bond already existing takes place. Thus the coordination site that was occupied by the ethyl group becomes available again and so the center is ready for another insertion step (Fig. 11.5). The chain migration during the insertion step leads to a *cis*-addition (this is known as the *chain migratory mechanism*).

Bi-metallic mechanisms, for example involving chloride bridged titanium–aluminum species, have also been described for chain growth but confirmation has proved difficult because of the heterogeneous nature of the catalytic species.

1 The reaction is not well defined and the products, because they are insoluble, are not easily analyzed. There are a number of different titanium species in various valence states. Because among the various species formed many catalyze the polymerization of ethene, this system is called a *multi-site* catalyst.

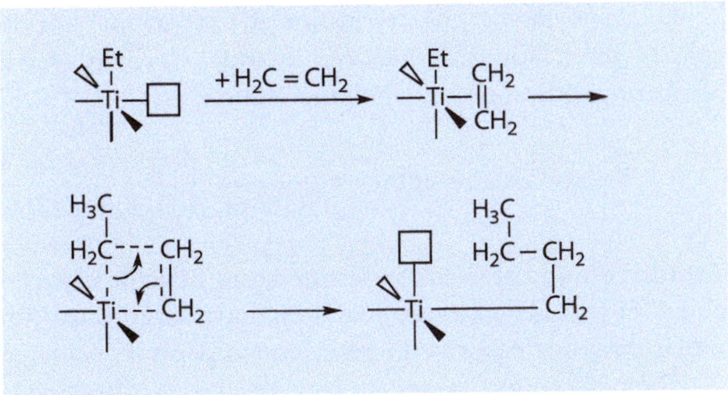

Fig. 11.5 Chain growth with Ziegler catalysts

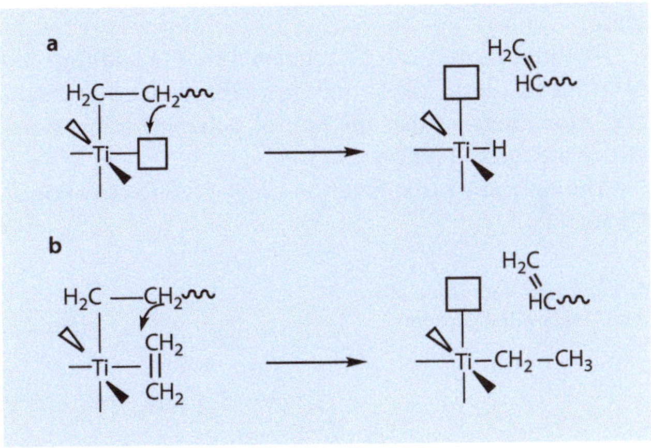

Fig. 11.6 Chain termination and transfer. (a) Via ß-hydride-elimination. (b) Via ß-hydride-transfer

11.2.3 Termination/Chain Transfer

Instead of an insertion, the β-carbon of the hydrocarbon chain can be transferred in the form of a hydride ion to the free coordination center or to a coordinated monomer molecule (Fig. 11.6). In both cases an unsaturated polymer is formed. These processes are called *β-hydride-elimination* or *β-hydride-transfer*, respectively. The titanium alkyl or titanium hydride species formed during these processes are still active polymerization catalysts and can coordinate with further ethene molecules to initiate new polymer chains.

It is possible to distinguish between the two mechanisms by varying the ethene concentration, i.e., the ethene pressure above the liquid phase. If termination takes place via β-hydride-elimination then an increase in the ethene concentration increases the rate of chain growth and the rate of termination remains the same. Thus, the molar mass increases with an increase in ethene pressure. If, however, a catalyst system is used, in which a β-H-transfer onto a coordinated monomer is favored, an increase in ethene concentration is also beneficial for chain growth and the molar mass is independent of the ethene pressure. Generally, however, both reactions occur concurrently.

Furthermore, termination of the catalytic process can take place because of contamination, by water for example.

During catalytic polymerization the molar mass of the polymer is controlled kinetically by the relationship between the rates of chain growth and termination, just as they are during radical or ionic polymerization:

$$\text{Kinetic chain length } \nu = \frac{\text{Olefin insertion rate}}{\text{Rate of } \beta\text{-H-elimination or -transfer}}$$

An increase in the reaction temperature benefits especially the rate of β-H-elimination and β-H-transfer as these reactions have large activation energies. Thus, the polymer molar mass decreases as the reaction temperature is increased.

The nickel effect, observed by Ziegler, is attributable to elimination taking place very quickly in the presence of nickel and therefore only a dimerization of the ethene takes place.

The molar mass can be regulated by the addition of hydrogen. Molecular hydrogen splits the titanium–alkyl bond (*hydrogenolysis*). The resulting hydride species can add further monomers so that the rate of polymerization stays the same but shorter polymer chains are formed (Fig. 11.7).

The different variants of the Ziegler "Aufbaureaktion" can be summarized as follows (Fig. 11.8):

Fig. 11.7 Regulation of molar mass with hydrogen

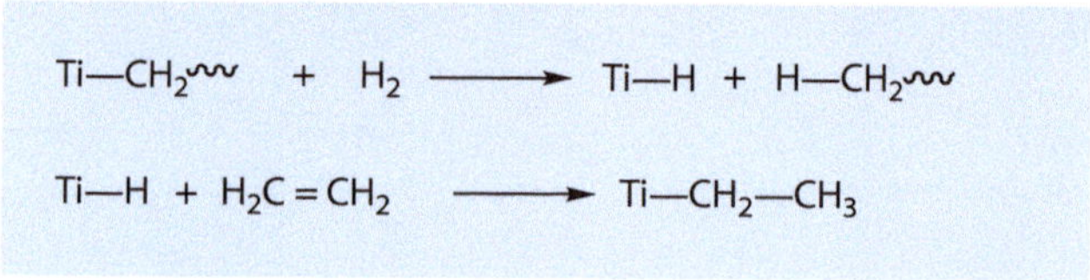

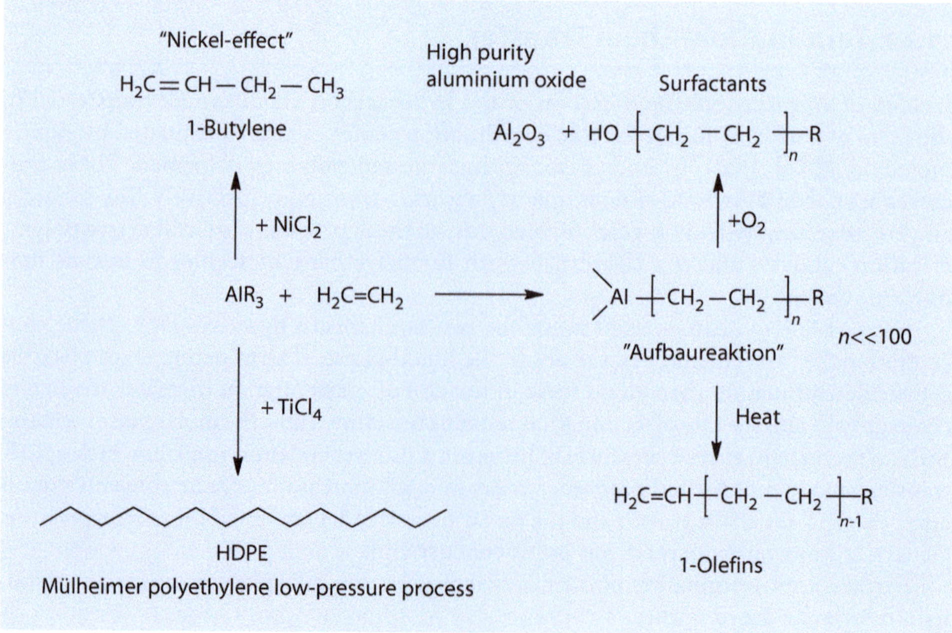

Fig. 11.8 Reactions of aluminum alkyls with ethene. *HDPE* high density polyethylene

- Oligomerization of ethene through the "Aufbaureaktion"
- Dimerization of ethene through the *nickel effect*
- Polymerization of ethene to polyethylene (*Muelheim polyethylene low-pressure process*)

The longer chain aluminum alkyls formed during the "Aufbaureaktion" have a narrow molar mass distribution (Poisson distribution, ▶ Chap. 10) and can be oxidized to alcohols with oxygen. This is a technically used route to valuable raw materials for surfactants. Very pure aluminum oxide is formed as a by-product. As an alternative, olefins can be produced by thermal elimination (◘ Fig. 11.8).

As well as the systems known as *Ziegler catalysts*, chrome-based catalysts have also found their way into technical polymerization processes. These systems, known as *Phillips catalysts,* consist of chromium trioxide on an SiO_2 carrier. Chrome(VI) is probably reduced down to chrome(II) during the reaction. Phillips systems are able to function without a cocatalyst, in contrast to Ziegler catalysts. The mechanism of polyinsertion is suggested to take place in an analogous manner to that described above for titanium-based catalysts. As a group, the reactivity of these systems is somewhat lower than that of Ziegler catalysts so that higher temperatures and higher olefin pressures are necessary and, usually, polymers of low molar mass are produced. Processes based on these catalysts are referred to as medium-pressure processes. The differences are summarized in ◘ Table 11.2.

As early as 1 year after their discovery, the applicability of the titanium catalysts discovered by Ziegler was substantially broadened by the Italian scientist Giulio Natta. Natta discovered that the systems described by Ziegler could polymerize not only ethene but also higher α-olefins. The polypropylene produced using the first generation of, not yet optimized, Ziegler catalysts was essentially atactic or stereo-irregular (▶ Sect. 1.3.2) and thus soluble in the reaction medium. No precipitated polymer could be observed during the reaction, in contrast to insoluble polyethylene, which is probably why this important discovery was overlooked in Karl Ziegler's laboratories.

If heterogeneous Ziegler catalysts are optimized by adding additional components, they are able to control the stereoregularity of the product (Natta 1964). The mechanism of the stereoregular polymerization of α-olefins is dealt with in the following paragraph (▶ Sect. 11.3) in combination with the polymerization with metallocene catalysts.

Karl Ziegler and Giulio Natta received the 1963 Nobel Prize for Chemistry for the discovery of the catalytic polymerization of α-olefins and the stereospecific polymerization of propene.

◘ **Table 11.2** A comparison of the process parameters for low and medium pressure processes

Process	Low pressure	Medium pressure
Catalyst	$TiCl_4/AlR_3$	CrO_3/SiO_2
p (bar)	1–10	35
T (°C)	70	150–180
M_w (kg/mol)	50–100	5–20

11.3 Homogeneous Polymerization Catalysts

The discovery of highly active and stereoselective metallocene catalysts by Brintzinger in 1995 opened up a further fascinating and, what is now, a widely studied research area of catalytic polymerization. These transition metal compounds are soluble in the reaction medium and are therefore referred to as homogeneous catalysts. Many of the mechanistic details of polyinsertion have been clarified with the aid of these well defined systems so that they have provided essential contributions to our understanding of catalytic polymerization.

Most, in particular the more important, homogeneous catalysts are metallocenes or half-sandwich compounds of the early transition metals titanium and zirconium. Examples are shown in ◘ Fig. 11.9. In principle, hafnium can also be used but the compounds based on this metal are generally less active than their zirconium counterparts—and hafnium is very expensive.

Similar to the heterogeneous Ziegler catalysts, most of the metallocenes are only active with a cocatalyst. Often, aluminum organic compounds are used as cocatalysts. However, the aluminum alkyls used in the heterogeneous systems have not proved to be very effective. Far more active catalyst systems can be obtained if methyl aluminoxane (MAO) is employed as a cocatalyst (◘ Fig. 11.10). This compound is the oligomeric product of the partial hydrolysis of trimethyl aluminum comprising approximately 20 aluminum atoms (Sinn and Kaminsky 1980). The structure of methyl aluminoxane has not been completely determined; it is probably not a linear oligomer, but rather contains cyclic and cage-like structures in which the oxygen atoms coordinate onto several aluminum atoms reflecting the Lewis acidity of aluminum.

However, it has been shown that systems not based on aluminum can also be employed as cocatalysts; only a high Lewis acidity is necessary. Thus, for example, triphenyl methyl cations $[C(C_6H_5)_3]^+$ with weakly coordinated anions such as tetrakis(pentafluorophenyl borane) $[B(C_6F_5)_4]^-$ can be used as cocatalysts.

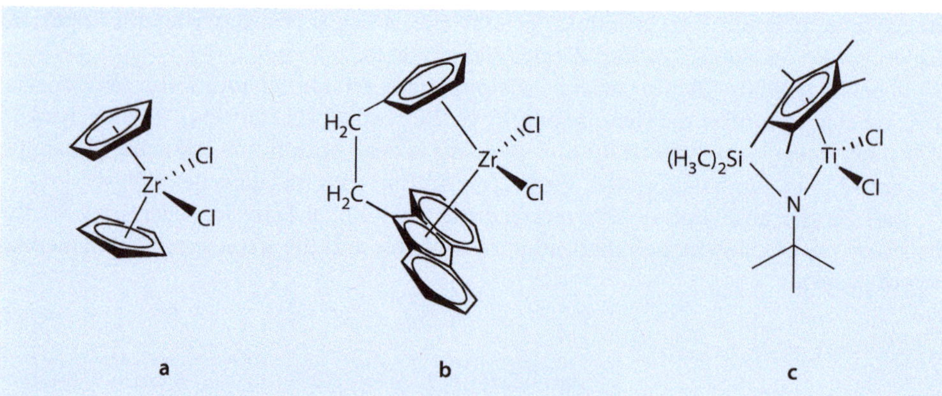

◘ **Fig. 11.9** Examples of (**a**, **b**) homogeneous metallocene catalysts and (**c**) half-sandwich complexes for the polymerization of olefins

◘ **Fig. 11.10** Synthesis of methyl aluminoxane via partial hydrolysis of trimethyl aluminum

$$Al(CH_3)_3 + H_2O \xrightarrow[\text{Stoich.}]{} \left[\begin{matrix} CH_3 \\ | \\ Al-O \end{matrix} \right] + 2\,CH_4 \uparrow$$

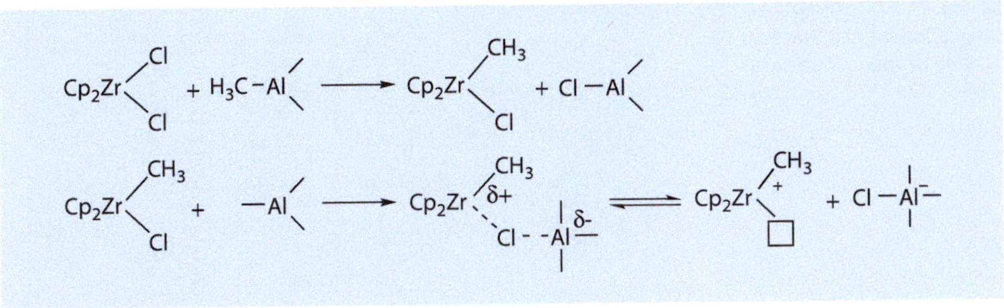

Fig. 11.11 Formation of the catalytically active moiety from a metallocene and a cocatalyst (here: methyl aluminoxane)

Fig. 11.12 Newman projection of propene along the C2–C3 bond

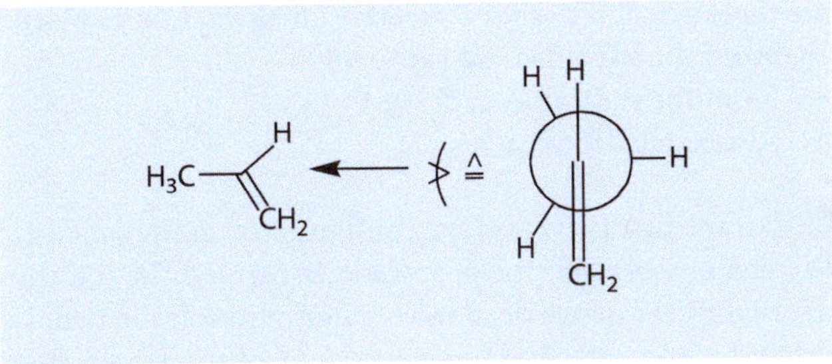

In contrast to the chemically relatively undefined heterogeneous systems, metallocene catalysts are well defined entities. Generally, there is only one, well defined, catalytically active component, which is why these systems are referred to as *single-site catalysts*. The uniform nature of the catalytic centers has considerably facilitated the mechanistic analyses of these catalyst systems.

For the polyinsertion of olefins, as a first step, the metallocene, which is mostly employed as a dichloride, is alkylated by the cocatalyst, such as methyl aluminoxane. In the next step, a further ligand is abstracted from the transition metal by the Lewis acid center of the cocatalyst. It is during this step that the active polymerization species moiety is formed, a cationic species which has the electron configuration d^0 (Fig. 11.11). As this is only formed by reaction of the metallocene with the cocatalyst, the metallocene is also referred to as a *precatalyst* or *precursor*. Coordination and insertion take place in the manner described for heterogeneous systems in ▶ Sect. 11.2.

The mechanism of stereo control of the polyinsertion reaction can be most easily understood using metallocene catalysts as an example.[2] It is assumed here that the different stereoisomers dealt with in ▶ Chap. 1 are understood.

Propene is a prochiral molecule (Fig. 11.12). This means that if an addition reaction takes place at the double bond of the C2 carbon atom, a chiral compound is produced (unless, coincidentally, after the addition step the two substituents at C2 are identical).

Whether the addition of the catalytically active metal alkyl complex takes place from the re or the si side of the propene[3] is decisive in determining the stereochemistry that results at C2 (Fig. 11.13).

---

2 The following discourse describes the basic mechanisms of stereo control. For a detailed discussion of stereo control during catalytic polymerization the reader is referred to the literature (e.g., Brintzinger et al. 1995; Angermund et al. 2000).

3 The re/si nomenclature denotes the sides of a planar, sp^2-hybridized center, from which, after a reaction in which a further substituent is added, a chiral center is formed.

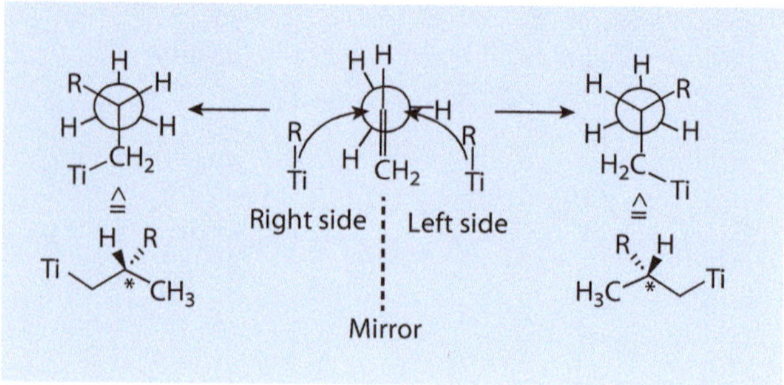

Fig. 11.13 Stereochemistry of an addition of a titanium alkyl to propene molecule

A stereoregular product is formed if addition takes place in a regular sequence. There are three possibilities for this; addition always takes place:

- From the left side in Fig. 11.13
- From the right side in Fig. 11.13
- Alternately from each side

For the first two cases the configuration at the prochiral carbon atom is the same along the chain; isotactic polypropylene is formed. In the third case, polypropylene with an alternating sequence of stereo configurations is formed, syndiotactic polypropylene. If the catalyst system used is unable to control the stereochemistry of the insertion reaction, then atactic polypropylene results. All stereochemical variants can be deliberately obtained using metallocene catalysts.

11

11.3.1 Catalysts for Isotactic Polypropylene

An example of an isoselective metallocene—a metallocene that predominantly catalyzes the formation of isotactic polymers—is shown in Fig. 11.14. The symmetry of the metallocene is identical to that of the point group C_2; the molecule has an axis of symmetry.[4] Because of this, the chlorine atoms of the metallocene, or both coordination points formed after the alkylization and formation of the d^0-cation, are identical in that they can be transferred into one another by rotating the molecule; such molecules are referred to as being *homotopic*.

There are four stereochemically different possibilities shown in Fig. 11.15 for the olefin to approach the catalytically active, tetrahedrally coordinated center. Complexes I and II are energetically unfavorable as the methyl group of the propene molecule points towards the ligand structure of the metallocene. This results in a sterically unfavorable interaction and causes insertion to take place in such a way that the transition metal adds to the C1 atom of the propene molecule and not the prochiral C2 atom. Option III is energetically disadvantaged in comparison to configuration IV as the ligand system forces the methyl group of the propene molecule into a position as far away as possible from the sterically demanding indenyl ligands. The most energetically favorable insertion option is the stereochemistry represented by IV. These effects can be modeled and quantitatively predicted using force field calculations (Angermund et al. 2000).

4 For a more detailed discussion of point groups and the elements of molecular symmetry, the interested reader is referred to the literature (Willock 2009)

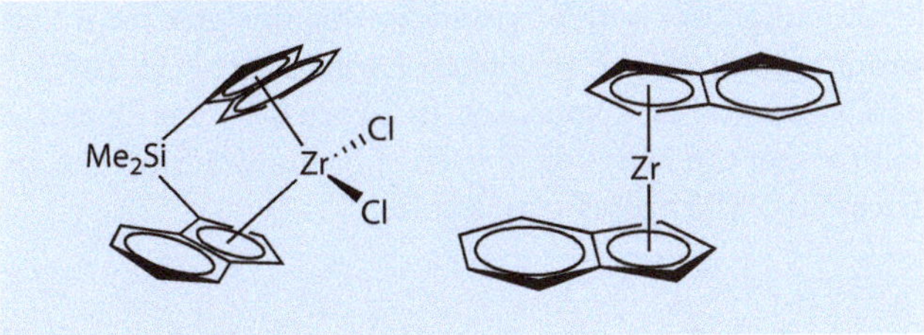

Fig. 11.14 Example of an isoselective metallocene and its simplified representation (in order to provide a better overview the chlorine atoms and the silyl bridges have been left out here and in the following figures)

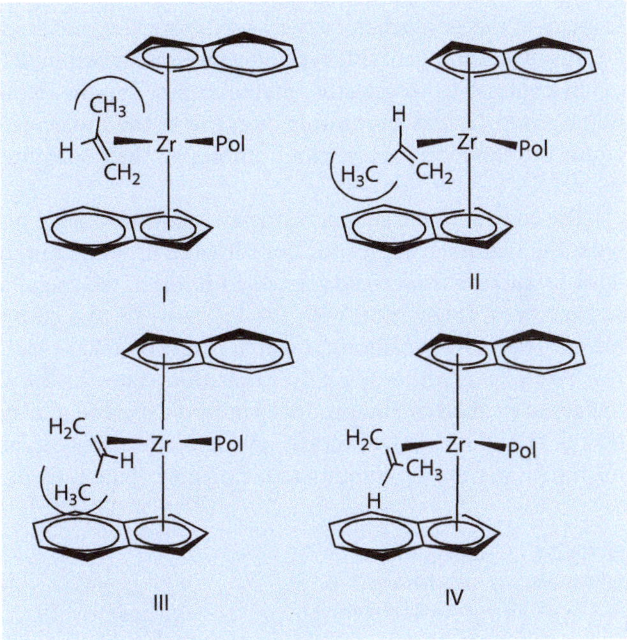

Fig. 11.15 Stereochemical options for the approach of a propene molecule to a metallocene with C_2-symmetry. *Pol*: the growing polymer chain

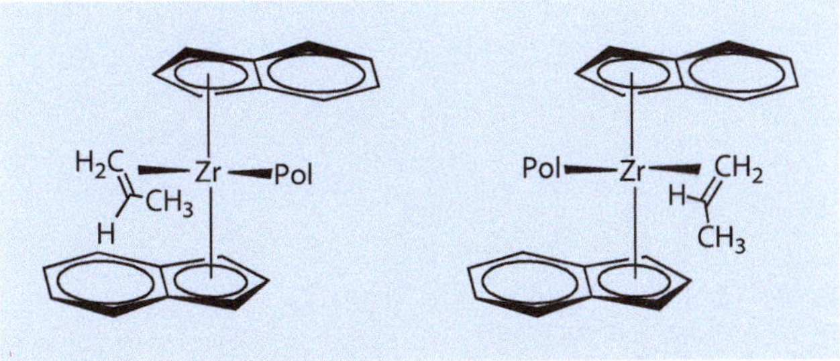

Fig. 11.16 Homotopy of the possible configurations when a propene molecule coordinates with an isoselective metallocene. *Pol*: the growing polymer chain

As has been explained above, both equatorial ligand positions are stereochemically identical at the transition metal (homotopic) in catalysts with C_2-symmetry. The configuration given by IV in Fig. 11.15 for a coordination of olefin with the 'left' coordination point is identical to the alternative coordination at the opposite ligand position (Fig. 11.16). Even though the polymer chain changes its position in the ligand structure every time an insertion step takes place (*chain migratory mechanism*, Sect. 11.2), every insertion takes place in a manner leading to the same configuration at the C2-atom of the inserted monomer unit. Isotactic polypropylene is formed (it-PP).

Metallocenes with a symmetry that deviates from that of the point group C_2 can also produce isotactic polypropylene (Brintzinger et al. 1995). It is only necessary that in every case the monomer approach is directed by the ligand structure of the metallocene to ensure that the insertion at both coordination sites always takes place from the same side (re or si) of the propene molecule.

11.3.2 Catalysts for Syndiotactic Polypropylene

Control of the stereochemistry by *syndioselective* metallocenes takes place in a very similar way. An example of this type of catalyst is shown in Fig. 11.17.

In contrast to isoselective metallocenes, syndioselective metallocenes belong to the point group C_S and have a mirror plane in the molecule. For this reason, both chlorine atoms are *enantiotopic* and not homotopic; their substitution results in an enantiomeric molecule.

The control of the stereochemistry follows the same principles as for isoselective catalysts: the methyl group of the coordinated propene monomer is directed away from the ligand structure (this results in the addition of the metal to the C1-atom) and positioned so that steric interaction with the ligands—in this case the fluorenyl system—is minimized. The ligand positions at which insertion takes place are, in contrast to the isoselective catalysts, enantiotopic. The transition states of the insertion steps are thus mirror images of each other and produce antipodal stereochemistry at the prochiral carbon atom (Fig. 11.18). As the two transition states alternate with one another because of the *chain migratory mechanism*, syndiotactic polypropylene is formed.

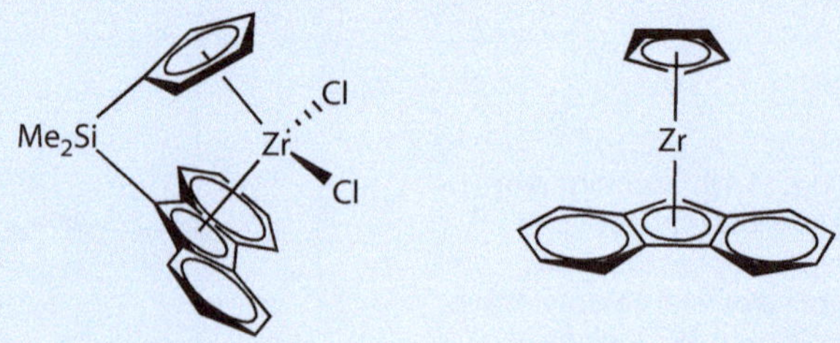

Fig. 11.17 Example for a syndioselective metallocene and its simplified representation

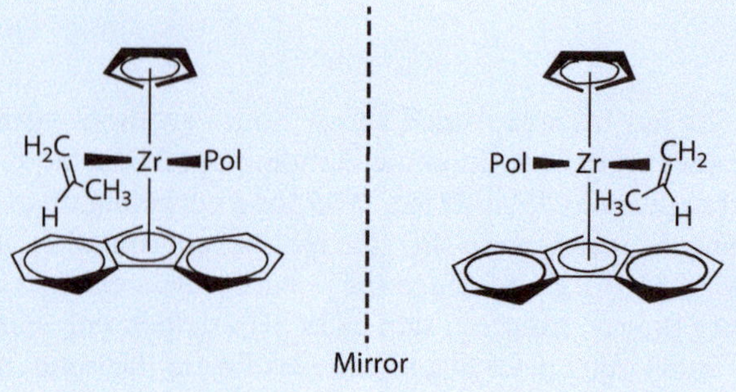

Fig. 11.18 Enantiotopicity of the possible configurations when a propene molecule coordinates with a syndioselective metallocene. *Pol*: the growing polymer chain

Considerable rigidity of the ligand structure is necessary to control the stereochemistry of both iso- and syndioselective metallocenes. This can be achieved by covalently linking both aromatic systems as shown in Figs. 11.14 and 11.17. Suitable bridges are ethylene or silyl moieties and, less commonly, a single carbon atom. The length of the bridges determines the angle of the aperture of the metallocene. As the bridge between the ligands is similar in form to a handle, these types of bridged metallocenes are referred to as *ansa*-metallocenes, after the Greek word for handle.

11.3.3 Catalysts for Atactic Polypropylene

Metallocenes that yield atactic polyolefins do not have an asymmetrical ligand structure. The simplest example of this type of catalyst is the metallocene of the point group C_{2v} shown in Fig. 11.19.

Because of the highly symmetrical nature of the molecule, neither of the stereochemically possible coordination options of the monomer and the active center is energetically favored. The configuration of the resulting stereocenter is therefore completely random and results in an atactic macromolecule.

Although the discovery of metallocene catalysts has led to a veritable flood of academic and industrial research work, the use of homogeneous catalysts is technically limited to niche markets. This is because of the high price of these catalysts, generated by the elaborate synthesis of tailor-made ligand structures. As well as this, there have been many patent litigations which have blocked a broad technical use of this new technology. The high price is only compensated for by a few tangible advantages in practice. Among them are a narrower molar mass distribution and the homogenous integration of comonomers. These aspects are briefly discussed in the following paragraphs.

Heterogeneous Ziegler catalyst systems have, as has already been mentioned, a number of different active centers. These centers are all different with respect to their insertion kinetics and their readiness to incorporate higher α-olefins as comonomers. As a consequence, a polymer is formed with a very broad molar mass distribution. It has also

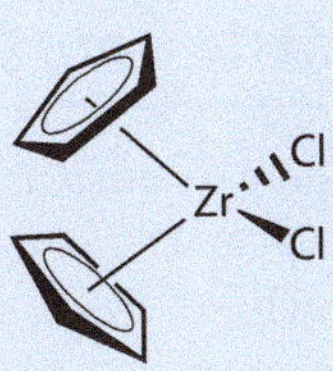

Fig. 11.19 Zirconocene dichloride as an example of a metallocene which is not stereoselective

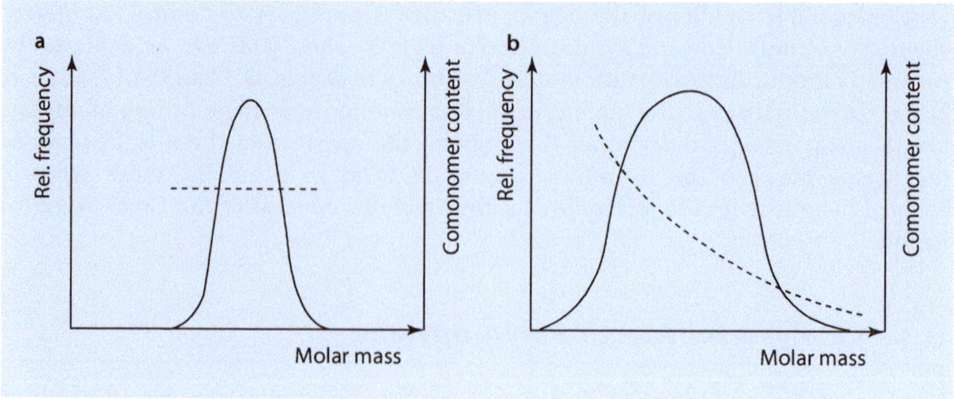

Fig. 11.20 (**a**) Homogeneous and (**b**) heterogeneous Ziegler catalysts. *Solid line:* molar mass distribution, *Dashed line*: comonomer fraction

been observed in practice that especially those active centers that incorporate α-olefins particularly readily tend to yield polymers of low molar mass, whereas the other centers tend to yield high molar mass homopolymers. Thus the product is a mixture of high polymers with a low concentration of comonomer and lower molar mass polymers with a higher concentration of comonomer (Fig. 11.20). This can result in extreme cases where the different macromolecules are immiscible and the product is inhomogeneous.

This, for the most part undesirable effect does not occur with homogeneous *single-site* catalysts. In practice, however, this rarely has an economic advantage. A technically relevant process in which polypropylene produced by metallocene catalysis is used is the high speed spinning of PP fibers. During the spinning process (▶ Chap. 17), the polymer molecules have to orientate themselves in the direction of the fiber axis. As is described in ▶ Chap. 17, this is facilitated by the polymer having a narrow molar mass distribution. Molecules with a too high degree of polymerization are difficult to orientate; molecules with a low molar mass negatively affect the mechanical properties of the fibers. Thus a narrow molar mass distribution is desirable for process optimization.

11.4 Catalysts Made from Late Transition Metals

In addition to the widely discussed catalysts made from early transition metals, which were dealt with in ▶ Sects. 11.2 and 11.3, in the last few years catalysts made from late transition metals such as iron, nickel, or palladium have also been studied. Figure 11.21 shows examples of nickel-based polymerization catalysts (Mülhaupt 2003). The catalysts shown at Fig. 11.21d and e can produce high molar mass polymers but the catalysts shown at Fig. 11.20a–c tend to result in the formation of oligomers. Nevertheless, this reaction has also become technically important for the commercial production of α-olefins from ethene and is known as the *Shell Higher Olefin Process (SHOP).*

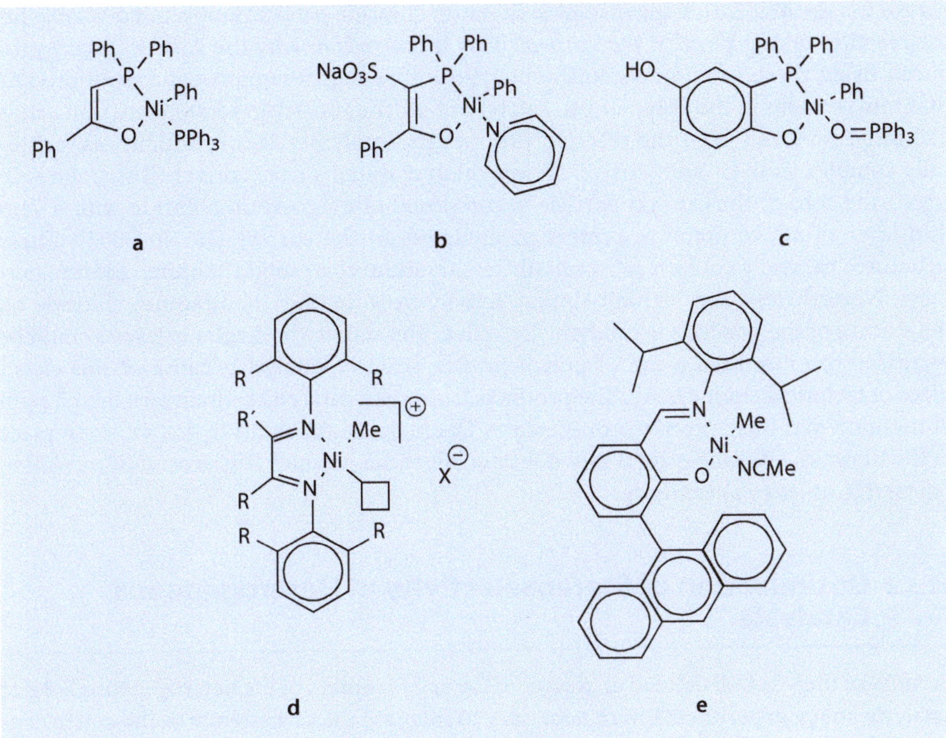

Fig. 11.21 Examples of nickel-based polymerization catalysts. Complexes described by (**a**) Keim, (**b**) Klabunde, (**c**) Ostoja-Starzewski, (**d**) Brookhardt, and (**e**) Grubbs

Catalysts based on late transition metals have the advantage of being less sensitive to the presence of substances containing hetero atoms, especially water, so that in principle a catalytic polymerization of α-olefins in aqueous emulsion is possible.

11.5 Technical Processes

Modern polymerization processes are often gas phase processes and do not involve solvents. The catalyst particles are brought into contact with the olefin in a fluidized bed reactor. Because of the special significance of the so-called gas phase polymerization, it is dealt with in more detail here.

A number of technical advances, which are discussed below, were necessary for the development of the gas phase polymerization of α-olefins, in which transition metals are used as catalysts.

11.5.1 Supports for Ziegler Catalysts

The activity of the first Ziegler catalysts—expressed as the amount (in grams) of polymer per amount of titanium (in grams)—was relatively low, because in these systems a large part of the metal is inside the insoluble titanium chloride particle and is not available for the reaction taking place at the surface. This is the reason why the first polymers produced using these catalyst systems contained rather high titanium concentrations. As titanium residues in the polymer can catalyze photochemical degradation reactions, they had to be removed from the reaction product after polymerization, which was technically complex as well as expensive. Newer catalyst systems have core-shell structures in which the core of the catalyst particle is composed of magnesium chloride with a very thin layer of active titanium centers precipitated on the surface (Böhm 2003). These structures naturally contain substantially less titanium than solid titanium chloride particles. Nevertheless, they exhibit similar activity because only the titanium chloride on the surface of the particles is catalytically active. The activity of Ziegler catalysts could be increased to more than a ton of polyolefin per gram of catalyst because of this clever piece of technical engineering. The products from such catalysts contain less than 1 ppm of titanium and its removal is unnecessary. The magnesium chloride that exists in place of the titanium chloride is inert and does not affect the product. These catalysts are often referred to as *leave-in catalysts*.

11.5.2 Optimization of Stereoselectivity of Heterogeneous Catalysts

Because of the less well defined character of the active centers of the heterogeneous Ziegler catalysts, many experiments were necessary to increase the consistency of the active centers. This can be achieved by the addition of Lewis bases such as alkyl benzoates, phthalates, or 1,3-diethers such as dimethoxy propane. The addition of these bases increases the stereoselectivity of the catalyst system. It is assumed that this is the result of the base reacting with non-stereoselective active polymerization centers on the surface of the catalyst particle which leads to their deactivation. This explanation is plausible if one considers that if a titanium atom is sterically so non-specific that even a relatively large Lewis base can coordinate with it, then it cannot be expected that the local geometry is so severely constricted that the catalytic center can stereoselectively direct the approach of the much smaller propene molecule.

By adding bases to the catalytic system the activity of the system decreases but the stereoselectivity of the system can be significantly increased. Because of this it was no longer necessary to remove the hitherto formed portion of (sticky!) atactic PP, which had been effected by elaborate extraction.

11.5.3 Control of Particle Morphology

Control of the morphology of the catalyst and polymer particles was also essential for the development of modern gas phase processes. In the most optimized processes, extremely porous agglomerates of very fine, catalytically active particles are employed. Because of this superstructure the catalyst particles can absorb the monomer as does a sponge and 'swell' as polymer is formed. After polymerization the polymer particles correspond to a significantly enlarged image of the original catalyst particles.

The use of such catalysts considerably simplifies the process; the product requires no further manufacturing, such as extrusion and pelletizing, to enable its easy handling (▸ Chap. 17).

The three developments dealt with in the Sects. 11.5.1–11.5.3 have significantly reduced the complexity of the production process of polyolefins. In the first few years of polyolefin manufacture the following process sequence was necessary:

- Polymerization reaction
- Catalyst deactivation
- Removal of the solvent
- Removal of catalyst residue
- Removal of atactic fraction
- Conversion into a manageable form, e.g., by extrusion

By application of the solvent-free gas phase process, this process sequence is simplified to one single production step and yields a dust-free product with acceptable levels of residual titanium and free of a troublesome, atactic fraction. Because polyolefins are large volume thermoplastics the technical optimization of the process was essential for the production of these products in a very competitive market.

11.6 Cycloolefin Copolymers

Cyclic olefins can also be polymerized by transition metal catalysts. Homo- and copolymers of norbornene and its derivatives are of special importance in this context. Ethene is the most often used comonomer (Fig. 11.22).

Cycloolefin copolymers (*COCs*) are engineering plastics and find application as medical packaging, optical fibers, adhesives and photo resists. As well as copolymers with ethene, terpolymers of ethene, propene, and ethylidene norbornene (Fig. 11.23) are a large volume elastomer. The market relevance of the thermoplastic cycloolefin copolymers is, however, not remotely comparable to that of polyolefins.

If cyclopentene is catalytically polymerized an anomaly occurs. The isomer, poly(1,3-cyclopentene) is obtained rather than the expected poly(1,2-cyclopentene) (Fig. 11.24).

The reason for this initially surprising result is that the metal alkyl compound formed by the insertion of cyclopentene is sterically too restricted to allow a further cyclopentene insertion. Thus, initially a β-H-elimination occurs and the product can then insert a further cyclopentene molecule in an alternative orientation (Fig. 11.25). This explains the 1,3-insertion.

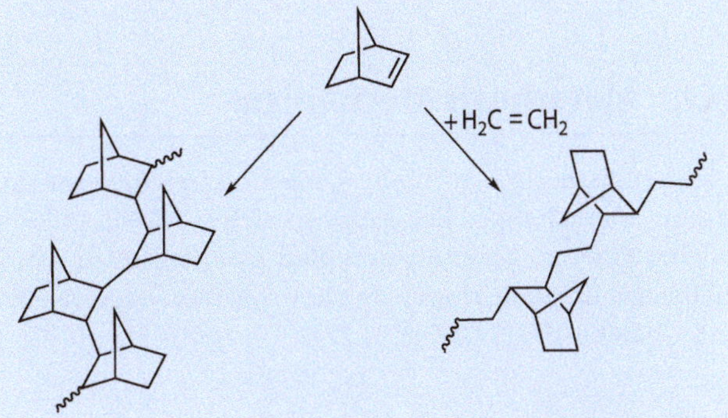

Fig. 11.22 Homopolymers and ethene copolymers from norbornene

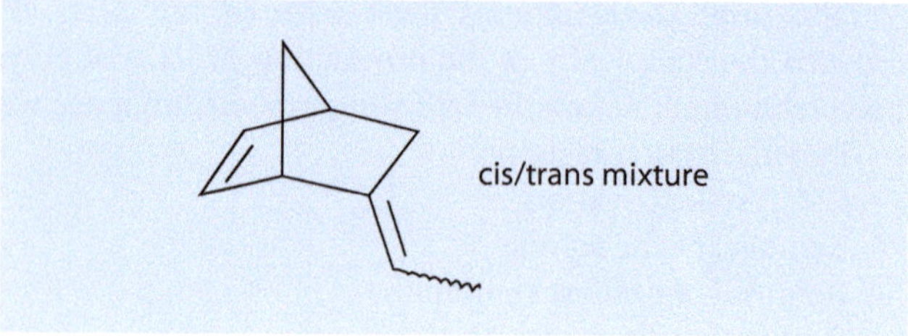

Fig. 11.23 Ethylidene norbornene

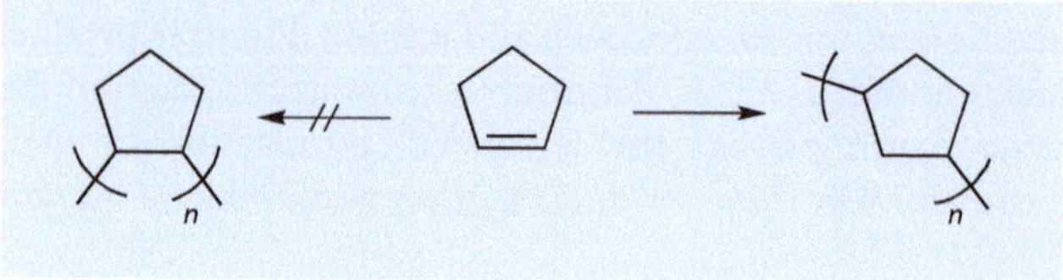

Fig. 11.24 Product of the catalytic polymerization of cyclopentene

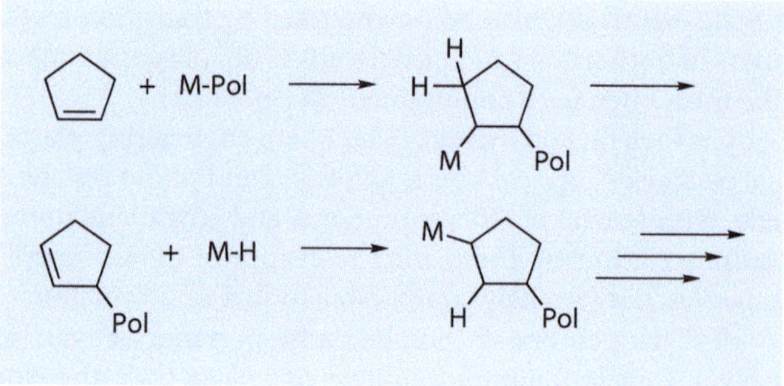

Fig. 11.25 Elimination and reinsertion during the catalytic polymerization of cyclopentene

11.7 Olefin Metathesis

One of the most promising developments in synthesis chemistry is the metathesis of olefins. The discoverers of this development (Grubbs, Schrock, and Chauvin) were awarded the Nobel Prize in 2005. Recently, because of well defined catalysts, even functionalized monomers can be employed in metathesis reactions.

11.7.1 Metathesis Mechanism

Olefin metathesis is, formally, a reaction in which the double bond is split and alkylidene groups are exchanged between two olefins (Fig. 11.26) (Calderon 1972).

The current, generally accepted mechanism in which metal carbene complexes and metallacyclobutane rings play a key role was suggested by Chauvin very early on (Hérisson and Chauvin 1971) (Fig. 11.27).

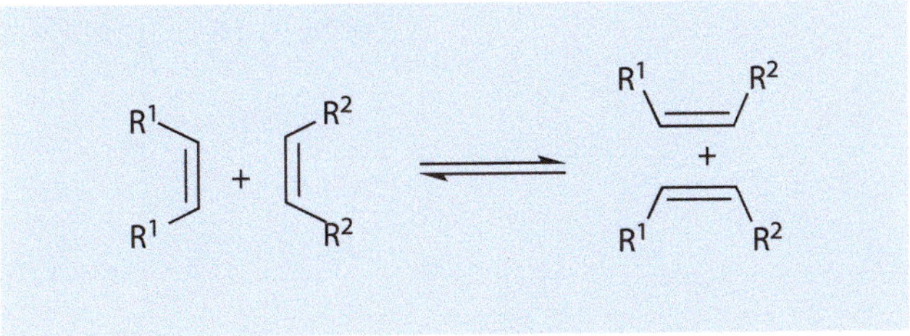

Fig. 11.26 Metathesis of two olefins

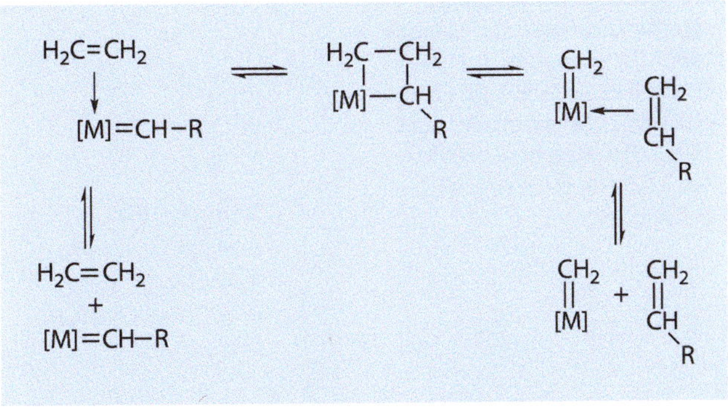

Fig. 11.27 Metallacyclobutane mechanism for olefin metathesis

Table 11.3 Heterogeneous metathesis catalysts

WO_3/Al_2O_3	MoO_3/Al_2O_3
WO_3/SiO_2	MoO_3/SiO_2
Re_2O_7/Al_2O_3	MoO_3/TiO_2
$Re_2O_7/SiO_2\text{-}Al_2O_3$	MoO_3/ZrO_2

11.7.2 Metathesis Catalysts

Metathesis catalysts can be divided into three groups: heterogeneous, homogeneous, and immobilized homogeneous catalysts.

11.7.2.1 Heterogeneous Catalysts

Classical heterogeneous metathesis catalysts are oxides of transition metals of the sixth group on Lewis acid oxide carriers. They are robust, durable, and can be regenerated. Typical examples are shown in Table 11.3.

These catalysts generally require high temperatures (for example 400 °C for WO_3/Al_2O_3). Some, however, are very effective even at room temperature (MoO_3/ZrO_2). Most of them do not tolerate other functional groups.

11.7.2.2 Homogeneous Catalysts

Homogeneous metathesis catalysts are nitrosyls, chlorides, or oxychlorides of tungsten or molybdenum in their higher oxidation states in combination with organotin compounds as cocatalysts (Table 11.4).

These systems can be used at room temperature. Their most serious disadvantage is that organometallic cocatalysts have to be used.

Table 11.4 Homogeneous metathesis catalysts

$WCl_6/EtOH/EtAlCl_2$	$MoCl_5/Ph_4Sn$
$WCl_6/BuLi$	$MoCl_3(NO)/EtAlCl_2$
$WOCl_4/EtAlCl_2$	

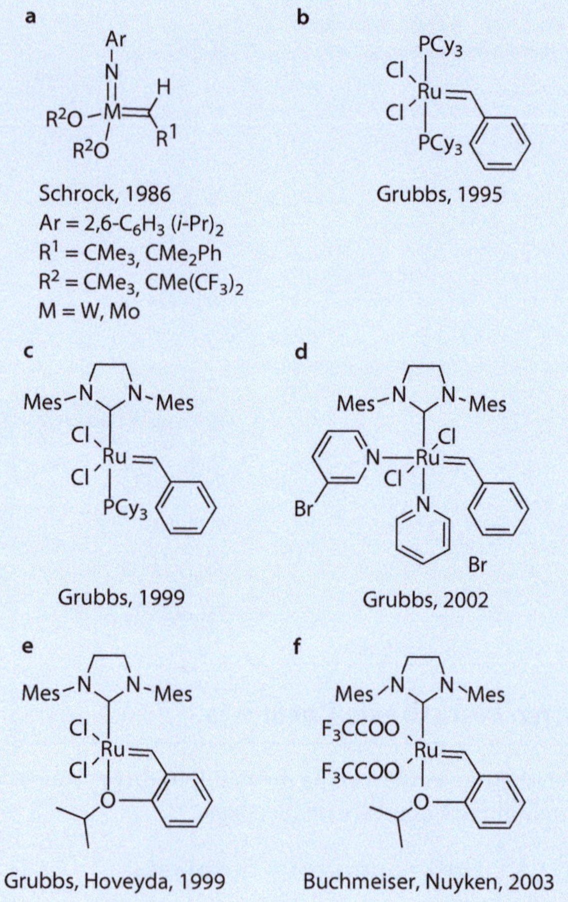

Fig. 11.28 A selection of well defined, homogeneous metathesis catalysts: (**a**) Schrock (1986), (**b**) Schwab et al. (1995), (**c**) Scholl et al. (1999), (**d**) Love et al. (2002), (**e**) Kingsbury et al. (1999), (**f**) Krause et al. (2003). *Mes*: 2,4,6-trimethyl phenyl, *Cy*: Cyclohexyl

The exceptionally active tungsten and molybdenum amido complexes, introduced by Schrock in 1986, were the starting molecules in the search for further, well defined catalysts (Schrock 1986). The ruthenium catalysts synthesized by Grubbs constituted a further breakthrough. Selected examples of well defined homogeneous catalysts are shown in Fig. 11.28.

Table 11.5 demonstrates how ruthenium-based metathesis catalysts triumph over the others. Although titanium-based catalysts do not tolerate any other functional groups, tungsten-based catalysts can tolerate esters and amides, molybdenum-based catalysts tolerate esters, amides, and ketones, and ruthenium-based catalysts are stable and active in the presence of nearly all functional groups. Ruthenium catalysts can even be used for metathesis in water.

Table 11.5 A comparison of the reactivity of metathesis complexes based on the most important metals with respect to olefins and other functional groups (decreasing reactivity from top to bottom) (Grubbs 1994)

Titanium	Tungsten	Molybdenum	Ruthenium
Alcohols, water	Alcohols, water	Alcohols, water	Olefins
Acids	Acids	Acids	Alcohols, water
Aldehydes	Aldehydes	Aldehydes	Acids
Ketones	Ketones	Olefins	Aldehydes
Esters, amides	Olefins	Ketones	Ketones
Olefins	Esters, amides	Esters, amides	Esters, amides

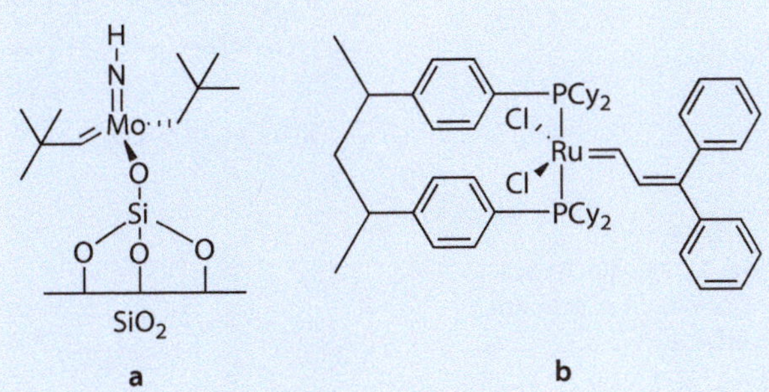

Fig. 11.29 The immobilization of well defined metathesis catalysts.
(**a**) Schrock-type on SiO_2.
(**b**) Grubbs-type on polystyrene

11.7.2.3 Immobilized Catalysts

The disadvantage of homogeneous metathesis catalysts is their insufficient reusability because the complex compounds are destroyed during the isolation of the products and undesirable transition metal traces remain in the product.

This problem can be overcome by immobilizing homogeneous catalysts on insoluble carriers. In this way the advantages of the homogeneous catalysts can be exploited, their disadvantages can be avoided, and the catalyst can be recovered by simple filtration for reuse. Two examples are shown in Fig. 11.29.

11.7.3 Metathesis Reactions

The most important metathesis reactions of olefins are summarized in Fig. 11.30.

As well as making unusual monomers accessible by disproportionation, the reciprocal transformation of monomers via metathesis is also of interest. Thus, for example, ethene reacts with 2-butene to give propene, a reaction which can also be reversed (Fig. 11.31).

Route b (Fig. 11.30) is most important for the synthesis of polymers by metathesis but they can also be synthesized using ADMET (route e, Fig. 11.30). By adding additional monomer the molar mass of the metathesis products can be reduced (i.e., reverse reaction of route e, Fig. 11.30). The ring closure metathesis (route c) and degradation

Fig. 11.30 Important metathesis reactions

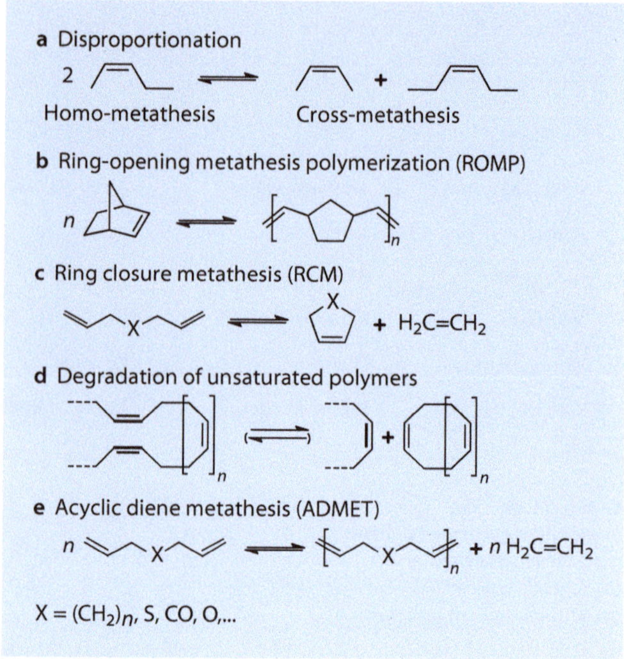

Fig. 11.31 Reciprocal transformation of propene and ethene

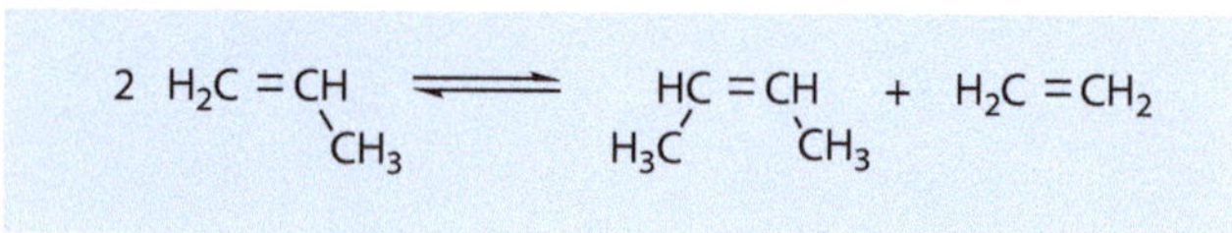

reactions (route d, Fig. 11.30) are especially well suited to the synthesis of some special organic molecules.

Important polymers produced by ROMP (Fig. 11.32) are polycyclopentene (Polypentenamer®), polycyclooctene (Vestenamer®), and polydicyclopentadiene, sold under several names.

Norbornenes can also be polymerized by ROMP. This reaction broadens the spectrum of interesting norbornene polymers that were already introduced in connection with polyinsertion (Fig. 11.22). Figure 11.33 gives an overview of the technically relevant structures. As ROMP-polymers have double bonds after polymerization, their resistance to atmospheric oxygen is limited (formation of allyl radicals!) and they are often hydrogenated to make them more weather-resistant.

A special case of a monomer that can be polymerized by metathesis catalysts is acetylene. Its polymerization (Fig. 11.34) is possible with classical metathesis catalysts as well as with Schrock systems whereas most of the ruthenium-based systems, such as the Grubbs' systems, are not suitable (Krause 2004).

This polymer is exceptional in that it has a continuous conjugated, pure sp^2-system along the polymer backbone; above and below the main polymer chain there is an unusually large, conjugated π-electron system. For this reason the polymer is black and electrically conductive when appropriately doped. Levels of conductivity can be reached that are

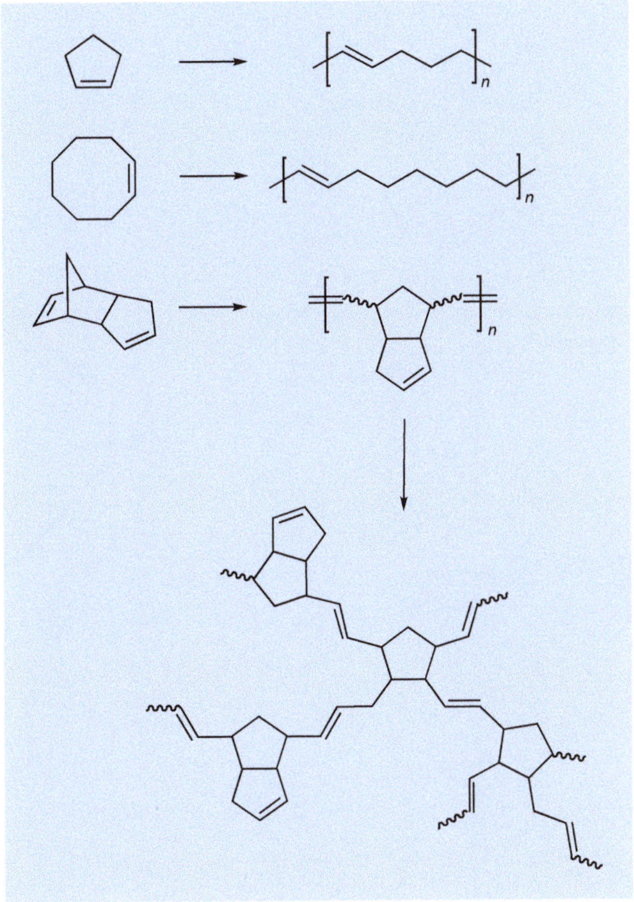

Fig. 11.32 Important polymers synthesized using ROMP

similar to those of conductive metals. However, two serious disadvantages have prevented the use of this fascinating material up to now. First, the very expansive π-electron system is very sensitive to oxidation so that polyacetylene (especially in its doped form) is sensitive to air; second, it is insoluble and infusible, which not only makes its processing but also its characterization difficult. Nevertheless, Heeger, MacDiarmid, and Shirakawa's work on polyacetylene was awarded the Nobel Prize for Chemistry in 2000.

Other monomers capable of metathesis are dialkynes, such as those shown in Fig. 11.35.

Dialkynes can be transformed using molybdenum catalysts of the Schrock kind or with special ruthenium catalysts (Fig. 11.28 (Buchmeiser, Nuyken)) into either perfect six-ring structures (molybdenum catalysts) or into five-ring structures (ruthenium systems) (Krause et al. 2004). These polymers are highly conjugated, similar to polyacetylene; they are however, significantly more stable and soluble, particularly in conventional solvents such as $CHCl_3$ and THF.

The living character of diyne-polymerizations can be exploited for the formation of block copolymers of the AB- and ABA-types. Monomers from Fig. 11.35 (forming the A-block) can, therefore, be combined with norbornenes (forming the B-block), for example.

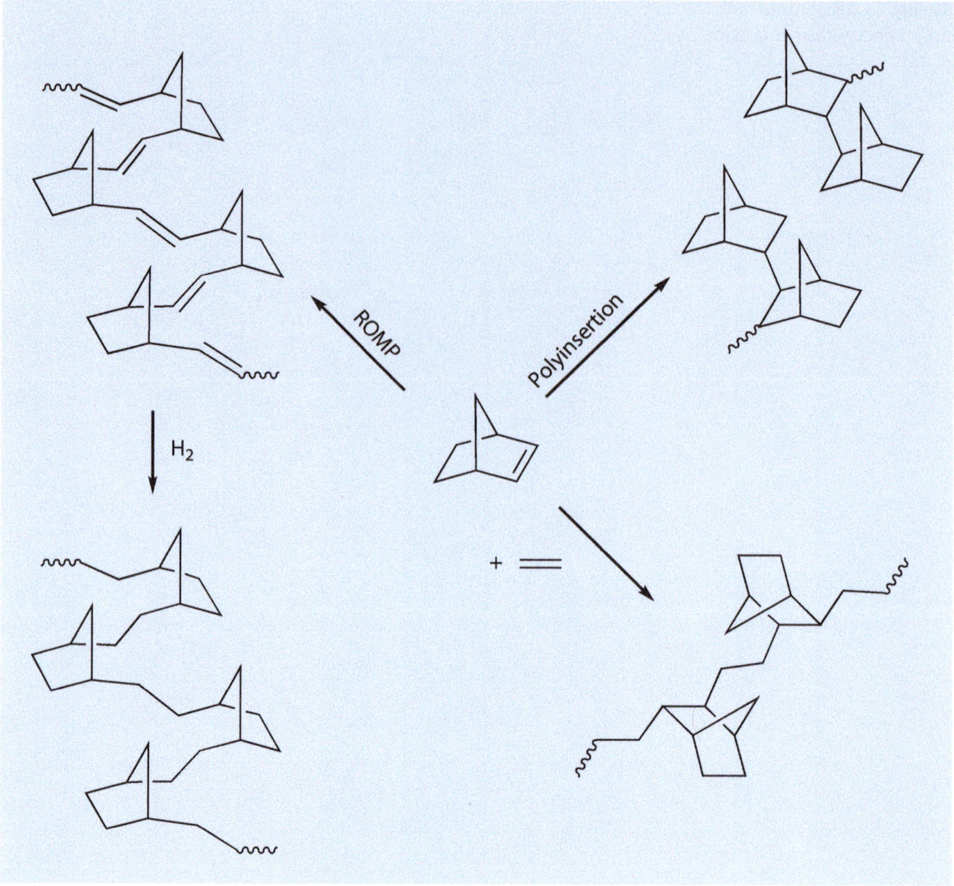

Fig. 11.33 Polymers based on norbornene

Fig. 11.34 Polymerization of acetylene with metathesis catalysts

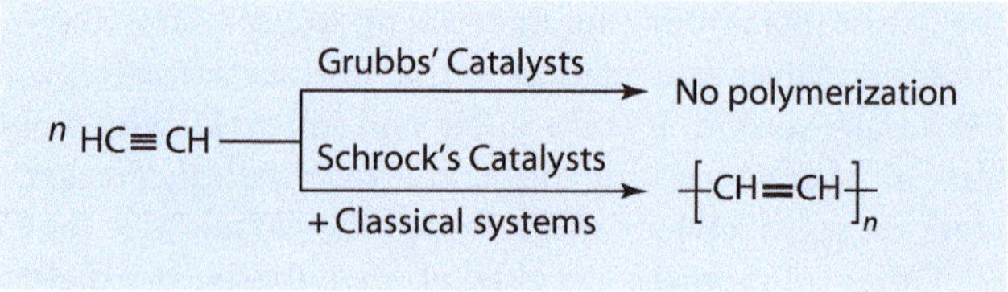

11.8 Copolymerization with Polar Comonomers

Most of the transition metal catalysts capable of polymerizing olefins are very sensitive to hetero atoms. This can be explained by the low nucleophilicity of the olefin double bond in comparison to most hetero atoms. This is why the copolymerization of olefins with monomers that are known from ionic or radical polymerizations has its limitations. Some cases exist, in which olefins or epoxides can be catalytically copolymerized with polar monomers and they are discussed in the following paragraphs.

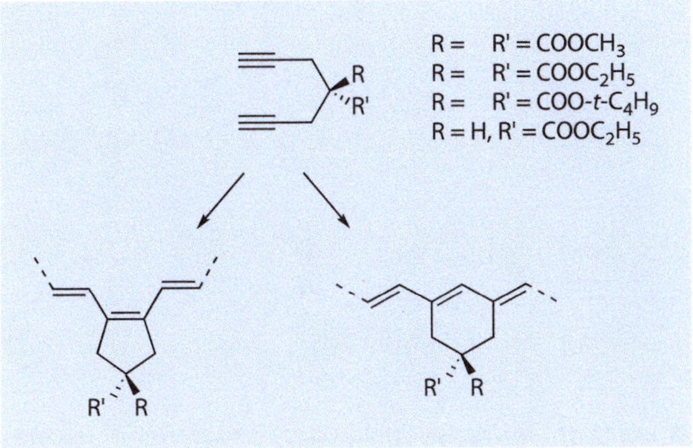

Fig. 11.35 The cyclopolymerization of 1,6-heptadiynes to five- and six-membered ring structures

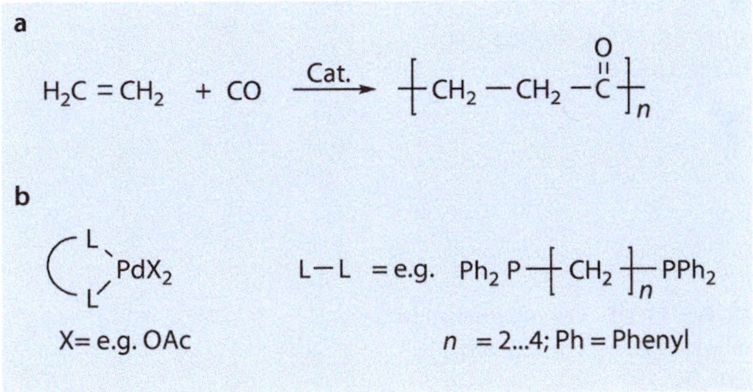

Fig. 11.36 (**a**) Copolymerization of ethene with carbon monoxide. (**b**) Possible catalysts

11.8.1 Olefin-CO Copolymers

A strictly alternating copolymer can be obtained by the copolymerization of ethene with carbon monoxide (Fig. 11.36). This copolymer is a polyketone. Palladium phosphine complexes are used as catalysts.

The mechanism (Fig. 11.37) is similar to that for olefin polymerization with metallocene catalysts. The monomers are coordinated and then inserted via chain migration.

The strictly alternating sequence of the two monomers is caused by the different stabilities and reactivities of the species formed during the catalytic cycle. The insertion of carbon monoxide into a palladium acyl compound is thermodynamically unfavorable. The insertion of an ethene molecule therefore always follows a CO-insertion. Conversely, the insertion of ethene is kinetically very inhibited. The rate of the insertion of a CO molecule is therefore significantly faster so that the insertion of carbon monoxide quickly follows the insertion of ethene and an alternating copolymer results.

As in the case of ethene, higher α-olefins can also be polymerized with CO. The generic polymer structure shown in Fig. 11.38 results. The insertion takes place in a regioselective manner, i.e., the substituent is bound to every third carbon atom along the polymer backbone in a regular fashion. The stereochemistry of the polymer can be controlled using catalysts with suitable symmetry.

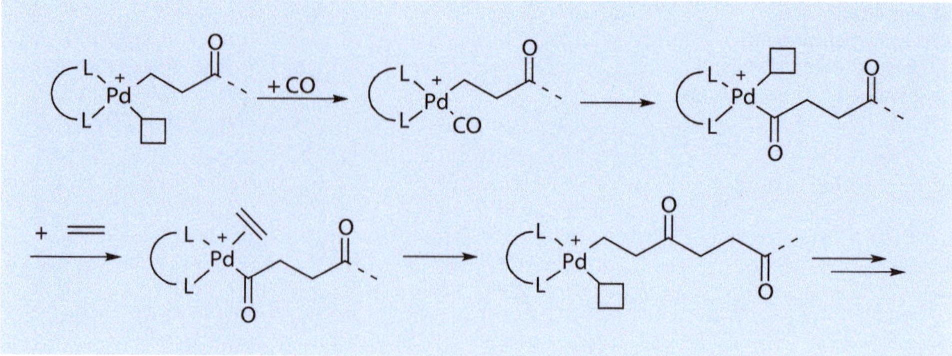

Fig. 11.37 Mechanism for the copolymerization of ethene and carbon monoxide

Fig. 11.38 Generic structure of copolymers from CO and α-olefins

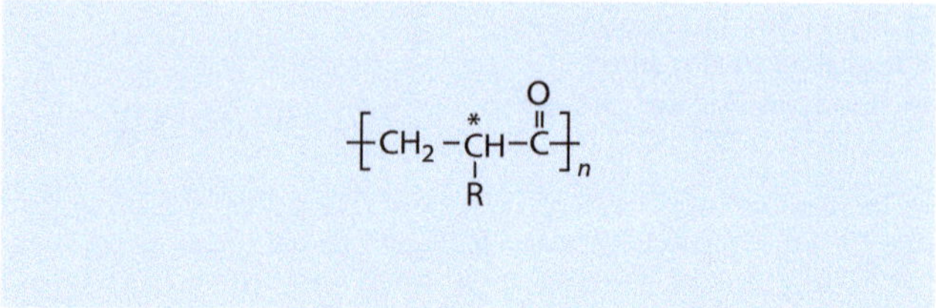

Fig. 11.39 Copolymerization of epoxides with (**a**) CO and (**b**) CO_2

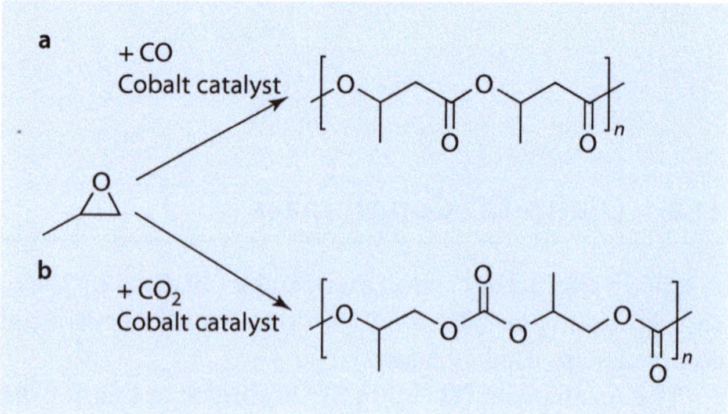

11.8.2 Copolymers From Epoxides and Carbon Oxides

Epoxides can be copolymerized with carbon monoxide as well as with carbon dioxide using cobalt catalysts to yield polyesters or polycarbonates, respectively (Fig. 11.39).

References

Angermund K, Fink G, Jensen V, Kleinschmidt R (2000) Towards quantitative prediction of stereospecificity of metallocene-based catalysts for alpha-olefins. Chem Rev 100:1457–1470

Böhm L (2003) The ethylene polymerization with Ziegler catalysts fifty years after the discovery. Angew Chem Int Ed 42:5010–5030

Brintzinger H-H, Fischer D, Mülhaupt R, Rieger B, Waymouth R (1995) Stereospezifische Olefinpolymerisation mit chiralen Metallocenkatalysatoren. Angew Chem 107:1255–1283

Calderon N (1972) Olefin metathesis reaction. Acc Chem Res 5:127–132

Grubbs RH (1994) The development of functional group tolerant ROMP catalysts. J Macromol Sci Chem A31:1829–1833

Hérisson JL, Chauvin Y (1971) Catalysis of olefin transformation by tungsten complexes, II. Telomerization of cyclic olefins in the presence of acyclic olefins. Makromol Chem 141:161–167

Kingsbury J, Harrity J, Bonnitatebus P, Hoveyda AH (1999) A recyclable Ru-based metathesis catalyst. J Am Chem Soc 121:791–799

Krause JO, Zarka MT, Anders U, Weberskirch R, Nuyken O, Buchmeiser MR (2003) Simple synthesis of poly(acetylene) latex particles in aqueous media. Angew Chem Int Ed 42:5965–5969

Krause J, Wurst K, Nuyken O, Buchmeiser MR (2004) Synthesis and reactivity of homogenous and heterogenous ruthenium-based metathesis catalysts containing electron withdrawing ligands. Chem Eur J 10:778–785

Love JA, Morgan JP, Trnka TM, Grubbs RH (2002) A practical and highly active ruthenium-based catalyst that effects the cross metathesis of acrylonitrile. Angew Chem Int Ed 41:4035–4037

Mülhaupt R (2003) Catalytic polymerization and post polymerization catalysis fifty years after the discovery of Ziegler catalysts. Macromol Chem Phys 204:289–327

Natta G (1964) Von der stereospezifischen Polymerisation zur asymmetrischen autokatalytischen Synthese von Makromolekülen. Angew Chem 76:553–566

Scholl M, Ding S, Lee CW, Grubbs RH (1999) Synthesis and activity of a new generation of ruthenium-based olefin metathesis catalysts coordinated with 1,3-dimesityl-4,5-dihydroxy-imidazol-2-ylidine ligands. Org Let 1:953–956

Schrock RR (1986) On the trail of metathesis catalysts. J Organomet Chem 300:249–262

Schwab P, France MB, Ziller JW, Grubbs RH (1995) A series of well-defined catalysts, synthesis and application of $RuCl_2(=CHR')(PR_3)_2$. Angew Chem Int Ed 34:2039

Sinn H, Kaminsky W (1980) Ziegler-Natta-catalysis. Adv Organomet Chem 18:99–149

Willock D (2009) Molecular symmetry. Wiley, West Sussex

Ziegler K, Holzkamp E, Breil H, Martin H (1955) Das Mülheimer Normaldruck-Polyäthylen-Verfahren. Angew Chem 67:541–547

Ring-Opening Polymerization

S. Koltzenburg et al., *Polymer Chemistry*, DOI 10.1007/978-3-662-49279-6_12

12.1 General Features

Ring-opening polymerization (ROP) is an important method of polymerization. It differs from radical polymerization (▶ Chap. 9), ionic polymerization (▶ Chap. 10), and step-growth polymerization (▶ Chap. 8). No low-molar-mass by-products are formed, except during the polymerization of Leuchs' anhydride (◘ Figs. 12.41 and 12.42). Furthermore, the driving force derived from the transformation of C=C double bonds into C–C single bonds, which offsets the loss of entropy during polymerization, is not available. A general feature of ROP is that the monomers are rings of varying size. Depending on the size and type of the ring structure, the ability to polymerize and the corresponding driving force varies. Small rings (three-, four-, or five-membered rings) can be polymerized because of the ring strain released when they are open. As an example, the enthalpy associated with the ring strain of oxirane is 116 kJ/mol. The release of enthalpy is also the driving force for the polymerization of seven- and eight-membered lactones and lactams, even though the ring strain is only about 16 kJ/mol for these monomers. Unstrained six-membered rings often do not polymerize via ROP. By contrast, the ROP of disulfides, silicones, and carbonates can be ascribed to the increase in the entropy that occurs during the polymerization of these monomers. This increase in entropy is based on the increase in the degrees of freedom of rotation gained when rings are transformed into open chains.

Many polymers that can be synthesized via ROP have obtained technical significance, such as polyethylene oxide, polyoxymethylene, polysiloxane, and polyphosphazene. Polyglycolides and polyactides have also become significant in the packaging sector because of their biocompatibility and degradability.

New metathesis catalysts provide new inspiration for the ROP of functionalized cyclic olefins and therefore create access to unusual polymer structures. ◘ Table 12.1 shows a selection of typical cyclic monomers.

In analogy with the structure of the other parts of this textbook, the following sections describe radical ROP (▶ Sect. 12.2), cationic ROP (▶ Sect. 12.3), and anionic ROP

◘ **Table 12.1** Typical cyclic monomers and the mode of their polymerization

Name	Structure	Ring size	Initiator type
Olefin	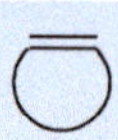	4, 5, 8	Metathesis
Ether		3, 4, 5, 7	Cationic, anionic
Thioether		3, 4	Cationic, anionic
Amine	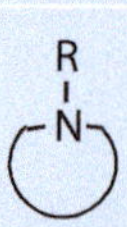	3, 4, 7	Cationic
Lactone		4, 6, 7, 8	Cationic, anionic

Table 12.1 (continued)

Name	Structure	Ring size	Initiator type
Thiolactone		4–8	Cationic, anionic
Lactam		≥4	Cationic, anionic
Disulfide		≥8	Radical
Anhydride		5 and ≥7	Anionic
Carbonate	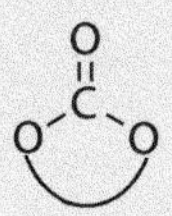	6,7,8 and ≥20	Anionic
Silicone	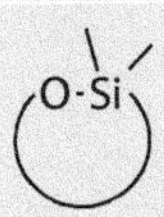	6, 8 and ≥10	Cationic, anionic
Phosphazene	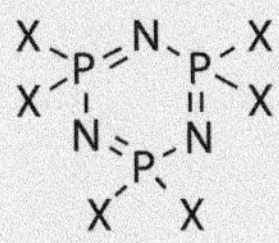	6	Cationic
Phosphonite	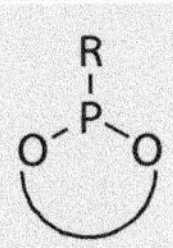	3, 5, 6, 7	Anionic
Oxazoline	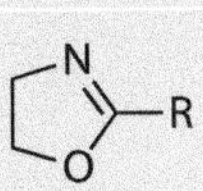	5	Cationic

(▶ Sect. 12.4). Finally, ring-opening metathesis polymerization (ROMP; ▶ Sect. 12.5) is discussed before covering a special case, the polymerization of phosphazenes (▶ Sect. 12.6).

12.2 Radical Ring-Opening Polymerization

The interest in ROP, and especially radical ROP (RROP), derives from the possibility that polymers can be obtained that have a similar or lower density than the corresponding monomers. These monomers are of interest for application as tooth fillings, coatings, molded electric and electronic tools, and equipment in which a constant volume during the polymerization process is important. Monomers that expand to give lower-density polymers can, for example, be polymerized with monomers, such as most vinyl monomers, which alone yield denser polymers (► Sect. 9.2.5, footnote 4). RROP creates new routes in which polyesters and polyketones can be obtained. Vinyl-substituted cyclic compounds are the systems that are most often described in this context (◘ Fig. 12.2). An unusual group of monomers that can undergo RROP are the bicyclobutanes and their derivates. The bridge across the four-membered ring of the monomer in ◘ Fig. 12.1 can be seen as a concealed C=C bond.

◘ **Fig. 12.1** Reaction of a radical with a bridged cyclobutane

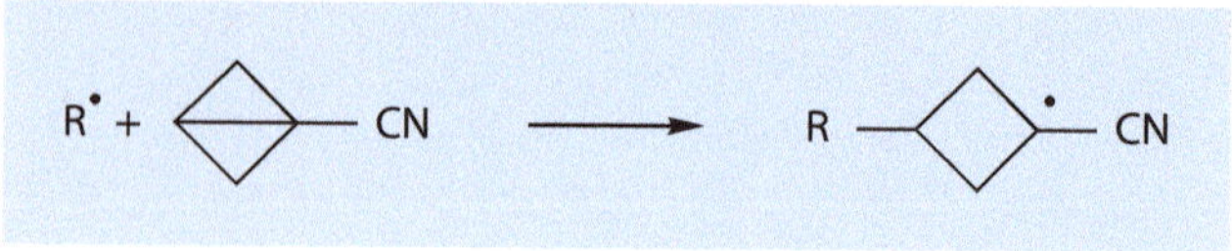

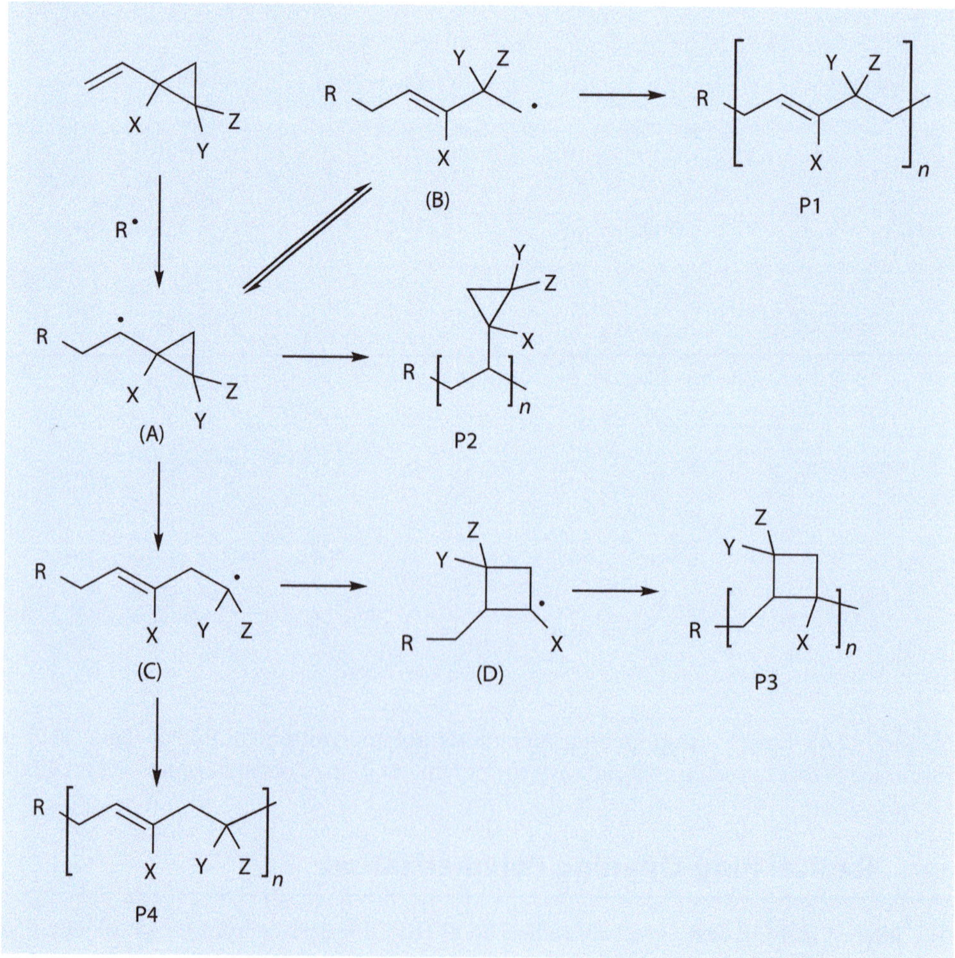

◘ **Fig. 12.2** Ring-opening polymerization of vinyl cyclopropane and the competing reactions

12.2.1 Vinyl Substituted Cyclic Compounds

Extensively researched examples of monomers that can be polymerized by RROP are vinyl cyclopropane and its derivatives. The three-membered ring of vinyl cyclopropane has a large ring strain so that the ring-opening competes with a normal radical vinyl polymerization so that polymerization takes place as shown in ◘ Fig. 12.2.

(A) is formed by the addition of R• to the C=C double bond. (A) can rearrange to (C) if the CX–CY bond is broken and (C) can then add a further monomer to form P4.

If the CX–CH_2 bond of (A) is broken, (B) is formed, and P1 results if (B) adds another monomer. It is, however, possible that (A) can polymerize to P2 via a vinyl polymerization without the three-membered ring being opened.

Another complication is the formation of a four-membered ring (D) from (C) which can give P3 by the addition of further monomer.

In general, the route to P4 is the favored one. Steric and electronic effects, and thus the substituents X, Y, and Z, influence to what extent, P1, P2, and/or P3, as well as P4, are formed.

Polyether structures can be formed from vinyl oxiranes (◘ Fig. 12.3).

Vinyl cyclobutane rings can also be polymerized using RROP (◘ Fig. 12.4).

12.2.2 Methylene Substituted Cyclic Compounds

The ring-opening of ketene acetals is a novel route to polyesters (see ◘ Fig. 12.5). However, because of their sensitivity to acid hydrolysis, the monomers are difficult to handle.

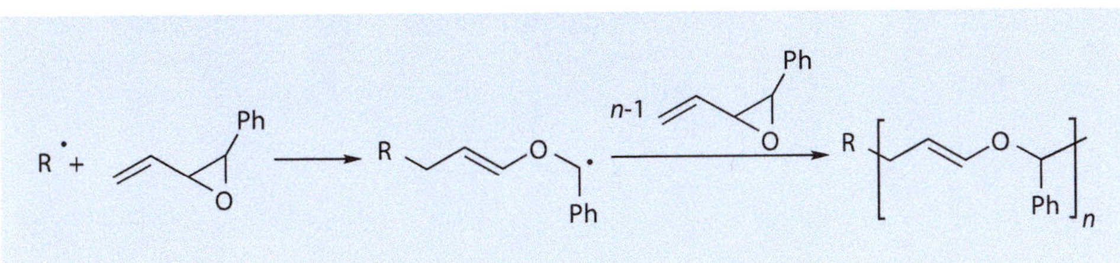

◘ **Fig. 12.3** RROP of vinyl oxiranes. *Ph*: phenyl

◘ **Fig. 12.4** RROP of a substituted vinyl cyclobutane

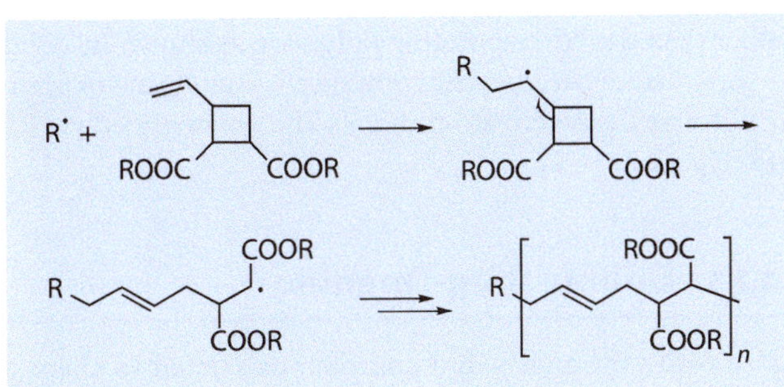

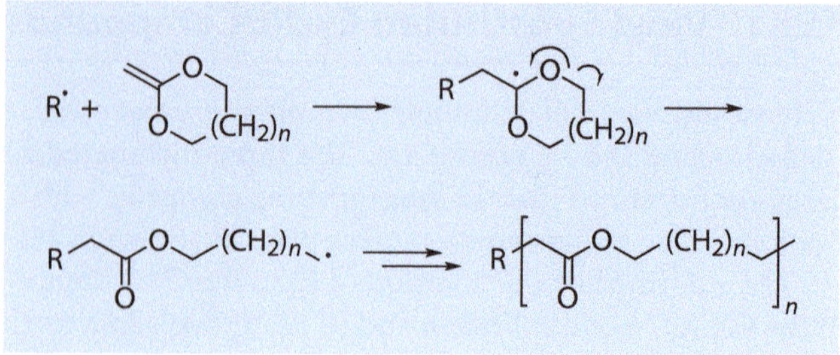

Fig. 12.5 RROP of ketene acetals

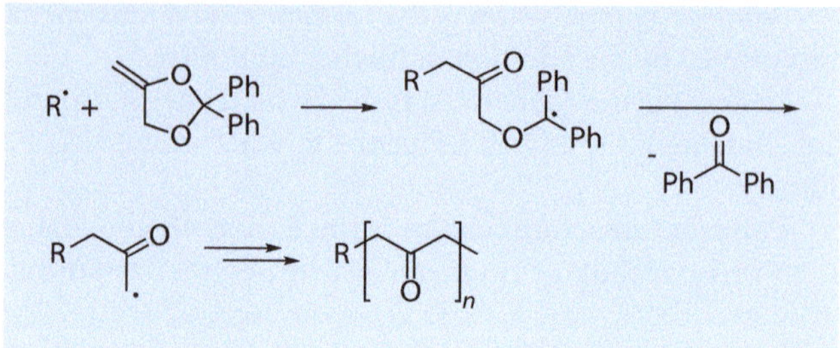

Fig. 12.6 Formation of a polyketone from 2,2-diphenyl-4-methylene-1,3-dioxolane

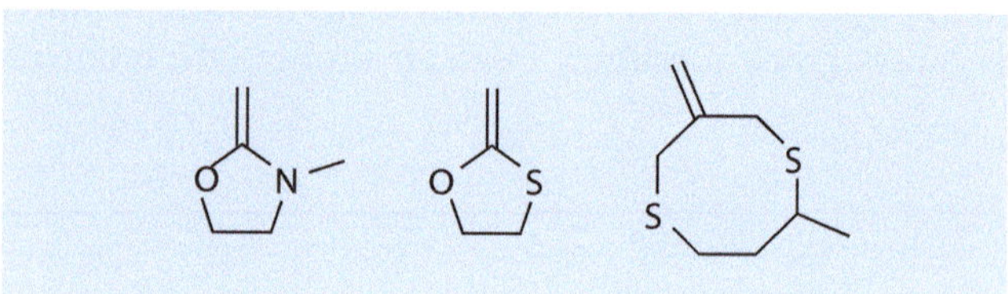

Fig. 12.7 Monomers which undergo RROP

An interesting secondary reaction yielding a polyketone is observed during the polymerization of 2,2-diphenyl-4-methylene-1,3-dioxolane (Fig. 12.6).

Because benzophenone is eliminated at every polymerization step, a polyketone results rather than the corresponding polyester, as shown in Fig. 12.5.

Apart from heterocyclic compounds containing oxygen in the ring, those with nitrogen or sulfur as the hetero atom can also be polymerized via RROP. Some examples are shown in Fig. 12.7.

12.2.3 Double Ring-Opening

Spiro orthocarbonates and compounds related to them are the subject of considerable interest, especially because of the volume expansion when the crystalline monomers are polymerized to an amorphous polymer. If, however, the monomer is first melted and then polymerized, a volume contraction takes place, typical for most polymerizations. The polymerization of a spiro orthocarbonate is shown in Fig. 12.8.

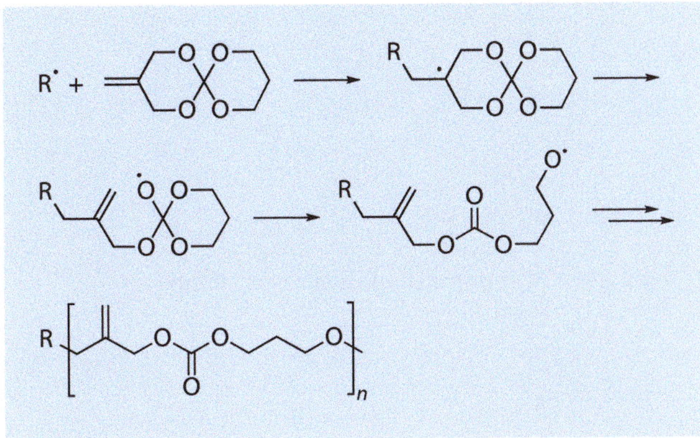

■ **Fig. 12.8** Ring-opening of a spirocarbonate

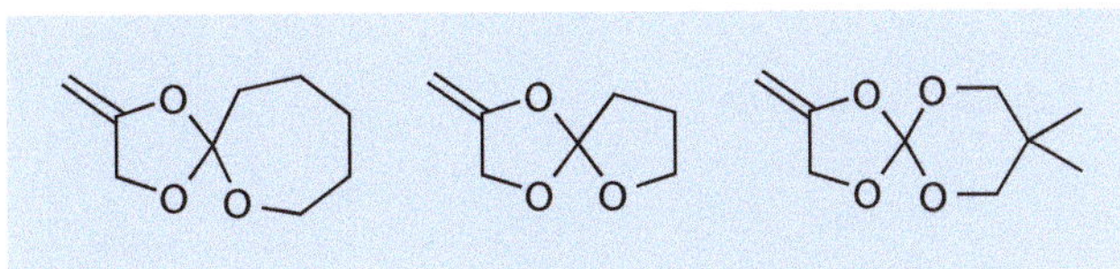

■ **Fig. 12.9** Spirocompounds which undergo RROP

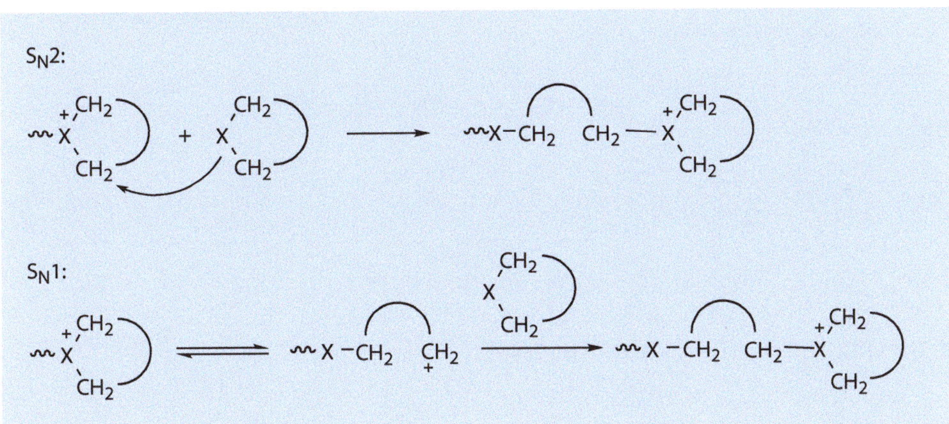

■ **Fig. 12.10** CROP as a nucleophilic reaction

Other spiro orthoesters (■ Fig. 12.9) can also be, essentially, quantitatively polymerized via ROP.

12.3 Cationic Ring-Opening Polymerization

Two different mechanisms are dealt with in the context of cationic ring-opening polymerization (CROP). First, the reaction of a monomer with an active, positively charged chain end by an S_N2- or S_N1-mechanism (■ Fig. 12.10). Second, a mechanism which takes place because of an *activated monomer* (■ Fig. 12.11).

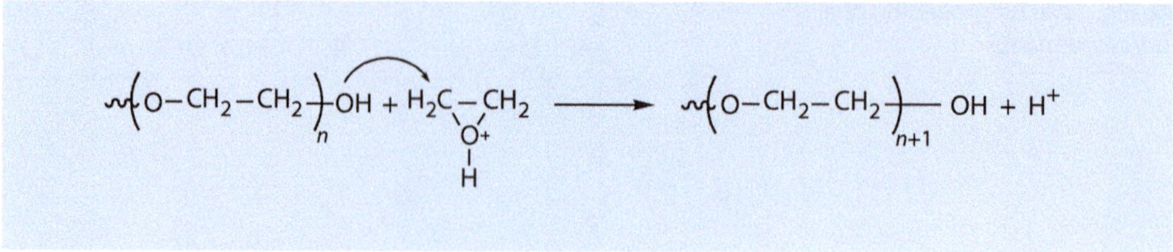

Fig. 12.11 CROP: *activated-monomer* mechanism

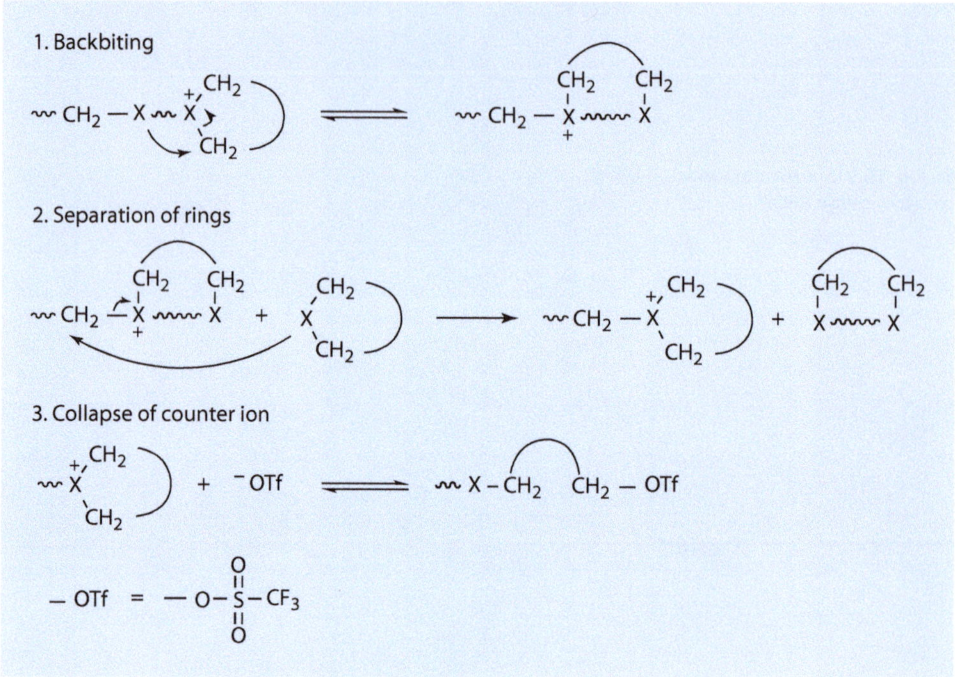

Fig. 12.12 Typical side reactions during a CROP

As can be seen in Fig. 12.10, the charge is always at the end of a growing chain for this first case, whereby X is the hetero atom such as oxygen, nitrogen, or sulfur.

In contrast to this, in the *activated-monomer* mechanism, protonated monomer molecules function as charge carriers (Fig. 12.11).

Several side reactions can take place during cationic ring-opening polymerization (Fig. 12.12).

12.3.1 Initiation

Cationic ring-opening polymerization can be initiated by Brønsted acids, carbenium ions, onium ions, covalent initiators, Lewis acids, and photo-initiators.

12

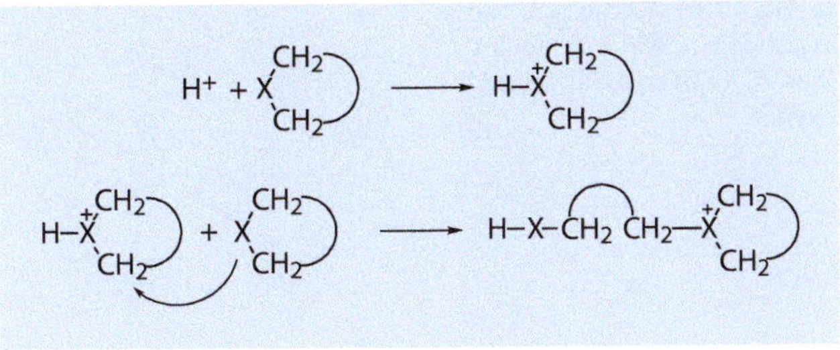

Fig. 12.13 Mechanism of proton initiation

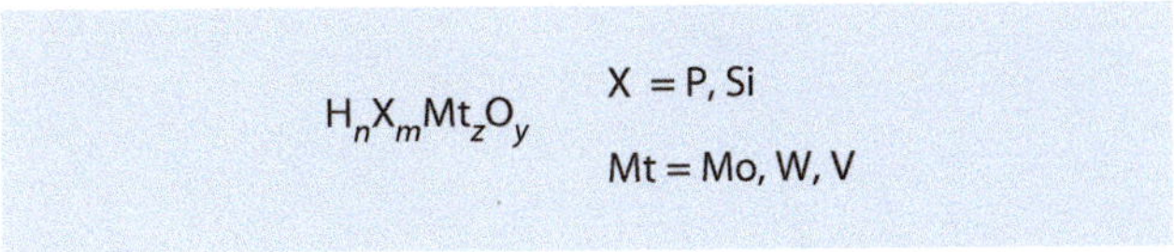

Fig. 12.14 General structure of a hetero poly-acid

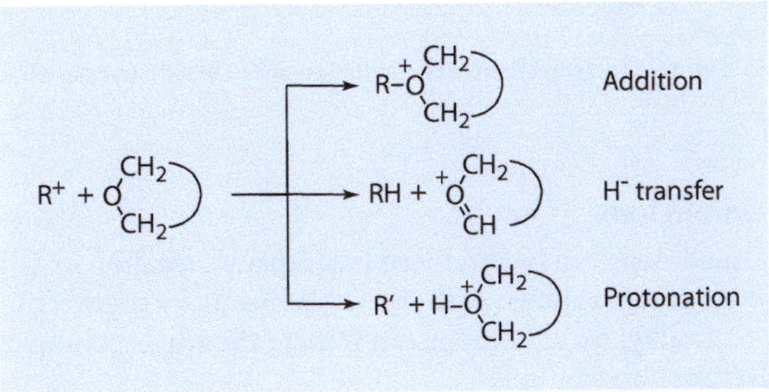

Fig. 12.15 Mechanism of the initiation of CROP with carbenium ions

Brønsted Acids

Protons can initiate the polymerization of nearly every heterocyclic monomer (Fig. 12.13). Anhydrous Brønsted acids such as HCl and H_2SO_4, but also more complex acids such as $HClO_4$ and $HOSO_2CF_3$, are often used as the proton source.

Hetero poly-acids with the general structure shown in Fig. 12.14 are a more recent addition to the group of Brønsted acid initiators for CROP.

Such acids are synthesized, for example, from the reaction of phosphoric acid with a corresponding metal oxide. These are readily available, stable in air, and soluble in numerous organic solvents. The best-known representatives of these proton sources are $H_3PMo_{12}O_{40}$ and $H_3PW_{12}O_{40}$.

Carbenium Ions

$R^+ = (C_6H_5)_3C^+$, $(C_6H_5)_2CH^+$, $(C_6H_5)CH_2^+$, $H_2C{=}CH{-}CH_2^+$, and $(CH_3)_3C^+$ have been extensively studied. They can react with cyclic monomers in a variety of different ways as shown in Fig. 12.15.

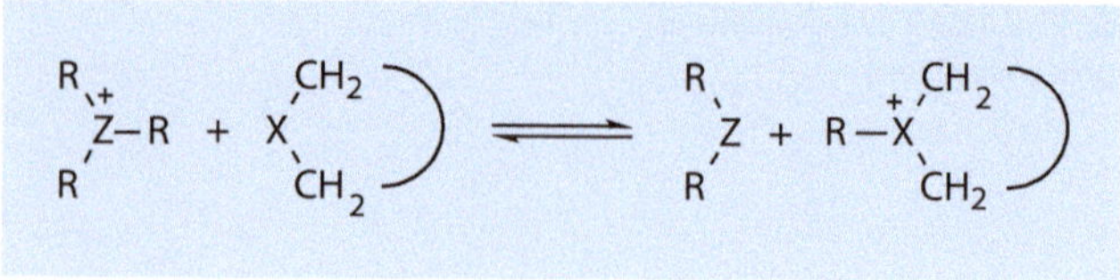

Fig. 12.16 Mechanism of the initiation of CROP with onium ions. *Z*, *X* can be identical, e.g., oxygen

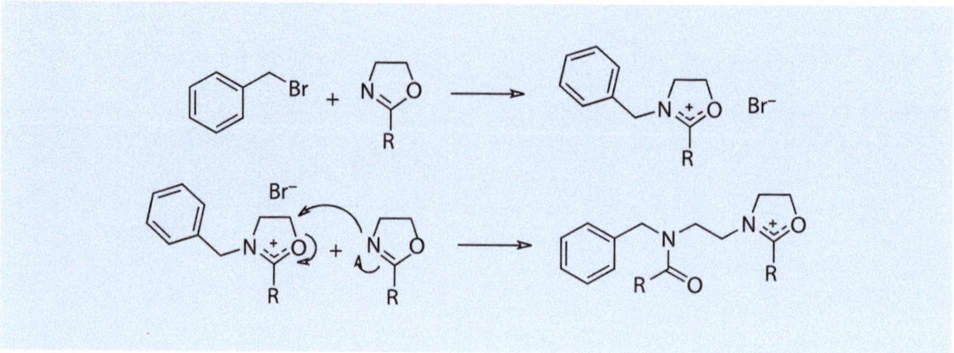

Fig. 12.17 Mechanism of the initiation of CROP with covalent initiators

Onium Ions

Onium ions can be produced in a separate reaction or in situ. Initiation with onium ions takes place as shown in Fig. 12.16, mostly by transferring an alkyl cation.

Trialkyl oxonium ions can initiate the ring-opening of cyclic acetals, ethers, sulfides, lactones, and amines.

Covalent Initiators

The polymerization of 2-alkyl-2-oxazolines can be initiated using typical alkylating agents such as benzyl bromide (Fig. 12.17).

However, $ROSO_2CF_3$ and $ROSO_2F$ are also suitable initiators.

Lewis Acids

The Lewis acids most often used for CROP are BF_3 and $BF_3 \cdot OEt_2$ but there are also in-depth studies with PF_5, for example for the polymerization of THF (Fig. 12.18).

Photoinitiators

Sulfonium and iodonium salts have achieved industrial significance as cationic photoinitiators. The reaction steps shown in Figs. 12.19 and 12.20 have been suggested for the formation of the cations that initiate the polymerization.

Heterolysis

The sulfonium salt is raised to a singlet state (Ph_3S^+) by light and this then relaxes to the ground state as Ph_2S and Ph^+. The Ph^+ species react with available RH moieties, which can be any molecule, such as solvent, polymer, or monomer, to give PhR and H^+.

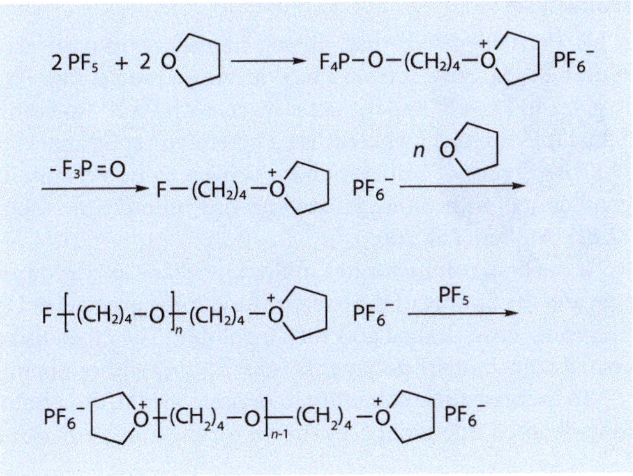

Fig. 12.18 CROP initiation of THF with PF_5, chain growth in two directions

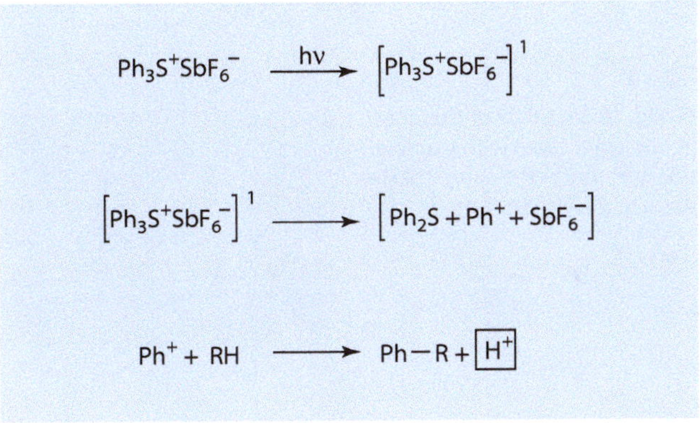

Fig. 12.19 Formation of the active initiator via heterolytic splitting of a photoinitiator. [1]Excited singlet state

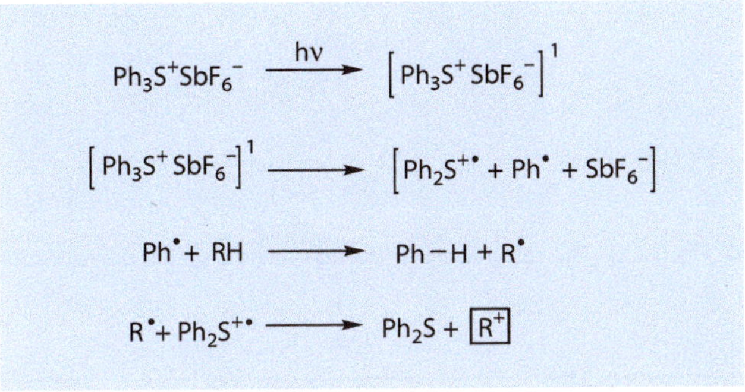

Fig. 12.20 Formation of the active initiator via homolytic splitting of a photoinitiator

Homolysis

It has been suggested that, during transition from an excited singlet state into a ground state, a Ph• is split-off from Ph_3S^+ leaving a radical ion, $Ph_2S^{+•}$. The Ph• can react with RH to give PhH and R• and the latter reacts with $Ph_2S^{+•}$ to become R^+ and Ph_2S. In this case, the actual initiator is R^+ whereas via a heterolytic splitting H^+ takes the role of the initiator.

These types of initiators have proven to be very useful for the cross-linking of hole conductors[1] with multiple oxetane ring side-chains used in OLEDs[2] (Figs. 12.21 and 12.22) (Müller et al. 2003).

If the hole conductor has multiple oxetane side-groups, the ring-opening polymerization and the linking of the oxetane functional groups lead to the polymeric hole conductor becoming cross-linked and thus insoluble. The cross-linked polymer can then simply be coated with another polymer by *spin coating* without being dissolved.

To increase their solubility in organic media, the photoinitiators can be modified with long-chained side groups, as shown for the iodonium salt in Fig. 12.23.

12.3.2 Chain Growth

Onium ions such as oxonium, sulfonium, ammonium, and phosphonium ions are the active species in cationic ring-opening polymerization (Fig. 12.24).

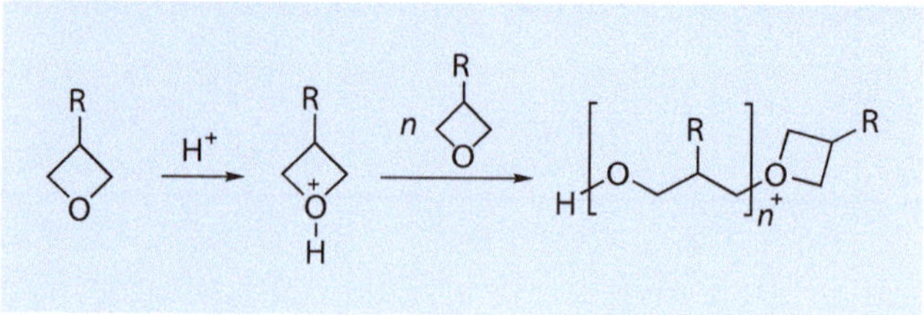

Fig. 12.21 CROP of oxetanes (R = polymer chain with oxetane side groups (for the source of H+; see, e.g., Fig. 12.19))

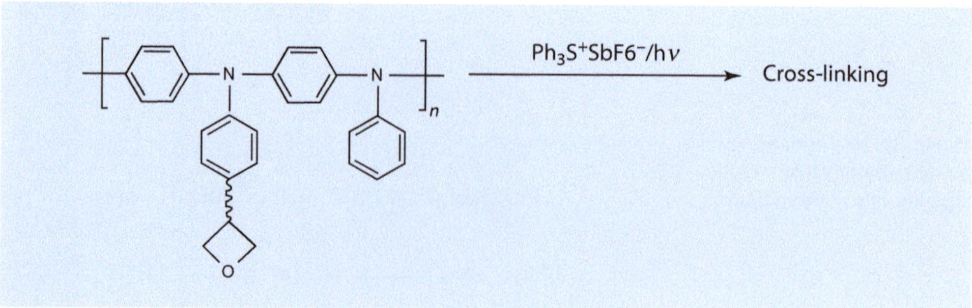

Fig. 12.22 The cationic cross-linking of a hole defect polyarylamine

---

1 Hole conductor: these can be aromatic amines, for example, which become radical cations after the transfer of one electron and therefore represent a 'hole' electronically speaking.

2 OLED: organic light emitting diode.

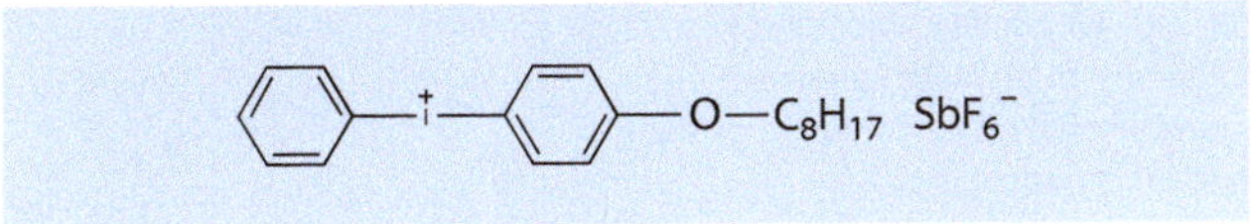

Fig. 12.23 Structure of an iodonium photoinitiator

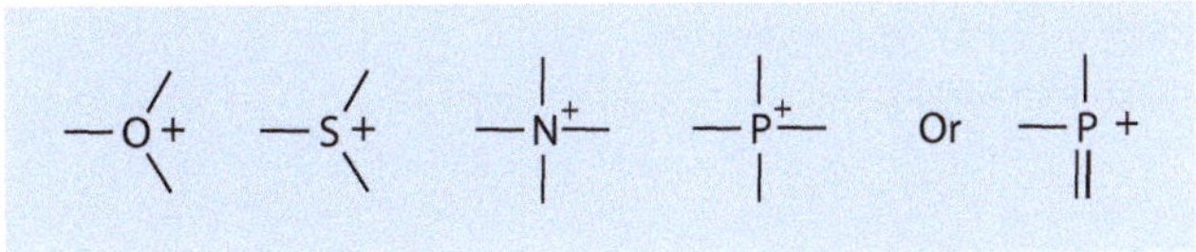

Fig. 12.24 Structure of typical growing chain ends during CROP

Table 12.2 Structure of the growing chain ends during CROP for a variety of monomers

Monomer	Structure of the growing chain end
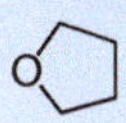	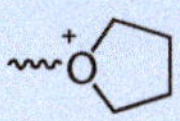
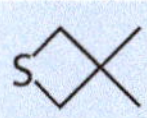	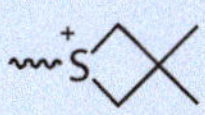
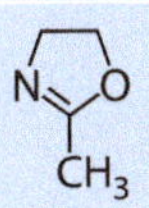	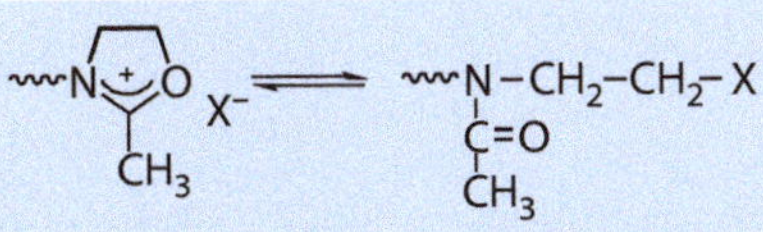
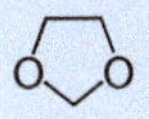	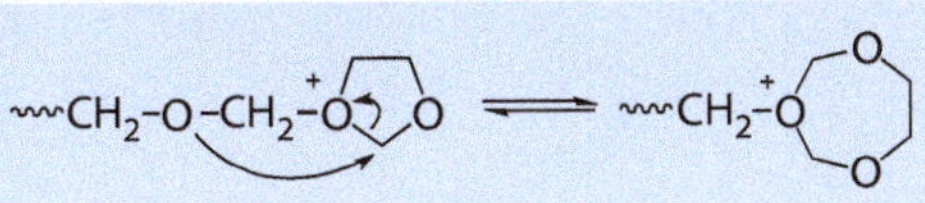

These can be determined by NMR spectroscopic analysis. Several examples are listed in Table 12.2.

Several reactions have been found to be useful for determining the concentration of the active chain ends (see, for example, Figs. 12.25 and 12.26).

The phenolate end group (Fig. 12.25) can be quantitatively identified using, for example, UV-spectroscopy. For this purpose the phenolate terminated polymer is reprecipitated several times and thus freed from excess phenolate reagent.

The reaction with PR_3 (Fig. 12.26) takes place quantitatively as PR_3 is more nucleophilic than the monomer. Quantitative determination can be made using $^{31}P\{^{1}H\}$-NMR spectroscopy.

The growth step of the *activated-monomer* mechanism can be visualized as shown in Fig. 12.27.

Chain growth via an activated monomer rather than an activated chain end can yield different polymer microstructures as shown in Fig. 12.28.

Because analysis has shown that structure type III predominates, it can be concluded that the polymerization of glycidol proceeds via an *activated-monomer* mechanism.

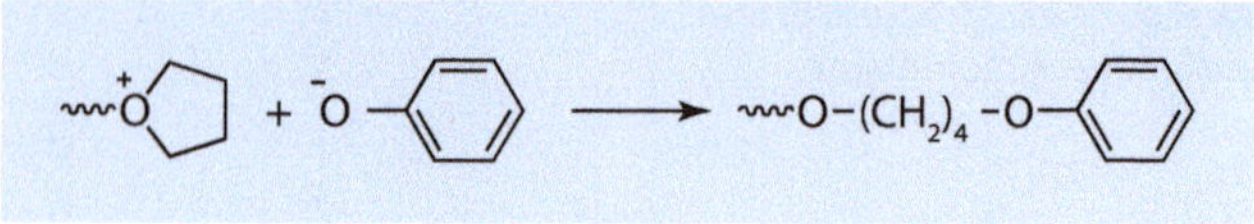

Fig. 12.25 Termination of CROP by phenolate to permit quantitative determination of the growing chains

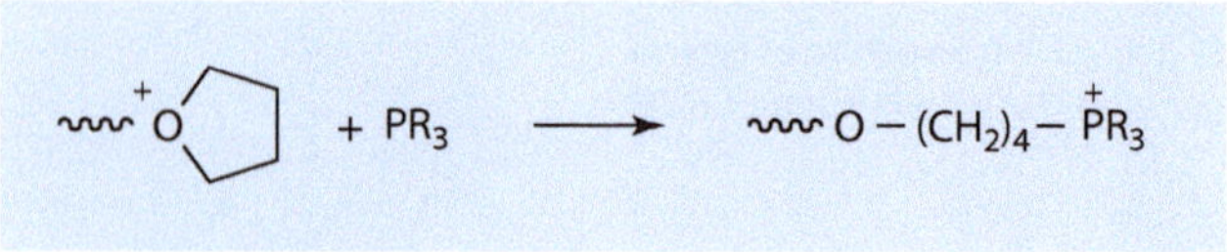

Fig. 12.26 Termination of a CROP with a phosphine to allow quantitative analysis

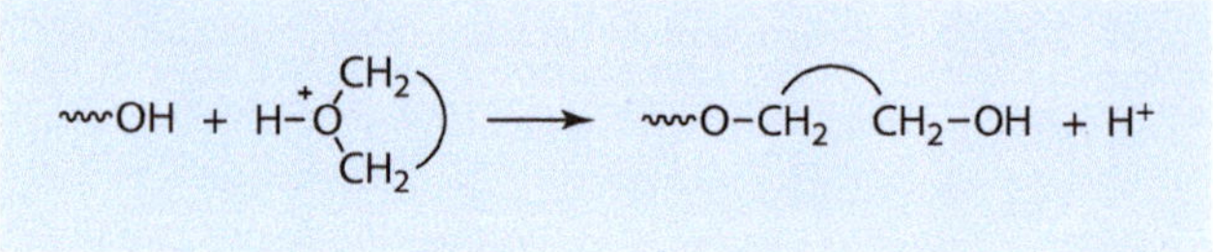

Fig. 12.27 Growth step in the *activated-monomer* mechanism

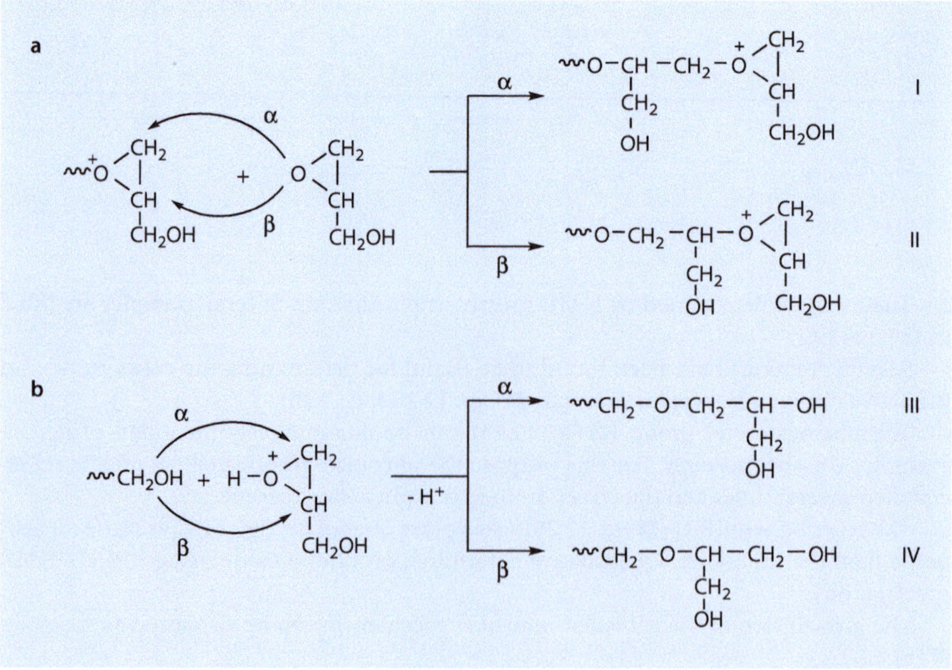

Fig. 12.28 The microstructure of poly-glycidol. (**a**) Growth via an active chain end. (**b**) Growth via an *activated-monomer* mechanism

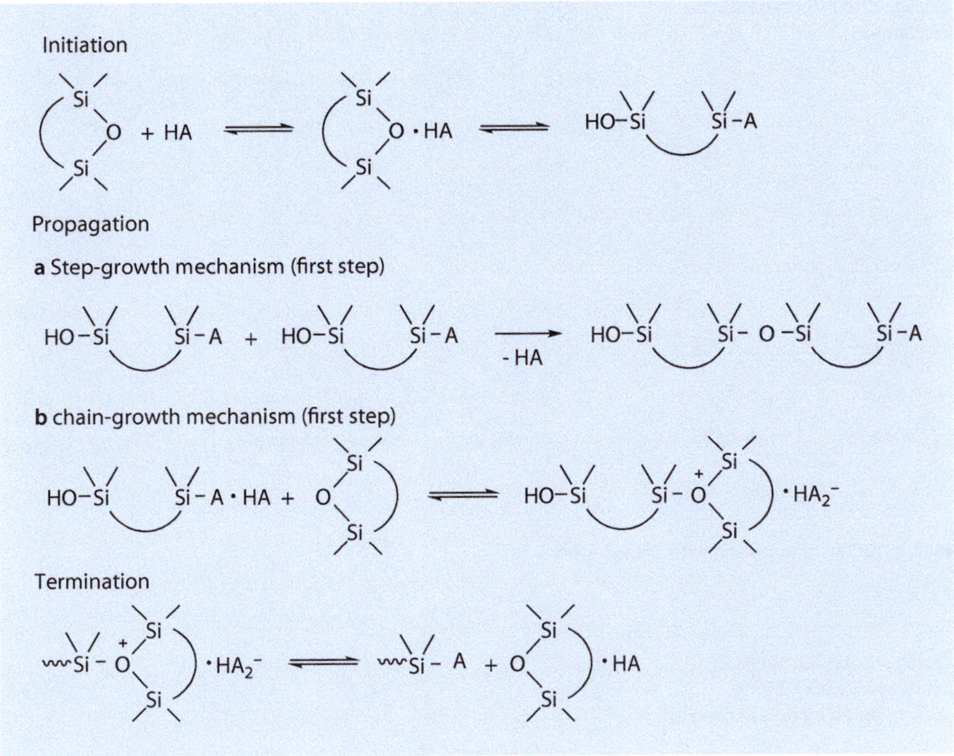

Fig. 12.29 Two alternative propagation routes for the ROP of cyclic siloxanes. (**a**) Step-growth. (**b**) Chain-growth

For the CROP of cyclic siloxanes two propagation reactions have been proposed (Fig. 12.29).

In contrast to radical chain growth polymerization, for example, in which a termination reaction results in chain growth being brought to a halt, the boundaries between termination and transfer during the polymerization of cyclic siloxanes are less clear. The termination reaction shown in Fig. 12.29 leads, in the first instance, to the growth of the chain being stopped. However, the resulting monomer formed can start the growth of a new chain as long as monomer is available. In this case it would be better to refer to the reaction as a transfer reaction rather that a termination reaction.

The exact mechanism of the polymerization of siloxanes is still a matter for discussion because the intermediate stages are very reactive and have proved difficult to identify.

12.3.3 Termination

Some ring-opening polymerizations are living but there is often an intrinsic termination reaction, namely an inter- or intramolecular alkylation (Fig. 12.30).

Such reactions become more dominant as the monomer is depleted. Close to complete conversion, these reactions become termination rather than transfer reactions because there is no monomer left to allow a new chain to grow.

Fig. 12.30 Termination via alkylation

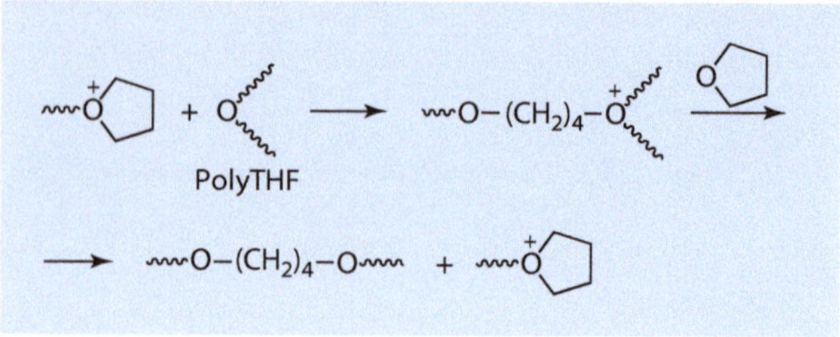

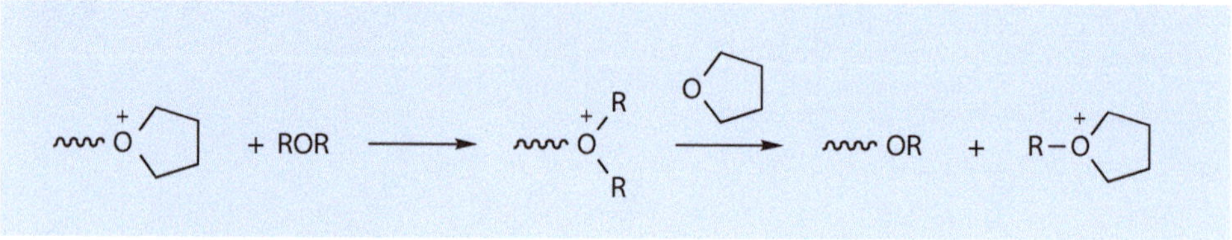

Fig. 12.31 Termination with dialkyl ether

Fig. 12.32 Termination produced by *back-biting* accompanied by ring expansion

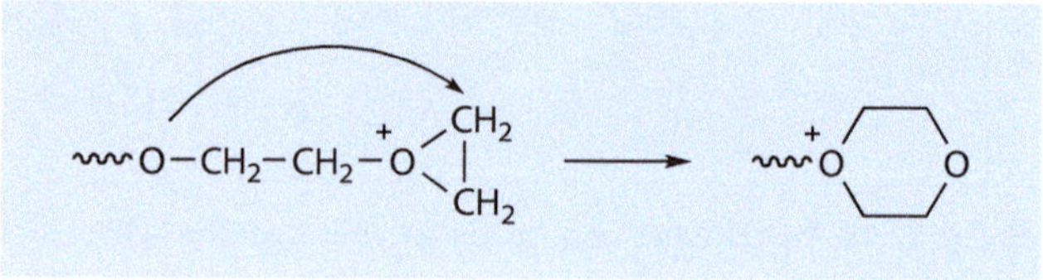

As well as being terminated by reaction with a polymer chain, polymerization at the active chain end can also be terminated by the addition of low molar mass compounds such as diethyl ether (Fig. 12.31).

The growth of an individual chain can also be terminated by intramolecular *back-biting* with ring expansion, an example of which – the polymerization of oxirane – is shown in Fig. 12.32. The resulting dioxonium ring, in this case, is inert to further chain growth.

A termination reaction that must always be considered is the collapse or decomposition of the counterion (Fig. 12.33).

12.4 Anionic Ring-Opening Polymerization

Anionic ring-opening polymerization (AROP) can be described as the nucleophilic reaction of an anionic chain end with a heterocyclic monomer (Fig. 12.34).

Several anionically polymerizable heterocycles and their corresponding chain ends are shown in Table 12.3.

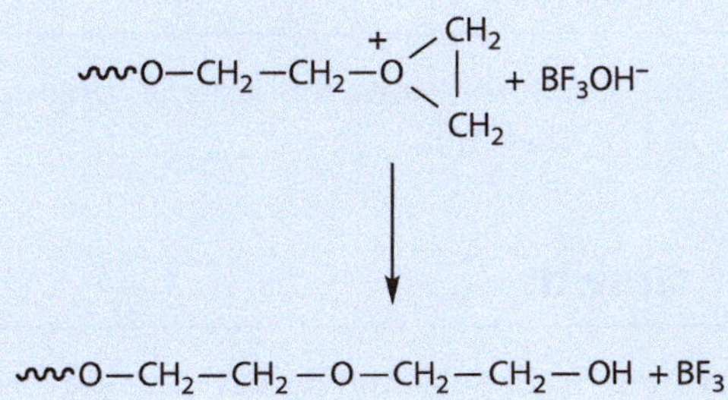

Fig. 12.33 Termination produced by collapse of the counterion

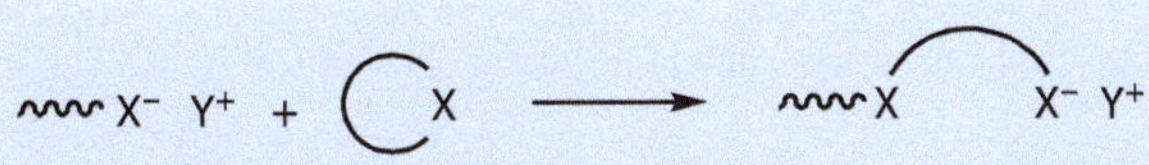

$X = O, S$; $Y^+ = Li^+, Na^+, K^+, Cs^+, R_4N^+, R_4P^+$

Fig. 12.34 AROP as a nucleophilic addition reaction

Table 12.3 Anionically polymerizable heterocycles and the corresponding active chain ends

Monomer	Structure of the growing chain end
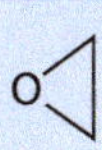	$\sim CH_2-CH_2-O^-$
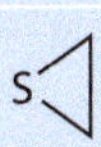	$\sim CH_2-CH_2-S^-$
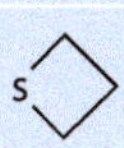	$\sim CH_2-CH_2-CH_2-S^-$
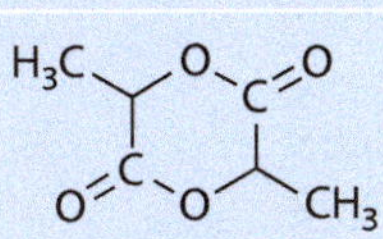	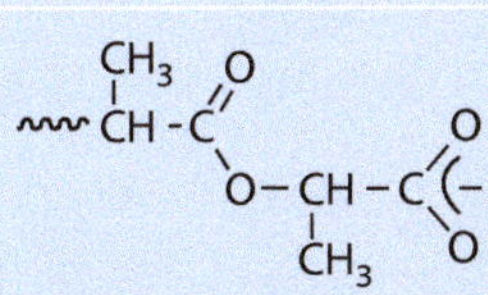
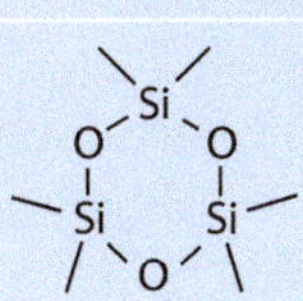	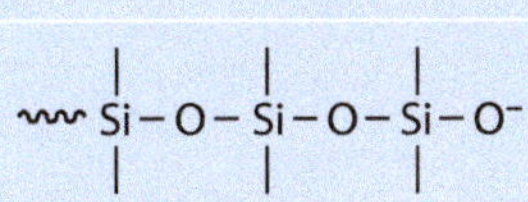

12.4.1 Initiators

Several important initiators for anionic ring-opening polymerization are listed in Table 12.4 (▸ Chap. 10).

12.4.2 Growth

For unsymmetrically substituted rings there are formally two ways the monomer ring can be opened as shown (Fig. 12.35).

Although not trivial, with careful microstructure and end group analysis the preferred reaction route can be identified. Additionally, simple consideration of the stability of the

Table 12.4 Selection of initiators for an AROP of heterocycles

Description	Structure	Monomer
Radical anion	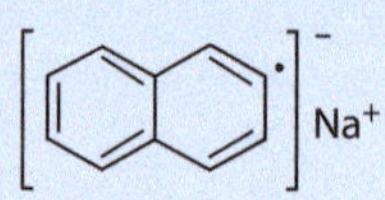	Ethylenoxide Propylene sulfide
Carbanion	$C_2H_5^-\ Li^+$ n-$C_4H_9^-\ Li^+$	Thietane Propylene sulfide Hexamethylcyclo-trisiloxane
	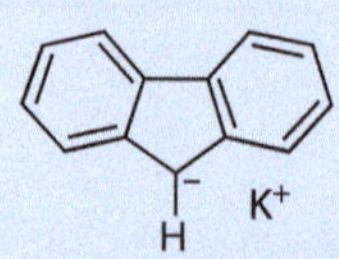	Ethylenoxide
Alcoholate Silanolate	$CH_3O^-\ K^+$ $(CH_3)_3SiO^-\ K^+$	Styrene oxide β-Propiolactone ε-Caprolactone
Carboxylate	$CH_3COO^-\ K^+$	β-Propiolactone Ethylene oxide Propylene oxide
Thiolate	$C_2H_5S^-\ K^+$	Propylene sulfide
Lactam anion	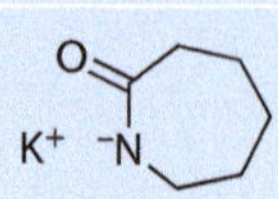	ε-Caprolactam
Amine	$(C_2H_5)_3N$	Propylene sulfide
Al-trialkoxide	$Al(O\text{-}t\text{-}C_3H_7)_3$	ε-Caprolactone
Al-monoalkoxide	$(C_2H_5)_2AlOCH_3$	Lactide ε-Caprolactone

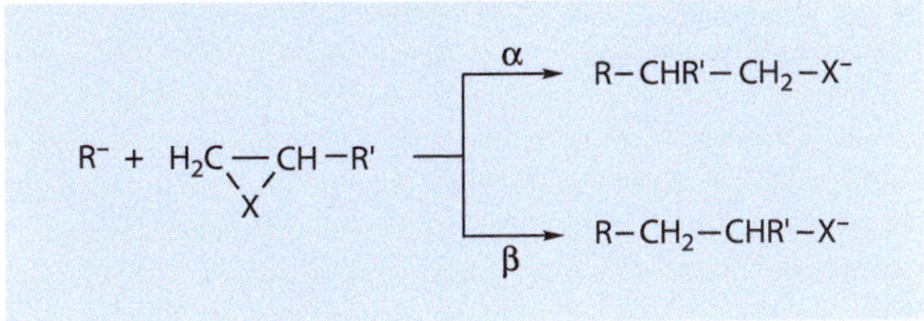

Fig. 12.35 Two possibilities of the nucleophilic ring-opening of asymmetrical heterocycles

reaction intermediates can prove helpful. Thus, the carboxylate ion (route β, Fig. 12.36) is energetically preferred, even for lactones whereby the nucleophilic attack takes place at the least substituted carbon atom (Fig. 12.36).

An additional possibility for determining whether route α or β (Fig. 12.35) is preferred is to short stop the polymerization, e.g., with diphenyl chlorophosphate (Fig. 12.37).

Valuable information about the neighborhood of the phosphorus atom can be obtained from a detailed analysis of the ^{31}P-NMR spectrum, which enables differentiation between the two alternative structures CH_2–X and CHR′–X.

The polymerization of lactams can be initiated by strong bases such as the alkali metals, metal hydrides, and metal amides, whereby a lactam anion is formed (Fig. 12.38).

The lactam ring is then opened in the propagation step shown in Fig. 12.39.

The very reactive, primary amine anion can also abstract a proton from another lactam to yield a new lactam anion (Fig. 12.40).

However, the so terminated chain can also react with a lactam anion and continue to grow.

The Leuchs' anhydride (*N*-carboxy-*α*-amino acid anhydride) is an interesting molecule. A CO_2 molecule is split off during every growth step (Fig. 12.41).

Primary amines can be used as initiators (Fig. 12.42).

As expected, the CO_2 originates solely from C2, as could be proven by using ^{14}C-marked Leuchs' anhydride. The free electron pair of the amine nitrogen atom of the initiator attacks the monomer at C5, the carbon atom with the lowest electron density. Then, one after the other, the C5——O1 and C2—N3 bonds are broken and CO_2 eliminated.

Alternatively, the base B^- abstracts a proton from the anhydride (Fig. 12.43) to form the nucleophilic anion which initiates chain growth.

It is also possible to polymerize cyclic siloxanes with AROP using alkali metal oxides or hydroxides as initiators. The polymerization reaction competes with depolymerizations so that chains and rings are in equilibrium (Fig. 12.44).

For this polymerization, $\Delta H \approx 0$ and the reaction is entropy driven because $\Delta S = +6.7$ J mol^{-1}. This gain in entropy arises from the increase in the degrees of freedom of the chains compared with that of the rings.

12.4.3 Transfer and Termination

Reactions that could strictly speaking be seen as transfer reactions have already been described in ▸ Sect. 12.4.2 (e.g., Fig. 12.40). The active centers of ring-opening

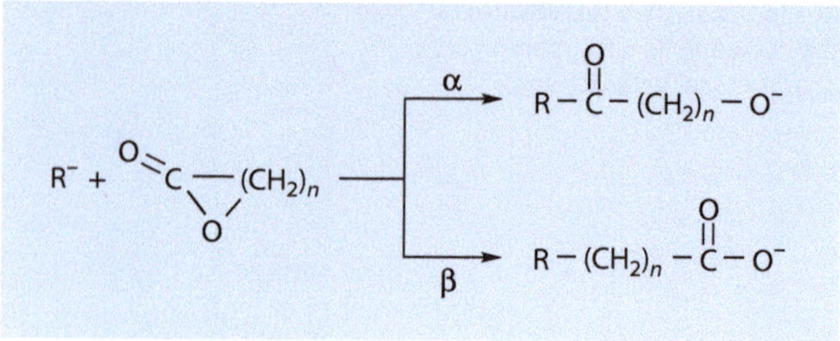

Fig. 12.36 Anionic ring-opening of a lactone

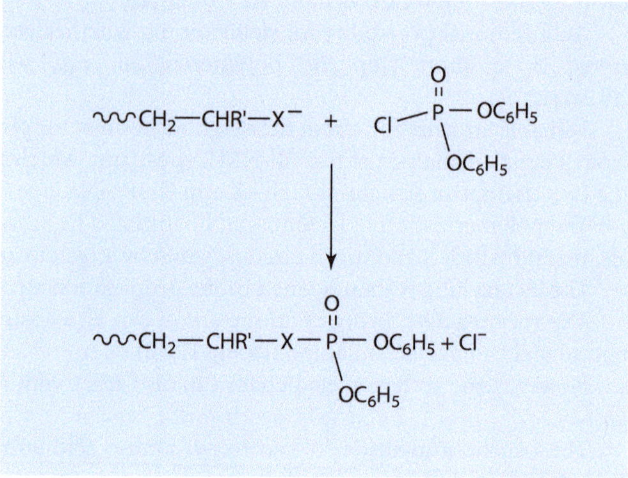

Fig. 12.37 Polymerization short stop with diphenyl chlorophosphate. Alcoholate X = O, thiolate X = S

12

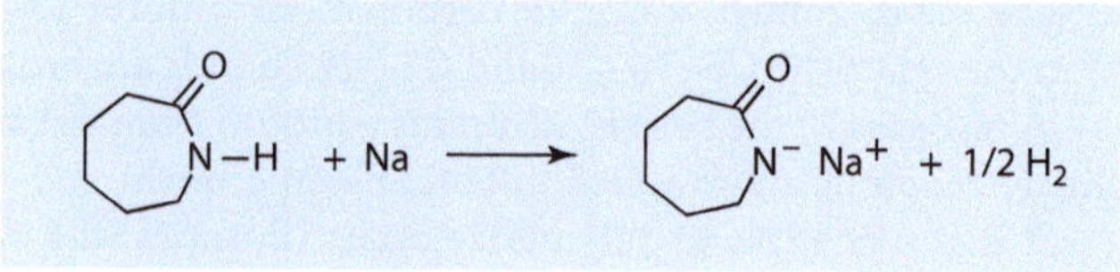

Fig. 12.38 Reaction of ε-caprolactam with sodium to yield a lactam anion

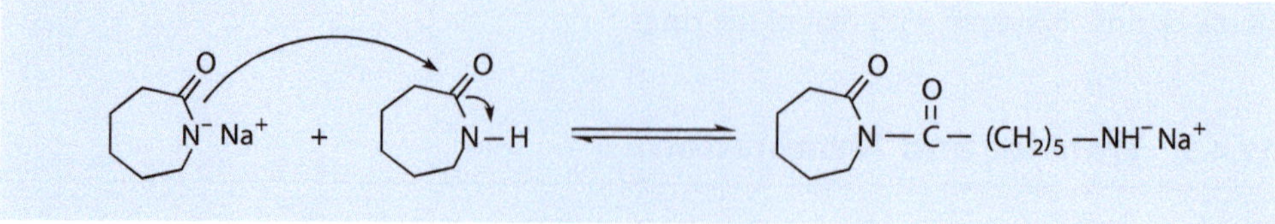

Fig. 12.39 Ring-opening of a lactam by a lactam anion

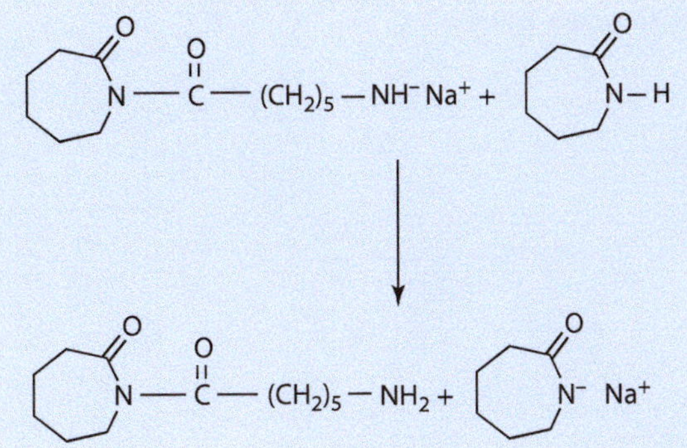

Fig. 12.40 Regeneration of the lactam anion

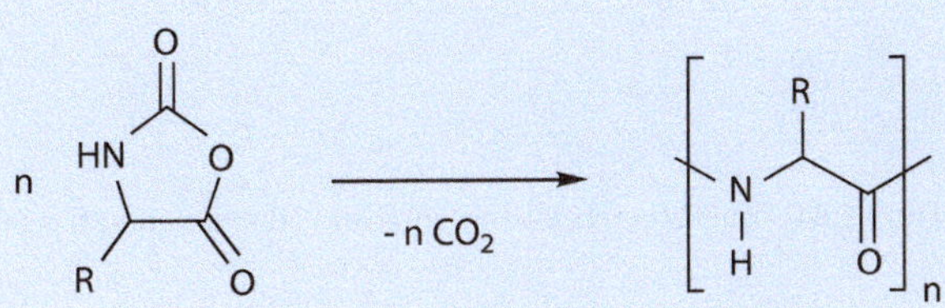

Fig. 12.41 Formation of Nylon-2 from Leuchs' anhydride

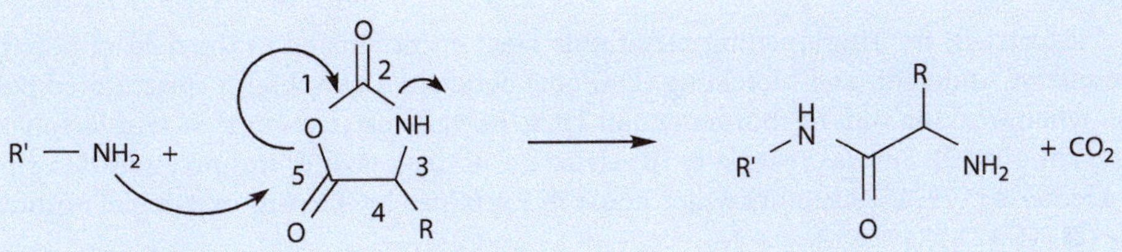

Fig. 12.42 Mechanism of CO_2 elimination from Leuchs' anhydride

polymerization, such as alcoholates and carboxylates, are not only nucleophilic but also alkaline. Thus, they can easily abstract protons from monomers. This reaction terminates the growing chain but an active anion is also generated so that this is a transfer and not a termination reaction.

This transfer reaction is the reason why the polymerization of propylene oxide with alkali metal alkoxides as initiators does not result in high molar mass products. However, the reaction can be limited by the addition of crown ethers as complexing agents for the counterions, which increases the concentration of free, growing chain ions. These seem to prefer adding on to the nearest monomer and are less inclined towards H-abstraction than ion pairs.

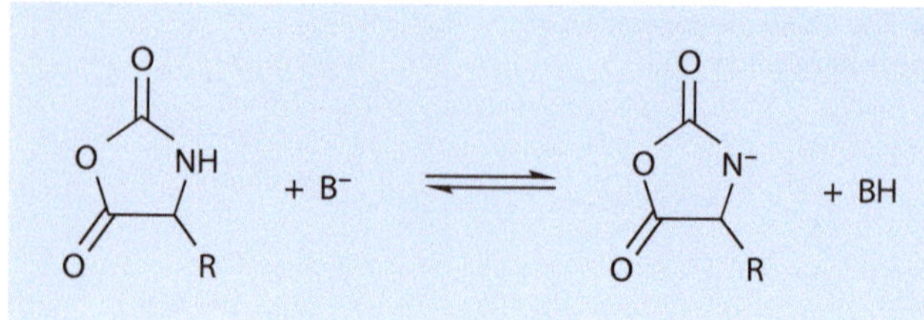

Fig. 12.43 Formation of the Leuchs' anhydride anion

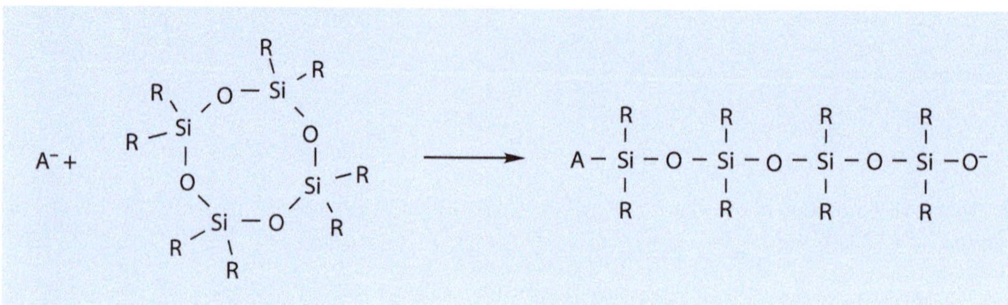

Fig. 12.44 Equilibrium between rings and chains during the AROP of cyclic siloxanes. *A*⁻: ~ $Si(R)_2O^-$

12.5 Ring-Opening Metathesis Polymerization

The fundamentals of metathesis as well as the catalysts used have already been discussed in ▶ Sect. 11.7.

Historically, the ring-opening metathesis reaction originated in the field of polymer chemistry: Anderson and Merckling (DuPont) obtained a new, highly unsaturated polymer when working with norbornenes and $TiCl_4$/EtMgBr-catalysts in 1955 (Anderson and Merckling 1955). Several years later the structure of this polymer was precisely described by Eleuterio (1957), a structure which could be explained by the ring-opening of norbornene (Fig. 12.45).

The driving force behind ring-opening metathesis polymerization is the opening of the strained ring. As the number of particles greatly decreases during polymerization, the positive ($-T\Delta S$) has to be compensated for by the negative ΔH, according to

$$\Delta G = \Delta H - T\Delta S \tag{12.1}$$

The ΔH, ΔS, and ΔG values of some cyclic olefins are listed in Table 12.5.

From this table it can also be seen why cyclohexene cannot be polymerized using a metathesis reaction at room temperature (positive ΔG^0).

The formation of macrocycles that is often observed during the polymerization of cycloolefins can be explained by the phenomenon of *back-biting* (Fig 12.46).

Polymers with diverse structures can be produced using ROMP. Selected examples are shown in Fig. 12.47.

The homopolymerization and copolymerization of various, functionalized cyclic olefins via ROMP has considerably increased the range of available polymers and completely

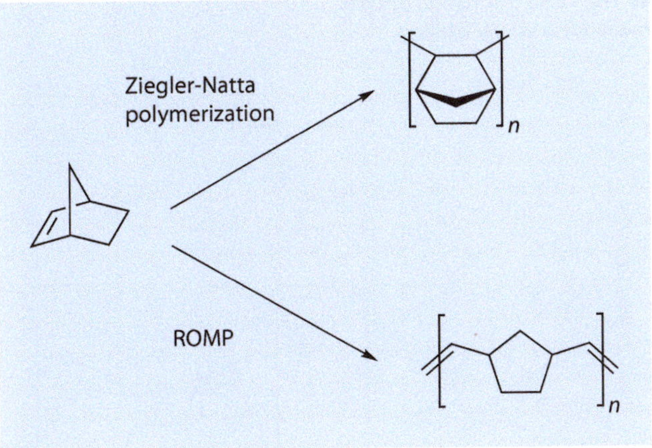

Fig. 12.45 Ziegler–Natta polymerization and ROMP of norbornene

Table 12.5 Thermodynamic parameters for the ROMP of a selection of liquid monomers to amorphous polymers at 25 °C

Monomer	Configuration of the double bond in the polymer	ΔH^0 (kJ/mol)	ΔS^0 (J/(mol K))	ΔG^0 (kJ/mol)
Cyclopentene	*cis* *trans*	−16 −20	−46 −46	−2.3 −6.3
Cyclohexene	*cis* *trans*	2 −2	−31 −28	11.2 6.3
Cycloheptene	*cis* *trans*	−16 −20	−20 −17	−10.0 −15.0
Cyclooctene	*cis* *trans*	−20 −22	−2 −2	−19.4 −21.5
1,5-Cyclooctadiene	*cis* *trans*	−25 −33	−5 −5	−23.5 −31.5

new, formally unknown structures have become accessible. Today, metathesis has become a standard tool for every synthesis and polymer chemist.

The best-known industrial uses are the Phillips-triolefin process, the Phillips-neohexene process, Norsorex (polynorbornenes) and Vestenamer (polycyclooctenes) (Figs. 11.32 and 11.33, ► Chap. 11).

12.6 Ring-Opening of Phosphazenes

The ring-opening polymerization of phosphazenes results in an interesting and promising substance class. More than 300 different polyphosphazenes are described in the literature (Allcock 1988). The most thoroughly examined reaction is the polymerization of $(PNCl_2)_3$ (Fig. 12.48).

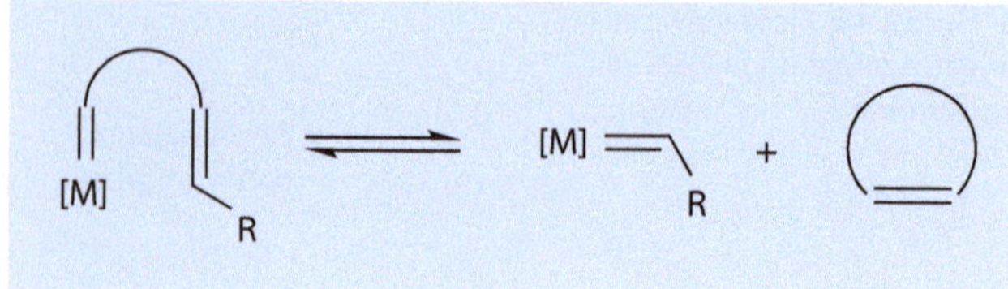

Fig. 12.46 Formation of cyclic byproducts via back-biting

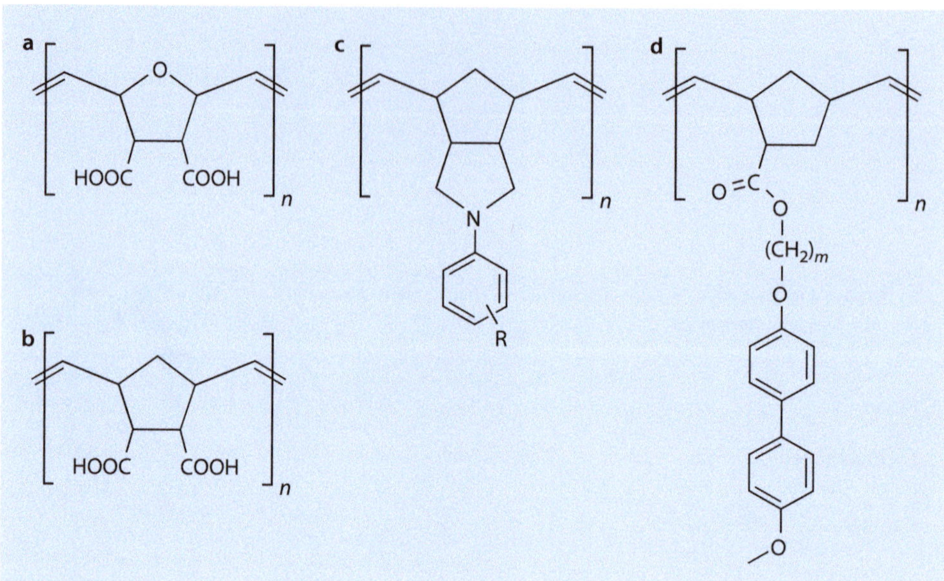

Fig. 12.47 Functional polymers via ROMP. (**a**) Super absorber, after crosslinking. (**b**) Ion exchange resin, flocculation aid. (**c**) High temperature thermoplastic. (**d**) Polymer with liquid crystal side chains ($m \geq 20$) (▶ Chap. 20)

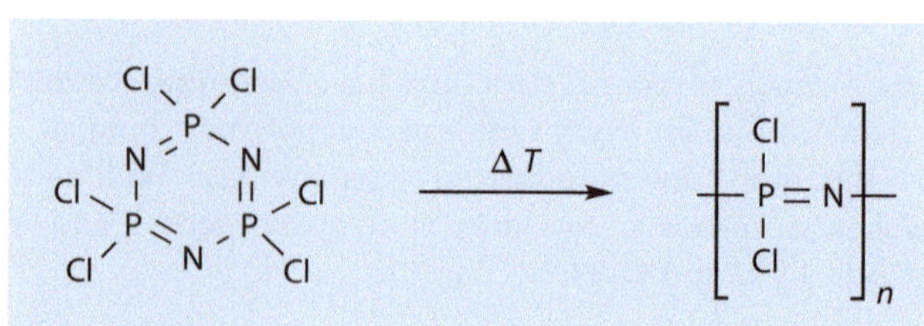

Fig. 12.48 Ring-opening of hexachloro triphosphazene at higher temperature

This reaction was first described by Liebig and Stokes in the nineteenth century (von Liebig 1834; Stokes 1897). Allcock's work has undoubtedly been responsible for generating renewed interest in this class of polymers (Allcock 1972, 1988).

Because polychlorophosphazene ($(PNCl_2)_n$) is sensitive to moisture, every Cl atom needs to be replaced by an alcoholate or amino group to obtain a material from $(PNCl_2)_n$

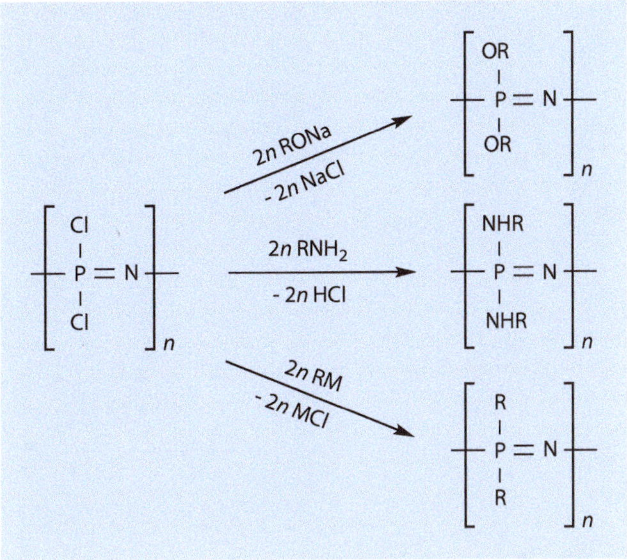

Fig. 12.49 Some substitution reactions with polychlorophosphazene

(referred to as "*Inorganic Rubber*," for which no technical use has been found up to now) that is stable under normal conditions and thus usable (Fig. 12.49).

To produce these materials, the cyclic monomer is heated up to roughly 250 °C in a closed system for several hours. Conversions of more than 50 % are avoided as these can result in insoluble products. In a second step, the stoichiometric reaction with alcoholates, amines, or organometallic compounds is conducted to give stable products. The polyphosphazenes obtained by substitution with fluorinated alcoholates have become technically relevant. They are used, for example, in seals which remain effective even at temperatures of −70 °C and do not swell in oil. Polyphosphazenes are also used in the medical field as carriers of medicines or as dental fillings.

With heat, hexachlorophosphazene dissociates to a cyclic cation and a chloride anion as a counterion. The cation reacts with the free electron pair of the nitrogen atom of a monomer. As a result, the P–N single bond of the nitrogen atom under attack is broken and the ring opens. An active chain with a $-P(Cl)_2^+$ end is formed. The formation of ions is catalyzed by BCl_3. With increasing conversion the thermally, or BCl_3 catalyzed, formation of cations also takes place along the chain and thus triggers the growth of side chains, which ultimately leads to the formation of insoluble products (Fig. 12.50).

Many attempts have been made to avoid the modification of the polymer (Fig. 12.49) by a corresponding substitution of the monomer. However, rings with methyl, OCH_2CF_3, OPh, or Ph instead of Cl simply decompose or undergo ring-expansion to rings with 8, 10, or 12 members instead of the original 6 when heated.

The ring strain is increased by replacing one of the PCl_2 groups in the ring with XCl and thus the temperature at which the ring can be opened is considerably decreased in comparison to $(PNCl_2)_n$ (Fig. 12.51) (Manners et al. 1989; Allcock et al. 1991).

However, it is just as important in this case, as it is with $(PNCl_2)_n$, to replace all the Cl with organic groups after polymerization to avoid hydrolysis of the polymer.

Another approach to increasing the ring strain in the monomer is shown in Fig. 12.52.

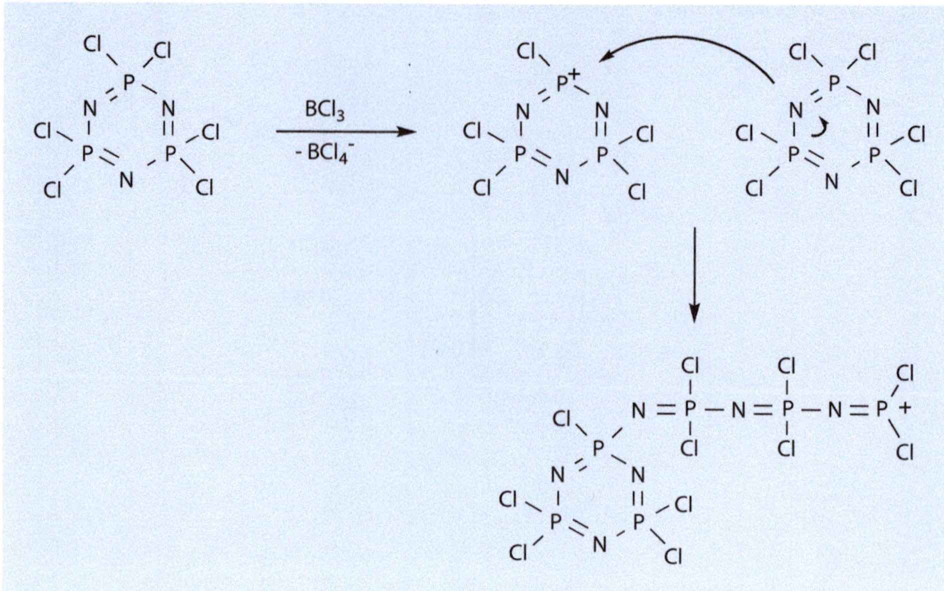

Fig. 12.50 Cationic polymerization of hexachloro triphosphazene

Fig. 12.51 Examples of phosphazene analog monomers with increased ring strain. X: C, S

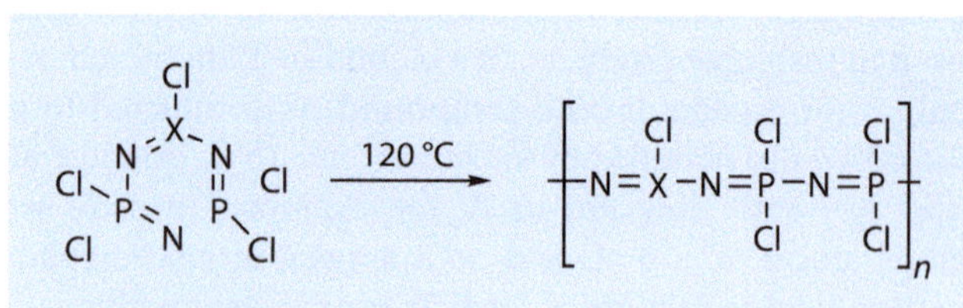

Fig. 12.52 Synthesis of a polyphosphazene with a ferrocene side chain

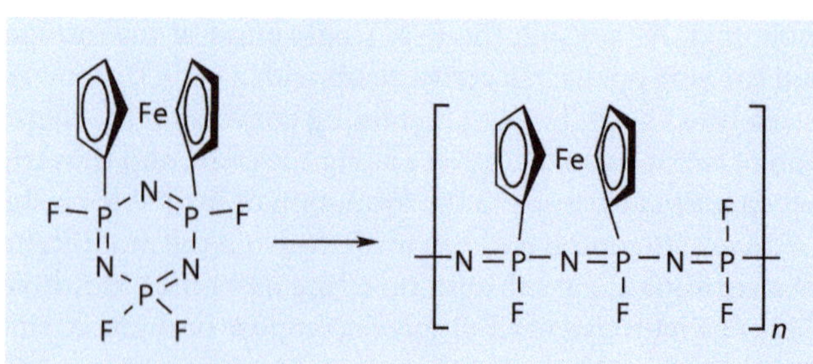

The substitution of the halogen with, for example, OCH_2CF_3, after polymerization is also necessary in this last case to obtain products that are stable with respect to hydrolysis.

References

Allcock HR (1972) Phosphorus-nitrogen-compounds. Academic Press, New York

Allcock HR (1988) Current status of polyphosphazene chemistry. ACS-Symp Ser 360:250–267

Allcock HR, Coley SM, Manners I, Renner G, Nuyken O (1991) Poly(aryloxycarbophosphazenes): synthesis, properties and thermal transition behavior. Macromolecules 24:2024–2028

Anderson AN, Merckling NG US 2721189 (1955) Du Pont de Nemours & Co. Chem Abstr (1956) 50:3008i

Eleuterio HS US 3074918 (1957) Du Pont de Nemours & Co. Chem Abstr (1961) 55:16005

Manners I, Renner G, Nuyken O, Allcock HR (1989) Poly(carbophosphazenes): a new class of inorganic-organic macromolecules. J Am Chem Soc 111:5478–5480

Müller CD, Falcon A, Reckefuss N, Rojahn M, Wiederhirn V, Rudati P, Frohne H, Nuyken O, Becker H, Meerholz K (2003) Multi-colour organic light-emitting displays by solution processing. Nature 421:829–832

Stokes HN (1897) Chloronitrides of phosphorus. Am Chem Soc 19:782–796

von Liebig J (1834) Über Phosphorstickstoff. J Liebigs Ann Chem 11:139

Copolymerization

S. Koltzenburg et al., *Polymer Chemistry*, DOI 10.1007/978-3-662-49279-6_13

If two different monomers, M_1 and M_2, are polymerized together, the structures shown in Fig. 13.1 result.

In the case of copolymers, we are not dealing with mixtures of the individual homopolymers but instead with products in which both monomers are covalently incorporated in the individual molecules. Although alternating and statistical copolymers can indeed be produced by polymerizing two different monomers together (simultaneously), the syntheses of block and graft copolymers require special methods. To differentiate the products, the following nomenclature has emerged:

- Statistical copolymer: Poly(M_1-*stat*-M_2)
- Alternating copolymer: Poly(M_1-*alt*-M_2)
- Block copolymer: Poly(M_1-*block*-M_2)
- Graft copolymer: Poly(M_1-*graft*-M_2)

If it is not intended to distinguish the specific copolymer architecture, the general term is poly(M_1-*co*-M_2).

Copolymerization allows us to synthesize a virtually infinite number of different products. If one assumes the existence of 50 technically interesting vinyl monomers, there are $(50 \cdot 49)/2 = 1225$ possible combinations of two different monomers. This number increases further by a large multiple if we vary the mixtures between 1 mol% and 99 mol% and take into account that for each monomer combination there are various possible copolymer architec-

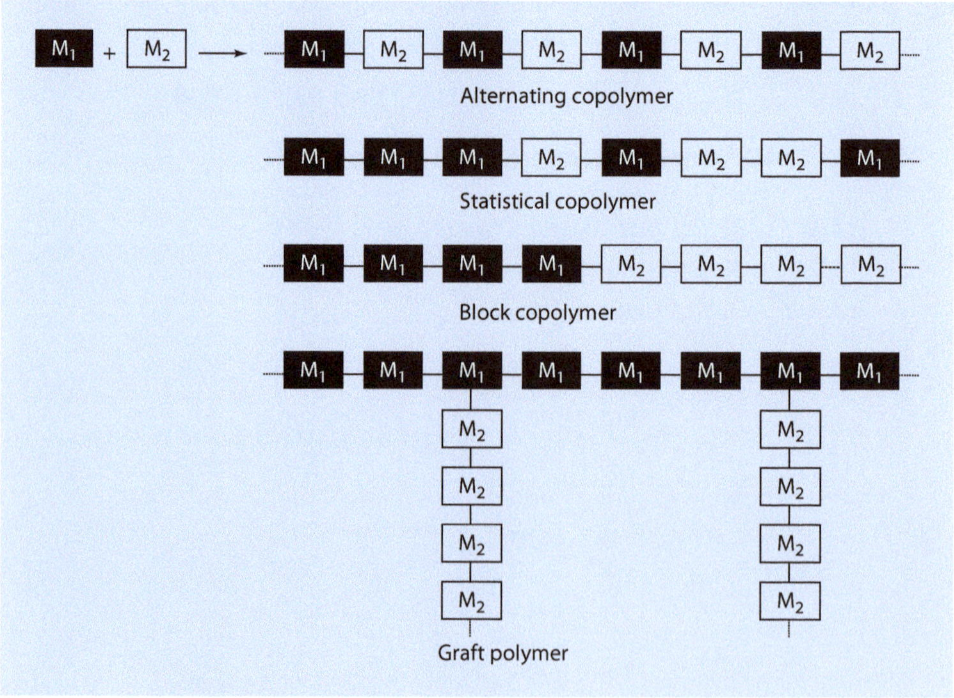

Fig. 13.1 Structures of alternating, statistical, block, and graft copolymers

tures. When copolymers are formed of more than two building blocks,[1] the number of combinations reaches vast proportions. An example of a group of copolymers that have become industrially very important is ABS—terpolymers of acrylonitrile, butadiene, and styrene.

Copolymerization has the great advantage that the attributes of the homopolymers can be deliberately combined. Taking the example of polystyrene, which is brittle and hard and has only moderate solvent resistance, it can be seen how, by introducing comonomers into the polymer, the service range of polymers based on styrene can be expanded. Copolymers with styrene as comonomer can be *thermoplasts* [e.g., poly(styrene-*co*-acrylonitrile), SAN] or *elastomers* [poly(styrene-*co*-butadiene) SBR,[2] and poly(styrene-*block*-butadiene-*block*-styrene), SBS] (▶ Sect. 14.5.3). For example, by incorporating acrylonitrile, the strength and ductility of SAN-polymers is improved and the resistance against aliphatic hydrocarbons and mineral oils is enhanced compared with polystyrene. SAN polymers with a wide variation in composition are available commercially.

13.1 Mayo-Lewis Equation of Copolymerization

The compositions of statistical copolymers are determined by the monomer composition and the relative reactivity of the comonomers. In the following, the corresponding rules are derived using the example of radical copolymerization.

13.1.1 Deduction of the Mayo-Lewis Equation of Copolymerization

If the radical polymerization is initiated in the presence of two different monomers, M_1 and M_2, the polymerization goes through all the individual phases of a radical polymerization; that is, the polymerization is initiated by the disintegration of the initiator, the primary radical reacts with the monomer (M_1 or M_2), and the chains grow until they terminate (▶ Sect. 9.2.1). Other than in the polymerization with only one monomer M_1, in the case of copolymerization, four different phases of growth, rather than one, need to be distinguished. A growing chain end derived from M_1 can react with either M_1 or M_2. Two further possibilities of growth emerge when the growing chain ends are derived from M_2 (◘ Fig. 13.2).

As the monomers are only consumed by chain growth, the following equations can be formulated for monomer consumption:

$$-\frac{d[M_1]}{dt} = k_{11}[\sim M_1^\bullet][M_1] + k_{21}[\sim M_2^\bullet][M_1] \tag{13.1}$$

$$-\frac{d[M_2]}{dt} = k_{22}[\sim M_2^\bullet][M_2] + k_{12}[\sim M_1^\bullet][M_2] \tag{13.2}$$

1 Polymers resulting from the copolymerization of more than two monomers are called *ter*polymers.
2 The abbreviation SBR derives from "styrene–butadiene rubber" reflecting the elastomeric properties of the copolymers.

Fig. 13.2 Chain growth during a radical polymerization of two monomers M_1 and M_2

$$\sim M_1^\bullet + \begin{cases} M_1 \xrightarrow{k_{11}} \sim M_1^\bullet & v_{11} = k_{11}[\sim M_1^\bullet][M_1] \\ M_2 \xrightarrow{k_{12}} \sim M_2^\bullet & v_{12} = k_{12}[\sim M_1^\bullet][M_2] \end{cases}$$

$$\sim M_2^\bullet + \begin{cases} M_1 \xrightarrow{k_{21}} \sim M_1^\bullet & v_{21} = k_{21}[\sim M_2^\bullet][M_1] \\ M_2 \xrightarrow{k_{22}} \sim M_2^\bullet & v_{22} = k_{22}[\sim M_2^\bullet][M_2] \end{cases}$$

Furthermore, assuming a stationary state with respect to the number of radicals (as described in ▶ Chap. 9), it follows that

$$k_{12}\left[\sim M_1^\bullet\right]\left[M_2\right] = k_{21}\left[\sim M_2^\bullet\right]\left[M_1\right] \tag{13.3}$$

and

$$\left[\sim M_1^\bullet\right] = \frac{k_{21}\left[\sim M_2^\bullet\right]\left[M_1\right]}{k_{12}\left[M_2\right]}t \tag{13.4}$$

The division of (13.1) by (13.2) yields

$$\frac{d[M_1]}{d[M_2]} = \frac{k_{11}\left[\sim M_1^\bullet\right][M_1] + k_{21}\left[\sim M_2^\bullet\right][M_1]}{k_{22}\left[\sim M_2^\bullet\right][M_2] + k_{12}\left[\sim M_1^\bullet\right][M_2]}$$

$$= \frac{[M_1]}{[M_2]} \cdot \frac{k_{11}\left[\sim M_1^\bullet\right] + k_{21}\left[\sim M_2^\bullet\right]}{k_{22}\left[\sim M_2^\bullet\right] + k_{12}\left[\sim M_1^\bullet\right]} \tag{13.5}$$

The expression $d[M_1]/d[M_2]$ is a measure of the rate at which the monomers M_1 and M_2 are incorporated into the growing polymer chains at any moment during the polymerization. Thus, this expression describes the chemical composition of the copolymers formed at any one moment in time. Introducing (13.4) into (13.5) gives

$$\frac{d\left[M_1\right]}{d\left[M_2\right]} = \frac{\left[M_1\right]}{\left[M_2\right]} \cdot \frac{k_{11} \cdot \dfrac{k_{21}\left[\sim M_2^\bullet\right]\left[M_1\right]}{k_{12}\left[M_2\right]} + k_{21}\left[\sim M_2^\bullet\right]}{k_{22}\left[\sim M_2^\bullet\right] + k_{12} \cdot \dfrac{k_{21}\left[\sim M_2^\bullet\right]\left[M_1\right]}{k_{12}\left[M_2\right]}} \tag{13.6}$$

which can be simplified to

$$\frac{d[M_1]}{d[M_2]} = \frac{[M_1]}{[M_2]} \cdot \frac{k_{11} \cdot \dfrac{k_{21}}{k_{12}} \cdot \dfrac{[M_1]}{[M_2]} + k_{21}}{k_{22} + k_{21} \cdot \dfrac{[M_1]}{[M_2]}} \tag{13.7}$$

By multiplying the second quotient in (13.7) by $[M_2]/k_{21}$ and simplifying one obtains

$$\frac{d[M_1]}{d[M_2]}=\frac{[M_1]}{[M_2]}\cdot\frac{k_{11}\cdot\dfrac{k_{21}[M_1]}{k_{12}[M_2]}\cdot\dfrac{[M_2]}{k_{21}}+k_{21}\dfrac{[M_2]}{k_{21}}}{k_{22}\dfrac{[M_2]}{k_{21}}+k_{21}\cdot\dfrac{[M_1]}{[M_2]}\cdot\dfrac{[M_2]}{k_{21}}}=$$

$$=\frac{[M_1]}{[M_2]}\cdot\frac{\dfrac{k_{11}}{k_{12}}\cdot[M_1]+[M_2]}{\dfrac{k_{22}}{k_{21}}[M_2]+[M_1]} \tag{13.8}$$

Equation (13.8) can be further simplified by introducing the so-called *copolymerization parameters* r_1 and r_2. These are defined as follows:

$$r_1=\frac{k_{11}}{k_{12}} \tag{13.9}$$

$$r_2=\frac{k_{22}}{k_{21}} \tag{13.10}$$

The copolymerization parameters, the reactivity ratios r_1 and r_2, are very important parameters. They describe the relative reactivity of the radicals towards the monomers M_1 and M_2. If $r_1>1$, then a polymer radical derived from M_1 preferably reacts with the monomer M_1. In this case, homopolymerization is favored over heteropolymerization. Moreover, only these two parameters are necessary to describe the copolymerization instead of the four rate constants.

By inserting the parameters r_1 and r_2 into (13.8) one obtains the Mayo–Lewis copolymerization equation:

$$\frac{d[M_1]}{d[M_2]}=\frac{m_1}{m_2}=\frac{[M_1]}{[M_2]}\cdot\frac{r_1[M_1]+[M_2]}{r_2[M_2]+[M_1]} \tag{13.11}$$

m_1	Concentration of M_1 in the newly formed polymer[3]
m_2	Concentration of M_2 in the newly formed polymer
$[M_1]$	Concentration of M_1 in the reaction mixture
$[M_2]$	Concentration of M_2 in the reaction mixture
r_1, r_2	Copolymerization parameters

Equation (13.11) describes the composition of the polymer formed at any one time as a function of the monomer composition at that time. During most copolymerizations, as the reaction progresses, and depending on the reactivity of the individual comonomers,

3 The proportion of M_1 and M_2 incorporated into the polymer at any one time cannot be determined simply from the composition of the polymer. However, if the copolymerization is prematurely stopped at conversions $p<5\,\%$, the difference between the composition at the time of stopping the reaction and the integral, experimentally obtained, copolymer is negligible. It should be noted that all the concentrations mentioned here are molar concentrations.

a more or less dramatic change in the proportion of the comonomers in the reaction mixture occurs. Thus, the copolymers produced after different reaction times, i.e., at different conversions, may be chemically different. Only at extremely low conversions of less than 5 % can the relative monomer concentrations be assumed to be approximately constant; at this stage, (13.11) gives a good first approximation of the copolymer composition.

If a copolymerization is allowed to continue to complete conversion, the average composition of all the polymer chains naturally corresponds to the initial monomer composition. However, in this case the product also contains a large number of chemically different macromolecules.

13.1.2 Example: Statistical Copolymerization of Styrene and Acrylonitrile

To copolymerize styrene with acrylonitrile, both monomers are mixed in different mole fractions (here: $0.1 \leq x_1 \leq 0.9$) and polymerized up to a conversion of max. 5 %. The polymer is separated from the residual monomer by precipitation and its chemical composition analyzed. The method of analysis of the copolymers is determined by the nature of the components. In the case of styrene and acrylonitrile, ^{1}H-NMR-spectroscopy, IR-spectroscopy (typical CN- and phenyl bands), UV-spectroscopy (absorption of the phenyl ring), and elementary analysis (acrylonitrile concentration via nitrogen analysis) are all appropriate.

A typical copolymerization of styrene (M_1) and acrylonitrile (M_2), conducted without a solvent, delivers the following data (Table 13.1).

A plot of the mole fraction of the monomer (M_1) in the polymer $m_1/(m_1+m_2)$ against the mole fraction of the monomer (M_1) in the initial monomer mixture shows clearly that rather than a simple linear relationship between $m_1/(m_1+m_2)$ and $[M_l]/[(M_l]+[M_2])$, their interdependence is more complicated (Fig. 13.3). This is discussed in more detail in the next section.

13

13.2 Copolymerization Diagrams and Copolymerization Parameters

A graph plotting the monomer composition in the polymer as a function of the initial[4] concentration of the monomers in the reaction mixture shown in Fig. 13.3 is referred to as a *copolymerization diagram*.

The intersection of the curve in Fig. 13.3 with the diagonal dotted line is referred to as the *azeotrope*. At this point the copolymer composition and the composition of the monomer mixture are identical. This is the only point at which the composition of the original monomer mixture does not change despite further conversion, i.e., chemically identical polymers are produced throughout the entire polymerization process. The mixture with this basic composition can therefore be fully polymerized without altering the

4 At very low conversion the actual monomer composition hardly differs from that of the initial composition so the initial mole fraction is used.

Table 13.1 Mole fractions of the monomers in the initial reaction mixture and in the polymer formed via a radical polymerization of styrene (M_1) and acrylonitrile (M_2), initiated with AIBN, $T=48$ °C[a]

$\frac{[M_1]}{[M_1]+[M_2]}$	$\frac{m_1}{m_1+m_2}$
0.1	0.30
0.2	0.40
0.3	0.43
0.4	0.46
0.5	0.47
0.6	0.49
0.7	0.54
0.8	0.60
0.9	0.70

[a]After a conversion of ca. 5 %, the copolymer is separated from the residual monomers by precipitation. This polymer composition was determined by ^{1}H-NMR-spectroscopy.

copolymer composition with increasing conversion. For all other monomer compositions, either styrene (in the example given here) is preferentially incorporated into the polymer (below or to the left of the azeotrope) or acrylonitrile (above or to the right of the azeotrope). This leads to the remaining monomer composition moving away from the azeotrope as the polymerization proceeds and a chemically inhomogeneous material is obtained. By making monomer additions of the monomer being preferentially incorporated during the polymerization, this effect can be corrected.

If the copolymerization parameters r_1 and r_2 are known, the azeotrope can be calculated using (13.11):

$$\frac{m_1}{m_2}=\frac{[M_1]}{[M_2]}\cdot\frac{r_1[M_1]+[M_2]}{r_2[M_2]+[M_1]} \tag{13.11}$$

At the azeotrope:

$$\frac{m_1}{m_2}=\frac{[M_1]}{[M_2]} \tag{13.12}$$

Inserting (13.12) into (13.11) gives

$$1=\frac{r_1[M_1]+[M_2]}{r_2[M_2]+[M_1]} \tag{13.13}$$

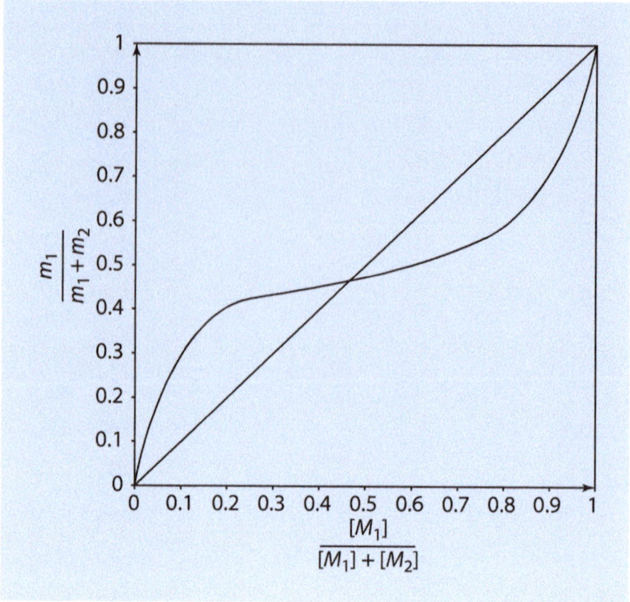

Fig. 13.3 Instantaneous copolymer composition as a function of the initial monomer composition using the example of the copolymerization of styrene (M_1) and acrylonitrile (M_2)

This can easily be transformed into

$$\frac{[M_1]}{[M_2]} = \frac{r_2 - 1}{r_1 - 1} \tag{13.14}$$

13

An azeotrope exists if r_1 and r_2 are both smaller than 1 (preferred heteropolymerization). This case is true for the radical copolymerization of styrene/acrylonitrile under the circumstances mentioned here ($r_1 = 0.40$; $r_2 = 0.17$). This is also referred to as a statistical copolymerization. Table 13.2 offers an overview of the combination of copolymerization parameters for each type of copolymerization.

In relation to the copolymerization parameters, we can distinguish between five different types of copolymerization (Table 13.2).

Typical curves analogous to Fig. 13.3 for the different cases given above are shown in Fig. 13.4. A detailed discussion of the individual cases can to be found in the subsequent sections of this chapter.

In Table 13.3 some examples of copolymerization parameters for selected monomer pairs are compiled.

13.3 Alternating Copolymerization

In some cases it is not possible for a monomer to add a growing radical derived from another molecule of the same monomer. Thus, the parameters r_1 and $r_2 = 0$, as well as k_{11} and $k_{22} = 0$ (Fig. 13.5).

Table 13.2 Combination of copolymerization parameters and the resulting class of copolymer

			Copolymerization class
a	$r_1 < 1$	$r_2 < 1$	Statistical (▶ Sect. 13.1)
b	$r_1 = 1$	$r_2 = 1$	Ideal (▶ Sect. 13.4)
c	$r_1 > 1$	$r_2 < 1$	Ideal, if $r_1 \cdot r_2 = 1$
	$r_1 < 1$	$r_2 > 1$	
d	$r_1 = 0$	$r_2 = 0$	Alternating (▶ Sect. 13.3)
e	$r_1 > 1$	$r_2 > 1$	unknown

(a) r_1 and r_2 are both less than 1. In this case, the copolymerization diagram has an azeotrope (Fig. 13.3).
(b) r_1 and r_2 both equal 1. It follows from the definition of the copolymerization parameters that both monomers are built into the growing polymer chain with the same rate constants and are therefore indistinguishable. In this case, the composition of the copolymers corresponds to that of the original monomer mixture. This copolymerization is referred to as ideal azeotropic copolymerization.
(c) r_1 is less than 1, r_2 is greater than 1 (or reversed). In this case one of the two monomers is always preferentially incorporated. If the product of the copolymerization parameters equals 1, this is referred to as an ideal non-azeotropic copolymerization. In this case the two polymer radicals react with the two monomers with identical ratios, i.e., the reactivity of the two radicals is the same in relation to the two monomers.
(d) r_1 and r_2 both equal 0. This situation implies that one active chain end subsequently always adds the other monomer. As a result, an alternating copolymer is formed.
(e) r_1 and r_2 are both greater than 1. In this case homopolymerization is kinetically favored over heteropolymerization. However, there are no examples of this case.

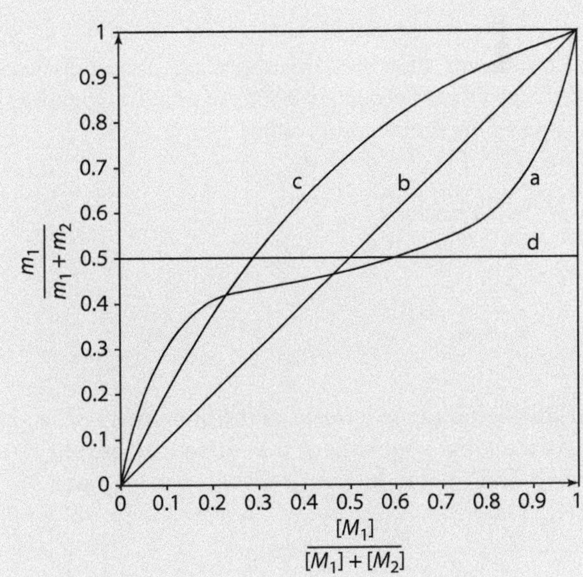

Fig. 13.4 Typical copolymerization curves (**a**) $r_1 < 1, r_2 < 1$, (**b**) $r_1 = 1, r_2 = 1$, (**c**) $r_1 > 1, r_2 < 1$, (**d**) $r_1 = r_2 = 0$

Table 13.3 Examples of copolymerization parameters[a]

M_1	r_1	M_2	r_2	Type
Methyl methacrylate	0.84	Vinylidene chloride	0.99	a, ~b
Styrene	0.58	Butadiene	1.35	c
Styrene	56	Vinyl acetate	0.01	c
Styrene	0.05	Maleic acid anhydride	0.005	~d
Acrylonitrile	5.5	Vinyl acetate	0.06	c
Styrene	0.40	Acrylonitrile	0.17	a

[a]The parameters are for radical polymerizations. These parameters are hardly affected by the temperature at which the polymerization takes place and, apart from a few exceptions (cf., for example ▶ Sect. 13.5), are almost independent of the solvent.

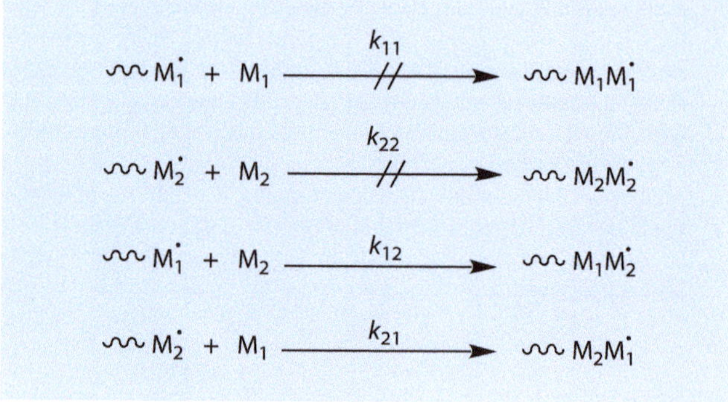

Fig. 13.5 Formation of an alternating copolymer from the exclusive addition of the growing chain end to a monomer different from the monomer from which it was derived

This means that the corresponding homopolymerization steps do not take place and the monomer sequence is alternating. Independent of the monomer composition, a polymer of the molar composition

$$\frac{m_1}{m_1+m_2}=0.5 \tag{13.15}$$

$$\frac{m_2}{m_1+m_2}=0.5 \tag{13.16}$$

with an alternating succession of the monomers M_1 and M_2 is produced.

Alternating copolymers are often the result if donor–acceptor complexes are formed from polar interactions between M_1 and M_2 and these polymerize in pairs (Fig. 13.6).

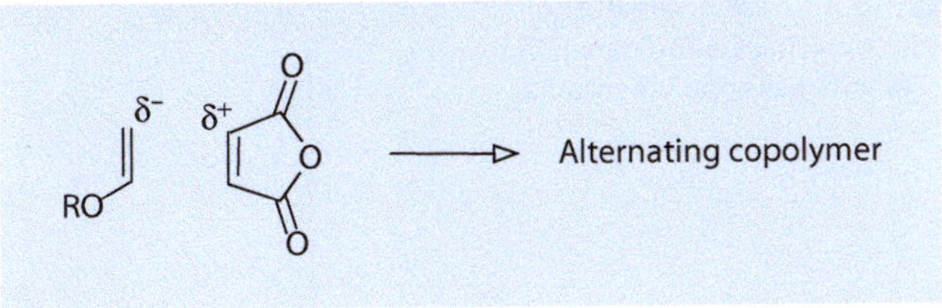

Fig. 13.6 Donor–acceptor complex from vinyl ether and maleic acid anhydride

13.4 Ideal Copolymerization

A copolymerization with $r_1 \cdot r_2 = 1$ is often referred to as an *ideal copolymerization*. If both parameters r_1 and $r_2 = 1$, this makes sense as it follows from (13.11) that for this case:

$$\frac{m_1}{m_2} = \frac{[M_1]}{[M_2]} \tag{13.17}$$

Such systems are rare but these monomer pairs can be quantitatively copolymerized to give copolymers having the same composition as that of the original monomer mixture. Such copolymerizations are thus easy to conduct—as opposed to statistical copolymerization, in which the disproportionately incorporated comonomer has to be continually replaced to obtain a homogeneous product.

Systems with the values $r_1 > 1$, $r_2 < 1$ (Table 13.2, Case c) can also meet the condition $r_1 \cdot r_2 = 1$. In this case, by inserting (13.9) and (13.10) and rearranging one obtains

$$\frac{k_{11}}{k_{12}} = \frac{k_{21}}{k_{22}} \tag{13.18}$$

According to the above, the relative rates at which M_1 and M_2 (second numeral of the subscript) react with the active chains ends $\sim M_1^{\bullet}$ and $\sim M_2^{\bullet}$ (first numeral of the subscript) are the same and independent of the nature of the active chain end. As a result, Eq. (13.11) can be simplified to

$$\frac{m_1}{m_2} = \frac{1}{r_2} \cdot \frac{[M_1]}{[M_2]} \tag{13.19}$$

For the experimental determination of r_2, various mixtures $[M_1]/[M_2]$ are polymerized up to a conversion of max. 5%. Thereafter, the copolymers are separated from the residual monomer, for example, by precipitation, and then analyzed. From a graph of $m_1/m_2 = f([M_1]/[M_2])$, $1/r_2$ or r_1 is given by the gradient of the straight line (Fig. 13.7).

The validity of the assumption $r_1 \cdot r_2 = 1$ should always be verified. To this end, the parameters should also be analyzed using the Mayo–Lewis and/or the Fineman–Ross equations (▶ Sect. 13.6).

The copolymerization diagrams of systems with $r_1 \cdot r_2 = 1$ can nevertheless assume extremely varied forms (Fig. 13.8).

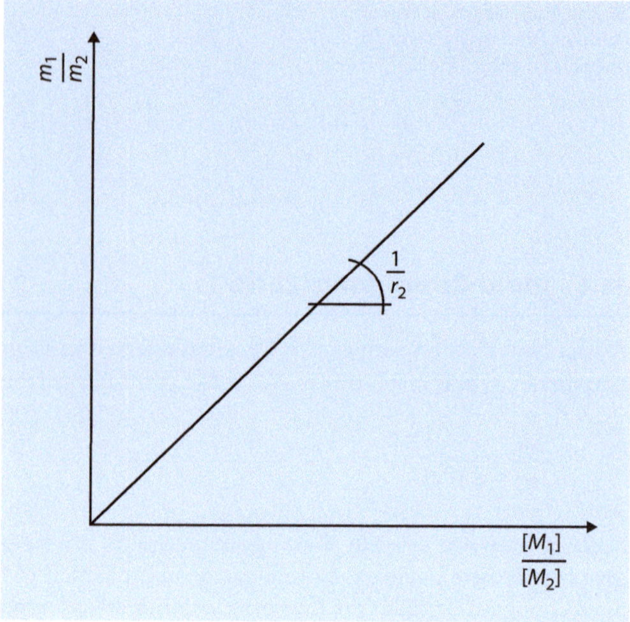

Fig. 13.7 Determination of the copolymerization parameter r_2 for an ideal copolymerization ($r_1 \cdot r_2 = 1$)

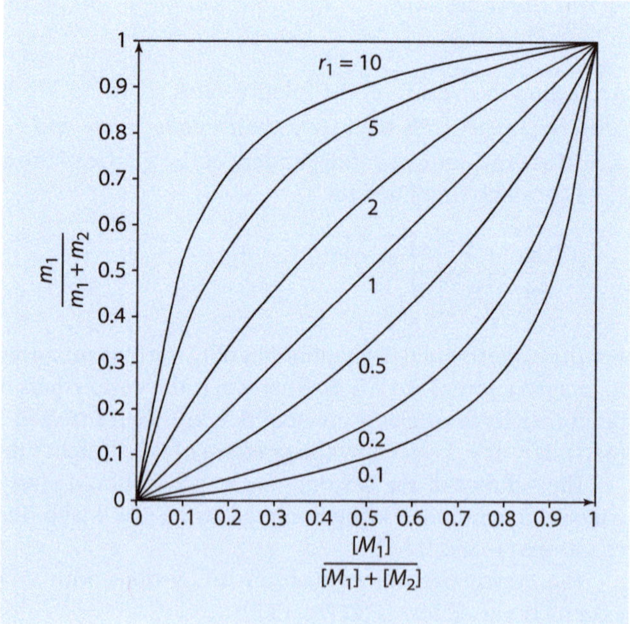

Fig. 13.8 Copolymerization diagrams for ideal copolymerizations with $r_1 = 0$, 1, ..., 10 ($r_1 \cdot r_2 = 1$)

Table 13.4 Copolymerization parameters for the radical copolymerization of acrylic acid (M_1) and MMA (M_2) in different solvents at 50 °C

Solvent	r_1	r_2
N,N-Dimethyl formamide	0.24	2.89
Tetrahydrofuran	0.25	2.46
Dioxan	0.30	2.32
Acetonitrile	0.33	2.17
Bulk (without solvent)	0.24	1.89

Polar solvents interfere to varying degrees with the formation of the hydrogen bonded dimers of acrylic acid.

13.5 Influencing the Inclusion Ratio of the Monomers

An influence of the solvent on the reactivity of the monomers can generally be neglected. Nevertheless, a systematic variation of the copolymerization parameters for the radical copolymerization of acrylic acid with **m**ethyl **m**eth**a**crylate (MMA) has been observed which depends on the propensity of the solvent to form hydrogen bonds (Table 13.4) (Glamann 1969).

Furthermore, the copolymerization behavior of monomers can be influenced by a change in the electron density of the C-C double bond. For example, if acrylic acid methyl ester is copolymerized instead of acrylic acid:

Styrene (M_1)/acrylic acid (M_2): $r_1 = 0.15$; $r_2 = 0.25$

Styrene (M_1)/acrylic acid methyl ester (M_2): $r_1 = 0.75$; $r_2 = 0.18$

The rate of propagation can also be drastically altered by complexation. Thus, the reactivity of acrylonitrile is changed considerably, e.g., by $ZnCl_2$ or $Al(CH_3)_3$ and is forced into an alternating copolymerization with ethylene. A change in mechanism from a radical to a cationic or anionic copolymerization has equally drastic consequences. For example, although the copolymerization of styrene and MMA results in statistical copolymers via a radical mechanism, homo-polystyrene is the sole product from a cationic initiation because MMA does not polymerize cationically. Via an anionic polymerization almost pure PMMA is produced initially before styrene is also incorporated in the polymer, i.e., in the ideal case poly(MMA-*block*-styrene) is produced.

13.6 Experimental Determination of the Copolymerization Parameters

As previously discussed in ▸ Sect. 13.1.2, the composition of the copolymers $m_1/(m_1 + m_2)$ is determined for various $[M_1]/([M_1] + [M_2])$ at a conversion of less than 5 % and, from the results, the copolymerization diagram drawn (Fig. 13.3).

Mathematically, according to Mayo and Lewis, (13.11) is transformed to give $r_2=f(r_1, [M_1], [M_2], m_1, m_2)$:

$$r_2 = \frac{[M_1]^2}{[M_2]^2} \cdot \frac{m_2}{m_1} \cdot r_1 + \frac{[M_1]}{[M_2]} \cdot \frac{m_2}{m_1} - \frac{[M_1]}{[M_2]} \tag{13.20}$$

With

$$a = \frac{[M_1]}{[M_2]} \tag{13.21}$$

and

$$b = \frac{m_2}{m_1} \tag{13.22}$$

it follows from (13.20) that

$$r_2 = r_1 \cdot a^2 \cdot b + a \cdot b - a \tag{13.23}$$

As $[M_1]$ and $[M_2]$ are given and m_1 and m_2 can be determined experimentally, an equation is obtained with two unknowns, r_1 and r_2. A second experiment with altered $[M_1]$ and $[M_2]$ and thus alternative values for m_1 and m_2 provides a second conditional equation. A graph of $r_2=f(r_1)$ results in straight lines for both experiments, the intersection of which yields the values for the required parameters r_1 and r_2. The reliability of the values determined increases with the number of experiments (Fig. 13.9).

13

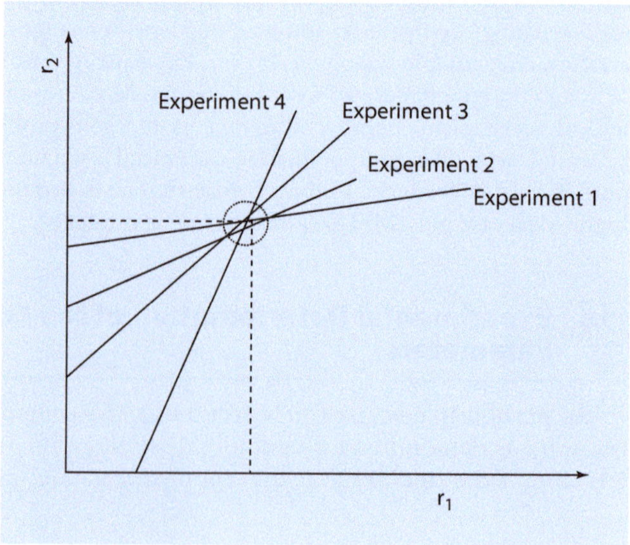

Fig. 13.9 Determining the copolymerization parameters using the method of Mayo and Lewis

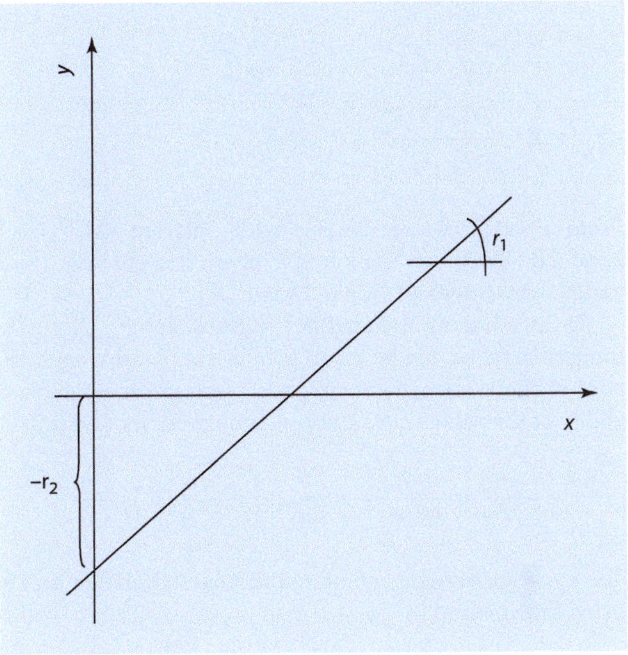

Fig. 13.10 Determining the copolymerization parameters using the method of Fineman and Ross

Depending on the accuracy with which m_1 and m_2 can be determined, there is often a larger or smaller deviation from the theoretically expected, common intersection. Alternatively, r_1 and r_2 can be determined from the gradient and the intercept of a graph of the so-called Fineman and Ross copolymerization equation (13.24); ▶ see Fig. 13.10.

$$y = r_1 x - r_2 \tag{13.24}$$

By extending the term on both sides of (13.11) with +1, the mole fraction variant of the Mayo–Lewis equation is obtained. Using definitions (13.25) to (13.30) and several transformations, the conditional equation for r_1 and r_2 in the form of equation (13.24) can be obtained.

$$f_1 = \frac{[M_1]}{[M_1]+[M_2]} \tag{13.25}$$

$$f_2 = \frac{[M_2]}{[M_1]+[M_2]} \tag{13.26}$$

$$F_1 = \frac{m_1}{m_1 + m_2} \tag{13.27}$$

$$F_2 = \frac{m_2}{m_1 + m_2} \tag{13.28}$$

and by combining f_1, f_2, F_1, and F_2 into x and y:

$$y = \frac{f_1(2F_1 - 1)}{f_2 \cdot F_1} \tag{13.29}$$

$$x = \frac{(1 - F_1) \cdot f_1^2}{F_1 \cdot f_2^2} \tag{13.30}$$

From a series of experiments with different $[M_1]/([M_1]+[M_2])$ a set of x, y pairs is obtained, which can be plotted to give a straight line. The gradient of the straight line is r_1 and r_2 the y-intercept (Fig. 13.10).

By introducing the auxiliary variable $H = F_{min}/F_{max}$ from (13.27) or (13.28), extreme monomer ratios can be given greater statistical weight (Kelen et al. 1980) and the influence of those measurements particularly prone to error can be reduced so that the reliability of the values of r_1 and r_2 is improved. By dividing (13.24) by $H+x$ one obtains

$$\frac{y}{H+x} = \frac{-r_2 + r_1 x}{H+x} = -\frac{r_2}{H} + \left(r_1 + \frac{r_2}{H}\right) \cdot \frac{x}{H+x} \tag{13.31}$$

The r_1-, r_2-parameters compiled in tables (Kelen et al. 1980) have often been optimized using this method.

13.7 The *Q,e* Schema of Alfrey and Price

The copolymerization parameters are specific to a particular monomer pair. As stated in the introduction to this chapter, n monomers produce $(n \cdot (n-1))/2$ combinations in binary systems (½, as $a \cdot b = b \cdot a$). For 100 technically interesting monomers, this leads to $(100 \cdot 99)/2 = 4950$ parameter combinations of which many are compiled in reference works such as the *Polymer Handbook* by J. Brandrup and E.H. Immergut (Brandrup and Immergut 1989).

13

Alfrey and Price developed a method to determine monomer rather than pair specific parameters using an empirical approach. In this method they describe the r-parameters in terms of four parameters for the reactivity and polarity of each individual monomer:

$$k_{12} = P_1 \cdot Q_2 \cdot e^{-e_1 e_2} \tag{13.32}$$

P_1	Reactivity of $\sim M_1^\bullet$
Q_2	Reactivity of M_2
e_1	Measurement for the polarity of $\sim M_1^\bullet$ and M_1
e_2	Measurement for the polarity of $\sim M_2^\bullet$ and M_2

By analogy:

$$k_{11} = P_1 \cdot Q_1 \cdot e^{-e_1 e_1} \tag{13.33}$$

Thus, for r_1 from Eq. (13.10):

$$r_1 = \frac{k_{11}}{k_{12}} = \frac{P_1 \cdot Q_1 \cdot e^{-e_1 e_1}}{P_1 \cdot Q_2 \cdot e^{-e_1 e_2}} = \frac{Q_1}{Q_2} \cdot e^{-e_1(e_1 - e_2)} \tag{13.34}$$

Again, by analogy, for r_2 from Eq. (13.11):

$$r_2 = \frac{k_{22}}{k_{21}} = \frac{P_2 \cdot Q_2 \cdot e^{-e_2 e_2}}{P_2 \cdot Q_1 \cdot e^{-e_2 e_1}} = \frac{Q_2}{Q_1} \cdot e^{-e_2 (e_2 - e_1)} \tag{13.35}$$

If r_1 and r_2 have been determined experimentally, only two conditional equations are available to provide values for the four unknowns Q_1, Q_2, e_1, and e_2.

To solve the problem, the reactivity Q and the polarity *e* of the monomer styrene are arbitrarily defined: $Q_{\text{styrene}} = 1.00$, $e_{\text{styrene}} = -0.80$.

With these numbers, the values for Q and *e* of other comonomers can be calculated. High values of *Q* represent high reactivity of the corresponding monomer. Strongly negative values of *e* indicate an electron-rich double bond.

The values of Q and *e* for selected monomers are compiled in Table 13.5.

In Fig. 13.11, the Q, *e* scheme is visualized as a "map".

This map should be interpreted as follows:

1. Copolymerization of monomers with extremely different values of *Q* is difficult (e.g., vinyl chloride/styrene)
2. Monomers with similar *Q* values (e.g., styrene/butadiene) can be copolymerized more easily
3. Systems with similar *Q* and extremely different *e* values tend to yield alternating copolymers

Table 13.5 *Q*- and e-values for a selection of monomers (Brandrup and Immergut 1989)

Monomer	*Q*	*e*
Styrene	1.00	−0.80
Acrylonitrile	0.48	≪23
1,3-Butadiene	1.70	−0.50
Isobutylene	0.023	−1.20
Ethylene	0.016	0.05
Isoprene	1.99	−0.55
Maleic acid anhydride	0.86	3,69
Methyl methacrylate	0.78	0.40
α-Methyl styrene	0.79	−0.81
Propylene	0.009	0.88
Vinyl acetate	0.026	0.88
Vinyl chloride	0.056	0.16

With these data, the copolymerization parameters for monomer pairs which haven't been experimentally determined can be predicted. However, although the *Q*, *e* scheme is a very useful, it is no more than an empirical concept so that a more thorough, theoretical interpretation is not worthwhile; these parameters should be regarded simply as semi-quantitative indicators for the corresponding monomer.

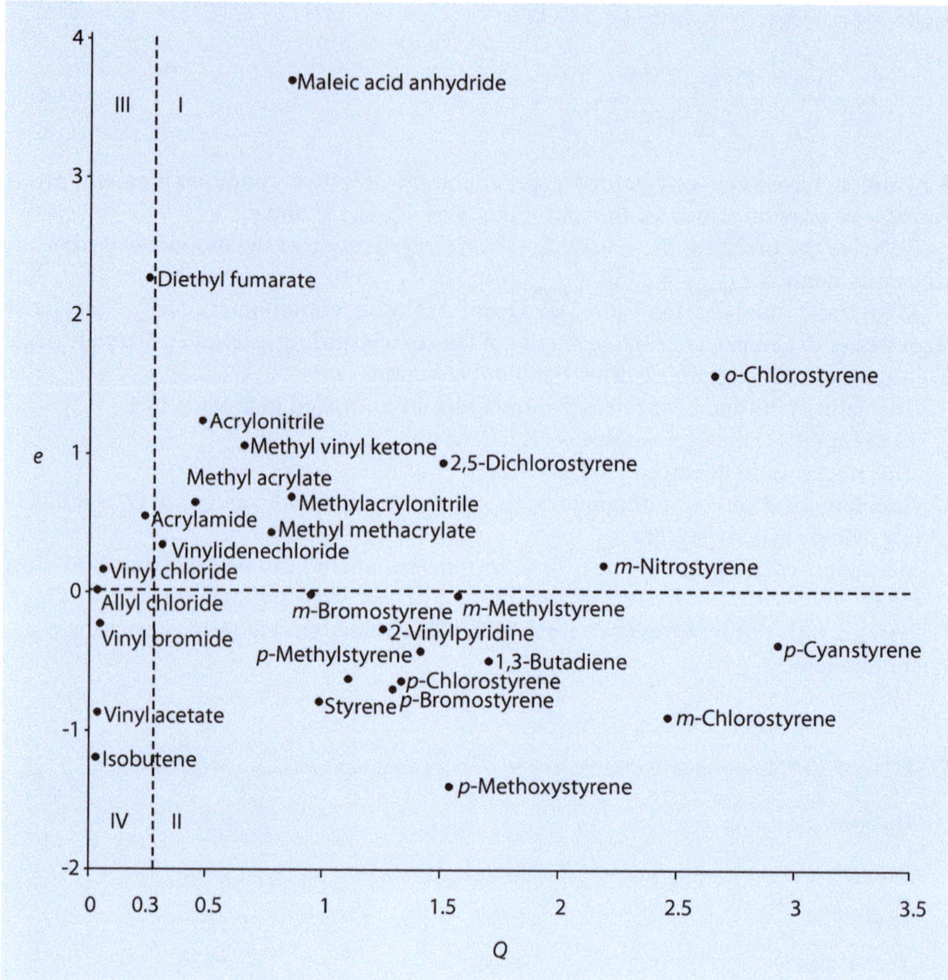

Fig. 13.11 *Q*, *e* scheme developed by Alfrey and Price

Copolymerization of monomers with different reactivity from quadrants III or IV with monomers from quadrants I or II is not favorable, whereas monomers from the same quadrant usually copolymerize very well. Copolymerization of monomers from quadrant I with monomers from quadrant II tends to yield alternating copolymers.

13.8 Copolymerization Rate

As for homopolymerization, chain initiation, growth, and termination are identified as the copolymerization steps. Here, too, the monomer is also consumed almost exclusively by the copolymer chain growth.

$$-\frac{d[M_1]+d[M_2]}{dt}=k_{11}\left[\sim M_1^\bullet\right][M_1]+k_{12}\left[\sim M_1^\bullet\right][M_2] \\ +k_{22}\left[\sim M_2^\bullet\right][M_2]+k_{21}\left[\sim M_2^\bullet\right][M_1] \quad (13.36)$$

With the termination reactions[5] given in Fig. 13.12 ($k_{ab12} \equiv k_{ab21}$) and assuming a quasi-stationary state for the radical concentration

$$k_{21}\left[\sim M_2^\bullet\right][M_1]=k_{12}\left[\sim M_1^\bullet\right][M_2] \quad (13.3)$$

If the rate of initiation is introduced using $v_i = v_{ab}$ (stationary state ⇒ the rate of initiation ~ rate of termination):

$$v_i = 2k_{ab11}\left[\sim M_1^\bullet\right]^2+2k_{ab12}\left[\sim M_1^\bullet\right]\left[\sim M_2^\bullet\right]+2k_{ab22}\left[\sim M_2^\bullet\right]^2 \quad (13.37)$$

The radical concentration in Eq. (13.16) can now be eliminated to give an equation for the overall rate of polymerization v_{br}[6]:

$$v_{br} = -\frac{d[M_1]+d[M_2]}{dt} \\ =\frac{\left(r_1[M_1]^2+2[M_1][M_2]+r_2[M_2]^2\right)\cdot\sqrt{v_i}}{\sqrt{r_1^2\delta_1^2[M_1]^2+2\cdot\Phi\cdot r_1 r_2\cdot\delta_1\delta_2[M_1][M_2]+r_2^2\delta_2^2[M_2]^2}} \quad (13.38)$$

In Eq. (13.8):

$$\delta_1=\sqrt{\frac{2k_{ab11}}{k_{11}^2}} \quad (13.39)$$

$$\delta_2=\sqrt{\frac{2k_{ab22}}{k_{22}^2}} \quad (13.40)$$

$$\Phi=\frac{k_{ab12}}{2\cdot\sqrt{k_{ab11}\cdot k_{ab22}}} \quad (13.41)$$

The δ-values are the reciprocals of the $k_w/\sqrt{k_{ab}}$ values known from the corresponding radical homopolymerizations (▸ Sect. 9.3.1).

The result of an evaluation using (13.38) is given in Fig. 13.13.

At $\Phi > 1$, cross-termination is more important than homo-termination ($v_{ab11}+v_{ab22} \ll v_{ab12}+v_{ab21}$) and with $\Phi = 6$ the experimental curve can be best described. It can be

5 As well as the terminations involving radical combination given in Fig. 13.12, termination via disproportionation or indeed other mechanisms is possible but, however, does not influence the kinetic analysis.

6 The symbol for the overall rate of the polymerization v_{br} has not been changed from the original text. The subscript "br" is derived from the German Brutto.

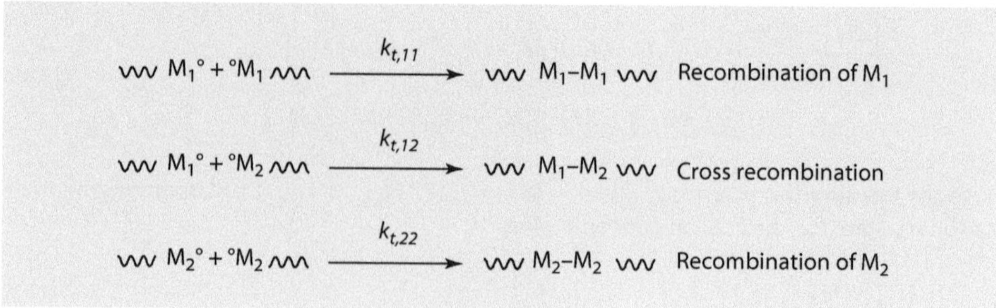

Fig. 13.12 Termination reactions during a copolymerization

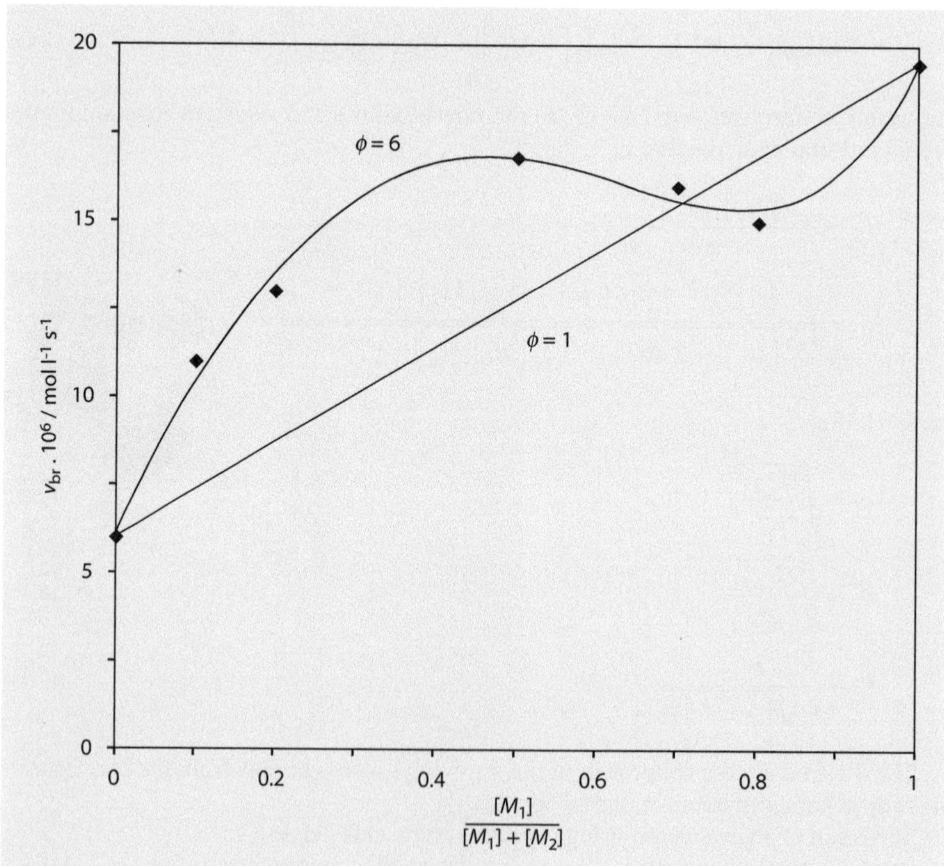

Fig. 13.13 Radical copolymerization of styrene (M_1) and methacrylonitrile (M_2) at 60 °C (Ito 1978)

concluded from the experimental data that an average rate for a copolymerization cannot, or at least can very rarely, be derived from the average rate of polymerization of the two homopolymerizations (Table 13.6).

Table 13.6 Φ-Values for a selection of monomer pairs

Monomer pair	Φ
Styrene–butyl acrylate	150
Styrene–methyl acrylate	50
Styrene–methyl methacrylate	13
Styrene–methacrylonitrile	6
Styrene–*p*-methoxy styrene	1

It is surprising that the fairly complex experimental curve in Fig. 13.13 can be described rather well by only five parameters r_1, r_2, δ_1, δ_2 and Φ

13.9 Block and Graft Copolymers

The main focus of this section is on methods of radical polymerization which lead to block and graft copolymers. Both polymer classes are not accessible via simultaneous radical polymerization of two or more monomers. Special methods are needed for such syntheses. Alternative methods for synthesizing block or graft polymers are described in ► Chaps. 9 and 10.

13.9.1 Block Copolymers

Block copolymers are distinguished from statistical copolymers by the monomer distribution. In the simplest block polymers, a longer sequence of repeating units ($-M_1M_1M_1-$) is joined with that of a sequence of different repeating units ($-M_2M_2M_2-$). Such products are also referred to as AB block copolymers (Fig. 13.1). If the sequences are repeated multiple times, they are called $(AB)_n$-block copolymers. An ABA block copolymer is shown in Fig. 10.58. In contrast to statistical copolymers, whose attributes, to a reasonable approximation, are averages of the corresponding homopolymer properties, block copolymers often have multi-phase morphologies and thus have the attributes of all elements from which they are formed. This makes them very interesting materials.

13.9.1.1 Block Copolymers by Controlled Radical Polymerization (CRP)

Methods such as ATRP, NMP, RAFT, and polymerization in the presence of 1,1 diphenyl ethylene have given strong impulses to the synthesis of block copolymers with radicals as chain transfer agents. These methods have already been discussed in ► Sect. 9.5 and therefore are not discussed here.

13.9.1.2 End Group Functionalized Polymers (Telechelics)

Oligomers or polymers with well-defined end groups (functional groups, subsequently denoted FG) can be joined together via these groups to build block copolymers (Fig. 13.14). This is why methods of radical synthesis which result in such components are presented here. The joining of the components is obviously a step-growth reaction (► Chap. 8).

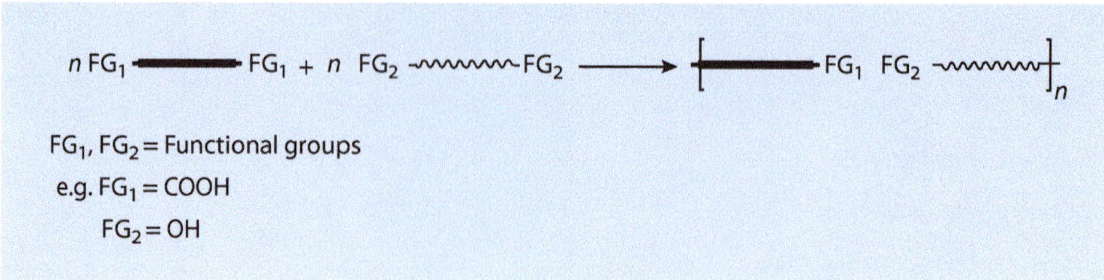

Fig. 13.14 Coupling oligomers or polymers with functional end groups to form block copolymers

Fig. 13.15 Synthesis of telechelics by *dead-end*-polymerization

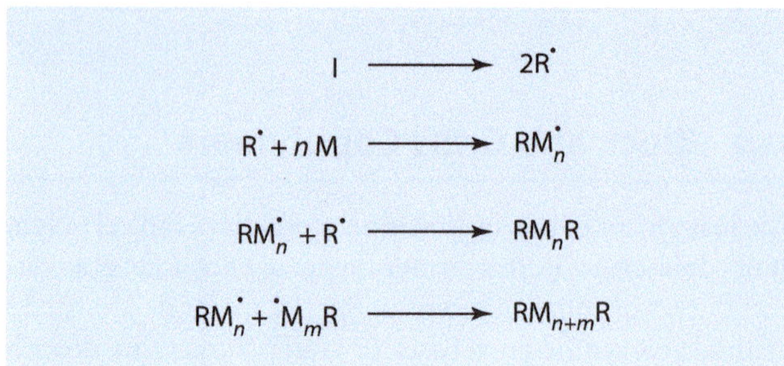

Fig. 13.16 Functionalized AIBN derivatives

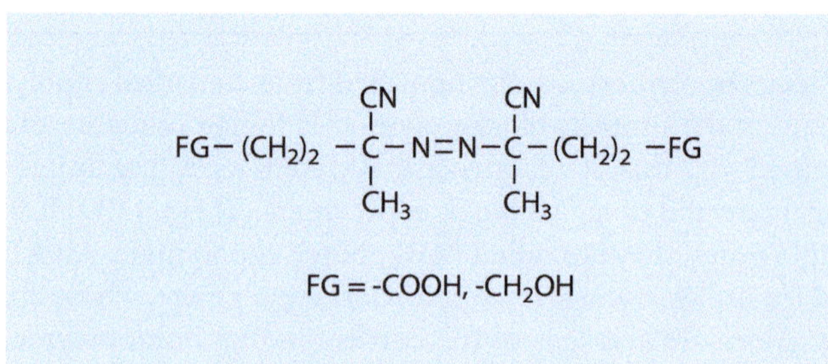

Functionalized Initiators

Radical polymerization in the presence of high initiator concentrations leads to the so-called *dead-end* polymerization and products which carry the functional group of the initiator fragment at their head and tail ends (Fig. 13.15).

The primary radicals $R^{\bullet}$ add to the monomer, but they can also terminate a growing chain. If no transfer occurs and, additionally, termination only involves primary radicals or two growing chains, polymer chains form with an *R*-group at each chain end. If a functional group is part of the *R* moiety, products are formed that can be coupled together to form block copolymers, as shown in Fig. 13.14.

Figure 13.16 shows typical functionalized initiators.

AIBN itself can also be seen as a functionalized initiator. However, the CN group is rather a potential functional group and needs to be transformed into more reactive groups after polymerization (Fig. 13.17).

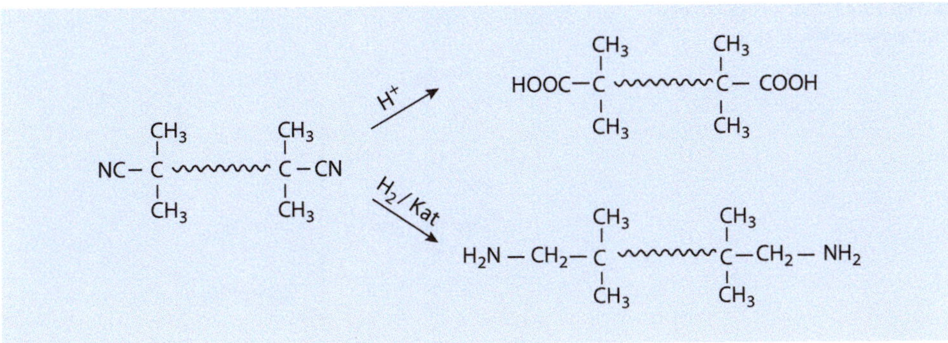

Fig. 13.17 Synthesis of acid- and amino-group terminated polymers starting from a nitrile group

Fig. 13.18 Introducing functional end groups employing a transfer reagent. *R* includes a functional group, e.g., COOH, CH_2OH, NH_2

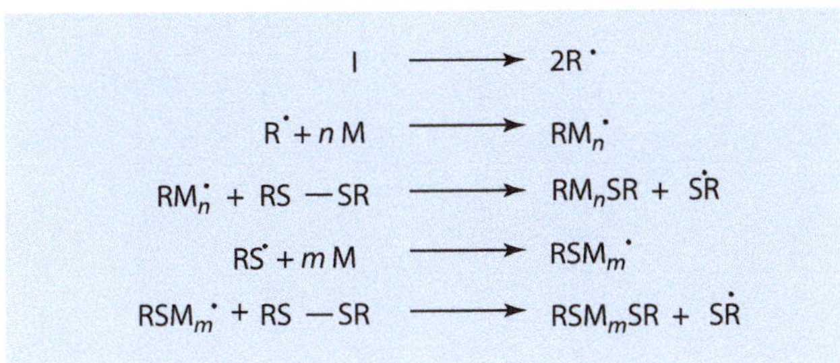

Functionalized Transfer Reagents

In Fig. 9.22 the reaction scheme leading to end group functionalized polymers from the reaction of the growing chain ends with a transfer reagent (CCl_4) is shown. An alternative route to such polymers is to carry out the polymerization in the presence of a disulfide derivative (Fig. 13.18).

The use of disulfides as chain transfer agents leads to bifunctional oligomers. These can then be converted into block copolymers of the type $(AB)_n$. Alternatively, if we use mercaptans for the chain transfer, monofunctional polymer segments result which can be transformed into AB-block copolymers.

13.9.1.3 Polymer Initiators

The concept of the synthesis of block copolymers from polymeric initiators (*macro initiators*) is shown in Fig. 13.19.

Depending on the type of chain termination, block copolymers of type AB or ABA are the result. Multi block copolymers are formed if initiators are used that contain multiple azo functionality as part of the polymer backbone. However, in most cases a mixture of AB, ABA, and $(AB)_n$ block copolymers is produced.

Figure 13.20 gives a typical synthesis of a polymeric initiator with N=N functionality as part of the polymer backbone.

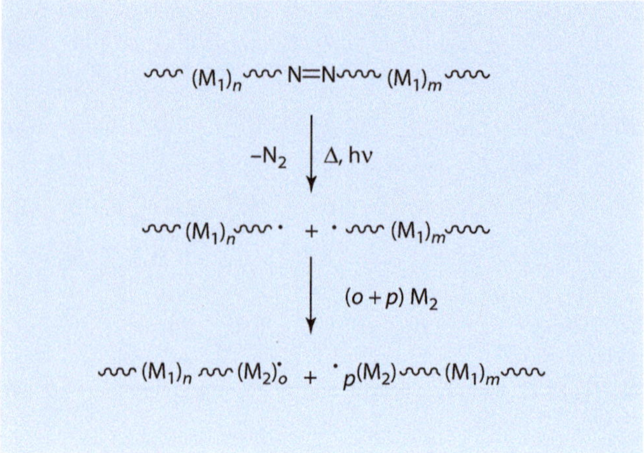

Fig. 13.19 Block copolymers from polymeric initiators

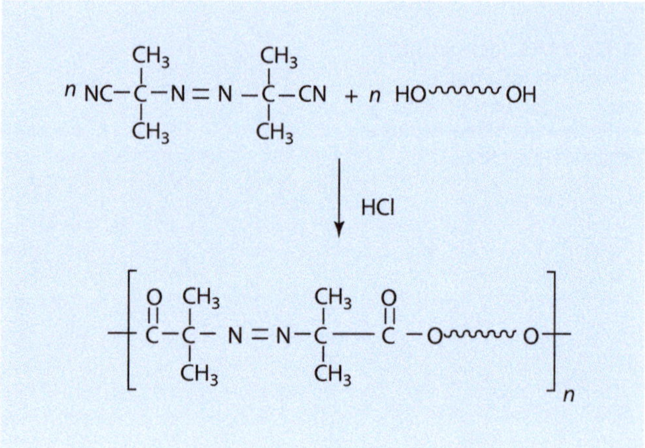

Fig. 13.20 Synthesis of a polymeric initiator

The nitrile groups of the AIBN are converted to acid groups via an acid catalyzed hydrolysis which in turn are esterified with a diol to give the polymeric initiator.

13.9.1.4 Change of Mechanism

A variant of the synthesis of block copolymers which, until now, has remained solely of academic interest involves changing the polymerization mechanism.

The reaction of $HgBr_2$ with a carbanion leads initially to an—HgBr-terminated polymer (Fig. 13.21).

If r_2Hg and RHgX (R = polystyrene) are heated in the presence of another monomer, for example in a solution of MMA in benzene at 80 °C, poly(styrene-*block*-methyl methacrylate) can be made.

Carbocations can also be transformed into radicals (Fig. 13.22).

13.9.2 Graft Polymers

Graft copolymers are branched polymers in which the main and side chains are chemically different (Fig. 13.1). As with block copolymers, they generally have a

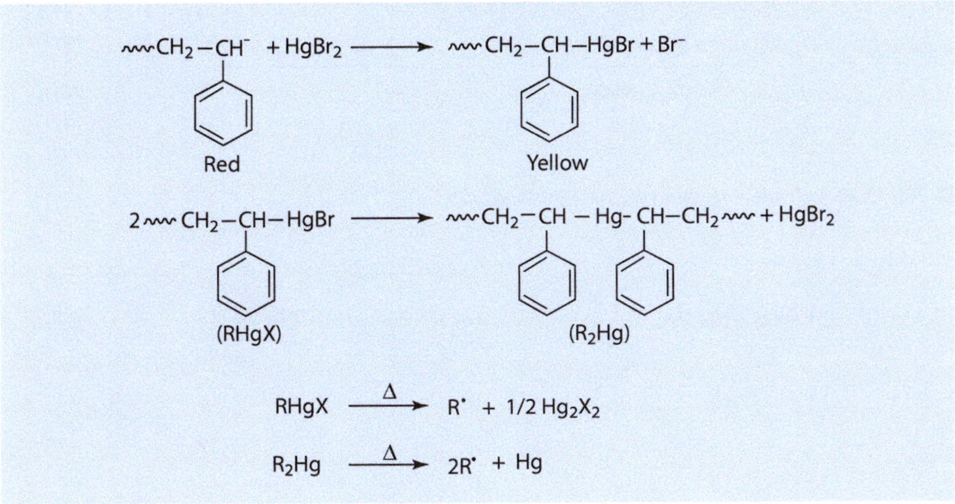

Fig. 13.21 Conversion of an anion into a radical

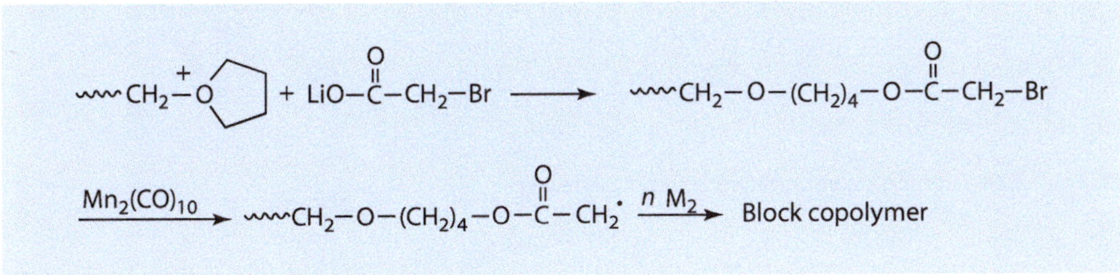

Fig. 13.22 Conversion of a cation into a radical

multi-phase morphology and thus exhibit the typical attributes of the homopolymers of the components from which they are made rather than average property values. This makes them interesting raw materials. They have considerable potential as compatibilizers to improve the mixing of incompatible homopolymers. Even small amounts of poly(M_1-*graft*-M_2) reduce the interfacial tension between homopoly(M_1) and homopoly(M_2).

13.9.2.1 Grafting Onto

Reactions which follow the scheme shown in Fig. 13.23 are referred to as *grafting onto*.

An actual example, is the radical grafting of polybutadiene with polystyrene (Fig. 13.24).

In this example polystyryl radicals are *grafted onto* the polybutadiene chain. Thus, polybutadiene is dissolved in a toluene/styrene mixture or in pure styrene and a radical polymerization of the styrene is initiated, e.g., with a peroxide or AIBN. The growing polystyrene chains can abstract H-radicals, from the polybutadiene chain in a transfer step to form allylic polybutadiene radicals. If other polystyrene radicals

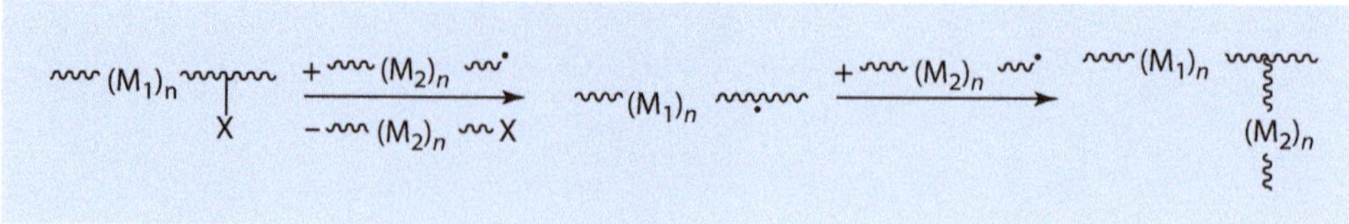

Fig. 13.23 Individual reaction steps for *grafting onto*

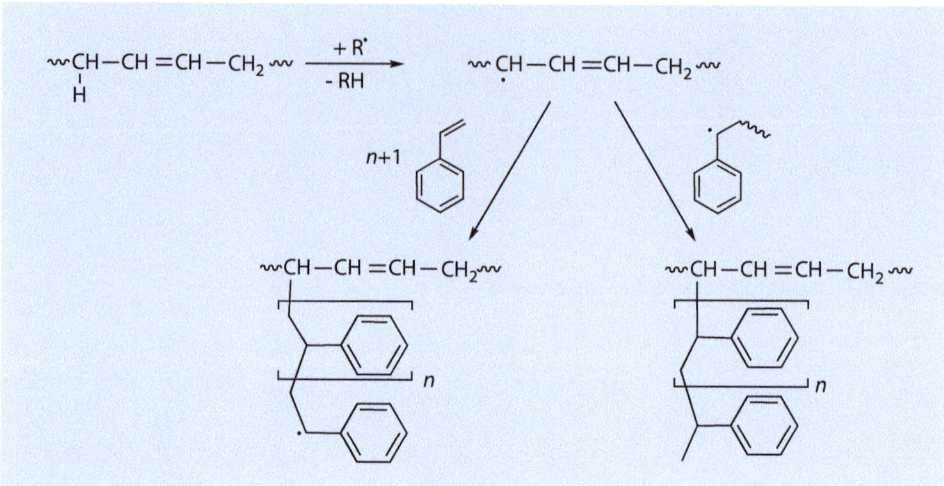

Fig. 13.24 Grafting polybutadiene with polystyrene

13

combine with these allyl radicals *grafting onto* occurs and the desired poly (butadiene-*graft*-styrene) is formed (Fig. 13.24, right). Alternatively, the allyl radicals can initiate a new polystyrene side chain (Fig. 13.24, left). This would be a *grafting from* reaction.

13.9.2.2 Grafting From

More elegant and efficient than *grafting onto* is the *grafting from* strategy. With this strategy polymers with initiator functions as side chains are used for the polymerization (grafting) (Figs. 13.25, 13.26 and 13.27). Such polymers are referred to as *polymeric initiators*, just as are those with initiator groups as part of the polymer backbone (▶ Sect. 13.9.1.3).

The polymeric initiators can be obtained from simple copolymerization (Fig. 13.26).

Furthermore, such polymeric initiators are also accessible via step-polymerization (Fig. 13.27).

As opposed to the polymeric initiators introduced in ▶ Sect. 13.9.1.3, which carry the initiating group as part of the polymer backbone and thus, following initiation yield block copolymers, the initiating groups of these polymeric initiators are side groups along the chain so that initiation leads to the growth of side chains from monomer M_2, graft copolymers.

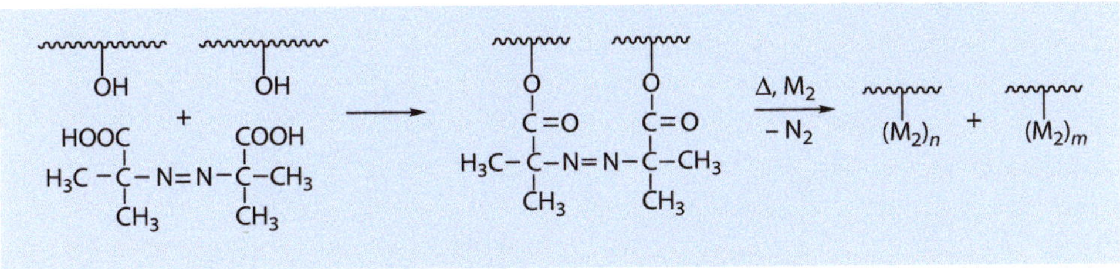

Fig. 13.25 Polymeric initiator for *grafting from* via polymer analog reaction

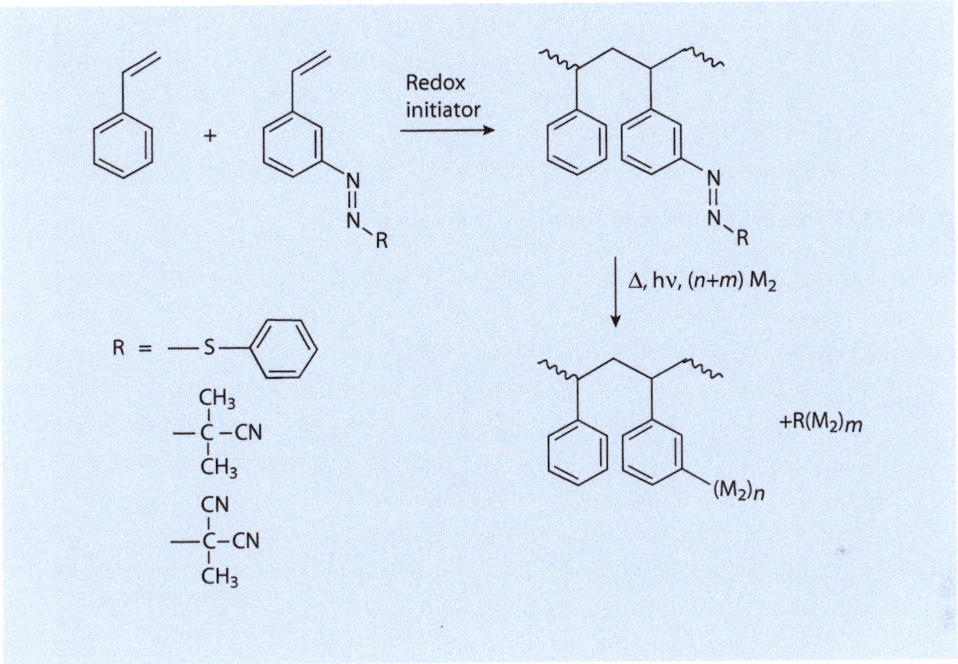

Fig. 13.26 Copolymerization to yield a polymeric initiator for *grafting from*

Using such polymeric initiators to synthesize graft copolymers results in several products: ungrafted backbone (A) (originally polymeric initiator), ungrafted side chain polymers (C) (initiated by R˙), and the target graft copolymer (B) (Fig. 13.28).

If, for example, methacrylonitrile is grafted from polystyrene, the products A, B, and C can be separated by selective extraction (Fig. 13.29).

During grafting, A and C are undesirable by-products and are avoided as much as possible. A can be minimized if the initiator has multiple initiator molecules per chain, since because every additional initiator function increases the probability of grafting. As soon as A has at least one side chain, it qualifies as grafted and thus becomes B. C can be minimized if *R*˙ is less reactive than the polymer radical so that, in the best case, it is not capable of initiating the polymerization.

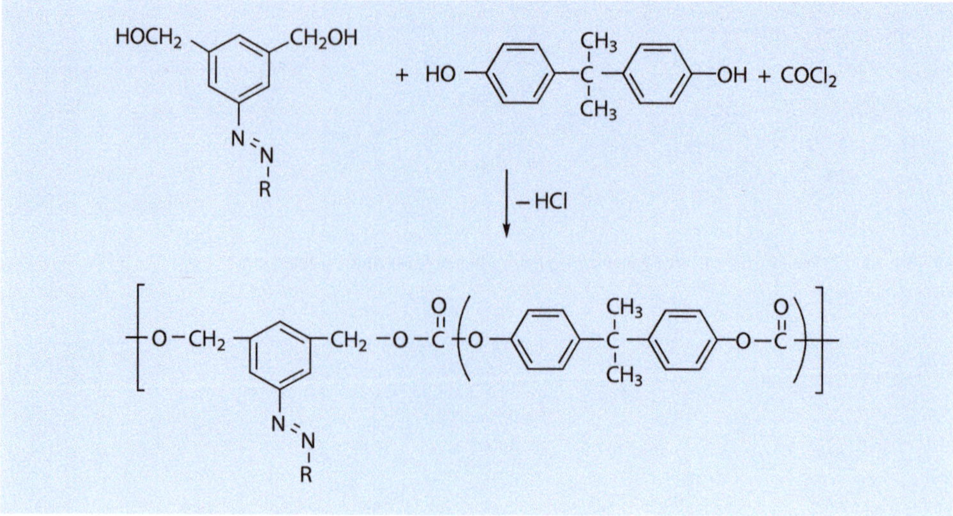

Fig. 13.27 Step-polymerization to yield a polymeric initiator

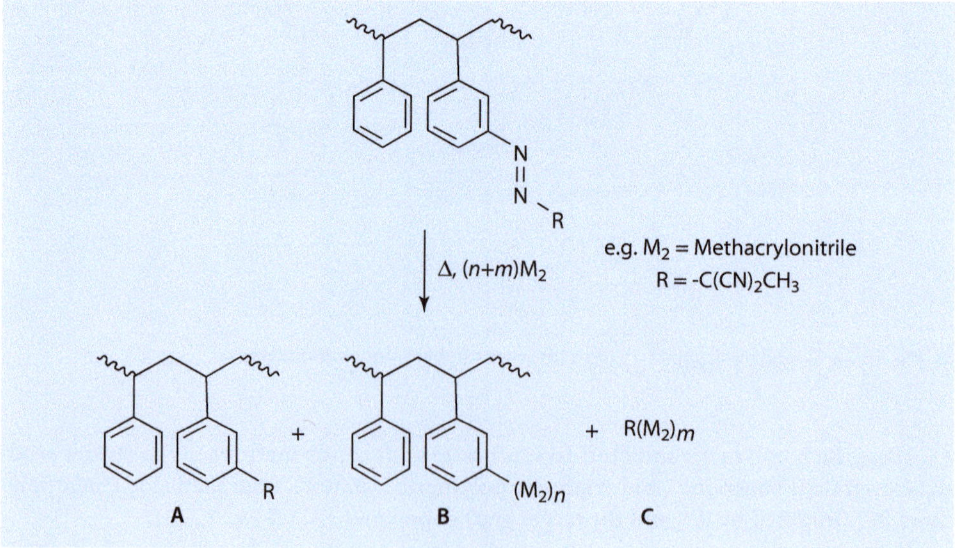

Fig. 13.28 The products from a graft polymerization

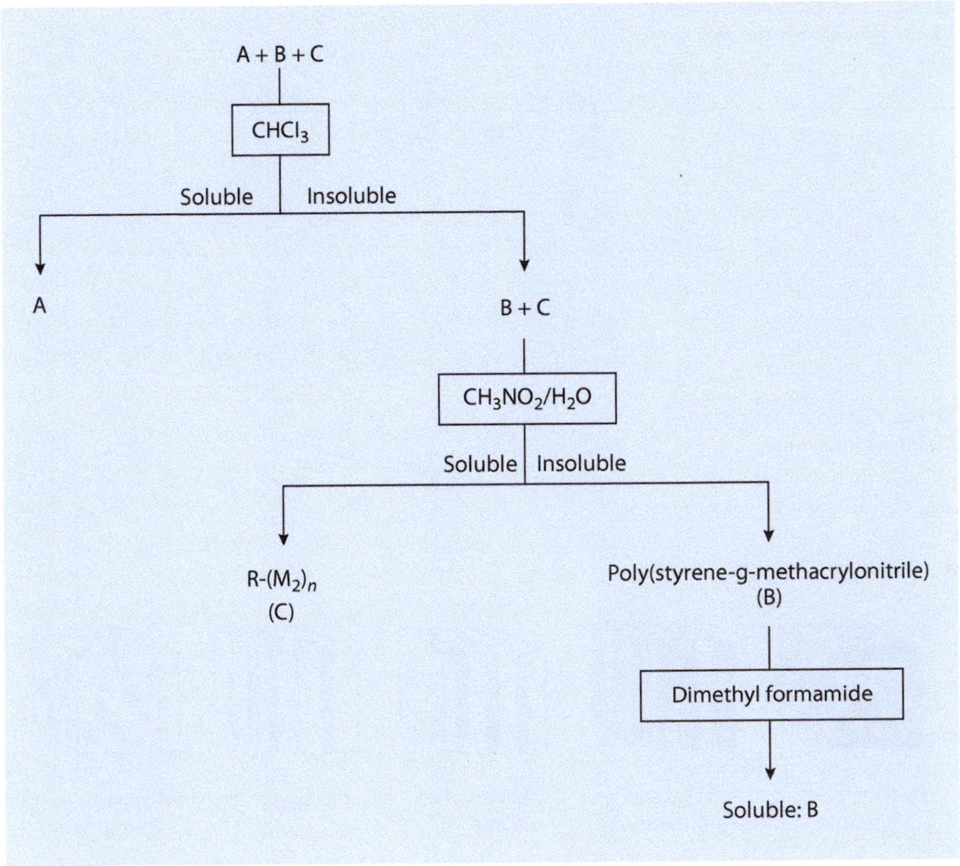

Fig. 13.29 Separating the products from grafting polystyrene with methacrylonitrile

13.9.2.3 Macromonomer Method

Macromonomers (also called macromers) are polymers (oligomers) with a polymerizable terminal functionality and thus are also suitable for copolymerization (Fig. 13.30). This is undoubtedly the most elegant synthesis route to graft copolymers.

The great advantage of this method is that unwanted homopolymers are almost completely avoided. There are a great many ways to synthesize macromonomers. The basic polymers can be prepared radically, anionically or cationically. The terminal functionality can be attached to the base polymer via the initiator, transfer reagent or the termination reagent. Alternatively, an end group can be reacted to yield the desired functionality. An example of the latter is the conversion of the terminal –OH of polyethylene oxide (PEO) into an acrylate macromonomer via esterification given in Fig. 13.31.

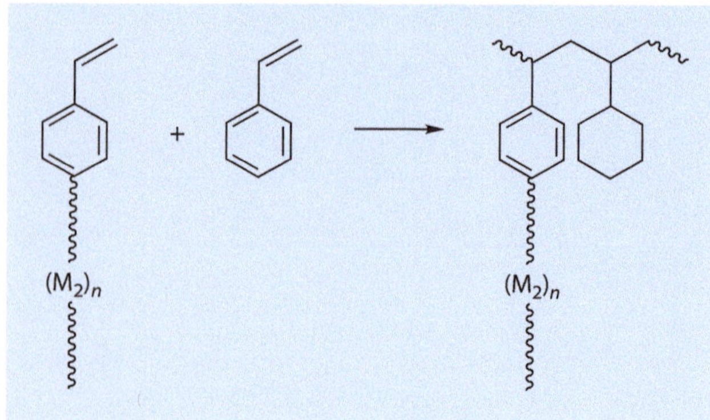

Fig. 13.30 Graft copolymers via the macromonomer method

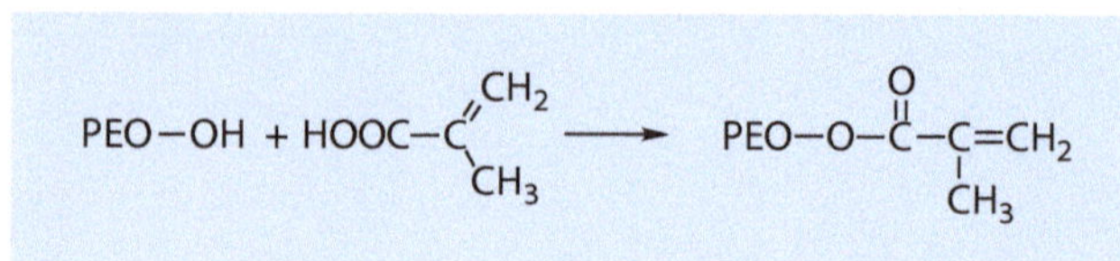

Fig. 13.31 Synthesis of a PEO-macromonomer

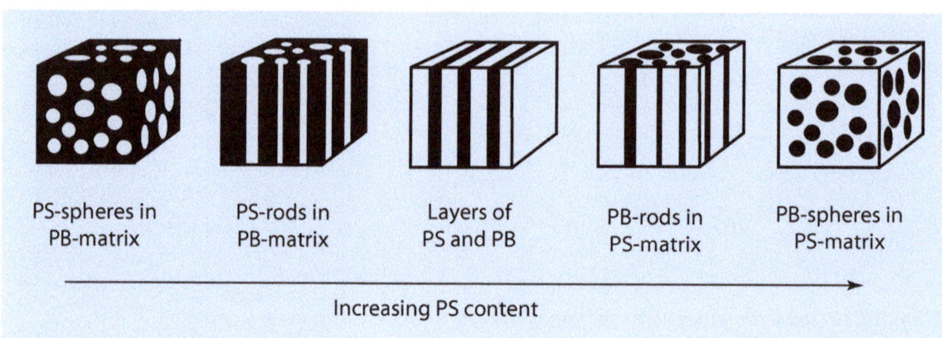

Fig. 13.32 Five basic morphologies for block- and graft-copolymers. *PS* polystyrene, *PB* polybutadiene

13.10 Technically Important Copolymers[7]

As mentioned above, statistical copolymers generally have properties reflecting the average values of the homopolymers, whereas block and graft copolymers usually have the attributes of both components. Poly(styrene-*block*-butadiene-*block*-styrene) is a typical and commercially very important example of these special polymers. The cause of this dualism of attributes is a microphase separation of the two distinct, incompatible polymers whereby the two phases are covalently bound together. The covalent bonding also inhibits a macroscopic phase separation. Taking the example of poly(styrene-block-butadiene-block-styrene) phases of polybutadiene coexist with phases of pure polystyrene. Depending on the ratio of the two polymers the morphologies shown in (Fig. 13.32) can be observed. These morphologies are described in more detail in ▶ Chap. 7.

7 The technically most important copolymers are described in ▶ Chap. 14 in subsections of the relevant homopolymers.

13.11 Structural Elucidation of Statistical Copolymers, Block and Graft Copolymers

To characterize statistical copolymers, the same methods can be used as those applied to homopolymers. As well as elementary analysis, ^{1}H-NMR-, ^{13}C-NMR-, UV-, and IR-spectroscopy, GPC (▶ Sect. 3.3.2) and DSC are used. One should not neglect simple solubility tests, by which, assuming a knowledge of the homopolymer solubilities, simple mixtures of the homopolymers can be ruled out. If the method used to synthesize the polymer is known, using the aforementioned analytical methods should allow a reliable determination of the total composition and indeed the monomer sequences. If the glass transition temperatures of the homopolymers are sufficiently different, the difference between a simple mixture of two homopolymers and a statistical copolymer is clear from a DSC; a mixture of two homopolymers has two glass transition temperatures, whereas a statistical copolymer exhibits only a single value.

To distinguish reliably between a homopolymer mixture and a block or graft copolymer, a selective extraction is essential. Mixtures are often easily separated with this method. Of course, the composition of the separate phases should always be analyzed. In this case, a DSC measurement is less appropriate, as a mixture of two homopolymers as well as block or graft copolymers all exhibit two glass transition temperatures.

The following methods of analysis have also proved suitable:

1. Solubility (see above)
2. Film transparency: block copolymers produce transparent films despite phase separation if the diameter of the microphases is less than $\lambda/20$ (λ = wave length of daylight)
3. Compatibility of polymer solutions (a solution of 10 % block copolymer usually remains transparent, whereas a mixture becomes turbid)
4. GPC: a block copolymer is generally monomodal whereas a mixture of two polymers with different molar masses is bimodal
5. Block copolymers tend to have high melt viscosities
6. Information about the morphology of the polymers can be obtained from REM, TEM, and staining, e.g., with OsO_4

The average sequence length and the order of the segments remain difficult questions as far as the complete structural description is concerned and require a knowledge of the synthetic method.

^{1}H- and ^{13}C-NMR spectroscopic measurements on statistical copolymers on the one hand and block and graft copolymers on the other are easily distinguishable because of the different sequences of the building blocks. Electron microscopy and small-angle scattering provide visual evidence for the presence of phase separation for block- and graft-copolymers whereas for statistical copolymers no special signals are observed. An increased melt viscosity can suggest that the material is a block- or graft-copolymer. If there is evidence of crystallization, then the sample is more likely to be a block- or graft-copolymer than a simple statistical copolymer.

Several methods for characterizing block- and graft-copolymers are summarized in ◻ Table 13.7.

Table 13.7 Methods for characterizing block- and graft-copolymers

Block- and graft-copolymers vs. homopolymer mixtures	
1.	Solubility, fractional extraction
2.	Film appearance
3.	Compatibility of polymer solutions
4.	Molar mass distribution, GPC
5.	Rheology
Block- and graft-copolymers vs. statistical copolymers	
1.	^{1}H- and ^{13}C-NMR spectroscopy
2.	IR spectroscopy
3.	DSC, glass transition temperature(s)
4.	Electron microscopy
5.	Small-angle scattering (SAS)
6.	Rheology
7.	Crystallization

As part of the structure analysis, all methods suitable for measuring mass are worthwhile. Additionally, selective polymer degradation can provide valuable information.

References

Brandrup J, Immergut EH (1989) Polymer handbook, 3rd edn. Wiley, New York, II:267

Glamann H (1969) Untersuchungen zum Einfluss des Reaktionsmediums auf die Copolymerisation von Acrylsäure und Methylmethacrylat. Dissertation, TU Berlin

Ito K (1978) An approach to the termination rate in radical copolymerization. J Polym Sci Polym Chem Ed 16:2725–2728

Kelen T, Tüdös F, Turcsanyi B (1980) Confidence intervals for copolymerization reactivity ratios determined by the Kelen-Tüdös-method. Polym Bull 2:71–76

13

Important Polymers Produced by Chain-Growth Polymerization

S. Koltzenburg et al., *Polymer Chemistry*, DOI 10.1007/978-3-662-49279-6_14

The technically most important polymers and copolymers produced by chain-growth polymerization are discussed in this chapter; those produced via step-growth polymerization are discussed in ▶ Sect. 8.5.

14.1 Polyethylene

Homopolymers of ethylene as well as ethylene copolymers with small amounts of other olefins of the type CH_2=CHR (R=CH_3, C_4H_9, C_6H_{13},...) are referred to generally in industrial circles as *polyethylenes* (PEs).

14.1.1 Properties

PE is a partially crystalline, nonpolar thermoplastic produced via radical or catalytic chain-growth polymerization. ◘ Table 14.1 provides an overview of the different PE types.

14.1.2 Manufacturing Processes

High-pressure processes are usually used for the production of low-density PE (LDPE), whereas high-density PE (HDPE) is produced using a low-pressure process (with either Ziegler or Phillips catalysts).

14.1.2.1 High-Pressure Processes

In the high-pressure process used to produce PE, high-purity ethene is heated to 100–200 °C in the presence of traces of oxygen (10–80 ppm) or peroxides at pressures of 1500–3800 bar. Under these conditions, ethene is in a supercritical state, so stirred autoclaves or

◘ **Table 14.1** Overview of the various polyethylene types

Name	Density	Structure
LDPE	Low	Highly branched
HDPE	High	Branched
MDPE	Medium	Linear
LLDPE	Low	Linear
VLDPE	Very low	Linear

HD high density, *LD* low density, *LLD* linear low density, *MD* medium density, *PE* polyethylene, *VLD* very low density
All PE types are outstanding thermal and electric insulators and chemically very stable. An important application for LDPE and LLDPE is plastic film, whereas HDPE is used to make containers for the transport of liquids, crates, pipes and plastic film. Lupolen (Basell), Dowlex (Dow), Paxon (Exxon) or Novex (BP) are some typical PE trade names

tube reactors are used. Polymerization proceeds by a radical mechanism. As a result of intra- and inter-molecular chain transfer reactions the molecules have numerous long- and short-chain branches which considerably decrease their ability to crystallize and thus reduce their density in comparison with linear PE grades. This can be of practical advantage in terms of weight and cost savings.

14.1.2.2 Low-Pressure Process

Low-pressure processes involve transition metal–catalyzed processes (▶ Chap. 11). These polymerizations can be carried out in solution, in suspension (slurry process), or in the gaseous phase. The reactions are conducted in stirred tank, fluid bed, or loop reactors.

Temperatures above 130 °C are used for the *solvent process*. At these temperatures PE is soluble in the various solvents such as, *n*-hexane, cyclohexane, and toluene. Depending on the chosen solvent, the pressure is as high as 70 bar.

The *suspension process* generally involves an aliphatic hydrocarbon or gasoline fractions as a non-solvent and is carried out at temperatures of 60–105 °C and pressures of between 8–40 bar. Ethene is soluble in such media, whereas the PE precipitates and is thus easily separated.

In the *gas phase process* ethene flows across a fluidized bed of catalyst at 85–100 °C and at a pressure of about 20 bar. Polymerization takes place on the catalyst particles.

The three processes mentioned above all have in common that very small amounts of catalyst (<1 ppm) are used so that these do not need to be removed after the reaction has taken place (so-called *leave-in* catalysts). As transfer reactions only take place to a lesser extent during ethene polymerization, the polymers formed are primarily linear and their degree of crystallinity, density, and melting point are higher than those of LDPE.

LLDPE (*l*inear LDPE) is a copolymer formed from ethene and an α-olefin such as 1-butene, 1-hexene, or 1-octene. These comonomers lead to polymers with short alkyl side chains; i.e., short chain branched polymers with lower degrees of crystallization and lower densities than pure PE are formed.

The molar mass of the products is controlled by the addition of hydrogen, which acts as a regulator in both low-pressure and high-pressure processes (◘ Fig. 14.1). The molar mass is generally between $2–5 \cdot 10^4$ g/mol for LDPE and $10^5–10^6$ g/mol for HDPE. PE grades with molar mass of up to $6.5 \cdot 10^6$ g/mol (UHMWPE; *u*ltra *h*igh *m*olecular *w*eight) have greater tensile strengths, better resistance to wear, and higher heat distortion temperatures than lower molar mass grades. However, they cannot be processed using conventional methods for thermoplastics; they are processed using so-called *RAM-Extrusion*

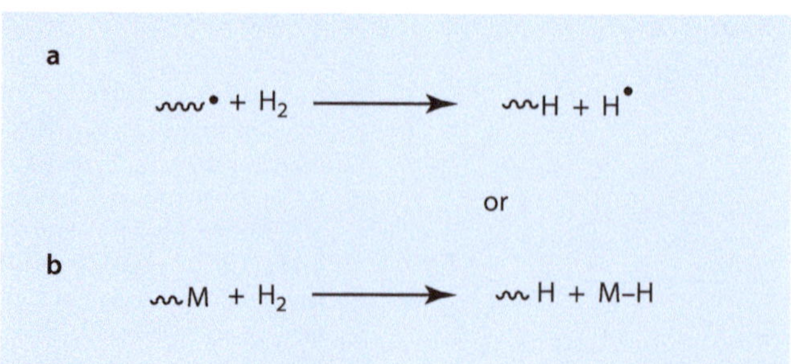

◘ **Fig. 14.1** Polymerization of ethene: chain transfer with hydrogen in (**a**) high pressure and (**b**) lower pressure processes. *M* metal of the catalyst (ram extrusion for the production of rods, tubes and bars) or by press-sintering. More recently grades which can be injection moulded have become available (▶ Chap. 17).

14.1.3 Modified Ethene Polymers

Copolymers of ethene (▶ Sects. 14.1.3.1 and 14.1.3.2) as well as branched polyethylenes (▶ Sect. 14.1.3.3) are both referred to as modified ethene polymers.

14.1.3.1 Ethene Copolymers with Nonpolar Comonomers

The copolymerization of ethene with 5–10 % α-olefin results in linear polymers with short chain branches (LLDPE types) that have lower densities and lower degrees of crystallization than HDPE.

If, instead of classical heterogeneous transition metal catalysts, stereospecific catalysts (metallocenes; ▶ Chap. 11) are employed for the copolymerization of, for example, ethene and 1-hexene, copolymers with special optical properties (crystal clear, low optical anisotropy) and exceptional ductility can be obtained. Copolymers of ethene and cycloolefins (e.g., norbornenes) are particularly suitable for optical data carriers, optical fibers and sterilizable medical apparatus. Such products have become well-known under the trade name Topas (Ticona/Mitsui).

14.1.3.2 Polar Ethene Copolymers

Copolymers made from ethene and vinyl acetate (Elvax (DuPont), Levapren (Bayer)) can be produced with all ratios of the monomers but using different technologies. Rigidity and stiffness decrease in comparison to PE with increasing vinyl acetate content, whereas tensile strength and shock resistance as well as flexibility and stretchability increase (◘ Table 14.2).

The partial hydrolysis of ethene–vinyl acetate copolymers (EVAc or EVM) with high vinyl acetate contents result in ethene–vinyl acetate–vinyl alcohol-terpolymers (EVAL) (◘ Fig. 14.2).

◘ **Table 14.2** Typical applications of ethene-vinyl acetate copolymers depending on their vinyl acetate content

Vinyl acetate content	Applications
1–10 %	Garden hose, film, glasshouse sheeting, pipes, profiles and sound dampening panels (mica filled)
10–30 %	Flame-resistant cable sheathing
30–40 %	Coatings and adhesives
40–50 %	Hot melt adhesive, additive for high impact PVC
40–90 %	Elastomers for, e.g., cable sheathing, conveyor belting

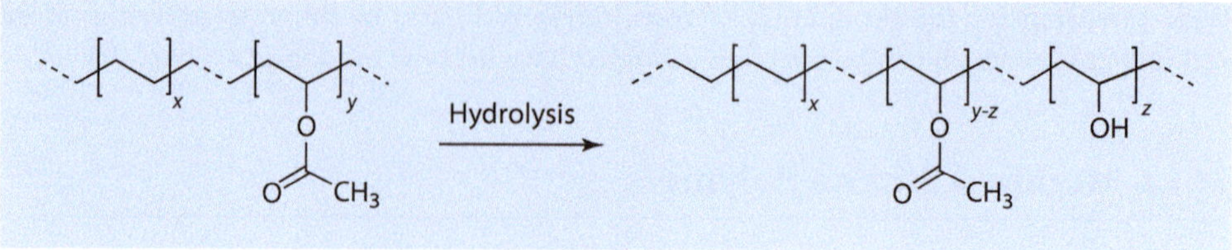

Fig. 14.2 Ethene-vinyl acetate-vinyl alcohol-terpolymer (EVAL) from the partial hydrolysis of ethene-vinyl acetate-copolymers (EVAc)

Fig. 14.3 Thermoreversible cross-link between two EAA-polymer molecules. M^{2+} metal ion with a double positive charge (► Sect. 15.2.5)

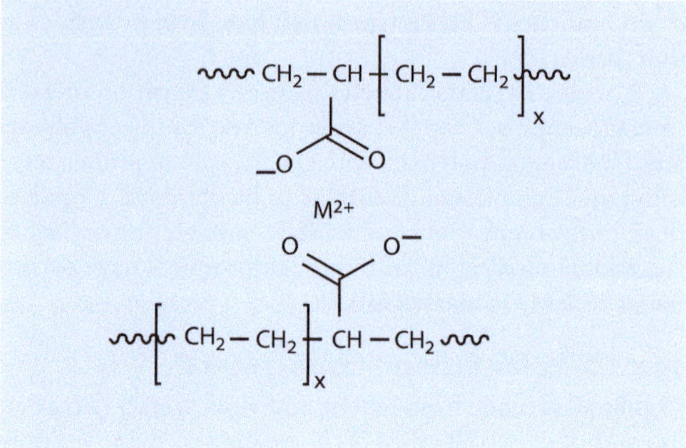

This copolymer (EVAL) is used, e.g., as a barrier layer in packages such as hose bottles and tubes together with PE, PP, PS, and PET, because of its excellent barrier properties with respect to gases such as O_2 and CO_2. The corrosion protection of street lights, traffic light posts, noise barriers, and motorway signs is often achieved using multi-layered polymers made from PE/EVAL/PE. Bonding agents are used to ensure that there is enough adhesion between layers.

14

Ethene acrylic acid copolymers (EAA) with between 5 wt% and 20 wt% % of the acid component adhere to metals very well; however, they are more sensitive to heat and oxidation than pure PE. These types of copolymers are often used as bonding agents in composite materials. A thermo-reversible interlinking of the copolymers is achieved using bivalent metal ions (M^{2+}) and the carboxylate groups (Fig. 14.3).

These "ionomers" are used as sealing films (tool handles, hammer heads, drink and pharmaceutical packaging).

14.1.3.3 Cross-Linking of Polyethylene (PE-X)

Cross-linking leads to polymers being insoluble and infusible. Because of this, the range of applications open for them is significantly expanded (► Sect. 15.2).

PE is cross-linked by high-energy radiation of prefabricated parts or by the addition of peroxides and subsequent heat treatment. During peroxide cross-linking hydrogen atoms

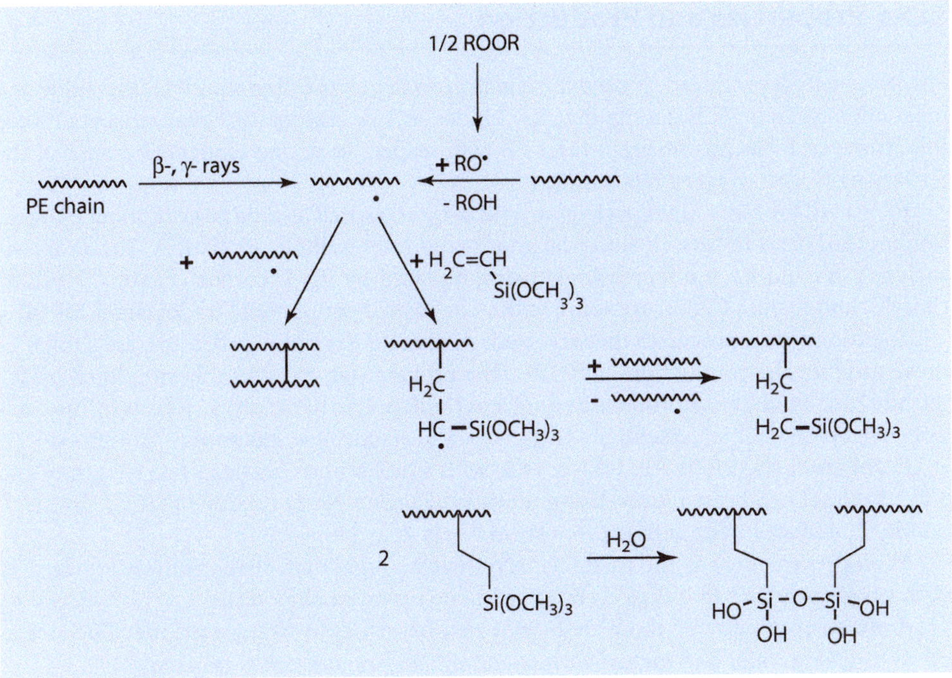

Fig. 14.4 Possibilities for cross-linking PE. *RO•* originates from ROOR

are abstracted from the polymer chain by peroxide radicals (RO•) to form polymer radicals. If these combine, cross-linking takes place.

If the radiation takes place in the presence of silanes (for example, trimethoxy vinyl silane) then the PE chain initially becomes Si-functionalized. The cross-linking via Si-O-Si-bridges then takes place by hydrolysis in a downstream water bath. Generally, it is the finished molded part which is cross-linked because, following cross-linking, further deformation is not possible. The possibilities for cross-linking PE are shown in Fig. 14.4.

PE-X is used in hot water pipes, for example, in underfloor heating systems and as the insulation layer of energy cables. PE-X has also been used for artificial knee joints but such medical prostheses are now more usually made from UHMWPE.

14.2 Polypropylene (PP)

Because of their mechanical properties, their availability, and the variability of their chemical structures, polypropylenes (propene homo- and copolymers) are very interesting from an industrial point of view and can be used for a wide range of applications.

14.2.1 Properties and Production

Polypropylene is produced by catalytic chain-growth polymerization. PP is crystalline and most often isotactic. It has a significantly higher service temperature than uncross-linked PE. However, PP is less resistant than PE with respect to strong oxidants because of the tertiary hydrogen of every repeat unit.

PP is used for films, rigid packaging, and carpets, as well as for a large number of injection molded parts in cars, in electrical goods, and in household appliances. Thermoplasts such as ABS and PA are increasingly being replaced by PP. Escorene (Exxon), Novolen (Basell), and Inspire (Dow) are some of the established commercial trade names for PP.

The production processes that are used, slurry and gas phase processes, are similar to those used for the production of HDPE. The process was considerably simplified by the introduction of liquefied propene as a solvent (bulk polymerization) as well as by the catalyst development which essentially eliminated the atactic, low molar mass products.

Highly isotactic PP (it-PP) (◘ Fig. 14.5) with a molar mass between $1.5 \cdot 10^5$ g/mol and $1.5 \cdot 10^6$ g/mol can be produced using modified Ziegler–Natta catalysts ($TiCl_4/Al(C_2H_5)_3$ on $MgCl_2$ surfaces). The atactic proportion is less than 5 %.

In all processes, as is the case for polyethylene, highly effective catalysts are used in such small quantities that they do not need to be removed afterwards.

A stereo specificity of more than 98 % can be achieved using *ansa* metallocenes as catalysts (◘ Fig. 14.6) and methyl alumoxane (MAO) as co-catalyst (▶ Chap. 11).

Syndiotactic polypropylene (st-PP) is characterized by its high transparency as the crystallites of this polymer are very small and therefore do not scatter light. It is

◘ **Fig. 14.5** Isotactic polypropylene

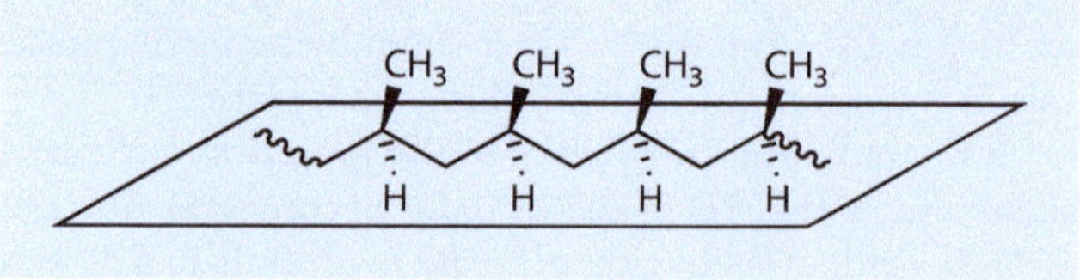

◘ **Fig. 14.6** Example metallocenes. (**a**) isoselective. (**b**) syndioselective

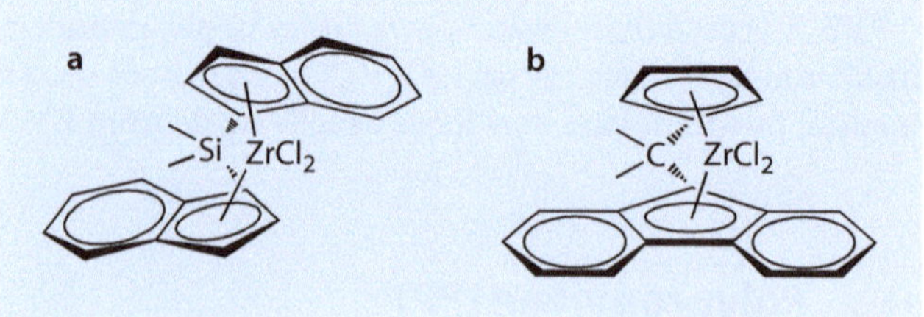

◘ **Fig. 14.7** Synthesis of ideally atactic polypropylene; 1,2- and 3,4-polymerization would disturb the ideal structure

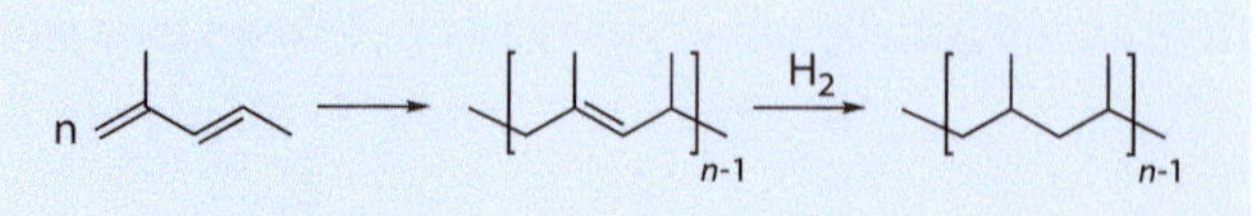

manufactured for use as films, non-woven fabrics, and special packaging in the cosmetic sector.

Atactic polypropylene (at-PP) is amorphous and has a soft rubbery consistency. It is used for coatings and for highly filled sealing compounds (for example, with kaolin or barite as a filler), as a sealant in construction work and as a coating for the backs of carpets. Originally, at-PP was a worthless by-product obtained during the polymerization of propene with Ziegler–Natta catalysts and was burnt or otherwise disposed of. Atactic polypropylene can also be obtained by the hydrogenation of poly(2-methyl-1,3-pentadiene) (◘ Fig. 14.7).

14.2.2 Propene Copolymers and Blends

Copolymerization of propene with 2–5 % ethene results in a decrease in crystallinity, brittleness, and rigidity; with increasing ethene the material becomes ever more flexible and impact resistant. Propene-ethene copolymers are used as shrink film, consoles, spoilers, radiator parts, car boot linings, bottle crates, adhesive tapes, chairs, and school bags. In addition to these applications, PP-PE copolymers are also used for toys and sports equipment.

Ethene-propene-diene elastomers (EPDM; in international elastomer nomenclature the M (*M*ethylene) indicates a rubber with a saturated backbone) are mostly terpolymers made out of ethene (E), propene (P), and 5-ethylidene-2-norbornene (ENB, D). The third monomer contains "internal" double bonds which polymerize considerably more slowly with both Ziegler–Natta and metallocene catalysts used to make these polymers. The remaining double bond can be used for cross-linking the final rubber article. Other dienes that have been used in EPDM are, 1,4-hexadiene (◘ Fig. 14.8), *cis,cis*-cyclooctadiene, and *exo*- and *endo*-dicyclo pentadiene (DCPD); only ENB- and DCPD-containing EPDM is commercially available today.

Technically important products have between three and seven C=C double bonds for every one thousand C-atoms that are not part of the main chain. A reaction of the double bonds remaining intact after cross-linking, with, for example, oxygen does not affect the main chain and thus has little effect on the properties of the material. EPDM is the basic component of, e.g., automotive profiles and roofing membranes.

SHIPP (*Super High Impact Polypropylene*) is produced by mixing PP with EPDM-rubber or the simple EPM copolymer (30–50 %). In these blends the rubber forms the dispersed phase and PP the matrix (▶ Chap. 7).

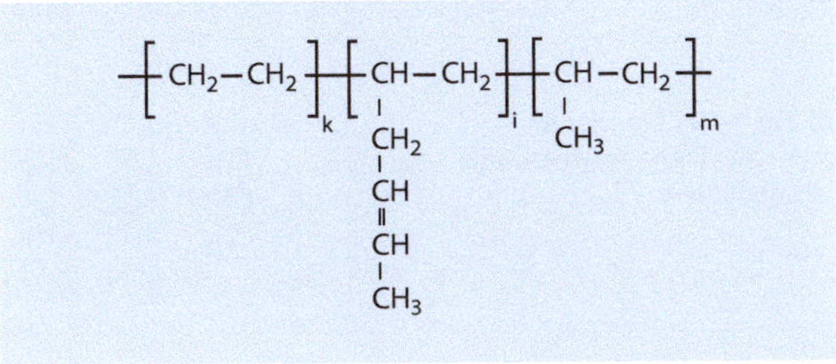

◘ **Fig. 14.8** Structure of EPDM with 1,4- hexadiene as the third monomer

14.3 Polyisobutylene (PIB)

Homopolymers of isobutylene (IB, isobutene or 2-methylpropene) and copolymers of IB with isoprene (butyl rubber, IIR) have become technically important products despite their technically elaborate synthesis by means of cationic chain growth polymerization.

14.3.1 Properties

Polyisobutylene is a thermoplastic material. Depending on its degree of polymerization, PIB is an oily liquid ($P_n = 10–40$), a sticky mass ($1,000 < P_n < 10,000$) or a rubbery product at room temperature. PIB has a low density (0.84 g/mL (amorphous) to 0.92 g/mL (crystalline)) as well as a high elongation at break and is resistant to acids, bases, and salts. PIB with low degrees of polymerization is used as an oil additive. PIB and IIR with an average molar mass form the basic material in chewing gum. PIB with a high degree of polymerization is used for sealants, corrosion-resistant pipe encasements, and case linings.

14.3.2 Production

IIR and PIB are the most important polymers that can be obtained by cationic chain-growth polymerization. Isobutene is polymerized in methylene chloride, propene, or ethene at temperatures between −100 °C and −30 °C. Propene or ethene do not polymerize cationically and therefore function solely as solvents.

Chain growth can be stopped by the addition of methanol. Temperature is an important control variable for the molar mass of PIB. Transfer to the monomer increases with increasing temperatures and so the chains become shorter (◘ Fig. 14.9).

The BASF band process is a procedurally interesting manufacturing process for PIB with an average molar mass (◘ Fig. 14.10). The manufacture of PIB according to the band process takes place at a temperature of about −100 °C. A mixture of isobutylene and ethene (volume ratio 1:1) is polymerized by the addition of an initiator solution of BF_3 in ethene. The reaction enthalpy is removed by the evaporation of the ethene, which can be reused after it has been compressed and cooled down. The polymer is transferred from the band reactor into the finishing extruder and from there into storage tanks.

Well-known trade names for PIB are Oppanol (BASF) and Vistanex (Exxon).

◘ **Fig. 14.9** Transfer to monomer in the polymerization of isobutylene

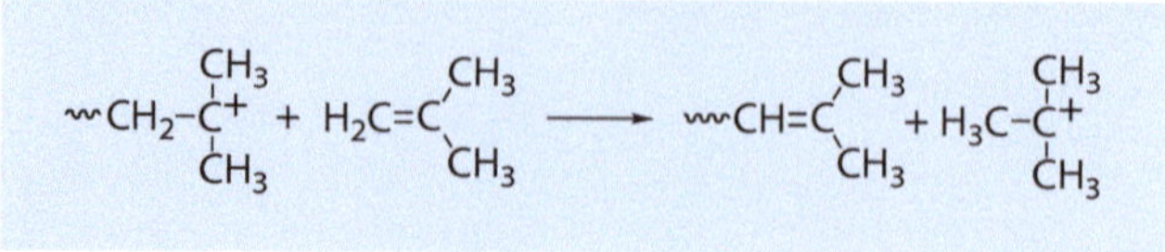

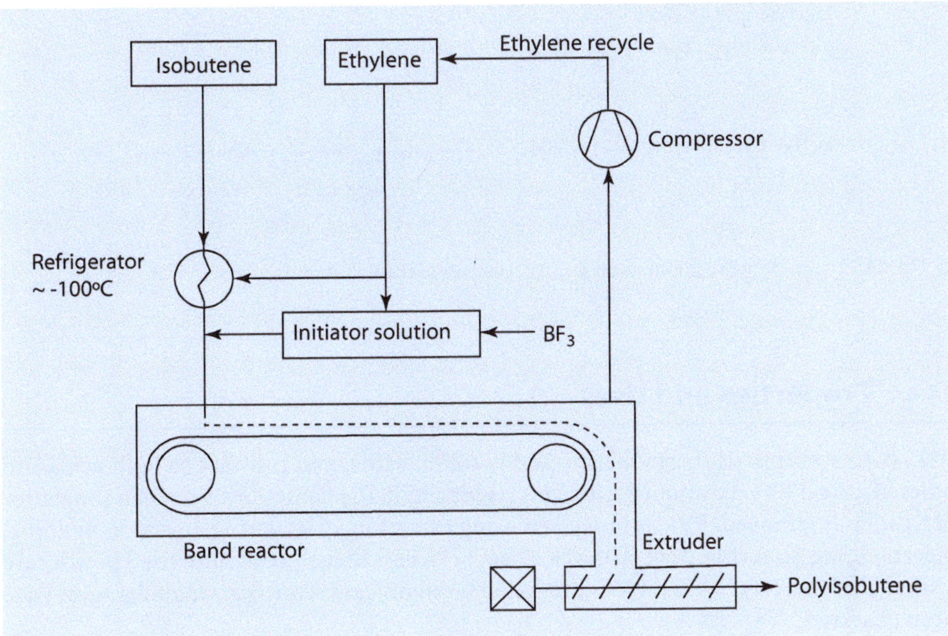

Fig. 14.10 BASF-band process for producing PIB with an average molar mass

14.3.3 Butyl Rubber

Butyl rubber (IIR) is a copolymer consisting of isobutylene (90–99.5 wt%) and isoprene (10–0.5 wt%), produced by cationic polymerization at temperatures between −100 °C and −40 °C by a solution process or a precipitation process in combination with $AlCl_3$ as an initiator. These polymers are very impervious to gas and are heat resistant as well as more resistant to aging than polymers with, e.g., double bonds in their main chains. These properties predestine them for use as membranes, bicycle tubes, rubbery cable insulations, and, most importantly, as the innerliner in tubeless tires. Copolymerization with isoprene allows the product to be cross-linked with sulfur via the remaining double bonds. Halogenation of the double bonds (with chlorine or bromine which allows cross-linking with metal oxides) leads to the technically most important product of this family of polymers (X-IIR). The major use of all polymers based on IB is as halogenated butyl rubber which is an essential part of a tubeless tire where it forms the innerliner to contain the air inside the tire.

14.4 Polyvinyl Chloride (PVC)

Homo- and copolymers of polyvinyl chloride belong to the group of the most technically important and versatile materials. With its chlorine content of 56.7 wt%, PVC represents a connective link between polymer chemistry and chlorine chemistry. In particular, because its properties can be modified over a wide range, PVC has conquered a very large market. Its application reflects its versatility.

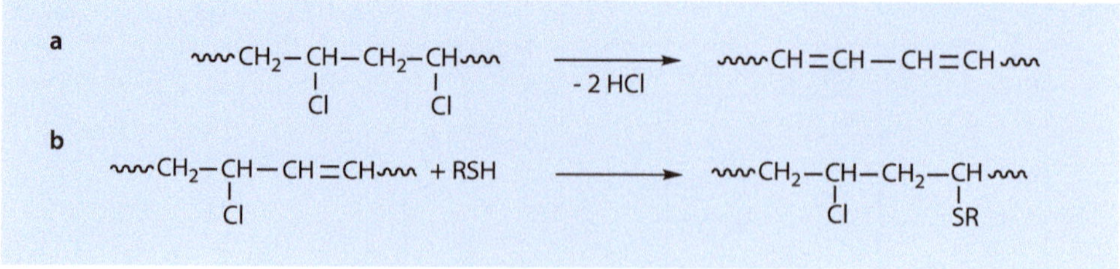

Fig. 14.11 (a) Dehydrochlorination and (b) "Healing" of the C=C-bonds

14.4.1 Properties and Uses

PVC is an amorphous thermoplastic and is solid, brittle, and resistant to both acids and bases. If a rigid PVC (without plasticizers) is set alight, the flame goes out when the source of ignition is removed. PVC is processed using extrusion, calendaring, injection molding, blow molding, sintering, or pressing (▶ Chap. 17). Glass fibers, chalk, and wood powder are typical fillers for PVC. When PVC is recycled no significant reductions in molar mass have been observed.

PVC can, however, not be processed without the addition of stabilizers because the polymer eliminates HCl when heated to form polyene and this reaction is autocatalytic. Additives such as sodium carbonate absorb the hydrochloric acid formed by dehydrochlorination whereas mercaptans add to the resulting C=C double bonds to "heal" them (Fig. 14.11).

PVC has become a workhorse for mechanical, construction, and electrical engineers as well as in the packaging industry. In engineering PVC is typically used for acid pumps, pressure pipes, and coatings. In construction, taps, pipes for fresh water, gas and drainage pipes, and window frames and shutters are all made out of PVC. Cable conduits and flush-mounted junction boxes are examples of electrical engineering uses of PVC and bottles and film are typical examples of PVC packaging. As well as all of the above, credit cards, adhesive tapes, drawing instruments, and toys are also often made from PVC.

Well-known trade names for PVC are Solvin (Solvay), Lacovyl (Atofina), and Vinnolit (Vinnolit).

14.4.2 Manufacturing Processes of PVC

Roughly 90% of all PVC is produced discontinuously by suspension polymerization. Vinyl chloride is dispersed in water using high pressure pumps or by high-speed stirrers (droplet diameter: 0.1–1 μm). Organic peroxides that are soluble in vinyl chloride are the most commonly used initiators. This process can be carried out using only small quantities of protective colloid (such as polyvinyl alcohol), which means that very pure PVC is obtained.

Emulsion polymerization is also used, which can be carried out either continuously or discontinuously. Vinyl chloride (boiling point: −14 °C) is liquefied at a pressure of 2.5 bar at room temperature and then dispersed in water. Water soluble peroxides such as $K_2S_2O_8$ or H_2O_2 are used as initiators, often in combination with reducing agents (such as sulfites, bisulfites, and ascorbic acid). After thoroughly removing any residual monomer, the milky polymer dispersion can be used directly as an adhesive or as a coating agent. If required, the polymer can be isolated by spray or roller drying. However, with these techniques the emulsifier and other polymerization aids remain in the polymer. Alternatively, the polymer can be separated by precipitation. The water soluble polymerization additives remain in the water to yield a polymer with a similar degree of purity to that obtained by suspension polymerization.

Polymerization in bulk takes place in two steps. In the first, vinyl chloride is polymerized, with organic peroxides as initiator, in a stirred autoclave with up to10% conversion. In a second autoclave, polymerization continues to ca. 80 % conversion. Subsequently, residual monomer is removed (to be recycled) by intensive degasification and a very pure polymer is obtained.

14.4.3 Copolymers

Copolymerization with vinyl acetate yields a product with a lower glass transition temperature (internal plastification), which can be more easily processed than the homopolyvinyl chloride.

The copolymerization of vinyl chloride with *N*-cyclohexyl maleinimide results in products with higher heat distortion temperatures (packaging for foodstuffs that are packaged when hot).

The grafting of vinyl chloride onto copolymers made from ethene and vinyl acetate results in products added to PVC to increase its impact resistance (▣ Fig. 14.12).

Blends of PVC with 5–12 wt% copolymer can considerably broaden its suitability for additional applications such as window profiles, shutters, and many other uses where it is exposed to the elements.

14.4.4 Soft PVC

Soft PVC, which is produced by adding large amounts (15–50 %) of plasticizers such as di-(2-ethylhexyl) phthalate to PVC, is a thermoplast with elastic, rubber-like properties. Because of the toxicity of the phthalate plasticizers, a wide range of non-toxic, often bio-based, plasticizers have been developed especially for PVC applications. The addition of low molar mass plasticizers is referred to as *external plasticization* emphasize the difference to the effect achieved via copolymerization. Soft PVC usually also

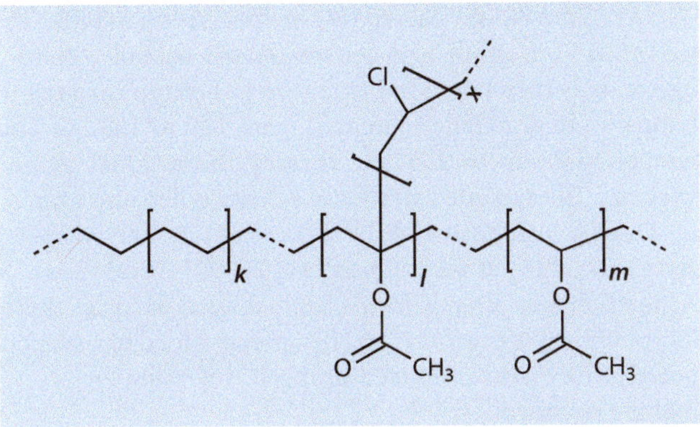

▣ **Fig. 14.12** Statistical ethene-vinyl acetate copolymer with polyvinyl chloride grafts

contains large quantities of fillers, sometimes simply to reduce the cost of the compound but also to achieve particular properties. Soft PVC is used in cables, hoses, coatings, automotive undercoating, flooring, rainproof clothing, and shower curtains to mention just a few examples.

14.5 Polystyrene (PS)

Polystyrene and its copolymers along with materials derived from them (often referred to as styrenics) are the third largest group of plastics after polyolefins and polyvinyl chloride.

14.5.1 Properties and Uses

Polystyrene is an amorphous thermoplastic and can be used to manufacture rigid, brittle, transparent, and non-conducting components. Polystyrene is resistant to acids and bases, alcohols, fats, and oils. It swells or partially dissolves, however, in hydrocarbons, halogenated hydrocarbons, esters, and ketones, restricting its range of application. It is not weather-resistant and burns, generating copious amounts of soot. As well as the homopolymer itself, copolymers, graft copolymers, and polymer mixtures containing styrene are of industrial importance. Such materials can be processed by, for example, injection molding, extrusion, or thermoforming. Especially important are polystyrene foams. Polystyrene-based materials find application in the electrical and electronics industries (radios and televisions), in lighting (light covers), in household goods (egg cups, butter dishes, washing pegs, shower walls, refrigerator lining), as packaging (as disposable packaging for foodstuffs, transparent films, cartons, bottles), and—in its foam form—as heat and sound insulation and for packaging. Some of the trade names are Edistir (Versalis), Empera (BP), Polystyrol (BASF), and Trinseo (Styrolution).

14.5.2 Manufacturing Processes for PS

14

Atactic Polystyrene

Atactic PS is produced by radical polymerization in bulk, suspension, or emulsion, although the polymerization in emulsion is only commercially used for copolymerization.

Very pure PS is obtained by polymerization in bulk. To begin with, styrene is polymerized at 80 °C to about 30 % conversion in a so-called continuous tower process. The syrup-like mass is then conveyed from top to bottom through the actual tower reactor and the temperature gradually increased from 140 to 180 °C. The product is finally extruded at temperatures up to 220 °C, whereby the residual monomer is removed by applying a vacuum. The extruded strands are then cooled and granulated.

During suspension polymerization, styrene, dispersed in water, is initiated with styrene soluble peroxides initially at 70–90 °C and then at 90–120 °C at a pressure of 5 bar. As the polymer directly forms bead-shaped particles there is no need for granulation. If a low boiling hydrocarbon (e.g., hexane) is added to a suspension polymerization, expanded polystyrene can be produced simply by releasing the pressure, a process not unlike that for making popcorn.

Stereo Regular Polystyrenes

Isotactic and syndiotactic PS are produced by the polymerization of styrene using Ziegler–Natta or metallocene catalysts. Isotactic PS is brittle and crystallizes very slowly and is therefore technically insignificant. In contrast, syndiotactic PS crystallizes rapidly and has a melting point of ca. 270 °C. Because of its thermal stability, it competes with polyamide, polybutylene terephthalate (PBT), and polyphenylene sulfide (PPS) for application in technical, structural components.

14.5.3 Copolymers and Blends

Around 60 % of the styrene that is produced is used for the production of thermoplastic styrene copolymers. Some of the most important modified styrene polymers are styrene-acrylonitrile copolymers (▶ Sect. 14.5.3.1), impact resistant, modified polystyrene (▶ Sect. 14.5.3.2), acrylonitrile-butadiene-styrene copolymers (▶ Sect. 14.5.3.3), and blends of polyphenylene oxide and polycarbonate (▶ Sect. 14.5.3.4). The commercially very important styrene butadiene copolymer, with 20–35 wt% styrene is an elastomer (styrene-butadiene rubber, SBR) and is discussed in ▶ Sect. 18.6.

14.5.3.1 Styrene-Acrylonitrile Copolymers (SAN)

Well known styrene-acrylonitrile copolymers are Luran (Styrolution), Lustran (Ineos), and Tyril (Dow). They are used as technical materials because of their improved toughness and tensile strength compared with simple PS, especially in combination with glass fibers. Weather resistant building components can be made from such materials. Typical applications are as film spools, video cassettes, radios, vehicle head-lamp housings, hazard triangles, control knobs in vehicles, and household appliances to name just a few.

These copolymers are produced by suspension polymerization and generally contain 76/24 parts by weight styrene/acrylonitrile. The monomer building blocks are distributed statistically along the polymer chain.

14.5.3.2 Impact Modified PS

Impact resistant PS (High Impact Polystyrene, HIPS) are multi-phase materials. They have a polystyrene matrix in which finely distributed rubber particles are embedded. Such materials can be obtained simply by mixing polystyrene with polybutadiene but the best properties can be achieved by dissolving polybutadiene in styrene monomer and then polymerizing the styrene. Suitable initiators are used to ensure that not only a homopolymerization of styrene takes place, but that some grafting of styrene onto the polybutadiene also takes place (◘ Fig. 14.13).

The morphology and therefore the properties of impact resistant polystyrene, are dependent on the ratio of the components, the structure of the matrix material, and how the rubber is attached to the matrix, allowing a wide variation of attainable properties.

SBS-block copolymers (SBS: segments of styrene (S) and butadiene (B)) represent a separate class of styrene copolymers which have become commercially very important under names such as Kraton (Kraton Performance Polymers, Inc) and Styrolux (BASF). SBS-block copolymers are produced by anionic block copolymerization (◘ Fig. 14.14; ▶ Sect. 10.2).

The PS and PB segments are incompatible with one another and tend to separate so that the morphology typical of thermoplastic elastomers (TPEs) and shown schematically in ◘ Fig. 14.15 develops.

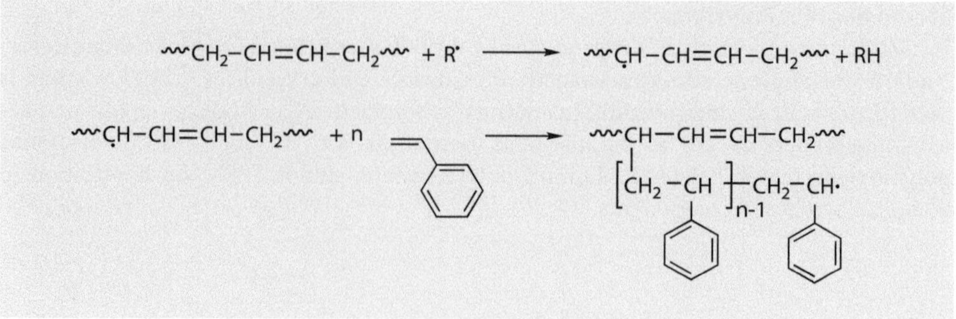

Fig. 14.13 Styrene–butadiene graft copolymers resulting from chain transfer to polybutadiene. The reaction between polystyrene and polybutadiene radicals also leads to graft copolymers. *R•:* can be an initiator or a polystyrene radical

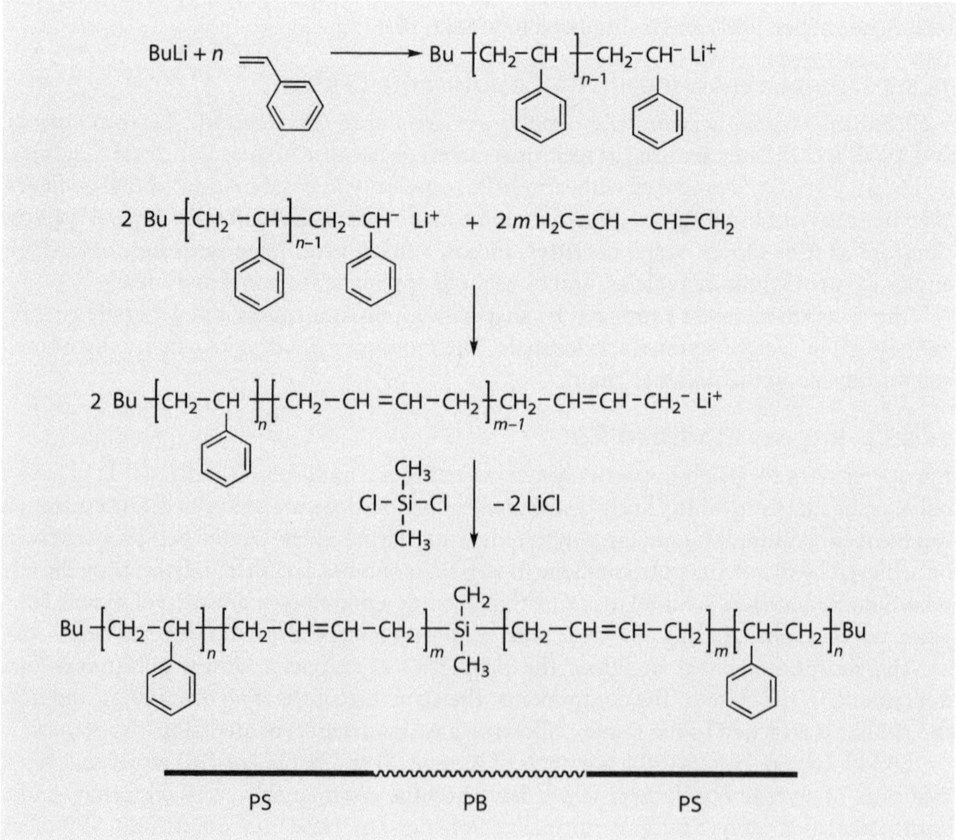

Fig. 14.14 Synthesis of SBS-block copolymers

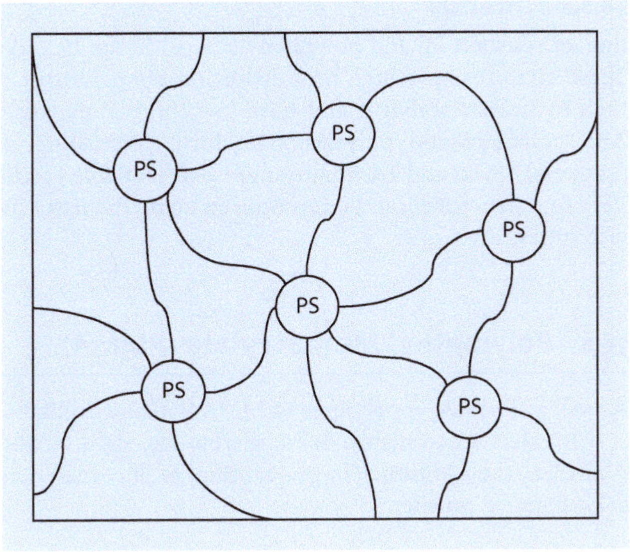

Fig. 14.15 Morphology of an SBS-block copolymer

14.5.3.3 Acrylonitrile-Butadiene-Styrene Copolymers (ABS)

ABS polymers are multi-phase materials just as are impact resistant PS. Soft particles of grafted polybutadiene rubber are embedded in a hard matrix of poly(styrene-co-acrylonitrile), i.e., it is a polymer blend. The grafting of polybutadiene increases the compatibility with the matrix and stabilizes the size of the polybutadiene domains.

Two technically important production processes exist for the production of ABS. In the so-called *bulk process*, polybutadiene is dissolved in a mixture of styrene and acrylonitrile and the monomers polymerized, whereby some grafting onto the polybutadiene takes place. The two polymer phases separate during polymerization as polybutadiene dissolves in the monomer mixture but not in poly(styrene-co-acrylonitrile). Thus, the soft, polybutadiene in the hard, styrene-acrylonitrile matrix morphology develops. The size of the grafted polybutadiene domains, which is of critical importance for the impact properties, is generally 1–4 μm.

An alternative method of production to this first production process is emulsion polymerization (► Chap. 16). In this process polybutadiene is polymerized in emulsion and then grafted with a mixture of styrene and acrylonitrile. After isolating the product from the emulsion it is fed to an extruder and mixed with additional poly(styrene-co-acrylonitrile) produced in a separate process. With this process the particle size of the grafted polybutadiene dispersion is determined by the emulsion process and is generally of the order of 300–600 nm.

ABS polymers are used for a wide variety articles such as toys (Lego), furniture edges, mobile phones, cameras, ventilation grilles, and armrests in vehicles, as well as being used in the sanitary sector (in a metallized form) for shower and bath fittings. ABS polymers are known, amongst others, under the following trade names: Cycolac (Sabic), Novodur (Styrolution), Magnum (Dow), and Terluran (BASF).

14.5.3.4 Blends

Impact resistant PS and polyphenylene oxide can be mixed in any ratio to one another. These mixtures have high heat distortion temperatures, absorb very little water, have a high hydrolytic stability, and have low thermal expansion coefficients (Noryl, Sabic). ABS polymers and polycarbonate form interesting multi-phase blends (Bayblend (Lanxess)). ABS and PA mixtures are also available (Terblend (BASF), Triax (Lanxess)). Very fine structures can be reproduced using the latter blends because of their low melt viscosities.

14.6 Polymethyl Methacrylate (PMMA)

As well as PMMA, copolymers of MMA with acrylonitrile, styrene, α-methyl styrene and acrylic esters are available. Solid, marble-like slabs (artificial marble) can be made from PMMA by the addition of large quantities of filler materials such as quartz, limestone, and color pigment powder.

14.6.1 Properties and Manufacture

PMMA is relatively brittle and one of the hardest thermoplastics. It is produced by radical polymerization. Well-known trade names are Plexiglas (Röhm), Degalan (Evonik), Altuglas (Arkema), and Acrylite (Evonik Cyro). This polymer is often used when light is an important factor because of its excellent transparency. PMMA exhibits excellent resistance to weathering and can be modified to improve its scratch resistance, which is not good when compared with inorganic glass. It can be extruded, injection molded, thermoformed, welded, and bonded.

It is typically used for spectacles and watch glasses, optical fibers, optical media, flat-panel displays, switch components, aircraft glazing, caravan windows, traffic signs, wash basins, baths, shower cubicles, and a casting resin for making delicate articles. It is also used for tooth fillings and prostheses.

Acrylic glass is mostly obtained by bulk polymerization. The pure monomer is pre-polymerized between two even sheets of glass and then fully polymerized very slowly (usually over several days). The PMMA objects (such as panes) produced in this manner have especially smooth surfaces. In contrast to this, suspension polymerization enables the monomer to be transformed into polymeric beads with diameters between 0.1 and 0.5 mm and also allows a controlled removal of the heat of reaction.

14.6.2 Copolymers of Methyl Methacrylate

Copolymers of methyl methacrylate with acrylonitrile are more impact resistant, heat resistant, and less susceptible to stress cracks than pure PMMA. Copolymers with up to 70 mol% acrylonitrile are used as barrier plastics for packaging. Copolymers that are sufficiently resistant to impact at low temperatures are made from methyl methacrylate, acrylonitrile, butadiene, and styrene. They are, however, not so weather resistant because of their butadiene content and the resulting backbone C=C double bonds.

An interesting variant are the impact resistant materials with a PMMA core and a cross-linked poly(butyl methacrylate) matrix. The refractive index of the matrix can be modified to be the same as that of the PMMA core by the addition of styrene to obtain a transparent, two-phase blend.

14.7 Polyacrylonitrile (PAN)

Homo- and copolymers can be made from acrylonitrile. Spun PAN (usually with small amounts of a second or third monomer) is the major raw material used for producing carbon fibers. SAN and ABS are important polymers, as well as copolymers with vinyl chloride (► Sect. 14.5.3.1 and ► Sect. 14.5.3.2).

14.7.1 Properties and Manufacture

PAN has a softening temperature of more than 200 °C because of very strong intermolecular forces between the nitrile groups (dipole-dipole interactions). As it disintegrates at temperatures above 200 °C, it cannot be thermoplastically processed without added plasticizer, which significantly limits its usefulness. It can, however, be spun into wool-like synthetic fibers from polar solvents such as dimethyl formamide or dimethyl sulfoxide as well as in concentrated aqueous salt solutions. Well-known trade names are Orlon (DuPont), Dralon (Dralon GmbH), and Dolan (Lenzing).

PAN is also used as a basic material in the production of high-quality carbon fibers (TorayCa (Toray), Tenax (Toho Tenax),Thornel (Cytec), Celion (Celanese), and Sigrafil C (SGL)).

The stiffness and strength of many plastics (ABS, SAN, ASA, vinyl chloride copolymers) can be improved by having acrylonitrile as a comonomer. Copolymers with up to 80 mol% acrylonitrile are used as so-called barrier plastics (they have good barrier properties for gases, aromas, and flavorings). These types of barrier plastics can be produced by precipitation or emulsion polymerization.

14.7.2 Carbon Fibers (C-Fibers)

Carbon fibers are formed from PAN fibers in a multi-step process.

In the first step, PAN is heated in an air atmosphere at 200–400 °C whereby the fibers are stretched and stabilized, the polymer is oxidized, and the side chains cyclize to form a ladder polymer of pyridine rings. This reaction is exothermic and its control is one reason why most carbon fiber precursors contain additional monomers such as itaconic acid. Subsequently the heat treatment continues in a second oven under nitrogen at temperatures up to 2,000 °C. In this phase the polymer loses both nitrogen and hydrogen (some 40 % of its original weight) to leave essentially only carbon—high strength (up to 8 GPa) and high modulus (up to 450 GPa) carbon fibers (◘ Fig. 14.16).

Carbon fibers can also be made from fibers derived from cellulose but the carbon yield is lower than that from PAN fibers and to date the fibers have not attained the high strengths or moduli of PAN-derived fibers. High modulus carbon fibers can also be

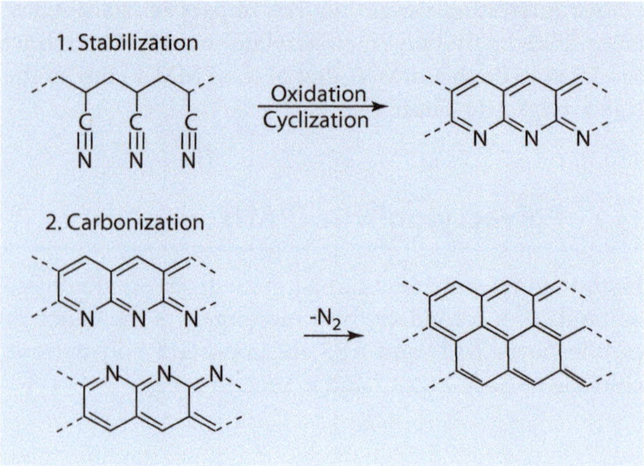

Fig. 14.16 Selected reactions from PAN to carbon fibers. C-fibers are >96 % carbon but their detailed structure is not completely defined

prepared from pitch but these fibers have relatively low tensile strengths and the pitch needs elaborate treatment before it can be used.

Carbon fibers are used as reinforcing material for fiber-reinforced thermoset and thermoplastic composites. These composites are much stronger than the unfilled matrix components. The thermoset composites require a longer manufacturing process and tend to be less impact resistant than those made with thermoplastics. The resulting materials are light, resilient, high-performance materials and are being increasingly used in the aerospace industry to replace metal components. They are also used in wind generators and high-end sport equipment and have begun to enter the automotive market as their price continues to be reduced.

14.8 Polyoxymethylene (POM)

Aliphatic polymers made from formaldehyde were studied by the famous polymer chemist Staudinger in the 1920s. However, it was not until the 1950s that they could be made stable enough to become useful. The breakthrough was a better understanding of depolymerization and ways to limit it taking place. Today POM is one of the recognized group of engineering thermoplastics.

14.8.1 Properties and Production

POM is one of the favored construction materials in the field of precision engineering because of its dimensional stability under changing environments, its excellent, low coefficient of friction, and its resistance to wear. It is a partially crystalline, linear polymer and is available both as a homopolymer (Delrin (DuPont), Tenac (Asahi)) and as a copolymer, with, e.g., small amounts of oxirane or dioxolane (Hostaform (Ticona), Ultraform (BASF)).

It is typically used for gear wheels, small gear boxes, valves, fittings, precision parts for telephones, faxes, and televisions, coupling parts for garden hoses, ski fittings, zips, and Playmobil toys.

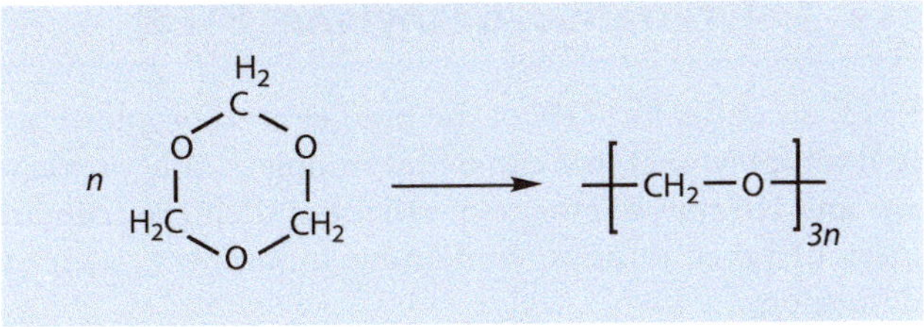

Fig. 14.17 Formation of POM from trioxane

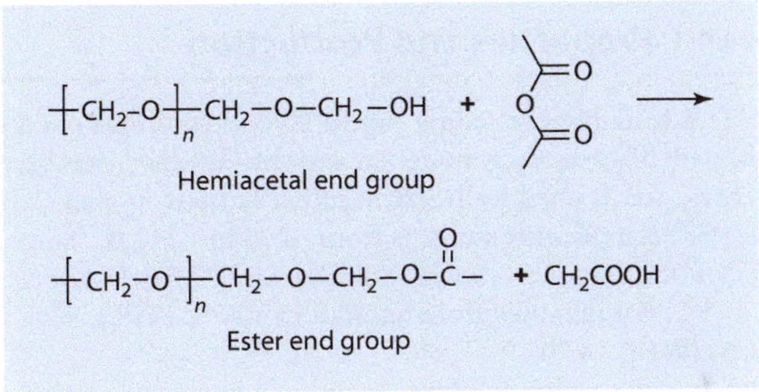

Fig. 14.18 Esterification of the terminal hemiacetal with acetic acid anhydride

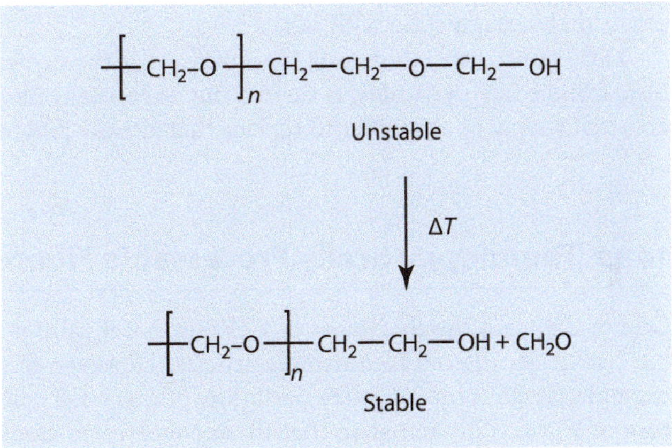

Fig. 14.19 Mechanism of the stabilization of POM by copolymerized oxirane monomers

The monomer for POM is either formaldehyde or its trimer trioxane (Fig. 14.17) and the polymerization is initiated either anionically with amines or phosphates, or cationically with BF_3/H_2O.

The zip-like degradation of the polymer is prevented by the reaction of the thermosensitive hemiacetal endgroup with an acid anhydride (Fig. 14.18).

Stabilization is also possible by copolymerizing trioxane with a small percentage oxirane, 1,3-dioxolane, or other cyclic ethers.

The degradation of the copolymer by depolymerization is halted at those points at which the comonomers, such as oxiranes, are incorporated into the main chain (Fig. 14.19). This degradation takes place during the production process so that everyday objects made out of POM do not eliminate formaldehyde during normal use.

14.9 Polytetrafluoro Ethylene (PTFE)

PTFE accounts for 60 % of the market of fluoroplastics. There is an increasing interest in fluoropolymers that can be thermoplastically processed, such as copolymers of ethene and tetrafluoro ethene or ethene and chloro trifluoro ethene as well as homopolymers of other fluorine-containing monomers, such as vinyl fluoride or vinylidene difluoride.

14.9.1 Properties and Production

PTFE is highly crystalline (up to 90 % crystallinity) and does not melt simply by being heated. It is extremely nonpolar and anti-adhesive, has excellent electrical insulating properties, and is chemically extraordinarily inert as well as being almost incombustible. Its service temperature range is from −200 to +250 °C. Some typical trade names are Teflon (DuPont), Algoflon (Solvay Solexis), and PTFE (Dyneon).

PTFE is obtained from tetrafluoro ethene (TFE) using an emulsion or suspension process (Fig. 14.20).

For an emulsion polymerization an aqueous solution of a peroxide initiator and special emulsifiers (such as ammonium perfluoro octanoate) is prepared and the monomer is then introduced and dispersed under high pressure. The polymer is isolated from the emulsion by coagulation with salt.

Little or no dispersing agent is employed for the suspension process. The polymerization, initiated by peroxides, is carried out at constant monomer pressure so that there is constant supply of monomer to replace that already polymerized.

14.9.2 Thermoplastically Processable Fluoroplastics

One of the main disadvantages of PTFE is that it cannot be thermoplastically processed but has to be sintered into useable articles. However, if TFE is copolymerized with, for example, hexafluoropropene or perfluoro(propyl vinyl ether), the regular molecular structure of PTFE is disrupted so that these copolymers can be processed using those techniques typical for thermoplastics. The monomer TFE is not used at all during the production of other thermoplastically processable fluoroplastics such as the homopolymers polyvinylidene difluoride or polyvinyl fluoride (Fig. 14.21).

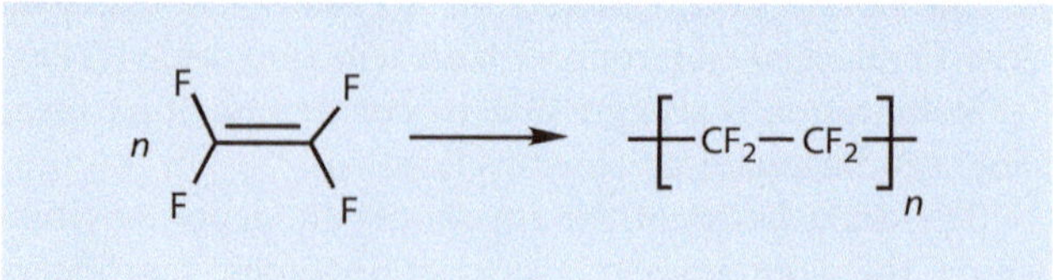

Fig. 14.20 Synthesis of PTFE from tetrafluor ethene

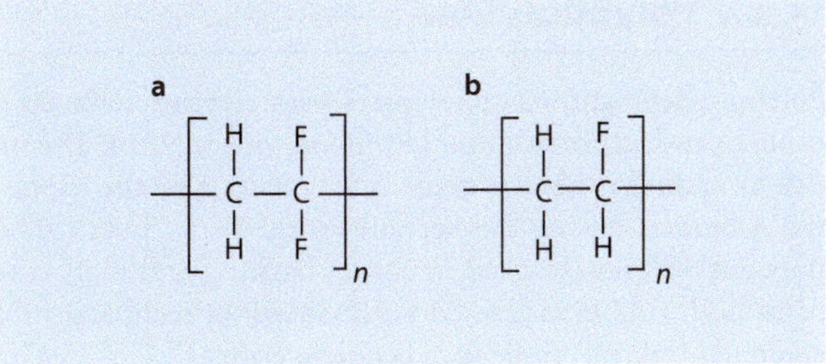

Fig. 14.21 Structure of (**a**) polyvinylidene difluoride and (**b**) polyvinyl fluoride

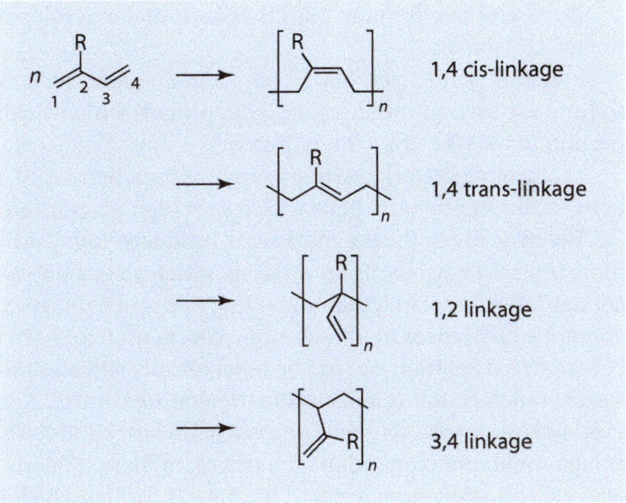

Fig. 14.22 Stereochemical possibilities for the incorporation of a 1,3-diene into a polymer

14.10 Polydiene

The expression *polydiene* is a general term for all diene homopolymers and copolymers. This includes polybutadiene, polyisoprene, polychloroprene, and all those corresponding copolymers which have obtained technical significance.

14.10.1 General Overview

1,3-Dienes with the general structural formula $CH_2{=}CR{-}CH{=}CH_2$ (R=H: butadiene, R=CH_3: isoprene, R=Cl: chloroprene) can be transformed into polymers by a chain growth mechanism. Fig. 14.22 shows the various ways that these monomers can be incorporated into the polymer backbone during the polymerization.

Depending on the initiator either 1,4-*cis*- or 1,4-*trans*- as well as 1,2-linked polymers are formed. For R$\neq$H the incorporation via the 3,4-C=C double bond is also possible.

14.10.2 Polybutadiene

Polybutadiene and its copolymers with styrene make up over 60 % of the total synthetic rubber production of some 16 million tons in 2014. The importance of SBR as an alternative to natural rubber is made most obvious by the explosive development of capacity in the Americas during the Second World War. These rubbers are primarily used to make tires but SBR is also used in carpet backing and mattresses and BR is used to make golf balls; both rubbers are used for a variety of technical rubber articles where temperature and chemical (oil) resistance is not so critical.

Butadiene can be anionically polymerized. Initially sodium dispersions in hydrocarbons were used for this purpose (butadiene-sodium (Natrium in German): Buna, a generic name in America).

The use of butyllithium (BuLi) as an initiator results in polymers with a high proportion of 1,2 isomers.

Initiators made from one alcohol and an olefin such as sodium isopropylate and allyl sodium are used in the so-called *Alfin* process which yields polybutadiene with high proportion (65–85 %) of 1,4-*trans*-isomers.

The heterogeneous polymerization of butadiene with $VCl_3/(C_2H_5)_2AlCl$ in hydrocarbons results in polybutadiene with a very high proportion of 1,4-*trans*-isomers.

The majority of the polybutadiene produced today is made using complex cobalt, titanium, nickel, or neodymium catalysts which yield a polymer with, predominantly, the 1,4 configuration with varying ratios of the *cis* and *trans* isomers along the backbone. There are subtle differences in the solution process used to make these polymers so that, e.g., the Nd-catalyzed reaction can be run adiabatically whereas those catalyzed by the other metal system require cooling and closer temperature control. Additionally, the different catalysts yield polymers with different degrees of linearity and have varying tendencies to form gel at high monomer conversion. The newest of these catalysts, the Nd family, yields polymers with exceptionally high levels of the 1,4-*cis*-isomer which have been shown to be particularly suitable for high-performance, so-called green tires.

The copolymers of butadiene with styrene can be synthesized using both radical and anionic initiators although, in terms of volume, radical processes are considerably more important. The copolymerization of 1,3-butadiene with styrene and acrylonitrile has already been discussed (▶ Sect. 14.5.3.2 and ▶ Sect. 14.5.3.3). Some 70 % of SBR is produced by emulsion (E-SBR) using redox initiating systems but the amount of SBR produced using an anionic solution (S-SBR) process is growing rapidly because of its use in so-called green tires. To facilitate the incorporation of the silica filler in such tires, S-SBR is functionalized, either by endcapping techniques or along the chain, either by the incorporation of additional monomers or by post polymerization modification.

Polybutadiene is covalently cross-linked for use in cable sheaths, vibration dampers, gaiters, vehicle tires, etc. (◘ Fig. 14.23; ▶ Sect. 15.2.2).

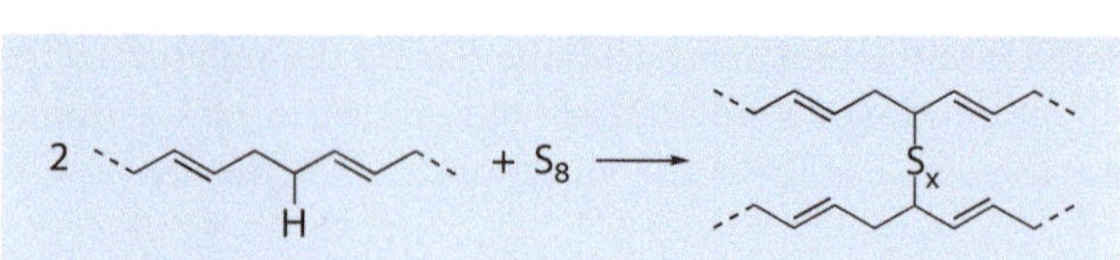

◘ **Fig. 14.23** Cross-linking polybutadiene with sulfur

Cross-linking (vulcanization) involves not only the primary reagent sulfur, but also a number of other additives which include: accelerators (amines), activation agents (such as ZnO), reinforcing fillers (such as carbon black, which also functions as a light and UV stabilizer), plasticizers (mineral or paraffin oils), and stabilizers to increase resistance to light and ozone so that their useful service life is extended.

14.10.3 Polyisoprene

Polyisoprene occurs naturally as *natural rubber* (100 % *cis*-1,4) and as *gutta percha* (*trans*-1,4). The world consumption of natural rubber was 12 million tons in 2014 whereas only about 1 million tons of synthetic polyisoprene was consumed. The synthetic material, although it is now possible to synthesize close to 100 % 1,4-*cis* polyisoprene using Nd catalysts similar to those used for BR (► Sect. 14.3.3), does not crystallize as fast or to the same extent when extended. This so-called strain crystallization is particularly important for the wear resistance of tires. Thus, the synthetic substitute is still not as good as the natural product.

cis-1,4-Polyisoprene is produced industrially from isoprene in solution using $TiCl_4$/$Al(C_2H_5)_3$ as a catalyst. The ratio of *trans*-1,4:*cis*-1,4 increases drastically with a decrease in temperature. Just as polybutadiene has to be cross-linked for its most important uses, so too, does polyisoprene. *cis*-1,4-Polyisoprene is typically used for car and truck tires (especially in the countries of the former Soviet Union because of their historical difficulties in obtaining natural rubber), shoe soles, safety gloves, and various specialized seals.

As a copolymer with isobutene, it is used for the inner liner of tubeless car tires (► Sect. 14.3.3).

14.10.4 Polychloroprene

Polychloroprene (CR) is produced on a large scale by the radically initiated emulsion polymerization of chloroprene. The polymer latex is destabilized by adjusting the pH to below 7 and then cooling the latex to form a solid sheet of polychloroprene. The sheet is then washed, dried, and rolled into a rope which is cut into the characteristic chip form in which it is sold. The commercial polymer contains considerable amounts of the emulsifiers used in the polymerization.

The polymer is largely the result of 1,4 addition and the *cis-trans* ratio is roughly 1:9. Allylic chlorine atoms, which are important for cross-linking, but only result from a 1,2-addition, are only present to a small extent. Vinyl chlorine atoms do not take part in the cross-linking process (vulcanization), which is not carried out using sulfur as is the general case for the other polydienes, but rather by reacting the polymer with ZnO and/or MgO as well as with ethylene thiourea as a sulfur source.

In the first step a nucleophilic attack of ethylene thiourea on the structural unit **I** occurs, causing the formation of the intermediate moiety **II**, which then reacts with, e.g., ZnO to form **III**. The reaction of **III** with **I** then causes a sulfur bridge to form and thus a cross-link site (◘ Fig. 14.24).

Cross-linked CR is used for hoses, cable sheathing, extruded profiles, windscreen wipers, and transmission belts, as well as for sportswear (the "Neoprene" wetsuits (DuPont trade name for CR)). The uncoagulated but concentrated latex is used for adhe-

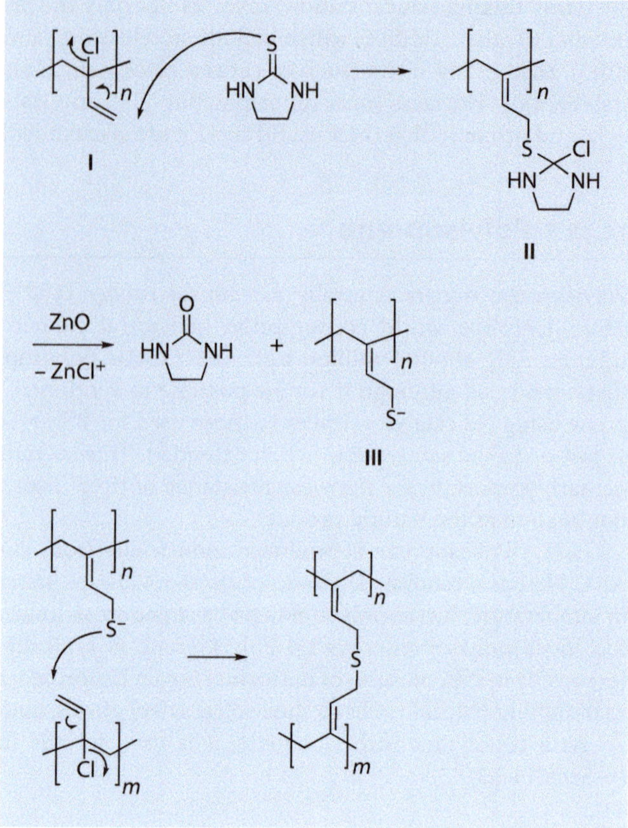

Fig. 14.24 Cross-linking of polychloroprene with ethylene thiourea/ZnO

sives, particularly for wood. Solvent (e.g., methyl ethyl ketone/toluene/heptane mixtures) based CR adhesives are also available.

Polychloroprene can be mixed with natural rubber and polybutadiene, which not only minimizes compound costs but also increases the flexibility at lower temperatures. Foamed CR (cellular rubber) is used for its excellent thermal insulation properties.

Chemistry with Polymers

S. Koltzenburg et al., *Polymer Chemistry*, DOI 10.1007/978-3-662-49279-6_15

Many macromolecules still contain chemically reactive groups even after polymerization has taken place; under certain conditions, macromolecules can also be reactive chemicals. They do, however, have some special characteristics compared with reagents with a low molar mass. This chapter describes such reactive macromolecules and their peculiarities.

It is of fundamental importance to differentiate between two types of chemical reactions that macromolecules can undergo: intentional chemical reactions or undesired degradation processes. Specific stabilization strategies have been developed to inhibit unwanted degradation processes, and these are also dealt with in this chapter.

15.1 Polymer Analogous Reactions

Reactions that involve parts of macromolecules, but that do not change their degree of polymerization, are known as polymer analogous reactions. The hydrogenation of unsaturated bonds in the polymer backbone or the hydrolysis of side groups are examples of these kinds of reactions. Because the molar mass of the repeat unit usually changes during these reactions, the molar mass of the macromolecule does not remain constant (in contrast to the degree of polymerization).

From a synthetic polymer chemist's point of view, polymer analogous reactions are fundamentally different to reactions involving only small molecules in several ways:

- By-products are difficult to separate from the desired products so that only reactions that do not generate significant amounts of unwanted by-products are suitable for polymer analogous reactions.
- The solubility of a chemically modified polymer can be very different from that of the base polymer. This is especially pronounced if charged moieties are added to an uncharged polymer, for example, the hydrolysis of ester to carboxylate groups. This can make polymer analogue reactions difficult as, because of its reduced solubility, the polymer precipitates during the reaction, resulting in the unconverted functional groups no longer being easily available for reaction; the reaction does not go to completion and a chemically inhomogeneous product is formed. Chemically inhomogeneous products are also obtained if a poorly soluble polymer (e.g., cellulose) is made more soluble by modification. Nevertheless, in many technical processes, despite chemical inhomogeneity, an acceptable product can be produced so that some polymer analogous reactions gain considerable technical significance. Some examples of these reactions are dealt with more thoroughly in ▶ Sect. 15.1.3. However, polymer analogous reactions are not so often used in academia, where the goal is usually to produce a well-defined material.
- Because of the high local concentration of the group taking part in the reaction in the interior of the polymer random coil, in contrast to the "empty" solvent phase in between the coils, special kinetic features occur:
  - A preference for intramolecular reactions
  - An increase or decrease in the reaction rate

These effects are discussed in more detail in ▶ Sects. 15.1.1 and 15.1.2.

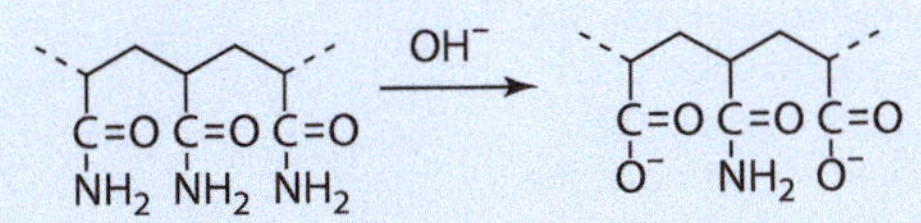

Fig. 15.1 Hydrolysis of polyacrylamide

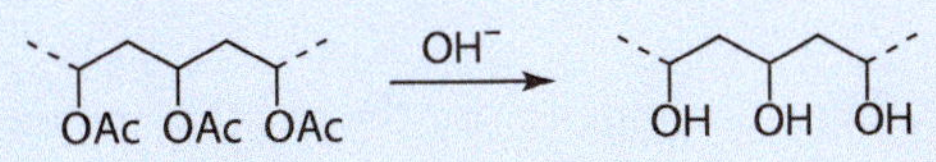

Fig. 15.2 Hydrolysis of polyvinyl acetate to polyvinyl alcohol. *Ac*: acetate

15.1.1 Special Kinetic Features of Polymer Analogous Reactions

In contrast to homogenous reactions of small molecules, the functional groups taking part in polymer analogous reactions are not evenly distributed throughout the reaction volume. Random polymer coils with high concentrations of functional groups are separated from one another by solvent. Because of the proximity of these functional groups there can be a strong impact on the kinetics of the reaction. This can result in the reaction being either retarded or accelerated.

An example of a self-retarding reaction discussed in the following is the (technically insignificant) saponification of polyacrylamide (Fig. 15.1). During alkaline saponification, negatively charged carboxylate groups are formed. These have an electrostatically repulsive effect on the attacking hydroxide ions so that the reaction of neighboring, unhydrolyzed groups is significantly slowed.

An amide group, such as the one shown in Fig. 15.1 which is flanked by two hydrolyzed, and thus negatively charged, groups, is electrostatically shielded from attack by a hydroxide ion because of the negative charges of these groups. This results in the reaction being slowed by a factor of 12 even at a degree of conversion as low as $p = 0.4$.

In contrast to this kind of retardation, an acceleration can also take place during polymer analogous reactions because of a high local concentration of functional groups. This is the case for the technically significant hydrolysis of polyvinyl acetate (Fig. 15.2).

The OH functional groups formed during this reaction can—in contrast to the less polar acetate group—form hydrogen bonds with the attacking hydroxide ion. This favors the attack of these ions. As a result, an acetate group flanked by two groups which have already been hydrolyzed is around 100 times more reactive than without these neighboring groups. This is why a strong acceleration in the rate of conversion as the reaction proceeds is observed. A further effect, which can lead to an acceleration of the reaction, is the formation of a hydrogen bond between an acetate group and a neighboring hydroxyl group, leading to a positive polarization of the carbonyl group (Fig. 15.3).

These kinetic effects have an important influence on the structure of the materials produced if the conversion is incomplete. During the hydrolysis of polyacrylamide, for

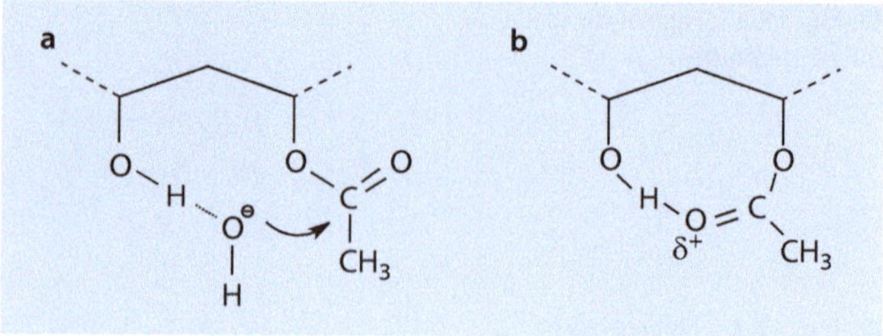

Fig. 15.3 Hydrogen bonds in partially hydrolyzed polyvinyl acetate. (**a**) Stabilization of an attacking hydroxide ion. (**b**) Intramolecular activation of the carbonyl function by hydrogen bonding. *Ac*: acetate

example, there is a tendency for the reaction to take place with groups that are as far apart as possible. This means that a partially hydrolyzed polymer tends to have a structure more akin to that of an alternating copolymer. On the other hand, partially hydrolyzed polyvinyl acetate can be regarded as a block copolymer.

15.1.2 Intra- Vs. Intermolecular Reactions

As already mentioned in ► Sect. 15.1.1, the functional groups attached to a polymer molecule are not distributed evenly throughout the polymer solution. This inhomogeneity is especially pronounced in dilute polymer solutions. In very dilute solutions the distance between the polymer coils (and therefore their functional groups) is especially large, whereas the distance between the functional groups within the polymer molecules is, at least initially, independent of the dilution. Thus, intramolecular reactions are more likely to take place during reactions involving a multifunctional reagent which can react with more than one of the functional groups of the polymer molecules.

At high polymer concentration, intermolecular reactions, i.e., a linking of two polymer molecules, can occur to a lesser extent. Formally, this no longer qualifies as a polymer analogous reaction because the degree of polymerization significantly increases. To avoid such reactions, polymer analogous reactions are best carried out in dilute solution.

A technically important example of this type of intramolecular polymer analogous reaction is the esterification of polyvinyl alcohol (Fig. 15.4).

15

The reaction is irreversible when carried out under alkaline conditions so that hydroxyl groups for which both neighboring groups have already reacted cannot react. For this reason, the reaction does not proceed to complete conversion with respect to the hydroxyl groups. It can be shown that the maximum conversion p_{max} is

$$p_{max} = 1 - \frac{1}{e^2} \approx 86.5\ \% \quad (15.1)$$

Complete conversion can only be achieved if the reaction is reversible and "reaction defects," such as those shown in Fig. 15.4, can be rearranged during the reaction process.

Polyvinyl butyral is widely used in the automobile industry to make laminated safety glass. Two glass panes are stuck to one another using a transparent film of polyvinyl butyral. The film is usually about 0.75 mm thick. In the event of mechanical damage the glass panes break but the glass splinters remain adhered to the polymer film. As a result, the material does not completely break, and the risk of injury from flying glass splinters is minimized.

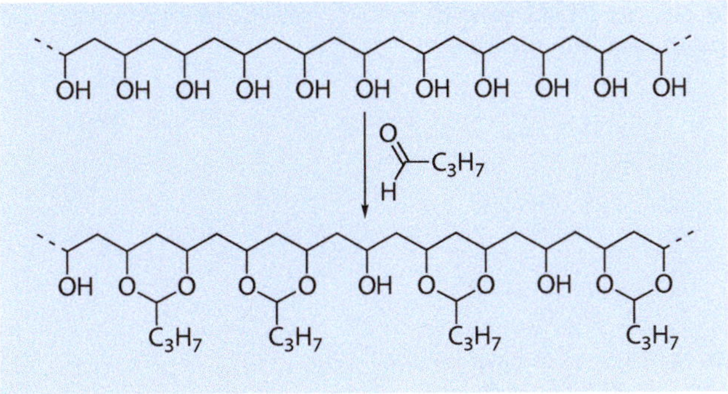

Fig. 15.4 Esterification of polyvinyl alcohol to polyvinyl butyral

15.1.3 Technically Significant Polymer Analogous Reactions

The aromatic side group of polystyrene can be chemically modified in many different ways. These reactions are often used commercially for polystyrene that has been cross-linked, for example, with divinyl benzene. As an example, the reaction of cross-linked polystyrene beads with sulfuric acid results in a sulfonic acid functionalized polymer network (Fig. 15.5) which can be used as a cation exchange resin.

Anion exchange resins can be made in a similar manner by chloromethylating polystyrene and subsequent reaction with a tertiary amine to yield quaternary ammonium functionalized polystyrene (Fig. 15.6).

The intermediate product, chloromethylated polystyrene, is also used as a resin for the Merrifield synthesis of peptides. The chloromethyl groups react with the amino groups of an amino acid to bind the peptide to the solid substrate.

Another macromolecule, often used for polymer analogous reactions, is cellulose, and cellulose derivatives have been technically significant for a long time because of its natural abundance. Cellulose is a polysaccharide consisting of β-1,4 linked D-glucose units (Fig. 15.7) leaving three hydroxyl groups available per monomer unit as reactive functional groups.

Cellulose Esters

The OH functional groups of cellulose can be esterified with a variety of different acids (Fig. 15.8).

The reaction of cellulose with nitrating acid gives so-called, nitrocellulose, which is well known as the main component of celluloid. Many everyday objects, in addition to film rolls, were made of celluloid in the nineteenth century and the early twentieth century. However, because of its flammability, it was soon replaced by other materials and now only has a few uses such as a raw material for paint.

Esterification of cellulose with acetic acid produces cellulose acetate. Textile fibers and cigarette filters are made from this biodegradable material.

The reaction of cellulose with carbon disulfide under alkaline conditions is of special interest. The resulting cellulose xanthate is sensitive to acid and can be transformed back into cellulose by lowering the pH. This permits cellulose xanthate to be spun from an acidic solution (► Chap. 17). During this process, the carbon disulfide is eliminated and cellulose fibers, known as rayon or artificial silk, are produced.

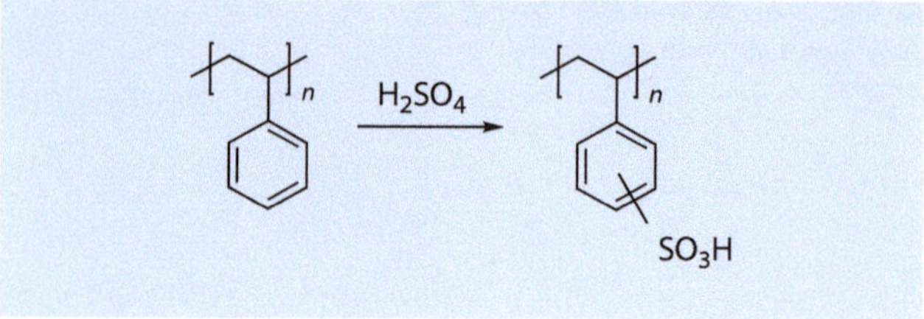

Fig. 15.5 Conversion of polystyrene with sulfuric acid

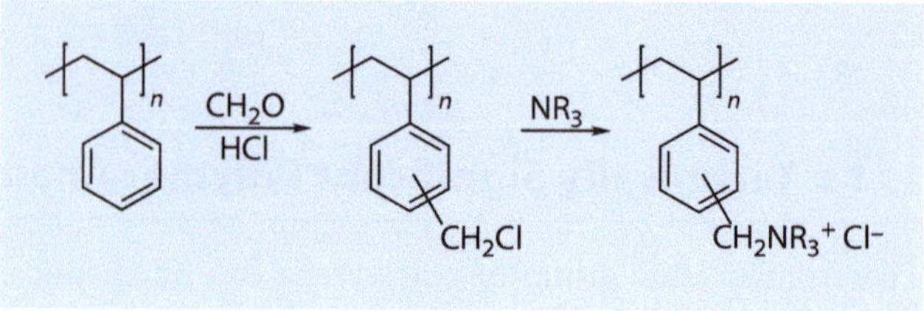

Fig. 15.6 Chloromethylation and amination of polystyrene

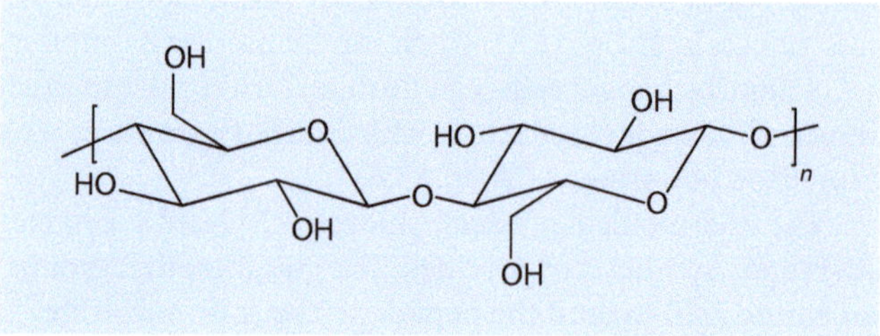

Fig. 15.7 Structure of cellulose and the D-glucose units

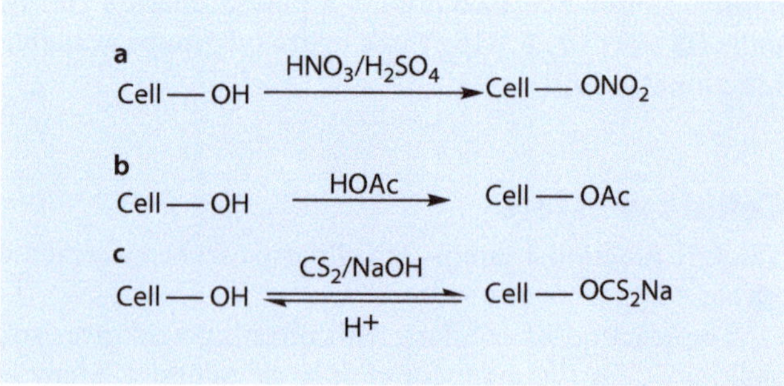

Fig. 15.8 Conversion of cellulose to (**a**) cellulose nitrate, (**b**) cellulose acetate, and (**c**) cellulose xanthate. *Cell-OH* cellulose

15

Cellulose Ethers

As well as from its reaction with acids, industrially important polymers also result from the etherification of cellulose with alkylating agents (Fig. 15.9). Thus the reaction of cellulose with alkyl halides yields a diverse selection of cellulose ethers. By employing ethylene oxide as reactant, hydroxyethyl cellulose is obtained. Another important reaction is the alkali catalyzed reaction with chloroacetic acid from which carboxy methyl cellulose (CMC) results.

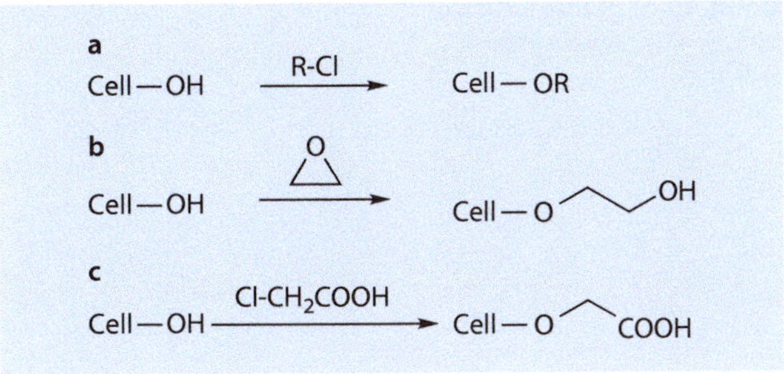

Fig. 15.9 Reaction of cellulose to (**a**) cellulose ethers, (**b**) hydroxyethyl cellulose (HEC), and (**c**) carboxymethyl cellulose (CMC)

These materials are functional polymers (► Chap. 19), for example as food thickeners (E466), stabilizers for suspensions of nano particles, glues, pharmaceutical additives, and flow improvers for concrete as well as for additives to textiles and paper. Cellulose ethers are also used as membrane substrates or, especially benzyl cellulose, as thermoplastic resins.

15.2 Cross-Linking Reactions

Cross-linking reactions result in covalent bonds between individual macromolecules and are thus, by definition, not polymer analogous reactions. However, the chemical cross-linking of polymers is also very important industrially, and has a great effect on their material properties, as mentioned briefly here:

- The glass transition temperature increases because of the decrease in the mobility of the polymer chains and the increasing molar mass which approaches an infinite value during cross-linking.
- The strength of the material increases.
- Because of the increase in molar mass, the solubility of the macromolecules decreases. When cross-linking is complete, the material is completely insoluble (► Chap. 1).

Industrially important cross-linking reactions of polymers are briefly discussed below.

15.2.1 Cross-Linking of Polyolefins

Polyolefins can be cross-linked in order to improve their material properties. This can be achieved chemically, thermally, or by radiation. These reactions usually take place radically (Fig. 15.10).

15.2.2 Vulcanization of Rubber

The vulcanization of rubber is probably the oldest chemical reaction intentionally carried out on polymers (Goodyear 1844). Natural rubber, which is sticky and thus not particularly useful in its original state, is transformed into an everyday, essential material by reacting it, e.g., with sulfur (Fig. 15.11). During this reaction the *cis*-polyisoprene chains are linked together by oligomeric sulfur bridges consisting of up to eight sulfur atoms.

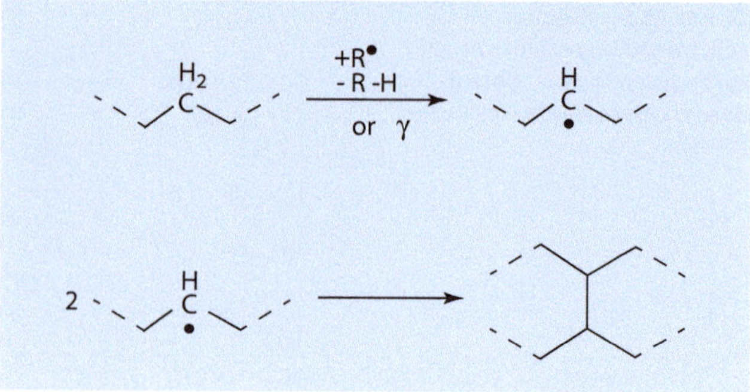

Fig. 15.10 Cross-linking of polymers by radicals or by radiation

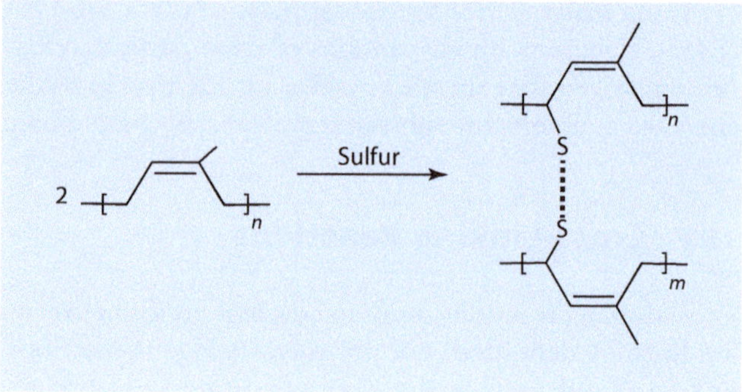

Fig. 15.11 Vulcanization of natural rubber (*cis*-polyisoprene)

15.2.3 Photoresists

The structuring of surfaces by light plays a crucial role in many industrial processes, especially in the electronics industry for the production of micro-structured electronic components. The structuring of polymeric surfaces can take place by a variety of means:

- Increasing the solubility by radiation
- Decreasing the solubility by radiation
- Direct removal with an intensive light beam

Polymeric surfaces, which interact with light in one of these ways, can be structured by simply exposing them to light through a mask. The structures accessible in this manner can be extremely small, which is an essential requirement for the production of microstructures. The processes of structuring and the materials used in these processes are now discussed in some detail.

15.2.3.1 Positive Resists

Polymers referred to as "positive resists" (*p-resists*) become more soluble on exposure to light of a suitable wavelength. An example is shown in Fig. 15.12.

The polarity of the molecule increases drastically as a result of the insertion of an ionic group. As a result, irradiated areas are soluble in polar solvents or water in contrast to those that have not been irradiated.

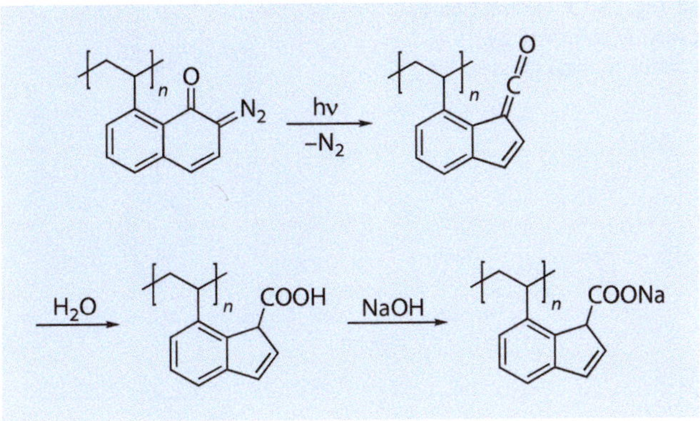

Fig. 15.12 Increasing polymer solubility by introducing a polar group with the aid of light

15.2.3.2 Negative Resists

So-called "negative resists" (*n-resists*) display exactly the opposite reaction when they are irradiated: they become insoluble. The easiest way to achieve this is by photochemical cross-linking (Fig. 15.13).

The material becomes insoluble in all solvents as a result of cross-linking so that the un-cross-linked macromolecules can be easily removed.

The basic procedure for the surface structuring of p- and n-resists is shown in Fig. 15.14. On the left a substrate is coated with an n-resin, which is soluble in its non-irradiated state. The irradiated regions are insoluble because of cross-linking after irradiation through a mask. In a further step, the regions which have not been irradiated, and thus remain soluble, are removed. A structured substrate remains. A similar procedure, but with a negative mask, can be employed for structuring using p-resins.

15.2.3.3 Polymers for Photoablation

By exposure to high-energy radiation, suitable polymers are broken down into small molecules. The low molar mass fragments are removed after irradiation. To obtain better defined structures, groups sensitive to radiation, such as triazene groups, can be included in the backbone of the macromolecules. When irradiated by a pulsed laser with a frequency in the nanosecond range, the surface becomes heated (in terms of atomic thermal motion) by the energy of the laser during each pulse. Because of the poor thermal conductivity of polymers, allowing only a slow energy transfer to the surrounding material, the energy from the laser is concentrated in a thin layer (around 1 µm for pulse lengths of 10 ns). The surface reaches very high temperatures during this process and instantaneous evaporation or disintegration of the material takes place. An example of a structure produced by laser ablation is shown in Fig. 15.15.

15.2.4 Paints

The paint on a modern automobile is about 110 µm thick so that per vehicle around 1 kg of paint is needed; nowadays, car paint comprises four layers (Goldschmidt and Streitberger 2002). The so-called *electrocoat* is galvanically applied directly onto the metal bodywork of the car. This electrocoat protects the car from corrosion. Then the *filler* is applied. The filler

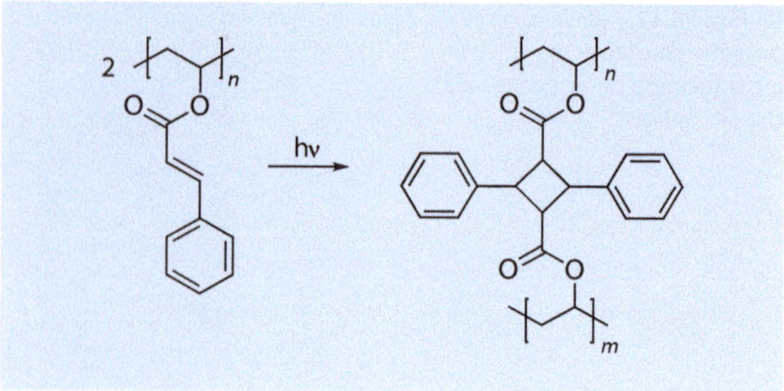

■ **Fig. 15.13** Reduction of polymer solubility by photochemical cross-linking

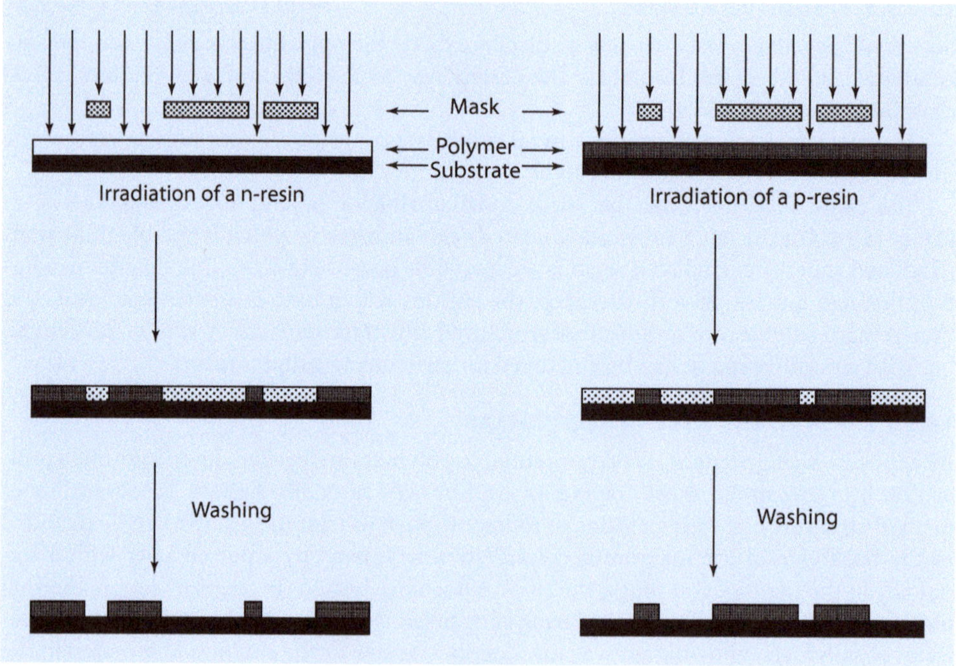

■ **Fig. 15.14** Structuring surfaces covered with n- or p-resists

levels out any microscopic unevenness of the bodywork and the electrocoat and protects the vehicle from the impact of small stones in use. The pigments that give the vehicle its color are contained in the next layer, the *base paint*. The top layer, made up of *clear coat*, is of special importance for the appearance of the vehicle. It is responsible for the glossiness and scratch resistance of the varnish system and is made up of a layer of polymers produced from their monomers and reactive oligomers (so-called pre-polymers) during the varnishing of the bodywork surface. This process is a challenging one for every polymer chemist, as the reaction mixture must form a perfect, absolutely even layer as soon as the varnish hardens, regardless of whether the surface is flat or curved or whether it is vertical

Fig. 15.15 Electron micrograph of a structure produced from a triazene functionalized polymer using pulsed laser ablation

Fig. 15.16 Pentaerythritol tetraacrylate as an example of a cross-linking monomer in clear coat varnishes

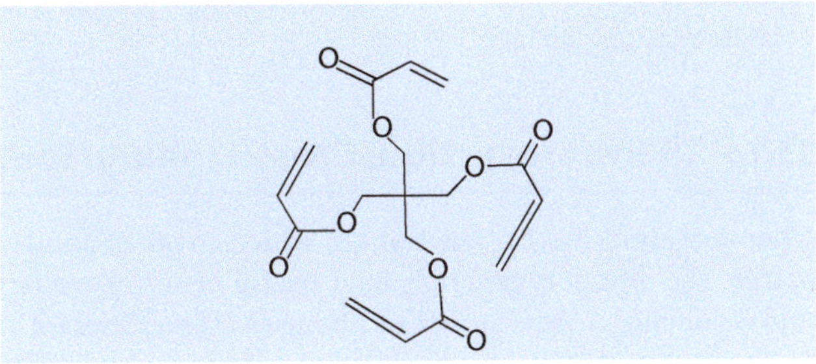

Fig. 15.17 Example photoinitiators and their decomposition products

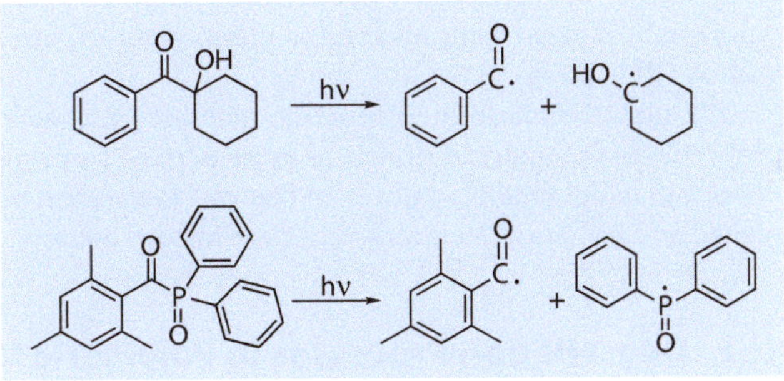

or horizontal in relation to the ground. The varnish mixtures used for these top coats are extremely complex. An example of a monomer used in clear coat varnish is pentaerythritol tetraacrylate shown in Fig. 15.16.

The system quickly reaches its gelling temperature as the molecule has four reactive double bonds (▶ Sect. 8.3.4). Even if the transformation of all double bonds is not complete, in practice every molecule is integrated into the network via at least one double bond. Hardening can be initiated photochemically after the surface has been coated with the clear coat. To do this a so-called photoinitiator is added to the varnish which disintegrates into radicals when irradiated with light of a suitable wavelength and these trigger polymerization, or rather cross-linking. Two examples of these kinds of initiators and their decomposition products when irradiated are shown in Fig. 15.17.

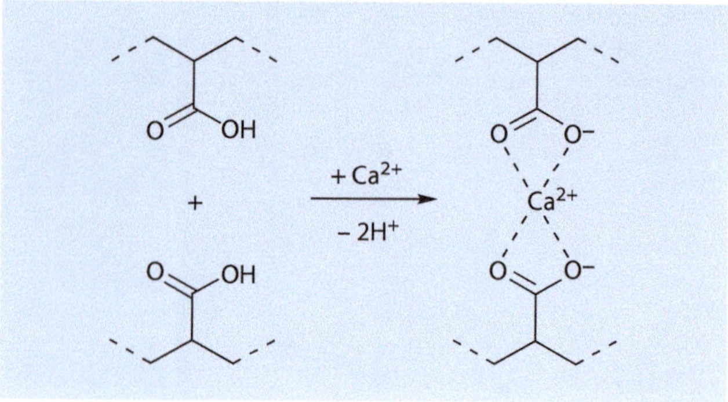

Fig. 15.18 Physical cross-linking of acid functionalized polymers with calcium ions

The advantage of the photochemical cross-linking of such, highly functionalized systems is the rapid speed of the process. Depending on the type of system, a short flash of light is enough to transform a viscous, sticky monomer mixture into a hard, glossy, and scratch resistant surface.

15.2.5 Thermoreversible Cross-Linking: Ionomers

Chemical cross-linking reactions in which covalent bonds are formed are irreversible processes. The products generally have useful material properties; however, they are infusible and insoluble so they cannot be changed into a different shape by melting or dissolution. In contrast to this, there are materials which are reversibly cross-linked by non-covalent, physical cross-links. One particular example is materials cross-linked via ionic interactions, such as the copolymers of ethylene with 5–10 wt% acrylic acid. These macromolecules can be ionized using alkaline solutions and then cross-linked using multivalent ions such as Ca^{2+} (Fig. 15.18).

The ionic bonds formed function as cross-links at lower temperatures and impart properties to the material similar to those derived from covalent cross-links. These bonds dissociate at higher temperatures, in contrast to covalent bonds, so that the material can be melted and reshaped. New ionic links are formed again as the material cools.

15.3 Degradation Processes in Polymeric Materials

Similar to non-polymeric materials, which age over the course of time (they become rusty or weathered), most polymeric materials also show signs of ageing to a greater or lesser extent when exposed to the environment. Ageing can be brought about by:
- Temperature stress
- Light or UV radiation
- Oxygen
- Water (hydrolysis)
- Microorganisms (biodegradation)
- Mechanical stress

Fig. 15.19 Depolymerization of a polymer

$$P_n \longrightarrow P_{n-1} + M$$

Fig. 15.20 Chain scission of a polymer

$$P_{n+m} \longrightarrow P_n + P_m$$

The ageing of polymers involves in particular the following processes:

- Depolymerization
- Chain scission
- Oxidation and cross-linking

Cross-linking processes have already been dealt with in ▶ Sect. 15.2.

Depolymerization is, formally, the reverse of polymerization according to the equation shown in Fig. 15.19.

This process is only thermodynamically possible if the ceiling temperature of the material is exceeded or if the monomer is continually removed from the system by, for example, evaporation.

Whereas depolymerization leads to the monomer as one of the products, *chain scission* refers to the process whereby the polymer backbone is split somewhere along its length to yield two, usually, polymeric fragments. This process can be described as in Fig. 15.20.

These two degradation processes have different effects on the development of the molar mass distribution. With the assumption that the monomer molecules or the fragments remain in the system, then the number average of the degree of polymerization develops along a hyperbolic curve that is identical for both degradation mechanisms. However, the weight average of the degree of polymerization decreases significantly more quickly with chain scission than it does during depolymerization as is shown in Fig. 15.21 for the degradation of an oligomer P_n with $n=13$. The reason for this is immediately apparent: if a monomer building block splits off from a polymer during depolymerization, then the weight average of the molar mass, or that of the degree of polymerization, is determined by the bigger molecule (▶ Chap. 3) and therefore only changes slightly. The weight average is therefore close to the original value for high degrees of polymerization. If, however, the same molecule is split in the middle because of a chain scission process, then two molecules that are roughly the same size are formed whose weight average degrees of polymerization are only half the size of that of the original macromolecule.

In general, the following applies: the closer the scission point is to the chain end, the smaller the effect on the weight average of the degree of polymerization. Conversely, if scission takes place in the middle of the chain, the difference between the various averages of the degree of polymerization is significantly smaller than that for depolymerization.

The courses of the different averages are shown in Fig. 15.21. In the case of chain scission, the curve for the weight average molar degree of polymerization can take on a slightly different course depending on how the process takes place as the position of scission is random during degradation. This, however, does not affect the validity of the statement in the previous paragraph.

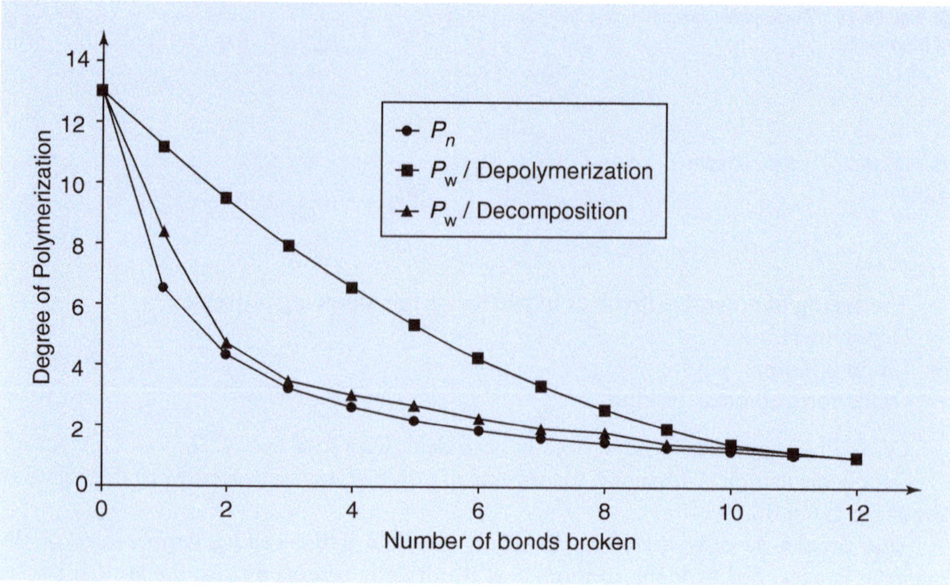

Fig. 15.21 Development of the number and weight average degrees of polymerization for a polymer with a degree of polymerization of 13 as a function of the mode of decomposition. The simulation assumes that all fragments remain in the system. P_n: number average degree of polymerization for both mechanisms (see text)

15.3.1 Degradation Mechanisms

Degradation processes that result in chain scission can take place with or without the participation of oxygen. In both cases, the first step of the process is the formation of radicals. This can take place thermally, photochemically, through the influence of gamma rays or electrons, and by oxidation. The polymer radical formed can then undergo a β-scission during which the degree of polymerization is reduced. During this process, the radical character of the system is not lost so that further radical reactions can take place. These processes are shown in Fig. 15.22.

A polymer radical can react in two different ways in the presence of oxygen, resulting in either an oxidative chain scission or peroxide formation.

The formation of peroxides occurs on the addition of a molecule of oxygen to the polymer radical. The peroxy radical formed in this process can then abstract a hydrogen atom from another polymer chain to form a hydroperoxide and a new polymer radical. This starts a chain reaction involving the absorption of oxygen and the formation of peroxides. This process is autocatalytic, and accelerates with time as the peroxides formed can generate new radicals. The peroxides produced can also rearrange into a ketone and an alcohol (Fig. 15.23).

The ketones formed during this process can take part in further scission reactions of the Norrish type when exposed to light. The Norrish-I-type reaction consists of an α-scission taking place and the generation of radicals which then initiate secondary reaction. In a Norrish-II-type reaction an excited carbonyl group abstracts a hydrogen atom from the γ-position (γ-H-abstraction) and the product decomposes to yield an α-olefin and a methyl ketone. These processes are shown in Fig. 15.24.

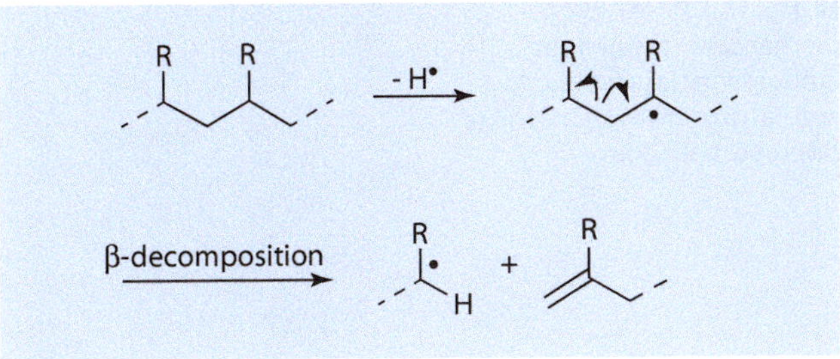

■ **Fig. 15.22** Radical chain scission without oxygen

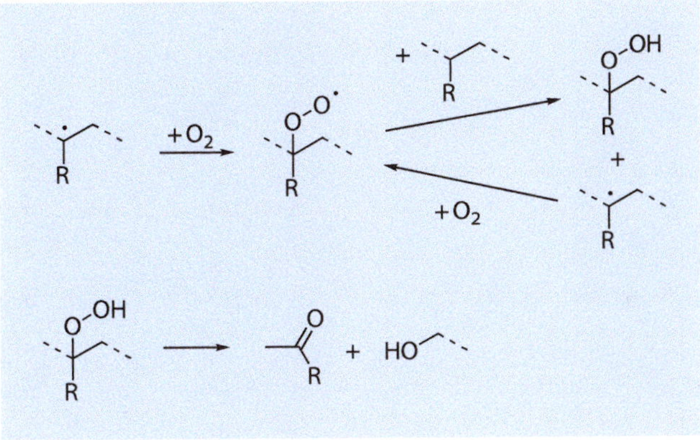

■ **Fig. 15.23** Oxidation of polymer radicals

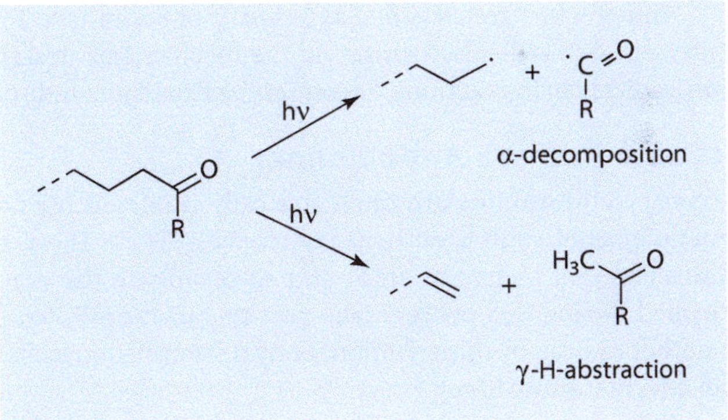

■ **Fig. 15.24** Photochemical decomposition of polyketones via Norrish reactions

15.3.2 Stabilization of Polymers

Polymer materials can be stabilized by basically two types of compounds: antioxidants and UV stabilizers. These compounds are added to the polymer as additives during processing. A prerequisite for effectiveness is the miscibility of the additive with the polymer (▶ Chap. 2). It is particularly important when choosing additives to prefer those with a low tendency to crystallize; the additive's enthalpy of crystallization competes with the free enthalpy of the mixing process.

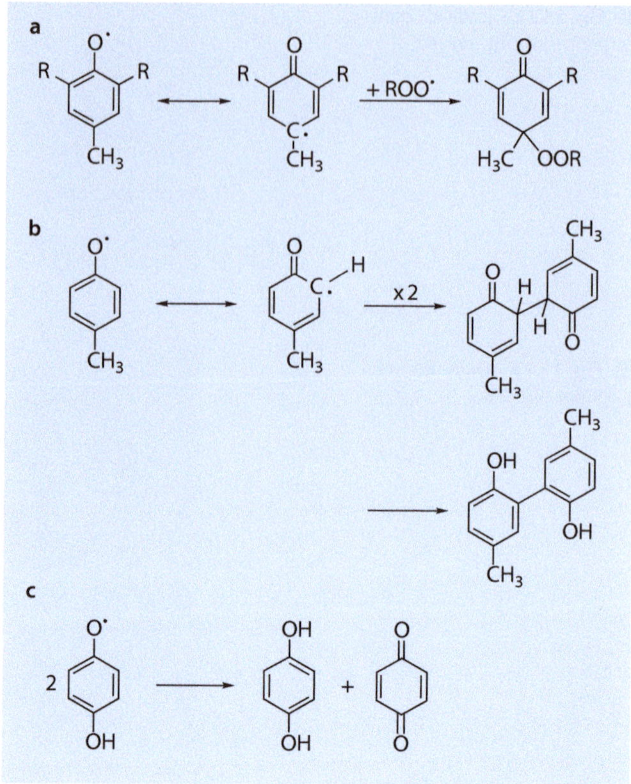

Fig. 15.25 Reaction mechanisms for phenolic antioxidants: (**a**) addition to a radical, (**b**) dimerization, and (**c**) disproportionation

Antioxidants are classified as primary or secondary. Primary antioxidants are used to intercept free radicals, formed via the mechanisms described above, and thus terminate any radical chain reactions. Secondary antioxidants reduce any peroxides that form.

15.3.2.1 Primary Antioxidants

Primary antioxidants are often sterically hindered phenols such as 2,6-di-*tert*-butyl-4-methylphenol (butylated hydroxytoluene, BHT). These substances saturate radicals by transferring a hydrogen atom and so terminate the reaction chain. The BHT radicals formed during this process take part in various follow-up reactions by combining with another radical, by dimerization, or by disproportionation (Fig. 15.25). The dimer is also an effective antioxidant.

15.3.2.2 Secondary Antioxidants

Sterically hindered amines, thioethers, or phosphites are used to reduce peroxides that have already been formed. A specific feature of amines is that they can react with peroxy radicals and with polymer radicals and can thus decrease the total radical concentration in the polymer. These processes are shown in Fig. 15.26.

As is evident from Figs. 15.25 and 15.26, antioxidants do not react catalytically, but stoichiometrically; i.e., they become completely consumed above a certain level of oxygen exposure. Once this level has been reached, the oxidation of the polymeric material takes place at the same rate as an unstabilized polymer. The period of use of the material is equivalent to the induction period of peroxide formation, i.e., the period of time in which

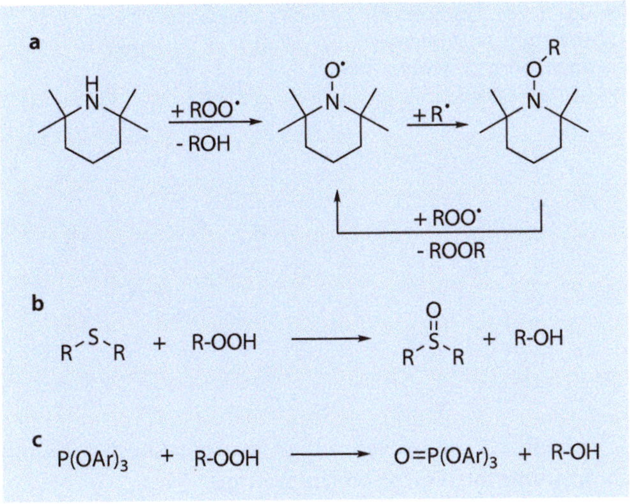

Fig. 15.26 Reduction of peroxides: (**a**) by tetramethyl piperidine (TMP) as an example of a sterically hindered amine, (**b**) by a thioether, and (**c**) by a triaryl phosphite

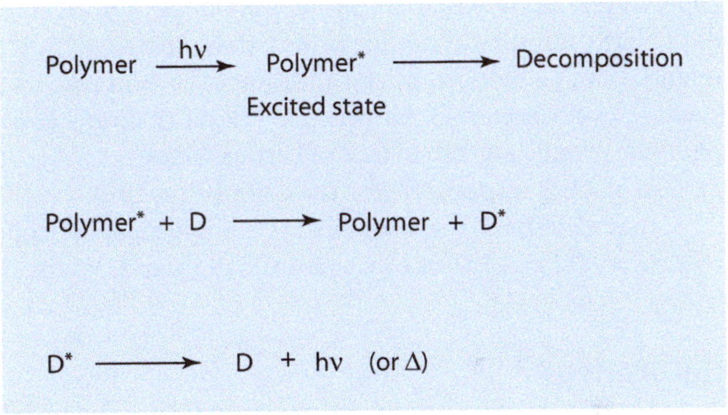

Fig. 15.27 Mode of action of UV-stabilizers. *D*: deactivator

the polymer is protected from oxidative stress by the remaining antioxidants. This time period can be lengthened by increasing the amount of additive. However, the amount is limited by the solubility of the additive and, if exceeded, the additive crystallizes or simply migrates out of the polymer. Too much additive can also, under certain circumstances, lead to detrimental side-effects, such as an increased absorption of water, oxygen, or other gases, together with changes in the mechanical or thermal properties (▶ Chap. 7).

15.3.2.3 UV Stabilizers

As light plays a significant role in many degradation processes, polymers can also be protected from degradation by UV-absorbing substances. How these so-called *UV stabilizers* function is shown in Fig. 15.27.

Initially, a chromophore in a polymer is excited by absorbing light. The excited chromophore then reacts, leading, for example, to the breaking of a bond. In the presence of a deactivator the excitation energy of the polymer can be transferred to the deactivator whereby the polymer reverts to its original state. The deactivation of the UV stabilizer can then take place either thermally or by the emission of light of a longer wavelength. In contrast to antioxidants, which are consumed by reaction with oxygen and radicals,

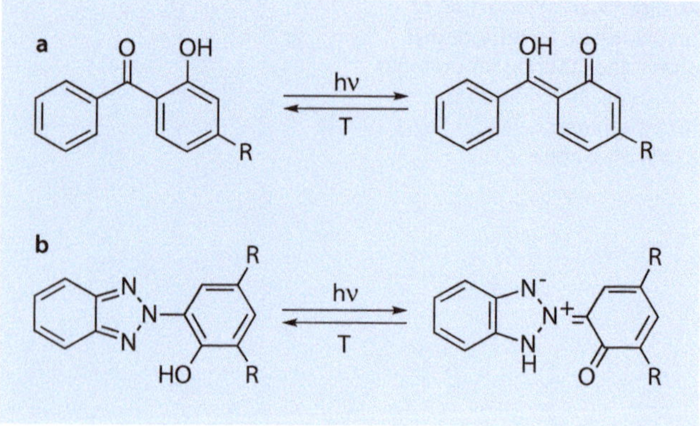

Fig. 15.28 Examples of UV-stabilizers. (**a**) 2-Hydroxy benzophenone. (**b**) Hydroxyphenyl benzotriazole

UV stabilizers can provide a high level of protection as long as no loss of stabilizer occurs, for example by desorption (migration).

Carbon black is the most common UV stabilizer. A good example of the use of carbon black is in car tires. These contain natural rubber, polybutadiene, and styrene-butadiene copolymers, all of which are susceptible to photoinduced, oxidative degradation because of the large number of double bonds along their polymer backbone. The UV stress of the material can be reduced by the addition of carbon black which provides for the long-term stability of the material. Absorbed UV light is simply converted into heat. As well as this, phenolic groups on the surface of carbon black act as radical traps (see above). The use of carbon black is responsible for the color of car tires.

Other examples of UV stabilizers are hydroxyl benzophenone or hydroxyphenyl benzotriazole. Fig. 15.28 shows examples of these UV absorbers.

References

Goldschmidt A, Streitberger HJ (2002) BASF Handbuch Lackiertechnik. Vincentz, Hannover

Goodyear C (1844) Improvement in India-rubber fabrics. U.S. Patent 3633

15

Industrially Relevant Polymerization Processes

S. Koltzenburg et al., *Polymer Chemistry*, DOI 10.1007/978-3-662-49279-6_16

In this chapter, industrially relevant polymerization methods are described in more detail. The focus is on chain growth polymerization, and special attention is paid to the heterogeneous methods—suspension and emulsion polymerization. Gas phase polymerization is dealt with separately in ▶ Sect. 11.5.

16.1 Overview

Polymerization can take place in bulk, from the gas phase, in suspension (monomer/water and, e.g., monomer/hexane), in solution, and in emulsion. Polymerization from the gas phase, but often from solution or bulk systems, is accompanied by the polymer precipitating and becoming a separate phase. The choice of method depends on the monomer, the resulting polymer, the conditions (e.g., temperature and pressure) necessary for the process, and the desired physical form of the product, so the industrial synthesis of polymers involves a broad range of methods (Table 16.1).

16.2 Polymerization in Mass

Polymerization by opening the C=C double bond is an exothermal process, so polymerization in substance ("in bulk") involves substantial challenges to remove the heat generated. Moreover, viscosity problems arise, which are caused by increasing the viscosity by multiple orders (transition of a liquid to a solid). Local overheating is possible as a result of the gel effect. In this case the molar mass distribution of the product can change or the polymer becomes discolored as a result of oxidation or decomposition. One solution is to interrupt the polymerization at a conversion of 40–60% and recycling the residual monomer.

Table 16.1 Examples of the preferred industrial processes for a selection of large volume polymers

Polymer	B	S	L	P	G	E	Mechanism
High density polyethylene		+	+	+	+		z, (m)[1]
Low density polyethylene	+	+					r
Polypropylene	+	+	+		+		z, (m)[1]
Polystyrene	+	+				+	r, (m)[1]
Polymethyl methacrylate	+	+	+			+	r
Butyl rubber				+			c
ABS-polymers			+			+	r
SBS-polymers				+			a
Polyacrylonitrile				+			r

B Bulk polymerization, *S* Suspension polymerization, *L* Solution polymerization, *P* Precipitation polymerization, *G* Gas phase polymerization, *E* Emulsion polymerization, *a* Anionic polymerization, *c* Cationic polymerization, *m* Metallocene Catalyst, *r* Radical polymerization, *z* Ziegler–Natta Catalyst[1]To date, less important

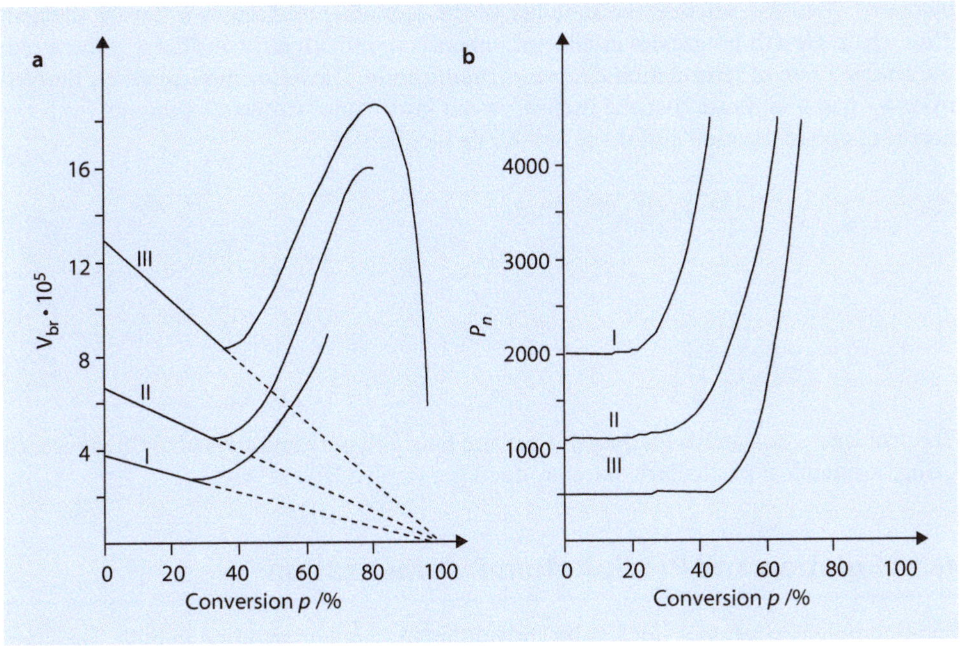

Fig. 16.1 Progress of (**a**) v_{br} and (**b**) P_n with time and the development of the gel effect ($[I]_I < [I]_{II} < [I]_{III}$). *Dashed line*: the bulk polymerization of styrene without a gel effect (Henrici-Olivé and Olivé 1958)

The *gel effect* (also referred to as the *Trommsdorff effect*) is particularly relevant for the polymerization of methyl methacrylate and styrene at high conversions, at which both the overall rate of reaction v_{br} and the average degree of polymerization P_n increase exponentially (Fig. 16.1), despite the fact that both these parameters should decrease because $v_{br} \sim [M]$, $P_n \sim [M]$, and $[M]$ decrease as conversion progresses (► Sect. 9.3.1).

The degree of polymerization does not increase as expected because the initiating radicals also react with the polymer as soon as polymerization begins, causing branching and higher degrees of polymerization (Fig. 16.1b). One possible explanation lies in the fact that at higher conversions during bulk polymerization the viscosity increases and, as a result, the heat dissipation becomes less efficient and the temperature increases. At the higher temperature, more radicals are formed per unit time. Thus, because

$$v_{br} \sim \left[P^\bullet\right] \tag{16.1}$$

the overall rate of reaction (► Sect. 9.2.2) increases.

Furthermore, from (9.28) and (9.31) follows:

$$P_n \sim \frac{1}{\left[P^\bullet\right]} \tag{16.2}$$

However, this suggests that the degree of polymerization would decrease (► Sect. 9.3) and is contrary to what is observed (Fig. 16.1b).

That a gel effect can also be observed when temperature control is very effective suggests it is the flexibility of the polymers that is most affected at higher conversion and increased viscosity, whereas the mobility of the monomer molecules is hardly affected. Thus, chain growth progresses unaffected, whereas termination is considerably impaired; the reaction rate of termination decreases significantly. The following equations, derived from ► Chap. 9, indicate that the molecules can grow "unobstructed," although both the degree of polymerization and the reaction rate increase:

$$v_{br} = k_p \cdot \sqrt{\frac{2k_d}{k_t} \cdot [I]} \cdot [M] \tag{9.5}$$

$$P_n = \frac{k_p [M]}{\sqrt{2k_t \cdot k_d \cdot [I]}} \tag{9.32}$$

The problems associated with the gel effect and bulk polymerizations lead to this approach being unsuitable if particularly pure products are wanted (PS, MMA).

16.3 Solution and Precipitation Polymerization

Solution polymerization is not significantly different to polymerization in bulk. The same kinetic laws apply. In both polymerization methods there are problems with viscosity and temperature control at higher conversion, although both problems are ameliorated by the added solvent. Formally, bulk polymerization can also be described as solution polymerization where the monomer is the solvent for the polymer rather than another, non-polymerizable solvent. Solution polymerization is a typical laboratory method.

Industrially it is used more for elastomer polymerizations (e.g., polybutadiene, EPDM recently for special SBS grades) than for thermoplasts; it is also often employed if the polymer is to be marketed as a polymer solution such as is often the case with functional water-soluble polymers (► Chap. 19).

In some cases the polymer is not or is only slightly soluble in its own monomer, so that with increasing conversion the polymer precipitates. This is therefore referred to as a precipitation polymerization. In most cases, the reason for the limited solubility is that the polymers are crystalline or semi-crystalline. The kinetics are not different to those in solution. However, different reaction rates can prevail in the liquid and solid phases. The most basic assumption that only different termination rate constants apply in the monomer-rich and the polymer-rich phases (impeded termination in the polymer phase) leads to a surprisingly good explanation of the experimental results. The chemical processes obviously overshadow the diffusion and mass-transfer processes. Popular examples are VC/PVC, VF/PVF, and ACN/PAN.

16.4 Suspension Polymerization

For radical polymerizations, suspension technology is very often employed. With this method a monomer which is only slightly soluble in water is dispersed as droplets or pearls with diameters of 0.01–3 mm. To stabilize this emulsion (oil in water), protective colloids or inorganic suspension aids are added which inhibit the coalescence of the droplets. The protective colloid in Fig. 16.2 is polyvinyl alcohol. Its hydrophobic backbone is positioned on the surface of the hydrophobic pearls and its polar functional groups, the –OH groups, face the water (Fig. 16.2).

With a monomer-soluble initiator, the polymerization is initiated by heating the reactor to the decomposition temperature of the initiator; a bulk polymerization takes place in the droplets. Each monomer droplet is a small, discontinuously operating batch reactor, optimally cooled by the surrounding water. Because of its two-phase nature, the reaction mixture—initially consisting of monomer droplets in water which becomes a mixture of polymer pearls in water with progressing conversion—retains a low viscosity similar to water even up to high monomer conversion. Thus, a considerably lower temperature gradient between the reactor contents and the coolant is observed than that existing for bulk polymerization (Fig. 16.3).

Whereas during a bulk polymerization (Fig. 16.3) the boundary layer (Fig. 16.3a) grows continuously with increasing conversion, the heat dissipation is aided significantly by carrying out the reaction as a suspension/pearl polymerization.

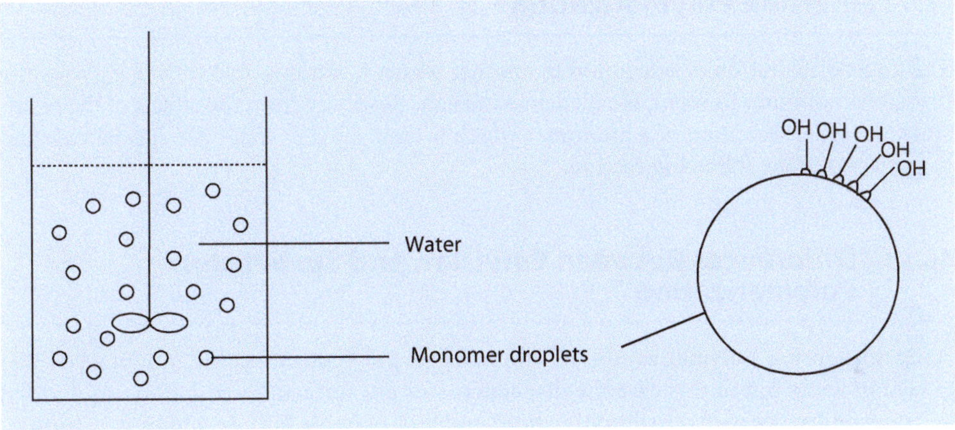

Fig. 16.2 Schematic diagram of a suspension polymerization and the mode of action of the stabilizer (polyvinyl alcohol)

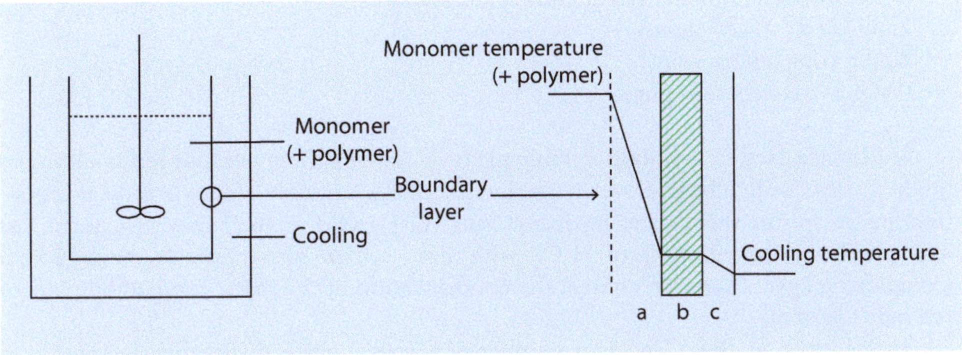

Fig. 16.3 Schematic diagram of a bulk polymerization and the temperature of the reactor zones. (**a**) Increasing temperature gradient in the monomer (and polymer) reaction mixture. (**b**) Reactor wall. (**c**) Boundary between the reactor wall and the coolant

The polymer forms as small, pearl-shaped particles which can be separated immediately. Suspension aids and protective colloids remain in the water or are easily removed from the polymer by washing. However, wastewater can, potentially, be a problem.

For a typical suspension polymerization, the following chemicals are required:

- Monomer(s) insoluble in water (styrene, methyl methacrylate, …)
- Initiator, soluble in the monomer
- Stabilizer (which preferably accumulates in the oil-water boundary layer)
- Water as dispersion agent (continuous phase)

Polyvinyl alcohol and partially saponified polyvinyl acetate, polyvinyl pyrrolidone, and methyl cellulose are typical protective colloids. As suspension agents, $BaSO_4$, $CaCO_3$, talcum, and hydroxyl apatite are often used.

The advantage of suspension polymerization over bulk or solution polymerization is emphasized in Figs. 16.2 and 16.3.

16.5 Emulsion Polymerization

The finest distribution of one liquid in another which is not miscible with it, e.g., a water insoluble monomer in water, is called an emulsion. Resulting from the effects of the emulsifier, the polymerization of a monomer which is emulsified in water is a special case and is discussed in the following section.

16.5.1 Differences Between Emulsion and Suspension Polymerization

As in suspension polymerization, in an emulsion polymerization the monomer is dispersed in water but in this case the dispersion aids are surfactants (e.g., sodium dodecyl sulfate—molecules with a hydrophilic head and hydrophobic tail). In addition, the initiators employed for the emulsion polymerization are soluble in water. The following chemicals are usually also included in a typical emulsion polymerization recipe:

- Monomer(s) not soluble in water
- Initiator soluble in water (frequently redox systems)
- Emulsifier(s) (surfactant(s))
- Buffer, complexing agent
- Water as a continuous phase

Buffers are used to maintain a stable pH level in the system because redox initiators are frequently only effective over a narrow pH range and only react via radical transition stages within this range. Hydroxyl ions (OH^-) (Fig. 16.7) may also form, for example from the reaction of H_2O_2 with Fe^{2+}, which also needs to be buffered. Complexing agents serve to control the concentration of the metal ion components of the redox system.

Recipes such as these lead to a number of features which distinguish the emulsion polymerization appreciably from a bulk or suspension polymerization.

One fundamental difference to the suspension polymerization is that the polymerization does not take place in the monomer drop but in micelles which are formed from surfactants above the so-called *critical micelle concentration* (CMC).

16.5.2 Mode of Action of a Surfactant

Surfactants are only partially soluble in water. If the threshold of molecular solubility is exceeded, the surfactant molecules occupy the air/water interface with the hydrophobic tail facing the air and the hydrophilic head in the water (▫ Fig. 16.4a, c_1). The surfactant reduces the surface tension of the water. If the interface is completely occupied by surfactant molecules, the surface tension is at a minimum and the CMC is reached. Any further increase in the surfactant concentration (▫ Fig. 16.4a, c_3) leads to the formation of micelles, structures of ca. 100 surfactant molecules made up of a hydrophobic middle and a hydrophilic surface, which are often spherical in shape.

Because of the relatively large range of emulsifier concentrations which have to be examined, it can prove more useful to plot a log-log-graph rather than the linear

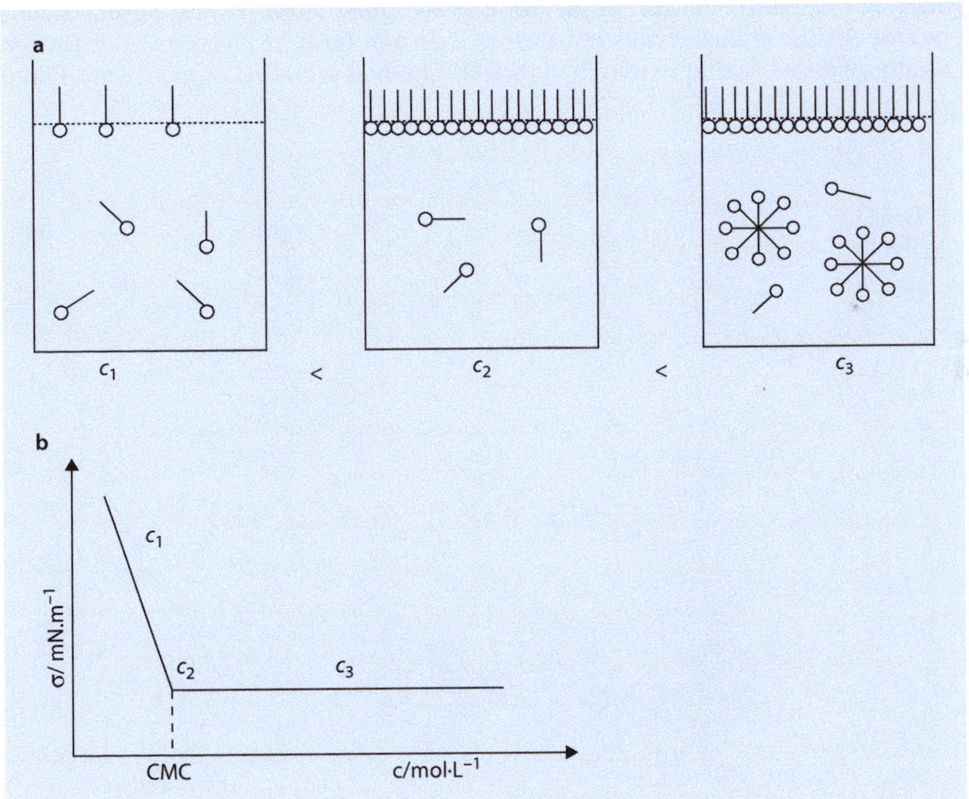

▫ **Fig. 16.4** (**a**) Localization of the surfactant at different concentrations c_1, c_2, and c_3. (**b**) Determining a surfactant's critical micelle concentration (CMC) by surface tension measurements

graph of the surface tension as a function of the emulsifier concentration shown in Fig. 16.4b.

16.5.3 Kinetics of Emulsion Polymerization

A snapshot of an emulsion polymerization is shown in Fig. 16.5. In this diagram all species, the monomer droplets, micelles, micelles filled with monomer, and micelles containing polymer are shown.

According to the generally accepted model (Smith and Ewart 1948) (▶ Sect. 16.5.5), a batch emulsion polymerization involves three stages: particle formation (I), particle growth (II), and monomer depletion (III). The changes in the overall reaction rate and surface tension as the reaction progresses through these three stages are shown in Fig. 16.6 and are explained in more detail below.

16.5.3.1 Particle Formation

The monomer molecules diffuse into the hydrophobic nucleus of the micelles and become solubilized. The generation of radicals from thermally induced decomposition of initiator molecules or redox initiators is the same as for conventional radical polymerization (▶ Chap. 9). The example redox initiators as shown in Fig. 16.7 play a significant role because of their solubility in water and their effectiveness at and below room temperature.

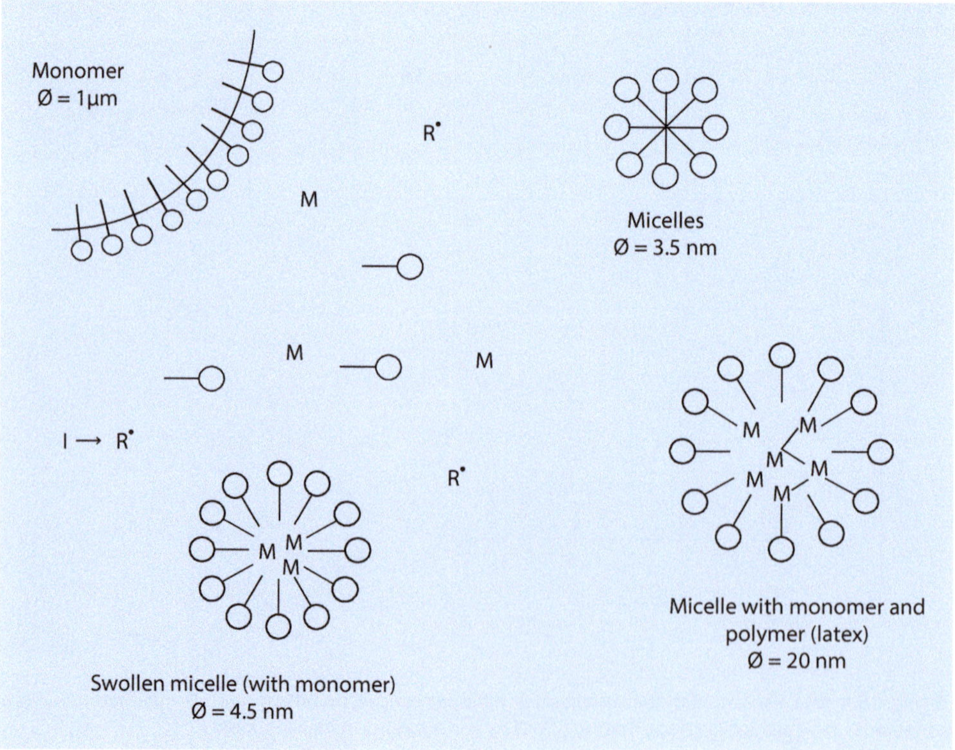

Fig. 16.5 Schematic snapshot of an emulsion polymerization

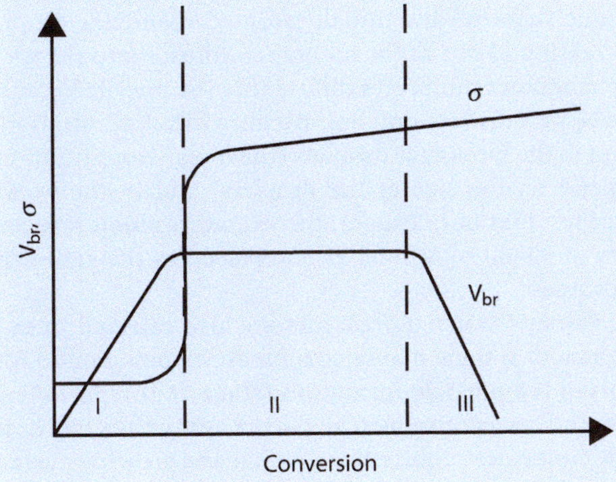

Fig. 16.6 Overall reaction rate v_{br} and the surface tension σ variation as a function of monomer conversion

$$H_2O_2 + Fe^{2+} \longrightarrow {}^{\bullet}OH + {}^{-}OH + Fe^{3+}$$

Fig. 16.7 Formation of initiator radicals by the oxidation of iron(II) with hydrogen peroxide

The –OH• radicals attack the monomers dissolved in water to form oligomer radicals. These oligomers now possess the same, and growing, impetus as a "surfactant" as the oligomer grows to diffuse into the micelle where it can continue to grow by consuming the monomer in the micelle (they become increasingly hydrophobic (the oligomer chain) and less hydrophilic (terminal –OH group)). An alternative theory is based on the concept that the growing oligomer radicals are no longer soluble in water as of a certain chain length and are stabilized by surfactant molecules which hinder their precipitation. Monomer then diffuses into the new micelle and can be polymerized. Oligomer radicals can also enter into the monomer droplets but, because the ratio of monomer droplets to micelles per cubic centimeter of emulsion is 10^{10}:10^{18}, this is highly unlikely.

The overall reaction rate steadily increases up to a total conversion of 10–15% (Fig. 16.6). According to the assumptions above, one after another the (nano-)reactors, the monomer-filled micelles, are "switched on" by the oligomer radicals. When 50% of all micelles are activated, the probability of an oligomer diffusing into an active or an inactive micelle is equal. A radical entering, an active micelle reacts with the radical present and the micelle is "switched off"; an inactive micelle is "switched on" by the entering radical. Thus, at this stage the overall reaction rate no longer changes[1] and the particle growth stage begins.

1 It is known from many studies that the assumption of merely one radical per micelle is extremely simplified and that multiple radicals can indeed coexist in the latex.

16.5.3.2 Particle Growth

In this stage, all the initially formed monomer droplets remain. They are, however, decreasing in size as the monomer diffuses into the water phase. From the water phase the monomer molecules diffuse into the micelle because of their limited solubility and the Nernst distribution law. Because the concentration of the monomer remains constant in the monomer droplets (the molar concentration of the pure monomer does not depend on the size of the droplets) and in the water phase (the number of active micelles does not change), the overall reaction rate remains constant up to a conversion of 50–60 %. Monomer is replaced in the active micelles at the same rate as it is consumed.

The micellar nanoreactors are also referred to as *latex particles*. In this stage of the reaction their diameter increases from about 3.6 nm (micelle) to 20 nm (latex). During the particle formation (stage I) and particle growth (stage II), polymerization and, as a consequence, particle growth occur in the active micelles. The surfactant molecules required to stabilize the growing surface area of the growing particles are initially supplied by the inactive micelles until the supply of these is depleted. Thereafter the surfactants in the air/water interface and finally those surfactant molecules dissolved in the water phase are used. As a result, the surfactant concentration becomes ever smaller, and the surface tension increases, until the CMC is reached (◘ Fig. 16.6).

16.5.3.3 Monomer Depletion

When the monomer droplets have all been consumed, fresh monomer is no longer available and the monomer concentration in the latex particles decreases, leading to a decrease in the rate of polymerization.

In an emulsion polymerization the polymerization takes place almost exclusively in the micelles and only in exceptional cases in the monomer droplets, as is the case in a suspension polymerization.

This can be verified by:

- The characteristic kinetics
- The latex particle size at the end of polymerization whereby the largest particles are smaller than the smallest monomer droplets
- Direct observation of the increasing size of the micelles during polymerization with the aid of electron microscopy, light scattering, or ultracentrifugation
- Statistical considerations, according to which it is highly unlikely that an oligomer radical formed in water encounters a monomer drop instead of a micelle
- Interrupting polymerization after a conversion of 5% and breaking the emulsion by adding a few drops of aqueous HCI so that it separates into two phases; almost all the polymer can be found not in the oil phase, but in the water phase

16.5.4 Polymerization Rate and Degree of Polymerization

As for bulk and solution polymerization, the polymerization speed can be written as

$$-\frac{d[M]}{dt} = k_w [M][P^\bullet] \tag{9.4}$$

Because the polymerizing radicals are present in the micelles, $[P^\bullet]$ is proportional to the number of micelles N_{Mic} and the radicals $\overline{n}$ per micelle:

$$[P^\bullet] = \overline{n} \cdot \frac{N_{Mic}}{N_L} \tag{16.3}$$

$\overline{n}$	Average number of radicals per micelle
N_{Mic}	Number of micelles per liter
N_L	Loschmidt's number

With (9.4) and (16.3) one obtains for the rate of polymerization:

$$-\frac{d[M]}{dt} = k_p [M] \cdot \frac{N_{Mic}}{N_L} \cdot \overline{n} \tag{16.4}$$

Assuming that only one radical can exist per micelle, because any additional radical terminates the growing chain by combination and only a further radical can re-start chain growth, $\overline{n}$ equals 1/2. This assumption is reasonable for the beginning of polymerization. However, with increasing particle diameter and the viscosity within the particles, it becomes likely that multiple radicals can coexist inside a single latex particle and do not directly encounter one another.

If transfer reactions are ignored, the degree of polymerization P_n is given by (► Sect. 9.3)

$$P_n = \frac{v_p}{v_t} = \frac{k_p [M] \cdot N_{Mic} \cdot \overline{n}}{N_L \cdot 2k_z \cdot f \cdot [I]} \tag{16.5}$$

If more surfactant is added, the number of micelles N_{Mic} increases, so that both rate and the degree of polymerization increase. The initiator concentration does not have any influence on the polymerization speed, unlike in bulk and suspension polymerizations. However, more initiator means more frequent "turning off and on" of the polymerization in the micelles so that the degree of polymerization is decreased.

From a comparison between a polymerization of butadiene in solution and in emulsion, it can be shown that the emulsion polymerization progresses two magnitudes faster and yields polymers of considerably higher molar mass (Echte 1993).

16.5.5 Quantitative Theory of Smith and Ewart

The first satisfactory quantitative description of an emulsion polymerization was achieved by Smith and Ewart (1948). It is based on the following assumptions.

Starting with the basic kinetics of radical polymerization:

$$v_{br} \cong v_p \tag{16.6}$$

$$-\frac{d[M]}{dt} = k_p [M][P^\bullet] \tag{9.4}$$

$$[P^\bullet] = \bar{n} \cdot \frac{N_{Mic}}{N_L} \tag{16.3}$$

$$-\frac{d[M]}{dt} = k_p [M] \cdot \frac{N_{Mic}}{N_L} \cdot \bar{n} \tag{16.4}$$

The radicals are formed in the water phase and enter *a single* latex particle with a rate ρ (number of radicals/(s · cm^3)):

$$\frac{dn}{dt} = \frac{\mathbf{r}}{N_{Mic}} \tag{16.7}$$

n	Number of radicals in the latex particle

The radicals can also exit the latex particle again (desorb):

$$-\frac{dn}{dt} = k_0 \cdot S \cdot \left(\frac{n}{\upsilon}\right) \tag{16.8}$$

S	Surface of the latex particle
υ	Volume of the latex particle
k_0	Desorption rate constant

Two radicals in one latex particle can combine and terminate chain growth:

$$-\frac{dn}{dt} = 2k_t \cdot \frac{n(n-1)}{\upsilon} \tag{16.9}$$

Each radical can now only react with $(n-1)$ radicals—it cannot react with itself.

If we look at the dispersion at the point in time t, it thus contains

N_0	Latex particles without radicals
N_1	Latex particles with one radical
N_2	Latex particles with two radicals
$\vdots$	
N_n	Latex particles with n radicals

The sum of all micelles (N_{Mic}) is then given by equation (16.10):

$$\sum_{n=0}^{n} N_n = N_{Mic} \tag{16.10}$$

Resulting from this, the average number of radicals equals

$$\bar{n} = \frac{0 \cdot N_0 + 1 \cdot N_1 + 2 \cdot N_2 + \cdots + n \cdot N_n}{N_0 + N_1 + N_2 + \cdots + N_n} = \frac{\sum_{n=0}^{n} n \cdot N_n}{\sum_{n=0}^{n} N_n} \tag{16.11}$$

How are the latex particles N_n formed that are filled with n radicals?

1. When each radical enters a latex particle N_{n-1} with $(n-1)$ radicals
2. When each radical exits a latex particle with $(n+1)$ radicals
3. At each termination between two radicals and a latex particle N_{n+2} with $(n+2)$ radicals

For the stationary state, the formation of N_n equals the decrease of N_n, so that

$$\begin{aligned} &N_{n-1} \cdot \frac{\rho}{N} + N_{n+1} \cdot \frac{k_0 \cdot S}{\upsilon} \cdot (n+1) + 2N_{n+2} \cdot \frac{k_t}{\upsilon} \cdot (n+2)(n+1) = \\ &N_n \cdot \frac{\rho}{N} + N_n \cdot \frac{k_0 \cdot S}{\upsilon} \cdot n + 2N_n \cdot \frac{k_t}{\upsilon} \cdot n(n-1) \end{aligned} \tag{16.12}$$

With the condition that radicals do not exit latex particles, i.e., $k_0 = 0$ and that two radicals combine very rapidly, i.e., $\rho/N \ll k_t/\upsilon$, it holds that

$$\begin{aligned} &N_0 \cdot \frac{\rho}{N} + \underbrace{N_2 \cdot \frac{k_0 \cdot S}{\upsilon} \cdot 2}_{=0,\text{since } k_0=0} + \underbrace{2 \cdot N_3 \cdot \frac{k_t}{\upsilon} \cdot 3 \cdot 2}_{=0,\text{since } N_3=0} = \\ &N_1 \cdot \frac{\rho}{N} + \underbrace{N_1 \cdot \frac{k_0 \cdot S}{\upsilon} \cdot 1}_{=0,\text{since } k_0=0} + \underbrace{2 \cdot N_1 \cdot \frac{k_t}{\upsilon} \cdot 1 \cdot 0}_{=0,\text{since one factor is } 0} \end{aligned} \tag{16.13}$$

If a desorption doesn't occur and the radicals combine rapidly, N_3 equals zero. Thus

$$N_0 \cdot \frac{\rho}{N} = N_1 \cdot \frac{\rho}{N} \tag{16.14}$$

$$N_0 = N_1 \tag{16.15}$$

$$\bar{n} = \frac{0 \cdot N_0 + 1 \cdot N_1}{N_0 + N_1} = \frac{N_1}{N_1 + N_2} = \frac{1}{2} \tag{16.16}$$

Equation (16.16) implies that a polymerization only takes place in every second micelle. A radical that now enters a micelle either terminates the growing chain or starts a new one. If the above requirements are not met, deviations from this simple result can be expected. A comprehensive solution of (16.12) was achieved by Stockmayer and O'Toole (Stockmayer 1957; O'Toole 1965).

16.5.6 Applications, Advantages, and Disadvantages of Emulsion Polymerization

Industrially, this polymerization method is of great importance; for example, polyvinyl chloride, polystyrene, polyacrylates, and polyvinyl acetate are produced with this method, among others. In many cases polymers are not used as such, but as dispersions (e.g., as emulsion paints, paper coatings, and glues). The polymers can also be separated from the water phase with spray driers, roller driers, or by coagulation, filtering, and drying. However, in most cases the emulsifier remains in the polymer and affects its properties.

The advantages of emulsion polymerization are (1) the process management in water is conducted without organic solvents and (2) the viscosity of the system remains low up to high solid contents so that, as with suspension polymerization, the heat of reaction can be removed without problems. The polymerized emulsion is, in many cases, ready for use. However, all the additives remain in the polymer.

16.6 Gas Phase Polymerization

Gas phase polymerization is most relevant in the catalytic polymerization of olefins such as ethylene and propene and is discussed in ▶ Sect. 11.5.

References

Echte A (1993) Handbuch der Technischen Polymerchemie. Chemie Weinheim, S 326

Henrici-Olivé G, Olivé S (1958) Kinetic measurements regarding Trommsdorff effect in polystyrene. Kunststoffe—Plastics 5:315–320

O'Toole JT (1965) Kinetics of emulsion polymerization. Appl Polym Sci 9:1291–1297

Smith WV, Ewart EH (1948) Kinetics of emulsion polymerization. J Phys Chem 16:592–599

Stockmayer WH (1957) The kinetics of emulsion polymerization. J Polym Sci 24:314–317

The Basics of Plastics Processing

S. Koltzenburg et al., *Polymer Chemistry*, DOI 10.1007/978-3-662-49279-6_17

With the exception of the functional polymers discussed in ▶ Chap. 19, polymers are typically used as solid materials for the widest variety of different objects, from plastic bags to prostheses. To fulfill each requirement, one needs to take the material, composed of macromolecules that have been produced in solution, bulk, or suspension, and give it form and precise geometry depending on the demands of the intended application. The exact conditions of normal use to which the object is exposed determine the choice of the chemistry of the material, for example, thermal, mechanical, and chemical properties; the actual use determines the geometry of the article. Many different processing technologies for shaping polymeric materials have been developed over the past few decades. However, all forms cannot be achieved with all processes, and all polymers cannot be processed using all techniques. This is why it is exceptionally important for a materials scientist to have a detailed knowledge of the interplay between shape, material, and processing technique. This chapter provides a short, introductory overview of the most important aspects of the basic concepts of the thermoforming of polymers. For further details that are beyond the scope of this book, the interested reader is referred to more comprehensive textbooks (Kaiser 2007; Michaeli 2010).

Generally, macromolecules are delivered to a processor or, as an intermediary, a *compounder* (7 Sect. 17.1.1) as powders or granulates. Polymers that require cross-linking before final use are also delivered as more or less viscous liquids. Elastomers (▶ Chap. 18) are generally supplied in bale (25 kg) form.

Processes for processing polymeric materials can be divided into two categories: *primary molding processes* and *reshaping processes.* Primary molding processes take a formless raw material, such as a powder or granulate, and turn it into a cohesive material as a more or less finished article or into continuous forms, such pipes and sheets. Particularly the sheets are often processed further into another final shape. These "half-finished" molded articles are simply raw materials to be processed further and reshaped, and are known as *semi-finished products*. The *reshaping processes* that are among the most important processing techniques for semi-finished products are those in which the shape of a part is changed without any material being added to or removed from it.

Further important processes for the processing of plastics are joining (welding) and separation processes as well as coating and finishing processes.

17.1 Primary Molding Processes

The most important processes for the manufacture of molded articles out of shapeless raw materials such as polymer powders or granulates are presented in the following paragraphs. Usually, additives (e.g., additional stabilizers or colorants) are added to the raw polymers before they are manufactured into articles. If these additions are made in a separate processing step, which may be necessary to ensure optimum dispersion, the process is referred to as *compounding*. Often the additives can simply be added during the primary molding stage as the polymer is being brought to a malleable temperature.

Shaping thermoplastic polymers is a purely physical process: the material is heated until its viscosity allows processing; it is then subjected to a shaping process and finally cooled down again. This process can be repeated as often as necessary as long as the material can withstand the thermal stress. By contrast, cross-linked systems such as duroplastics and elastomers can be shaped only once. The cross-linking takes place during or after the shaping of these materials so that they are then transformed into insoluble and infusible material. No further reshaping can take place. Of course, cross-linked articles can be joined together by gluing or cut (ground) into their final shape for use.

Some of the most important molding processes are:

- Extrusion
- Injection molding
- Blow molding
- Calendering
- Pressing

The processes for the production of fiber-reinforced plastics as well as polymer foams and fibers are also discussed below.

17.1.1 From the Raw Polymer to an Article

Macromolecules are often mixed with other substances during processing. These can be dyes, pigments, plasticizers, fillers, flame retardants, stabilizers (▶ Sect. 15.3.2), or processing aids such as lubricants. If the preparation is not trivial (such as simply adding a colorant or additional stabilizer), it is often completely separated from the shaping process and may be carried out by specialized companies, known as *compounders*, between the manufacturers of the polymer and the article in the supply chain.

Generally, simply mixing solid additives with solid polymer powder or granulates is not enough to obtain a homogeneous mixture of the components. Thus, the mixing of additives and macromolecules usually takes place in the melt. This cannot be done in a conventional stirred vessel because of the high viscosity of polymer melts (▶ Chap. 4), so a so-called *extruder* is usually used. An extruder is similar, in principle, to a meat mincer. It is basically made up of a massive, cylindrical case with a screw rotating inside (◘ Fig. 17.1).

Heating and cooling elements are integrated in the extruder housing. Initially, the cold extruder generally needs to be heated to ensure that the polymer melts and becomes malleable. When the extruder is running the friction between the screw and the viscous polymer melt generates so much heat that external heating can be considerably reduced or, under some circumstances, cooling may even be necessary.

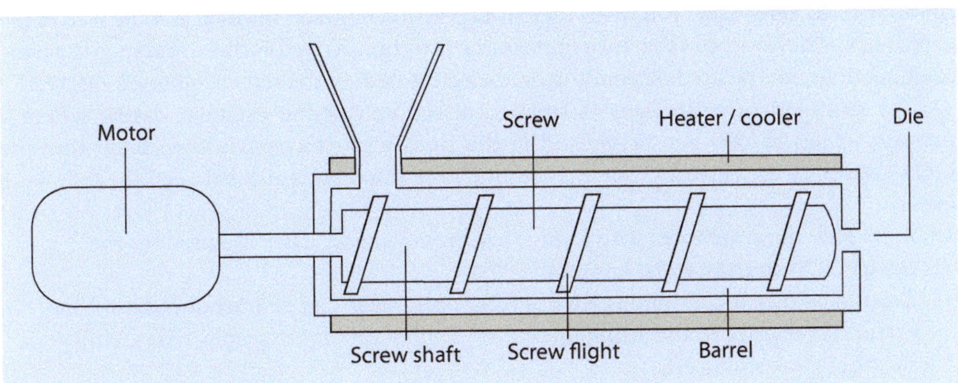

◘ **Fig. 17.1** Simplified diagram of an extruder

Once the polymer mass has reached the end of the extruder it exits the system through an orifice that is generally called a die. By using different shaped dies the cross-section of the extruded material can be varied. If the extruder is being used to mix a compound for further processing, the polymer strand is usually granulated. To this end, the hot polymer strand can be cut by knives rotating close to the die plate. This is referred to as *hot granulation*. Usually numerous strands exit a single die plate. Depending on how soon the polymer stops being sticky, this may be done in a closed system filled with flowing water (*underwater granulation*). An alternative for polymers with greater melt strength is that the polymer strand is first drawn through a water bath after exiting the die and cooled until it is solid (*cold pelletizing*). Other molding processes used after the polymer mass has left the extruder die are discussed in detail in ▶ Sect. 17.1.2 (extrusion) and ▶ Sect. 17.4 (fibers).

Other raw polymer compounding steps may involve reducing the size of the polymer particles or removing certain components such as water or residual solvent left over from the polymerization process. The latter can be achieved using dry, warm air. This can, however, be detrimental for polymers with functional groups sensitive to oxidation, such as double bonds, as it can lead to premature aging of these materials. In these cases drying in vacuo is preferable.

17.1.2 Extrusion

The basic mode of operation of an extruder was just discussed in the previous paragraph (▶ Sect. 17.1.1). An extruder takes up a polymer mass in a solid or (less frequently) liquid state, melts it, compacts it, and forces it through a small outlet—the die. In addition, other process steps can be carried out in an extruder, such as degassing and mixing with additives. Chemical reactions involving polymers can also be performed in an extruder. Industrial extruders can process several tons of material per hour. The most important part of an extruder is the conveying screw. By varying the geometry of the screw, the extrusion process can be optimized for each required process step, such as compaction, mixing, or degassing.

In simple extruders, the conveying effect is brought about by the friction between the polymer melt and the walls of the extruder housing. If this friction is too little, however, then, in the extreme case, the material would simply rotate with the conveyer screw without being moving toward the exit die. This effect can be prevented by using *twin-screw extruders*. In such extruders the polymer melt is carried down the extruder by two parallel screws. Either the two screws rotate tangentially or they intermesh and force the material to move toward the extruder exit, even when there is no friction between the melt and the wall of the extruder. Conveyer screws that turn in the same direction, as well as those that turn in opposite directions, can be used, depending on the orientation of the screw flights (■ Fig. 17.2).

The so-called *extruder head* is situated at the end of the extruder casing where a pressure of up to 600 bar prevails. The die itself and its construction determine the cross-section of the extruded polymer strand. One can distinguish between the following groups:

- *Axial dies*: these are tools with a spherical cross-section. They are used for the extrusion of strands, rods, or pipes.
- *Cross-head dies*: a continuous wire or fiber optic cable can be introduced into the extruded polymer strand using these dies. This is how, for example, wires with sheathing are produced.
- *Profile dies*: these are dies with non-spherical cross-sections. Profiles, sheets, and films can be made with these dies.

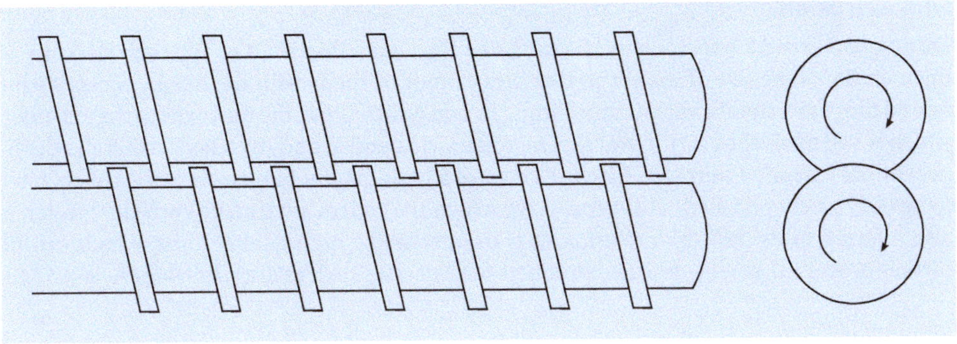

Fig. 17.2 Principle of a twin-screw extruder

Usually, further process steps take place after the material has left the extruder die:

- The dimensions of the part that has been produced are checked and, if necessary, corrected during the *calibration* process. This takes place in a vacuum or using compressed air. At the same time, the polymer is cooled down to a temperature at which it is no longer malleable. The extrudate is normally cooled using either water or cool air.
- The calibrated workpiece is carried away from the extruder mechanically. This can influence or control the dimensions of the workpiece if it is still malleable. If the speed with which the extrudate is transported is greater than the speed with which it exits the extruder, the material is *drawn* and its cross-section becomes thinner than the diameter of the die (▶ Sect. 17.4). By removing the material at a suitable speed, one can compensate for so-called *die swell*, which is a result of the loss of pressure of the polymer melt when it leaves the extruder.
- At the end of the process, the extrudate is either cut *(to length)* and the pieces stacked or wound into coils (continuous products, e.g., cables) for storage.

Primarily, the following articles are produced by extrusion:

Pipes

These are typically made out of PVC, polyethylene, polypropylene, polyamide, or polyoxymethylene. One of the largest volume applications for these pipes is in construction where they are used for transporting both clean water and sewage.

Sheets

PMMA, PC, PVC, polyolefins, or ABS sheets are usually produced in thicknesses of 1–10 mm and are mainly used as semi-finished products which go on to be reshaped to their final form. An exception to this rule is so-called *flat films* which are often less than 1 mm thick.

Film Extrusion

During *film extrusion* the polymer is extruded through a thin slit die. The high viscosity of the melt limits the size of the slit so that dies cannot be made with the height necessary for typical films, i.e., significantly under 1 mm. To make such films, the film exiting the extruder, which is still malleable, is run over several rollers, a calender, and stretched so that the thickness of the extruded film is reduced. To ensure a high quality film surface, the rollers have to be elaborately polished. This process, in which the rollers are usually cooled by water, is also referred to as *chill-roll-extrusion*. It is used to make high quality films, which can be further processed as semi-finished products by processes such as *deep drawing* (► Sect. 17.5).

Blown Film Extrusion

This process is used to manufacture very thin films, e.g., cling film. The diameter of these films, used, e.g., in packaging applications or as plastic bags, is in the order of tens of micrometers. Evidently, it is not possible to make the profile of the extruder die as thin as desired without the back pressure becoming too high so that these very thin films cannot be extruded directly. For *blown film extrusion* (◘ Fig. 17.3) thin tubes with a wall thickness as thin as possible are initially extruded through a ring die. As soon as the film leaves the tool it is inflated by pressurized air flowing through the center of the die into the tube (◘ Fig. 17.4). Depending on the speed of the extrusion and the pressure of the air flow, the tube is stretched and its walls become thinner. The process is similar to inflating a balloon. The tubular film is then cooled while maintaining a blown form until it reaches the so-called *frost line*, at which point it is no longer plastic. Finally, the tubes are laid together and rolled for storage.

Blown film extruder equipment can exceed heights of 15 m. Frequently, different materials are extruded one over the other by *coextrusion*. The tubes produced by blown film extrusion can be either cut to give thin continuous films or, for example, cut, sealed at one end, and printed to become bags or bin liners.

Wire Sheathing

A diagram of a wire sheathing installation is shown in ◘ Fig. 17.5.

The central machine is a simple extruder. A special tool, a cross-head die, forms the head of the extruder. This tool allows a wire or fiber to be inserted into the extruding polymer which then passes through a ring-shaped die. The exact dimensions of the tool determine to what extent the polymer is pressed onto the fiber or wire. Modern systems can produce wire at speeds considerably above 100 km/h.

17.1.3 Injection Molding and Reaction Injection Molding

The process of injection molding is the most important shaping process for plastics. It can be used for thermoplastic materials and is also used for duroplastics and elastomers. In contrast to extrusion, injection molding is a batch process; it is shown schematically in ◘ Fig. 17.6.

During this process a given shape (the *form*) is filled with a precise amount of polymer melt from a short reciprocating extruder. The amount of material injected, the so-called *shot weight*, varies and can weigh from a few milligrams to several kilograms. For thermoplastic polymers the injected form is cooled until it can no longer be plastically

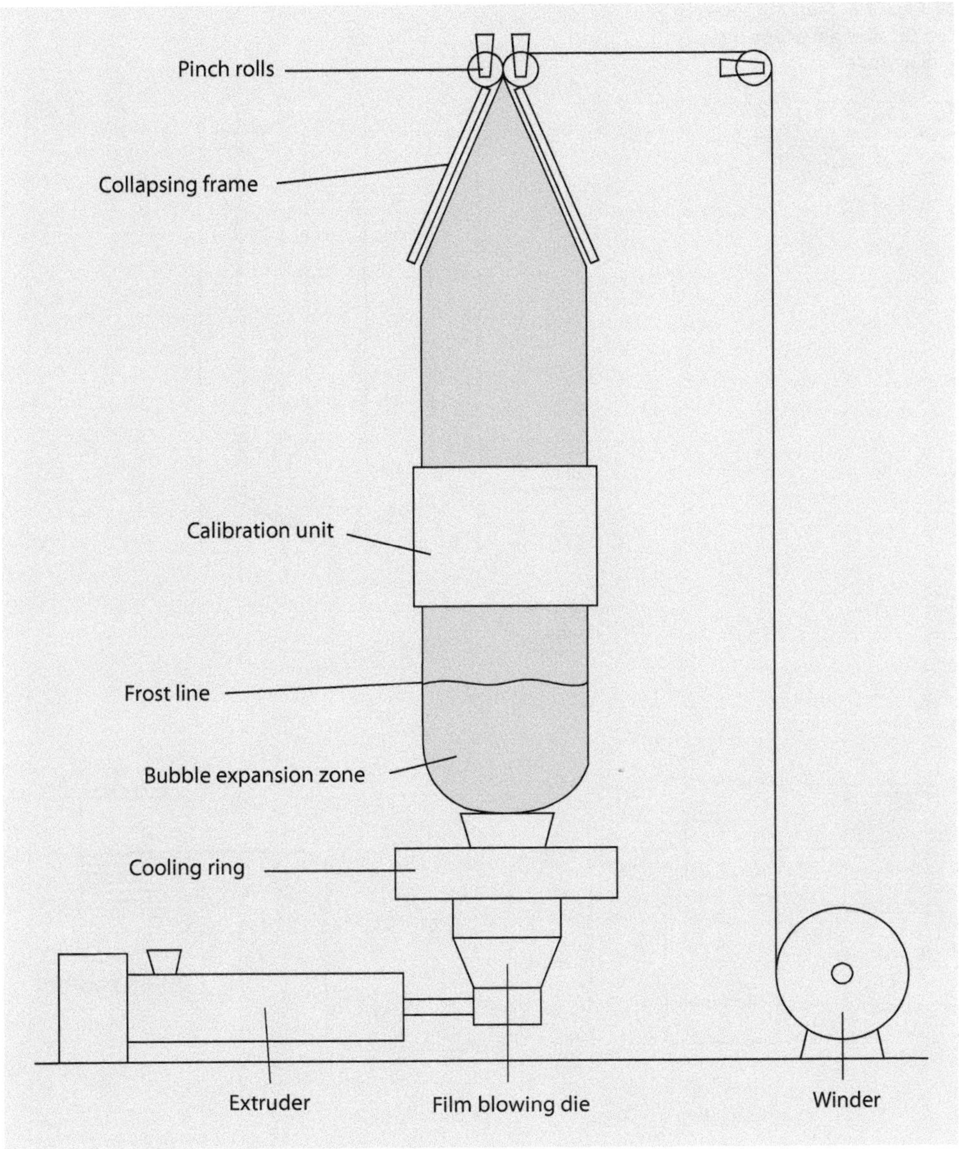

Fig. 17.3 Simplified sketch of a blown film apparatus

deformed. Thus, amorphous polymers are cooled to below their glass transition temperatures and partially crystalline polymers to temperatures below their melting temperatures. In contrast to this, duroplastic and elastomer molded parts, which at the time of injection are not cross-linked, are heated after injection whereby the chemical cross-linking reaction is initiated. The cross-linked material produced is insoluble and infusible. This process is referred to as *reactive injection molding*. The molded part can then be taken out of the mold (*demolded*) after complete cross-linking while it is still hotter than the glass transition temperature of the polymer.

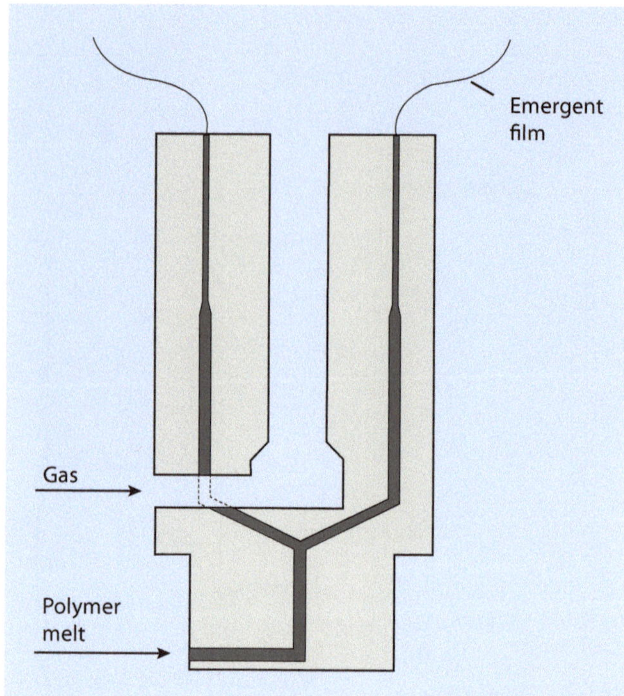

■ **Fig. 17.4** Diagram showing the detail of a die for a blown film extruder

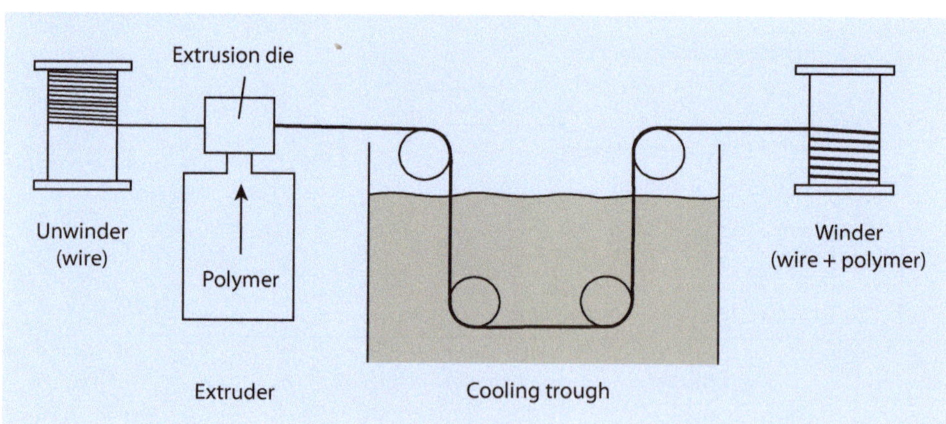

■ **Fig. 17.5** Simplified diagram of a wire sheathing machine

17

The hardening process, i.e., the period of time between the injection of the polymer melt and the demolding of the finished part, is dependent on the form used and the geometry of the part, the shot weight, and (for partially crystalline thermoplastics) the rate of crystallization. This process can take anything from under 1 s to several minutes. If high quality products are required it is especially important not to demold too early. The cooling down period thus significantly determines the cycle time and is an important component of the process cost. Very fast cooling leads to the best surface appearance and can reduce cycle times and thus cost.

To compensate for the loss of volume that takes place during the cooling and solidifying processes, the process is carried out at even higher pressures than for continuous

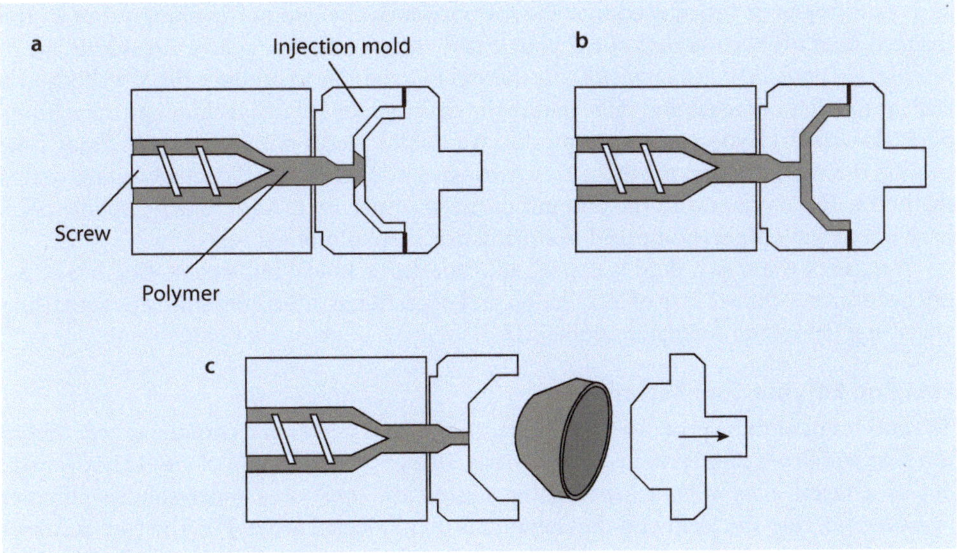

Fig. 17.6 Simplified injection molding process. (a) Start of the polymer melt injection. (b) End of the injection. (c) Demolding and ejection

extrusion, in some cases, even at pressures above 1000 bar. As a result, molded parts with very precise geometry (such as molded parts for precision engineering) can be produced. Another advantage of this technology is that injection processes can be automated. However, the injection tools required are very expensive so this technique is only economically viable for larger scale production processes. More recently, the forms can be made using 3D printing techniques. These forms are much cheaper and can be used for prototype production but are generally not suitable for large series products.

Typical materials manufactured using injection molding techniques are polypropylene, PBT, polyether sulfone, PMMA, polyamide, polyoxymethylene, polycarbonate, polystyrene, and styrenic copolymers. These materials are used in car manufacturing (dashboards, bumpers, air intake manifolds, lighting parts) or in the production of toys (such as Lego building blocks made from ABS).

Based on the principle of injection molding, process variations have been developed which are discussed in the following sections.

Multi-component Injection Molding

In this variant, several different materials are introduced into the injection molding tool simultaneously or sequentially. This can be done using a single gate or several gates. This process enables the production of complex workpieces which have, e.g., sandwich structures. Components made out of different materials with pronounced mechanical differences such as sealing units with integrated gaskets can also be made in this way.

Alternatively, two differently colored batches of the same polymer can be processed together into multicolored parts, such as the rear light covers for cars, using multi-component injection molding.

Fluid Injection Technique

It is possible to produce hollow parts by fluid injection molding in which a gas or a liquid is injected into the form together with the polymer melt. The fluid, mostly an inert gas such as nitrogen, is introduced into the form towards the end of the filling process. The injected fluid displaces a part of the plastic and creates a hollow space. The advantage of *internal gas pressure injection molding* is the ability to be able to produce thick-walled parts such as handles or ledges and rails. Before the development of this technology these kinds of thick-walled plastic parts were produced as solid pieces with long cycle times (low process throughput), making them very expensive. A further, significant advantage of this method is the reduction in the amount of material per part. Additionally, hollow parts have improved stiffness compared to equally heavy, solid molded parts.

The use of water as a fluid is conceivable but the technical implementation has so far not been successful because of installation and operational difficulties inherent with using water (e.g., leaks and corrosion problems).

Lost and Soluble Core Techniques

The end result of these processes are similar to the fluid injection technique except that in this case a hollow space is created in the plastic mass by the insertion of a solid body made out of a metal alloy with a low melting point (*lost core*) or a water-soluble polymer (*soluble core)* into the form and the polymer is then injected around it. The core is subsequently melted or dissolved out. For example, tin/bismuth alloys are used for lost core technology. The advantage of using this technique instead of fluid injection is that hollow spaces with complex geometry, not possible with a simple fluid core, can be produced. An example of the use of this kind of manufacturing process is the production of automotive intake manifolds. Alternatively, for less complex components, the production of two half-shells, which are then welded or glued together, or blow molding (► Sect. 17.1.4) are often less complicated, more cost-effective, methods.

Injection Stamping

During this process the tool is not closed completely so the polymer melt is injected essentially without excess pressure. Only after injection is complete is a higher pressure created by closing the form. The advantage of this technique is the ability to produce parts with relatively low levels of internal stress as the polymer chains undergo less shearing and are less deformed during this process than during injection into a closed form. Thus, this process is well suited for the production of CDs or DVDs as well as optical components of a high quality such as glass panes or lenses.

Powder Injection Molding

Powder injection molding (PIM) is used to produce molded parts made out of metal or ceramics. This is done by mixing a metal or ceramic powder with a plastic granulate to obtain a mix that can be processed by injection molding. The mixture is prepared in such a way that the very abrasive metal or ceramic particles are completely encapsulated by the less abrasive polymer melt so that premature wear of the injection apparatus can minimized. A molded part made out of metal or ceramic particles bound together by the polymer is formed. In a subsequent *bonding agent removal step* the polymer is removed. This can be achieved by dissolving the polymer in an organic solvent or in water (for polyglycols, for example). Alternatively, a chemically induced depolymerization is also possible. This can be achieved by the addition of acid if polyoxymethylene is employed as the bonding agent; this polymer decomposes to gaseous formaldehyde as a result of its low ceiling temperature (► Sect. 10.2.5).

After removing the polymeric binder, a metal or ceramic part is obtained with only a minimum mechanical stability—a "green body." The green body is then sintered to produce the final article.

This process enables technically challenging and complex metal and ceramic parts (such as very small or filigree structures) to be molded in large numbers in an efficient manufacturing process with excellent tolerance limits. Even though shrinkage during PIM is often ca. 20 % (depending on the proportion of bonding agent and the particle size distribution of the polymer used), very precise and delicate components can be produced with excellent reproducibility.

17.1.4 Blow Molding

Blow molding, also known as *extrusion blowing or blow molding of hollow objects*, is a method used for the production of hollow parts from thermoplastic polymers. This two-step process is similar to a combination of extrusion and injection molding.

In the first step a tube-like preform, the so-called *parison*, is produced using a conventional extrusion process with a ring die (see above, blown film extrusion). In the second step this preform is then transferred to a tool whose inner contours correspond to the outer contours of the molded part being produced. A hollow mandrel or needle is inserted into the top end of the hot and still malleable tube. The tool is closed, resulting in the other end of the tube being pressed, welded together, and sealed. Compressed air is forced into the tube through the mandrel or hollow needle, forcing the polymer onto the contours of the form (Fig. 17.7).

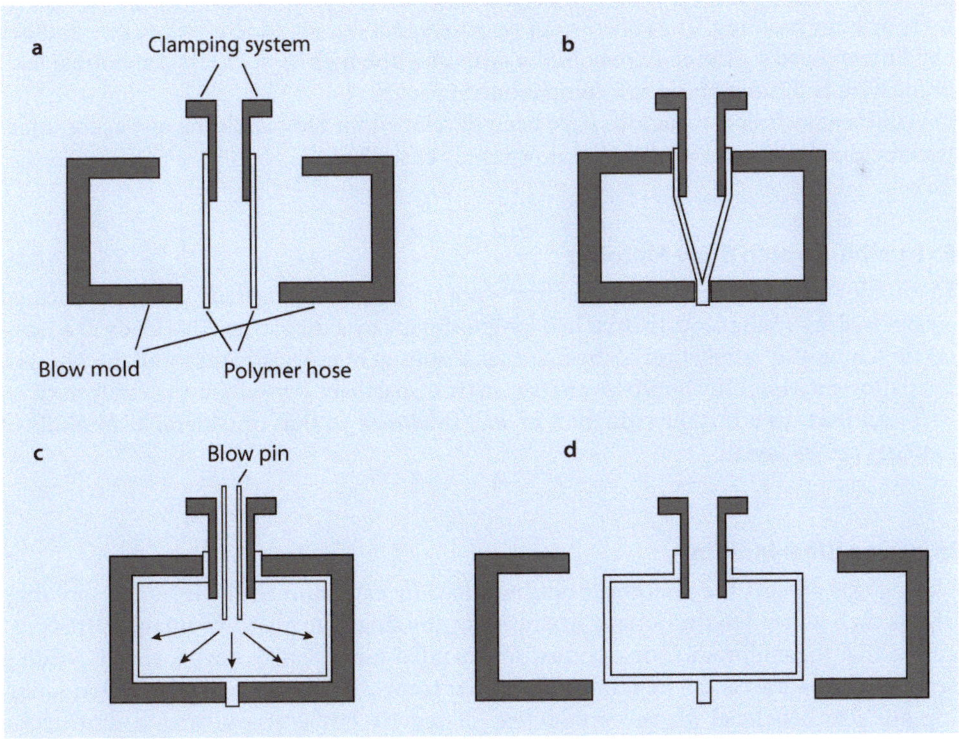

Fig. 17.7 Process steps during blow molding hollow objects. (**a**) Introducing the preform into the form and onto the mandrel. (**b**) Form closure. (**c**) Inflation of the preform to the final form. (**d**) Opening the blowing mold and demolding

An analogous process is used to give tires their profile using inflatable "bladders" inside the unvulcanized tires. After cooling one obtains a hollow part. Hollow objects with volumes as small as several milliliters or as large as several cubic meters can be produced using this process.

Control of wall thickness is subject to a certain degree of process-related fluctuation so tolerances are greater than those of injection molded parts. However, it is possible to vary continuously the wall thickness of the preform during extrusion so that the preform can have different wall thicknesses along its long axis. This enables the hollow part produced to have walls that have intentionally been made thicker in critical areas which need greater strength. As the cooling process for hollow parts is slow (one-sided wall contact) the form is usually cooled actively by circulation of a cooling agent within the tool.

Blow molding is traditionally widely used in the packaging sector to make, for example, plastic bottles. Technical objects such as drums, fuel tanks, ventilation ducts, boots (CVJ-boots), or expansion joints can also be produced in this way. Polyolefins, PET, PVC, polyamide, polycarbonate, PMMA, ABS, thermoplastic polyurethane, and thermoplastic elastomers are typical materials. The process can also be modified by inserting labels into the form which are then firmly attached to the hollow part during the blow molding process (so-called *in-mold labeling*). Handles can also be inserted into special holding brackets in the form and thus welded onto the molded object.

In the same way as different polymers can be coextruded simultaneously (► Sect. 17.1.2), it is also possible to produce objects comprising layers of different polymers by blow molding. The cohesion of polymers that do not adhere well to one another can be improved by the coextrusion of a layer of a polymer (*a tie-layer*) compatible with both layers and functioning as a compatibilizing layer.

Different technical variations have been developed for blow molding just as for injection molding, and some of these are now briefly described.

Extrusion-Stretch Blow Molding

In this blow molding variant the preform is not only stretched laterally using compressed air but is also simultaneously stretched longitudinally by a die stamp. This leads to a biaxial stretching and orientation of the material, resulting in especially thin wall thicknesses. Biaxial orientation also improves the strength of partially crystalline materials such as PET and leads to a further reduction of wall thickness so that considerable amounts of material can be saved.

Injection Blow Molding

During this process the preform is not produced by extrusion but by injection molding. This is then inflated to the desired size by blow molding. The main advantage of this process is that the endproduct, in contrast to extruded tubes, which have a visible welding seam on the base area, can be produced without seams. The advantages of injection molding are also beneficial for the production of precise bottle screw threads and necks, although the process involves extra investment cost.

Injection-Stretch Blow Molding

This process is a combination of extrusion-stretch blow molding and injection blow molding. A preform produced by injection molding is stretched through a die stamp in its plastic state and is then inflated and laterally stretched by compressed air. The advantage of this process is that the preforms are especially compact and can thus, if their manufacture does not take place on-site, be easily and cost-effectively transported elsewhere. This too is a process generally only used for small number production runs.

17.1.5 Calendering

Calendering is the process whereby polymers are rolled out into broad strips (films or sheets) at temperatures above their melting temperature for partially crystalline plastics or above their glass transition temperature for amorphous materials. The calendered sheets are often semi-finished products for further process steps. Calendering is similar to many thermoforming processes in that it is basically rather simple: the plastic material is first passed through an extruder or kneading machine to heat it to a temperature at which it is malleable and then passed over a sequence of heated roller pairs through narrow gaps (*nips*). As a result the material is rolled out into a continuous flat shape of well-defined thickness (which can be varied by adjusting the nip between the rollers). In the simplest case the roll axes are arranged in a single plane but other arrangements are often employed (◘ Fig. 17.8).

Usually the amount of material initially added to the calender is greater than that which can immediately pass between the first pair of rollers. The addition of material needs to be adjusted to the speed at which material is processed. The backlog made up of some of the polymer melt forms what is called the *bank* in which, because of the rotating motion of the rollers, intensive mixing occurs and from which the surface is continually removed. Guides at the sides of the rollers ensure that all the material stays on the rollers.

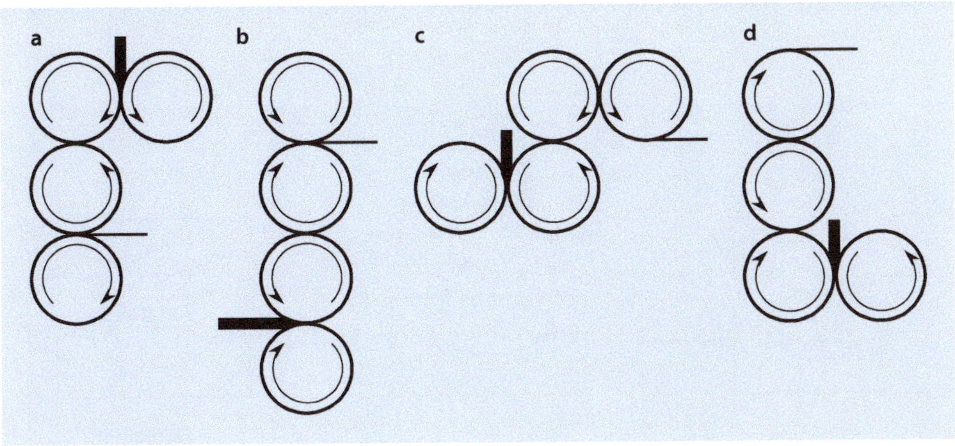

◘ **Fig. 17.8** Roller configurations for a four-roll calender. (**a**) F-Calender. (**b**) I-Calender. (**c**) Z-Calender. (**d**) L-Calender

After the material has passed through the nip of the first pair of rollers, the material tends to stay on the hotter roller or that with a faster speed of rotation and this roller transports the material to the next pair. After the material has been conveyed through the last nip it is picked up by suitable tools and either wound to a roll or cut to size and stacked.

Calendering is a technically important process for making thick films or sheets from PVC used for flooring or curtains. Rubbery materials (*elastomers*, ▶ Chap. 18) are often compounded and calendered before being cross-linked. Calendered polyethylene and polystyrene are also common but thinner films are usually made using blow extrusion processes (▶ Sect. 17.1.4).

The high viscosity of the polymer melt exercises considerable pressure on the rollers at the nip and this can lead to their deformation. This would result in the calendered sheet being thicker in the middle and must therefore be prevented. This is especially difficult for very wide rollers (some rollers, such as those used for roofing membranes, are several meters wide) so that calendering facilities that can accurately produce larger width sheets or films are extremely expensive. Thus, this process is only viable for mass production.

17.1.6 Pressing

Pressing is, in principle, also a simple technique for thermoforming polymers. It is of great importance for the processing of polymers whose final manufacturing stage involves cross-linking. The un-cross-linked polymers are placed in a heated, open form. At this stage the polymer can be a powder, a granulate, or in lump or sheet form. The form is then closed. As a result of the pressure on the closed form, the material is evenly distributed throughout the mold. The chemical cross-linking of the material is initiated by the high temperature of the form (◘ Fig. 17.9). After allowing the material to adjust to the form, the article can either be removed from the press hot or cooled before removal. This technique can also be used for making foamed articles.

Although apparently very simple, this process requires considerable attention to detail in practice; the material properties, geometry of the mold, temperature variation, and duration of the cycle are all important parameters. In particular, the onset of cross-linking

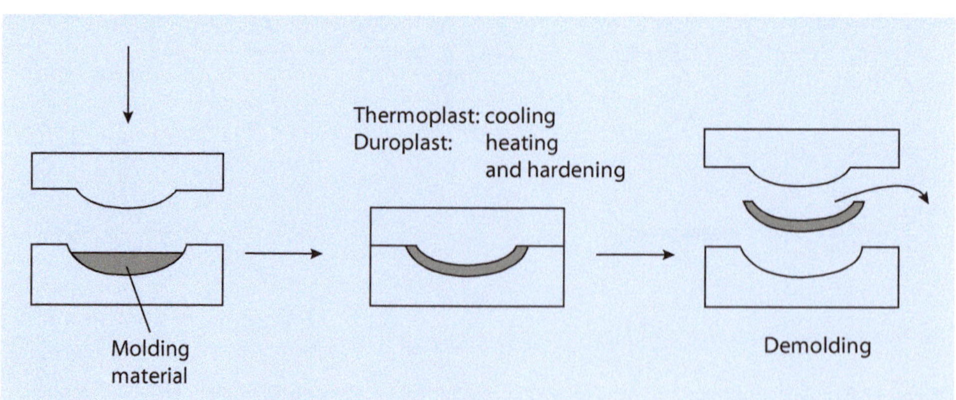

◘ **Fig. 17.9** Pressing process steps

needs to be avoided before the mold is completely filled by the malleable polymer. It is also important not to use excessive temperature or cycle times to keep the heat history of the material (its premature ageing) to a minimum.

As well as for elastomers (▶ Chap. 18), pressing is often employed for forming duroplasts such as melamine, phenol, or formaldehyde resins and for polyester and epoxide resins (▶ Sect. 8.6). It is also of considerable technical significance for the molding of fiber-reinforced composites (▶ Sect. 17.2).

17.1.7 Other Primary Molding Processes

As well as the processes described so far, there are also other molding processes whose detailed description are beyond the scope of this book. Examples of these are:
- Casting
- Dip molding
- Rapid prototyping
- Stereolithography

For further information, the interested reader is referred to specialized textbooks discussing the processing of thermoplastics.

17.2 Fiber-Reinforced Plastics

Fiber-reinforced plastics (FRP) are composite materials in which fibrous reinforcing materials are enclosed in a plastic matrix. The fibers lend the material a higher modulus and an increased tensile strength. Any stress applied to the material, such as when it is stretched, is transferred to and borne by the reinforcing fibers. Obviously, these effects depend on the physical properties of the fiber, such as E-modulus, elongation at break, and tensile strength, being greater than those of the matrix. On the other hand, the matrix polymer stabilizes the fibers so that, for example, they do not break when folded across their long axis. Load transfer between the matrix and the fibers can only take place if there is sufficient adhesion between the fibers and the matrix material. Adequate adhesion is thus essential for the improvement of the material properties. To achieve this, most fibers are treated before being used for composite manufacture by coating them with a matrix-specific layer—the sizing.

Reinforcing fibers can be either inorganic fibers such as glass or basalt fibers or organic fibers such as polyaramide or carbon fibers. Natural and thus renewable fibers made out of wood, flax, hemp, or sisal are also used. However, the latter induce only moderate improvements in the mechanical properties of the material, although they do have advantages with regard to their cost and potential recyclability.

Reinforcing fibers can be categorized into three groups according to their length:
- *Short fibers* are between 0.1 and 1 mm long. Because of the short fiber length, polymers containing short fibers can be processed by, for example, the typical injection molding or extrusion processes described above. Composite materials reinforced with short fibers usually have a fiber content of 20–50 wt%.
  The material can be recycled if it is very selectively collected or sorted (▶ Chap. 21).

- *Long fibers* are between 1 and 50 mm long. Thermoplastics containing long fibers can also be extruded. Long fibers are also often used to reinforce duroplasts. The proportion of long fiber in composites is usually 60–70 vol% and thus higher than the usual level for short fibers. Because of their outstanding material properties, such as their high specific modulus, these materials can be used as high-performance materials in lightweight construction applications such as the rotor blades of helicopters. However, although materials reinforced with long fibers have such excellent material properties, they are not easily recycled.
- *Continuous fibers* are even longer fibers. Their use in composite materials results in materials with the highest strengths and moduli.

To be able to handle these fibers more easily they are often bundled or made into woven or non-woven materials.

Matrix polymers can be thermoplastics or cross-linked thermosets. In the class of thermoplastic materials it is possible to reinforce nearly all polymers with fibers. Examples are polyamides, PE, PP, polyamide, PS, PBT, polyether ether ketone (PEEK), and polysulfone. Epoxide, phenol and formaldehyde, melamine, and urea resins and unsaturated polyester resins are the dominant duroplast matrices (▶ Sect. 8.6) but cyclic polyolefins such as polynorbornene and cyclopentadiene are also used as thermoset matrices. FRP materials made from duroplasts often have the best material properties although their impact resistance is often not as good as that achieved by thermoplastic composites. Furthermore, once cross-linked, the articles can no longer be re-formed so that recycling is essentially impossible. Lastly, the process of making thermoset composites is generally time-consuming and expensive.

In practice, the fiber bundles are often impregnated with the matrix material and processed to semi-finished products called *prepregs*.

Glass mat reinforced thermoplastics (GMT) and long fiber reinforced thermoplastics (LFT) are especially important for thermoplastic FRPs. GMT semi-finished products are usually transformed into the desired final form by pressing and heating. LFT materials can be extruded into a press mold, for example (▶ Sect. 17.1.6).

Duroplastic fiber-reinforced materials are categorized into three groups: *sheet molding compounds* (SMC), *bulk molding compounds* (BMC), and *prepregs*. SMC semi-finished products are non-cross-linked prepolymer and reinforcement, in the form of short or long fibers, in sheet form. Bulk molding compounds are similar to SMC in composition; however, they are doughy, viscous, shapeless masses. Other names for BMC are *dough molding compounds* (DMC) or *premixes*. SMC are always processed using hot presses. BMC can be processed either by pressing or by reactive injection molding (▶ Sect. 17.1.3). Prepregs containing continuous fibers are usually used in a roll or sheet form. They are cured (hardened) in autoclaves under pressure at higher temperatures.

17

Besides the processes described in the previous paragraphs for the processing of short and long fiber-reinforced FRP, the following special processes have also been developed for the processing of fiber-reinforced composite materials.

Layer Pressing

With this process SMC are layered one above another and processed into multi-layer laminates. As an alternative to this, non-planar workpieces such as pipes can be produced by wrapping layers of the impregnated SMC onto an inert core and then hardening them.

Fiber Spraying

This process is used for the production of molded parts with a matrix of unsaturated polyester resins (UP-resins) and for polyurethanes with glass fiber reinforcements. The process consists of spraying comparatively short glass fibers and the UP resin simultaneously onto a surface where they are thermally cured. For this process, continuous glass fibers are cut and sprayed onto a surface using compressed air with a fiber spraying gun. Wood and other materials can be used for the molds. Because of the relatively simple and cost-efficient nature of this process, it is suitable, in contrast to many other plastic processing processes, for short and medium production runs as well as for coatings.

Pultrusion

Pultrusion involves fiber bundles or mats being impregnated with resin and then being given the desired form by pulling the material through a die. In contrast to an extrusion which works by pushing or squeezing the polymer melt through an orifice, a pultrusion machine pulls the material through the die. If the die is heated, the cross-linking of duroplasts is initiated; in the case of thermoplastic materials the fiber strands are welded together. The cross-linking process for duroplasts is completed on a heated hardening line. Pulling the material through the die requires significant force because of the frictional forces that are created, so hydraulic systems are used. Pultrusion is typically used for producing long and continuous fiber-reinforced composites.

Resin Transfer Molding

For resin transfer molding (RTM), flat reinforcing materials such as glass fiber mats are inserted into the mold. The form is then closed and evacuated. The resin (usually the prepolymers of a duroplast) is injected under pressure, impregnates the inserted fibrous material, and is then cured.

Hand Lay-Up

As the name of this process suggests, this process requires a lot of manual labor and is therefore only cost-efficient for high-price, small scale production of large molded parts. On the other hand, little investment is required.

Hollow objects of large volume or open molded parts can be produced using a hand lay-up process. For this process an inert carrier (such as wood or plaster), on whose surface the FRP molded part is produced, is covered with a release agent so that it can be separated from the FRP part at the end of the process. The matrix resin and the reinforcing material are alternately applied onto the carrier manually, mostly using either a spray gun or a brush, until the desired material thickness is reached and the object can be hardened.

This process is used for producing items such as silos or other big containers, boats, or bodywork for small-scale vehicle production series (e.g., prototypes). Unsaturated polyester resins (UP-resins) or epoxide resins are the typical matrix materials.

17.3 Foams

Foams are heterogeneous materials comprising a gaseous phase and a solid or liquid phase. Nearly all polymers can be made into a foam just as can other materials, such as metals, ceramic materials, and, as is well known, watery systems such as soap solutions.

17.3.1 Properties of Foamed Materials

Foams have a low density because a portion of their volume is gaseous. Their density (weight/volume) is often given as part of their specification and can be as low as ~6 kg/m^3 (for comparison, air has a density of slightly over 1 kg/m^3 at ambient temperature and pressure). For so-called *microcell* or *micro-cellular* foams the pores or cells have diameters below 100 µm. *Fine-celled* foams have cell diameters of 100 µm to 1 mm, whereas the cells of the so-called *large cell* foams have diameters larger than 1 mm. These values are only a rough guide. In the last few years *nanocell* foams have been made whose pore size is considerably less than 1 µm and which have great potential in the area of thermal isolation (▶ Sect. 17.3.2, aerogels).

There are three basic types of foam which can be categorized depending on the shape of the cells or pores and their distribution within the foam:

- The pores of the so-called *closed-cell* foams are not connected but are isolated from one another. This is why these materials are not permeable to gases or liquids. They also cannot be compressed as easily as *open-cell* structures. Such materials are very good thermal insulators. They can also be used for applications where shock and impact need to be absorbed.
- In contrast to closed-cell foams, in *open-cell* foams the individual cells are connected to each other. A liquid or gaseous medium can therefore freely circulate within the foam as long as the pore size is not too small. Thus, aided by capillary action, open-cell foams always adsorb liquid. They are very useful, for example, as cleaning sponges. However, the extent to which they absorb liquid does depend on the wettability of the pore surfaces by the particular fluid. In addition, in micro-cellular foams with their very small pore size, free convection of the trapped fluid is largely impeded (Hagen–Poiseuille law). Additionally, such foams also absorb sound waves very well as these "play themselves out" within the extremely complex inner structures. Thus, they are also used for sound insulation. Open-cell foams are also very good thermal insulators so they form thermal insulation components in construction as well as in refrigerators and freezers. Open-cell foams (provided the appropriate polymer is used) can also be very soft and easily deformable as the air contained within the foam can readily escape when the foam is compressed. They therefore often form the cushions in seating.
- As well as the above foam types, foams that belong somewhere between the two groups also exist and these are referred to as *mixed-cell* foams.

If the pore size distribution is not uniform throughout the component but is subject to a regular gradient, then the material is referred to as an *integral foam*. The variation in density along a cross-section of a block of this sort of material is shown in ◘ Fig. 17.10. The weight of a component can be reduced by using integral foam without having to forego a stiff outer shell.

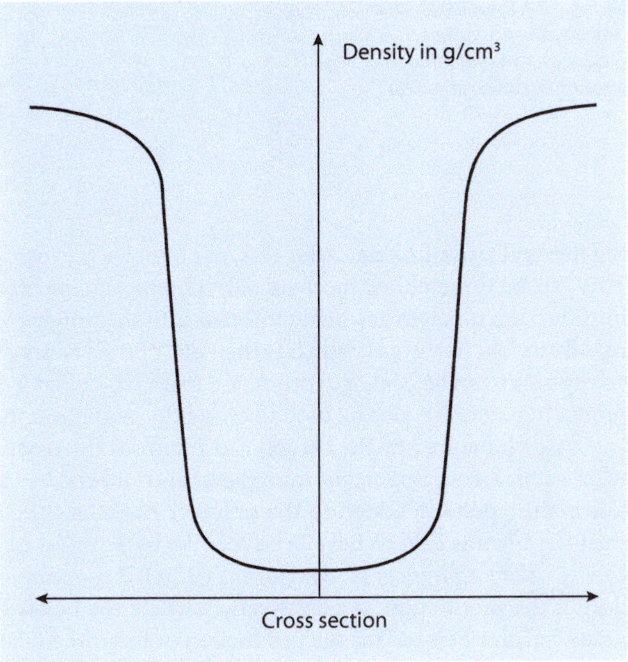

Fig. 17.10 Diagram showing the variation in density along the cross-section of an integral foam

The material properties of foams are numerous and varied. These properties are not only linked to the specific gravity, pore size, and type but also depend on the processing conditions and, of course, to a large extent on the chemistry and physical properties of the polymer from which they are made, such as its glass transition temperature. Thus, foams made from duromers such as urea- or phenol-formaldehyde resins are brittle and are referred to as *hard foams*. They are not markedly deformed under compression; however, they do break if the force is too large. In contrast, tough hard foams exist, such as those made from polyurethane or polystyrene, which can be more compressed and can, at least partly, take on their original shape again when the compressive force is released. *Flexible foams* can be easily deformed and take on their original shape again once the applied force is removed.

The main uses of foams are as follows:

- Heat insulation, e.g., in buildings or refrigerators and freezers
- Sound insulation, e.g., for rooms or vehicles
- Lightweight constructions, e.g., vehicle and aircraft construction
- Cushions, e.g., in mattresses or furniture upholstery, but also in shoe soles or packaging
- Safety, e.g., in car bumpers or motorbike helmets

17.3.2 Foam Production

To make a polymer foam using a thermoplastic, the polymer must first be heated to transform it into its plastic state. If the foam is to be made from duromers, the prepolymers are usually already capable of flowing. If not, they are gently warmed to avoid initiat-

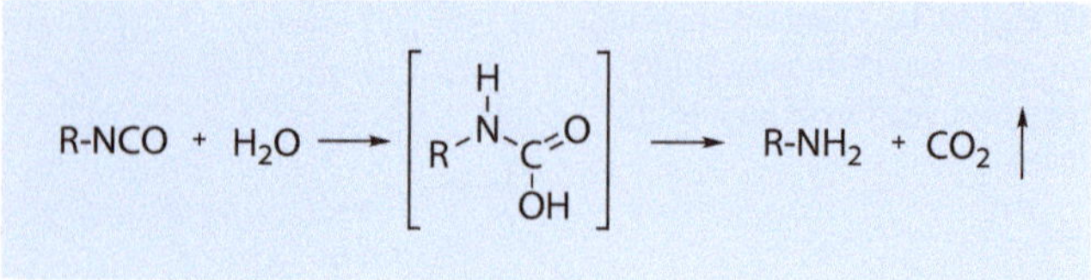

Fig. 17.11 Reaction of an isocyanate and water to produce carbon dioxide (foam forming in polyurethanes)

ing thermal cross-linking. After this, gas bubbles are introduced into the plastic material. This can be done either mechanically, chemically, or physically. Mechanical gas bubble introduction involves gas being injected into the polymer melt or the melt is stirred very rapidly to fold in the gas, which is then distributed as smaller bubbles, in the same way as an egg white can be beaten to produce a foam (so-called *blown foams*). Chemical reactions producing a gas can also be used to create the gas within the polymer. Polyurethane foams (see below), foamed by the carbon dioxide from the reaction of the isocyanate monomer with water, are an important example of such a process (Fig. 17.11). If a solvent with a low boiling point is added to the polymer mass and the solvent evaporated by heating, then the foam is said to have been formed by physical means. The blowing agent in this case is called a *physical blowing agent* (PBA). An example is the evaporation of pentane during the production of polystyrene particle (or bead) foam (see below). Supercritical fluids can also be used for the production of micro-cellular foams.

The foam structure of thermoplastics is frozen by cooling the polymer below its melting or glass transition temperature. In the case of duroplasts or elastomers, cross-linking, initiated thermally, proceeds until the foam hardens.

Making open-cell foams is more challenging from a technical point of view than making closed-cell systems. The reason for this is that the rate of formation of the blowing agent has to be precisely controlled to ensure that it is fast enough to make the cells connect with one another and to break the polymer film between them, but not so fast that the foam collapses.

In principle, many of the typical thermoplastic manufacturing techniques discussed in the previous paragraphs can also be used for the production of polymer foams. However, several important special techniques exist, especially for the production of polystyrene or polyurethane foams, which are discussed in the following.

17.3.2.1 Foam Extrusion

During this process, which is mostly used for the production of polypropylene foams, but also for the production of polyethylene, polystyrene, or PET foams, the melted polymer and a propellant (often pentane or carbon dioxide) under high pressure are extruded. To achieve an even distribution of the gas bubbles and to control the extrusion, two extruders in a *cascade* or *tandem* arrangement are sometimes used. In the first extruder the polymer is melted and the blowing agent (the *propellant*) is added. As long as the pressure in the extruder is above the vapor pressure of the propellant, it remains as a liquid or becomes dissolved in the melt. Because the dissolved propellant acts as a plasticizer, the viscosity of the system is significantly decreased so that it is often necessary to cool the system to ensure that the melt does not spray out of the extruder in an uncontrolled fashion. The easiest way to achieve this technically is to transfer the melt into a second extruder; the cooling extruder. This arrangement is shown schematically in Fig. 17.12.

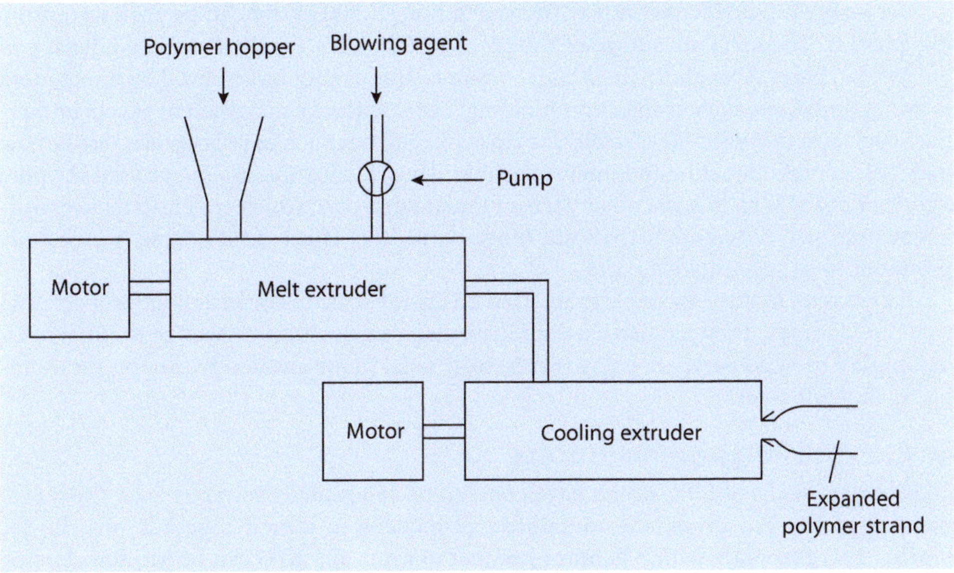

Fig. 17.12 Tandem extruder arrangement for foam extrusion

After leaving the cooling extruder's die, the pressure decreases dramatically and the polymer mass expands by a factor of between 20 and 50. This process is used to produce foamed molded parts such as foamed polyethylene or polypropylene sheets which are used, for example, as under-floor sound-proofing. Foamed polystyrene (e.g., ceiling) tiles can also be made in this way. Foamed granules can also be obtained, which are then joined together with the desired molded part in a subsequent step, for example, by welding (▶ Sect. 17.6).

As a variation on this process, smaller strands can be extruded into a pressure vessel filled with water. The extruded strands are cooled by the water and then cut into granules by rotating knives. The water pressure hinders an evaporation of the propellant and this remains in the polymer. After the polymer has been cooled down, an expansion is no longer possible so one obtains a polymer granulate containing propellant. This can be foamed using the *molded part process*, a discussion of which follows.

17.3.2.2 Molded Part Process

This technique is used to make, for example, molded parts from polystyrene (EPS, expanded polystyrene) and polypropylene (EPP, expanded polypropylene). Foamed granules, produced by processes such as foam extrusion, described above, are manufactured into larger molded parts. The foam granules, after extrusion, generally contain up to half the original amount of propellant either dissolved in the polymer or in a liquid state. These granules are placed in a form and heated in an autoclave, for example using steam, to a temperature above the glass temperature or melting point of the polymer. When this temperature is reached, two processes take place. The propellant, because of its vapor pressure, causes the granules to expand further, whereby these fill the form and are pressed against another. Parallel to this, the malleable particle surfaces are fused together to give the molded part a degree of cohesiveness despite the original individual granules remaining discernible. For this reason, such parts are said to be made of *particulate foam*.

Very large molded foam parts with densities of 12–300 kg/m^3 can be produced using this process. The main advantage of the process is the very evenly distributed density of the molded parts throughout their entire volume. This cannot be achieved by direct mold foaming processes, such as injection molding, because the foam's inherent excellent thermal insulation properties (especially for closed-cell foams such as polystyrene) lead to the materials cooling in a slow and uneven fashion. Thus, during the cooling process the pore structure changes so that the outer parts of the molded part, which cool faster, have a different pore size distribution than the inner parts. This effect is, of course, particularly pronounced in large molded parts.

Polystyrene molded foam parts are used on a large scale in construction for insulating roofs and facades. They are also used for insulating refrigerators. As well as thermal insulation, EPP molded parts are being increasingly used in the automotive sector, for example, as bumper cores.

17.3.2.3 The Polystyrene Process

To produce blocks of EPS, which have come to be colloquially referred to by the tradename Styropor®, a granulate containing propellant is joined together into bigger blocks. This granulate with a pentane content of around 5 wt% can be produced using pearl polymerization. In this process, a suspension polymerization (► Sect. 16.4) is carried out in a pressure vessel and pentane is added, which then dissolves into the evolving PS-pearls under pressure. Before the vessel is depressurized the system is cooled down to below the glass transition temperature of the pentane-containing polymer (plasticizer effect!) so that the pentane is trapped in the polymer and the pearls cannot expand when the vessel is opened. Alternatively, one can also produce polymer granulate containing propellant by foam extrusion followed by underwater granulation under pressure (see above). Both processes are used as large-scale manufacturing processes.

In a first step, this semolina-shaped granulate with particle sizes of between 0.5 and 2 mm is pre-expanded by heating, for example using steam, to temperatures above the glass transition temperature of polystyrene (100 °C). Pentane boils at about 30 °C (*n*-pentane 36 °C, iso-pentane 28 °C) and the polymer granules expand to up to 40 times their original volume; their diameters are ca. 3–4 times larger. The density of the foam produced is determined by duration and the temperature of the pre-expansion process; longer pre-expansion times and higher temperatures lead to increased expansion of the pearls and, consequently, a lower density.

After pre-expansion, the pre-expanded product is stored for 1–2 days. During this process air diffuses into the pores in which reduced pressure exists because of the pentane having cooled and condensed in the individual pores. Additionally, some pentane escapes so that a propellant content of roughly 3% results. To foam mold the material it is, as described above, placed in a porous pressure mold heated to around 130 °C by pressurized water steam, and further expanded whereby the individual granules fuse together to form molded parts. This process is summarized in ◘ Fig. 17.13. The shape-giving autoclave molds are nearly completely filled with pre-expanded granules, so that the density of the final part is largely determined by the density of the pre-expanded material. The final parts are stored for a longer period to ensure that the (highly flammable!) pentane has completely evaporated.

17

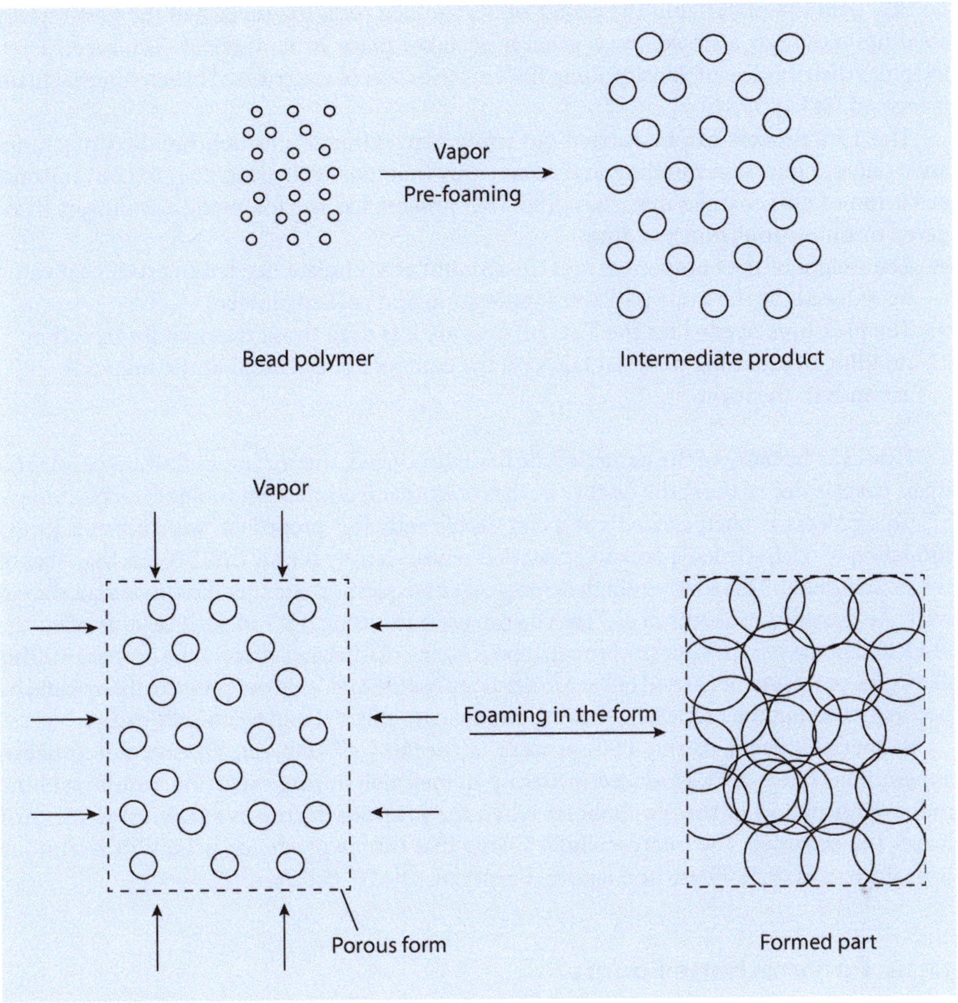

Fig. 17.13 Molded parts from foam granules, e.g., Styropor®

17.3.2.4 Thermoplastic Foam Injection Molding (TSG)[1]

With this variant of the injection molding process described in ▶ Sect. 17.1.3, polymers containing propellants are the starting materials. The propellant forms a gas by evaporation or chemical reaction during the melting process in the extruder which then results in an expansion of the material once it leaves the extruder and enters the mold. As the solubility of most propellants used for polymers is limited and there is a danger that physical propellants outgas during storage, only a limited amount of propellant (generally less than 10 %) can be introduced into the granulate in this manner. If larger quantities are required (lower density articles), additional propellant can be introduced under pressure into the extruder of the injection molding machine in the same way as described above for foam extrusion.

After the material has left the extruder it is injected into a mold, gas develops, and the material expands as a foam. The material next to the walls of the form cools down more

1 The abbreviation TSG is derived from the German: **T**hermoplastschaum-**S**pritz**g**ießen

quickly than the material in the center of the molded part; the surface of the part rapidly solidifies although a slower pore generation takes place in its interior. The result is an irregular distribution of density along the cross-section of the part so that an integral foam is formed (Fig. 17.10).

The TSG process can be carried out using conventional injection molding machines and relates to injection molding in the same way that foam extrusion does to conventional extrusion of non-foamed materials. The TSG process has the following advantages compared to simple injection molding:

- The weight of the component and the amount of granulate needed to produce it can be reduced for the same part size (application and cost advantage)
- The pressures needed for the TSG process are less than those required for injection molding because the material takes on the contours of the mold as the material expands in the form

However, because of the nature of the foam produced, the surface definition of integral foam parts is not of the same quality as that of unfoamed injection molded parts.

This process is often carried out using significantly less propellant, which results in the formation of relatively few pores and materials whose density is only ca. 10 % less than that of the solid material. This relatively high density calls into question whether these materials should really be classified as foams at all. The main reason for using TSG in such cases is often the reduction of the system viscosity brought about by the plasticizing effect of the propellant. This allows the process to be carried out at a lower temperature and/or lower pressure than would be necessary without the propellant, making it more cost-effective and technically easier.

A special variant of the TSG process is the *MuCell®-Process*. During this process, supercritical fluids are introduced into the polymer melt during extrusion, resulting in the massive formation of tiny gas bubbles when the pressure decreases suddenly as the melt leaves the extruder. The micro-cellular foams that can be produced using this technique have pore sizes of 5–50 μm and a pore density of 10^7–10^9 cells/cm^3.

17.3.3 Polyurethane Foams

Polyurethanes are formed from a reaction of a diisocyanate with a diol as shown in Fig. 17.14.

The properties of polyurethanes can be varied enormously by employing different starting materials as the diol and isocyanate components. As well as non-cross-linked and linear polyurethanes, which are processed and used as typical thermoplastics, the foamed and cross-linked (duroplastic) polyurethane foams (PU foams) are important representatives of this class of polymers.

The production of polyurethane foams differs from the processes discussed in the previous paragraphs in several important aspects:

- As is usual for duroplasts, cross-linking takes place in the course of the primary molding process. However, as the reaction between isocyanate and alcohol

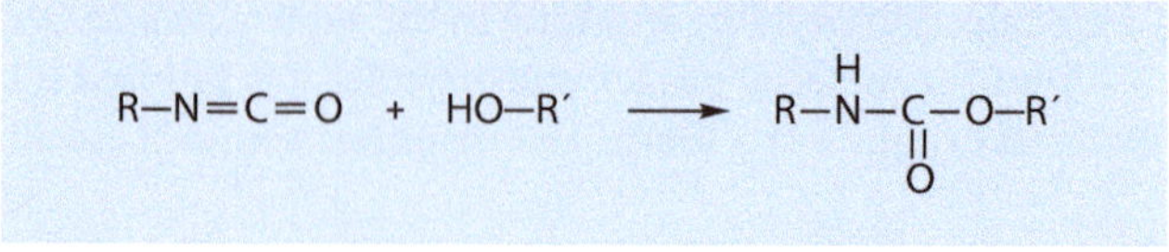

Fig. 17.14 Reaction of an isocyanate with an alcohol to form a urethane

components takes place at room temperature, these monomers cannot be stored together. Thus, the individual components are introduced into the process individually and are mixed only just before entering the mold.

- The starting materials are small molecules with low viscosities so they can be dosed with simple pumps and mixed in static mixers. In contrast, duroplasts from higher viscosity prepolymers generally require extruders or gear pumps to be used.
- Water is generally used as the propellant because it generates carbon dioxide by reaction with the surplus isocyanate functions present as shown in Fig. 17.11. The amine also formed by this process can react with another isocyanate moiety to form a urea derivative. Thus, the reaction does not lead to network defects. The use of water as a propellant can cause problems because of the relatively large reaction enthalpy. As well as this, the urea groups formed tend to make the material more brittle. Pentane or cyclopentane can be used as alternative propellants, and foams made with these propellants have been shown to have better thermal insulation characteristics.
- The character of the product network can be achieved using isocyanates with a functionality greater than two and/or oligo- or polyalcohols. The network density can be controlled by the functionality and the molar proportions of the various monomers. Thus, with the wide range of possibilities for the variation of the monomer components available, PU foams are a particularly versatile class of materials.

Processing starts with the feeding of the individual components into a blender head in which the starting materials are intensively mixed, either by a stirrer or in a static mixer. This process has to be carried out quickly as the reaction between the components takes place as soon as they have been mixed. After mixing, the reacting mass can , analogous to extrusion processes, either be continuously cast on a conveyer belt (so-called *reaction foam casting*) where the mixture cures to form a semi-finished product, such as blocks or, in a similar process to that of TSG, be injected into shaped molds. In contrast to the TSG process, during this process, referred to as *reaction injection molding* (*RIM*), pressures of only about 10 bar are needed because the viscosity of the liquid starting materials is low. Integral foams can also be produced as with the TSG process by varying the mold temperature.

Fiber-reinforced PU foams are produced using a variant of the RIM process. In this variant, reinforcing materials such as glass fibers are added during the production of these foams.

The material properties of PU foams can be varied over a wide range by choosing the appropriate starting materials. Thus, both soft foams and rigid foams can be produced. Soft foams are often produced using polymeric or oligomeric diols, such as polyethylene glycol. Soft and elastically deformable, open-celled structures, which are chemically not densely cross-linked and have a blocky character, are formed if the process is correctly managed. If shorter diol starting materials are used, the degree of cross-linking is often greater and hard, generally closed-cell, foams can be made.

The diverse properties of PU foams leads to them being employed for a myriad of applications. They are typically used in mattresses, upholstery, shock-absorbent packaging, and shoe soles, as well as in vehicle components, such as bumpers, auxiliary springs, and dashboard coatings. In addition, open-cell PU foams are used as sound insulation whereas closed-cell PU foams find application in the thermal insulation of refrigerators, roofs, and refrigerated warehouses.

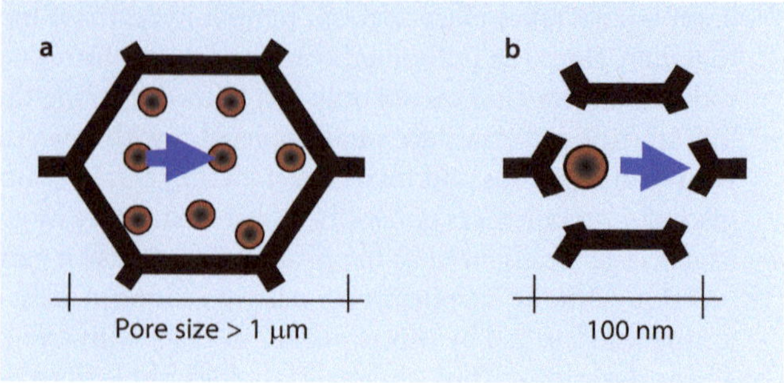

Fig. 17.15 Diagram to explain the Knudsen effect in nanocell foams (**b**). If, in contrast to larger pores (**a**), the pore diameter is smaller than the mean free path between gas molecules, the gas molecules collide more frequently with the walls of the pores than with each other

17.3.4 Aerogels

As mentioned in ▶ Sect. 17.3.1, foams have been developed in recent years having pores with diameters in the nanometer range. Such foams have especially low thermal conductivities and thus represent excellent thermal insulation materials.

The thermal conductivity of a foam λ results from the sum of three components:

- The thermal conductivity of the polymer λ_{pol}
- The thermal conductivity of the gas contained in the pores of the foam λ_{gas}
- A contribution from the thermal radiation in the infrared λ_{IR}

Thus:

$$\lambda = \lambda_{pol} + \lambda_{gas} + \lambda_{IR} \tag{17.1}$$

For conventional foams the thermal conductivity of the enclosed gas contributes some 60 % to the total conductivity of the system, whereas the contribution from the thermal radiation in the IR is generally smaller than the other two terms.

The thermal conductivity of the enclosed gas can be considerably reduced if the foam's pore size is smaller than the mean free path of the gas. This effect is known as the Knudsen effect (Notario et al. 2015) and is explained schematically in Fig. 17.15.

Practically, this effect starts to become important when the pore size is considerably less than 1 μm; it does not suddenly become effective but gradually does so as the pore size decreases. Interestingly, the Knudsen effect is not restricted to closed cell foams but, as indicated in Fig. 17.15b, is also observed with open cell foams.

Materials with nanosized pores based on silica, known as *aerogels*, have been known since 1930 and have been commercially available since the turn of the century. They are made using supercritical liquids, in particular carbon dioxide.

The synthesis of aerogels involves the following process steps:

1. A solvent swollen gel, i.e., a network swollen with a suitable solvent, is prepared.
2. The temperature and pressure are adjusted to make the solvent supercritical or the solvent is first exchanged for another solvent with more accessible supercritical conditions, and the appropriate conditions then established. This generally involves using an autoclave because of the high pressures involved.
3. The supercritical fluid is allowed to escape the autoclave to leave the dry aerogel network.

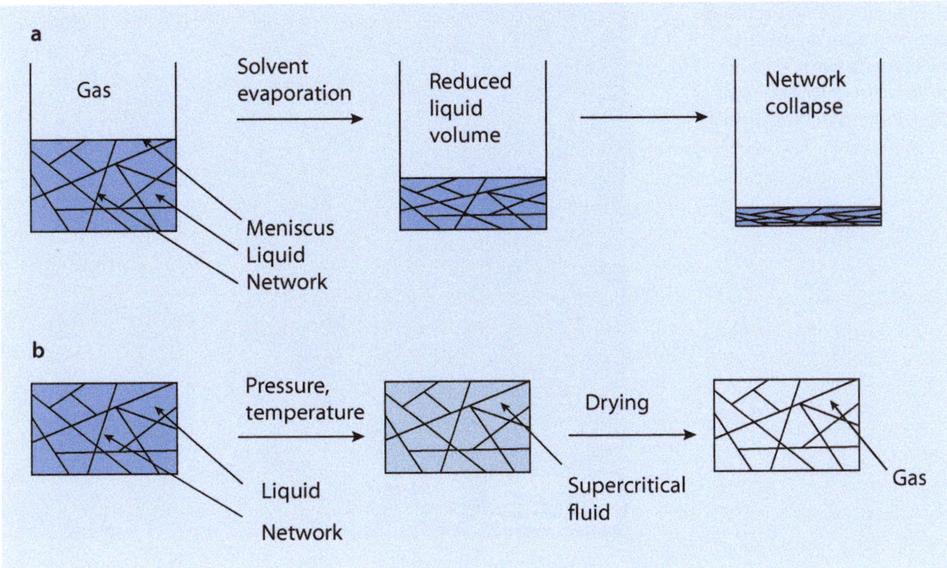

Fig. 17.16 Comparison of a "normal" (**a**) and a supercritical (**b**) drying process

During this so-called supercritical drying process, it is important to ensure that the temperature and pressure in the autoclave maintains the remaining fluid in a supercritical state. The process must be controlled so that only a dry network containing gas molecules remains. A "normal" drying, whereby the liquid is simply evaporated, would lead to a collapse of the network structure which would be confined to the ever diminishing liquid volume (Fig. 17.16).

Aerogels prepared in this way have some fascinating properties. Optically, silica aerogels resemble frozen smoke. They have an extremely low density but, depending on the detailed chemistry, can exhibit considerable mechanical stability. As expected from a consideration of the Knudsen effect, they are excellent as thermal insulation materials and have thermal conductivities as low as 15 mW/(m·K).

Aerogels based on polymers have only become commercially available since 2015 (from BASF Polyurethanes GmbH, based on BASF research carried out at ISIS in Strassburg and with support from the Technical University, Hamburg). These materials are polyurethanes. At the time of writing, further details regarding the chemistry of the material have not been published. These aerogels are dried using supercritical carbon dioxide. Under an electron microscope these materials resemble a fine spider's web with extremely small pores (Fig. 17.17).

Particularly challenging when developing and producing such aerogels is making them suitable for practical use. The materials need to be mechanically stable so they can be safely handled without breaking up or releasing a fine powder. The mechanical properties need to be adjusted to make them suitable for the particular application, such as in the construction industry. The exceptional variability of polyurethane chemistry, with respect to the many isocyanates and polyols available, ensures that the properties can be successfully optimized.

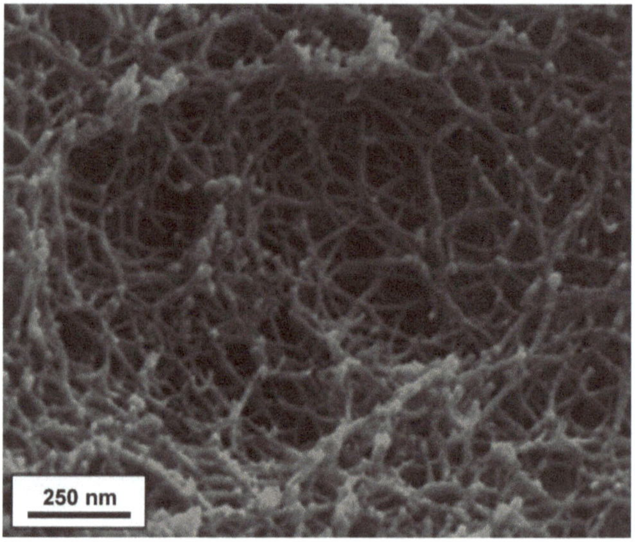

Fig. 17.17 Polyurethane aerogel under an electron microscope (with kind permission of Dr. M. Fricke, BASF Polyurethanes GmbH)

17.4 Fibers

About half the world-wide fiber production is based on synthetic polymers. The global demand for fibers, especially for textiles, is steadily increasing because of the continuously growing world population and increasing industrialization. The development of new agricultural areas for the production of natural fibers, such as wool or cotton, cannot keep pace with the increasing demand. Thus, an increase in the proportion of synthetic fibers is to be expected. As well as in the textile industry, polymer fibers are also frequently used for carpets and for fiber-reinforced polymers (▶ Sect. 17.2). Fibrous fleece is also used as an alternative to foam for sound and thermal insulation.

Chemical fibers are stretched polymer filaments and they all have in common that the main chain of the molecule is orientated, more or less, in the direction of the fibers during the so-called *spinning process*. This orientation has two significant effects on the properties of the materials produced:

- The strength of the material increases in the direction of the fibers as the ability of the molecules to react to macroscopic forces by simple rotation around the chain's σ-bonds is considerably impaired because of the deviation from a coiled structure (▶ Chap. 2).
- Additionally, the chains aligned parallel to each other can more readily interact. The crystallinity of partially crystalline polymers (PET, PA) increases, partly because of *strain crystallization*, whereby the crystallites also tend to be orientated parallel to the length of the fiber.

The fiber's material properties are anisotropic as a result of the orientation of the polymer chains. Strong interactions between the chains stabilize and maintain this orientation. Therefore, typical polymer fibers are unbranched and have symmetrical structures, such as PE or it-PP. Nevertheless, atactic polymers can also be suitable for making fibers if they have a large cohesive energy density, i.e., a large number of interactions involving a favorable enthalpy per unit of volume. This is the case, for example, for (atactic) polyacrylamide, polyacrylonitrile or PVC. It is, however, even more advantageous if the macromolecules can form hydrogen bonds between their chains. This is especially the case for polyamides, polyurethanes, and polyester urethanes. Polyethylene terephthalate is also widely used for textile fibers.

Depending on the detailed chemistry of the base polymer, the optimum molar mass for the best fiber properties varies. If the molar mass of the polymer is too small, interactions between the chains will not be strong enough—the number of interactions per chain is proportional to the length of the chain—and the properties of fibers made from comparatively low molar mass polymers are usually unsatisfactory. Polymers with very large molar mass, however, cannot be orientated well enough during the spinning process, which also affects the properties of the obtained fibers negatively. Additionally, for solution spinning, the solution viscosity increases with molar mass and economic spinning conditions limit the viscosity of the solution and thus the molar mass of the polymer. As a rule of thumb, nonpolar polymers, which lack interactions between their chains, require a larger molar mass to yield satisfactory fiber properties than polymers for which dipole–dipole-interactions or even hydrogen bonds between the chains are possible.

In general, the chemistry of the material and the processing steps are coordinated with one another in such a way as to ensure that the fibers have an optimum tensile strength—i.e., the maximum tensile strength at break—in the direction of their fibers, are flexible, and have a high abrasion resistance. Examples for further properties of polymer fibers, which have to be optimized depending on their intended application, are elasticity, moisture absorption, and dyability.

A strong moisture absorbancy of fibers is generally not desired as the absorbed water acts as a plasticizer which can have a negative effect on the properties of the material.[2] Also, wearing “wet” clothes is not comfortable. Hydrophobic textiles that hardly absorb any water, such as PP fibers, are thus more comfortable, particularly as sportswear. However, polyolefin fibers, with their simple aliphatic structures, are especially difficult to dye as they are practically unable to form any energetically favorable interactions with dye molecules.

The production of polymer fibers is made up of two important process steps: the actual spinning process and the stretching or drawing of the fibers. The most important spinning processes are discussed in the following.

Melt Spinning

This method is relatively simple and is similar to extrusion. As with extrusion, the material being processed is melted, additives are added if necessary, and the melt is forced through a shape-giving tool—the spinneret. The spinneret is simply a perforated plate, whereby the holes are generally more or less conical behind the front of the die plate to minimize shear and “dead zones” behind the plate. Spinnerets with as many as several thousand holes with diameters between 0.2 and 2 mm are not unusual. The polymer melt is extruded through this spinneret at high pressure because of the high viscosity of the polymer melt. After the

---

2 This rule is not valid when the melting or glass transition temperature of the polymer is at a temperature where water absorption does not shift it either into or out of the service temperature range. Thus, polyamide always contains a few percent water, which has no effect on the technical properties at room temperature or other “normal” service temperatures.

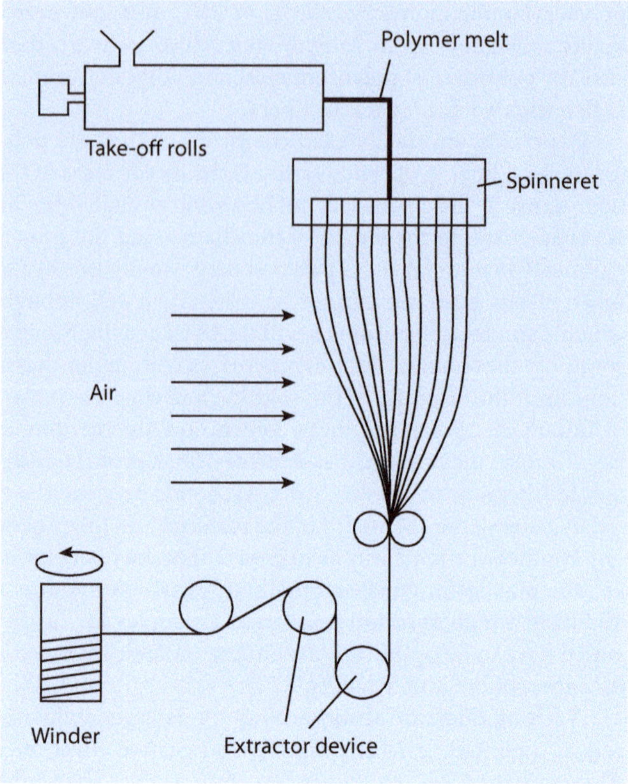

Fig. 17.18 Diagram of the melt spinning process

melt leaves the die plate, the fibers are pulled (drawn) and become thinner than the die holes. Thus, the diameter of the fibers can be determined by the speed of the fiber take-up. The ratio of the diameter of the holes in the spinneret to the diameter of the final fiber is a measure of the extent of elongation of the fiber and affects the degree of orientation of the fiber molecules. The speeds of fiber take-up are often of the order of 1000 to 6000 (!) m/min. Cooling and hardening is effected by cool air. As a basic rule it can be stated that polymers with relatively large molar mass, which are therefore also highly viscous, have to be spun at slower speeds to avoid the tensile stress exceeding that of the fiber. A scheme of this process, which is an important method for polypropylenes, polyesters, and polyamides, is shown diagrammatically in Fig. 17.18.

17

Dry Spinning

In this process the polymer is not melted as for melt spinning but is dissolved in an appropriate, generally volatile, solvent such as acetone or methylene chloride. Suitable polymers are, e.g., acrylic polymers. The solutions are heated and then forced through a spinneret. Because of the relatively low viscosity, compared to the solvent-free melt, it is possible to have significantly smaller spinneret holes (diameters <250 μm). After leaving the spinneret the solvent is evaporated from the fibers using warm air. The solvent is recovered in a separate manufacturing step. Apart from this, the set-up is principally the same as for melt spinning. The speed of take up is generally of the order of 1,000 m/min.

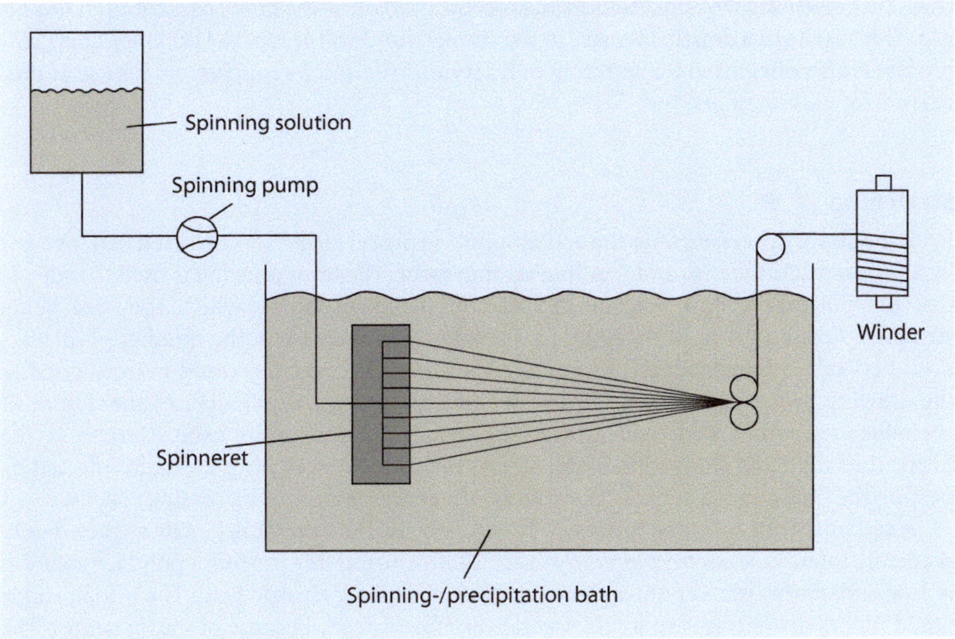

Fig. 17.19 Production of polymer fibers by wet spinning

Wet Spinning

The polymer to be spun is also used in a dissolved form during this process. However, in contrast to dry spinning, the fiber is fed into a coagulation bath after it leaves the spinneret (Fig. 17.19). This contains a non-solvent for the polymer, which is precipitated and the solvent used for spinning is washed out. To facilitate rapid precipitation and complete solvent removal, the two solvents should be miscible.

This process is significantly more complex than dry spinning as the fibers have to be subsequently dried and (if organic solvents are used) the solvent mixture from the coagulation bath must be separated for reuse. However, it is of considerable importance for the production of cellulose fibers by the "viscose process" (▶ Chap. 15) as cellulose cannot be processed either from the melt or from solution because it is infusible and insoluble in suitable solvents. Thus, cellulose is converted into the soluble cellulose xanthogenate at high pH and then spun into an acid bath. The xanthogenate dissociates by the action of the acid and (the now orientated) cellulose is reformed.[3]

The extraction speeds for wet spinning are around 50–200 m/min and thus considerably slower than those for dry spinning.

A variant of wet spinning is *air gap spinning*. This spinning process is mostly used for liquid-crystalline polymers (▶ Chap. 20), particularly polyaramides. Air gap spinning is also a wet spinning process; however, in contrast to the process described above, an air gap is left between the spinneret and the coagulation bath. The mesogens of the polymer

3 A process does exist in which cellulose is dissolved in *N*-methylmorpholine-*N*-oxide and then spun from this solution. This process can be carried out without carbon disulfide, although it is not used very often because of the toxicity of the morpholine derivate.

chains are orientated to constitute an anisotropic domain as the fibers pass through this air gap. This results in a drastic increase in the strength and stiffness of the polymer fiber. This process is also often used for spinning polyacrylonitrile and its copolymers for use as precursors for carbon fibers.

Stretching

In all the above processes, after the actual spinning process the fibers are stretched (drawn) by an appropriate take-up and winding arrangement. The term drawing is used to indicate that the polymer chains become oriented in the direction in which they are being stretched—the long axis of the fiber. This results in an increase in the number of interactions between neighboring chains because the chains become more and more aligned as the drawing process progresses. In many cases the process also increases the degree of crystallization which contributes to an increased modulus and tensile strength of the fibers, thus reducing elongation under stress. Technically, stretching is usually effected by passing the fibers over a sequence of rollers which are rotating at increasing speeds.

A rather special spinning process, which, however, until now has largely attracted only academic interest, is *electrospinning* (▶ Sect. 22.2). During this process a polymer solution is accelerated towards a counter electrode within a strong electric field. The initial, single liquid thread that ejects is split into several very fine threads because of static charges. The solvent is evaporated and *nanofibers*, with diameters well below 1 µm, are deposited on the counter electrode. To date, the main disadvantage with this method is that it has not been found possible to achieve a cost-efficient scale-up.

17.5 Forming Processes

In contrast to primary molding processes in which a solid part is formed from particulate material, such as a granulate, the starting material for forming processes is an already solid part, usually the product of a primary molding process, the form of which is to be altered. The weight and the internal cohesion of the parts employed do not change. Typical forming processes, which are dealt with in more detail below, are:

- Bending
- Tensile forming
- Compressive forming
- Tensile-compressive forming ("deep drawing")

Combined processes are also possible.

Generally, forming processes require the material to be heated to a plastic state in which it possesses a degree of flowability. Because cross-linked polymers do not exhibit any plasticity, they cannot be reformed by forming processes. Only thermoplastic molded parts can be thermally reshaped.

The temperature at which reshaping is possible is above the glass transition temperature, T_G for amorphous polymers and close to or above the crystal melting point, T_m for partially crystalline polymers to achieve adequate ductility. Heating can take place using convection, contact with heating media (hot gas, heat transfer oil, or solid heating elements), or by IR radiation.

Polymers exhibit maximum ductility (*maximum stretchability temperature*) at temperatures just above their glass transition or melting temperatures. This can be explained by the chains becoming more and more mobile with increasing temperature so that the material can be reshaped. However, the chain–chain interactions are still strong enough to maintain cohesiveness. The thermal energy and the mobility of the chains are not sufficient to overcome the chain entanglements so that the material doesn't flow in the absence of any applied force. If a force is applied, the material can be shaped but does not break. If, however, the processing temperature is significantly above T_G or T_m, the chains disentangle as a result of their increasing mobility relative to one another and the material breaks when force is applied. This effect is called *thermal brittleness*.

Shaping at temperatures close to the ductility maximum avoids the problem of thermal brittleness. However, because the macromolecules deform but do not disentangle to any significant extent, the shaping has the effect of creating a certain amount of internal stress. This can result in the reformed shape undergoing a spontaneous reversal of the deformation if it is reheated to above T_G or T_m, the so-called *memory effect*.

The most important forming processes are briefly described below.

Bending

Bending of polymeric materials is, in principle, trivial: the material, such as a pipe or sheet, is heated and then bent over an edge or an axle (Often a special mandrel is inserted for bending pipe structures to avoid buckling.). The wall thickness of the molded part does not changed significantly during this process.

Tensile Forming

During tensile forming a flat molded part is clamped at its edges and then stretched by force being exerted on its middle in one direction either by pressurized air, vacuum, or a punch. The wall thickness of the molded part decreases proportional to the size of the final part because of the edges being fixed.

Compressive Forming

During compressive forming, the semi-finished part is placed on a solid support and pressure applied in the form of a stamp or punch. Material is displaced from the area under pressure. A typical example for this kind of shaping process is the embossing of patterns.

Tensile-Compressive Forming

Similar to the tensile forming process, tensile-compressive forming involves applying a force to the middle of a molded part using a punch or stamp. In contrast to pure tensile forming, the edges of the material are not rigidly clamped so the material can follow the movement of the punch. Therefore, considerably less stretching occurs in the area being deformed. The advantage of this process, also known as *deep drawing*, is that the wall thickness of the molded part remains almost constant. This process is widely used, for example for the production of rotationally symmetric parts such as buckets.

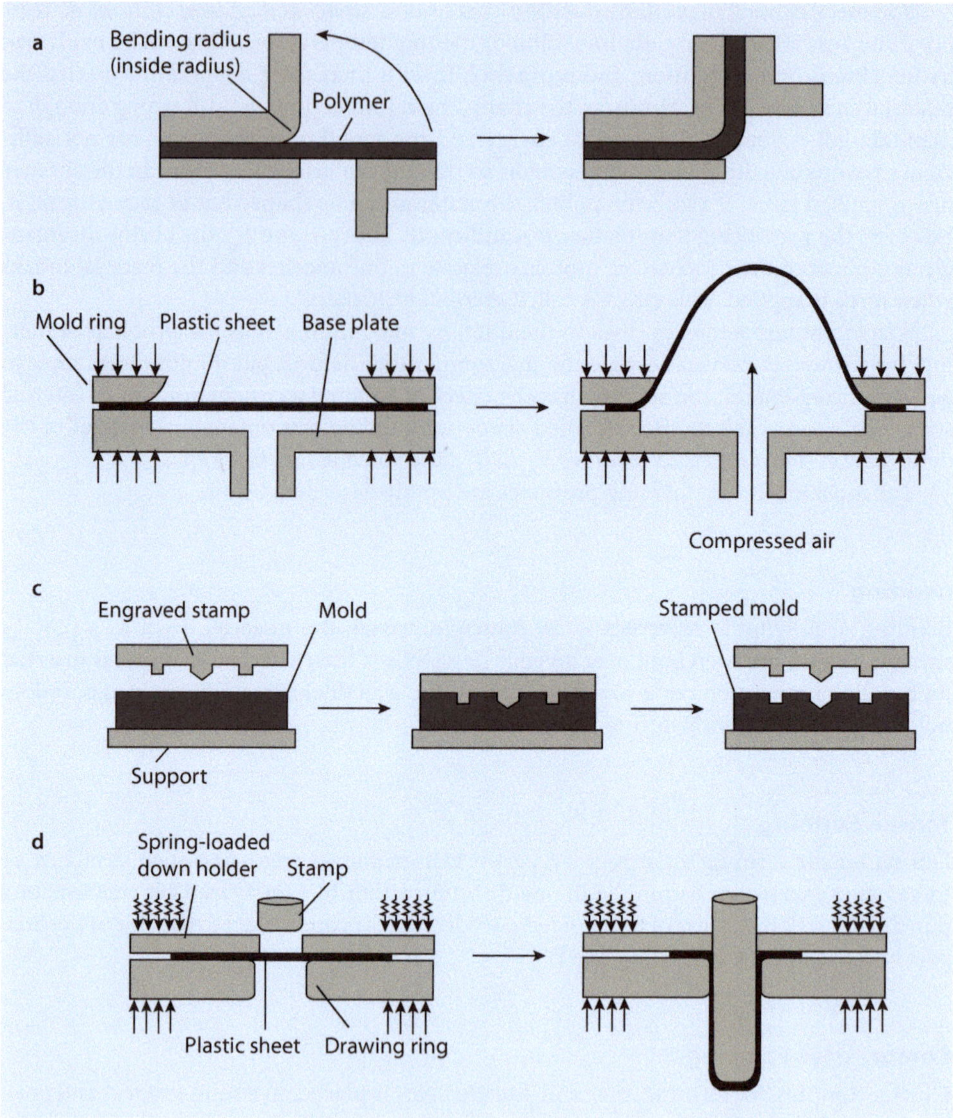

Fig. 17.20 Most important forming processes. (**a**) Bending. (**b**) Tensile forming. (**c**) Compressive forming. (**d**) Tensile-compressive forming

An overview of these most important forming processes is given in Fig. 17.20.

17.6 Joining Processes

Joining processes are processes in which two molded parts are joined together by creating a cohesive bond between them. Significant processes in this context are welding (▶ Sect. 17.6.1) and gluing (▶ Sect. 17.6.2) plastic molded parts.

17.6.1 Welding Processes

The welding of polymers basically consists of locally melting the molded parts at the points at which they should be joined and then pressing them together. A locally limited melting without melting the whole part is usually easy as plastics are generally poor heat conductors. Because of the mobility of the molecules at higher temperatures, the macromolecules or parts of these diffuse out of their entanglements and intertwine with other molecular sections within the other molded part according to the reptation model (▶ Chap. 6). The new entanglements become fixed on cooling and the two parts are bound together. As this process requires the formation of the melt state it can only be used for thermoplastic materials and not for cross-linked polymers. Furthermore, the rate of diffusion of whole molecules decreases drastically with increasing molar mass so that high-molar mass polymers are difficult to weld.

Another precondition for the welding mechanism requires that the polymers of the two parts are miscible. Because macromolecules with different chemistry are principally poorly miscible (▶ Chap. 2), successful welding of different polymers is only possible in exceptional cases.

When welding reshaped parts it is necessary to consider that locally melting the material may induce deformation to the original shape because of the memory effect.

The energy required for melting the part can be supplied in various ways as identified below:

Hot Plate Welding

The energy for melting the material is provided by thermal conduction from a heating element, generally heated electrically.

Hot-Gas Welding

The transition into the melted state is achieved by contact with hot gas. Inert gases are used if the polymer is sensitive to oxidation.

Radiation Welding

The heat is introduced from light or laser beams. Laser welding has become especially important. Suitable materials are chosen in such a way that one of the polymers contains an appropriate absorber. This additive absorbs the radiation from the laser beam and transforms it into thermal energy. The welding process can then be carried out in a very elegant fashion with the radiation beam passing through the polymer not containing an additive to reach the surface to be welded. The component, which is transparent to the wavelength range applied, is hardly heated at all. However, when the laser beam meets the component containing the additive, the energy is absorbed, whereby the melted area, and thus the welded area, can be very precisely controlled (◻ Fig. 17.21). This process, referred to as *transmission welding*, is mostly carried out using IR radiation.

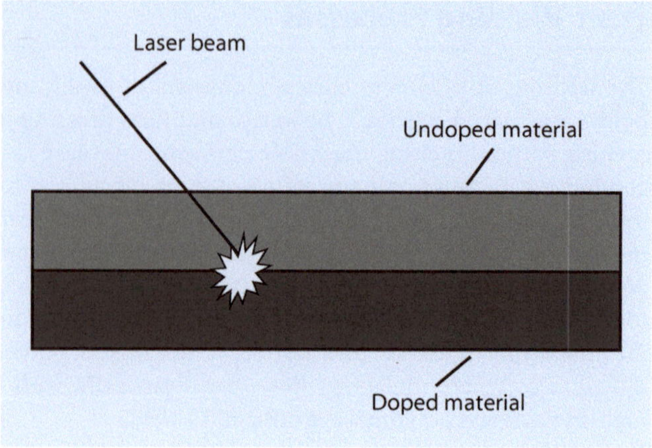

Fig. 17.21 Principle of transmission welding

Friction or Vibration Welding

The components to be welded together are moved relative to one another. As a result of the frictional heat produced on the component surfaces, the material is heated and eventually melts. This process is particularly suitable for the welding of components with large surface areas; e.g., it is used in the automotive industry.

As an alternative to these processes, welding can also be achieved by ultrasound and electric or magnetic alternating fields (high frequency or induction welding).

17.6.2 Gluing

In contrast to welding processes, which necessitate melting at least some part of the components, thermosets, i.e., cross-linked and infusible materials, can also be glued together with an appropriate adhesive. Gluing parts together has become much more important in recent years, for example, in automotive and medical engineering.

The principle of adhesion involves inserting an intermediate layer between the surfaces of the two parts to be joined. The intermediate layer must have sufficient internal mechanical strength (cohesion) and must be able to bond with both the surfaces to be joined (adhesion). These principles are summarized in Fig. 17.22. If the level of adhesion is inadequate, the glue layer separates from the surface it is adhered to under stress; if the level of cohesion is insufficient then the fracture occurs in the adhesive layer. If two parts have been perfectly glued together, then under stress, fracture occurs within one of the parts rather than at the adhesive joint.

Essentially, two effects are involved in an adhesive bond to a surface, which are referred to as *mechanical adhesion* and *specific surface adhesion*. The basic principle of mechanical adhesion is the mechanical interlocking of the adhesive layer with the microstructural surface roughness. Specific adhesion terms describe the non-mechanical interactions between the adhesive and the surface such as hydrogen bonds, dipole–dipole, or van-der-Waals interactions. Polyolefins generally have a low level of specific adhesion because of their nonpolar nature and are therefore difficult to glue. For this reason they are often oxidized (e.g., by treating the surface with plasma or a corona discharge) before being glued (or before painting) to achieve adequate adhesion. Covalent chemical bonds considerably improve specific adhesion between the surface and the adhesive if the surface has the

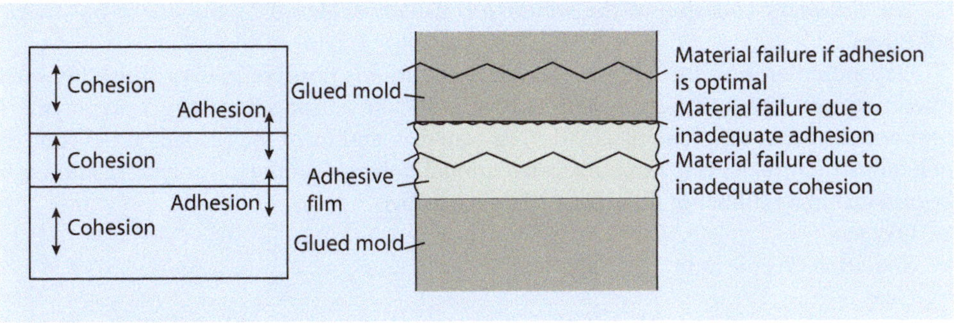

Fig. 17.22 Adhesion and cohesion effects and the modes of fracture at an adhesive joint between two materials

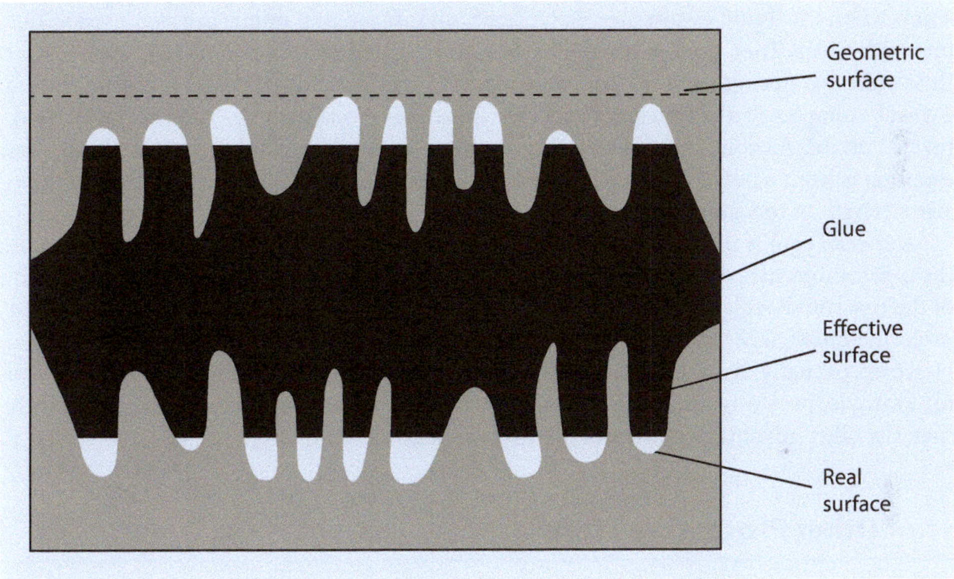

Fig. 17.23 Surface effects of glued surfaces

appropriate chemical functionality. Both adhesion effects are increased by an increase in the surface area so roughening a surface to be glued before gluing strongly improves the result of the process (Fig. 17.23).

To achieve an adequate specific adhesion of the adhesive to the surface, an optimum wetting of the surface by the adhesive is required. Surface wetting is improved for surfaces with high surface energies. The surface energy is defined for a surface in contact with air and is generally dramatically decreased by contact with a compatible liquid. This effect facilitates the wetting of the surface with an adhesive. Very polar materials with high surface energies, such as metals, glass, or ceramics, are thus wetted well and they can be more successfully glued than less polar plastics. Indeed, extremely nonpolar polymers, such as Teflon® or polypropylene, are extremely difficult to glue without prior surface treatment. A surface treatment using plasma or wet chemical processes creates the necessary surface functionality and is often used to prepare such surfaces for a gluing or painting process.

The necessary cohesion of the adhesive is usually achieved by the use of polymeric adhesives.

Depending on how the adhesive sets or hardens, it is possible to distinguish between physically and chemically setting adhesives. Chemically setting adhesives are prepolymers or monomers that react chemically during adhesion and form the (usually cross-linked) polymer structure. As is the case for conventional polymer syntheses, polymerization can be initiated in a number of different ways, such as by:

- Oxygen
- Radiation (visible light, UV radiation)
- Heat
- Humidity (i.e., cyanoacrylates, ▶ Sect. 10.2.2)
- Hardeners/catalysts

Physically setting adhesives are preformed macromolecules that are, for example, dissolved in a solvent or used in dispersed (emulsion or suspension) form and then create a film when the liquid evaporates. Polymers with low melting points are also used as hot-melt adhesives. They bond with the surface to be glued as they are melted and develop their cohesive strength as they cool and solidify, which usually involves a crystallization. Typical examples of a hot melt adhesive are those based on ethylene–vinyl acetate copolymers. Polychloroprene is an interesting example of an adhesive because as the solvent evaporates from a polychloroprene solution the polymer crystallizes to improve the cohesive strength of the adhesive layer.

A special case is the gluing of plastic surfaces using pure solvents. During this process the macromolecules on the surface are locally partially dissolved to facilitate the diffusion of the macromolecules between the surfaces. After the solvent has evaporated the macromolecule entanglements remain and the surfaces are effectively welded to one another. However, partially dissolving materials which have considerable internal stress because of, for example, previous shaping processes with a solvent can lead to stress cracking as the material takes advantage of the possibility to reduce its stress.

17.7 Other Processing Steps

As well as the techniques described in the previous sections, there are numerous other processes important in processing polymeric materials, for example:

- Separating processes (sawing, cutting)
- Coating processes (e.g., with polymer dispersions)
- Finishing processes (e.g.. printing)

For a more detailed discussion of these processes the interested reader is referred to the specialized literature.

References

Kaiser W (2007) Kunststoffchemie für Ingenieure. Hanser, München

Michaeli W (2010) Einführung in die Kunststoffverarbeitung. Hanser, München

Notario B, Pinto J, Solorzano E, de Saja JA, Dumon M, Rodríguez-Pérez MA (2015) Experimental validation of the Knudsen effect in nanocellular polymeric foams. Polymer 56:57–67

Elastomers

S. Koltzenburg et al., *Polymer Chemistry*, DOI 10.1007/978-3-662-49279-6_18

Characteristic of elastomers is that they have a glass temperature below ambient temperature; however, in contrast to thermoplastics, the most obvious property of cross-linked elastomers is their almost completely reversible elastic behavior. Some cross-linked elastomers can be stretched to up to seven times their original length and still return almost immediately to their original length. On account of this, this chapter is dedicated to this class of materials.

On a molecular level, this reversible elasticity requires that the polymer chains in these materials have a high degree of mobility. Elastomers are therefore exclusively amorphous materials with low glass transition temperatures that lie below their service temperature range. To prevent the material flowing at these temperatures, the polymer chains need to be cross-linked. However, the cross-linking density must not be so high that the mobility of the chains becomes severely restricted. Elastomers are therefore wide-meshed, cross-linked macromolecular materials and can be distinguished from thermoplastics, which are linear or branched, and duromers, which are densely cross-linked. Elastomers are not fusible because of their cross-links and, if exposed to excess thermal stress, they decompose without melting.

18.1 Permanently Cross-Linked Elastomers

As is well known, polymers can be divided into different groups according to their architecture as linear, branched, or cross-linked structures (Fig. 18.1). In addition, linear polymers, whose chains are longer than a certain (*critical*) length, tend to form entanglements.

Permanently cross-linked elastomers can be obtained by irreversibly chemically linking neighboring chains to one another. A wide-meshed, cross-linked elastomer is also referred to as a *rubber*. It is important to distinguish between *vulcanized* (cross-linked) and non-cross-linked (*unvulcanized*) elastomers.

The naming of elastomers distinguishes the various chemical structure of the main chain. An elastomer with a C=C double bond along the main chain is an *R-elastomer*, whereas an *M-elastomer* has a main chain of methylene units. Elastomers having both C and O atoms along the main chain are designated as *O-elastomers* and those having a siloxane main chain as *Q-elastomers*. *T-elastomers* are those with sulfur and carbon in the main chain and *U-elastomers* those with a carbon, oxygen, and nitrogen backbone. Typical examples of these different kinds of elastomers are shown in Table 18.1.

Some commercial products are listed in Table 18.2.

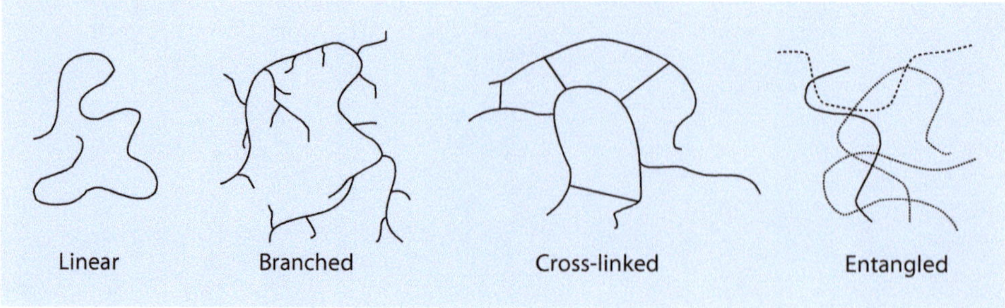

Fig. 18.1 Various polymer structures

Table 18.1 Name, type and chemical structure of typical elastomers

Name	Type	Repeating unit
Chloroprene rubber	R	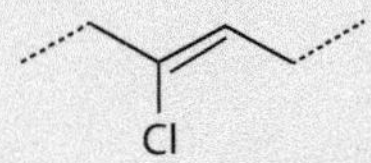
Natural rubber (Polyisoprene)	R	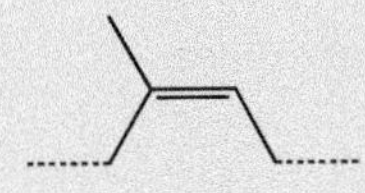
Copolymer of ethylene and propylene (EPM)	M	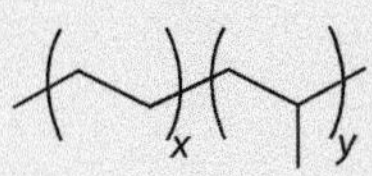
Copolymer of epichlorhydrin and oxirane	O	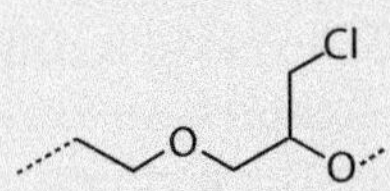
Silicone rubber	Q	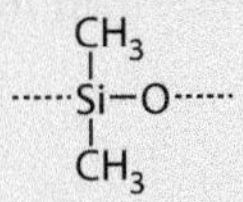
Polysulfide rubber	T	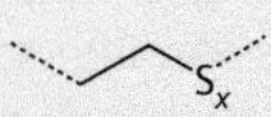
Polyether-urethane rubber	U	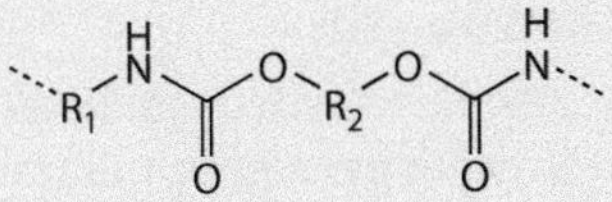

18.2 Properties

To use elastomers optimally they should be extensively tested and evaluated for the particular application. Some of the more critical properties and how to test these quantitatively are presented in this section.

Table 18.2 Selected elastomers, trade names, and suppliers

Elastomer	Trade name	Supplier
Styrene-butadiene rubber	Buna S	(Several)
Acrylonitrile-butadiene rubber	Perbunan N	Lanxess
Polychloroprene rubber	Baypren Neoprene	Lanxess DuPont
Acrylonitrile-butadiene-styrene rubber	Terluran	BASF
Isobutene-isoprene rubber	X_Butyl Exxon-Butyl	Lanxess Exxon Chemicals
Ethylene-propylene-diene rubber	Keltan	Lanxess
Methyl-fluoro-silicone rubber	Silastic Silopren	Dow Corning Momentive performance materials

Hardness

The hardness of a material is defined as the resistance of the material to permanent shape change when a compressive force is applied by a hard body of a certain dimension with a defined strength. The hardness of elastomers is measured on the *Shore A* scale using a truncated cone with a mass of 1 kg, which is pressed onto the test material. The hardness value is measured on a scale of 0 (very soft, no resistance) to 100 (very hard, no penetration). The method is described in detail in the DIN 53505 norm.

Rebound Elasticity

In this test, a swing hammer is dropped onto the test material from a horizontal 90° position. The proportion of the distance that the pendulum rebounds is calculated. A value of 0 signifies that the material is not elastic; there is no rebound. A test material that is completely elastic would yield a value of 100 (DIN 53512).

Tensile Strength

A test sample with defined dimensions is pulled apart using a testing machine. The maximum strength per square millimeter is the tensile strength in MPa (N/mm^2). The method is described in DIN 53504.

Ultimate Elongation

This is the maximum elongation that can be achieved during tensile testing and is given as a percentage of the original length of the sample. This can also be referred to as *elongation at break*. The measuring technique for this is described in DIN 53504 (Fig. 18.2).

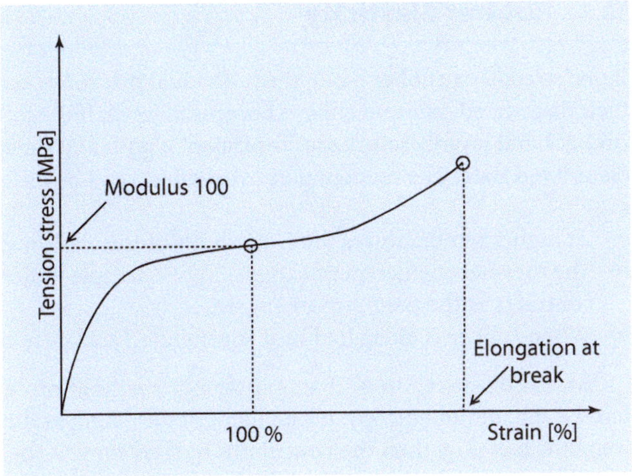

Fig. 18.2 Stress-strain diagram for a cross-linked elastomer

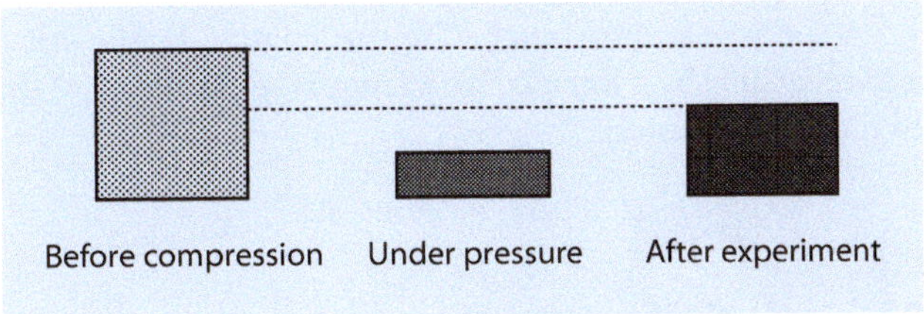

Fig. 18.3 Determining the compression set

Modulus of Elasticity

The modulus of elasticity (or e-modulus or Young's modulus) is defined as the gradient (tensile stress/extensional strain) of the linear (initial) part of a stress-strain-diagram. In practice, the stress values at 100% or 300% elongation are used to compare elastomer samples. These values are also referred to as Modulus 100 or Modulus 300. The values are given in MPa (N/mm^2), i.e., the force divided by the cross section of the (original) sample (DIN 53504; Fig. 18.2).

Compression Set

An important criterion for evaluating the residual deformation after the material has been put under compressive stress and the stress removed is the compression set. A sample of a defined height is placed under compression by a certain amount according to DIN 53517. After removal of the stress and a certain amount of time at a given temperature, the height of the material is determined (Fig. 18.3).

Further properties of elastomers that are often determined are tension set (DIN 53518), tear resistance (DIN 53516), and swell resistance (DIN 53521). As for all testing, it is important to control all the relevant test parameters very exactly to obtain repeatable and meaningful results. For cross-linked elastomers, the time and the temperature are particularly important variables.

In many material tests, cross-linked natural rubber compounds exhibit better values than synthetic elastomer compounds so natural rubber still represents some 40% of the total world market for rubber.

18.3 Rubber Elasticity

If one stretches a rubber band, the individual macromolecules are also stretched and leave their disordered, random coils to become aligned. If the force being exerted is removed the rubber band rapidly returns to its original length and the molecules revert to their coiled, disordered state. The following unusual effects can be observed:

- At higher temperatures the E-modulus of the elastomer increases
- The thermal coefficient of expansion of a sample under stress is negative; the sample contracts as the temperature increases
- When rubber is elongated heat is generated and its temperature increases

As was discussed in ▶ Chap. 4, polymer chains at rest exist as random coils. This form has the maximum entropy. If one increases the temperature of a rubber band on which a weight is hanging, then the contribution of entropy to the whole free enthalpy of the system increases and the macromolecule coils contract. The rubber band becomes shorter, which explains the second observations mentioned above.

A rubber band is stretched to the length L by the force F. If it is strained isothermally and reversibly by a force ΔF to a length ΔL (◘ Fig. 18.4) then, according to the second law of thermodynamics:

$$\Delta G = \Delta F \Delta L + p\Delta V = \Delta H - T\Delta S \tag{18.1}$$

ΔG	Change in free enthalpy
ΔH	Change in enthalpy
ΔS	Change in entropy
T	Absolute temperature
$p\,\Delta V$	Applied pressure
ΔF	Force required to extend the rubber band by ΔL

Thus, for the applied force ΔF:

$$\Delta F = -p\frac{\Delta V}{\Delta L} + \frac{\Delta H}{\Delta L} - T\frac{\Delta S}{\Delta L} \tag{18.2}$$

◘ **Fig. 18.4** Stretching a rubber band

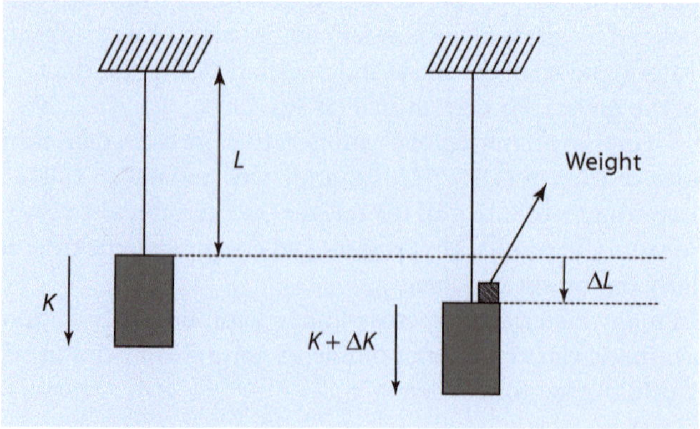

During elongation of the rubber band, only the volume can be neglected ($\Delta V \approx 0$). Additionally, the change in enthalpy caused by the elongation is insignificant because all possible conformations have essentially the same enthalpy ($\Delta H \approx 0$). Thus, (18.2) can be simplified to

$$\Delta F = -T\frac{\Delta S}{\Delta L} \tag{18.3}$$

From this equation it is clear that the force ΔF necessary to extend the rubber band by a length ΔL is proportional to the absolute temperature, explaining the first observation listed above. Furthermore, because the order increases as the band is stretched, ΔS is negative. This equation (with ΔF and ΔS constant) also explains why the rubber band contracts if warmed; additional force is needed to keep the length ΔL constant (a more rigorous explanation follows). This finding can be verified using the experimental arrangement shown in Fig. 18.5.

If, at room temperature, a rubber band has a force F exerted on it and is extended to a length L and is then heated from T to T_1 ($T_1 > T$), the rubber band contracts and the additional force ΔF needed to elongate the rubber band to the original length L can be determined.

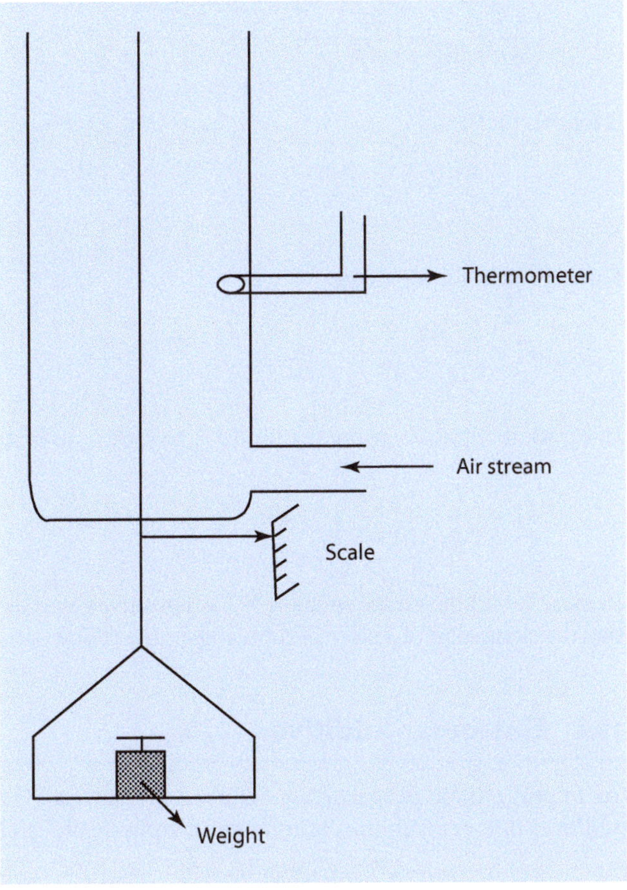

Fig. 18.5 Experimental set-up to verify Eqns. (18.3) and (18.11)

For an adiabatic reversible change of state of an elastic body it holds that

$$dQ = TdS = 0 \tag{18.4}$$

Hence

$$S = 0 \tag{18.5}$$

If S is perceived as a function of T, p, and L, then the following is true at constant pressure:

$$\left(\frac{\partial S}{\partial T}\right)_{p,L} dT + \left(\frac{\partial S}{\partial L}\right)_{p.T} dL = 0 \tag{18.6}$$

In combination with the following equations borrowed from general physical chemistry ((18.7) and (18.8)):

$$\left(\frac{\partial S}{\delta L}\right)_{p,L} = c_{p,L} \cdot \frac{1}{T} \tag{18.7}$$

$c_{p,L}$ Thermal capacity at constant length and pressure

$$\left(\frac{\partial S}{\delta L}\right)_{p,T} = -\left(\frac{\partial F}{\delta T}\right)_{p,L} \tag{18.8}$$

it follows that

$$c_{p,L}\left(\frac{\partial T}{\delta L}\right)_{p} = T\left(\frac{\partial F}{\delta T}\right)_{p,L} \tag{18.9}$$

Solving (18.9) for dT yields

$$dT = \frac{T}{c_{p,L}} \cdot \left(\frac{\partial F}{\delta T}\right)_{p,L} \cdot dL \tag{18.10}$$

Integration of (18.10) using the limits T_0 to T or L_0 to L results in

$$\Delta T = \frac{T}{c_{p,L}} \cdot \int_{L_0}^{L} \left(\frac{\partial F}{\delta T}\right)_{p,L} \cdot dL \tag{18.11}$$

Because for rubber elastic solids $\partial F/\partial T$ is a positive value, according to (18.11) it must follow that the elongation of a piece of rubber heats the rubber—the third observation listed above.

18

18.4 Elastomer Additives

To improve their properties, a number of materials are added to elastomers. These additives interact with the elastomer either physically or chemically to varying degrees.

Fillers

Generally, fillers increase the hardness of the rubber compound and its volume. If the filler is cheaper than the elastomer, rubber parts can be manufactured more cost effectively. If the filler interacts with the elastomer (active fillers, especially carbon black and silica), drastic improvements, e.g., to the strength and abrasion resistance of the resulting part, can be achieved. Other fillers (inactive fillers, e.g., kaolin and chalk) simply reduce the cost of the part or change its color. In addition to the chemistry of the filler particles, the particle size and the shape of the filler particles are important factors which affect how the filler changes the properties of the final rubber.

Plasticizers

These additives decrease the hardness of the rubber, influence its swelling properties, and improve its low-temperature flexibility by decreasing the glass transition temperature. Plasticizers also decrease the viscosity of the compounds, making their processing easier. Synthetic or mineral oils are most often used.

Cross-Linki ng Agents

The most important cross-linking agent is sulfur (▶ Sect. 14.10.2). Peroxides can also be used, as can phenol-formaldehyde resins. In special cases cross-linking can also be initiated using high-energy radiation (▶ Sect. 14.1.3).

Antioxidants

Aging processes, i.e., interactions of the elastomers with light and oxygen, result in a deterioration of the performance characteristics over time. Antioxidants and light stabilizers can delay light-related and oxygen-related aging processes (▶ Sect. 15.3).

Processing Aids

A broad spectrum of additives is available to ensure the optimal use and safe processing of elastomers. Examples are resins to increase the tackiness to facilitate the manufacture of blank disks, viscosity reducing agents, and so-called masticators (such as aromatic disulfides) which help break down very high molar mass material to shorten mixing times and improve filler dispersion. Additionally, colorings, propellants, fragrances, and flame retardants can all be added to elastomers depending on the intended application.

18.5 Processing

Part of the processing of elastomers involves their cross-linking (vulcanization). The processing steps can be divided into a compounding step, the forming of these compounds, and, finally, the vulcanization process.

A detailed description of these processes would go beyond the scope of this book. However, some of the processes described in ▶ Chap. 17 can also be used for elastomers.

18.6 Technically Important Elastomer Types

Rubbers based on natural rubber (NR, natural *cis*-1,4 polyisoprene) are especially used in objects whose dynamic properties are critical. Natural rubber is a common component of commercial vehicle and aircraft tires as well as being used for conveyer belts, transmission belts, and gloves (these are dipped using the latex).

A major portion of the styrene-butadiene rubber (SBR) produced is used to make tires where it can substitute, at least in part, natural rubber. Other applications for SBR are in shoe soles, seals, flooring, and facade paints (latex).

Butadiene rubber (BR) is especially useful because of its excellent elasticity; it also has good abrasion resistance and its low T_G is responsible for its excellent low-temperature flexibility. A large amount of BR is employed in tires as blends with NR and SBR.

Butyl rubber (IIR) consists generally of >95 wt% isobutene (2-methylpropene) and <5 wt% isoprene (2-methyl-1,3-butadiene). Its primary use (>80 %) is as an inner liner for tubeless tires because of its low gas permeability. As well as this, IIR exhibits good aging resistance because of its low content of C=C double bonds.

Elastomers made from ethylene, propylene, and a diene monomer (EPDM) are very resistant to aging because of their saturated main chain; they are also resistant to many chemicals. EPDM is used for window and door seals, particularly in the automotive industry, and for belting, damping elements, and cable sheathing.

Acrylonitrile-butadiene rubber (NBR) is used when resistance to petrol, oils, and fat is required. The properties of NBR vary greatly depending on the acrylonitrile content. Grades with low acrylonitrile contents (18 mol%) are more elastic and have a lower T_G but they are less resistant to oils than grades with high acrylonitrile contents (40–50 mol%). Grades with 38 mol% ACN are compatible with PVC and PVC-NBR blends are an important application for this type of elastomer. Other important uses of NBR are seals, O-rings, roller coverings, hoses, and linings where oil resistance is a critical property requirement.

The distinguishing features of chloroprene rubber (CR) articles are their weather and ozone resistance, their good dynamic properties, and their good flame resistance produced by the high chlorine content. Some CR grades crystallize and are used as adhesives. Probably the best known use of CR is as diving suits which are often referred to as neoprene suits.

Ethene-vinyl acetate elastomers (EVM) are particularly noted for their ability to be very highly filled with less active fillers such as magnesium and aluminum hydroxides, typical cost effective, non-halogen flame retardants. EVM is used for seals, cable sheathing (flame resistant non-corrosive, FRNC), and particularly for hot melt, adhesives. Their elastic nature increases with their vinyl acetate content, up to about 80 % VA.

Silicones (VMQ) demonstrate high flexibility at low temperatures as low as −60 °C. They are suitable for service temperatures up to ca. 180 °C. They are often used where food contact approval is necessary and for insulators, seals in the electrical industry, and medical products.

18.7 Thermoplastic Elastomers

Thermoplastic elastomers are elastomers that, in contrast to covalently (and therefore permanently) cross-linked elastomers, involve thermally reversible cross-links. Thermoplastic elastomers can be made from block copolymers for example. As polymers are usually incompatible, most block copolymers separate into at least two phases (▶ Sect. 13.9.1). One phase functions as the “soft” elastomer phase and the other as the “hard” phase and has a glass transition temperature or melting point above the service temperature. The “hard” phase forms the cross-links in place of covalent network points. These hinder the material flowing and ensure that the material reverts back to its original shape after the release of any applied stress. However, at higher temperature, above the T_G or melting range of the “hard” phase, the latter becomes softer and the material can be processed as a thermoplastic without decomposition

Multi-phase graft copolymers and polymer blends can also form thermoplastic elastomers. A special group of thermoplastic elastomers are the thermoplastic vulcanizates (TPV) in which small rubber particles are vulcanized as the rubber is being blended with a thermoplastic which forms the matrix phase. The best known examples of such systems are the TPV based on EPDM and PP. Notwithstanding, the largest volume thermoplastic elastomers are those made from block copolymers and these are described in more detail here.

Multi-phase block copolymers with TPE (thermoplastic elastomer) properties are available in many different variations (◘ Table 18.3). Characteristic properties of thermoplastic elastomers are compiled in ◘ Table 18.4.

All these grades exhibit excellent tensile strength and elasticity. These properties are especially pronounced in thermoplastic elastomers based on ether-esters or ester-urethanes wherein the polyurethane or polyester segments can crystallize. Polyether-esters and polyester-urethanes have better high-temperature characteristics than TPEs based on styrene-diene. The latter are more suited to low- or ambient-temperature applications because of the low glass transition temperature of the elastomer blocks.

SBS and their hydrogenated analogs are a better choice as electrical insulators because of their low polarity. Styrenic TPEs are widely used in the automotive industry as motor parts, O-rings, seals, hoses, and shock absorbers. They are also used for shoe soles, bitumen modifiers, and adhesives.

The use of thermoplastic elastomers in the shoe industry, and especially the sport shoe industry, is becoming ever more important. The TPE fiber *Spandex*®, which is based on polyurethane, has become known world-wide.

◘ **Table 18.3** Types of thermoplastic elastomers

Building blocks	Type	Trade name/supplier
Styrene-butadiene-styrene	ABA	Kraton/Kraton; Styroflex/BASF
Ether-ester	$(AB)_n$	Hytrel/DuPont; Solpren/Dynasol
Urethane-ester/ether	$(AB)_n$	Pellethane, Estane/Lubrizol, Desmopan/Bayer

Table 18.4 Typical properties of thermoplastic elastomers (+: especially favorable)

Property	Styrene-butadiene-styrene (SBS)	Hydrogenated styrene-butadiene-styrene (SEBS)	Ester-ether	Urethane-ester/ether
Ductility			+	+
Reusability	+	+		
Top service temperature			+	+
Lowest service temperature	+	+		
Aging resistance		+		
Acid-base resistance	+	+		
Oil resistance			+	+
Electrical resistance	+	+		
Abrasion resistance				+
Processability			+	
Price	+			

In spite of the enormous application potential of thermoplastic elastomers, their synthesis, their characterization, and modifying them to achieve optimum performance are challenges. The styrenic TPE are generally synthesized via anionic polymerization in dilute solutions. The polyurethane TPE and the copolyester ethers are manufactured in bulk in continuous or semi-continuous step growth processes. The introduction of controlled radical polymerizations can be expected to widen the possibilities for block thermoplastic elastomers.

To characterize thermoplastic elastomers adequately, knowledge of their history is required. Then statements about whether they are block copolymers, statistical copolymers, or mixtures of homopolymers can be made with certainty. Taking polyether amide as an example, ▶ Sect. 18.8 demonstrates how the properties and thus the possible applications of a thermoplastic elastomer vary as its structure changes.

18.8 Polyether Amide-12 Elastomers

18

Polyether amide-12 elastomers are ABA-block copolymers made from PA-12 and a polyether such as polyTHF (polytetrahydrofuran) (Fig. 18.6).[1]

The polyether amide-12 elastomer is synthesized using individual building blocks, PA-12 and α,ω-hydroxy-telechelic of THF. If PA-12 is first converted using a dicarboxylic

1 Most of the figures/illustrations shown in this section are taken from information on Vestamid 12® produced by Evonik. The authors are grateful for permission to use these.

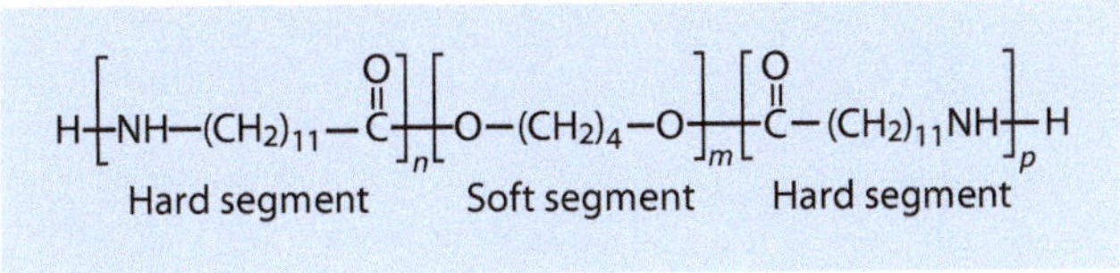

Fig. 18.6 Structure of polyether amide-12 elastomers

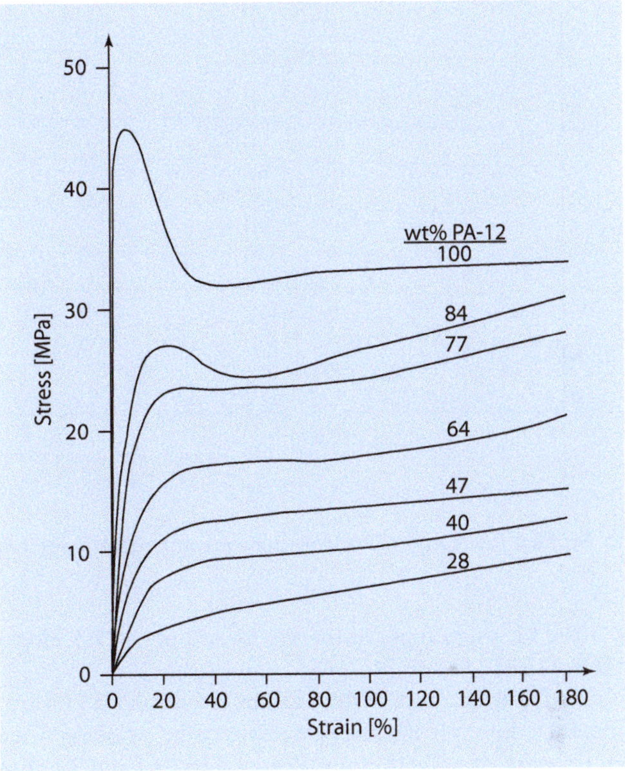

Fig. 18.7 Stress-strain curves of PA-12 elastomers with polytetrahydrofuran soft blocks

acid, then building blocks form that have COOH-functions at both ends. Their reaction with α,ω-polytetrahydrofuran diol results in $(AB)_n$-block copolymers.

As the polyether content increases the products become more elastomeric; they become more flexible and their impact-resistance improves. Additional typical characteristics of this family of TPE are:

- Low density
- Resistance to chemicals and solvents
- Wide variation of hardness and flexibility
- Good elasticity and ductility

These products are employed for a wide range of applications such as low-noise gear units, seals, the functional elements of sport shoes, ski boots, and conveyer belts. PA-12-rich TPEs have properties more typical of crystalline thermoplastics but the elastomeric character increases as the amount of polyether increases, as can be seen from the stress-strain curves shown in Fig. 18.7.

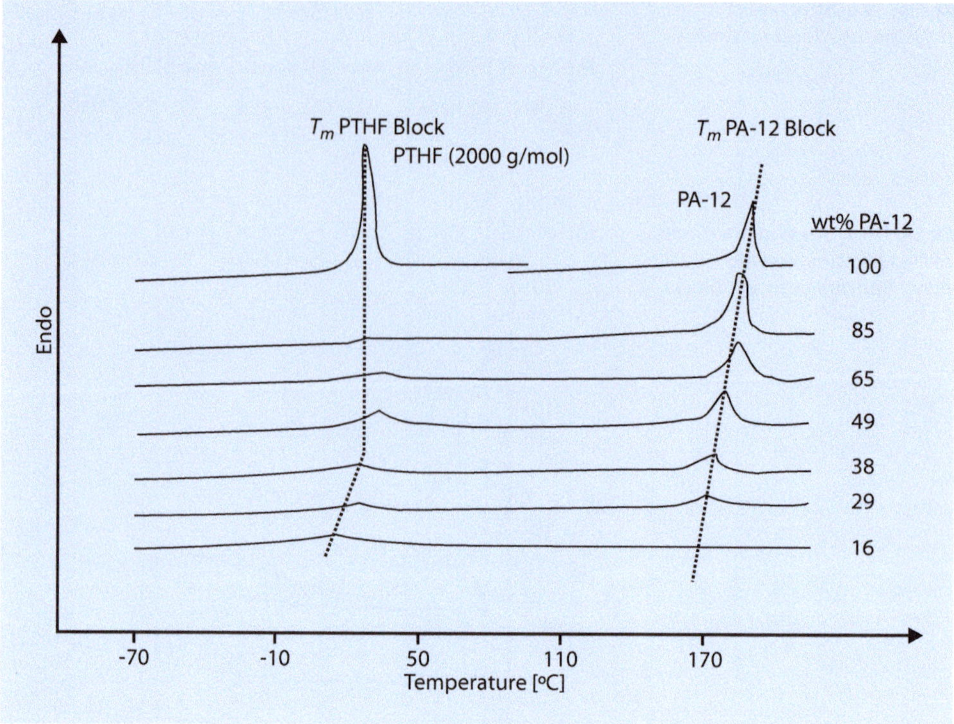

Fig. 18.8 DSC curves of PA-12 elastomers with polytetrahydrofuran soft blocks (PTHF)

The melting ranges of the PA-12 and poly-THF blocks can also be observed by DSC (Fig. 18.8).

As for elastomers with covalent crosslinks, TPEs are also characterized, in the first instance, by their hardness, measured by pressing a truncated cone into the sample. In the case of polyether amides, materials from 30 Shore D (more rubber-like) to 68 Shore D can be synthesized by varying the hard/soft block ratio (Fig. 18.9).

The relatively low temperature dependency of the mechanical properties of PA-12 elastomers, produced by the relatively high T_M of the polyamide blocks, is of crucial importance for many applications. As well as this, the favorable abrasion resistance, low water absorption, and good dimensional stability of PA-12 elastomers make them interesting materials.

18.9 Liquid Crystalline Elastomers

Networks in which relatively freely mobile and thus orientable mesogens are incorporated are gaining ever-increasing importance (▸ Sect. 20.6.3). These kinds of networks have, on the one hand, the characteristic anisotropic phase behavior of liquid crystals and, on the other, properties of cross-linked elastomers such as form stability and rubber elasticity.

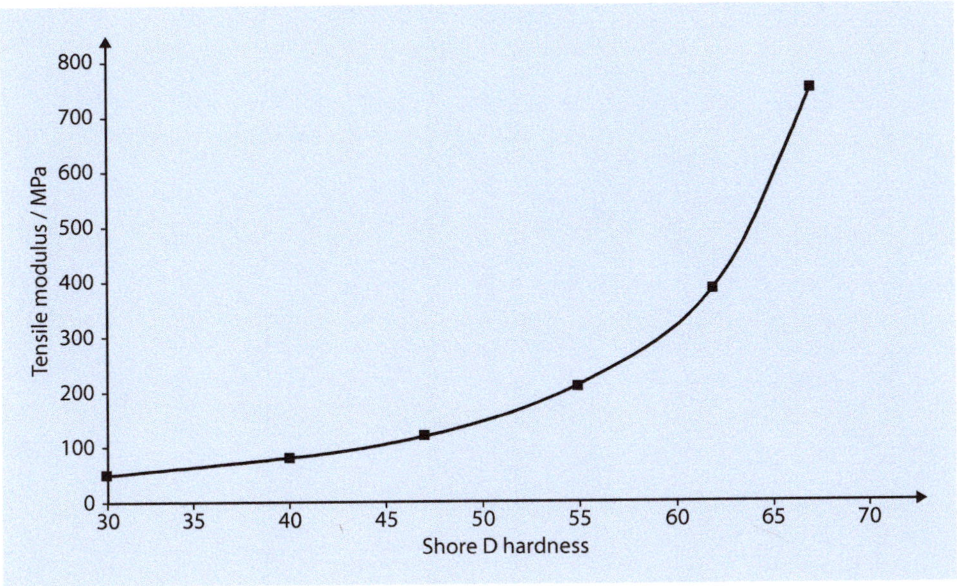

Fig. 18.9 Plot of Young's modulus against Shore D hardness for vestamid elastomers. *E40* Elastomer has a Shore D Hardness of 40

The mechanical deformation of elastomers has the same influence on the orientation of the liquid crystalline molecules as electric and magnetic fields have on the orientation of low molar mass liquid crystals and non-cross-linked liquid crystalline polymers (▶ Chap. 20).

Functional Polymers

S. Koltzenburg et al., *Polymer Chemistry*, DOI 10.1007/978-3-662-49279-6_19

Classical polymer materials such as polyethylene and polyamide are familiar to all of us because of the multiplicity of uses in everyday life. Equally as omnipresent, but less well known, are the so-called functional polymers. As opposed to the structures previously discussed in ▶ Chaps. 13, 14, and 15, these polymers do not generally belong to the group of solid materials. Instead, these materials are mostly used in solution, where they induce a certain physical effect. Thus, most frequently it is not the polymer itself that is recognized but rather its effect or function—hence the name *functional polymers*. They are also referred to as *polymeric materials* or *effect substances* (Göthlich et al. 2005). Despite being so inconspicuous, many areas of our daily lives would be very different without them. For example, they are used in detergents, pharmaceuticals, and cosmetics.

As with most systems with which we come into contact in our daily lives, almost all environments in which functional polymers are used are aqueous, so most of these polymers are soluble in water. As demonstrated in ▶ Chap. 2, when a polymer dissolves in a solvent only a relatively small entropy gain is achieved. To reach the desired solubility in water, only relatively hydrophilic monomers can be considered; these interact, for example, via hydrogen bonds with water and are thus responsible for a favorable (negative) contribution to the free enthalpy of the solution. Such monomers are often electrically charged. The ionic polymers made from them are referred to as *polyelectrolytes*.

In principle, four different types of monomers are potential elements of functional polymers (◻ Fig. 19.1):

- Anionically charged, that is, acidic monomers such as acrylic acid
- Cationically charged, that is, alkaline monomers such as monomers carrying amino groups
- Electrically neutral, highly polar monomers, for example, vinyl pyrrolidone
- Zwitterionic monomers

Some examples of such monomers can be found in ◻ Fig. 19.1.

A substantial amount of the commercially available functional polymers are homo- and copolymers based on acrylic acid, vinyl pyrrolidone, and vinyl acetate. In addition, acrylate and methacrylate monomers play a vital role. Depending on their substituent, they can be hydrophilic (e.g., hydroxyethyl acrylate) or hydrophobic (e.g., lauryl acrylate). Cationic acrylates, based on amino functionalized acrylates, are common.

Zwitterionic monomers, such as the betaine presented in ◻ Fig. 19.1, have found only limited use to date. However, because of their unusual solubility behavior, they remain of interest. As opposed to most polymers, which are soluble in water and display an *upper critical solution temperature* and thus become insoluble above a certain temperature (▶ Chap. 2), zwitterionic monomers can give polymers that have a *lower critical solution temperature*, that is, a minimum temperature at which the polymer is soluble.

As well as the unsaturated monomers mentioned above, which are mostly radically polymerized in aqueous solution, some monomers are also used to make functional polymers, which are converted via a ring-opening polymerization (▶ Chap. 12), particularly ethylene oxide, ethylene imine, and oxazolines. Ethylene oxide can be polymerized both cationically and anionically and delivers a linear polymer that is soluble in water (◻ Fig. 19.2). Particularly important are polyethylene oxides with a lower molar mass, which are also referred to as polyethylene glycols. Industrially, anionic polymerization has achieved the greater relevance. In the polymerization of ethylene imine, the reaction is complicated because both the nitrogen atom of the monomer and that in the polymer are able to react with the active species (ammonium ion). The latter possibility leads to branched products, as demonstrated in ◻ Fig. 19.3.

19

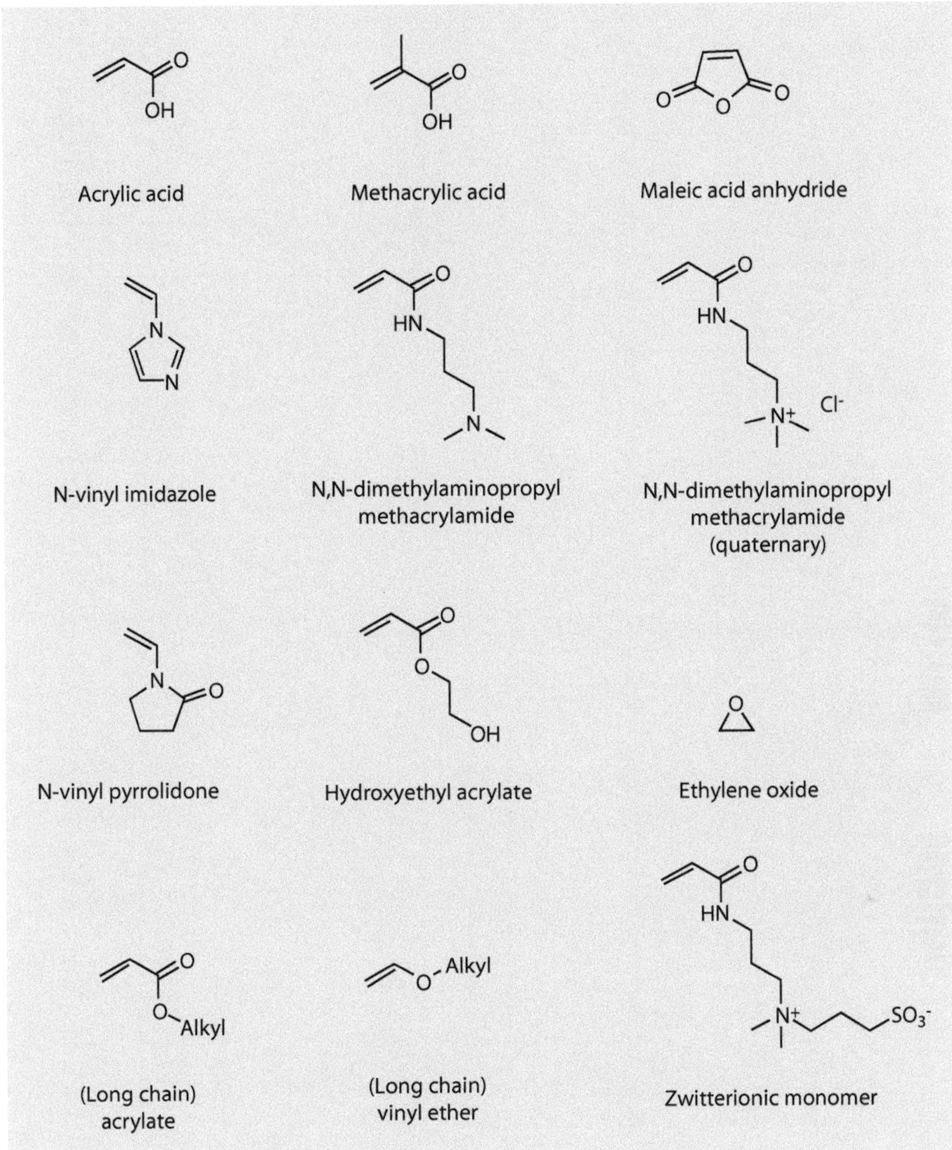

Fig. 19.1 Examples of monomers of functional polymers

Fig. 19.2 Polymerization of ethylene oxide to polyethylene oxide

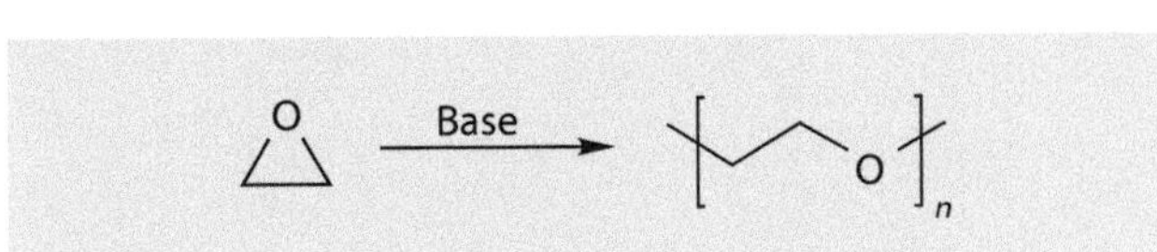

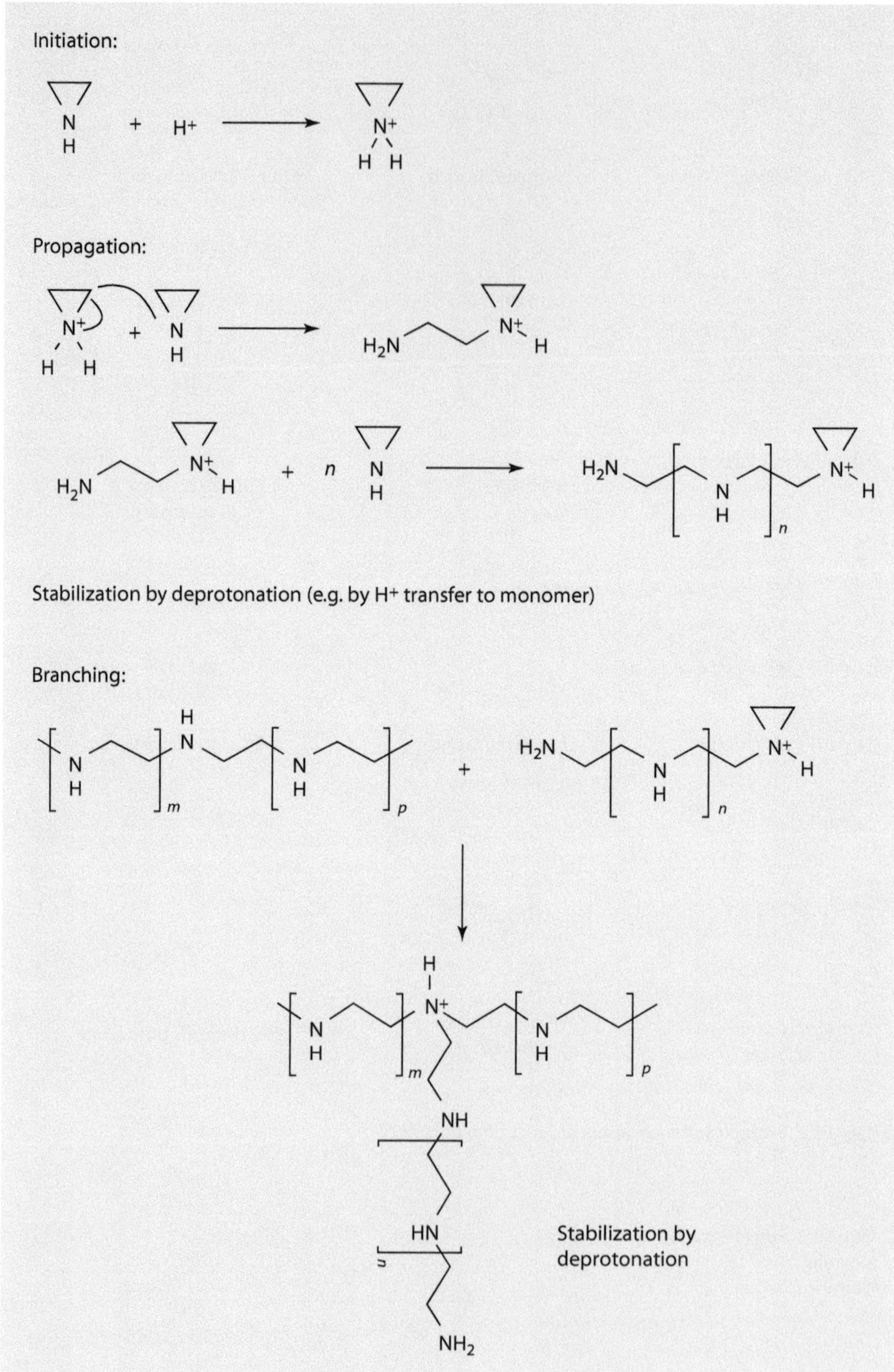

Fig. 19.3 Polymerization of ethylene imine to polyethylene imine

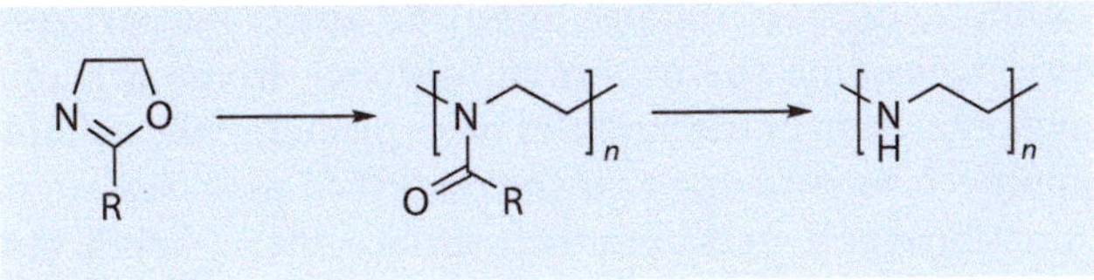

Fig. 19.4 Synthesis of linear poly(ethylene imine) by hydrolysis of a polyoxazoline

Linear polyethylene imine can be synthesized via a polymer analog reaction whereby an oxazoline monomer is cationically polymerized via ring opening (▶ Sect. 12.3) and subsequently hydrolyzed (Fig. 19.4).

By copolymerizing the monomers shown in Fig. 19.1 with other monomers, a vast number of chemically varied polymer structures are accessible. The technical profile of these polymers can be additionally fine-tuned by selectively varying their molar mass.

In addition to the synthetic polymers discussed here, natural polymers or modified biopolymers, such as, for example, derivatives of cellulose (▶ Chap. 14), play an important role as functional polymers.

19.1 Polymer Dispersion Agents

One of the most important applications of functional polymers is the stabilizing of so-called *colloid systems*, which can also be referred to as dispersions. Here we are dealing with systems of two or more immiscible substances in which one material is finely distributed in the form of minute particles or droplets (dispersed phase) in the other (continuous phase) material. For reasons of clarity, the dispersed phase is often referred to as particles, although the models may also include emulsions.

Systems in which a solid is dispersed in a liquid are referred to as *suspensions*. If a liquid is homogenously and finely distributed in another liquid with which it is immiscible, it is referred as an *emulsion*.

In most cases, the only systems referred to as *colloid systems* are those whose dispersed particle diameters are less than 1 μm. Milk, in which small fat droplets are dispersed in water, is probably the most well-known example of a natural colloid system. Colloids are also technically of great importance. Thus, for example, all polymer dispersions which are produced via an emulsion polymerization and used, for example, as coating agents fall into this classification. It is also necessary to find a suitable method to disperse plant protection substances, which are often water insoluble, into stable, sprayable aqueous dispersions for convenient use by farmers.

Because of the large surface area of the colloid nanoparticles, the simple dispersion is, in most cases, thermodynamically unstable and requires stabilization. Without sufficient stabilization, the dispersion segregates into two macroscopic phases; for example, the oil droplets of an emulsion tend to float (cream) to the top of the system and finally coalesce into a separate phase. In general, the stabilization of a colloid system can occur in two ways:

- By electrostatic stabilization
- By steric stabilization

Electrostatic stabilization involves all particles being identical so that they repel each other electrostatically. The simplest approach is to adsorb charged molecules, for example, dodecyl sulfates, onto the dispersed particles. Once charged, the particles tend to assume the maximum distance from each other so that no flocculation or agglomeration of the particles takes place and the colloid stability is increased. Another approach to ensuring that the dispersed particles are mutually repulsive is to adsorb functional polymers onto their surface. Steric stabilization is discussed in detail below. It is also sometimes practical to stabilize colloids by the simultaneous adsorption of both polymers and ions—or of charged polymers. This approach is called *electrosteric stabilization*. For a more detailed study of colloid systems, the interested reader is referred to Tadros (2005).

19.1.1 Steric Stabilizing of Colloid Systems: Protective Colloids

In Fig. 19.5 two colloid particles (not to scale) are shown which have been sterically stabilized by the adsorption of a polymer layer. For the steric stabilization of a particle with a diameter of 1 μm, the adsorbed polymer layer is ideally 5–10 nm thick. When two stabilized colloid particles approach each other, two unfavorable energy effects occur which impede or prevent close approach:

- The chains overlap (Fig. 19.5b) and/or
- The adsorbed polymer layer is compressed (Fig. 19.5c)

In both cases, the segment density between the particles increases. Both processes lead to a strong repulsion of the particles for two reasons:

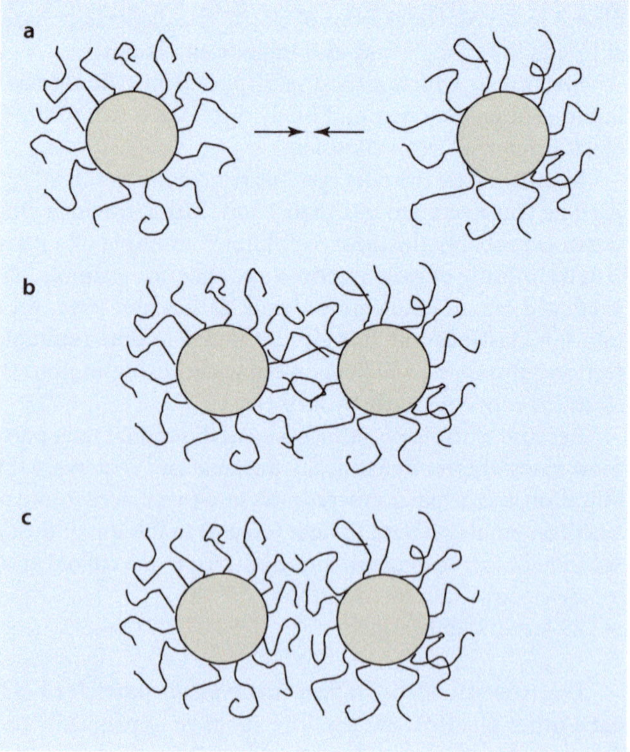

Fig. 19.5 Scheme showing the processes occurring when two sterically stabilized colloid particles approach each other. (**a**) Starting condition. (**b**) Approach involving an entanglement of the polymer chains. (**c**) Approach involving a compression of the polymer chains

- The increased segment density between the particles leads to an osmotic pressure, which in turn results in the diffusion of solvent into the region between the particles and thus counteracts any further approach
- The configuration entropy of the chains decreases because of their reduced flexibility

These repulsion effects are especially strong if the chains protruding into the continuous phase are very soluble in it so that they are highly solvated. On the other hand, the adsorption and thus the stabilization is obviously only effective if at least some subsection of the polymer also shows an affinity to the surface of the particle. Thus, *amphiphilic block copolymers* consisting of hydrophilic and hydrophobic blocks have proven to be particularly effective colloid stabilizers. In a similar manner, comb copolymers with a hydrophobic polymer backbone and hydrophilic side chains exhibit a marked tendency to adsorb onto the surface of particles or droplets.

The adsorption processes are shown schematically in ◻ Fig. 19.6. The thermodynamic driving force of the adsorption is both enthalpic and entropic in nature:

- The absorption of a polymer onto a surface involves a favorable entropy change, because *one* polymer chain at the surface of the particle supplants a larger number of solvent molecules
- The favorable enthalpy results, in the case of hydrophobic particles in water, for example, are that the hydrophobic areas of the polymer accumulate onto an equally hydrophobic surface or, in an emulsion, they can dissolve in an oil phase

Such amphiphilic steric stabilizers are also referred to as *protective colloids*. Technically, block copolymers consisting of ethylene oxide and propylene oxide are commonly used (◻ Fig. 19.7). In these polymers the hydrophobic propylene oxide middle section functions as an anchor group for the adsorption of the polymer on the surface, whereas the hydrophilic polyethylene oxide segments protrude into the water phase leading to steric stabilization.

Concrete plasticizers, such as those discussed in ▸ Sect. 19.1.2.3, are examples of amphiphilic comb polymers.

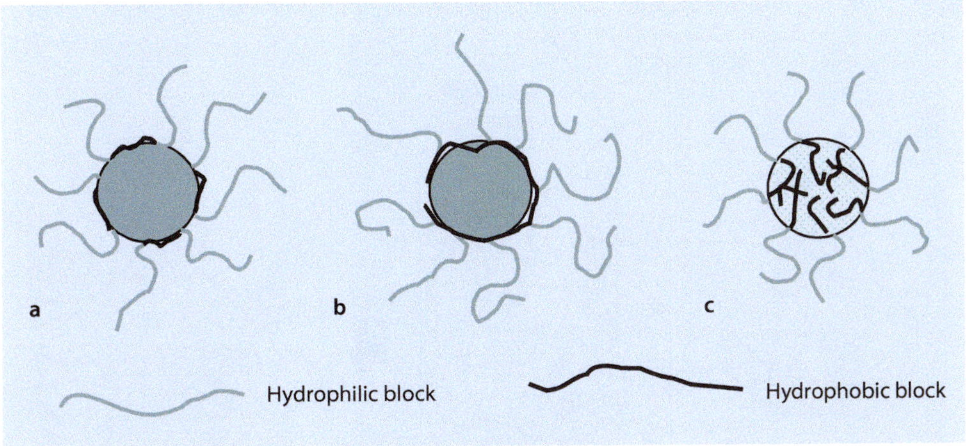

◻ **Fig. 19.6** Possibilities for adsorbed amphiphilic polymers. (**a**) Adsorption of an ABA-triblock copolymer on a particle. (**b**) Adsorption of a comb copolymer on a particle. (**c**) Adsorption of an AB-block copolymer to an oil droplet

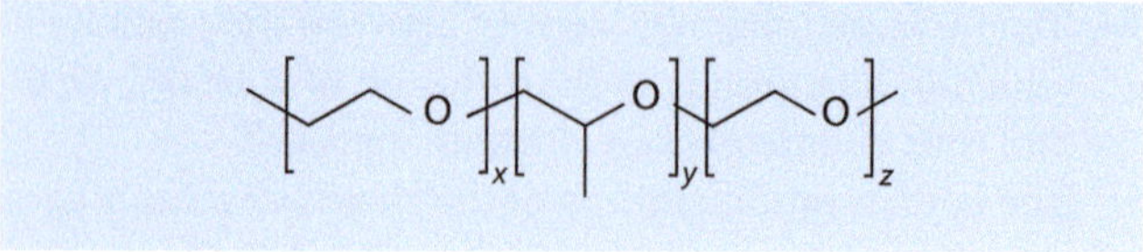

Fig. 19.7 ABA-triblock copolymer of ethylene oxide (forming the A-block) and propylene oxide (forming the B-block) as an example of a typical protective colloid

19.1.2 Applications of Polymer Dispersion Agents

In this section some important applications of polymer dispersion agents are discussed in more detail.

19.1.2.1 Paper Coatings

The surface of most paper one encounters in daily life is modified with a surface coating. This coating fulfills various functions, such as to generate gloss (common in illustrated magazines) or to reduce absorbency. The latter is an essential characteristic of writing and printing paper to prevent the ink or printing paste from running as they do on blotting paper so that a sharp printing image can be achieved.

A material often used for paper coating is so-called *China Clay*, a layered silicate with negatively charged surfaces and positively charged platelet edges. Thus the platelets tend to form a "house-of-cards" type structure (Fig. 19.8b). An extremely voluminous but mechanically fairly stable superstructure emerges. To destroy it and to transfer the layered silicate into a fluid form of low viscosity which can be spread onto the paper, a large amount of water must be added. Even at a water content of 40 wt% the mixture is still pasty and not fluid at room temperature. To obtain a spreadable mass using only water would require considerably higher amounts of water but these would then have to be evaporated in the finishing process which would be energetically unfavorable, time consuming, and expensive.

An alternative to using a dilute spreading mass is to add a small amount of polyacrylic acid with a molar mass M_w of ca. 2000 g/mol. Polyacrylic acid is largely in its dissociated form at a neutral pH and is thus negatively charged. For this reason it preferably adsorbs onto the positively charged platelet edges, neutralizing their positive charge. This leads to

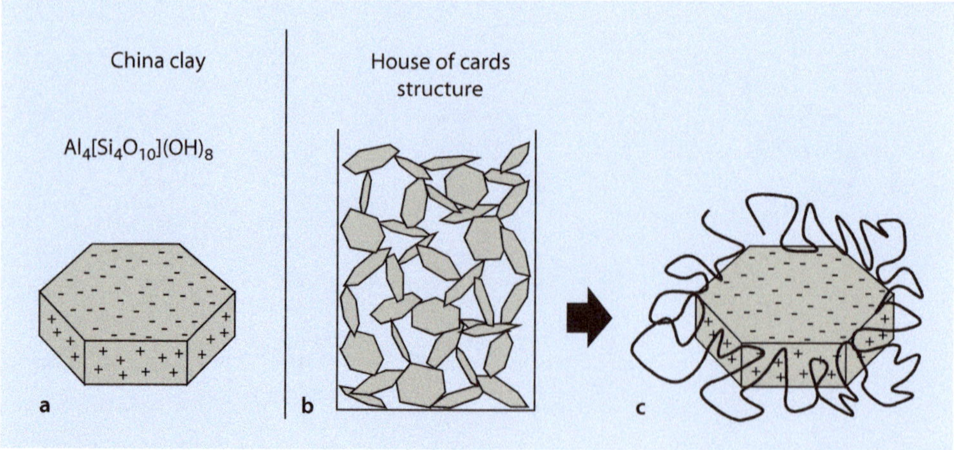

Fig. 19.8 (**a**) Surface charge distribution. (**b**) "House of cards" structure of China Clay. (**c**) Interaction of China Clay with polyacrylic acid

the collapse of the stable "house-of-cards" structure. The viscosity decreases rapidly to yield a relatively thin, spreadable paste, even at high solid contents. The mechanisms described here are shown schematically in ◘ Fig. 19.8.

19.1.2.2 Additives for Seawater Desalination

Clean drinking water is becoming an ever more valuable resource in view of an increasing world population and ecological and climatic changes. Especially in the countries of the Middle East, seawater desalination is employed to supply much of the water demand. The majority of desalination plants operate via reversed osmosis and large plants can desalt several hundred thousands of cubic meters of seawater per day. The remaining concentrated brine must be effectively prevented from crystallizing and clogging up the pipes. These residue dispersions can be stabilized by adding polymer dispersion agents to prevent agglomeration or precipitation on the walls. In this case too, negatively charged polyelectrolytes based on acrylic acid are commonly used. The working principal of these dispersion agents is similar to that shown in ◘ Fig. 19.8.

19.1.2.3 Plasticizers

Concrete is the largest volume industrially produced product in the world. Chemically, it is a mixture of cement, sand, gravel, and water. Cement is a heterogeneous mixture of various minerals based on the calcium aluminates, silicates, and sulfates. With water, this mixture forms hydrates in a strongly exothermic reaction whereby the mineral particles are joined to each other, leading to the increasingly solid material as the concrete sets. Because of the heterogeneity of the materials and the complexity of the various processes, the exact details have not yet been identified.

Notwithstanding, the addition of water to the above mixture has two effects: it makes the water available for forming the hydrates essential for the setting process and it turns the solids into a liquid facilitating rapid mixing and forming processes, for example, it can be poured into a shape before it solidifies—the so-called *hardening*. The amount of water added is critical for both the processability and the strength of the concrete. Although the flowability of the concrete increases with the amount of water added, only approximately 30 % water (relative to the mass of the concrete) is chemically bound as hydrate water in the end product; the rest evaporates over time. Too much water leads to an increased porosity and a decrease in the strength of the resulting concrete. For the production of high-performance concrete, necessary, for example, for the construction of skyscrapers, it is beneficial to limit the amount of added water to a minimum. To this end, polymer plasticizers are a good choice.

Functional polymers have been used as plasticizers for concrete for a long time. Older products of this type are sulfonated lignin, sulfonated melamine-formaldehyde-resins, or sulfonated naphthalene-formaldehyde-resins. Lignin sulfonate is a structurally ill-defined polymer that is negatively charged by its sulfonic acid groups. It is a by-product of paper production. Formaldehyde condensates are discussed in ▶ Chap. 8. These products are inexpensive but achieve only a limited effect. For high-performance *superplasticizers* for cases where high demands are placed on the hardened concrete, polycarboxylate ethers are commonly used. Polycarboxylate ethers are polymers based on carboxylic acids, for example, acrylic acid or methacrylic acid. These are copolymerized with a comonomer containing a hydrophilic polyethylene chain, a so-called *macromonomer* (◘ Fig. 19.9). A typical example of a macromonomer is polyethylene glycol, which has a functionalized group, such as an acrylate function, at one end. The resulting polymers have the structure of graft copolymers.

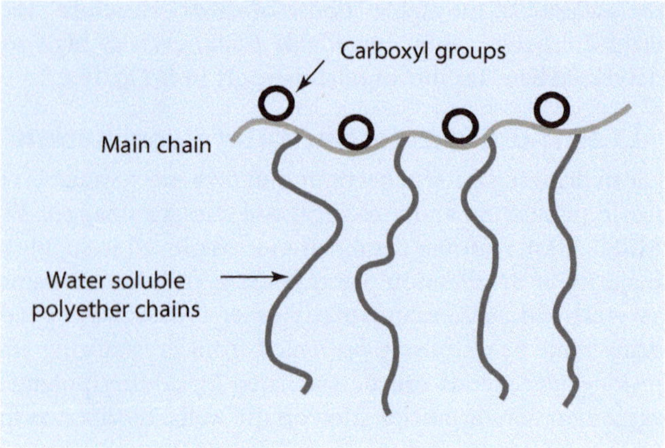

Fig. 19.9 Sketch showing the structure of a polycarboxylate ethers

This generic structure is extremely variable in detail. By varying the molar mass of the comb polymer and the number of polyether side chains and their length, the effect of the polymer on the hardening and setting process of the concrete can be influenced considerably and hence controlled. The ability to control the process is important in practice, as the demands on the hardening and setting kinetics vary considerably depending on the application. Whereas a manufacturer of ready-set concrete parts, who mixes the concrete and pours it into molds on site, requires the quickest possible setting in order to finish the product and to fill his forms anew, a producer of ready-mix concrete, who needs to transport his product to the desired site in a concrete transport vehicle, aims to achieve the opposite: obviously, the concrete should not harden in the transport vehicle. By varying the different parameters, the hardening and setting process kinetics can be controlled over a wide range, depending on the application.

As a mechanistic model of the liquefying effect, it is assumed that the polymer with its polycarboxylate backbone adsorbs onto the surface of the mineral particles of the concrete and the polyether side chains protrude into the aqueous phase. Because of the steric dispersion effect of the long polyether side chains, the cement particles are effectively held apart. This postpones the hardening of the concrete and enables a prolonged processing time (cf. Fig. 19.6b).

19.1.2.4 Scale and Color Transfer Inhibitors

Functional polymers also play a large role in detergents. Modern heavy-duty detergents contain about 7 % of various polymers, which predominantly inhibit color transfer and scale development and should protect the textile becoming soiled. In the following, these functions are briefly discussed in more detail.

Scale Inhibitors

As in a seawater desalination plant, in the washing machine there is a danger that, over time, calcium and magnesium sediments build up in the machine components—especially on the heating elements—and compromise the performance of the equipment.

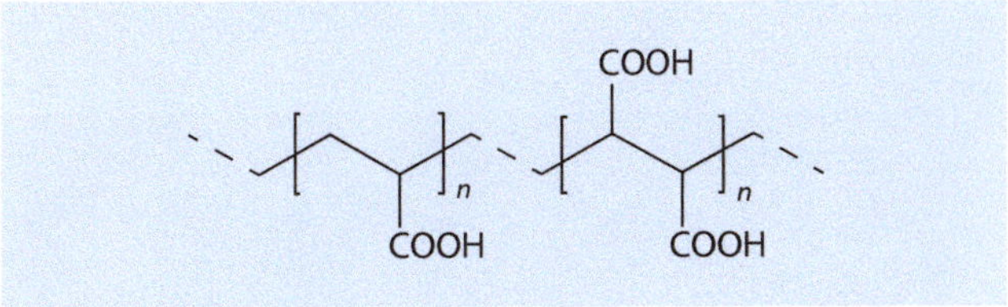

Fig. 19.10 Structure of acrylic acid-maleic acid copolymers

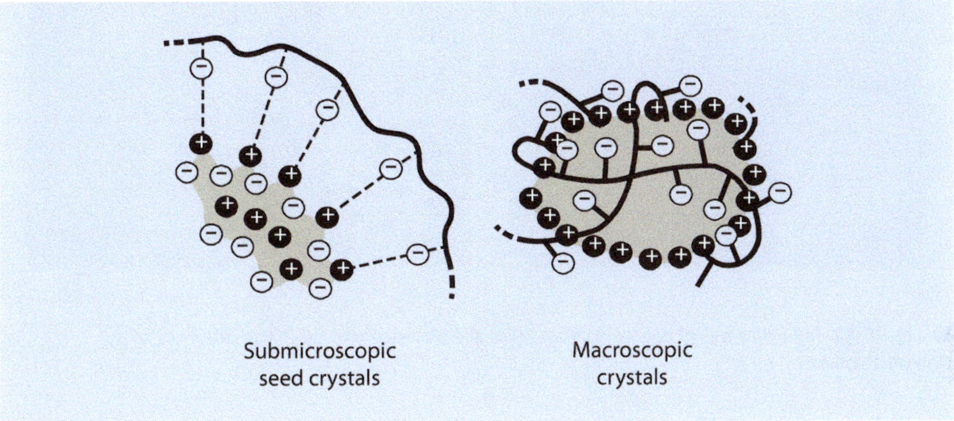

Fig. 19.11 Scheme showing the interaction of polycarboxylates with calcium or magnesium crystallites

Moreover, such fine precipitates can also settle on the textile material, reducing wear comfort. Adding negatively charged polyelectrolytes, such as those described in ▶ Sects. 19.1.2.1, 19.1.2.2, and 19.1.2.3, to the wash process can also prevent the development of scale. In detergents, statistical copolymers of acrylic acid with maleic acid are often used (Fig. 19.10), which (after neutralizing the carboxylic acid groups) have a particularly high charge density. Generally, the molar mass of the polymers used in detergents are a little higher than those used in paper coating.

The mode of action of these polymers presumably relies on several effects. As with all carboxylic acid groups, the carboxylate functions in the polyelectrolytes being discussed have a high affinity toward calcium and magnesium ions and so reduce the activity of these ions in the wash water (the so-called wash liquor). The carboxyl-functionalized polymers also effectively adsorb onto the surface of alkaline earth minerals. The absorbed polymer chains prevent the crystals from further growth. At the same time, as discussed above, the polymer serves as a dispersion agent and thus prevents the precipitation of sediment on surfaces (Fig. 19.11).

Two REM pictures of calcite crystals formed in the presence or absence of a polycarboxylate but under the same conditions are shown in Fig. 19.12; the effects are clearly visible. Because the polymer does not react stoichiometrically but rather in a comparatively small amount with the surface of the crystals, amounts of less than 1 ppm of polymer

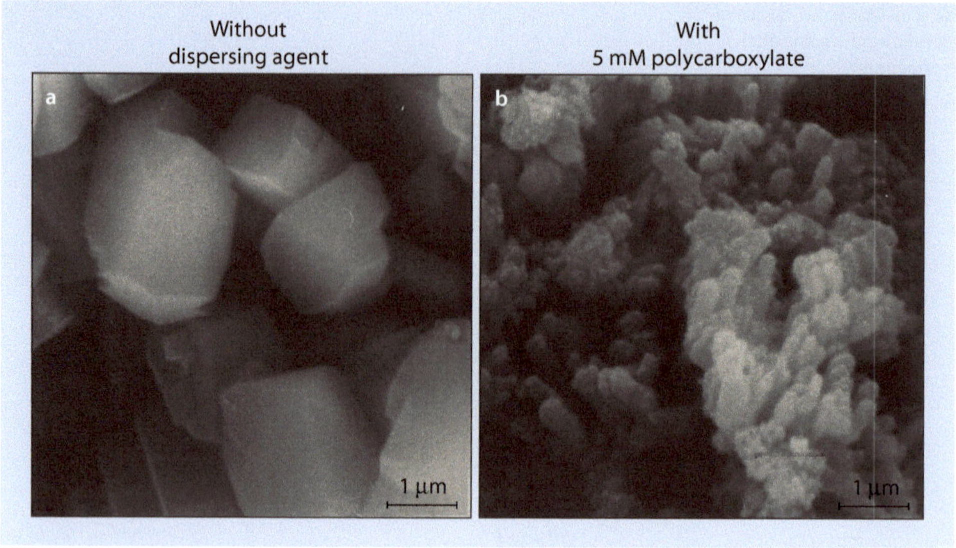

Fig. 19.12 REM-pictures of calcite crystals (**a**) in the absence and (**b**) in the presence of a polycarboxylate

are sufficient for achieving a massive effect. In total, the crystallization of inorganic minerals can often be influenced or controlled considerably by adding dissolved polymers to crystallizing solutions. This has been discussed in numerous scientific papers (e.g., Cölfen 2003; Wegner et al. 2007; Huang et al. 2008).

Color Transfer Inhibitors

Another problem arising when washing colored textiles is color transfer, commonly referred to as "color running." This undesired effect arises because color molecules that have not adsorbed onto the surface of the textile permanently enough dissolve into the wash water during the washing process and re-adsorb onto differently colored or even white textiles in an equilibrium process. These processes can also be inhibited by using functional polymers; in this case, homo- or copolymers based on polyvinyl pyrrolidone (PVP) are effective (Fig. 19.13).

Despite lacking an electric charge, PVP is an extremely polar polymer because of the high dipole moment of the pyrrolidone group. This dipole moment enables a relatively stable complex formation with any color molecules in the wash water, reducing their concentration and their tendency to re-adsorb. Figure 19.14 shows a model of such a colorant–PVP-complex. With its electroneutrality, the pyrrolidone group can interact with both ionic and neutral molecules so that a large number of colorants can be prevented from "color running" with this polymer.

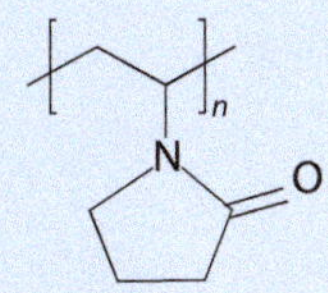

Fig. 19.13 Structure of PVP

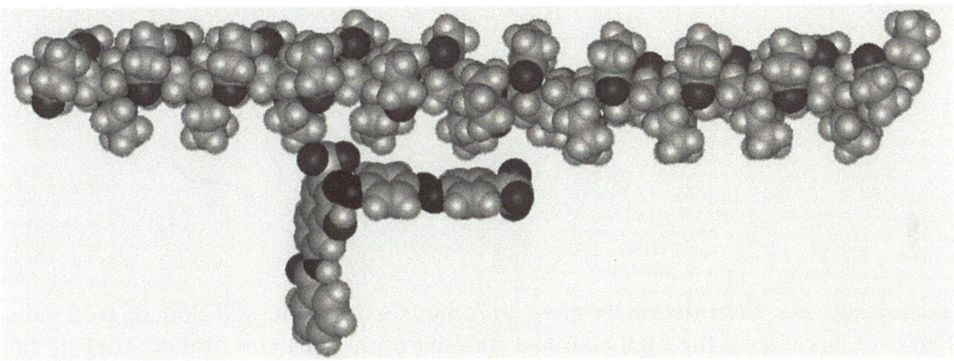

Fig. 19.14 Model for a PVP-complex with the Dye Direktrot 81

19.2 Flocculants

Polymer dispersion agents have been discussed in ▶ Sect. 19.1.2 but if the basic polymer structure is changed, functional polymers which have the exact opposite effect can be made. Such polymers are designed to cause colloidal systems to flocculate. To this end, the fact that such systems are usually thermodynamically unstable and sterically or electrostatically prohibited from coagulating by absorbed materials is exploited. If the activation barrier associated with this kinetic stabilization can be overcome, the system flocculates and the particles agglomerate.

Flocculation is of great importance in the field of wastewater treatment. A large portion of the pollutants in wastewater is present in the form of particles or droplets, which are colloidal and highly stabilized by negatively charged surfaces. These often extremely small particles are hard to remove from the wastewater. Filtering techniques can be ruled out because of the vast amounts that have to be treated. Techniques which precipitate the pollutant particles are more effective and result in purified water and a so-called sludge. The latter can be disposed of, for example, by incineration.

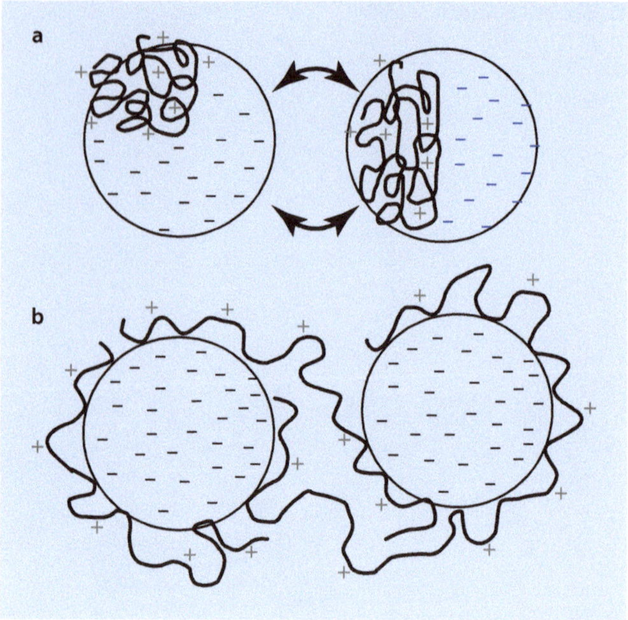

Fig. 19.15 Mechanisms of the flocculation employing high molar mass cationic polymers. (**a**) Mosaic adhesion. (**b**) Bridge building

Technically, the flocculation is often accomplished by the addition of iron salts. However, because of the high colloidal stability of the particles formed, they do not agglomerate effectively and the accumulating sludge still contains up to 90% water. For this reason, further polymer flocculating agents are added to induce improved aggregation. The goal of this second flocculation is to produce larger structures by destabilizing the system, which then sediments or can be separated more easily. For this application, copolymers based on acrylamide with cationic comonomers are used. Because of their positive charge these polymers adsorb onto the negatively charged surfaces of the pollutant particles very effectively. However, in contrast to the charged dispersion agents discussed in ► Sect. 19.1.2, for this application high molar mass polymers are generally employed.

The mechanism by which these polymers are believed to work is shown in Fig. 19.15. In principle, two mechanisms are discussed:

- With the so-called *mosaic adhesion* the polymers adsorb to only some areas of the particle surface. In these areas, the originally negative charge of the particles is locally reversed, which results in the system containing particles whose surfaces have both positively and negatively charged areas. The oppositely charged areas of two or more particles attract each other so that the particles—in analogy to the "house of cards" structure in Fig. 19.8b—agglomerate.
- For the so-called *bridging precipitation* it is assumed that a large molar mass polymer is adsorbed onto more than one particle to form "bridges". Obviously, a relatively high molar mass is required for this to occur.

A great advantage of the second precipitation is the markedly reduced water content in the resulting sludge, which requires considerably less energy by incineration.

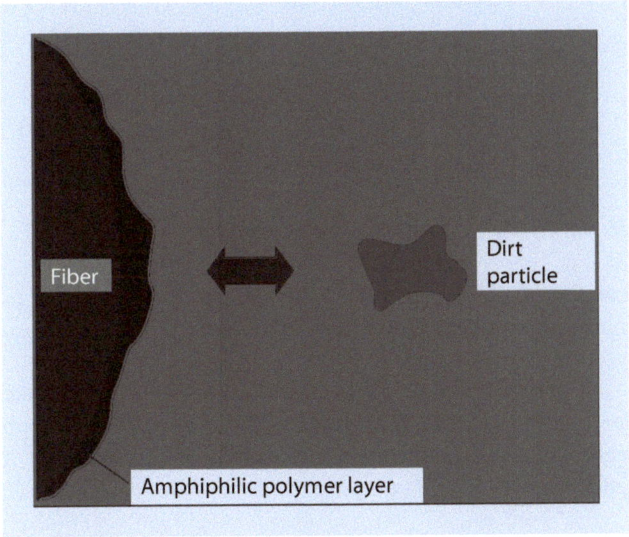

Fig. 19.16 Schematic diagram of the impregnation of textiles with amphiphilic polymers to protect against the adsorption of hydrophobic grime

19.3 Amphiphilic Systems for Surface Functionalization

Amphiphilic polymers, polymers which have both hydrophilic and hydrophobic molecular parts, can be synthesized with different polymer structures. Statistical copolymers made of hydrophilic and hydrophobic monomers exhibit a relatively narrow sequence of these elements, whereas block copolymers are made up of a few, relatively large domains of different polarity. Graft copolymers have a similar structure, in which the hydrophilic branches are grafted onto a hydrophobic backbone or the other way round.

If such a polymer is adsorbed onto a surface with, for example, a hydrophobic character, then the hydrophobic parts of the polymer are orientated predominantly toward this surface whereas the hydrophilic parts face outward. Macroscopically, this leads to an alteration of the polarity of the surface, which is expressed, for example, in an altered wettability by water or organic solvents. Moreover, the tendency of further hydrophobic molecules or particles to adsorb onto the surface with reversed polarity is significantly reduced.

This effect is used technically, on a large scale, for textile impregnation. For this purpose, amphiphilic polymers can be used as additives in detergents. During the washing process they are organized onto the hydrophobic textile fibers and render them hydrophilic (Fig. 19.16). The adsorbed polymers have the effect that hydrophobic grime particles, such as oil droplets, which soil the textile as it is being worn, cannot bind onto the fabric as strongly because the adsorption of these hydrophobic grime particles onto the hydrophilized surfaces is energetically less favorable. Subsequently, they can be washed out considerably more effectively in the washing processes. For this reason, these macromolecules are referred to as *soil release polymers.*

Chemically, these polymers are generally copolyesters based on terephthalic acid, glycol, and polyethylene glycol (Fig. 19.17). By incorporating polyethylene glycol, the polyester becomes a block-structure in which the hydrophilic polyethylene glycol blocks are separated by hydrophobic polyethylene terephthalate structures. The block structure

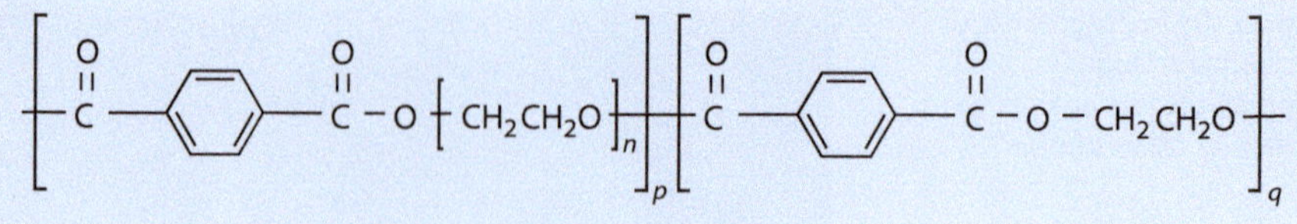

Fig. 19.17 Structure of soil release polyesters

can be effectively controlled by the molar mass of the polyethylene glycol used for the polycondensation and the molar ratio of glycol and terephthalic acid to polyethylene glycol.

The hydrophobic blocks adsorb as anchor groups effectively onto polyester fibers and polyester mix textiles, whereas the polyethylene terephthalate and glycol blocks are responsible for making the surface hydrophilic and thus impregnating the fabric. Similar effects are less strong on pure cotton because of its different surface structure.

19.4 Thickeners

As well as those applications already discussed, which rely on an interaction between polymers and surfaces, dissolved functional polymers can also have a strong physical effect on the total system. One profound effect is the polymer-typical increase of the solution viscosity (▶ Chap. 3). This thickening effect is used in many domestic products, for example, cosmetics, in which the system is required to have a viscous or gelatinous consistency.

In principle, every high molar mass, water-soluble polymer could be used as a thickener. However, it is often desired that the viscosity of the system is switchable, that is, it can be influenced by changing the intrinsic system parameters. With polyelectrolytes this can be achieved by varying the pH or the salt concentration. This is illustrated in Fig. 19.18 with a polycarboxylic acid as an example.

At low pH values, the carboxyl functions of the polyelectrolyte are protonated and the polymer entanglements are electrically neutral. Thus, they are present simply as solvent swollen polymer coil (▶ Chap. 2). If the carboxyl functions are neutralized by adding a base, for example, NaOH, an increasing accumulation of negative charge builds up along the polymer chain. As these repel each other, the polymer chains spread out to reduce the interaction. At a certain point the chains overlap and the viscosity of the system increases. This effect is reversible by adding acid. Alkali salts, for example, sodium chloride lead to a similar effect which can be explained as follows.

The strength of the electrical field generated by an ion in solution is, intuitively, inversely proportional to the distance from the charge. Mathematically, this decrease can be described by an exponential function. Here the distance in which the potential of the electrical field falls by the factor of Euler's number (approx. 2,718, this means a decrease of approx. 63.2 %) is given by the so-called *Debye length* λ_d. This length is given by

$$\lambda_d = \sqrt{\frac{\varepsilon \cdot kT}{2N_A \cdot e^2 \cdot I}} \tag{19.1}$$

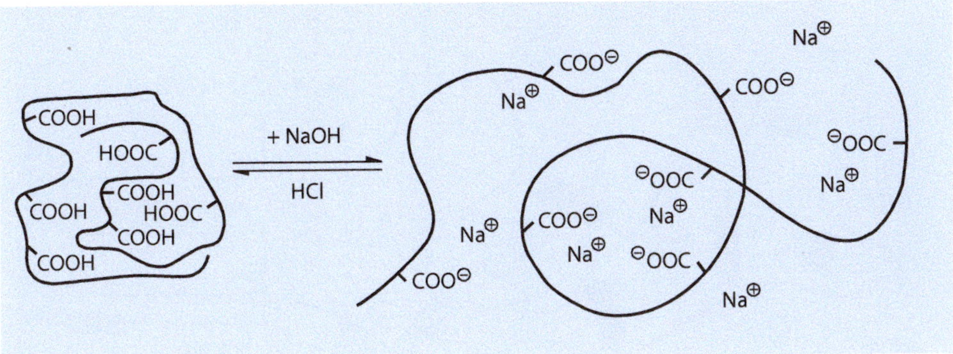

Fig. 19.18 Spreading of a polyelectrolyte coil by varying the pH

ε	Electrical permittivity
k	Boltzmann constant
N_A	Avogadro constant
e	Elementary charge
I	Ionic strength of the solution

For singly charged salts such as sodium chloride, the ionic strength equals the activity, and thus, approximately, the concentration of the salt. Equation (19.1) means that the electrical field of an ion falls at a higher salt concentration in a solvent. Thus the Debye length in a 0.001 M NaCl solution at 25 °C is ~10 nm and shortens to ca. 1 nm in a 0.1 M solution. Thus, the effective electrical field of the ion "shrinks." This effect is often termed a *shielding* of the electrical charges. As a result, the charges along the polymer backbone repel each other less if NaCl is added to the solution. This leads to the chains contracting and the viscosity decreasing. Many thickeners are therefore sensitive to the salt concentration of the solution.

The so-called *associative thickeners* are particularly effective. These are statistical copolymers consisting of a hydrophilic monomer, for example, acrylic acid, with a small amount of a non-polar comonomer, for example, a long-chain acrylate. If such polymers are dissolved in water, the alkyl moieties tend to agglomerate. If such an association occurs between various polymer chains, this results in a physical bridging between these chains and eventually leads to gel formation and a large increase in the viscosity of the system. Because of the extreme growth of the polyelectrolyte, polymer entanglements and the strength of such hydrophobic interactions in aqueous systems, minute amounts of thickeners (a few percent, sometimes even less) are enough to gel the complete system.

19.5 Super-Absorbents

In contrast to the soluble polymer systems discussed thus far, with the so-called *super-absorbents* we are dealing with cross-linked insoluble polymers. Such polymers can only swell in contact with a solvent. Although this effect is not pronounced for most polymers, super-absorbers are optimized to absorb a maximum amount of water. To achieve this goal, they are made from cross-linked polyacrylic acid which has been

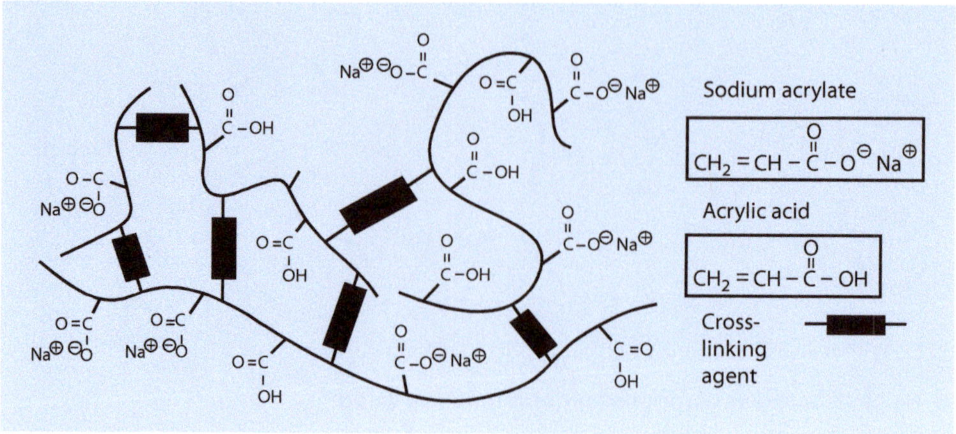

Fig. 19.19 Structure of cross-linked sodium polyacrylate (super-absorber)

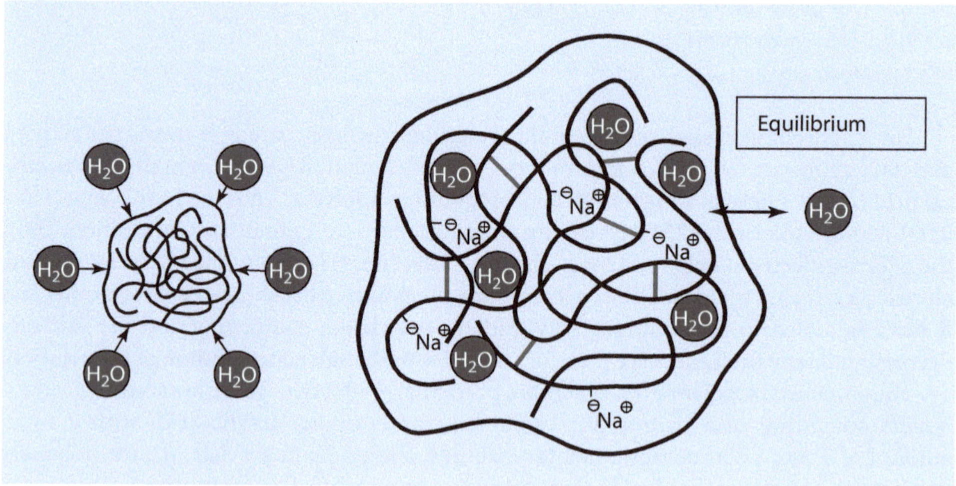

Fig. 19.20 Swelling of a super-absorber produced by osmosis

partially neutralized with sodium ions (Fig. 19.19). Because of their high internal salt concentration, a considerable osmotic pressure develops when they come into contact with water and the water is "sucked into" the gel particles. The gel particles swell dramatically. The water absorption is only halted when the restoring force built up by the chains becoming stretched between the cross-linking points is equal to the osmotic driving force (Fig. 19.20).

The main application of super-absorbers is diapers and other hygienic products. For these products, super-absorber particles are wrapped in a so-called *fluff* pulp (bleached, non-woven cellulose fibers). This textile ensures a rapid distribution of, in most cases, the suddenly increasing amounts of liquid, and the super-absorbers take care of its storage, making the article appear dry. Osmotically functioning super-absorbers show notable advantages in practice:

- The amount of liquid that can be absorbed is extremely large. Some super-absorbers can absorb up to a thousand times their own weight before they cease to swell. This results in a gel which consists of up to 99.9 % water but which feels dry on the outer surface.
- Because of the osmotic effect, the liquid is bound much more strongly to the absorber than is the case, for example, for absorption from the capillary action of a sponge. Thus, water absorbed by a super-absorber cannot be released by simply "squeezing," as it can with a sponge. For use in diapers this is a crucial factor, as the bound urine is not released as the baby moves about.

Further, volume-wise less important, applications of super-absorbers are their use as soil improvers in crop fields (preservation of moisture for dry periods), for moisture control in tunnel construction, and to prevent contact sweat in furniture seats.

19.6 Polymers for the Formulation of Active Ingredients

To conclude this chapter, the use of functional polymers in the area of formulation of active ingredients, particularly in pharmaceutical applications, is highlighted

In a typical tablet, that is, in a solid oral dosage form, functional polymers are used in the following applications.

19.6.1 Matrix Materials

Many pharmaceutical ingredients are already active in milligram amounts. As these minute amounts cannot be handled easily, polymers are used in which the active ingredients dissolve and become homogeneously distributed. Hereby the active ingredient is transformed into a form which a patient can handle. Such a "solid solution" can be imagined as a conventional solution of an organic material in a solvent, in which the solvent has been replaced with a polymer. Thus the active ingredient is no longer present in crystalline but in molecular form. An example for a suitable matrix material is the PVP mentioned above.

19.6.2 Coatings

Tablets covered in a smooth polymer coating can be swallowed more easily. Moreover, the coating can be colored. Colored tablets are easier for the patient to recognize, which reduces the danger of confusion. Furthermore, the often unpleasant taste of the active ingredients can be masked with a suitable coating. An example of a polymer suitable for pharmaceutical coating is a graft copolymer of vinyl acetate on polyethylene oxide, whereby the grafts have been hydrolyzed to vinyl alcohol (◘ Fig. 19.21 and ► Chap. 15).

Many oral pharmaceutically active ingredients are sensitive toward acid and should only be released after passing through the stomach. In these cases a coating is used which is resistant with respect to the stomach acid. Such polymers are usually insoluble in acid but soluble above pH 5.5. Such a pH-dependent solubility can be achieved by building carboxyl groups into the polymer. These are ionized at high pH levels and markedly

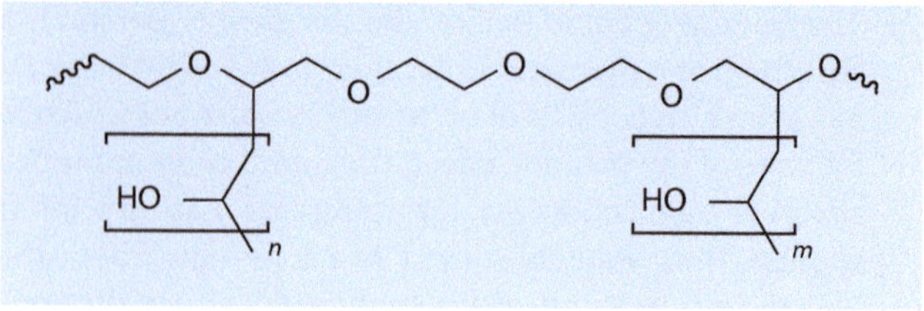

Fig. 19.21 Graft copolymer of vinyl alcohol (from vinyl acetate) and polyethylene oxide

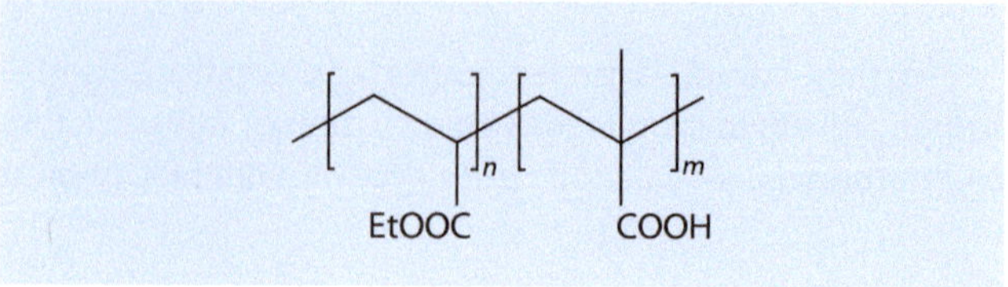

Fig. 19.22 Copolymer of methacrylic acid and ethyl acrylate as an example for a stomach acid-resistant tablet coating

increase the solubility of the polymer in water. Thus, for example, copolymers consisting of methacrylic acid and ethyl acrylate are employed (Fig. 19.22). Another area of application for such pH-sensitive coatings is the formulation of active ingredients which would otherwise damage the stomach lining.

19.6.3 Exploders

Pharmaceutical tablets can also include so-called *exploders*. These can be, for example, cross-linked PVP. Analogous to the super-absorbers discussed in ▶ Sect. 19.5, these swell in contact with water and cause the tablet to disintegrate so that its contents can dissolve more rapidly.

19.6.4 Solubilizers

Many active pharmaceutical ingredients—especially those which have only recently been discovered—have an unfavorable solubility in water and/or only dissolve slowly. Because an oral formulation must be taken up by the body within 24 h or risk being excreted, the speed with which an active ingredient dissolves can, under certain circumstances, limit its bioavailability. In many cases this problem can be solved with polymers which assist the dissolution process. In the same way as color transfer inhibitors, discussed in ▶ Sect. 19.1.2.4, solubilizers are capable of building water-soluble complexes with the active ingredient and thus increase its solubility and dissolution rate in aqueous media. A common solubilizing polymer certified for pharmaceutical use is polyethylene glycol esterified with a (monomeric or dimeric) stearic acid (Fig. 19.23). In addition, polymers have the advantage that they are usually relatively non-toxic, obviously a precondition for pharmaceutical use.

With an appropriate choice of formulation aids, the way the active ingredient is released can be significantly influenced. Thus, on the one hand a sudden, rapid release is

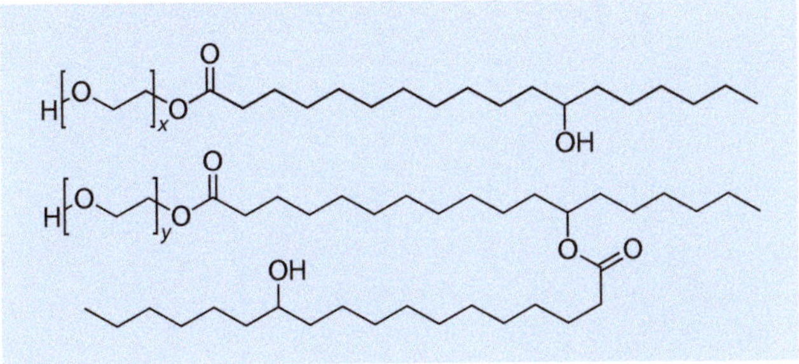

◘ **Fig. 19.23** Structures of the oligomeric main components of the pharmaceutical solubilizer Solutol HS 15®

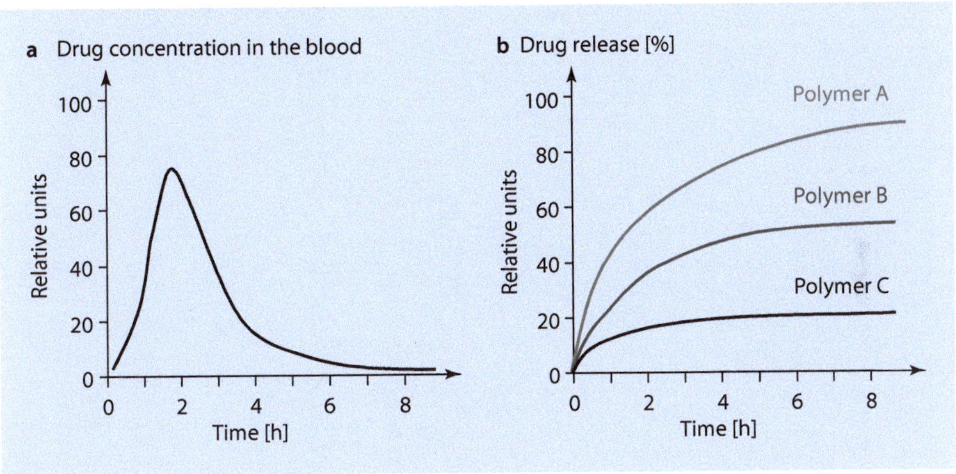

◘ **Fig. 19.24** Graphs of different release profiles. (**a**) Instant release. (**b**) Retard formulation

possible. This is always desirable when symptoms need to be reduced rapidly and effectively. Examples are painkillers for acute pain or medication to treat allergic or epileptic attacks. Such formulations are referred to as *instant release* formulations. On the other hand, a steady release over the course of the day may be more appropriate. So-called *retard* formulations ensure a constant level of active ingredient in the body and thus reduce any side effects of any excess active ingredient shortly after intake. Both release profiles are displayed schematically in ◘ Fig. 19.24.

References

Cölfen H (2003) Precipitation of carbonates: recent progress in controlled production of complex shapes. Curr Opin Colloid Interface Sci 8:23–31

Göthlich A, Koltzenburg S, Schornick G (2005) Funktionale polymere im Alltag. Chem Unserer Zeit 39:262–273

Huang S-C, Naka K, Chujo Y (2008) Polym J 40:154–162

Tadros T (2005) Applied surfactants. Wiley, Weinheim

Wegner G, Demir M, Faatz M, Gorna K, Munoz-Espi R, Guillemet B, Gröhn F (2007) Macromol Res 15:95–99

Liquid Crystalline Polymers

S. Koltzenburg et al., *Polymer Chemistry*, DOI 10.1007/978-3-662-49279-6_20

In this chapter, liquid crystalline polymers are defined, methods for their characterization are described, and some examples of liquid crystalline polymers are discussed.

20.1 The Liquid Crystalline State

Solid bodies in which elementary building blocks (atoms, molecules, ions) are arranged regularly and at geometrically precisely defined points in a lattice are called crystals. The defined structure prevails throughout the whole crystal and is also referred to as *long-range order*, referring not only to the position of the crystal building blocks but also to their orientation in relation to the crystal lattice (long-range position and orientation order). This order means that the crystal has a regular structure at a macroscopic level and generally anisotropic mechanical, electrical, and optical properties. By contrast, most liquids are only locally ordered. Thus, a degree of order can only be considered in terms of one or two molecular diameters at most, and the physical properties are independent of direction; they are *isotropic*.

The transition from an ordered solid body to a disordered liquid can be achieved either by melting the body or by dissolving it in a suitable solvent. Most crystalline solid materials transition directly from an ordered state into a disordered, liquid state when heated above their melting point or dissolved. However, a number of substances, under certain conditions, form an additional state—that of an anisotropic liquid. On the one hand, this phase exhibits the typical characteristics of a liquid, that is, it has a more or less viscous flow behavior. On the other hand, its physical properties are anisotropic, typical of crystals. As the physical properties of this phase lie somewhere between those of a crystal and those of an isotropic liquid, it is referred to as *liquid crystalline* or *mesophase*. A substance that can form a mesophase is called *mesogenic*. The three-dimensional position and orientation long-range-order present in the crystal is partially lost in this phase. The formation of this kind of liquid crystalline phase is also known as *mesomorphism*.

If the transition of a pure substance from a crystalline state to a mesophase is caused by a change in temperature, then the mesophase formed is referred to as a *thermotropic* mesophase. Liquid crystalline phases, which are transformed into a mesogen by adding a solvent, are called *lyotropic* mesophases. *Amphotropic* substances can form both kinds of mesophases.

20.2 Lyotropic Liquid Crystals

So-called lyotropic phases are two- or multi-component systems made up of a solvent, amphiphilic compounds, and, often, water. Lyotropic mesogens are often typical soaps, such as the alkaline salts of long-chain fatty acids or quaternary alkyl ammonium compounds with at least one long alkyl chain, but can also be polyhydroxy compounds and carbohydrates that have been modified to make them, in part, water repellant.

The dissolved amphiphiles form micelles in certain temperature and concentration ranges. The shape of the micelles is strongly dependent on the solvent and the shape of the amphiphilic molecules and can vary greatly, with the resulting shapes ranging from spheres and cylinders to double-layered shapes (▪ Fig. 20.1).

Micelles are usually able to move around freely when the surfactant concentration is low. As long as this is the case, these liquids are not referred to as liquid crystals. If the

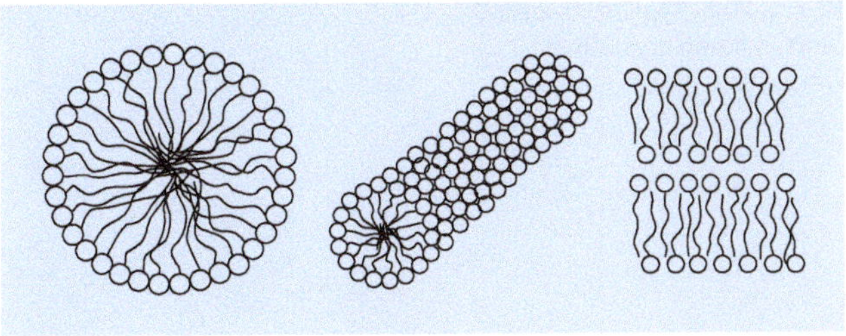

■ **Fig. 20.1** Examples of the different shapes formed by amphiphilic molecules. (**a**) Spherical micelle (*left*). (**b**) Cylindrical micelle (*middle*). (**c**) Double layer lamella (a mesophase)

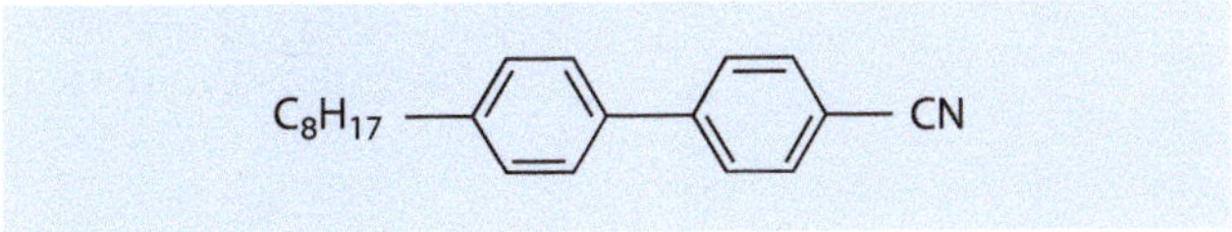

■ **Fig. 20.2** 4′-Octyl biphenyl-4-carbonitrile—an example of a thermotropic liquid crystalline material

concentration of amphiphile is greater the micelles become organized into more highly ordered liquid crystalline phases and superstructures.

With changes in temperature or concentration the mesophases can transform into one another. Such transformations are reversible.

20.3 Thermotropic Liquid Crystals

Thermotropic liquid crystals do not directly transition from a crystalline into a liquid state when heated. Instead they pass through one or several additional phases within limited and well-defined temperature ranges. These phases often have anisotropic physical properties, such as those that can occur in crystals; however, they exhibit flow behavior. A well-known example is 4′-octyl biphenyl-4-carbonitrile (■ Fig. 20.2) which undergoes transition into a liquid crystalline state at 22 °C and then transforms into an even less ordered mesophase when heated further to 34 °C. The transition into an isotropic melt only takes place at 41 °C.

The thermotropic liquid crystalline state typically, but not exclusively, occurs in compounds with pronounced molecular anisotropy. This means that this state often occurs in compounds made up of molecular rods (calamatic mesogens) or discs (discotic mesogens). Several selected structural varieties are shown in ■ Fig. 20.3.

20.4 Liquid Crystalline Structures

Typical liquid crystals are characterized by an orientational order of their anisometric molecules and a partial or complete destruction of the positional foci of their molecules. In the simplest case the orientational order is maintained, such as the orientation of the long axes of rod-shaped *calamatic* molecules in small regions, so-called *domains*. The molecular orientation found in *nematic* or *cholesteric* calamatic phases are shown schematically in ■ Fig. 20.4.

The rod molecules' centers of gravity are statistically distributed in the nematic phase. However, all molecules are orientated in one preferential direction, the *director*.

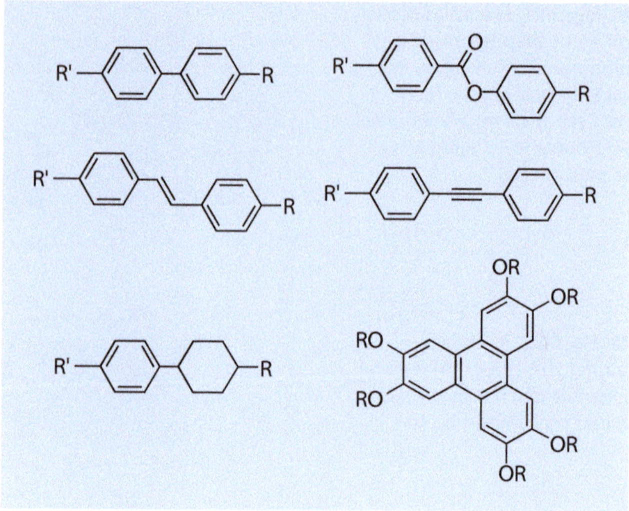

Fig. 20.3 Typical structural units in liquid crystalline substances

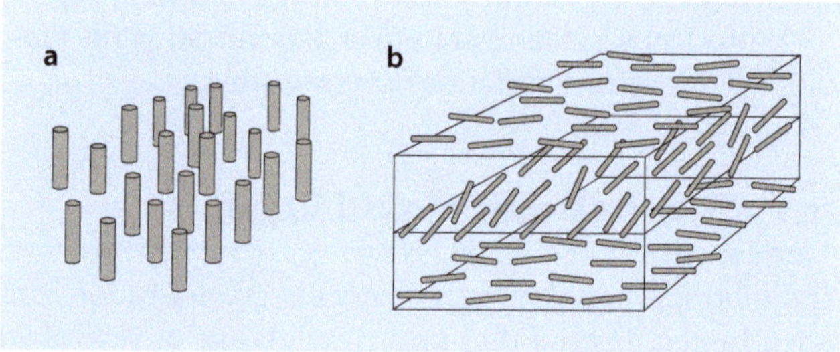

Fig. 20.4 Visualization of (**a**) nematic and (**b**) cholesteric phase (the rotation of one layer with respect to another by 90° is fortuitous)

The degree of orientation in relation to the direction field is a result of the thermal movement of the molecules. This value is designated by the *order parameter S*:

$$S = \frac{3}{2}\left(\cos^2\Theta - \frac{1}{3}\right) \tag{20.1}$$

Θ is the deviation of the long axis of a molecule from the preferential direction. If all the molecules are oriented exactly in parallel, $S=1$. For an isotropic melt, $S=0$.

The cholesteric phase is closely related to the nematic phase. One can view the orientation shown in Fig. 20.4a as a level on which a vertical axis can be placed. If, along this axis, the preferential direction of orientation changes continually, the result is a nematic phase with a helical superstructure (cholesteric phase, Fig. 20.4b).

Cholesteric phases usually only occur in chiral mesogens or mixtures of achiral mesogens with optically active substances. This is why these phases are also called *chiral-nematic*. They are of great technical importance for appliances with displays.

The so-called smectic phases have a higher degree of order than nematic liquid crystals. In smectic phases the mesogens are ordered because of intermolecular interactions in equidistant lamellae, that is, with positional orientation (Fig. 20.5). At present a number of smectic phases are known which have been labeled from S_A to S_K depending on the date of their discovery. The individual phases can be distinguished by the orientation of the

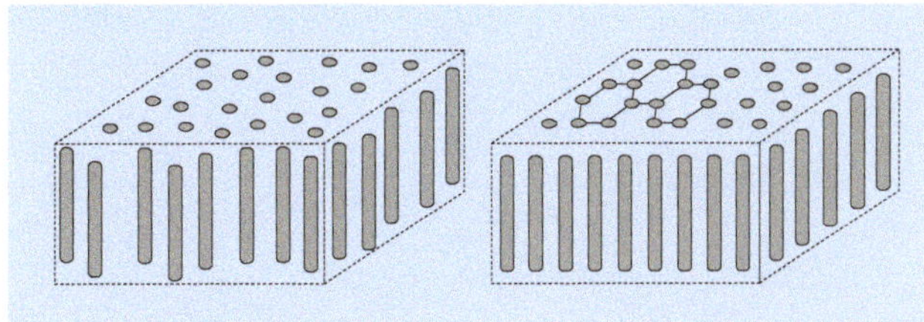

Fig. 20.5 Examples of smectic phases with varying degrees of order

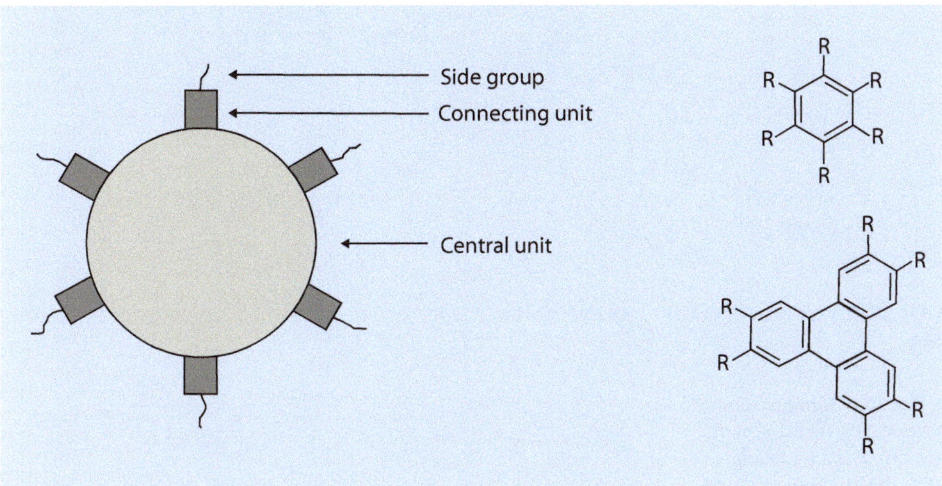

Fig. 20.6 Structural principle and examples of discotic mesogen with rigid central cores

mesogens in each layer and the deviation of the director from a right angle to the layers. For more detail the interested reader is referred to the specialist literature.

In contrast to the situation in crystals, the individual layers in a smectic phase remain mobile with respect to one another. The viscosity of smectic phases is higher than that of nematic and cholesteric phases as a result of the higher degree of order.

So-called *polymorphic* mesogens pass through several mesophases when heated. As the temperature increases the phase with the highest energy (lowest order) is the most stable. Discotic mesophases can form from disc-shaped molecules. The general structural principle of this class of compounds is shown in Fig. 20.6. Discotic mesomorphism can occur if optimal area coverage is achieved by the side chains in the peripheral region and perfect space filling in the central area is achieved by the disc-shaped core of the mesogen.

Classical discots are made up of a rigid, flat, disc-shaped central moiety onto which flexible side chains are attached via connecting groups. The central core can be an aromatic, heteroaromatic, or metal-complex system, onto which flexible side chains with ether, thioether, ester, or amide connecting groups are attached.

Scyllo-inositol hexaesters (Fig. 20.7) are examples of non-rigid central cores that form discotic mesophases. Mesogens with regional cores and pyramidal mesogens are also systems that deviate from the usual basic design of discotic mesogens. Their more ordered phases are thus better referred to as *columnar* or *stacked mesophases*.

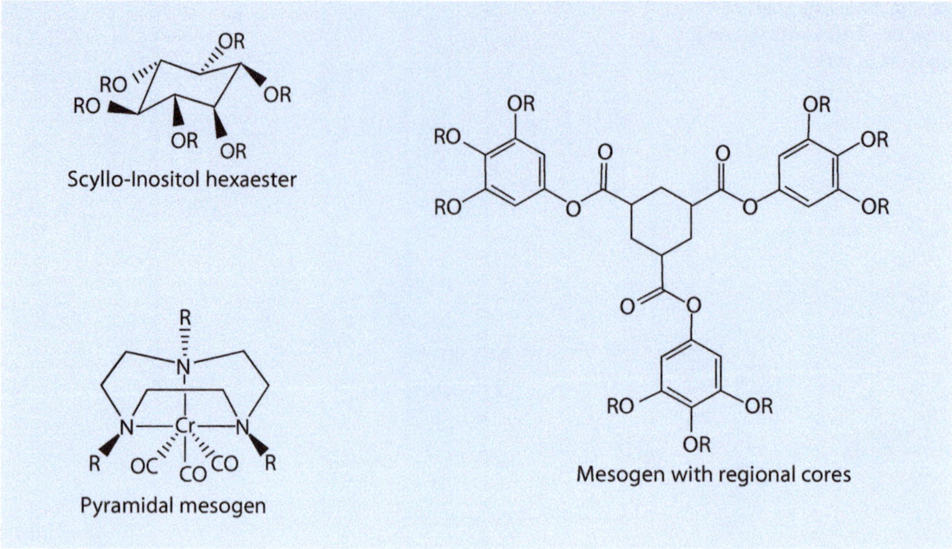

Fig. 20.7 Examples of discotic mesogens without rigid, flat cores and without pronounced disc-like form

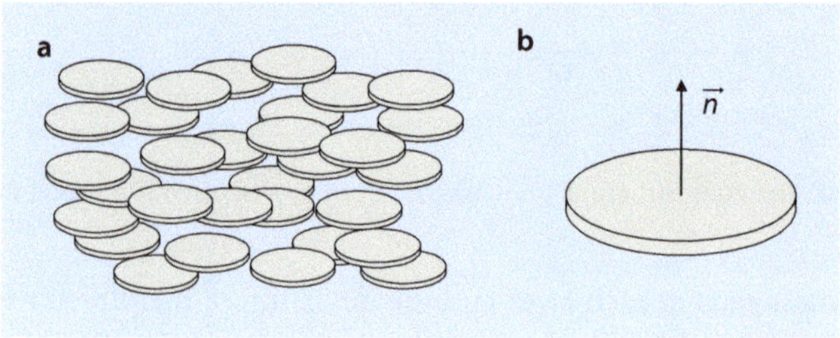

Fig. 20.8 Nematic-discotic mesophases. (**a**) Schematic diagram of the molecular order. (**b**) Orientation of the director

The nematic phase is also the least ordered phase in discotic systems. As in the calamatic-nematic phase the centers of gravity of the molecules are distributed isotropically in space and the mesogens have only a one-directional order. Although the director is parallel to the molecule's long axis in rod-shaped molecules, the preferential direction for discots is normal to the disc plane (Fig. 20.8).

If discotic mesogens are stacked on top of one another, and thus form columns, then structures with two-dimensional order, such as hexagonal-columnar phases, are formed (Fig. 20.9). The discotic molecules can have differing degrees of order along the column axis.

Some substances are able to form mesogens by association. This association can be caused by columbic-, dispersion-, van-der-Waals-, charge-transfer-, or hydrogen bond interactions. An example of this is shown in Fig. 20.10.

20.5 Characterization of Mesophases

A characterization of mesophases can be made using differential scanning calorimetry (DSC), polarization microscopy, and X-ray diffraction.

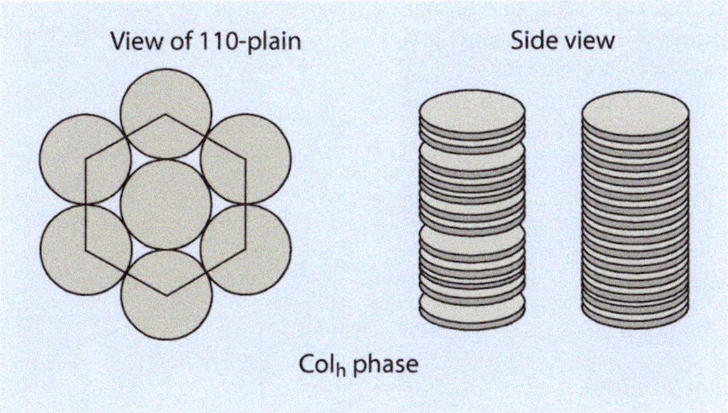

Fig. 20.9 Schematic diagram of a hexagonal columnar phase

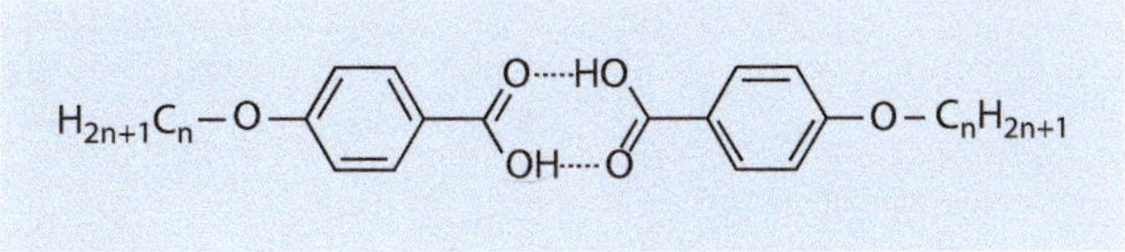

Fig. 20.10 Mesogenic cyclic dimer of a 4-alkoxy benzoic acid

20.5.1 Heat Flow Calorimetry

This method of DSC is one of the most important methods used in modern polymer analysis and in the analysis of mesophases for determining the thermal characteristics of new materials.

The temperature of the phase transitions and the associated enthalpy and entropy can be determined using DSC. With this method the first-order nature of the transitions between the crystalline, liquid crystalline, and disordered phases is exploited. Thus, the transition appears as a peak in the DSC thermograph, whose integral gives a measure of the heat associated with the transition. Figure 20.11 shows a typical thermograph with two transitions.

Such thermographs provide information about the temperature of the phase transitions and the degree of order. Highly-ordered phases generally exhibit strong hysteresis. In addition, the enthalpy of the transition increases with the difference in the degree of order across the transition (Table 20.1). The difference in temperature between the heating and the cooling cycle peaks for the same phase transition is caused by the relationship between the respective rates of phase transition.

20.5.2 Polarization Microscopy

Whereas DSC provides only indications of the existence of a mesophase, using a polarization microscope equipped with a thermostat-controlled stage enables one to determine exactly whether a mesophase is indeed present. A polarization microscope has a polarizer between the source of light and the condenser lens as well as an analyzer beneath the eyepiece. The sample can therefore be viewed in linear, polarized light. All (liquid) crystals that are not cubically symmetrical demonstrate birefringence.

When the light ray hits the coplanar birefringent sample it is split into two separate beams whose polarization planes are perpendicular to one another. These separate beams move in two different directions with different refractive indices and therefore have different speeds. This leads to a path difference when the light exits the sample, which is linearly

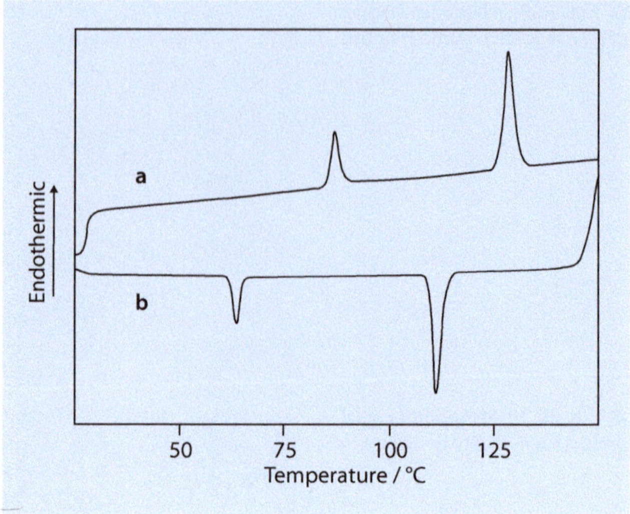

Fig. 20.11 DSC-curves of poly(8-(4′-methoxy biphenyl-4yl oxy)-octyl methacrylate) showing two phase transitions. (**a**) Heating. (**b**) Cooling

Table 20.1 Phase transition enthalpies (Kelker and Hatz 1980)

Nematic → isotropic	0.8–9.6 kJ/mol
Smectic → isotropic	2.9–12.9 kJ/mol
Crystalline → isotropic	7.0–117.0 kJ/mol

dependent on the thickness of the layer and the difference between the refractive indices. Because of this path difference, interferences occur, which result in a rotation of the oscillation plane of the light reaching the analyzer.

During the analysis, a thin layer of the sample is placed between two glass sheets. Subject to the arrangement of the mesogens in the different phases, characteristic defect structures are formed on the surface of the glass sheets when they come into contact with domains having different orientations, and when different domains come in contact with one another. The sum of these defects is referred to as texture. Examples of smectic, nematic, and cholesteric textures are shown in Fig. 20.12.

20.5.3 X-Ray Diffraction

X-ray diffraction is the most important method for characterizing the different mesophases in liquid crystals. The analysis of the position and intensity of the X-ray reflections allows an essentially complete description of the mesophase structure. Using monochromatic X-rays with a wavelength of λ, the direction of constructive interference of the beams diffracted by the structures with a periodicity d below the angle 2Θ can be determined using the Bragg equation:

$$n \cdot \lambda = 2 \cdot d_{hkl} \cdot \sin\left(\Theta_{hkl}\right) \tag{20.2}$$

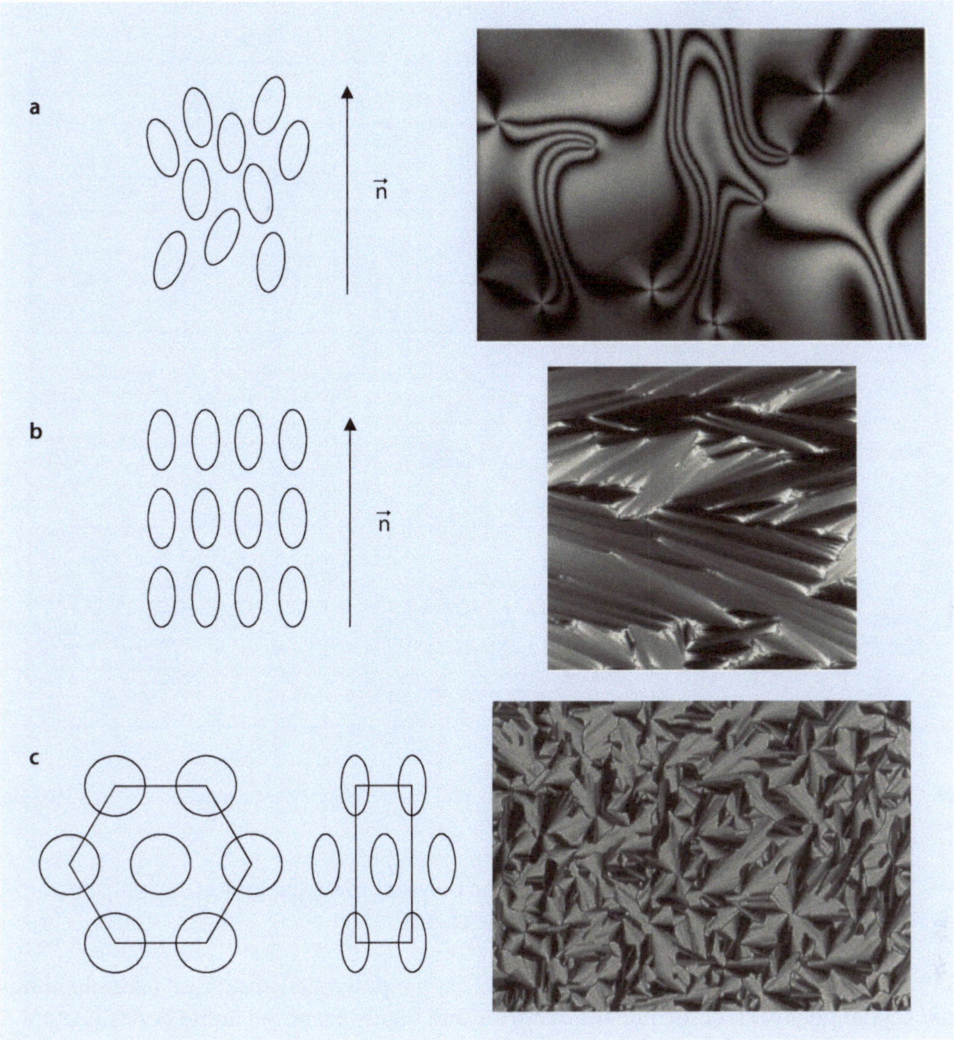

Fig. 20.12 Examples of (**a**) nematic, (**b**) smectic, and (**c**) cholesteric textures (Lavrentovich 2011)

A detailed discussion of the X-ray diffraction patterns of liquid crystals is rather complex and lengthy and is not given here.

Additional analytical methods for characterizing mesophases are miscibility studies with known liquid crystals, examination of the electro-optical behavior of mesophases, dielectric spectroscopy, atomic force microscopy (AFM), and rheology.

20.6 Liquid Crystal Polymers

A large number of structures are conceivable for polymers that have typical anisometric mesogen groups as repeating units (Fig. 20.13). In the first instance, one can distinguish between two basic types:

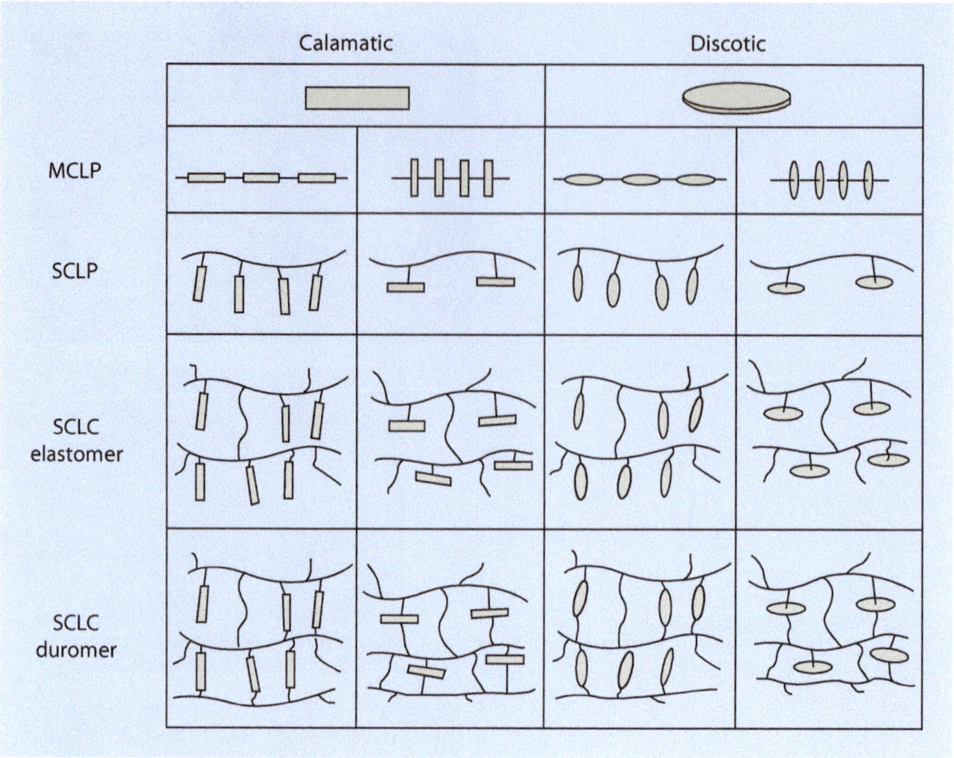

Fig. 20.13 Structures of polymeric liquid crystals. MCLP main chain liquid crystal polymers, SCLP side chain liquid crystal polymers

- Those with mesogen groups as part of the polymer main chain
- Those mesogen groups as part of the side chains

The subgroups MCLP (main chain liquid crystal polymers) and SCLP (side chain liquid crystal polymers) constitute the dendritic and highly branched liquid crystalline polymers described in ► Sect. 20.6.4.

20.6.1 Liquid Crystalline Main Chain Polymers

As can be seen in Fig. 20.13, anisometric mesogens are linearly or laterally connected with respect to their long axes. For this reason, rotations around the horizontal or vertical axes of the mesogens are inhibited. Most MCLP are linked linearly and the whole spectrum of reactions described in ► Chap. 8 can be used to synthesize them (Fig. 20.14).

However, the direct linking of mesogens often leads to problems, especially if, as a result of polymerization, the melting point changes and becomes higher than the decomposition temperature. A typical example is 4-hydroxybenzoic acid. Benzoic acid phenyl esters with low molecular weights are well-known liquid crystals. The analog polymer is, however, no longer fusible. The possibilities shown in Fig. 20.15 can be used to achieve a liquid crystalline state:

20

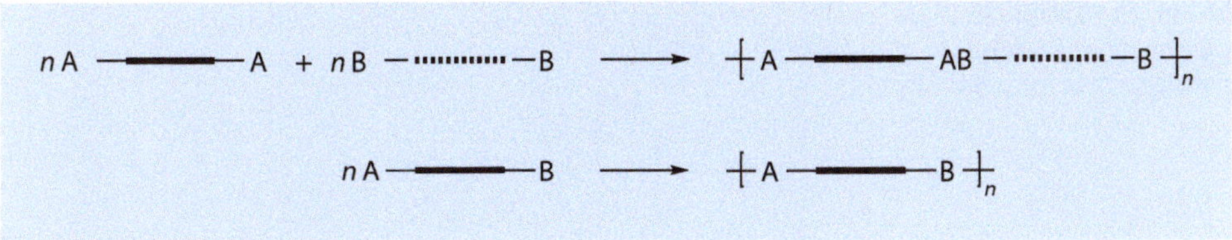

Fig. 20.14 Possibilities for combining monomeric mesogene to form MCLP

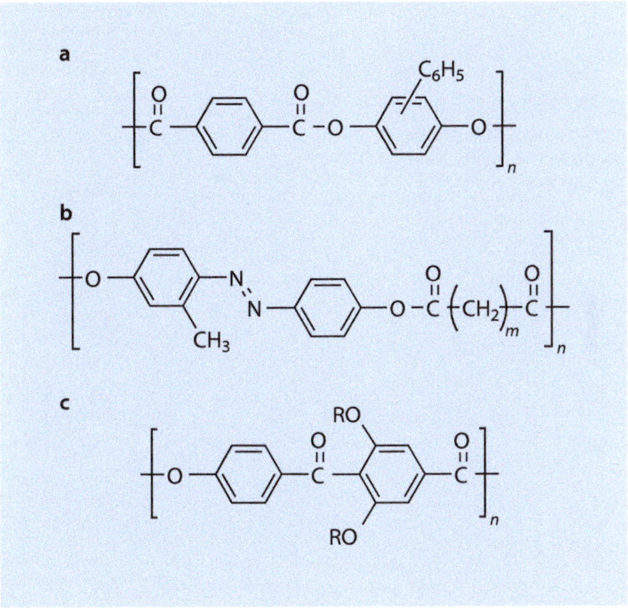

Fig. 20.15 Liquid crystalline polymers (**a**) with aromatic side groups, (**b**) with flexible main chain segments, and (**c**) with flexible side chains. *R*–$(CH_2)_nCH_3$

1. Comonomer building blocks with aromatic side groups are incorporated into the main chain (Fig. 20.15a)
2. Rigid mesogen groups are linked with one another using flexible spacers (Fig. 20.15b)
3. Flexible chains are laterally inserted into the mesogens as substituents in linear macromolecules (Fig. 20.15c)

Liquid crystalline main chain polymers often form nematic phases above their melting temperature. If a polymer is made up of two or more different monomer building blocks then smectic order is possible. The formation of the more highly ordered smectic phase is facilitated by the insertion of spacers between the liquid crystalline building blocks.

The melting point of rigid, rod-shaped polymers can be decreased by the addition of a suitable solvent to such an extent that a liquid crystalline state can be achieved. Thus, the aromatic polyamide shown in Fig. 20.16 can form a nematic phase in concentrated sulfuric acid.

Figure 20.17 shows an example of a lateral arrangement of a mesogen group in a main chain.

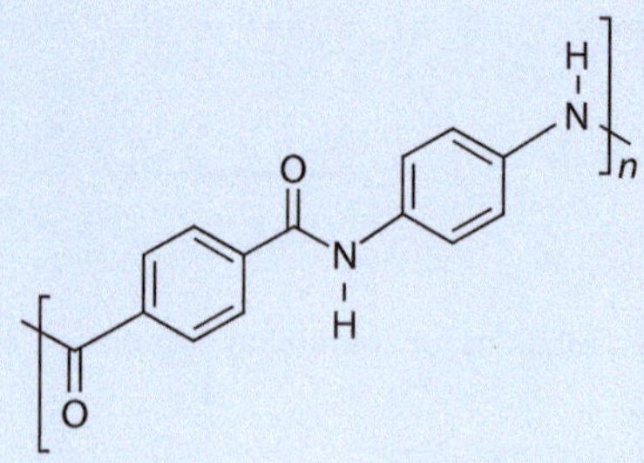

Fig. 20.16 Aromatic polyamide which forms a nematic phase in sulfuric acid

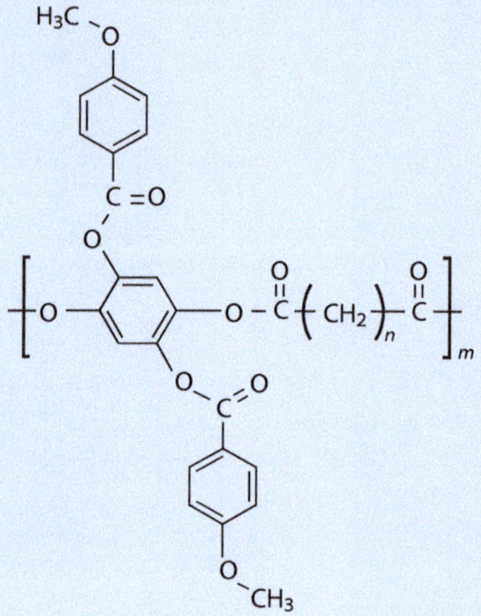

Fig. 20.17 Lateral arrangement of a mesogen group in the main chain. $n = 10$, 12, 14, 20

By spinning the polymers shown in Figs. 20.16 and 20.17 through a die, a macroscopically uniform orientation along the fiber can be imposed (Fig. 20.18). The ordered nematic structure is retained after cooling the fiber to a glass. The mechanical strength in the direction of the long axis is produced by the covalent bonds along this axis, and is thus very high, whereas horizontally van-der-Waals interactions, that is, significantly weaker forces, determine the tensile strength. Thus, polymers processed in this way are often used as high performance fibers.

Main chain polymers with liquid crystalline properties have gained industrial importance as fibers and in composite materials because of their exceptional mechanical properties, such as their high tensile strength after spinning, or their dimensional stability and low thermal expansion coefficients.

20.6.2 Liquid Crystalline Side Chain Polymers

The mesogen groups in SCLP are generally attached randomly along the polymer backbone by flexible spacers (Figs. 20.19 and 20.20). Radical polymerization and polymer analogous reactions are the most commonly used methods for synthesizing SCLP.

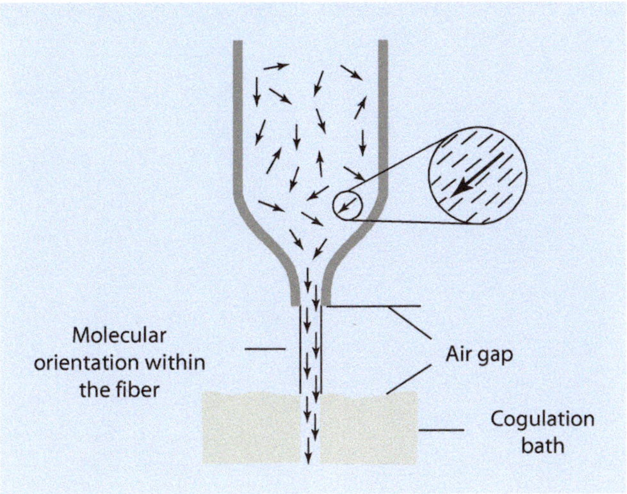

Fig. 20.18 Spinning an MCLP. → Director

In contrast to the analogous small molecule mesogens, the corresponding SCLP generally have lower transition temperatures and the liquid crystal phase transitions are dependent on the degree of polymerization; increasing with increasing degree of polymerization, usually, up to a degree of polymerization P_n of about 100.

The flexible linkage between the mesogens and the polymer backbone facilitates the formation of more highly ordered smectic structures if the side chain lengths are longer. The flexible spacer needs to be at least four atoms long. The rotation around the long axis of the spacer is only possible if the mesogen has a linear as opposed to a lateral arrangement.

If one cools LC- (liquid crystal-) polymers below the temperature at which liquid crystallinity first appears, phases with greater degrees of order can sometimes be obtained, just as for small molecular liquid crystals. However, the structure does not continue to change below the glass transition temperature of the polymer. This combination of glass state and anisotropic physical properties enables such polymers to be used as optical storage elements in the fields of linear and nonlinear optics. In modern electro-optical display elements, the position of the director is influenced by the application of external electrical or magnetic fields and its ordered state is then frozen by cooling the polymer to below its glass temperature.

These materials can also be used for holography. An ordered polymer, which contains a dye exhibiting *cis-trans*-isomerism, can be heated by radiation and transformed into a liquid crystalline state. If isomerization of the linear *trans*-dye in the angled *cis*-dye takes place at the same time, then the state of order of the phase changes (Fig. 20.21).

The disorder created by the laser beam is frozen by cooling the polymer to below the glass temperature. Thus, one can 'inscribe' something, such as an image, into a polymer layer.

20.6.3 Liquid Crystalline Elastomers (LCE)

If SCLP with flexible main chains are transformed into networks, the translational diffusion of the main chain is inhibited. However, the chain segments remain mobile. The movement of the mesogens linked to the network is less inhibited, and these can form liquid crystalline phases. Such systems display the behavior of small molecule liquid crystals and are dimensionally stable and rubber elastic, two typical properties of elastomers

Fig. 20.19 Synthetic route to SCLP

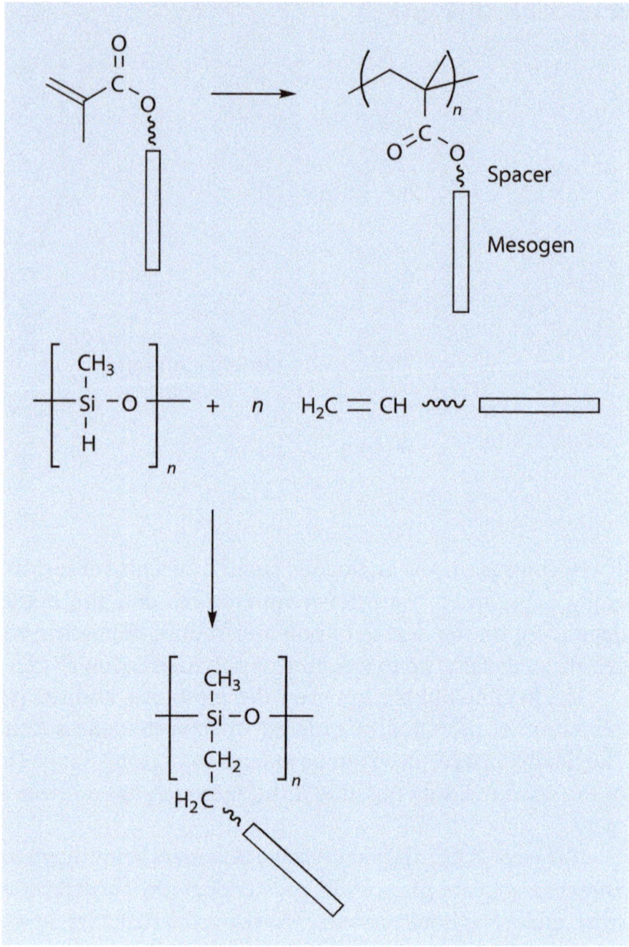

(► Sect. 18.8). The synthesis route shown in Fig. 20.22 has proved to be successful for these kinds of polymers.

It involves the copolymerization of mesogen monomers with monomers having an additional reactive functionality (A) that can be transformed into networks by reacting with B. Mechanical deformation of the rubber leads to a macroscopic orientation of the mesogens. These materials can be used, for example, as separating membranes.

A highly promising application of LCEs is as actuators. The incorporation of photochromic moieties into LCEs leads to materials which shrink when irradiated with UV light because of the photochemical reactions of the photochromic groups which can result in a decrease in the nematic order.

20.6.4 Dendritic and Hyper-Branched Polymers

Dendritic polymers can be distinguished from linear polymers by their multiple branches. These can be either structurally perfect with centrally-symmetric architectures (dendrimers) or star polymers without radial symmetry and randomly distributed

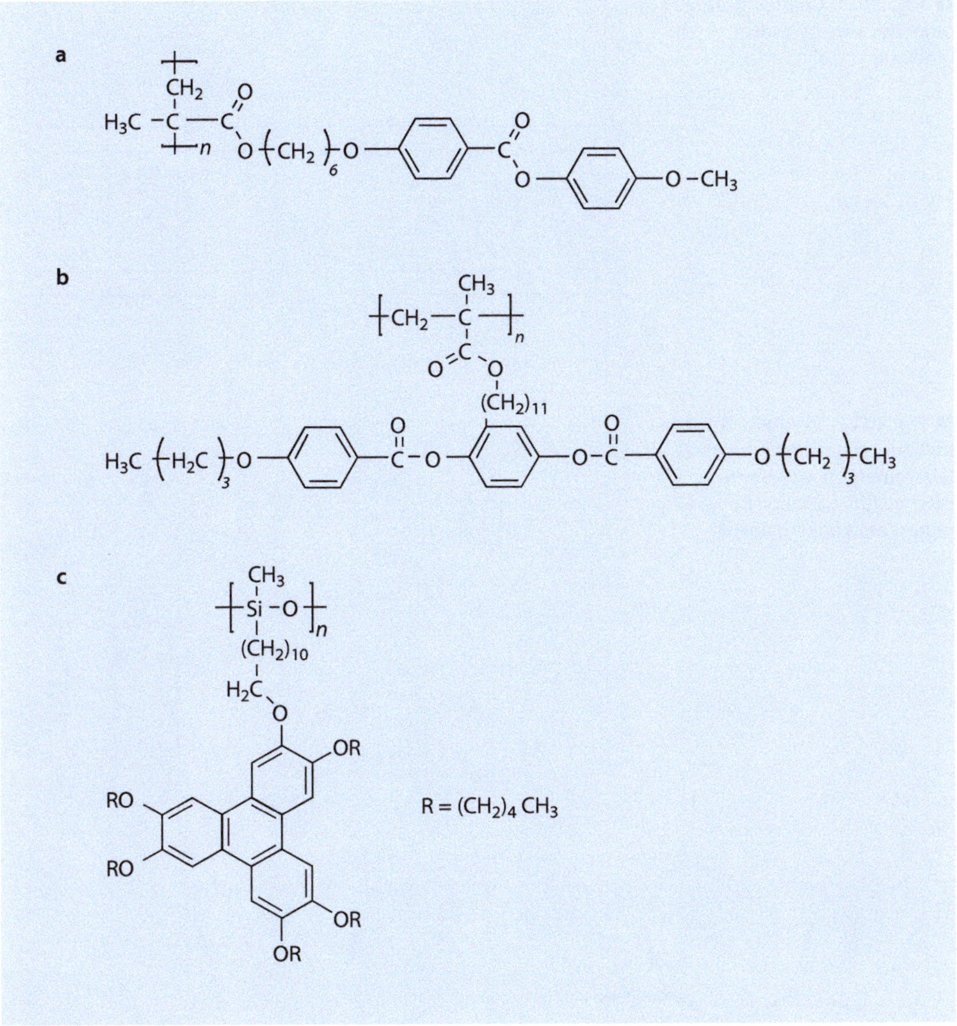

■ **Fig. 20.20** Selected examples of SCLP. (**a**) Linearly bound calamitic mesogens. (**b**) Laterally bound calamitic mesogens. (**c**) Discotic mesogens

branches (hyper-branched polymers ▶ Sect. 8.3.3). Irrespective of structural differences, by linking mesogenic units to these branched structures, liquid crystalline phases can result.

As for linear LC-polymers, a distinction is made between those in which the mesogens are linked to the polymer structure via flexible alkyl units (■ Fig. 20.23a) and those in which the mesogens form the terminal sections of the chains (■ Fig. 20.23b).

Hyper-branched polymers are more easily synthesized than dendrimers. In spite of their less perfect architecture, they should thus be considered interesting alternatives to dendrimers as carriers of mesogens. Hyper-branched MCLPs were first reported in 1992

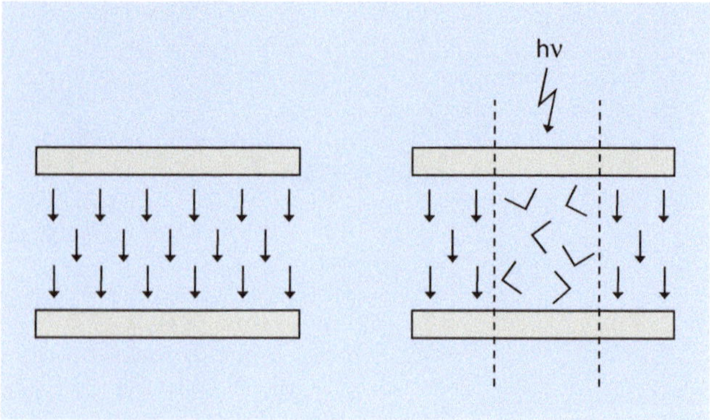

Fig. 20.21 Changing state of order because of laser irradiation

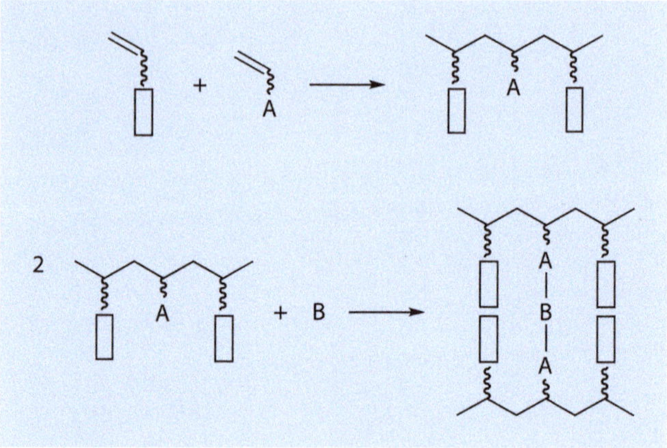

Fig. 20.22 Synthetic strategy for liquid crystalline networks (elastomers). *A* represents a reactive functionality, *B* represents a linking agent

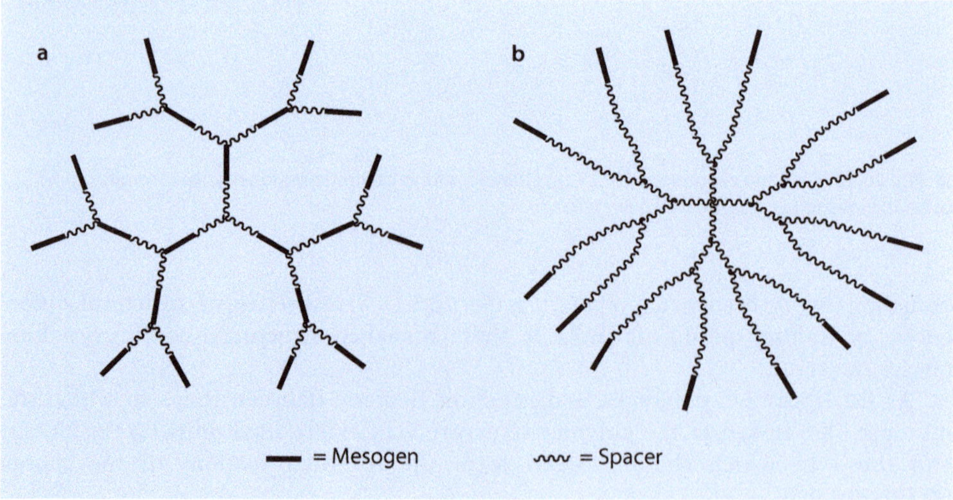

Fig. 20.23 Structures of dendritic LC-polymers. (**a**) Mesogens in the main chain. (**b**) Mesogens in the side chains

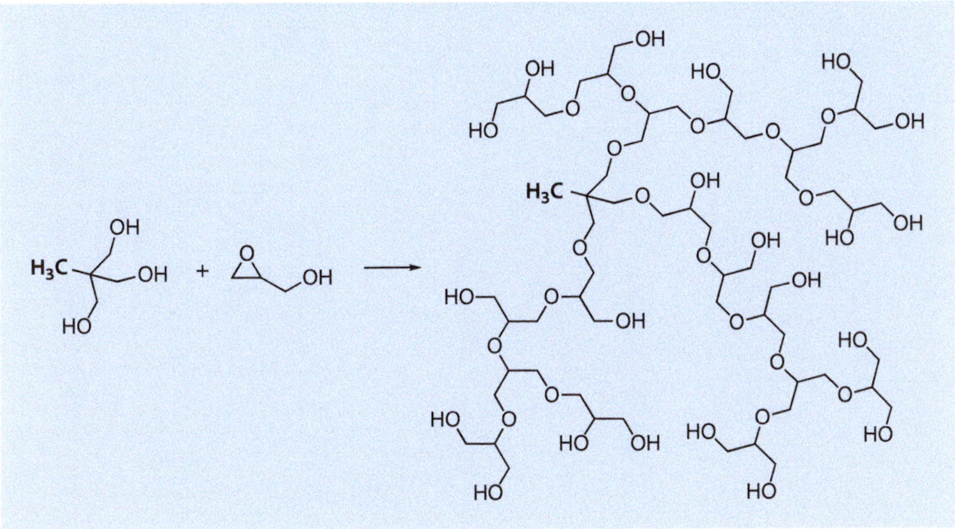

■ **Fig. 20.24** Synthesis of polyglycerine

■ **Fig. 20.25** Schematic visualization of (**a**) lateral and (**b**) terminal addition of mesogens to a dendrimer or a hyper-branched polymer. *M* Mesogen, *D* Dendrimer, *H* Hyper-branched polymer

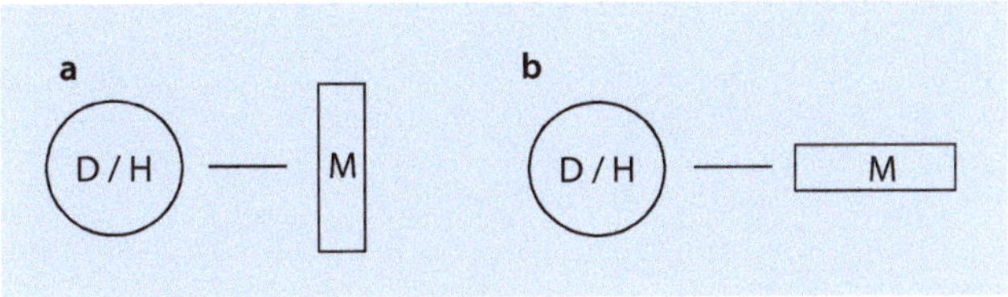

(Percec and Kawasumi 1992). A suitable structure for end group functionalization with mesogens is the highly flexible and water-soluble polyglycerine (■ Fig. 20.24).

Thermotropic, hyper-branched LC-polymers can be obtained by functionalization with cyanobiphenyl (Sunder et al. 1999). The mesogens can either be linked laterally or at a terminal position to the dendritic or hyper-branched polymer (■ Fig. 20.25).

References

Kelker H, Hatz R (1980) Handbook of liquid crystals. Verlag Chemie, Weinheim, S 36

Lavrentovich OD (2011) Liquid crystal institute. Kent State University, Ohio, OH

Percec V, Kawasumi M (1992) Synthesis and characterization of a thermotropic nematic liquid crystalline dendrimer polymer. Macromolecules 25:3843–3850

Sunder A, Quincy M-F, Mülhaupt R, Frey H (1999) Hyperbranched polyether polyols with liquid crystalline properties. Angew Chem Int Ed 38:2928–2930

Polymers and the Environment

S. Koltzenburg et al., *Polymer Chemistry*, DOI 10.1007/978-3-662-49279-6_21

Around 280,000,000 (280 million!) tons of synthetic polymers were produced for subsequent use in 2011. An annual increase in per-capita consumption of around 5% is expected through 2015. In light of these numbers, one must question the impact this consumption and use have on the environment. This concerns not only the resulting high volumes of waste produced but also issues such as recycling, energy, renewable raw materials, and sustainability. Close relationships and interactions exist between all these fields, which should not be ignored but are, nevertheless, all too often insufficiently considered in current discussions in these fields.

An all-encompassing solution to the problems that exist in these fields cannot be provided within the scope of this book; indeed, such a solution remains to be found. The aim of this chapter is to explain the basic concepts and terms that are relevant in relation to the above-mentioned interdependencies.

21.1 Introduction and Definitions

Some of the basic terms used when dealing with the recycling of plastics are defined in this section.

Reuse

The general term *reuse* encompasses all the options available for the processing of plastics, regardless of whether the waste products are directly processed into new plastics or whether they are reused in a different manner. This term also includes so-called *energy utilization* in particular, that is, the use of plastic waste for the purpose of producing energy.

Recycling

The recycling of plastic waste refers to the processing of old waste into new plastics, or their being processed to be put to another use, with the exception of energy recovery. Recycling thus refers to all types of *material utilization.*

The relationship between these terms can thus be described by the following equation:

$$\text{Reuse} = \text{Recycling} + \text{Energy utilization}$$

21.2 Options Available for Recycling Plastics

Plastic waste occurs at every stage along the chain of production and use of plastics (◘ Fig. 21.1). Fundamentally, the waste can be recycled in four different ways, which are explained briefly in the following. The options available differ from one another according to the stage in the value chain where the waste occurs and at which stage the plastics can be reintroduced into the chain. Each of the four principle routes is used—for reasons that are easily understood but are explained—to considerably different extents.

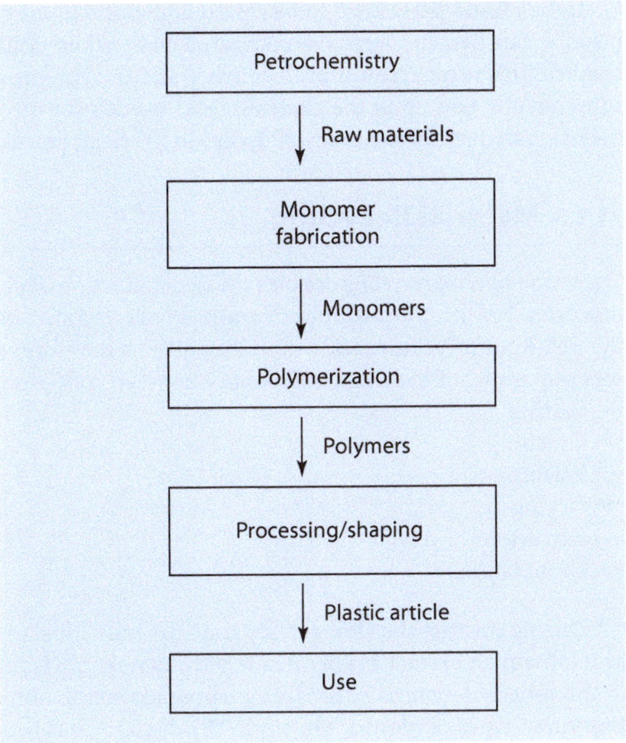

Fig. 21.1 Synthetic polymer value chain

21.2.1 In-Plant Recycling

In-plant recycling refers to the reuse of plastic waste created during processing. Plastic waste products such as stamping remnants (*Chad*), *deburred* remnants from injection molded parts, or products not meeting the product specifications are used in this context. These waste products are collected, and, often simply after grinding, reused for the same process. The advantages are obvious. The accumulated waste not only consists of a single polymer, it is also clean. Thus, it is ideal for reuse in the same process, although it is often necessary to blend the recycled waste with virgin material to maintain the required mechanical properties; for this reason, this recycling option can be considered as general practice.

21.2.2 Direct Reuse

Many plastic parts can, generally, be used multiple times. Requirements for this are that they can be detached from other components to which they may be attached without damage, and that no wear and tear has occurred. The classic example is plastic shopping bags that are used multiple times. The above-mentioned requirements are, however, often only partly valid for plastic parts that are components in complex systems, such as materials used in automobile parts and electronic appliances. As a rule, using a plastic bag multiple times is not problematic; however, packaging that has come into contact with perishable goods, such as yogurt or fresh meat, needs to be carefully cleaned before being reused.

If the plastic parts need to be sorted and cleaned, the overall ecological and economic balance can become negative compared with other available options examined in this chapter. This is on account of the energy and the volume of cleaning agents (which often subsequently end up in the environment) needed for the cleaning stage. Reuse is, nevertheless, a frequently used recycling option for clean plastic parts.

21.2.3 Material Recycling

The term *material recycling* denotes the reintroduction of plastic waste into a manufacturing processes. For this purpose, plastic waste is collected after use—in Germany, for example, in the well-known "yellow sack"—and then given a new form (▶ see Chap. 17). However, complex processing of the waste is necessary between collection and forming. This includes:

- Sorting
- Crushing
- Washing
- Drying
- Extrusion
- Granulation

During sorting, the aim is to separate the individual plastic types from one another so as to obtain an extract as chemically pure as possible from the range of materials present in the collected materials. For this purpose a general, numerical classification of the most important types of plastics, shown in Table 21.1, has been introduced.

The classification can be found on the molded part in a triangle shaped out of arrows (Fig. 21.2). The first number zero is occasionally omitted.

Table 21.1 Classification of polymers for recycling

Class	Material
01	Polyethylene terephthalate (PET)
02	High density polyethylene (HDPE)
03	Polyvinylchloride (PVC)
04	Low density polyethylene (LDPE)
05	Polypropylene (PP)
06	Polystyrene (PS)
07	Others

Fig. 21.2 Polymer labeling; here LDPE

As most polymers are immiscible with one another (▶ Chap. 2), good segregation is necessary to obtain a high-quality material for further processing.

However, as well as the obvious benefits in reducing waste, the materials recycling option also has significant disadvantages:

- The material history is unknown. During reprocessing, it can be expected that the material may have been decomposed and/or contains peroxide or keto groups from oxidation processes (▶ Chap. 15) or unsaturated double bonds (from elimination reactions). Likewise, polymers such as polybutadiene, which per se contain double bonds, may have further cross-linked.
- Additionally, as mentioned above, thorough cleaning of the waste is necessary which is costly.
- As mentioned in ▶ Chaps. 15 and 17, many synthetic materials contain additives such as pigments/dyes, fillers and reinforcing materials, plasticizers, and reacted or unreacted stabilizers. All of these cannot, generally, be separated from the polymers. As these accompanying substances can vary vastly from one batch of waste to another, it is exceptionally difficult to produce a consistent, high-quality product from the resulting product.

For these reasons, the plastics obtained from material recycling are generally significantly inferior with respect to their quality when compared with non-recycled ("virgin") polymers.

Furthermore, it should be noted that simple material recycling cannot be carried out on the following, widely used, polymer classes:

- Copolymers and polymer blends, which cannot be separated into single grades
- Duroplasts which, as a result of their cross-linked state, are insoluble and infusible and therefore cannot simply be reprocessed

All in all, material recycling only makes sense if the material available is clean and well-characterized. This is, for example, often the case for plastic drink bottles (largely poly(ethylene terephthalate)). However, in most other cases, the inferior material properties of the recycled material, the high costs of segregating the different plastic types, and the negative energy balance of the many complex individual steps all question the efficacy of material recycling of finished and used polymer articles.

21.2.4 Raw Material Recycling

So-called *raw material recycling* of synthetic material involves, for example, a degradation of the polymers via pyrolysis or solvolysis, in contrast to material recycling, discussed above, during which the macromolecules are not chemically altered. During this process, for example, the main chain of a polyolefin is thermally split into different olefins. Polymers with groups that can be hydrolyzed, and especially monomer polyesters, can be regenerated hydrolytically or by exposure to particular solvents. Mixtures of substances of low molecular weight form which, according to the starting material used, can be re-used as monomers, as basic chemicals, or as fuel. The processes used are usually relatively harsh and non-selective. However, they have the advantage that they can be used for synthetic waste that is not well separated and sorted according to type and for contaminated and/or composite materials. They are also suitable for copolymers and polymer blends. A typical example of a new industry that is developing based on such techniques is tire recycling. The high costs and energy required by such processes are problematic. The latter in particular is dealt with in more detail in the next section.

21.3 Plastics and Energy

The plastics value chain and the different recycling options that exist are considered in relation to the whole material flow of the petrochemical industry in this chapter. As well as this, the energy balance for polymeric materials is also discussed.

21.3.1 Energy Recovery from Synthetic Materials

Against the background of depleting amounts of fossil fuel, it is worth briefly considering the continuing role of crude oil as the most important raw material in the chemical and synthetic materials industries.

Worldwide, roughly 90 % of the oil produced is used in approximately equal proportions for both transport/mobility and heating/heat production, that is, uses in which this raw material is fed into oil-fired heating systems or vehicle motors. Barely 10 % of the oil produced is used for the production of chemical products. Around half of this amount becomes the raw material for synthetic materials. ◘ Figure 21.3 shows the flow of polymeric materials taking the different recycling options, discussed in ▶ Sect. 21.2, into consideration.

Consideration of ◘ Fig. 21.3 begs the question as to whether it would not make more sense first to use a large proportion of crude oil in the form of a chemical product and then to use it for the production of energy, rather than simply feeding it into a combustion process without any interim use. Naturally this only makes sense for chemical products that have an energy value comparable to that of crude oil. However, with regard to this last point, polymers are especially attractive, as ◘ Fig. 21.4 demonstrates. This figure clearly shows that polyolefins and polystyrene in particular, both have energy values corresponding almost exactly to those of fossil energy sources. Why this is the case becomes clear immediately when viewing the chemistry of these macromolecules (pure hydrocarbons). It was for polyolefins, in particular, that the term "firm crude oil" was coined.

Polymeric materials thus have a high potential with regard to their use as environmentally friendly energy sources. Their significance remains insignificant, probably because of a lack of political or societal acceptance rather than a rational or scientific reason. Although it is not intended to become involved in an environmental and political debate in this text, it must be clearly stated that it is neither ecologically, nor economically worthwhile to invest more energy in the recycling of a plastic than would be needed for its production. On the other hand, it is immediately clear that it makes good sense first to produce a yoghurt carton from crude oil and then to use the yoghurt carton for the production of energy rather than simply burning the crude oil.

There is no single recycling method for polymer waste better than any other. To recycle polymers in the best way from both economic and ecological points of view, a mix of

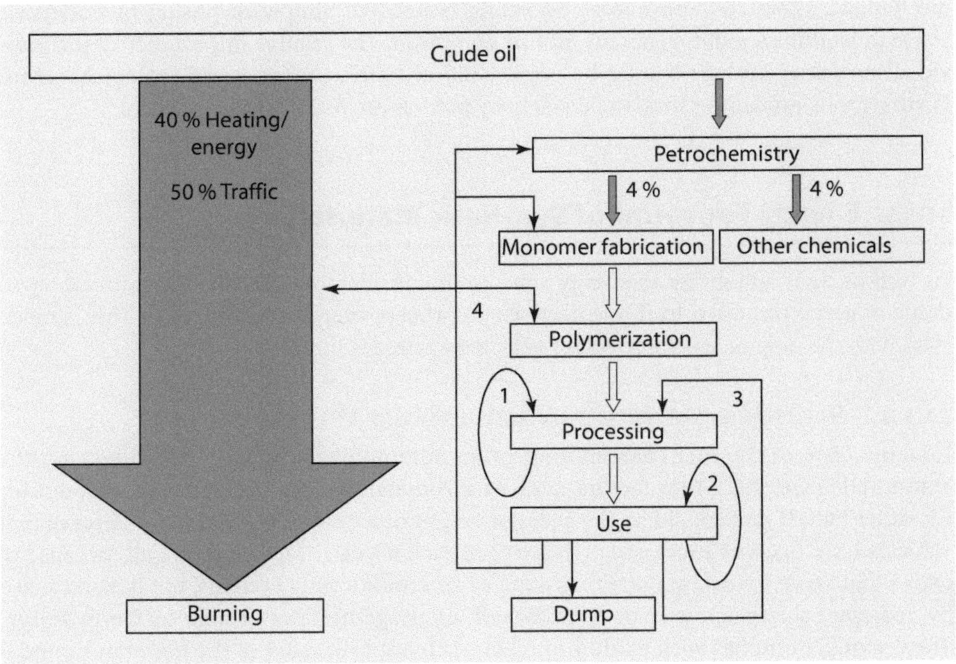

Fig. 21.3 Crude oil material flow including polymer recycling. (1) In-plant recycling. (2) Direct reuse. (3) Material recycling. (4) Raw material recycling. *Numbers* are approximations

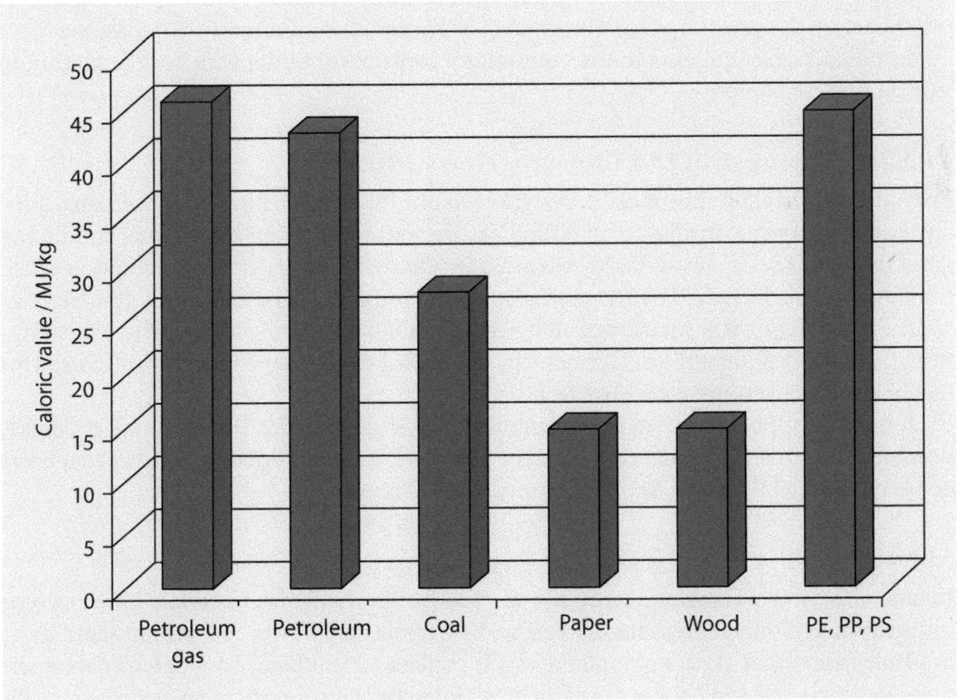

Fig. 21.4 Calorific value of selected materials

mechanical, chemical, and energy recycling is needed. Simply disposing of polymeric waste in landfills should, generally, not be an option. The relative importance of the individual recycling options should be considered in both economic and ecological terms. Furthermore, guidelines motivated solely by politics are usually not expedient.

21.3.2 Energy Footprint of Synthetic Materials

As well as their suitability as energy sources of considerable specific energy, waste synthetic material can often lead to a decrease in process energy consumption. This is made clear with the help of several specially selected examples in the following.

21.3.2.1 Synthetic Materials in Automobile Manufacturing

Roughly 50 % of the fuel consumption of an automobile is linearly dependent on the automobile's weight. Other factors, such as aerodynamic drag, for example, account for the other half. If one considers the average weight of a new vehicle, one can observe that vehicles have become increasingly heavier in recent years. This is, above all, because of extras that serve to increase comfort, such as air conditioning systems, but is also caused by necessary safety extras such as airbags or an altogether more stable car body design. The use of light metals such as aluminum is not feasible because of the high raw material costs. The use of plastics in engine and bodywork parts, for example, has made an important contribution to reducing weight increase. Prominent examples of the replacement of metals with plastics are in induction pipes or wheel trim caps made from polyamide. Bumpers have also been made using plastics for many years now. The use of plastics in other parts of the exterior of car bodies such as car roofs or mudguards is becoming more common and developments in tire technology contribute significantly to a reduction in vehicle fuel consumption.

21.3.2.2 Polymers in the Construction Industry

From the time of their invention, polystyrene foams (Styropor®) have been used on a massive scale for thermal insulation in buildings. The extraordinarily low thermal conductivity of the closed cell foams is being exploited (▶ Chap. 17). The energy consumption of old buildings can be decreased by 70 % by the introduction of modern insulation. Calculations show that subsequent insulation of all housing could decrease the total energy consumption of the Federal Republic of Germany by 20–30 %. Polyurethane foams are also used for insulation purposes in the construction sector.

If one takes the energy-saving potential of polystyrene foams into account, it quickly becomes clear that these materials have the potential, if used properly, to ensure that more crude oil is saved than was necessary for their production.

21.3.2.3 Synthetic Materials as Packaging

In many sorts of packaging, especially for foodstuffs, synthetic materials have become indispensable, if only for reasons of hygiene. In several cases, however, there are alternatives available from other classes of materials such as glass. Thus, the use of yoghurt containers that can be used multiple times are being demanded by some groups as an environmentally friendly alternative to plastic packaging. Such demands, however, often ignore the following considerations:

- The transport from production to point of consumption. For example, the transport of a 150-g portion of yoghurt requires the transport of 85 g of packaging material if the packaging is made of glass with an aluminum lid compared with ca. 5.5 g for the corresponding plastic package. This increases enormously the energy consumption of transportation in a refrigerated vehicle. The proportion of packaging for a plastic yoghurt container is 3.5 % in contrast to 36.2 % for glass packaging. In other words, the energy required by three lorries laden with yogurts in plastic packaging would be sufficient for only two lorries carrying yogurts in glass containers. It is evident that this is not beneficial for the environment.
- As has already been mentioned, reusable packaging, especially in the foodstuff sector, must be cleaned exceptionally thoroughly, which in turn consumes energy which could just as well be used for the production of new packaging. Furthermore, a significant amount of the cleaning agents ends up in the environment.

With this in mind, it is clear that the careful use of polymers can save significantly more energy than is necessary for their production. Even in uses for which this is not the case, the use of polymeric materials is frequently energetically and ecologically more favorable than using alternative materials.

21.4 Biopolymers

The term "biopolymers" is today often mentioned in discussions. It is often ill-defined, as are the terms "biodegradable" and "bio-based".

Biologically degradable or *biodegradable* polymers are materials that can be degraded by microorganisms or enzymes within a reasonable length of time under ambient conditions. Materials that can be biologically degraded within a composting plant are referred to as *compostable* polymers. It should be obvious that the degradation products are non-toxic. Biodegradability is a function of the chemical structure of a polymer.

Polymers based on renewable materials are by definition macromolecules that are produced from renewable raw material sourced from nature. Biogenic polymers such as cellulose or chitin, which can be directly isolated from renewable sources, belong to this group, as do polymers that can be produced by chemical synthesis from renewable sources. The counterpart to these bio-based polymers are polymers based on typical raw materials, primarily crude oil, which is referred to as a *fossil resource*. The transition is not always clearly defined. Thus, for example, there are polymer blends made from polymers based on fossil resources mixed with bio-based polymers. Cellulose, for instance, is often modified using reagents from fossil sources.

Whereas the expression "bio-based" refers to the origin of the polymer or the raw material from which it was produced, the expression "biodegradable" defines the application or recycling options of the polymer at the end of its period of use. Because biological degradability is a function of the polymers structure but not of its raw material, polymers based on fossil resources can also be biodegradable. The opposite is also possible—polymers based on renewable resources that are not biodegradable.

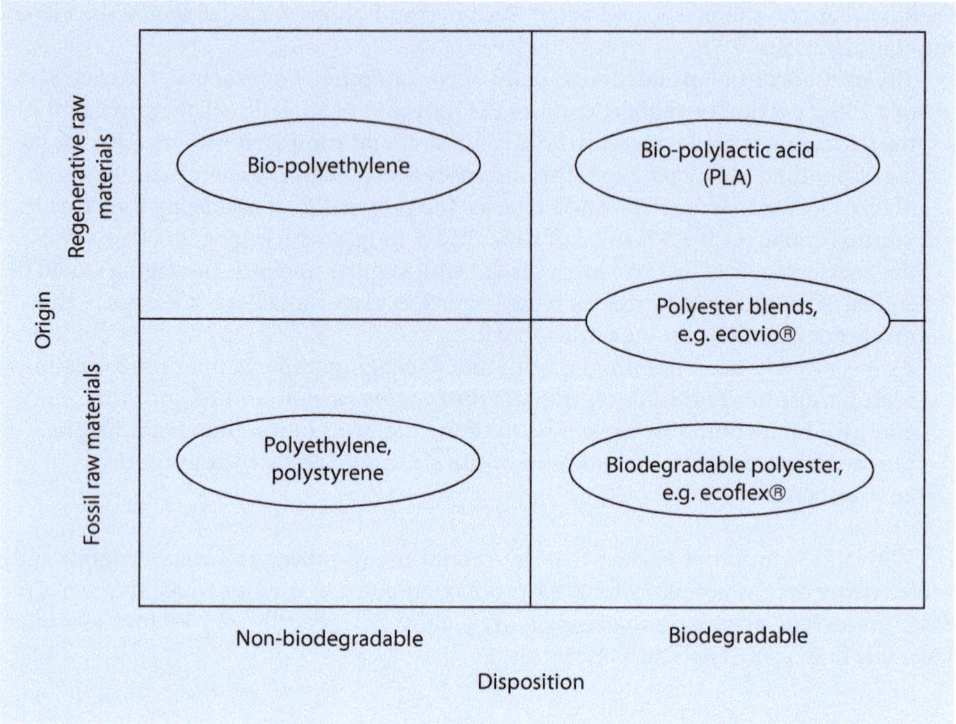

Fig. 21.5 Biopolymer classification; the named biopolymers are referred to in the text

Both properties are therefore in no way connected to one another and should be carefully differentiated. Figure 21.5 graphically emphasizes this.

In general language use, it has unfortunately become commonplace to refer universally to both polymers of biological origins and biodegradable polymers as *biopolymers*.[1] When using this term it should always be made clear whether, in the particular context, biodegradable or bio-based materials are being referred to.

The following sections deal with bio-based and biologically degradable polymers in more detail.

21.4.1 Polymers Based on Renewable Raw Materials

At a first glance, the use of renewable resources as raw materials appears to be an attractive alternative to crude oil. The enormous diversity of organic reactions and especially "white biotechnology",[2] which has become increasingly efficient in recent years, enables many chemical products—such as the monomers that have been dealt with in previous

1 Other literature limits the use of the term "biopolymer" to unmodified naturally occurring polymers such as proteins or natural rubber.

2 The term *white biotechnology* is used to describe biotechnology used for industrial purposes. With this technology, basic raw materials, pharmaceuticals, fine chemicals, and monomers are produced using microorganisms or enzymes in so-called bioreactors.

chapters—to be produced not only from crude oil but also from renewable raw materials. At the time of writing, the proportion of renewable resources in the German chemical industry makes up around 15% of the total raw material demand. A large proportion of this percentage is natural fats and oils, which can be processed into chemical products. In addition, starch is hydrolyzed into glucose and then processed further into ethanol. The ethanol obtained can be dehydrated to become ethene, and this polymerized to polyethylene. Via this and other routes, many of the polymers currently made from petrochemical raw materials can be made from renewable resources. Moreover, as well as the polymers that can be produced by classical, chemical means from bio-based raw materials, a large number of polymers are directly available in nature. Starch, cellulose, lignin, chitin, proteins, natural rubber, and polymeric nucleic acids can be mentioned in this context.

There are, therefore, two fundamentally different approaches to producing polymers that are either completely or to a large extent bio-based:

1. The use or modification of naturally occurring macromolecules
2. The synthesis of conventional or even novel polymers based on low molecular weight building blocks derived from natural sources

In the following, both of these options are discussed in more detail.

21.4.1.1 Naturally Occurring Polymers and Their Derivatives

In the field of water soluble, so-called functional polymers (▶ Chap. 19), many of the polymers used are already bio-based. Although DNA does not play a role in technical applications, for obvious reasons, proteins, polysaccharides, and their derivatives are indispensable for many of applications. In contrast, natural polymers and their derivatives play only a minor role as structural polymers—classical materials—when compared with the use of fully synthetic materials such as polyethylene or polystyrene.

Proteins

Naturally occurring proteins or those obtained from natural sources have great significance for the food industry. This is not least because of the strict regulations for non-natural products in the foodstuff sector; synthetic polymers based on fossil resources therefore have barely any significance in this sector.

One material used in large amounts is gelatine. Gelatine is basically made up of the protein collagen and is obtained from the connective tissue of different livestock species, particularly from the skin and bones of pigs and cattle. Because of the continuing concern of the public as well as religious dietary rules, special types of gelatine are produced from fish.

Gelatine is, as the name already implies, a typical gelling agent, which is used as a thickening agent in cooking. Well-known examples are wine gums, jelly, brawn, and aspic. Gelatine is also used on a large scale in the pharmaceutical industry to coat medicinal capsules or as a thickener for liquid medicinal products. Furthermore, gelatine is used as a protective colloid for the stabilization of dispersions (▶ Sect. 19.1) of vitamins and carotenoids, for example.

Apart from these functional uses, proteins have not gained any great relevance in contrast to their synthetic analogues, the polyamides. The reason for this is that the degradation of the protein is undesirable for many of these applications. Moreover, the biological degradation of proteins often occurs in combination with the formation of degradation products with an unpleasant smell. An exception is the use of silk in the textile field.

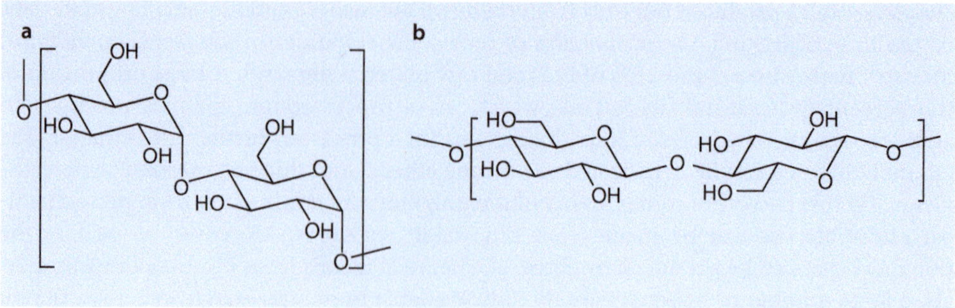

Fig. 21.6 Segments of the chemical structures of starch (**a**) and cellulose (**b**)

Spider silk proteins have an even better tensile strength than silk. However, so far it has not been possible to produce them on a technically relevant scale because of problems related to spider husbandry and breeding. Biotechnological processes that reproduce the amino acid sequence of spider silk have not produced fibers whose mechanical properties are comparable to spider silk with its more complex structure. The difference can be attributed to variations in the secondary and tertiary structures.

Polysaccharides and Their Derivates

Of the polysaccharides and their derivatives, starch and cellulose play the major role (Fig. 21.6). Both polymers are isomeric forms of poly-D-glucose, and differ from one another in the stereochemistry of the links between the glucose units. In starch, the glucose units are α-linked, whereas the building blocks of cellulose are linked to one another by β-links.

Starch is often used in its original form, that is, without chemical functionalization of its OH-functions, albeit the natural degree of polymerization is reduced to meet the specifications of the particular application by hydrolysis. Starch with a low degree of polymerization is also referred to as *maltodextrin*. Complete hydrolysis of starch leads to D-glucose, which can be further processed into other bio-based substances, such as bioethanol.

Starch is almost exclusively obtained from plant sources where it is part of the energy storage and is usually present in the form of fine granules.

Naturally occurring starch is generally a mixture of *amylose* and *amylopectin*. The difference between these two starch isomers is that amylopectin chains are highly branched whereas amylose chains are, largely, linear. The ratio of amylose to amylopectin depends to a large extent on the plant from which the starch has been obtained. Industrially, starch is mainly produced from potatoes, rice, cereals, and cassava.

Similar to gelatine, starch is widely used in the food industry, and is an essential component of noodles, flour, bread, and other baked goods. In addition, starch is also used in the paper industry as a binding or coating agent and in the pharmaceutical industry as an ingredient in pills, for example, as a filler, disintegrant, or binding agent (▶ Sect. 19.6).

Thus, the main applications of starch, as is the case for gelatine, are different from the classical uses of synthetic materials. From a technical point of view, its especially strong tendency to absorb water has proven to be a big problem; the absorption of water has considerable influence on the melting and glass transition temperatures and on the moduli. Thus the material properties and the temperature range in which the product can be used are dependent on its water content. The use of starch as a molded part is therefore limited to relatively undemanding applications such as loose filler materials for the packaging of sensitive goods. However, even this use is decreasing. Custom-fit molded parts made of polystyrene or polyethylene foam are becoming more popular. Blends can be made from starch with other polymers, for instance, biodegradable synthetic polymers, which are less prone to absorbing water and are thus more stable and whose properties are less affected by the material's history.

In addition to these two major polysaccharide representatives, a large number of other structures such as guar (E-412), xanthan (E-415), and alginate can also occur naturally. Many of these polysaccharides are used as thickeners, for example, in cosmetics, but also in the crude oil industry. A comprehensive discussion of the extraordinarily diverse range of polysaccharides would, however, go beyond the scope of this book. For further information the interested reader is referred to the specialist literature (e.g., Ramawat and Mérillon 2015; Aspinall 1983; Habibi and Lucia 2012).

21.4.1.2 Bio-based Polymers from Renewable Building Blocks

As has become clear from the previous section, a large number of monomers can be produced from renewable resources:

- Ethene (glucose → ethanol → ethene)
- Ethylene oxide (from bio-ethene)
- Diacids (e.g., for polyesters or polyamides)
- Polyols (e.g., for polyurethanes)

The properties of the polymers obtained are not dependent on whether the building blocks are derived from renewable or fossil raw materials (as long as the educts have identical degrees of purity) and the biodegradability of these materials is just as limited as their synthetic analogs. However, the "most important of all product properties"—production cost—is often critical. The production processes for products based on renewable resources are often more complex, and thus more costly, than the typical fossil-based processes, particularly in the field of large-volume, basic chemical products.

Before biodegradable and bio-based polymers are discussed, the term "biodegradable" is first explained in more detail.

21.4.2 Biologically Degradable Polymers

As has already been explained above, the biological degradability of a polymer is dependent on its chemical structure. Well-known examples of biologically degradable polymers are the naturally occurring polymers, such as starch, cellulose, lignin, chitin, and proteins mentioned above. As these are continuously produced by living organisms, but do not accumulate in nature, some form of degradation must take place. These polymers are used by microorganisms (bacteria, yeasts, fungi) as energy sources and degraded into biomass, water, carbon dioxide, and (depending on the ambient conditions) methane. In an oxygen-rich environment, the main products of so-called *aerobic* decomposition are water and CO_2. This is typically the case during controlled composting or degradation in soil. The carbon produced is then fed back into the biological cycle during photosynthesis. Alternatively, under *anaerobic* conditions, where little oxygen is available, methane as well as carbon dioxide is produced. Examples of anaerobic environments are deep water and rubbish tips. Anaerobic decomposition can also be used technologically to obtain biogas from biomass.

The microorganisms responsible for degradation have to be able to absorb the organic molecules and to metabolize them in their cells in order to be able to use them as a source of food. To facilitate this, two processes outside the microbial cell are necessary.

As a prerequisite, the polymer to be degraded must contain flexible structural units. In particular, any crystalline domains must be dissolved; this step is of course unnecessary for amorphous polymers or for the amorphous domains of partially crystalline materials.

The flexible chain segments must be chemically split into smaller molecules. Microorganisms secrete so-called *extracellular enzymes* that can split the polymer chains, often via hydrolysis. For this to take place, the chains have to be flexible enough to be able to penetrate the catalytic center of the enzyme. The fragments formed have to be water-soluble and able to pass through the cell walls or membranes of the microorganisms. After being taken up by the microorganism the fragmented parts are metabolized.

Not all polymer chains can be split enzymatically. In particular, polymers with simple, saturated hydrocarbon backbones are deemed to be either not very degradable or not degradable at all. An exception is polyvinyl alcohol.

Temperature, humidity, and oxygen content of the medium have a significant influence on the biological degradation. Of course, as well as the microbiological degradation pathways described here, thermal or photochemical degradation mechanisms mentioned in ► Chap. 15 also play a part in the degradation of polymers in the environment.

To establish whether a given polymer is biodegradable or not, there are different, partly national, standards. The most important is the European Standard EN 13432, the American ASTM D 6400-04, and the Japanese GreenPla. All of these standards define basic requirements for packaging materials which can decompose in industrial composting plants and are thus compostable. The most important test of aerobic degradability is the so-called *controlled composting test* according to ISO 14855. This test involves incubating the polymer with mature, defined compost at 58 °C in a compost container. Polymer decomposition is followed by measuring the amount of carbon dioxide formed.

Relatively large numbers of the polymers produced commercially in large volumes are biodegradable. This is true both for polymers based on fossil raw materials and for bio-based polymers. The most important examples are discussed in the following.

Biodegradable Polymers that are not Bio-based

An important class of fossil-based biodegradable polymers are the aliphatic or partially aliphatic polyesters. Examples are poly(ε-caprolactone) and a polyester based on adipic and terephthalic acids and butane diol, marketed under the name ecoflex®. Further examples of biodegradable polymers based on fossil fuels are polyvinyl alcohol, polyethylene oxide (depending on molar mass and which end groups are present), and polyaspartic acid. The latter is produced as shown in ◘ Fig. 21.7 from maleic anhydride, water, and ammonia.

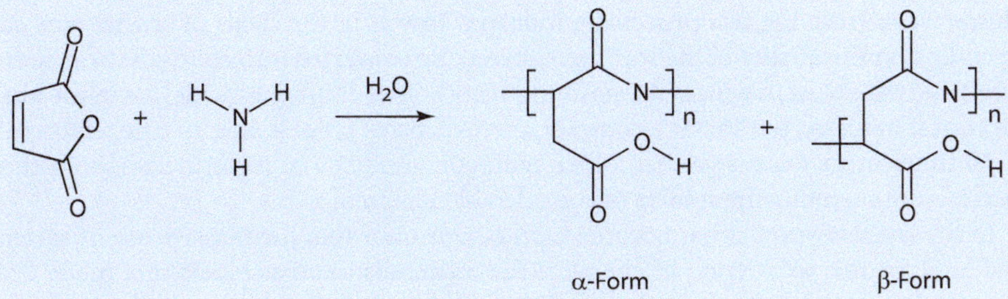

Fig. 21.7 Synthesis of polyaspartic acid from maleic acid anhydride, ammonia, and water

Biodegradable Polymers Based on Renewable Raw Materials

An example of a technologically manufactured, bio-based, and biodegradable polymer is polylactic acid. The lactic acid starting material is obtained by fermenting molasses or by the fermentation of glucose with the aid of different bacteria. The polymer is most often synthesized by ring-opening polymerization of the cyclic diester of lactic acid, lactide, the dimer of lactic acid.

Biodegradable Polymers that are Partially Bio-based

As has already been explained in the introduction to ▶ Sect. 21.4, one can, of course, mix biodegradable polymers that are based on different raw materials. The biological degradability of the polymer blend formed is not affected by this. An example of this kind of partially bio-based, completely biologically degradable polymer is marketed under the name ecovio®. It is a mixture of ecoflex® with polylactic acid.

21.4.3 Discussion

21.4.3.1 Bio-based Materials

Naturally occurring polymers such as lignin and chitin are often described in both the academic and industrial literature as materials available in abundance as waste from, for example, the paper and food industries. However, the details are often misrepresented. Thus, although lignin (a complex cross-linked polyphenol which makes up the bulk of plant cell walls) is a by-product of the paper industry, it is often used to provide a good proportion of the energy required in the paper manufacturing process because of its substantial fuel value. Some paper manufacturing companies (so-called paper mills) cover 80–100 % of their energy needs by burning lignin. Lignin is not by any means the primary product of a paper mill, but it is certainly not a valueless waste product as often suggested Furthermore, the lignin produced in the papermaking process tends to be of a dark color which makes any material recycling more complex. Chitin, a polymer of *N*-acetyl glucosamine, which makes up the bulk of the shells of crustaceans, can be considered as a large volume waste from the fish processing industry. However, the shells of crustaceans can, depending on the quality of the starting material, be converted into chitosan (by deacetylation) and then used in winemaking (fining agent), agriculture (pesticide), or in the pharmaceutical industry, but this is a complex and expensive process and, to date, chitosan is 10–20 times more expensive than other synthetic alternatives. Nevertheless, both these materials are currently the subject of considerable research.

In the last few years, it has become increasingly clear that the massive use of agricultural land for the production of chemical raw materials, such as bioethanol made from maize or sugar cane, has meant that the following problem has occurred. The amount of agriculturally useful land available worldwide is decreasing because of climate change and increasing soil erosion. At the same time, the world population is continuously increasing. Furthermore, the demand for high value foodstuffs, such as meat, whose production requires a particularly large area of arable land, is also increasing.[3] This means that, mid- and long term, it will become even more difficult than it already is to meet the nutritional needs of the world's population. Against this background, the use of arable land for the production of raw materials rather than food is controversial. The area of arable land needed to replace conventional polymers with polymers based on renewable raw materials to a significant extent is therefore not going to be available in the foreseeable future. The use of renewable resources for the production of plastics or other chemicals therefore has to be restricted to sources of raw materials not useful to humans either as food or for other purposes. Examples of such raw materials are wood waste or plants with low nutritional value which thrive in soil or areas where the production of high quality crops is not possible. The progress mentioned at the outset of this chapter in the fields of white genetic engineering and biorefineries gives reason to hope that useful and sustainable access to bio-based chemical building blocks that are not in competition with food production may be achievable.

21.4.3.2 Biodegradable Materials

For technological materials, for which as long a life span as possible and therefore long-term retention of mechanical and other properties is desirable, biodegradation is, of course, unwanted. This is, for example, the case for most of the uses in the automotive industry, in construction, and for the vast majority of electronic applications.

Thus, the use of biologically degradable polymers is limited to uses in which a long product life span is unnecessary. Examples include:

- Biodegradable garbage bags
- Disposable crockery
- Fast-food packaging
- Agricultural sheeting

The use of bio-based and/or biodegradable polymers would also often be sensible and technically possible in the case of functional polymers (▶ Chap. 19). The limited success of such materials can often be ascribed to their high production costs in comparison to fossil-based polymers.

---

3 During its life, a cow eats, on average, 16 kg maize per kilogram of beef produced from the animal after slaughter.

Biological degradation is, generally, a slow process, especially under anaerobic conditions, such as in landfill sites. Additionally, as explained above, the climate-damaging greenhouse gas methane is released during anaerobic degradation. Ultimately, the use of biodegradable polymers, does not bring any benefits at all if the product eventually ends up in a landfill site. A comprehensive solution cannot be achieved simply by providing consumers with biodegradable products; practical and socially acceptable solutions for the product life time, including the management of its waste, need to be found.

The final disposal of waste in landfill sites is no longer permitted in Germany but is still accepted in many countries. The highest priority must therefore be to avoid the disposal of waste at landfill sites.

Conclusion

An all-encompassing solution to the challenges in all aspects of waste disposal, such as recycling, energy, and food requirements, cannot be provided within the scope of this book. However, it is important to ensure that any debate about polymer recycling is carried out within the broader context of the debate on energy and food. A practical solution can only be achieved if energy resources and agricultural land are employed to optimum advantage and the final disposal of waste in landfill sites is avoided wherever possible. Furthermore, a debate on the recycling options available for synthetic materials, especially in the areas of thermal recycling and bioplastics, based on well-founded information and free of political and ideological prejudices, is urgently needed.

References

Aspinall GO (ed) (1983) The polysaccharides. Academic, New York

Habibi Y, Lucia LA (2012) Polysaccharide building blocks: a sustainable approach to the development of renewable materials. Wiley, Hoboken, NJ

Ramawat KG, Mérillon JM (2015) Polysaccharides: bioactivity and biotechnology. Springer, Heidelberg

Current Trends in Polymer Science

S. Koltzenburg et al., *Polymer Chemistry*, DOI 10.1007/978-3-662-49279-6_22

In this chapter some developments are discussed that have caught our attention over the past few years for a variety of reasons. The selection is, of course, subjective, and no attempt is made to suggest whether these developments are likely to become widespread technology or remain laboratory curiosities. Nevertheless, we believe the examples presented are interesting in their originality and show some of the directions being followed in current polymer science.

22.1 Nanocomposites

Nanocomposites consist of two or more types of organic or inorganic macromolecules that form structures in the dimension of nanometers, or are even molecularly mixed. Such materials are playing increasingly important roles as surface coatings, reinforcements, and barrier materials. Monolithic nanocomponents are interesting precursors for highly porous oxides and polymer networks. Because of their large internal surface, such materials have potential, for example, for gas storage.

There are various strategies for synthesizing nanocomposites, for example, *simultaneous polymerization* whereby two monomers A and B are polymerized side by side (simultaneously) (▫ Fig. 22.1).

Furfuryl alcohol is polymerized cationically; the water produced induces the sol–gel process, in which $Si(OR)_4$ is converted into SiO_2. This process is by no means trivial; the phase separation during and after the synthesis can negatively influence the dimensions of the structural elements formed.

Consecutive polymerization is an alternative to simultaneous polymerization. As shown in ▫ Fig. 22.2, the functions A and B of a monomer are activated successively.

However, the most promising method is *twin polymerization*, which occurs as shown in ▫ Fig. 22.3 (Grund et al. 2007; Spange and Grund 2009).

If the polymerization of blocks A and B can be made to occur undisturbed side by side—without interruptions by phase separation or copolymerization—one can achieve products with a large internal surface area (800–1000 $m^2\,g^{-1}$) after removing one of the components.

▫ Figure 22.4 gives a selection of A, B monomers that have been successfully used in twin polymerization.

The ionic polymerization of tetrafurfuryloxysilane results in a nanostructured network. It can be shown by ^{13}C and ^{29}Si-solid nuclear magnetic resonance, as well as by chemical analyses that form in the process two independent networks that are intertwined but not chemically connected with one another, called interpenetrating networks.

▫ **Fig. 22.1** Simultaneous polymerization. (**a**) Schematically. (**b**) Example: polymerization of furfuryl alcohol and tetraalkoxysilane

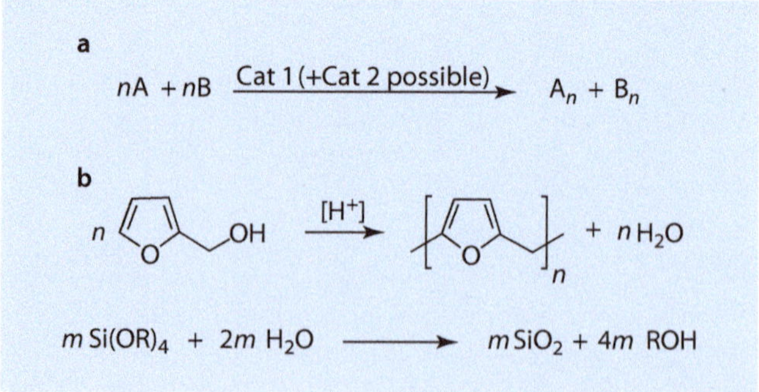

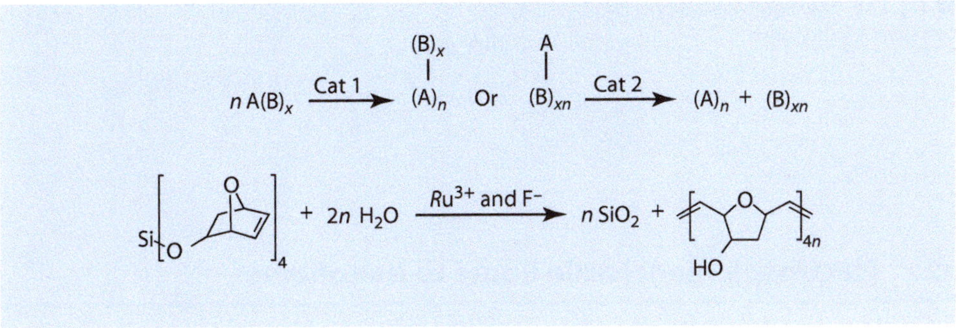

Fig. 22.2 Scheme of consecutive polymerization

Fig. 22.3 Scheme of a twin-polymerization

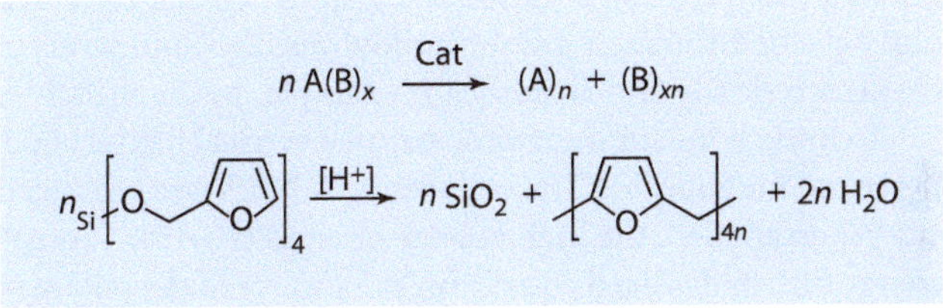

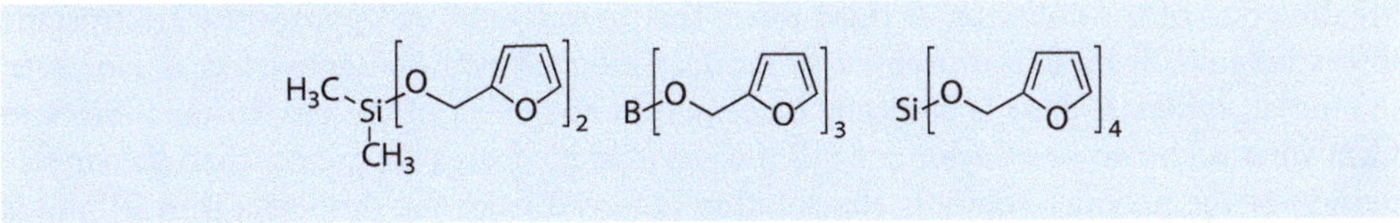

Fig. 22.4 Examples of bi-, tri-, and tetra-functional molecules that are suitable as monomers for twin-polymerization

After burning the organic phase of the nanocomposite of polyfurfuryl alcohol/SiO_2, one obtains monolithic SiO_2 particles with a surface area of 1000 $m^2\ g^{-1}$. If the inorganic polymer is selectively and successfully removed, one obtains an organic network. In the case of the nanocomposite composed of polyfurfuryl alcohol/SiO_2, however, this is impossible because the organic network is damaged by removing the SiO_2 (depolymerization).

The cationic polymerization of 2.2-spiro[4*H*-1,2,3-benzodioxasilane] has been shown to be more successful. The subsequent selective removal of the inorganic or the organic network results in stable carbon or SiO_2 monoliths with surface areas in the region of 1000 $m^2\ g^{-1}$.

The possible variations of the inorganic part of the A, B monomer is evident. The search for suitable organic ligands has proven more difficult. With the discovery of spiro molecules (Fig. 22.5), an interesting and promising way to produce materials with very small particle sizes in the range of 0.5–2 nm and extremely small and uniform pore diameters seems to have been found.

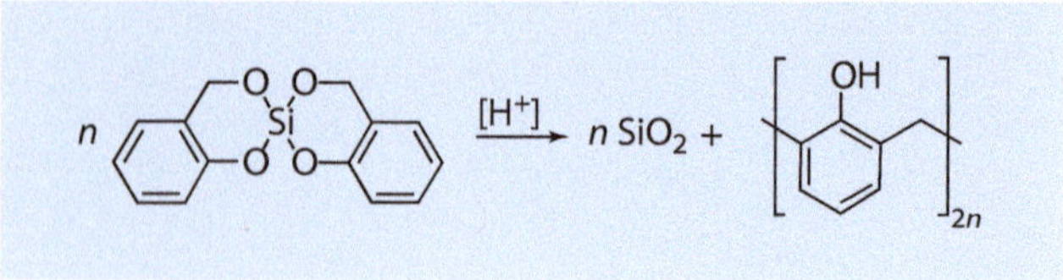

Fig. 22.5 Twin-polymerization of a spiro-A,B-monomer

22.2 Electrospinning: Facile Route to Nanofibers

Electrospinning is an easy and elegant method for producing very fine polymer fibers (Reneker and Chun 1996). In Fig. 22.6 an electrospinning apparatus is shown schematically. It consists of relatively few components; a syringe, a nozzle (usually a hollow needle), a high voltage power supply, and a collector.

First, the polymer solution to be spun is placed in a syringe out of which it is continuously pushed through a tight hollow needle (spinneret) as a thin stream. The syringe is depressed slowly and steadily, for example, by an electric motor with precision gearing.

To initiate the spinning process, an electrical field is generated between the metallic spinneret and the collector. The voltages of 1–30 kV are produced by a high voltage power supply.

When applied, the high voltage charges the drops conveyed from the spinneret, and the charge distributes itself equally on the surface of the polymer solution. Only two electrostatic forces have an effect on these drops—an electrostatic repulsion between the individual charges on the surface of the material exiting the spinneret and the Columbic attraction in the direction of the collector. A third force, the surface tension, opposes the electrostatic forces and initially inhibits movement of the drop induced by the other two forces. However, the initial, spherical drop is conically deformed by the effect of the electrostatic forces to form what is known as a Taylor cone. If the electrostatic forces are greater than the surface tension of the polymer solution, the solution is forced from the drop as a thin jet and is accelerated in the direction of the collector; velocities of several meters per second can develop. The solvent continuously evaporates, and the thread of polymer solution is stretched so that it arrives at the collector solution-free and with a drastically reduced diameter of the order of nanometers. There, over time, it forms a disorderly fiber mat. More modern techniques can produce an ordered, for example, parallel, deposition of fibers.

The morphology of the electrically spun fibers is defined by the intrinsic qualities of the polymer solution. The key parameters are summarized in Table 22.1.

The literature is inconsistent as to the effect of the various parameters given in Table 22.1. For example, although a decrease of the fiber diameter with increasing field strength has been widely discussed, the opposite has also been reported. Nevertheless, this technology has already been used successfully for many different polymers (Table 22.2).

The most important application of traditional fibers is the reinforcement of a large variety of materials. Nanofibers should interact much more successfully with the matrix than traditional fibers because of their high aspect ratio (length/diameter).

A further use for these fibers is as filter elements. With the filter mats made from nanofibers, particles with $d < 0.5$ µm can be filtered. In the field of air filtration, such systems are said to have great potential.

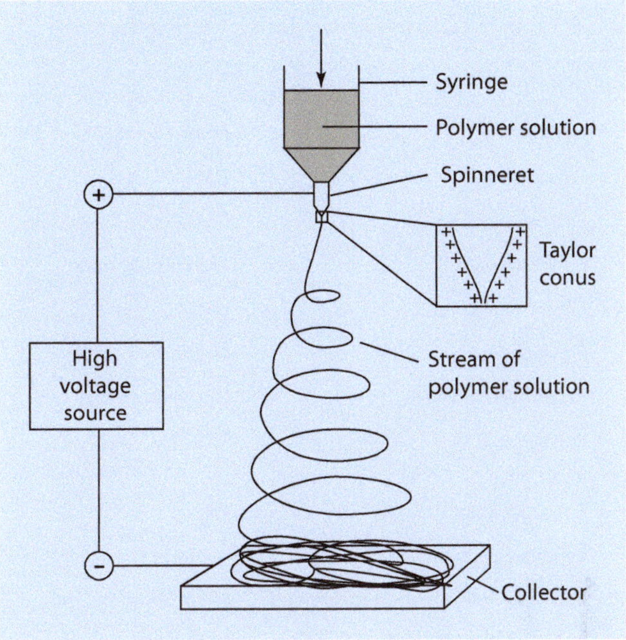

Fig. 22.6 Scheme for the electrospinning process

Table 22.1 Key parameters for the electrospinning process (Li and Xia 2004)

Intrinsic polymer solution parameter	Process parameter
Polymer type	Field strength
Conformation of the polymer chain	Distance between spinneret and collector
Viscosity or concentration of the polymer solution	Flow rate of the polymer solution Ambient humidity and temperature
Elasticity	
Electrical conductivity	
Polarity and surface tension of the solvent	

Because of their large specific surface area, nanofibers are ideal substrates for catalysts.

A broad and highly promising area of use of these fibers is in medicine, for example, as a support for soft tissue and blood vessels. They have also been proposed as a coating for hard prostheses. Moreover, their use as a support material for *tissue engineering*, the specified growth of cellular tissue on a carrier material, is also plausible because the similarity of these fibers to the native extracellular matrix encourages the colonization of cells.

A further application is in the treatment of wounds, in which fibers (possibly infused with medication) are spun directly onto the skin and can thus protect injured areas against viruses and bacteria.

22

Table 22.2 Electrospun polymers—selected examples

Polymer	Solvent	Possible application
Nylon 6,6	Formic acid	Safety clothes
Polyurethane	DMF	Safety clothes, filter elements
Polybenzimidazole	DMAc	Safety clothes
Polycarbonate	Dichloromethane	Sensors, filter elements
Polyacrylonitrile	DMF	Carbon fibers
Polylactide	Dichloromethane	Medical applications
Polyethylene oxide	Isopropanol/H_2O	Filter elements
Polystyrene	THF	Catalyst support, filter elements
Polyvinyl phenol	THF	Antimicrobial action
Cellulose acetate	Acetone	Membranes
Poly(2-hydroxyethyl-methacrylate)	Ethanol/formic acid	Flat tapes
Polyvinylidene fluoride	DMF/DMAc	Flat tapes

DMF Dimethyl formamide, *DMAc* Dimethyl acetamide

22.3 Spinning-Disc Reactor (SDR)

The success of a polymerization process depends not only on its kinetics but also on the way it is carried out. Polymerization in continuous or discontinuous processes in stirred tanks, stirred cascades, loop reactors, or flow tubes determines the morphology, molar mass, molar mass distribution, branching, or, if applicable, the cross-linking of the product and thus also determines the useful characteristics of the polymers produced. All reactor types can be traced back to three basic types (Fig. 22.7).

These three basic types—stirred tanks, continuously stirred tanks and flow reactors—are used in many variations and also combined with each other (Echte 1993). Nowadays such reactors are well understood. However, new challenges in the field of process control continue to arise. The term *process intensification* summarizes a response to the demands of modern process technology to reduce energy usage, required space, inventory of product and raw materials, number of process steps, emissions and noise and to increase efficiency and safety. The SDR (Fig. 22.8) is one solution to all of the above (Ramshaw 1983; Cai 2006; Pask et al. 2012).

In the SDR, a very thin liquid film is formed on a thermostatted rotating disc from a centrally added mixture of liquids or separated educts added at the center of the disc. The reaction mixture is tangentially accelerated by the shear stress acting at the interface between the disc and the liquid. The desired reaction occurs on the thermostatted disc. The product is spun onto the cooled outer wall, and collected at an outlet. The continuous operation, short retention time, and excellent heat transfer and mixing render this reactor of interest for polymerization reactions as long as the viscosity of the resulting product solution is not too high. It has been shown, for example, that from polyols and diisocyanates, polyurethane-pre-polymers of varying compositions and molar mass can be produced.

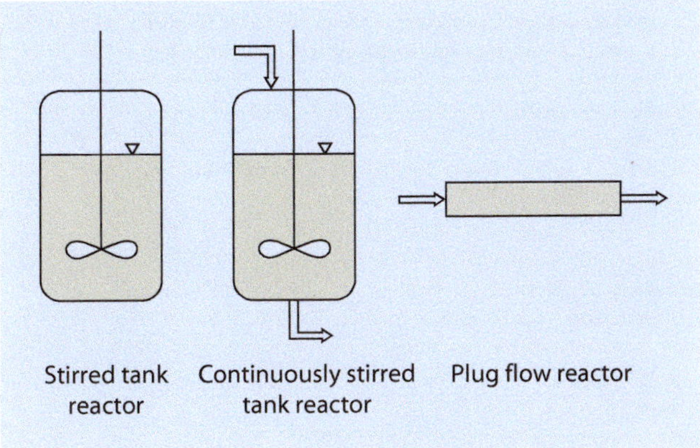

Fig. 22.7 Basic types of chemical reactors

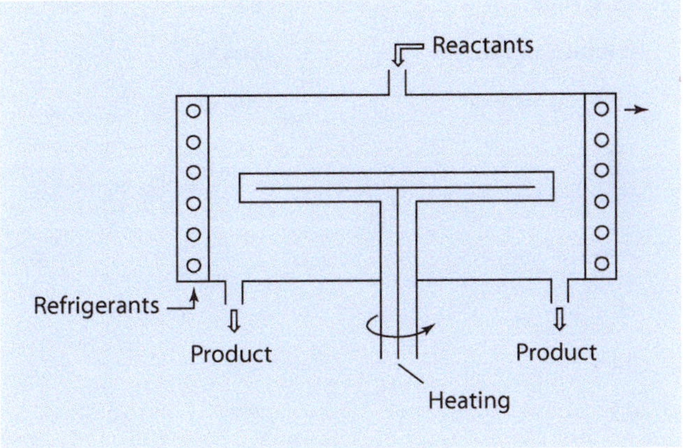

Fig. 22.8 Sketch of a spinning-disc reactor (SDR)

Polyreactions can be optimized in a short time and with comparatively small amounts of material by optimizing the disc temperature, the speed of the disc's rotation, and the flow rate. This reactor is also very suitable for photopolymerizations because light can easily and completely penetrate the ultrathin coating (from nanometers to micrometers, depending on the viscosity) on the rotating disc.

The advantages of the SDR compared to the stirred tank reactor are summarized in Table 22.3. In order to produce 3.6 tons of product requires 16 h in the stirred tank but in an SDR with a disc diameter of 1 m only 10 h.

In the SDR, the amount of substance per unit time is very small so that, in the event of an incident, only milliliters have to be discarded as compared to the total batch in the stirrer tank. Further advantages are its flexibility, for example, product change, but also the small amount of cleaning materials required, the relatively low investment costs, and the small size of the reactor.

In terms of process engineering, the SDR is particularly advantageous, compared to the discontinuous stirred tank reactor, when scaling-up a reaction (turning a laboratory experiment into an industrial process) because with the SDR this can be done without the use of characteristic numbers such as Reynolds-Prantl number and others (Table 22.4).

The tenfold multiplication of the diameter requires only one action: increasing the flow by a factor of 100. Thus, the film thickness and residence time remain identical. In a stirred tank reactor the transition from laboratory to the ton scale is much more laborious

Table 22.3 Comparison between an SDR and a stirred tank reactor using the example of the synthesis of polyurethane prepolymers (Cai 2006)

Reactor type	Stirred tank	SDR
Size	4 m^3—volume	100 cm—disc diameter
Charging	1.5 h	100 mL s^{-1}
Heating	2.5 h	27 min
Reaction	7 h	3 s
Cooling	1.5 h	20 min
Emptying	1.5 h	100 mL s^{-1}
Cleaning	2 h	100 mL s^{-1}
Total time	**16 h**	**10 h**
Production volume	**3600 kg**	**3600 kg**
Cleaning material	1000 kg	12.5 kg

Table 22.4 Parameters for the scale-up of an SDR process

Disc diameter (m)	Flow rate (mL s^{-1})	Film thickness (μm)	Residence time (s)	Production (t/8000 h)[a]
0.1	2	68.9	0.27	58
0.3	18	68.9	0.27	518
1.0	200	68.9	0.27	2880

[a]Assumption: annual available production time = 8000 h

and requires attention to, for example, the decrease of surface area to volume (which leads to different cooling characteristics) and the increase in the size of the stirrer (which leads to different shear and mixing parameters), to name just a few of the challenges.

22.4 Polymers for Organic Light-Emitting Diodes (OLEDs)

With the discovery that some organic substances display electroluminescence when voltages of less than 10 V are applied to them (Tang and van Slyke 1987) and that certain polymers also display this characteristic (Friend et al. 1999), a rapid development started, resulting in the multicolored light-emitting diodes that now cover the entire visible spectrum (Müller et al. 2003). The structure of a multilayer light-emitting diode is shown in Fig. 22.9.

The organic material (polymer or small molecule) is present in one or more layers of varying composition between a cathode and an anode on a glass substrate. Electrons flow into the organic phase at the cathode (Ba, Ca, or Al) and migrate through the electron conducting layer (I) to reach the emitter layer (II). There they encounter electron holes, which are formed

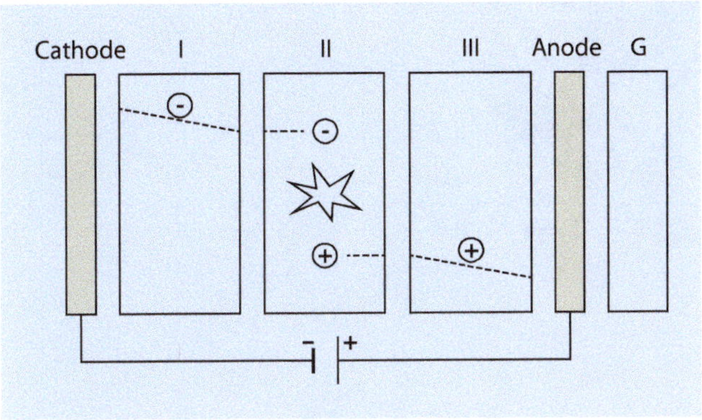

Fig. 22.9 Schematic construction of a three-layer OLED. *I* Electron Conductor, *II* Emitter, *III* Hole conductor, *G* Glass support, *Cathode* Ba, Ca, Al, *Anode* ITO (indium tin oxide)

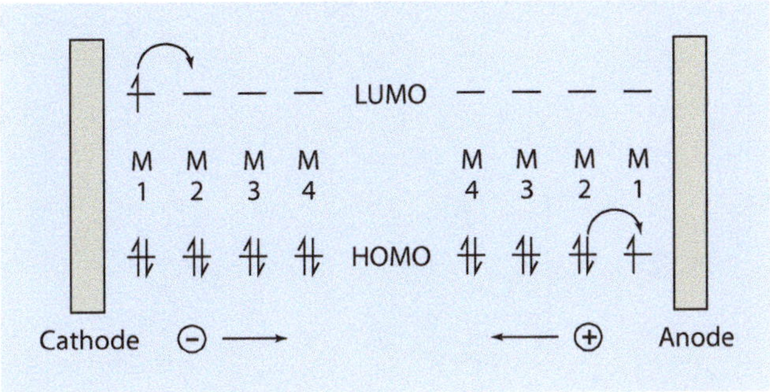

Fig. 22.10 Schematic representation of the electron and hole conduction through the organic layers of an OLED

at the anode by electrons being removed from the hole conducting layer (III). During the collision, they form an excited singlet state which relaxes to the ground state, emitting light. The emitted light passes through the transparent indium tin oxide (ITO) anode and then through the glass surface layer. The passage of the electrons and holes is outlined in Fig. 22.10.

The electron conduction occurs in the lowest unoccupied molecular orbit (LUMO). Without mass transport, the electron jumps from molecule to molecule toward the anode. Hole conduction occurs in the highest occupied molecular orbit (HOMO). The hole which is formed by "sucking up" an electron at the anode (electron deficient) is filled under the influence of the electrical field of its neighbor, leading to a hole there. From a chemical perspective, electron and hole conduction are a sequence of redox reactions driven by the electrical field. The reactions take place via radical cations (holes) and radical anions. The negative charge travels, without mass transport, from left to right and the positive charge in the opposite direction. Typical examples of electron and hole conductors are given in Fig. 22.11.

The basic processes for producing OLEDs are sublimation (*vapor deposition*) and preparation from solution (e.g., by so-called *spin coating*, whereby a coating is formed on a rotating surface by distributing the solution over the surface by centrifugal force). Sublimation technology is expensive and unsuitable for polymers. For smaller, high-performance OLEDs, sublimation is the method of choice because of the additional cleaning step involved in spin coating and the possibility of generating multilayered OLEDs without damaging the previously applied layer.

Polymers are applied exclusively from solution. Therefore the application of second and third polymer layers in the multilayered construction presents a challenge during the application process, as previously formed layers can redissolve. The problem can be simply solved by

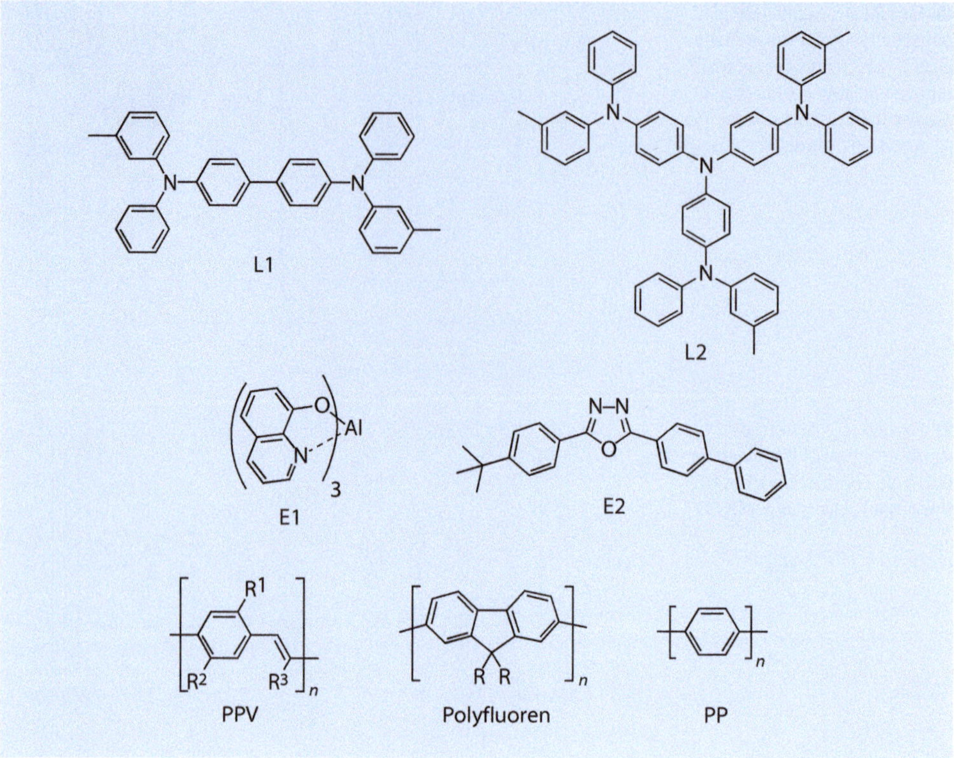

Fig. 22.11 Typical structures of monomeric hole and electron conductors and multifunctional polymers which can also function as emitters. *L1*, *L2* Hole conductors, *E1*, *E2* Electron conductors *PPV Polyfluorene*, *PP* Polymers

photo-cross-linking each layer as it is applied, thereby rendering it insoluble and preventing damage by the application of further solution layers. This procedure is illustrated in Fig. 22.12.

A polymer or a small molecule building block with at least two photochemically linkable side groups per molecule (e.g., oxetane groups) is applied onto a glass carrier with an ITO surface. After removing the solvent, the layer is cross-linked via the oxetane groups in the presence of a photoinitiator (▶ Chap. 10, Fig. 10.17) with the aid of UV light. Thereafter, the cycle can be repeated with a second and third polymer solution to produce a multilayer diode. A cathode is then applied and the whole construction is "encapsulated" and thus protected by a transparent, inert polymer layer.

Aromatic polyamines have proved successful as hole conductors as have oxetanes as cross-linking moieties. The basic structure of this type of system is shown in Fig. 22.13.

Both small molecule and polymeric aromatic amines can be elegantly synthesized by a Hartwig–Buchwald coupling (Louie and Hartwig 1995; Guram et al. 1995).

The required functional (electroluminescence and cross-linking) groups can also be incorporated into the polymer chain by copolymerization. The synthesis route, shown in Fig. 22.14, undoubtedly increases the possibilities for variation in the structure of

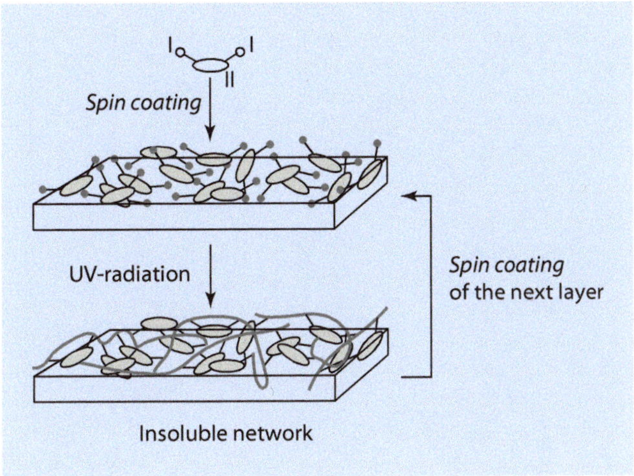

Fig. 22.12 Manufacture of a multi-layer OLED. *I* Cross-linkable group, *II* Electroluminescent monomer or polymer

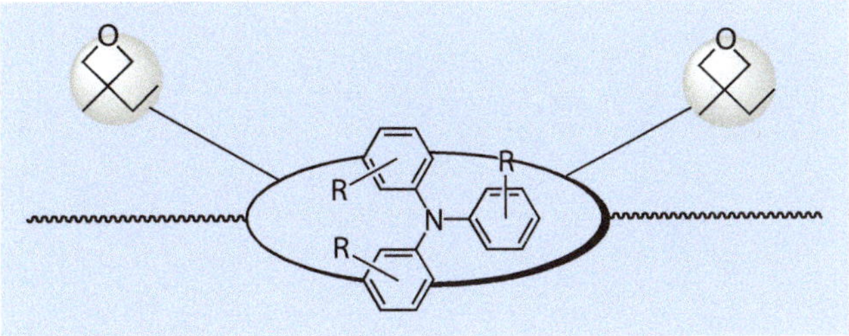

Fig. 22.13 Basic structure of an aromatic amine with cross-linkable side groups

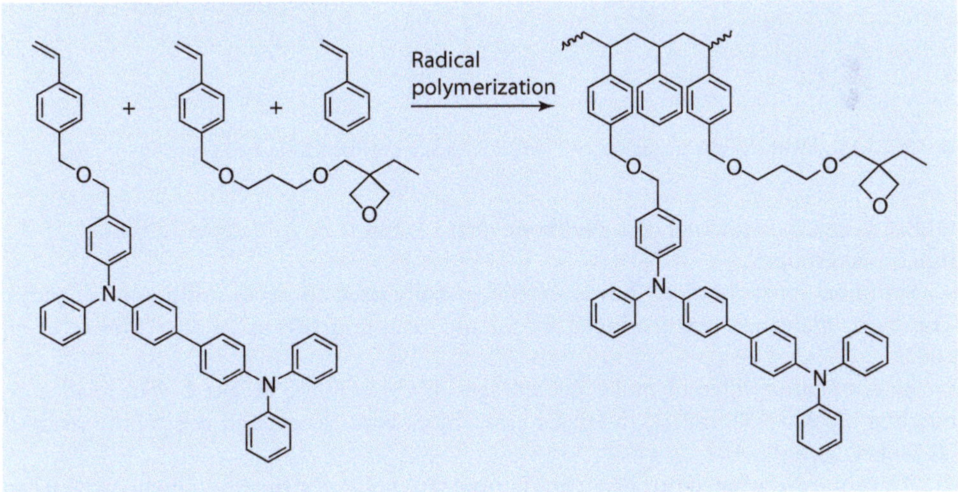

Fig. 22.14 Cross-linkable OLED-polymer via copolymerization

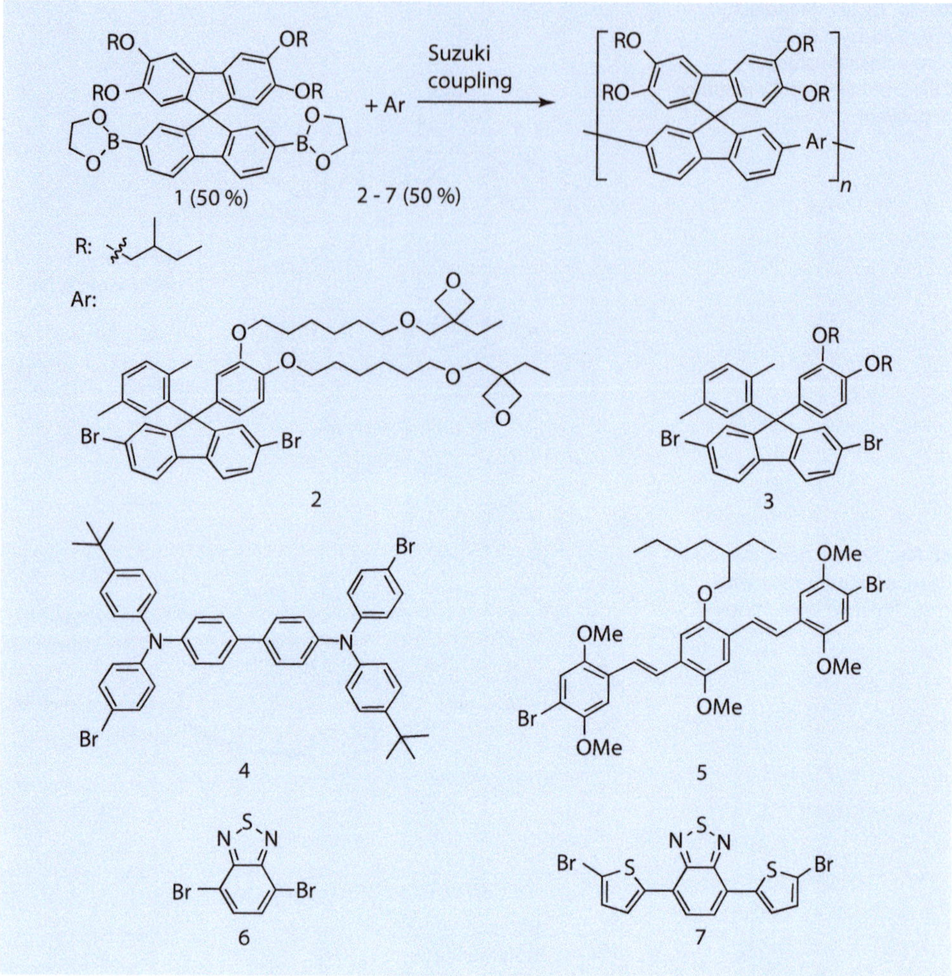

■ **Fig. 22.15** Synthesis of blue, green, and red luminescent polymers by Suzuki-coupling

OLED materials. However, the backbone chain exhibits an undesired "dilution" of the functional groups.

Spin coating technology has been successfully used to make multicolor displays. The basic materials are produced by Suzuki coupling (■ Fig. 22.15) (Müller et al. 2003).

By combining different molar percentages of the building blocks **2–7** in relation to building block **1** (50 mol%), the three base colors blue, green, and red can be realized (■ Table 22.5).

If layers are irradiated through a mask, superfine colored structures remain, which can be addressed individually.

One important innovation is the Baytron®-initiated cross-linking of OLED polymers with oxetane side groups (Köhnen et al. 2009). Baytron® is a mixture of poly(3,4-ethylene dioxythiophene) (PEDOT) and polystyrene sulfonate (PPS). To produce a multilayer

Table 22.5 Proportions of the building blocks **1–7** from Fig. 22.15 in the OLED-polymers P-Blue, P-Green, and P-Red

Building block	P-blue	P-green	P-red
1	50	50	50
2	25	25	25
3	15	–	–
4	10	10	10
5	–	15	–
6	–	–	10
7	–	–	5

display, a PEDOT/PPS layer is covered with an oxetane-containing OLED polymer. Experience has shown that this layer can be "arbitrarily" thick. If the layer is heated to the glass transition temperature of the polymer to be cross-linked, it is completely cross-linked with the reaction starting at the interface with the PEDOT/PPS layer. Subsequently, a further cross-linkable layer of a different OLED polymer can be applied. By creating suitable conditions for cross-linking ($T > T_G$, t), this layer can also be cross-linked without any problem. Thus a multilayer display can be constructed very easily without the necessity of adding another (photo-) initiator.

The oxetane moiety in contact with PEDOT/PPS at the interface is apparently activated by a proton, initiating a cationic ring opening of the oxetane moieties and thus cross-linking, which propagates throughout the whole layer. This layer remains active on its surface, so its activity can be transferred to the subsequent layer. Again, complete cross-linking without the loss of activity is triggered, which can be transfer to the subsequent layer. Because of its size, PPS definitely cannot be considered as a counterion. An alternative is that, under curing conditions, PPS eliminates HSO_4-ions which accompany the active species on its journey through the layers.

Thus it comes as no surprise that OLED technology is said to have a bright future. In particular, it can be expected that large-surface, freely formable light sources and displays are possible using polymer building blocks. One challenge is to increase the efficiency and life expectancy of white light-emitting OLEDs. In addition, the hermetic encapsulation of displays remains another as yet unsolved challenge.

22.5 Polymer Membranes for Fuel Cells

Given the increasing scarcity of energy resources and the increasing environmental problems caused by traditional energy sources, it is not surprising that the field of fuel cells is attracting increased attention (Scherer 2008).

Fuel cells, electrochemical devices that directly transfer the chemical energy of fuel (hydrogen, methane, propane, methanol, etc.) into electricity using oxygen from the air without resorting to a conventional combustion process have been known for a long time (Fig. 22.16).

The reduction and oxidation stages of the combustion reaction take place separately in the fuel cell. The two reactions take place in different parts of the fuel cell separated by an electrolyte membrane. Depending on the structure (Fig. 22.17 and Table 22.6), and especially on the electrolyte, positive or negative ions are transported through the membrane.

The fuel cell consists of a porous anode and a likewise porous cathode (e.g., of carbon fibers covered with activated carbon or carbon black doped with noble metals) which are separated by a membrane. High demands are placed on the membrane in terms of stability and efficiency.

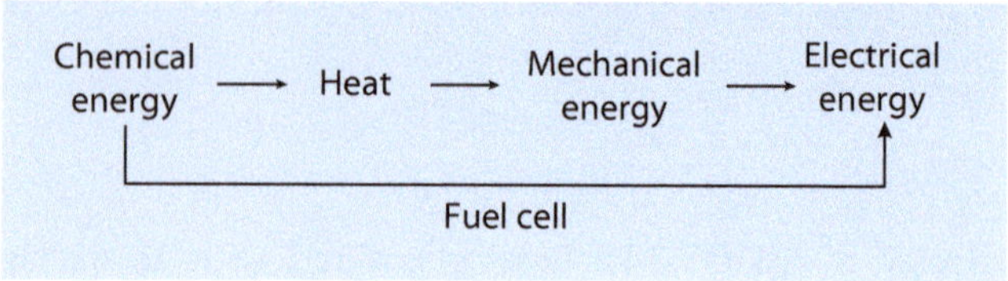

Fig. 22.16 Comparison of conventional conversion of chemical in electrical energy with fuel cells

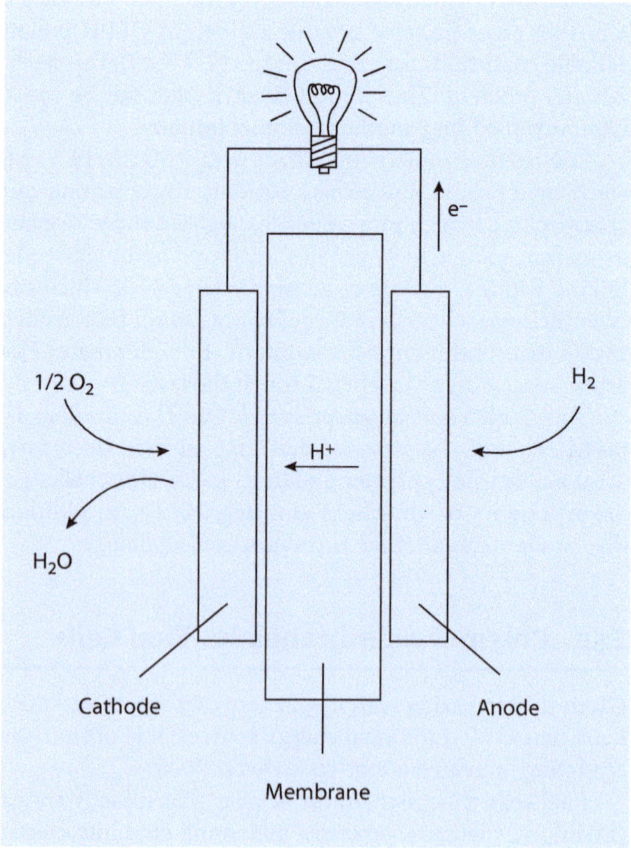

Fig. 22.17 Construction and mode of operation of a fuel cell. In this example protons are transported through the electrolyte, in this case the polyelectrolyte membrane

Table 22.6 Fuel cell types

Type	Electrolyte	Working temperature	Charge carrier	Application
PEFC	Ion exchange membrane	80 °C	H^+	Automobile
AFC	KOH	60–120 °C	OH^-	Space travel
PAFC	Immobilized phosphoric acid	200 °C	H^+	Power stations
MCFC	Molten immobilized carbonate	650 °C	CO_3^{2-}	Power stations
SOFC	Ceramic	800–1000 °C	O^{2-}	Power stations

PEFC Polymer electrolyte fuel cell, *AFC* Alkaline fuel cell, *PAVC* Phosphoric acid fuel cell, *MCFC* Molten carbonate fuel cell, *SOFC* Solid oxide fuel cell

Fig. 22.18 Chemical reaction at the anode in a fuel cell

$$H_2 \longrightarrow 2\,H^+ + 2\,e^-$$

Fig. 22.19 (**a**) Chemical reaction at the cathode. (**b**) Complete fuel cell process

a

$$1/2\,O_2 + 2\,H^+ + 2\,e^- \longrightarrow H_2O$$

b

$$1/2\,O_2 + H_2 \longrightarrow H_2O$$

Hydrogen is separated into protons and electrons at the anode (Fig. 22.18). The electrons created on the anode can be used to produce electricity—they migrate to the cathode via an electrical circuit. The protons move through the polyelectrolyte membrane toward the cathode, driven by the electrical field. There the protons combine with oxygen and the electrons to form water (Fig. 22.19).

The following fuel cell types, depending on the type of membrane, can be distinguished (Table 22.6).

The membranes have special relevance. In the case of PEFC and PAFC, polymers are employed. Depending on the application, very different chemical structures have been suggested and indeed used. *Nafion*® (DuPont) with the structure shown in Fig. 22.20 is one of the most widely known polymer membrane materials.

Nafion® is a proton conductor whose conductivity can be attributed to the sulfonic acid groups in the side chains of the polymer but water is essential to the process. Because

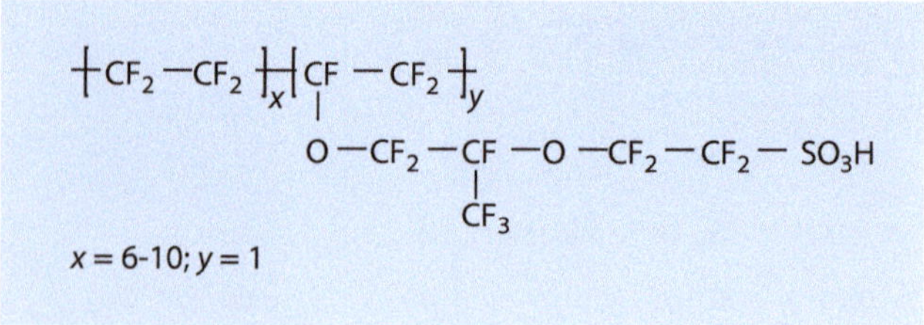

Fig. 22.20 Basic structure of Nafion®, a proton conductor

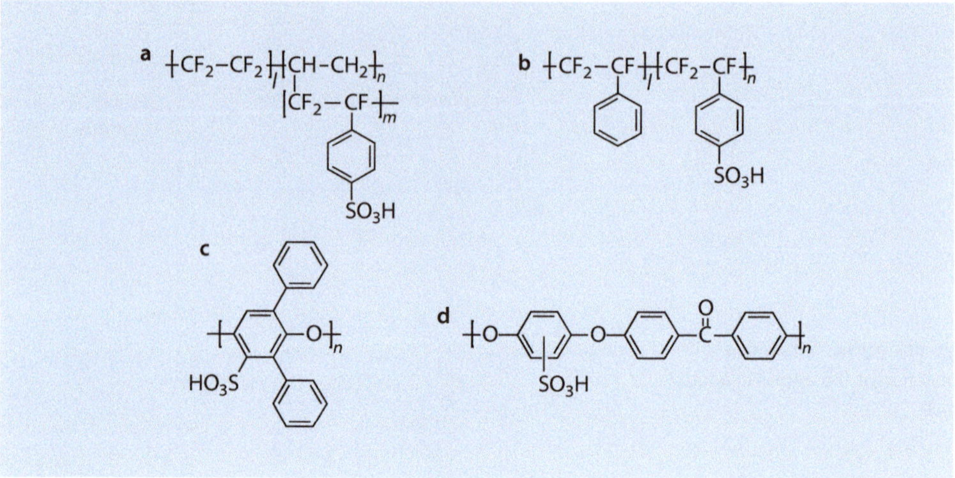

Fig. 22.21 Selected examples of proton conductors. (**a**) Scherer (1990). (**b, c**) Ballard Power Systems. (**d**) Maier and Meier-Haack (2008)

of the incompatibility of the sulfonic acid groups and the fluorinated polymer backbone, channels form in the polymer into which the sulfonic acid groups protrude. The transport of the protons occurs in these channels.

In Fig. 22.21, various alternative candidates for proton conductors are shown.

Ion transport depends heavily on water absorption. For example, Nafion® is an isolator in its dry state. Therefore, for most polymers, operating temperatures are not above 100 °C as loss of water by evaporation must be avoided.

Higher application temperatures can be reached with polybenzimidazole (PBI) doped with phosphoric acid (Fig. 22.22). Sulfonated PBI is also used as a proton transporter (Fig. 22.23).

Generally, the requirements of the membranes differ greatly depending on their application. For example, the use of PBI membranes in the car industry is problematic because of starting temperatures below 25 °C (the proton conductivity is too weak) and the formation of phosphoric acid when the car stops. However, for continuous use, for example, in power stations, these PBI membranes soaked in phosphoric acid, particularly because of their heat stability, provide an interesting solution.

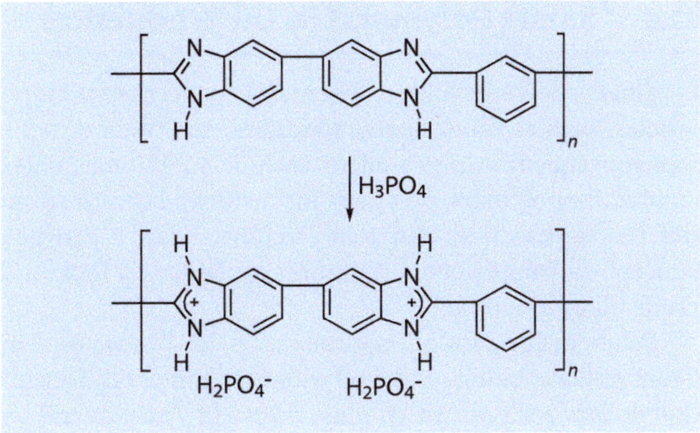

Fig. 22.22 Polybenzimidazole (PBI) -membrane modified with phosphoric acid (Samms et al. 1996)

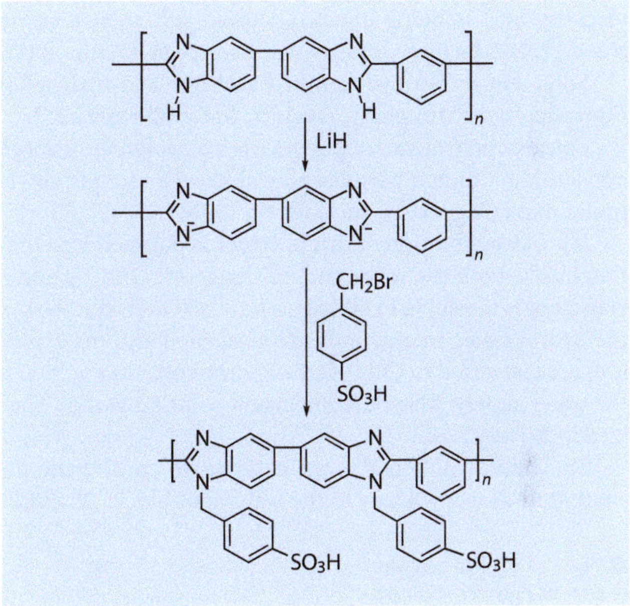

Fig. 22.23 Sulfonation of PBI (Glipa et al. 1997)

For future research, several major challenges remain to improve:

- Proton transport
- Mechanical stability, particularly the flexural fatigue strength necessary because of cyclic differences in the degree of swelling
- Thermal stability
- Chemical stability
- Permeability of reactants
- Cost efficiency

22.6 "Smart Polymers" in Oil Production

Originally, polymer science was mainly concerned with polymers as materials for making articles, such as films, fibers, adhesives, and other structural components. The spectrum was subsequently increased to include functional materials, such as super absorbers, conductive polymers, polymers for medicinal uses, and optics. When a polymer can fulfill different tasks under different circumstances, it can be called a "smart polymer." The concept of "smart polymers" is best explained by looking at an example, here their use in crude oil production.

When drilling for oil, not only oil fields but water veins are struck, resulting in about 3 tons of water being produced with each ton of oil. This results in 3 tons of waste or 40 billion dollars per year which must be used to extract and separation of the water. An intelligent solution would be to find a material that inhibits the water entering into the conveying shaft of the oil well but allows the oil to flow without restriction. Such a polymer might find and block the water (Shashkina et al. 2003). This material should exhibit a low viscosity as it is being injected, should gel when it comes into contact with water, but should retain its low-viscosity when in contact with oil (Fig. 22.24).

Polymers with a hydrophilic backbone and hydrophobic side chains are suitable for this task. They form physical gels in water (Fig. 22.25).

One example of such a polymer is a terpolymer of acrylamide, sodium acrylate, and an acrylamide, which is partially alkylated with a longer alkyl chain ($n = 8, 11$) and has a large molar mass ($M_w \approx 10^6$ g/mol) (Fig. 22.26).

If a watery solution of this polymer is pumped into the conveying shaft, the gel formation blocks both the water and oil transport. Only by adding an inhibitor that dissolves in water, but is insoluble in oil (e.g., a few wt% KCl) can the system be prohibited from gelling; the hydrophobic interactions are weakened and the chain is extended. If the carboxylate ions are converted to COOH, the hydrophobic interaction increases and a plug of gel closes the water supply. Thus, the intelligent solution works. The effect of the inhibitor is visualized in Fig. 22.27.

This "smart polymer" system serves to control the flow of water into the conveying shaft. It finds its own way to the water and blocks its influx by forming a gel.

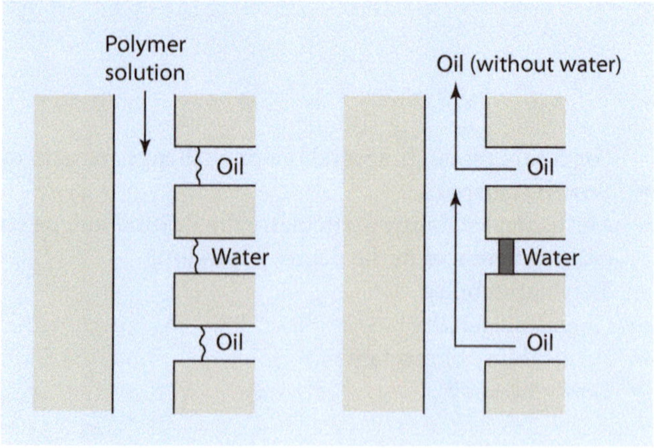

Fig. 22.24 Ideal behavior of a "smart polymer" during the production of crude oil

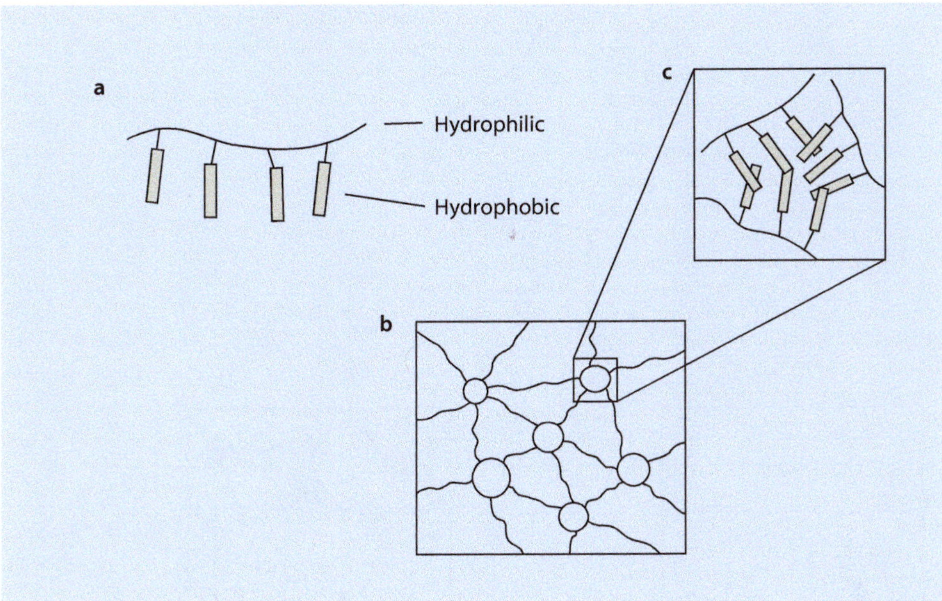

Fig. 22.25 (**a**) Schematic construction and (**b**) effect of a gelling polymer in water. (**c**) Detail showing a single cross-linked area

Fig. 22.26 Chemical composition of a "smart polymer" for the production of crude oil

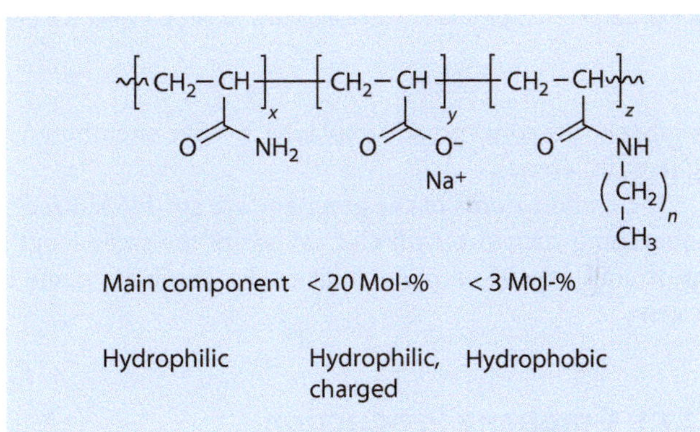

22.7 Graphene as Nanofiller

Until 2004, single-layered carbon coatings (graphene) were deemed thermodynamically unstable. The report on the preparation of graphene monolayers was for this reason all the more remarkable (Novoselov et al. 2004). For their work on graphene the main authors, A. Geim and K. Novoselev, were awarded the 2010 Nobel Prize for Physics.

If the single-layered carbon coatings are imagined in a convoluted form, then they form carbon nanotubes (CNT). If a portion of the six-membered carbon rings of the

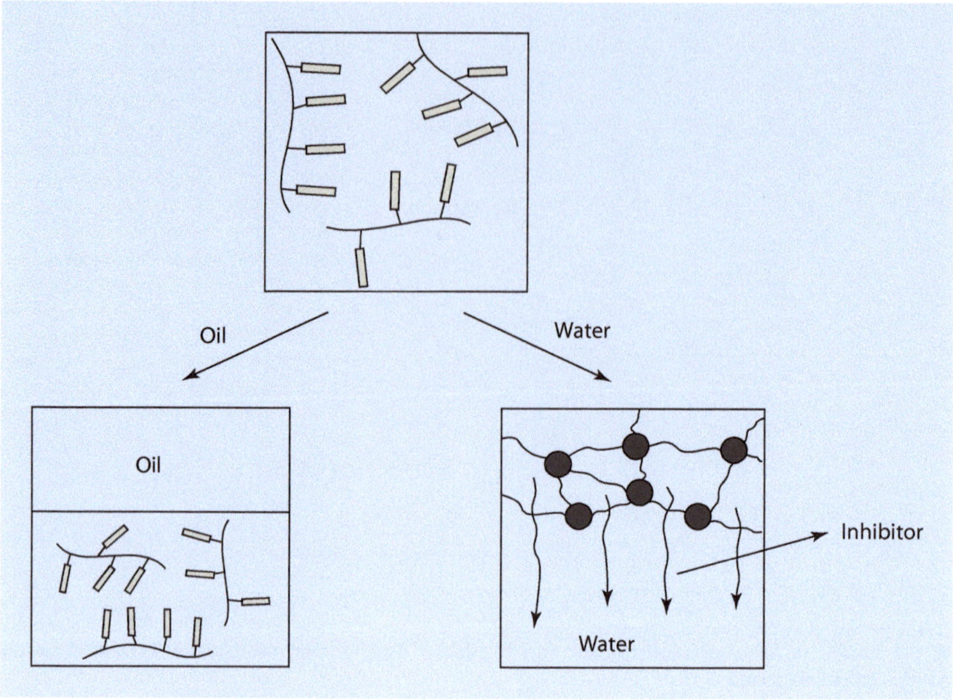

Fig. 22.27 Visualization of the effect of a "smart polymer" vis-a-vis water and oil

graphene are conceptually replaced by five-membered rings, the surface rolls into a sphere (fullerene).

The carbon atoms of the graphene are sp^2-hybridized, resulting in a two-dimensional honeycomb structure with C–C bonds of the same length (1.42 Å). The non-hybridized 2p-orbitals extend perpendicular to the graphene plane and form a delocalized π-bond system.

22.7.1 Graphene Production

The *exfoliation* of graphite—the mechanical separation of individual layers—was implemented by the aforementioned Nobel laureates. In this process, adhesive tape is pressed onto a graphite block and rapidly removed. The graphite monolayers—graphene—stuck to the adhesive tape are then transferred onto a wafer coated with photoresist as the adhesive tape is pressed onto the varnish and then removed. The graphene remains attached to the photoresist. After dissolving the photoresist, the graphene remains on the wafer.

An alternative route to producing graphene is a *top-down strategy*, which begins with graphite being oxidized to graphite oxide (GO). This is then neutralized with KOH, and then reduced (Fig. 22.28).

Alternatively, graphene structures can be synthesized following a *bottom-up strategy* with various preliminary stages. As an example, the synthesis of hexa-*peri*-hexabenzocoronene (HBC) is visualized in Fig. 22.29 (Clar and Ironside 1958).

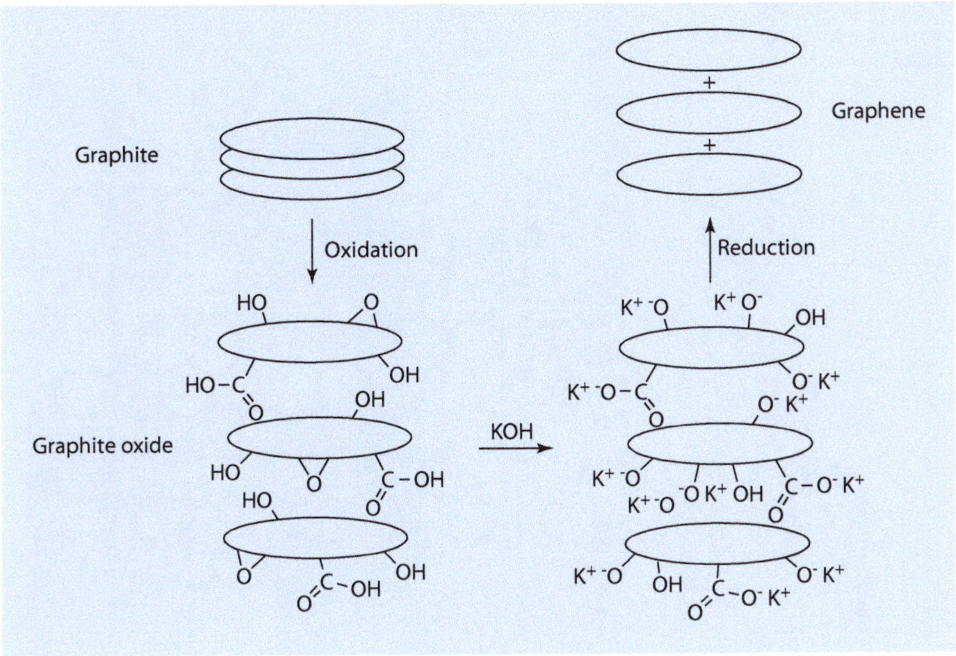

Fig. 22.28 Scheme showing the conversion of graphite to graphene via graphite oxide

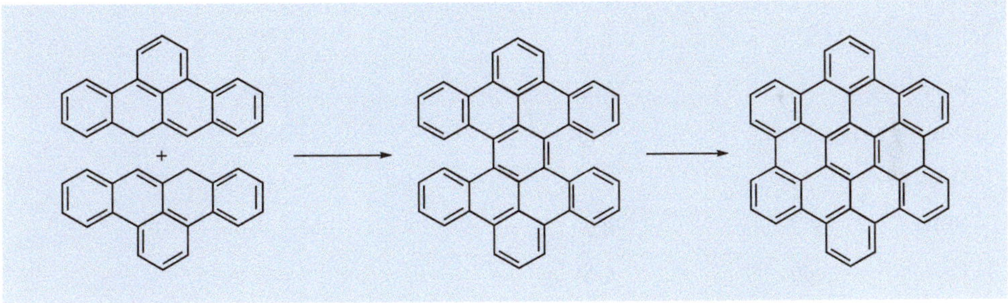

Fig. 22.29 Synthesis of hexa-*peri*-hexabenzocoronene (HBC), a precursor for graphene

The molecule presented in Fig. 22.30 is the largest polycyclic hydrocarbon known to date (Simpson et al. 2002). By modifying HBC it has been possible to produce a large number of well-defined, graphene-related molecules of various shapes and sizes (Wu et al. 2007).

Additionally, graphene can grow epitactically on metallic substituents, for example, by decomposition of ethylene on iridium. A further method for synthesizing individual graphene layers involves the thermal decomposition of hexagonal SiC layers. Graphene is a result of chemical vapor deposition (CVD) of carbon on inert substrates such as copper, which are subsequently dissolved.

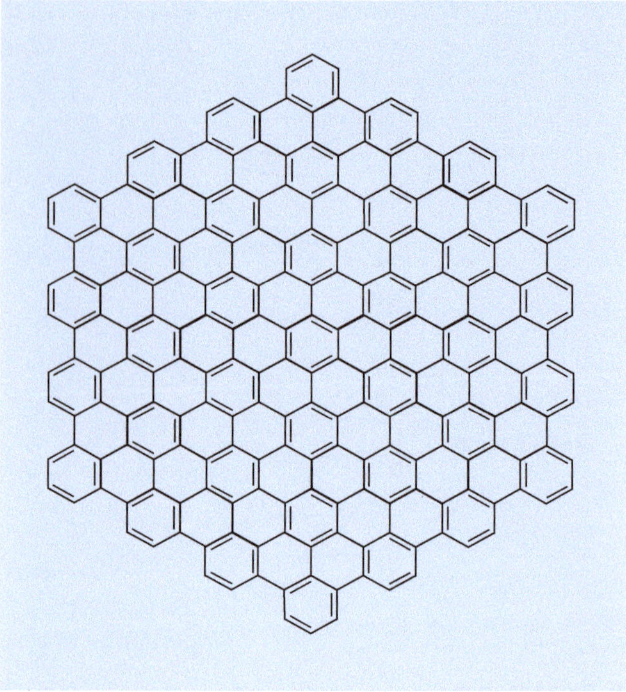

Fig. 22.30 Largest polycyclic hydrocarbon known to date

Table 22.7 Properties of selected fillers for polymers (Kuilla et al. 2010)

Material	Tensile strength (GPa)	Thermal conductivity (W m^{-1} K^{-1})	Electrical conductivity (S m^{-1})
Graphene	130	5000	7200
CNT	60–150	3500	4000
Steel	1770	5	135,000
HDPE	15	0.5	Isolator
Kevlar®	3600	0.04	Isolator

22.7.2 Properties and Use as a Nanofiller

Because graphene is two-dimensional, and consists of planar surfaces of sp^2-hybridized carbon atoms with a thickness of one atom, it is the thinnest material known to date.

Because of its unusual properties, such as high tensile strength and high thermal and electrical conductivity (Table 22.7), it is predestined as a filler for polymers suitable for use in "high-tech" processes such as in the field of electronic sensors, for flexible, transparent electrodes in displays, and in solar cells.

Polymer/graphene nanocomposites display excellent mechanical, thermal, and electrical properties when compared to their unfilled matrix polymers. For example, they have exceptional gas-barrier properties and flame resistance.

The thermal stability of a polymer can be improved by to 100 °C by filling it with graphene and even a small percentage of added graphene renders the composite electrically conductive. However, achieving an adequate dispersion of the strongly associated graphene particles is not trivial.

References

Cai Z (2006) Herstellung von Polymeren und Partikeln in einem. Spinning disk reactor. Dissertation, TU München

Clar E, Ironside CT (1958) Hexabenzocoronene. Proc Chem Soc, p 150

Echte A (1993) Handbuch der Technischen Polymerchemie. VCH Wiley, Weinheim, S 355

Friend RH, Gymer RW, Holmes AB, Burroughes JH, Marks RN, Taliani C, Bradley DDC, Dos Santos DA, Brédas JL, Lögdlund M, Salaneck WR (1999) Electroluminescence in conjugated polymers. Nature 397:121–128

Glipa X, El Haddad M, Jones DJ, Roziére J (1997) Synthesis and characterisation of sulfonated polybenzimidazole, a highly conducting proton exchange polymer. Solid State Ionics 97:323–331

Grund S, Kempe P, Baumann G, Seifert A, Spange S (2007) Zwillingspolymerisation: ein Weg zur Synthese von Nanokompositen. Angew Chem 119:636–640

Guram AS, Rennels RA, Buchwald SL (1995) Eine einfache katalytische Methode zur Synthese von Arylaminen aus Arylbromiden. Angew Chem 107:1456–1459

Köhnen A, Riegel N, Kremer JHWM, Lademann H, Müller DC, Meerholz K (2009) The simple way to solution-processed multilayer OLEDs—layered blockcopolymer networks by living cationic polymerization. Adv Mater 21:879–884

Kuilla T, Bhadra S, Yao D, Kim NH, Bose S, Lee JH (2010) Recent advances in graphene based polymer composites. Prog Polym Sci 35:1350–1375

Li D, Xia Y (2004) Adv Mater 16:1151

Louie J, Hartwig JF (1995) Palladium catalyzed synthesis of aryl amines from aryl halides. Mechanistic studies lead to coupling in the absence of tin reagents. Tetrahedron Lett 36:3609–3612

Maier G, Meier-Haack J (2008) Sulfonated aromatic polymers for fuel cell membranes. Adv Polym Sci 216:1–62

Müller CD, Falcou A, Reckefuss N, Rojahn M, Wiederhirn V, Rudati P, Frohne H, Nuyken O, Becker H, Meerholz K (2003) Multi-colour organic light-emitting displays by solution processing. Nature 421:829–833

Novoselov KS, Geim AK, Movozov SV, Jiang D, Zhang Y, Dubonos SV, Grigorieva IV, Firsov AA (2004) Electric field effect in atomically thin carbon films. Science 306:666–669

Pask SD, Nuyken O, Cai Z (2012) The spinning disk reactor: an example of process intensification technology for polymers and particles. Polym Chem 3:2698–2707

Ramshaw C (1983) Higee distillation—an example of process intensification. Chem Eng 389:13–14

Reneker DH, Chun I (1996) Nanometre diameter fibres of polymers, produced by electrospinning. Adv Mater 16:216–223

Samms SR, Wasmus S, Savinell RF (1996) Thermal stability of proton conducting acid doped polybenzimidazole in simulated fuel cell enviroments. J Electrochem Soc 143:1225–1232

Scherer GG (1990) Polymer membranes for fuel cells. Ber Bunsenges Phys Chem 94:1008–1013

Scherer GG (ed) (2008) Fuel cells I/II. Adv Polym Sci 215:1–268 and 216:1–261

Shashkina YA, Zaroslov YD, Smirnov VA, Phillippova QE, Khokhlov AR, Pryakhina TA, Churochkina NA (2003) Hydrophobic aggregation in aqueous solutions of hydrophobically modified polyacrylamide in the vicinity of overlap concentration. Polymer 44:2289–2293

Simpson CD, Brand JD, Berresheim JD, Przybilla L, Räder HJ, Müllen K (2002) Synthesis of a giant 222 carbon graphite sheet. Chem Eur J 8:1424–1429

Spange S, Grund S (2009) Nanostructured organic-inorganic composites by twin-polymerisation of hybrid-monomers. Adv Mater 21:2111–2116

Tang CW, VanSlyke SA (1987) Organic electroluminescent diodes. Appl Phys Lett 51:913–915

Wu J, Pisula W, Müllen K (2007) Graphene as potential material for electronics. Chem Rev 107:718–747

Supplementary Information

S. Koltzenburg et al., *Polymer Chemistry*, DOI 10.1007/978-3-662-49279-6

Index

D

Q

R

FSC
www.fsc.org
MIX
Papier | Fördert gute Waldnutzung
FSC® C083411

Zeitfracht Medien GmbH
Ferdinand-Jühlke-Straße 7
99095 Erfurt, Deutschland
produktsicherheit@kolibri360.de